Deep-Sea Pycnogonids and Crustaceans of the Americas

Michel E. Hendrickx
Editor

Deep-Sea Pycnogonids and Crustaceans of the Americas

Springer

Editor
Michel E. Hendrickx
Inst Ciencias del Mar y Limnología
National Autonomous University of Mexico
Mazatlán, Mexico

ISBN 978-3-030-58409-2 ISBN 978-3-030-58410-8 (eBook)
https://doi.org/10.1007/978-3-030-58410-8

This Springer imprint is published by the registered company Springer Nature Switzerland AG
The registered company address is: Gewerbestrasse 11, 6330 Cham, Switzerland

This book is the result of a large effort by many American scientists and some of their overseas colleagues. Participation of academics from Latin America is particularly important as it throws new light on the deep-sea communities occurring off their countries' coasts or, in some cases, in much larger areas within the geographic limits of the Americas. In this context, we wish to dedicate this book to two of our colleagues who actively participated in the early edition of this book but passed away while this process was taking place.

Our Chilean colleague, Marco Antonio Retamal, was certainly a founding architect for the investigation in marine biology in Chile. Local expert in crustaceans, Marco

Antonio was involved in a wide series of projects in his homeland, successfully forming and advising many students and young scientists during his career.

Jorge López, from El Salvador, shared his interest between the fishery activities in Central America and a genuine, strong desire of studying the deep-water fauna in the region. An expert in fisheries and conservation issues, he often participated in regional meetings and local advisory panels.

Marco Antonio and Jorge were not only distinguished academics with a profound interest in the biology and ecology of crustaceans and deep-water communities in Latin America, they were also dear friends to many of us. Their contribution to this book will remain forever a contribution to our knowledge of the deep-sea fauna of the Americas. Both true gentlemen, they will be remembered with respect, affection, and admiration.

Preface

Biogeography is a synthesis science integrating all the knowledge from geography, climatology, paleontology, zoology, and phylogenetic. The first step to obtaining all this knowledge is a good sampling. Taxonomy is also essential to compare species between geographies. Nowadays, the knowledge acquired from diverse sources is collected in computer databases, and the largest ones in zoology and marine ecology are OBIS and WoRMS. However, before it can be integrated inside these databases, regional revisions are necessary to clarify the data.

During the last decades, knowledge about the Arthropoda from the deep sea, especially Pycnogonida and Crustacea, increased considerably. The increase in marine deep-sea cruises with a mixture of boats, manned submersibles, or even ROVs brought several new species to light and more information about the way of life of these animals. Surprisingly, some areas on the planet remained very poorly known. It is the case for the coast of the Americas, more precisely South America, where a large gradient of latitudes offers a great diversity of ecosystems.

The present volume is a very successful attempt to fill the gap in the knowledge on America's crustacea. Crustaceans are present in all ecosystems and at any depths, from the pelagic to the benthic zones, and play a critical role in the food web. There are scavengers such as amphipods or isopods, plankton eaters like copepods, suspension feeders like cirripeds, and even carnivores such as lobsters or hermit crabs. At any given place in the oceans, these species they are essential to the functions of the ecosystems.

The American continent, which extends from the Arctic to the Antarctic waters, separates the fauna between the Atlantic Ocean and the Pacific Ocean. Very few benthic species, if any, can be found on both sides of the Americas. Furthermore, it is still unclear why the west American coast fauna is so different from the Indo-Pacific one. The closing of the Isthmus of Panama, only 2.8 million years ago, explains partially this point as well as the long distances and strong currents separating the East from the West Pacific Ocean. Surprisingly, only 120 deep-sea species of decapods are known from the pacific coast of the Americas considering that we find more than ten times this number in the Western Pacific. Perhaps, the hypoxic conditions found along a large part of America's coastlines, with the oxygen minimum zone found within a large bathymetric range, could well be responsible for this poor fauna diversity.

Besides the scientific value, this kind of "regional" cataloguing of crustacea is important for the history of marine biology, drawing up a list of oceanographic vessels and expeditions undertaken along the American coasts. It is also an opportunity to acknowledge the dedicated work of community experts such as taxonomists or ecologists.

Some articles are very original and precious considering that they describe very light and fragile animals difficult to discover. They were collected in the deep sea, near the bottom or inside of the upper layer of the sediments, which required the use of sledge-dredges with micrometric meshes or box-corer.

The catalogue of Pycnogonida from the Uruguayan continental slope has been enriched with several new deep-sea species. The groups of peracarid crustacea are insufficiently sampled anywhere in the world. It is well known that the harpacticoid copepods are very rich in species, present in all the benthic habitats and sometimes forming huge biomass. However, surprisingly, the review of the species from the Americas is only comprised of 22 families and 169 species. For the larger peracarids animals, the knowledge is a little bit better: 27 families and 118 species of amphipods for western Mexico only.

The biology of the pelagic species living in the largest ecosystems of the world is still poorly known. Therefore, the observations about the reproduction and bathymetric repartition of *Nematocarcinu* species in the Gulf of California are very useful. This group is present in all the oceans, characterized by their very long legs. For the first time, the sex ratio has also been evaluated.

A large sampling in the Gulf of Mexico by numerous oceanographic cruises brings good data on the 4 families and 14 species of Penaeoid shrimps. On the other side of South America, different cruises have collected 78 species of pelagic shrimps, Dendrobranchiata and Pleocyemata, between Peru and Tierra del Fuego. A large part of these is associated with the Nazca and Salas y Gomez Ridge seamounts. The lobsters are also included with the Nephropidae and Polychelidae of the Gulf of Mexico. The very large group of Galathoidea, one of the best-studied group of crustaceans in the world, has been studied on the Brazilian coast with the help of molecular tools.

Fisheries in the deep sea are problematic. Generally, the growth rate of the targeted species is unknown as well as the generations turnover. In the case of crustacea, there is another difficulty: the halieutic models are validated for fish, with linear growth, and not for crustacea with several moults. Studies on Lithodids from the Peruvian coast are bringing new useful data on species of *Parolomis* and *Lithodes*. In several places, the deep-sea populations of crustacea are already being overexploited.

Other chapters bring interesting comments on decapod fisheries concerning 79 species in Uruguay and 181 species along the Brazilian coast, mainly on shrimps and geryonid crabs. For each species, an updated synonymy is given. The sampling of deep-sea crustacea along the Chilean coast is showing a change in the diversity between the warm waters in the north and sub-Antarctic waters south of Chiloé

Island brought by the Humbolt Current. The pacific coast of Central America was also sampled from Guatemala to Panama.

Conservation strategies are proposed to keep an equilibrium between healthy ecosystems and sustainable fisheries in the deep sea. In these environments, the biological conditions are different from the shallow-water fisheries, and the stocks are easily overexploited because the species have extreme longevity, slow growth rates, late maturity, and low fecundity.

This volume highlights the ambitious exploration sampling programs in the deep sea to increase our knowledge of the fauna from North, Central, and South America. This knowledge could help to protect the deep-sea environment essential for the sustainability of our planet.

MNHN-Nouméa
Paris, France

Bertrand Richer de Forges

Prologue

Although investigation in the Latin America part of the Americas' deep waters has significantly increased in the last decades, there is still a profound gap between what we know about this environment in most of the northern hemisphere and in the southern hemisphere. Exploratory cruises financed by local governments are few, and occasional bilateral projects, although often well financed and organized, have not been sufficient to significantly increase the rate at which deep-water communities are sampled and studied. On the other hand, the number of scientists who have found opportunities to get involved in this kind of research is extremely reduced due to a certain lack of priority for this category of science. Globally, the number of large, modern sea-going vessels available for research is also very reduced in several countries in the region. Operational costs are extremely high and, in some cases, finding resources to support offshore surveys is near to impossible. Some research vessels are quite old and should ideally be replaced by more efficient vessels, but, again, there are often other priorities and critical investments that are needed and focused on more urgent situations. Equipment is another issue, and prices of modern sampling gear and recording equipment are most of the time too high for local budgets. In some countries, a strong interest has arisen for deep-water fishing and some of our colleagues have had the opportunity of taking advantage of this sampling effort to gather valuable information on deep-water communities that were previously mostly unknown. This, of course, raises the question of deciding whether the deep-water fishing industry should be given a "carte blanche" or be supervised in order to limit its impact on the poorly known, slow-growing species that might prove to be commercially exploitable on the short term but might be profoundly affected on the long term. Remote equipment have been intensively deployed in regions where institutions or government agencies can afford the acquisition of these expensive equipment and support their offshore operation. Long-term video recording from deep-water observatories is certainly a friendly way to study the local fauna, provided we do have a fair knowledge of what species inhabit the area. These low-impact techniques will certainly proliferate in the near future, but elevated cost to build and maintain these infrastructures obviously represent a huge limitation for many countries. One of the purposes of this book is to call the attention to the extraordinary richness of the deep-sea fauna in our region. Simultaneously, it is a warning that what we know about this fauna, its distribution, its basic biology, and its ecology is far from being sufficient in order to understand and manage all the factors that are already impacting negatively on these communities. In fact, they could very well vanish before we are able to study all its components and their interaction. The challenge is enormous and tantalizing. To be successful, it will take a gigantic effort, a great deal of investment, and, above all, comprehensive multinational collaboration programs aimed at sharing our present experience and our future goals.

Contents

Chapter 1
The Deep-Water Crustacean and Pycnogonid Fauna of the Americas in a Global Context

K. E. Schnabel, S. T. Ahyong, A. J. Gomez, M. E. Hendrickx, R. A. Peart, and J. N. J. Weston

Abstract Uniting arthropods and the deep sea as a research focus allows us to combine the largest habitat on Earth with the most abundant and successful animal group on our planet. To set the stage for this volume, we briefly introduce the deep-sea environment, the history of its exploration and gaps in our knowledge in terms of global distributions of biodiversity, before we discuss the current understanding of the Crustacea and Pycnogonida within the phylum Arthropoda. The ecological context in the deep sea of the Americas is addressed and briefly outlined, including the geological history, biogeographic boundaries, and the effect of a large oxygen minimum zone in the region. Deep-water (>200 m) areas of the Americas represent roughly 72% of the national exclusive economic zones (EEZ) in the west Atlantic Ocean and 86% in the east Pacific Ocean. As in other regions of the world,

K. E. Schnabel (✉) · R. A. Peart
Coasts and Oceans Centre, National Institute of Water & Atmospheric Research, Wellington, New Zealand
e-mail: Kareen.Schnabel@niwa.co.nz; Rachael.Peart@niwa.co.nz

S. T. Ahyong
Australian Museum, Sydney, NSW, Australia

School of Biological, Earth & Environmental Sciences, University of New South Wales, Kensington, NSW, Australia
e-mail: shane.ahyong@austmus.gov.au

A. J. Gomez
Climate Change Analysis, Ministry for the Environment, Wellington, New Zealand
e-mail: Antony.Gomez@mfe.govt.nz

M. E. Hendrickx
Instituto de Ciencias del Mar y Limnología, Universidad Nacional Autónoma de México, Unidad Académica Mazatlán, Mazatlán, Mexico
e-mail: michel@ola.icmyl.unam.mx

J. N. J. Weston
School of Natural and Environmental Sciences, Newcastle University, Newcastle Upon Tyne, UK
e-mail: j.weston2@ncl.ac.uk

M. E. Hendrickx (ed.), *Deep-Sea Pycnogonids and Crustaceans of the Americas*, https://doi.org/10.1007/978-3-030-58410-8_1

our knowledge of the diversity and ecology of deep-water arthropod communities of the Americas is still very limited. Comprehensive research programs aimed at exploring the vast offshore areas off the Americas must increase rapidly to study the deep-water fauna before it is substantially, and possibly, irreversibly impacted by human activities and global climate change.

Keywords Pancrustacea · Multicrustacea · Astrobiology · Biogeography · Oxygen minimum zone · Atlantic ocean · Pacific ocean · Caribbean sea · Pelagic · Benthic · Bathyal · Abyssal · Exclusive economic zone

1.1 Introduction

1.1.1 Our Watery World and the Beginnings of Life

> Look out across the ocean on a calm day, from the shore or from the deck of a ship. The vista is daunting in its scale yet innocuous in its features. But beneath this tranquil skin lies a teeming horde of organisms, from the tiniest of viruses to the mightiest whales, all of which are continually influenced by the physical features of the seawater within which they move—and by which they are moved. (Herring 2002)

Pycnogonids and crustaceans play important biological roles in all the Earth's oceans, and in many habitats they dominate in numbers and/or biomass. Considering the deep sea, however, large gaps in our knowledge remain, and before a range of chapters present diversity, distribution, and significance of crustaceans in the American regions, let us expand our focus: How much do we know about our ocean? Do we still "know less about the bottom of our ocean than of the surface of the Moon"? What do we consider the deep sea to be and how diverse is this environment? And of course, the complexity of the question: What is an arthropod, and what place do crustaceans and pycnogonids hold within it?

The deep sea, with its unique chemical properties, remains a hot contender for the origin of all life on Earth around 4 billion years ago (Dodd et al. 2017). Most recently, Jordan et al. (2019) added to evidence that the complex organic compounds necessary for basic cell formation can be formed around deep-sea hydrothermal vents. This prompted the headline news "Life on Earth probably originated in deep-sea vents and aliens could be growing the same way now, scientists suggest" (Cockburn 2019). Could complex life have similarly evolved in oceans of other solar system bodies?

The Earth may not be the only world in our solar system to contain deep oceans. Many of its celestial bodies show the presence of water either as traces of vapor in their atmospheres or as ice on the surface. The Moon itself has strong indications of ice at its South Polar regions, making it a site considered for a permanently occupied base. Despite a long distance from the Sun in the coldness of space, deep, salty liquid oceans are suspected to exist beneath the icy surface of three of the Galilean moons of Jupiter, Europa, Ganymede, and Callisto, and similarly on Enceladus, a moon of Saturn (Paganini et al. 2020; Schmidt 2020; Cable et al. 2020). Tidal flexing from the gravitational pull of the nearby planet generates sufficient internal heat, and presumably hydrothermal vents, on these moons to form liquid oceans tens of kilometers deep below an icy surface, itself several kilometers thick. Despite their relatively small size in comparison to the Earth, there are strong suggestions that there is significantly more water in liquid oceans of some of these worlds than exists on Earth. Naturally, the question that comes to mind is: Are these ocean worlds hospitable for life to exist?

Proposed astrobiology programs are set to explore the conditions in and under the ice with melt probe missions and deployment of ocean gliders, using the Arctic and Antarctic sea ice as analogs (Schmidt 2020). Thus, before extrapolation to any extraterrestrial conditions can be made, it is first critical to have a sound knowledge of our own watery worlds. As such, the contributions to our understanding of ice-ocean interactions on Earth and the life it can support, provided by the initiatives of the planetary analog research, are significant. Together, they represent progress toward a better understanding of our own oceans under the ice here on Earth while pushing the boundaries of how data from other ocean worlds can be interpreted.

1.2 The Deep Sea

1.2.1 How Little Do We Know of Earth's Deep Ocean?

In 1818, British explorer John Ross reported that he had collected a basket star from a depth of 1600 m during his search of the Northwest Passage – the first true deep-sea faunal record. Despite 200 years of subsequent intensive deep-sea research, a common quote persists: "95 % of the ocean remains unexplored and we know more about the surface of the Moon than the ocean floor." Does this statement still hold in the twenty-first century?

The exploration of both the deep sea and space is relatively new, but true deep-sea discoveries predate the Russian and US space programs by nearly 140 years with the first satellite *Sputnik 1* launched into Earth's orbit by the Soviet Union only in 1957. Also, the surface of the Moon is about one-tenth of the area covered by ocean on Earth, i.e., 38 compared to 362 million km^2 (Gregersen 2010; Charette and Smith 2010), and the combined total of physical samples available from the Moon are about 380 kg (Zeigler et al. 2019), compared to the swathes of deep-sea geological samples and biological specimens collected over the last two centuries. Twelve men have walked on the Moon, and, as of July 2020, thirteen people have visited the deepest point on Earth, the Challenger Deep of the Mariana Trench (Lobner 2020). Notably, the dive of the *Bathyscaphe Trieste* took Jacques Piccard and US Navy Lieutenant Don Walsh to a depth of around 10,900 m in 1960, just one year before Soviet Air Force Pilot Yuri Gagarin was the first person in space in 1961.

Yet, the deep sea is more than the deepest point on Earth. Every year, hundreds of manned submersible dives are conducted into the deep sea, taking scientists to bathyal and abyssal depths to study underwater features and faunal communities. In fact, several manned submersibles are rated to dive more than 1000 m, including the 6500 m rated American *DSV Alvin* and Japanese *DSV Shinkai* 6500 and the 7000 m rated Chinese *DSV Jiaolong* (Kelley et al. 2016). At even greater depths, the ability to research at hadal depths has substantially increased in 2018 and 2019 with the Five Deeps Expedition, which completed 39 dives with the submersible DSV *Limiting Factor* to the deepest point in every ocean (Jamieson 2020). In addition to manned submersibles, many research expeditions every year deploy a variety of sampling equipment and sensors, such as CTDs, sediment traps, baited landers, epibenthic sleds, and remotely operating and autonomous vehicles. They survey, image, and collect specimens in order to study the deep sea. There are also long-term research stations, like the Monterey Bay Aquarium Research Institute's "Station M" which has been sampling the abyssal zone since 1989 (Smith et al. 2017). Arguably, there is a greater continued presence of scientists in the deep sea than in space, and the International Space Station (ISS) has been continuously occupied only since November 2000, with a total of 240 individuals from 19 countries who have visited the ISS since then (Garcia 2020).

The metrics that are applied to compare what we know of the bottom of the ocean and the surface of the Moon, however, are those of remote sensing and mapping. Detailed maps of the surfaces of the Moon, Mars, and Venus so far exceed the resolution available for our ocean floor. A near-global map of the lunar surface was compiled in 2011 by the NASA's Lunar Reconnaissance Orbiter (LRO) to a scale of 100 m, which included the second ever image of the far side of the Moon. In contrast, the entire ocean floor has now been mapped to a maximum resolution of around 5 km (Sandwell et al. 2014), which allows for the identification of large features such as underwater mountains, ridges, or trenches. However, less than 18% of the seafloor has actually been measured directly (Mayer et al. 2018). While indeed the surface of our nearest celestial neighbors is better mapped than the deep ocean, this might change in the near future.

With an ambitious target, the Nippon Foundation-GEBCO Seabed 2030 Project is a focused global initiative for mapping the entire ocean floor to a resolution of about 140 m (at 4000 m depth), similar to those provided by the LRO, by the year 2030 (Mayer et al. 2018). Of course, accurate bathymetry maps are crucial for understanding the geomorphology that impacts many seafloor processes and habitat heterogeneity and, in turn, their biological communities (Stewart and Jamieson 2018). However, in situ images and specimens are what many scientists are interested in, but considering the global scale and the cost and effort to sample the deep sea (Clark et al. 2016), the distribution of deep-sea biological specimens remains very scarce globally. This is exemplified using the Ocean Biogeographic Information System (OBIS) database, the world's largest repository of marine biological data. The distribution of over 4.4 million sample records from depths ≥200 m (per degree latitude and longitude) across the surface of all oceans is shown (Fig. 1.1). It is immediately evident that some areas (e.g., parts of continental Americas, Europe,

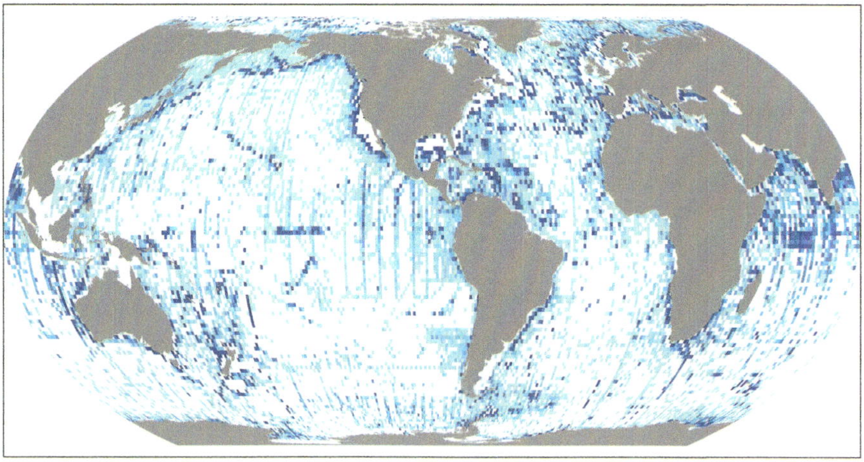

Fig. 1.1 OBIS records for all Animalia, depth ≥200 m, 4,440,609 records, 1° grid colored by counts of records from 0 (white) to 1055 (dark blue). Projection: Equal Earth (sphere) Americas, ArcGIS Pro 2.4.2. (The Esri Third-Party Software Acknowledgments are available online at https://links.esri.com/open-source-acknowledgments.) (Extracted 28 November 2019)

and Asia) are relatively densely sampled, but the majority, particularly of the oceanic areas, have not been sampled at all.

The same pattern emerges when one examines the global distribution of samples with depth. Webb et al. (2010) revealed "biodiversity's big wet secret," using OBIS records to clearly show that the number of biological records sharply declined from a depth of >200 m and then again between 5000 and 6000 m (Fig. 1.2). More than 50% of all OBIS records are from the continental shelf, which constitutes <10% of the ocean surface. The global distribution of records through the water column further highlights the paucity of records from the deep pelagic ocean. Thus, we can put to rest the oversimplistic and nonsensical comparison between our presumed knowledge of the deep ocean and the Moon. Over the past century, much work, time, and innovation have allowed us to reach deeper toward a fuller understanding of the deep ocean. Yet, there is much work in the next century to be done.

1.2.2 The Deep-Sea Environment

"Ninety per cent of the two-thirds of the surface of the Earth covered by the sea lies beneath the shallow margins of the continents; and most lies under 2 km or more of water. We may, therefore, with some justification, speak of the deep-sea bottom as constituting the most typical environment, and its inhabitants as the typical life-forms of the solid face of our planet. Yet, because of the remoteness of this habitat and the difficulties in observing and sampling these organisms, they are known to only few scientists; and as living rather than pickled specimens, to less than a

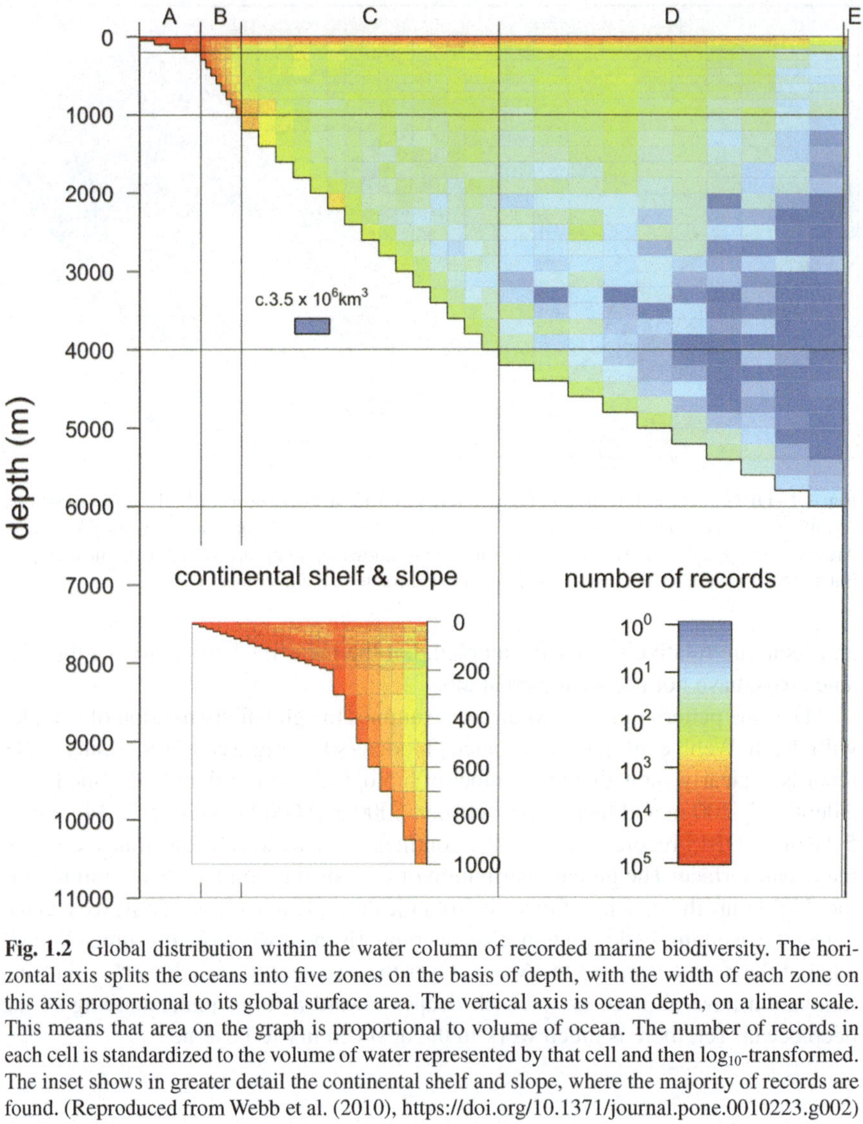

Fig. 1.2 Global distribution within the water column of recorded marine biodiversity. The horizontal axis splits the oceans into five zones on the basis of depth, with the width of each zone on this axis proportional to its global surface area. The vertical axis is ocean depth, on a linear scale. This means that area on the graph is proportional to volume of ocean. The number of records in each cell is standardized to the volume of water represented by that cell and then $\log_{10}$-transformed. The inset shows in greater detail the continental shelf and slope, where the majority of records are found. (Reproduced from Webb et al. (2010), https://doi.org/10.1371/journal.pone.0010223.g002)

handful. Yet the possibility of life existing at these great depths, and a curiosity about the nature of these life forms in what appears to be one of the most 'difficult' of environments has fascinated Man since the early days of oceanic exploration" Gage and Tyler (1991: xi).

This Preface of what has become a standard textbook on deep-sea biology still rings true. The sheer scale of the three-dimensional space, that is, the open ocean, remains as daunting today as it was three decades ago. The average depth of the ocean is around 3700 m and covers an estimated area of 362 million km^2 (Charette and Smith 2010), holding nearly all the space that can be inhabited by life on Earth. The deep sea as defined herein is considered as the area deeper than the continental shelf (on average to 200 m depth), reaching to nearly 11,000 m at the Challenger Deep, Mariana Trench, Pacific Ocean. This includes the world's largest mountain range (the Mid-Atlantic Ridge) and the deepest canyons. Within the deep sea, there are three zones: the bathyal extending down the continental slope from 200 to 3000 m; the abyssal, from 3000 to 6000 m, covering 54% of the Earth's surface with a vast network of abyssal plains that are punctuated by ocean ridges and trenches (Smith et al. 2008); and the hadal, from 6000 to 11,000 m, accounting for the deepest 45% of the ocean and consisting of subduction trenches, fracture zone, troughs, and depressions (Jamieson et al. 2009).

The world's five oceans are connected by powerful global currents. Paradoxically, however, instead of being a single, uniform system, the ocean is complex and heterogeneous. There are myriad communities with changing interactions and relationships, and many of the species that inhabit these communities are adapted to the extreme environmental conditions of increasing hydrostatic pressure with depth, no light, low temperature, and limited food inputs (Perrone et al. 2003; Downing et al. 2018). Some deep-sea communities are highly specialized, such as the hydrothermal vent assemblages. These hydrothermally active habitats have only been recently discovered, the first in 1977 around the Galapagos Rift (Lonsdale 1977). Other distinct deep-sea communities, including prominent crustaceans, inhabit only the extreme depths of the hadal subduction trenches (Beliaev 1989; Jamieson 2015). In contrast, some species inhabit all oceans, with cosmopolitan species among pelagic crustaceans not uncommon (see review by Halsband et al. 2020). Increasingly powerful genetic tools, however, have revealed that many supposedly widespread species are mosaics of several cryptic or pseudocryptic species, in both pelagic (e.g., Andrews et al. 2014; Cornils and Held 2014; Cornils et al. 2017) and benthic shallow-water and deep-sea crustaceans (e.g., Garlitska et al. 2012; Havermans et al. 2013). In most cases, though, some general boundaries around regional assemblages can be drawn based on submarine topography, stable environmental characteristics, and geotectonic history. For example, Longhurst (2007), refined by Reygondeau and Dunn (2019), proposed nested global biogeographic partitions for pelagic ecosystems. Spalding et al. (2007), with Marine Ecoregions of the World (MEOW), did the same for the world's coastal and shelf areas (depths <200 m), and Watling et al. (2013) refined the UNESCO's (2009) Global Open Oceans and Deep Seabed (GOODS) classification with a delineation of bathyal, abyssal, and hadal provinces. While these marine biogeographic boundaries are rarely strict, the classifications in all cases are proposed to be of use as management tools and for analyses (such as predictive modelling) or to guide establishment of marine protected areas and minimize the negative impact of commercial activities in marine environments. The delineations of these global classifications were substantially guided by knowledge of deep-sea fauna distributions, so a common limitation is that of the fauna itself, which continues to bring surprises.

1.2.3 *The Deep Sea of the Americas*

1.2.3.1 The Ecological Context

The Americas stretch in a continuous landmass from the Arctic Ocean in the north to the Antarctic Ocean in the south and covers nearly 30% of the Earth's land area. The continent delimits the Atlantic Ocean from the Pacific Ocean and entirely encompasses the Caribbean Sea. To the north, narrow and shallow passages between Alaska and Russia (~80 km distance) and Canadian Ellesmere Island and Greenland (16 km distance) limit the present movement of water between deep oceans (Fig. 1.3). Instead, the only connection of deep oceanic waters remains the relatively small gap of the Drake Passage, between the southern tip of South America (56°S) and the northern tip of Palmer Peninsula in west Antarctica (62°S, Fig. 1.4). Here, waters join the Antarctic Circumpolar Current that fuels the thermohaline circulation or the global conveyor belt (Iudicone et al. 2008). "Engines" of this global circulation, where warm surface water sinks downward to join the cold bottom waters, are located near the Americas in the northern Atlantic Ocean (Labrador Sea) and on either side of South America (Ross Sea and Weddell Sea; Kuhlbrodt et al. 2007).

In geological history, the continental movements changed underwater topography and exchange of water and associated fauna, with biological signals of these events still detectable in regional faunas (e.g., Feldmann and Schweitzer 2006). The breakup of Gondwana and Laurasia about 200 million years ago (mya) resulted in the formation of the separate Pacific, Indian, and Atlantic oceans which was completed during the Eocene Epoch (~50 mya). The formerly widespread Tethyan fauna began to regionally differentiate, facilitating lineage divergence. The eastern Pacific Ocean fauna remained connected to that of the Atlantic until relatively recently, being separated by the formation of the Isthmus of Panama during the Pliocene; strong evidence indicates a progressive shoaling and that a deep-water connection was shut off around 10 mya (Bacon et al. 2015; O'Dea et al. 2016). The linking of North and South America by the Isthmus of Panama had major impacts on global climate, oceanic and atmospheric currents, and biodiversity, yet the timing of this critical event remains contentious. In 1914, a new connection between the Pacific Ocean and the Caribbean Sea had been established by the Panama Canal with a number of channels and locks transiting through the freshwater Gatun Lake. Surprisingly little is known about the effect of the Panama Canal on the distribution of marine biota, but crustaceans appear to be some of the most successful hitchhikers that have been transported in both hull-fouling and ballast waters (Cohen 2006; Ros et al. 2014). The discussions typically consider shallow-water species, but it is probable that some vertically migrating planktonic stages of deep-water taxa are transported in ballast waters and survive the interoceanic transit, particularly in light of a significant expansion of the capacity of the Panama Canal (Brierley 2014; Muirhead et al. 2015).

Fig. 1.3 Geophysical and bathymetric map of North America and surrounding shelf and seas, showing the 200m bathymetric contour line (dark lines) and the national exclusive economic zones (white lines). The Gulf of Mexico is abbreviated (GOM). Projection: North America Lambert Conformal Conic. Credit: Esri, Garmin, GEBCO (General Bathymetric Chart of the Oceans), NOAA (National Oceanic and Atmospheric Administration) National Geophysical Data Center, and other contributors

The modern distributions of marine organisms around the Americas generally follow global latitudinal clines. The range boundaries of crustaceans and other marine invertebrates generally coincide with transitions between oceanic domains with distinct biogeochemical properties and planktonic and benthic communities (Watling et al. 2013; Reygondeau and Dunn 2019). In general, little is known about distribution pattern of deep-water pycnogonids of the Americas. Some species of *Colossendeis* feature a cosmopolitan or very wide distribution (Staples 2007; Hendrickx this volume). American species of Pycnogonida have been sparsely

Fig. 1.4 Geophysical and bathymetric map of South America and surrounding shelf and seas showing the 200 m bathymetric contour line (dark lines) and the national exclusive economic zones (white lines). Projection: South America Lambert Conformal Conic. Credit: Esri, Garmin, GEBCO (General Bathymetric Chart of the Oceans), NOAA (National Oceanic and Atmospheric Administration) National Geophysical Data Center, and other contributors

studied, and this volume contributes to their general knowledge (Soler et al. this volume).

In the western Atlantic Ocean, the biogeographic boundaries for marine species are often related to convergence regions of currents, where the cold Labrador Current meets the Gulf Stream in the north or the Falkland Current meets the Brazil Current in the south. These regions often show marked changes in regional species assemblages in both shallow and deep water (e.g., see Boschi (2000) for shallow-water crabs and Schnabel et al. (2011) for deep-water squat lobsters). While the Caribbean and the Gulf of Mexico communities appear to be distinct and diverse based on continental shelf species (Boschi 2000; Briggs 2003), these distinctions are less pronounced at pelagic bathyal and abyssal depths (Watling et al. 2013; Sutton et al. 2017). As a result, no single pattern is evident.

The separation between biogeographical provinces in the eastern Pacific Ocean is similarly related to major oceanographic processes. This includes areas of upwelling off Peru, Chile, and California and the presence of an extensive oxygen minimum zone that creates a large median wedge between a generally northern and southern fauna (see below). These biogeographic regions have been reported for shallow-water decapods (Wicksten 1989), bathyal squat lobsters (Macpherson et al. 2010), and mesopelagic plankton (Sutton et al. 2017). In all cases, however, studies point to extensive caveats, sampling gaps, and general patchy knowledge that often confound and preclude extrapolations to large regional and global scales.

1.2.3.2 The Oxygen Minimum Zone

Striking features affecting the composition and abundance of the deep-water pelagic and benthic communities worldwide are oxygen minimum zones (OMZ). While many small coastal areas are occasionally affected by short-term or smaller-scale oxygen depletions (Rabalais and Turner 2001; Gooday et al. 2009), at the oceanic level, OMZs are particularly ubiquitous and persistent in the East Pacific, the northern part of the Indian Ocean, and the East Atlantic (Diaz and Rosenberg 1995; Levin et al. 2000) (Fig. 1.5). Of these, the East Pacific OMZ is by far the widest of its kind (Fig. 1.5), ranging almost uninterrupted from Alaska to central Chile (Helly and Levin 2004), over about 90° of latitude and up to about 7000 km offshore (Diaz and Rosenberg 1995). The widest OMZ cores are found off western Mexico, Central America, and Peru. The core narrows considerably toward higher latitudes (the northern part of the Baja California Peninsula to Oregon and further north) as well as south of Peru (Helly and Levin 2004). In these areas, dissolved oxygen concentration rapidly decreases with depth and reaches severe hypoxic values or even close to anoxic conditions in what is known as the OMZ core (Helly and Levin 2004, Fig. 1.6). The depth at which the upper and lower limits of the OMZ core occur and the degree of intensity of the OMZ depend on geographic location. Thus, the bathymetric range (interval) corresponding to hypoxic (either mild or severe) or anoxic conditions will vary significantly from one locality to another (Helly and Levin 2004; Serrano 2012; Papiol et al. 2016).

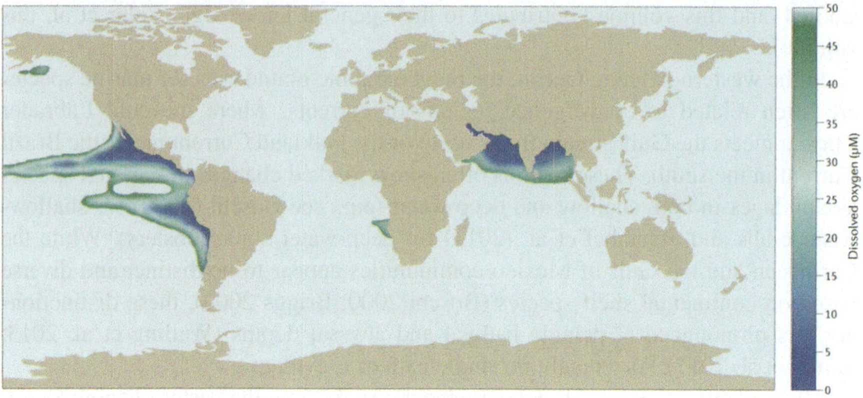

Fig. 1.5 Worldwide distribution of oxygen minimum zones. Data from the World Ocean Atlas

Oxygen depletion in the water column strongly affects pelagic communities not adapted to hypoxic or near-anoxic conditions. Likewise, in areas where the OMZ intercepts the continental shelf and slope, the composition and size of benthic communities will depend upon their tolerance to oxygen depletion. Additionally, the amount of organic matter available for consumption by detritus feeders is another critical driver (Levin et al. 2000, 2010; Papiol et al. 2016). Of paramount importance for local fisheries is the shallowest depth at which the OMZ core occurs, and shallow habitat compression has been predicted for hypoxia-intolerant taxa (Stramma et al. 2010). Indeed, the shallowest the depletion of dissolved oxygen starts, the highest the impact on fishery activities. In most of the East Pacific Ocean, fishing activities have traditionally taken place on the continental shelf (Méndez 1982; Hendrickx 1995; Rojas-Morales et al. 2000; Álvarez-León 2002; Tabash Blanco 2007; Hendrickx and Serrano 2010; Díaz Merlano et al. 2011; Mendo and Wosnitza-Mendo 2014; Reyes et al. 2014; Villalobos-Rojas et al. 2014; Vieira et al. 2016). Considering the entire East Pacific Ocean and with a few local exceptions, the upper boundary of the OMZ found between 40 and 60°N is located much deeper (between 650 and 550 m depth) than in other areas. Indeed, it is very shallow between 25°N and 5°N (200 m depth or less) and between 5°S and about 20°S (100 m depth or less; Fig. 1.6). Similarly, the depth at which the OMZ core lower boundary is found on average varies with latitude. In the higher latitudes (40–60°N), it is located at 1100–1200 m depth; in the intermediate latitudes (30–5°N), it occurs at 800–1000 m; and in the lower latitudes (0–25°S), it is much shallower (400–500 m; Helly and Levin 2004). Consequently, as for other groups of invertebrates, deep-water crustaceans tolerant of (some) oxygen depletion, or living permanently below the OMZ core, will be found at different depths depending on the latitudinal range they inhabit (Papiol et al. 2016).

Along the OMZ, the deep-water fauna occurring below the lower boundary of the core is almost entirely distinct from the fauna found on the shelf or above the upper boundary of the core (Kameya et al. 1997; Quiroga et al. 2009; Hendrickx and

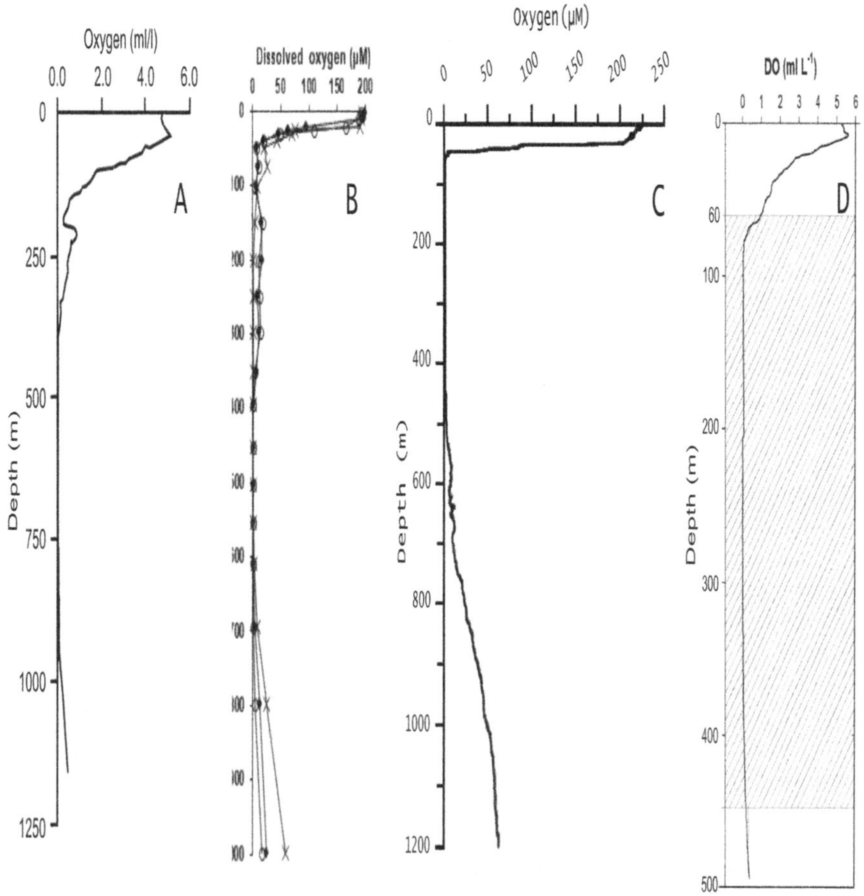

Fig. 1.6 Dissolved oxygen profiles obtained offshore in four localities in the East Pacific. (**a**) Off western Mexico. (**b**) Off Costa Rica. (**c**) Off Peru. (**d**) Off Chile. (Sources or by courtesy of (**a**) Hendrickx and Serrano (2010), (**b**) Lu et al. (2019), (**c**) Mosch et al. (2012), (**d**) Escribano et al. (2009))

Serrano 2010). Migration of adult stages of benthic species through the OMZ, from shallow to deep water or inversely, appears to be strongly limited due to their physiological intolerance to oxygen depletion (Quiroga et al. 2009; Hendrickx and Serrano 2014). In a similar manner, dispersal and vertical migrations of pelagic forms (including the larval stages of benthic species) are also perturbed or limited (Judkins 1980; Saltzman and Wishner 1997; Hidalgo et al. 2005), particularly in localities where the upper boundary of the OMZ is very shallow (e.g., less than 25 m depth; Criales-Hernández et al. 2008). Consequently, this might interfere with latitudinal and longitudinal distribution and dispersion patterns of many planktonic and nektonic species in areas where the OMZ is wide and strong (Judkins 1980; Morales et al. 1999; Escribano et al. 2000).

1.2.3.3 Political Boundaries and Deep Territorial Seas

All but two of the 35 countries of the Americas (namely, Bolivia and Paraguay) adjoin at least one of the three major oceans. Combined, in the Atlantic Ocean, the total area covered by national exclusive economic zones (EEZ), the territorial sea extending from the coastline to 200 nautical miles offshore (outlined in white in Figs. 1.3 and 1.4), covers about 14.6 M km^2, of which 10.5 M km^2 (72%) are deeper than 200 m. Comparatively, in the Pacific Ocean, the EEZs cover about 13.9 M km^2, of which 11.9 M km^2 (86%) are deeper than 200 m (Figs. 1.3 and 1.4; FMI 2018, 2019, GEBCO 2020). Higher proportion of deep sea in the Pacific is linked to the presence of several faraway offshore islands, e.g., Clarion, Malpelo, Galápagos, and Easter Islands.

The extent of the deep sea (>200 m depth) in these American countries is extraordinarily variable, depending on two factors: (1) the size of their EEZ and (2) the steepness of the continental margin. For the Americas, the largest EEZ is that of the USA (not including Hawaii), but this includes the Alaska portion which makes up nearly half (about 2.5 M km^2 of a total of 5.5 M km^2). In contrast, the smallest EEZ is found around San Cristobal and Nevis of the Lesser Antilles with 9,974 km^2; however, this has a proportionally very large deep-water extension, with 93.5% of its EEZ deeper than 200 m. The country with the highest proportion of water deeper than 200 m is Barbados, with 99.8%. By comparison, Argentina with the smallest portion of the EEZ deeper than 200 m (26.1%) is a country with a proportionally large continental shelf. The deepest point in the Americas is Milwaukee Deep, at 8376 m, of the Puerto Rico Trench off the Island of Puerto Rico (Stewart and Jamieson 2019). Given the significant proportion of deep-water habitat lying in the EEZ of countries of the Americas, national tasks of exploring and understanding the structure and functions of these large ecosystems remain colossal. Intensive studies of both the deep-water benthic and pelagic communities that are recognized as diverse and abundant should be considered, as well as monitoring in a permanent manner the environmental conditions prevailing in these ecosystems, particularly deploying deep-water observatories in key areas.

1.2.4 Marine Arthropoda

The phylum Arthropoda comprises insects, spiders, pycnogonids, scorpions, millipedes, and crustaceans. They all share segmented, chitinous exoskeletons and need to molt regularly to grow. Arthropods have a body form so versatile they have been compared to Swiss Army knives and are the most diverse metazoan form of life; with more than 1.3 million known species, they encompass nearly 80% of all known animals (Zhang 2013). However, the relationships among these groups remain hotly debated. Aria and Caron (2017) recently observed that retracing the evolutionary history of arthropods has been one of the greatest challenges in biology. The contributions herein primarily cover Crustacea, as the dominant marine arthropods, but

Pycnogonida (sea spiders) are also included as they are one of only two other relatively diverse groups of marine arthropods (the other being the marine mites, e.g., Bartsch 2006). The exclusively marine Pycnogonida are currently placed within the subphylum Chelicerata, which includes the terrestrial spiders and ticks, but the exact phylogenetic placement of pycnogonids remains unclear. Recent genomic-scale analysis of the chelicerates indicated that the marine forms (pycnogonids and the horseshoe crabs) form successive sister groups to terrestrial chelicerates, suggesting a single colonization of land within Chelicerata (Lozano-Fernandez et al. 2019b).

Progressing from the historical view, the Crustacea are now accepted to not be monophyletic inasmuch as insects (Hexapoda) are derived from within the crustaceans (Regier et al. 2010; Schwentner et al. 2017; Lozano-Fernandez et al. 2019a). The crustacean sister group to insects is still debated, but the consistent consensus based on major molecular and morphological datasets points to the remipedes, a small group of highly derived cave dwellers (Lozano-Fernandez et al. 2019a; Ahyong 2020) that were thought by some to be the most "primitive" crustacean form (Schram 1983). The crustaceans and insects are now variously grouped together as Pancrustacea, Tetraconata, or simply Crustacea sensu lato, in which the hexapods are understood as terrestrial crustaceans.

In this volume, the chapters addressing Crustacea are primarily focused on the class Malacostraca. This highly diverse class includes Decapoda (e.g., crabs, prawns, and lobsters), Peracarida (e.g., amphipods, isopods, and cumaceans), and Stomatopoda (mantis shrimps). Other chapters present aspects of the Copepoda, which are united with the Malacostraca in the superclass Multicrustacea (Schwentner et al. 2018; Lozano-Fernandez et al. 2019a). It comprises over 57,100 species, about 85% of all known Crustacea (~67,000) worldwide (Ahyong et al. 2011). Multicrustacea are primarily marine and can be found throughout the world's oceans and depths. They range in size from a fraction of a millimeter in length for the minute parasitic Tantulocarida Boxshall and Lincoln, 1983 to the Japanese spider crab, *Macrocheira kaempferi* (Temminck, 1836), with a leg span that can reach 3.8 m (Davie et al. 2015). Moreover, Multicrustacea can be described by a number of superlatives: the aforementioned Japanese spider crab is the largest living arthropod; the coconut crab, *Birgus latro* (Linnaeus, 1767) is the largest land invertebrate with a weight of up to 4 kg and a leg span of up to 1 m (Drew et al. 2010, Fig. 1.7 center); the deep-sea amphipod of the genus *Hirondellea* Chevreux, 1889 dominates the scavenging community at extreme depths in the world's trenches including Challenger Deep (Dahl 1959; Kobayashi et al. 2012; Lacey et al. 2016); planktonic copepods are the dominant mesozooplankton in the marine environment, comprising as much as 80% of its total biomass (Kiørboe 1998). With the diversity in form and number of marine arthropods, it is clear that distributions (vertical and horizontal), lifestyles, and abundances are not uniform throughout the deep sea but differ according to both geological and evolutionary histories (Wilson and Ahyong 2015). Deepwater decapods are most diverse and abundant in the bathyal zone, substantially giving way to peracarids at abyssal depths and to almost entirely relinquish to peracarids at hadal depths. Just as the habitats of the deep sea are not uniform,

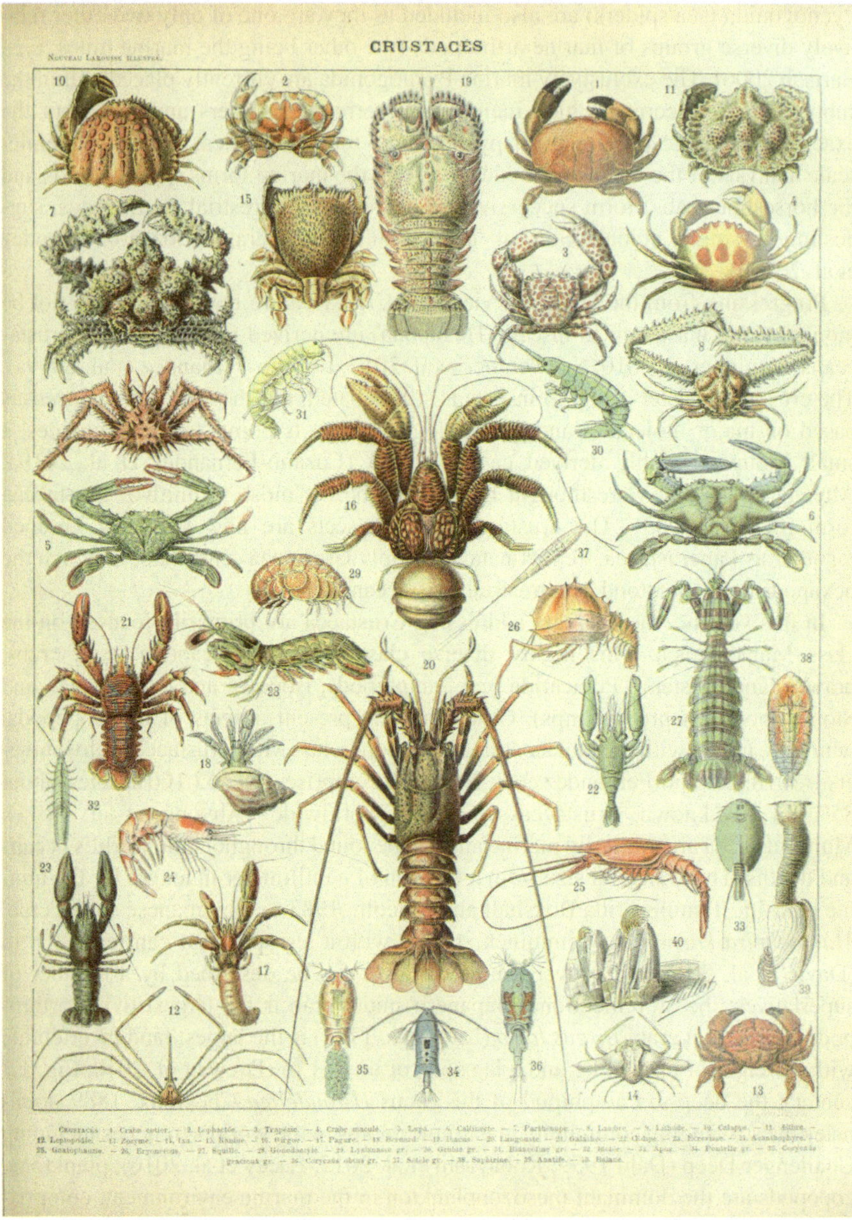

Fig. 1.7 The diversity of shape and form illustrated in the Nouveau Larousse Illustré Ocean Educational Chart by Adolphe Millot (1857–1921), showing examples of various decapods such as crabs, shrimps, and lobsters (1–26), stomatopods (27–28), amphipods and isopods (29–32), notostracan (33), copepods (34–38), and barnacles (39–40)

neither are the faunas, but nearly universally, the ecological significance of Multicrustacea in the marine realm cannot be overstated. They play a fundamental role in exporting, redistributing, and repackaging carbon and nutrients (e.g., Longhurst 1995; Smith et al. 2008; Jamieson et al. 2009), and many species are economically important.

As marine arthropods constitute a significant portion of the marine biodiversity in nearly all regions and across all depths, they present an excellent model for our understanding of large-scale patterns at global scales. Yet, large areas of the deep sea remain unknown, at the same time as the global climate is undeniably changing.

1.3 Conclusions

The deep ocean surrounding the North and South American continents hosts a wide diversity of arthropods, from microscopic peracarids and benthic copepods to some of the largest decapods such as king crabs. Contributions herein address a wide range of biological aspects related to deep sea crustaceans and pycnogonids, ranging from taxonomic inventories, phylogenetics, and biogeography to community analysis, ecology, reproduction, and fisheries (see Preface).

As in many other regions of the world, our knowledge of the composition and ecology of the deep-water arthropod communities of the Americas is still very limited. Most countries of the Americas, in particular in Central and South America, are yet to implement comprehensive research programs aimed at exploring the vast offshore areas extending off their coastlines. In many cases, limitations are directly linked to the high cost of managing offshore research. Nevertheless, it is noteworthy that despite very limited means, there has been a consistent effort in the last decades to explore the deep-water resources with a view to identify potentially lucrative deep-water fishing grounds or mineral deposits. While this effort is understandable, there is obviously a very high risk that these marine communities are adversely affected when we lack detailed knowledge of the composition and ecology. Returning to the earlier claim, we know more about the surface of the moon than the ocean floor – is it still that case? We certainly have more data, more images, and more samples from the deep sea than the surface of the Moon. However, the sheer scale and complexity of the deep sea and its living systems mean that even in the twenty-first century, the claim remains arguably true.

Unfortunately, it is becoming apparent that depth does not shield the deep ocean and its inhabitants from anthropogenic disturbance (Jamieson et al. 2017, 2019). The state of impact is symbolized by the recent description of *Eurythenes plasticus Weston, 2020*, a scavenging amphipod from hadal depth of the Mariana Trench named for the presence of microplastic in its gut (Weston et al. 2020). At a time where the global community is increasingly aware of the challenges and risks due to human disturbances and climate change, the immense effects of the deep ocean upon our climate remain a significant piece in the puzzle that is ill-understood (Bindoff et al. in press). Indications are that climate change might threaten the biodiversity of the deep sea proportionately more compared to shallow-water ecosystems (Brito-Morales et al. 2020). Let us remind ourselves that research and the regional and global initiatives to tackle these challenges, through technical advances and protection measures such as marine protected areas and sustainable management of fishing and mining, have to be considered with optimism for the future. A prerequisite for any of these initiatives is detailed knowledge of the fauna and communities in the deep sea. A challenge to all of these initiatives in the Americas is the significant knowledge gap that exists for many parts of the region. Contributions in the present volume are intended to fill some of these gaps.

Acknowledgments Our thanks to Pieter Provoost (OBIS Data Manager and IOC Software Engineer) for a complete OBIS database export and Arne Pallentin and Sally Watson (NIWA) for GIS mapping support. We are grateful to Britt Lonneville, Vlaams Instituut voor de Zee, Oostende, Belgium, for providing data related to the EEZ extension in the Americas. We are grateful to Yanhong Lu (Department of Ocean Science, Hong Kong University of Sciences and Technology) for generously providing original data of Fig. 1.6b. We also thank Ruben Escribano (Departamento de Oceanografía, Universidad de Concepción) and Thomas Mosch (Helmholtz-Zentrum für Ozeanforschung Kiel, GEOMAR) for their assistance with the elaboration of Fig. 1.6c. Our gratitude to Danièle Guinot (Muséum national d' Histoire naturelle, Paris) and Christine Carrau (Bibliothèque Théodore Carrau, Paris) for tracing the source of the Nouveau Larousse Illustration plate of crustaceans (Fig. 1.7). KS and RP were funded by a New Zealand Strategic Science Investment Fund to the National Institute of Water and Atmospheric Research (NIWA) Coasts and Oceans Research Programme 2 (2019/2020 SCI).

References

Ahyong ST (2020) Evolution and radiation of Crustacea. In: Thiel M, Poore GCB (eds) The natural history of the Crustacea: evolution and biogeography, vol 8. Oxford University Press, New York, pp 53–79

Ahyong ST, Lowry JK, Alonso M, Bamber RN, Boxshall GA, Castro P, Gerken S, Karaman GS, Goy JW, Jones DS, Meland K, Rogers DC, Svavarsson J (2011) Subphylum Crustacea Brünnich, 1772. In: Zhang Z-Q (ed) Animal biodiversity: An outline of higher-level classification and survey of taxonomic richness. Zootaxa 3148:165–191

Andrews KR, Norton EL, Fernandez-Silva I, Portner E, Goetze E (2014) Multilocus evidence for globally distributed cryptic species and distinct populations across ocean gyres in a mesopelagic copepod. Mol Ecol 23:5462–5479. https://doi.org/10.1111/mec.12950

Aria C, Caron J-B (2017) Burgess Shale fossils illustrate the origin of the mandibulate body plan. Nature 545(7652):89–92. https://doi.org/10.1038/nature22080

Álvarez-León R (2002) Aprovechamiento de los crustáceos de importancia comercial del Pacífico colombiano. In: Hendrickx ME (ed) Contributions to the Study of East Pacific Crustaceans 1. [Contribuciones al Estudio de los Crustáceos del Pacífico Este 1]. Instituto de Ciencias del Mar y Limnología, UNAM, Mexico City, pp 1–35

Bacon CD, Silvestro D, Jaramillo C, Smith BT, Chakrabarty P, Antonelli A (2015) Biological evidence supports an early and complex emergence of the Isthmus of Panama. Proc Nat Acad Sci 112(19):6110–6115. https://doi.org/10.1073/pnas.1423853112

Bartsch I (2006) Halacaroidea (Acari): A guide to marine genera. Organ Divers Evol:1–104

Beliaev GM (1989) Deep sea ocean trenches and their fauna. Scripps Institution of Oceanography, San Diego

Bindoff NL, Cheung WWL, Kairo JG, Arístegui J, Guinder VA, Hallberg R, Hilmi N, Jiao N, Karim MS, Levin L, O'Donoghue S, Purca Cuicapusa SR, Rinkevich B, Suga T, Tagliabue A, Williamson P (in press) Changing ocean, marine ecosystems, and dependent communities. In: Pörtner H-O, Roberts DC, Masson-Delmotte V, Zhai P, Tignor M, Poloczanska E, Mintenbeck K, Alegría A, Nicolai M, Okem A, Petzold J, Rama B, Weyer NM (eds) IPCC special report on the ocean and cryosphere in a changing climate

Boschi EE (2000) Species of decapod crustaceans and their distribution in the American marine zoogeographic provinces. Rev Invest Desarr Pesq 13:1–136

Boxshall GA, Lincoln RJ (1983) Tantulocarida, a New Class of Crustacea Ectoparasitic on other Crustaceans. J Crust Biol (1):1–16. https://doi.org/10.2307/1547849

Brierley AS (2014) Diel vertical migration. Current Biol 24(22):R1074–R1076

Briggs JC (2003) Marine centres of origin as evolutionary engines. J Biogeogr 30:1–18

Brito-Morales I, Schoeman DS, Molinos JG, Burrows MT, Klein CJ, Arafeh-Dalmau N, Kaschner K, Garilao C, Kesner-Reyes K, Richardson AJ (2020) Climate velocity reveals increasing exposure of deep-ocean biodiversity to future warming. Nat Clim Chang 10:576–581. https://doi.org/10.1038/s41558-020-0773-5

Cable ML, Neveu M, Hsu H-W, Hoehler TM (2020) Enceladus. In: Meadows V, Des Marais DJ, Arney G, Schmidt B (eds) Planetary astrobiology. University of Arizona Press, Tuscon, pp 217–246

Charette MA, Smith WHF (2010) The volume of Earth's ocean. Oceanography 23(2):112–114. https://doi.org/10.5670/oceanog.2010.51

Chevreux E (1889) Amphipodes nouveaux provenant des campagnes de "l'Hirondelle" 1887–1888. Bull Soc Zool France 14:283–289

Clark MR, Consalvey M, Rowden AA (2016) Biological Sampling in the Deep Sea. Wiley Blackwell, West Sussex

Cockburn H (2019) Life on Earth probably originated in deep-sea vents and aliens could be growing the same way now, scientists suggest. In: The Independent Digital News & Media Limited. https://www.independent.co.uk/news/science/origin-life-deep-sea-theory-hydrothermal-vents-aliens-planets-oceans-a9185991.html. Accessed 20 May 2020

Cohen AN (2006) Species introductions and the Panama Canal. In: Gollasch S, Galil BS, Cohen AN (eds) Bridging divides: Maritime Canals as invasion corridors. Springer, Dordrecht, p 127e206

Cornils A, Held C (2014) Evidence of cryptic and pseudocryptic speciation in the *Paracalanus parvus* species complex (Crustacea, Copepoda, Calanoida). Front Zool 11:1–17. https://doi.org/10.1186/1742-9994-11-19

Cornils A, Wend-Heckmann B, Held C (2017) Global phylogeography of *Oithona similis* s.l. (Crustacea, Copepoda, Oithonidae) – a cosmopolitan plankton species or a complex of cryptic lineages? Mol Phylog Evol 107:473–485. https://doi.org/10.1016/j.ympev.2016.12.019

Criales-Hernández MI, Schwamborn R, Graco M, Ayón P, Hirche HJ, Wolff M (2008) Zooplankton vertical distribution and migration off Central Peru in relation to the oxygen minimum layer. Helgol Mar Res 62(1):85–100

Dahl E (1959) Amphipoda from depths exceeding 6000 meters. Galathea Rep 1:211–241

Davie PJF, Guinot D, Ng PKL (2015) Anatomy and functional morphology of Brachyura. In: Castro P, Davie PJF, Guinot D, Schram FR, Von Vaupel Klein C (eds) Treatise on zoology – anatomy, taxonomy, biology – the Crustacea, complementary to the volumes translated from the French of the Traité de Zoologie, vol 9(C) (I), Decapoda: Brachyura (Part 1). pp 11–163

Díaz Merlano JM, Vieira Betancourt C, Melo Saldarriaga G (2011) Diagnóstico de las principales pesquerías del Pacífico colombiano. Fundación Mar Viva, Bogotá, Colombia

Diaz RJ, Rosenberg R (1995) Marine benthic hypoxia: A review of its ecological effects and the behavioural response of benthic macrofauna. Oceanogr Mar Biol An Rev 33:245–303

Dodd MS, Papineau D, Grenne T, Slack JF, Rittner M, Pirajno F, O'Neil J, Little CTS (2017) Evidence for early life in Earth's oldest hydrothermal vent precipitates. Nature 543(7643):60–64. https://doi.org/10.1038/nature21377

Downing AB, Wallace GT, Yancey PH (2018) Organic osmolytes of amphipods from littoral to hadal zones: Increases with depth in trimethylamine N-oxide, scyllo-inositol and other potential pressure counteractants. Deep-Sea Res Part I Oceanogr Res Pap 138:1–10. https://doi.org/10.1016/j.dsr.2018.05.008

Drew MM, Harzsch S, Erland S, Hansson BS (2010) A review of the biology and ecology of the Robber Crab, *Birgus latro* (Linnaeus, 1767) (Anomura: Coenobitidae). Zool Anz 249:45–67. https://doi.org/10.1016/j.jcz.2010.03.001

Escribano R, Marín V, Irribarren C (2000) Distribution of *Euphausia mucronata* at the upwelling area of Peninsula Mejillones, northern Chile: the influence of the oxygen minimum layer. Sci Mar 64(1):69–77

Escribano R, Hidalgo P, Krautz C (2009) Zooplankton associated with the oxygen minimum zone system in the northern upwelling region of Chile during March 2000. Deep Sea Res Part II Top Stud Oceanogr 56(16):1083–1094

Feldmann RM, Schweitzer CE (2006) Paleobiogeography of Southern Hemisphere decapod Crustacea. J Paleontol 80(1):83–103. https://doi.org/10.1666/0022-3360(2006)080[0083:POSHDC]2.0.CO;2

FMI (2018) Flanders Marine Institute. IHO Sea Areas, version 3. Available online at https://www.marineregions.org/ doi:https://doi.org/10.14284/323

FMI (2019) Flanders Marine Institute. Maritime boundaries geodatabase: Maritime boundaries and exclusive economic zones (200NM), version 11. Available online at https://www.marineregions.org/. https://doi.org/10.14284/386

Gage JD, Tyler PA (1991) Deep-sea biology: a natural history of organisms at the deep-sea floor. Cambridge University Press, Cambridge

Garcia M (2020) International Space Station facts and figures. https://www.nasa.gov/feature/facts-and-figures. Accessed 20 May 2020

Garlitska L, Neretina T, Schepetov D, Mugue N, De Troch M, Baguley JG, Azovsky A (2012) Cryptic diversity of the 'cosmopolitan' harpacticoid copepod *Nannopus palustris*: genetic and morphological evidence. Mol Ecol 21(21):5336–5347. https://doi.org/10.1111/mec.12016

GEBCO (2020) GEBCO Compilation Group 2020 Grid. https://doi.org/10.5285/a29c5465-b138-234d-e053-6c86abc040b9

Gooday A, Jorissen F, Levin L, Middelburg J, Naqvi S, Rabalais N, Scranton M, Zhang J (2009) Historical records of coastal eutrophication-induced hypoxia. Biogeosciences 6:1707–1745

Gregersen E (2010) The inner solar system: the Sun, Mercury, Venus, Earth, and Mars. The Rosen Publishing Group, Inc, New York

Halsband C, Ahyong ST, Brandt A, Kosobokova K, Ward P, Goodall-Copestake WP, Macpherson E (2020) Chapter 6. Biogeography of the oceans. In: Thiel M, Poore GCB (eds) The natural history of the Crustacea: evolution and biogeography of the Crustacea. Oxford University Press, New York, pp 121–154

Havermans C, Sonet G, d'Udekem d'Acoz C, Nagy ZT, Martin P, Brix S, Riehl T, Agrawal S, Held C (2013) Genetic and morphological divergences in the cosmopolitan deep-sea amphipod *Eurythenes gryllus* reveal a diverse abyss and a bipolar species. PLoS One 8(9):e74218. https://doi.org/10.1371/journal.pone.0074218

Helly J, Levin L (2004) Global distribution of naturally occurring marine hypoxia on continental margins. Deep-Sea Res Part I Oceanogr Res Pap 51:1159–1168

Hendrickx M (1995) Introducción. In: Fischer W, Krupp F, Schneider W, Sommer C, Carpenter K, Niem V (eds) Guía FAO para la identificación de especies para los fines de la pesca. Pacífico centro-oriental Vol. I, Plantas e Invertebrados. FAO, Roma, pp 1–7

Hendrickx ME (this volume) The deepwater Colossendeis tenera Hilton, 1943 (Pycnogonida, Pantopoda, Colossendeidae) Off Western Mexico. In: Hendrickx ME (ed) Deep-sea pycnogonids and crustaceans of the Americas. Springer, Cham

Hendrickx M, Serrano D (2010) Impacto de la zona de mínimo de oxígeno sobre los corredores pesqueros en el Pacífico mexicano. Interciencia 35:12–18

Hendrickx M, Serrano D (2014) Effects of the oxygen minimum zone on squat lobsters distribution in the Gulf of California, Mexico. Central Eur J Biol 9:92–103. https://doi.org/10.2478/s11535-013-0165-6

Herring P (2002) The biology of the deep ocean. Biology of habitats. Oxford University Press, New York

Hidalgo P, Escribano R, Morales C (2005) Ontogenetic vertical distribution and diel migration of the copepod *Eucalanus inermis* in the oxygen minimum zone off northern Chile (20–21 S). J Plank Res 27(6):519–529

Iudicone D, Speich S, Madec G, Blanke B (2008) The Global Conveyor Belt from a Southern Ocean Perspective. J Phys Oceanogr 38(7):1401–1425. https://doi.org/10.1175/2007jpo3525.1

Jamieson A (2015) The Hadal Zone: life in the Deepest Oceans. Cambridge University Press, Cambridge

Jamieson AJ (2020) The Five Deeps Expeditions and an update of full ocean depth exploration and explorers. Mar Technol Soc J 54(1):6–12. https://doi.org/10.4031/MTSJ.54.1.1

Jamieson AJ, Fujii T, Mayor DM, Solan M, Priede IG (2009) Hadal trenches: the ecology of the deepest places on Earth. Trends Ecol Evol 25(3):190–197. https://doi.org/10.1016/j.tree.2009.09.009

Jamieson AJ, Malkocs T, Piertney SB, Fujii T, Zhang Z-Q (2017) Bioaccumulation of persistent organic pollutants in the deepest ocean fauna. Nat Ecol Evol 1(3):51. https://doi.org/10.1038/s41559-016-0051

Jamieson AJ, Brooks LSR, Reid WDK, Piertney SB, Narayanaswamy BE, Linley TD (2019) Microplastics and synthetic particles ingested by deep-sea amphipods in six of the deepest marine ecosystems on Earth. R Soc open sci 6(2):180667. https://doi.org/10.1098/rsos.180667

Jordan SF, Rammu H, Zheludev IN, Hartley AM, Maréchal A, Lane N (2019) Promotion of protocell self-assembly from mixed amphiphiles at the origin of life. Nat Ecol Evol 3(12):1705–1714. https://doi.org/10.1038/s41559-019-1015-y

Judkins D (1980) Vertical distribution of zooplankton in relation to the oxygen minimum off Peru. Deep-Sea Res Part I Oceanogr Res Pap 27:475–487

Kameya A, Castillo R, Escudero L, Tello E, Blaskovic V, Cordova J, Hooker Y, Gutiérrez M, Mayor S (1997) Localización, distribución y concentración de langostinos rojos de profundidad. Bol IMARPE, Spec Pub (1997):1–47

Kiorboe T (1998) Population regulation and role of mesozooplankton in shaping marine pelagic food webs. Hydrobiologia 363:13–27. https://doi.org/10.1023/a:1003173721751

Kobayashi H, Hatada Y, Tsubouchi T, Nagahama T, Takami H (2012) The hadal amphipod *Hirondellea gigas* possessing a unique cellulase for digesting wooden debris buried in the deepest seafloor. Plos One 7:e42727

Kuhlbrodt T, Griesel A, Montoya M, Levermann A, Hofmann M, Rahmstorf S (2007) On the driving processes of the Atlantic meridional overturning circulation. Rev Geophys 45(2). https://doi.org/10.1029/2004rg000166

Lacey NC, Rowden AA, Clark MR, Kilgallen NM, Mayor DJ, Linley TD, Jamieson AJ (2016) Community structure and diversity of scavenging amphipods from bathyal to hadal depths in three South Pacific trenches. Deep Sea Res Part I Oceanogr Res Pap 111:121–137. https://doi.org/10.1016/j.dsr.2016.02.014

Levin L, Gage J, Martin C, Lamont P (2000) Macrobenthic community structure within and beneath the oxygen minimum zone, NW Arabian Sea. Deep-sea Res Part II Top Stud Oceanogr 47:189–226

Levin LA, Sibuet M, Gooday AJ, Smith CR, Vanreusel A (2010) The roles of habitat heterogeneity in generating and maintaining biodiversity on continental margins: an introduction. Mar Ecol 31(1):1–5

Lobner P (2020) 60th Anniversary of the First Visit by Humans to the Deepest Point in the Ocean – the Challenger Deep. https://lynceans.org/all-posts/60th-anniversary-of-the-first-visit-by-humans-to-the-deepest-point-in-the-ocean-the-challenger-deep/. Accessed 20 May 2020

Longhurst A (1995) Seasonal cycles of pelagic production and consumption. Progress Oceanogr 36(2):77–167. https://doi.org/10.1016/0079-6611(95)00015-1

Longhurst A (2007) The ecological geography of the sea, 2nd edn. Elsevier Inc, Amsterdam/Heidelberg. https://doi.org/10.1016/B978-0-12-455521-1.X5000-1

Lonsdale P (1977) Clustering of suspension-feeding macrobenthos near abyssal hydrothermal vents at oceanic spreading centers. Deep Sea Res 24(9):857–863. https://doi.org/10.1016/0146-6291(77)90478-7

Lozano-Fernandez J, Giacomelli M, Fleming JF, Chen A, Vinther J, Thomsen PF, Glenner H, Palero F, Legg D, Iliffe T, Pisani D, Olesen J (2019a) Pancrustacean evolution illuminated by taxon-rich genomic-scale data sets with an expanded Remipede Sampling. Genome Biol Evol 11(8):2055–2070. https://doi.org/10.1093/gbe/evz097

Lozano-Fernandez J, Tanner AR, Giacomelli M, Carton R, Vinther J, Edgecombe GD, Pisani D (2019b) Increasing species sampling in chelicerate genomic-scale datasets provides support for monophyly of Acari and Arachnida. Nature Comm 10(1):2295. https://doi.org/10.1038/s41467-019-10244-7

Lu Y, Xia X, Cheung S, Jing H, Liu H (2019) Differential distribution and determinants of ammonia oxidizing Archaea sublineages in the Oxygen Minimum Zone off Costa Rica. Microorganisms 7(10):2–14. (453)

Mayer L, Jakobsson M, Allen G, Dorschel B, Falconer R, Ferrini V, Lamarche G, Snaith H, Weatherall P (2018) The Nippon Foundation—GEBCO Seabed 2030 Project: the quest to see the World's Oceans Completely Mapped by 2030. Geosciences 8(2):63

Méndez M (1982) Crustáceos comerciales y otras especies comunes en el litoral peruano. Bol Lima 20:1–20

Mendo J, Wosnitza-Mendo C (2014) Reconstruction of total marine fisheries catches for Peru: 1950–2010. Univ British Columbia Working Pap Ser 2014-21, pp 1–23

Morales C, Hormazábal S, Blanco J (1999) Interannual variability in the mesoscale distribution of the depth of the upper boundary of the oxygen minimum layer off northern Chile (18–24S): Implications for the pelagic system and biogeochemical cycling. J Mar Res 57(6):909–932

Mosch T, Sommer S, Dengler M, Noffke A, Bohlen L, Pfannkuche O, Liebetrau V, Wallmann K (2012) Factors influencing the distribution of epibenthic megafauna across the Peruvian oxygen minimum zone. Deep Sea Res Part I Oceanogr Res Pap 68:123–135

Paganini L, Villanueva GL, Roth L, Mandell AM, Hurford TA, Retherford KD, Mumma MJ (2020) A measurement of water vapour amid a largely quiescent environment on Europa. Nat Astron 4:266–272. https://doi.org/10.1038/s41550-019-0933-6

Papiol V, Hendrickx ME, Serrano D (2016) Effects of latitudinal changes in the oxygen minimum zone of the northeast Pacific on the distribution of bathyal benthic decapod crustaceans. Deep-sea Res Part II Top Stud Oceanogr 137:113–130. https://doi.org/10.3897/BDJ.4.e8661

Perrone FM, Della Croce N, Dell'Anno A (2003) Biochemical composition and trophic strategies of the amphipod *Eurythenes gryllus* at hadal depths (Atacama Trench, South Pacific). Chem Ecol 9(6):441–449. https://doi.org/10.1080/02757540310000095723

Quiroga E, Sellanes J, Arntz W, Gerdes D, Gallardo V, Hebbeln D (2009) Benthic megafaunal and demersal fish assemblages on the Chilean continental margin: the influence of the oxygen minimum zone on bathymetric distribution. Deep-sea Res Part II Top Stud Oceanogr 56(16):1112–1123

Rabalais NN, Turner RE (2001) Hypoxia in the northern Gulf of Mexico: Description, causes and changes. In: Rabalais NN, Turner RE (eds) Coastal hypoxia: consequences for living resources and ecosystems, coastal and estuarine studies, vol 58. American Geophysical Union, Washington, DC, pp 1–36

Regier JC, Shultz JW, Zwick A, Hussey A, Ball B, Wetzer R, Martin JW, Cunningham CW (2010) Arthropod relationships revealed by phylogenomic analysis of nuclear protein-coding sequences. Nature 463(7284):1079–1098. https://doi.org/10.1038/nature08742

Reyes C, Rodríguez C, León R (2014) Análisis bioeconómico de las pesquerías de camarón blanco del Pacífico colombiano. Rev Luna Azul 39:153–164

Reygondeau G, Dunn D (2019) Pelagic biogeography. In: Cochran JK, Bokuniewicz HJ, Yager PL (eds) Encyclopedia of ocean sciences, 3rd edn. Academic, Oxford, pp 588–598. https://doi.org/10.1016/B978-0-12-409548-9.11633-1

Rojas-Morales J, Campos-Montero J, Segura-Hernández A, Mug-Villanueva M, Campos R, Rodríguez-Solano O (2000) Shark fisheries in Central America: A review and update. Pesquería de tiburón en América Central: Una revisión y actualización. Uniciencia 17:49–56

Ros M, Ashton GV, Lacerda MB, Carlton JT, Vázquez-Luis M, Guerra-García JM, Ruiz GM (2014) The Panama Canal and the transoceanic dispersal of marine invertebrates: Evaluation of the introduced amphipod *Paracaprella pusilla* Mayer, 1890 in the Pacific Ocean. Mar Environ Res 99:204–211. https://doi.org/10.1016/j.marenvres.2014.07.001

Saltzman J, Wishner K (1997) Zooplankton ecology in the eastern tropical Pacific oxygen minimum zone above a seamount: 2. Vertical distribution of copepods. Deep-Sea Res Part I Oceanogr Res Pap 44:931–954

Sandwell DT, Müller RD, Smith WHF, Garcia E, Francis R (2014) New global marine gravity model from CryoSat-2 and Jason-1 reveals buried tectonic structure. Science 346(6205):65–67. https://doi.org/10.1126/science.1258213

Schmidt B (2020) The astrobiology of Europa and the Jovian System. In: Planetary astrobiology, pp 185–216

Schnabel KE, Cabezas Padilla P, McCallum A, Macpherson E, Ahyong S, Baba K (2011) Worldwide distribution patterns of squat lobsters. In: Poore GCB, Ahyong ST, Taylor J (eds) The biology of squat lobsters. CSIRO Publishing/CRC Press, Melbourne/Boca Raton, pp 149–182

Schram FR (1983) Remipedia and crustacean phylogeny. In: Schram FR (ed) Crustacean phylogeny, Crustacean Issues Vol 1. A.A. Balkema, Rotterdam, pp 23–28

Schwentner M, Combosch DJ, Nelson JP, Giribet G (2017) A phylogenomic solution to the origin of insects by resolving crustacean-hexapod relationships. Current Biol 27(12):1818–1824.e5. https://doi.org/10.1016/j.cub.2017.05.040

Schwentner M, Richter S, Rogers DC, Giribet G (2018) Tetraconatan phylogeny with special focus on Malacostraca and Branchiopoda: highlighting the strength of taxon-specific matrices in phylogenomics. Proc R Soc B: Biol Sci 285(1885):20181524. https://doi.org/10.1098/rspb.2018.1524

Serrano D (2012) La zona de mínimo oxígeno en el Pacífico mexicano. In: Zamorano P, Hendrickx M, Caso M (eds) Biodiversidad y comunidades del talud continental del Pacífico mexicano. Secretaría del Medio Ambiente y Recursos Naturales (SEMARNAT), Instituto Nacional de Ecología (INE), Mexico, pp 105–119

Smith CR, De Leo FC, Bernadino AG, Sweetman AK, Martinex Arbizu P (2008) Abyssal food limitation, ecosystem structure and climate change. Trends Ecol Evol 23(9):518–528. https://doi.org/10.1016/j.tree.2008.05.002

Smith KLJ, Sherman AD, McGill PR, Henthorn RG, Ferreira J, Huffard CL (2017) Evolution of monitoring an abyssal time- series station in the northeast Pacific over 28 years. Oceanogr 30(4):72–81. https://doi.org/10.5670/oceanog.2017.425

Soler-Membrives A, Lucena RA, Company JB, Rotllant G (this volume) Deep-sea Pycnogonids from Uruguay: every deep cruise adds valuable information. In: Hendrickx ME (ed) Deep-sea pycnogonids and crustaceans of the Americas. Springer, Cham

Spalding MD, Fox HE, Allen GR, Davidson N, Ferdaña ZA, Finlayson M, Halpern BS, Jorge MA, Lombana A, Lourie SA, Martin KD, McManus E, Molnar J, Recchia CA, Robertson J (2007) Marine ecoregions of the world: a bioregionalization of coastal and shelf areas. BioScience 57(7):573–583. https://doi.org/10.1641/b570707

Staples DA (2007) A new species of *Colossendeis* (Pycnogonida: Colossendeidae) together with records from Australian and New Zealand waters. Memoirs of Museum Victoria 64:79–94

Stewart HA, Jamieson AJ (2018) Habitat heterogeneity of hadal trenches: Considerations and implications for future studies. Prog Oceanogr 161:47–65. https://doi.org/10.1016/j.pocean.2018.01.007

Stewart HA, Jamieson AJ (2019) The five deeps: The location and depth of the deepest place in each of the world's oceans. Earth-Sci Rev:102896. https://doi.org/10.1016/j.earscirev.2019.102896

Stramma L, Schmidtko S, Levin L, Johnson G (2010) Ocean oxygen minima expansions and their biological impacts. Deep-Sea Res Part I Oceanogr Res Pap 57(4):587–595

Sutton TT, Clark MR, Dunn DC, Halpin PN, Rogers AD, Guinotte J, Bograd SJ, Angel MV, Perez JAA, Wishner K, Haedrich RL, Lindsay DJ, Drazen JC, Vereshchaka A, Piatkowski U, Morato T, Błachowiak-Samołyk K, Robison BH, Gjerde KM, Pierrot-Bults A, Bernal P, Reygondeau G, Heino M (2017) A global biogeographic classification of the mesopelagic zone. Deep Sea Res Part I Oceanogr Res Pap 126:85–102. https://doi.org/10.1016/j.dsr.2017.05.006

Tabash Blanco F (2007) Explotación de la pesquería de arrastre de camarón durante el período 1991–1999 en el Golfo de Nicoya, Costa Rica. Revista de biologia tropical 55(1):207–218

Temminck CJ (1836) Coup-d'oeil sur la faune des iles de la Sonde et de l'empire du Japon: discours préliminaire, destiné à servir d'introduction à la Faune du Japon. https://doi.org/10.5962/bhl.title.119899

UNESCO (2009) Global Open Oceans and Deep Seabed (GOODS) – biogeographic classification. vol IOC Technical Series. UNESCO-IOC, Paris

von Linné C (1767) Systema naturae: per regna tria natura, secundum classes, ordines, genera, species, cum characteribus, differentiis, synonymis, locis 12.2 edn. Laurentii Salvii, Holmiae [Stockholm]. https://doi.org/10.5962/bhl.title.68927

Vieira C, Diaz MC, Díaz J (2016) Ordenamiento y manejo pesquero en la costa norte del Pacífico colombiano. In: La pesca artesanal en la costa norte del Pacífico colombiano: un horizonte ambivalente. Fundación MarViva, Bogotá, pp 45–57

Villalobos-Rojas F, Herrera-Correal J, Garita-Alvarado C, Clarke T, Beita-Jiménez A (2014) Actividades pesqueras dependientes de la ictiofauna en el Pacífico Norte de Costa Rica. Rev Biol Trop 62(4):119–137

Watling L, Guinotte JM, Clark MR, Smith C (2013) A proposed biogeography of the deep ocean floor. Prog Oceanogr 111:91–112

Webb TJ, Vanden Berghe E, O'Dor R (2010) Biodiversity's big wet secret: the global distribution of marine biological records reveals chronic under-exploration of the deep pelagic ocean. Plos One 5(8):e10223. https://doi.org/10.1371/journal.pone.0010223

Weston JNJ, Carrillo-Barragan P, Linley TD, Reid WDK, Jamieson AJ (2020) New species of *Eurythenes* from hadal depths of the Mariana Trench, Pacific Ocean (Crustacea: Amphipoda). Zootaxa 4748(1):163–181. https://doi.org/10.11646/zootaxa.4748.1.9

Wilson GDF, Ahyong ST (2015) Lifestyles of the species-rich and fabulous: the deep-sea crustaceans. In: Watling L, Thiel M (eds) The natural history of the Crustacea: lifestyles and feeding biology. Oxford University Press, New York, pp 279–298

Zeigler RA, Mosie AB, Kent JJ, Krysher CH, Watts LA, McCubbin FM (2019). The Apollo sample collection: 50 years of solar system insight. Presented at the 50th Lunar and Planetary Science Conference, The Woodlands, Texas

Zhang Z-Q (2013) Animal biodiversity: an update of classification and diversity in 2013. Zootaxa 3703(1):005–011

Chapter 2
Deep-Sea Pycnogonids from Uruguay: Every Deep Cruise Adds Valuable Information

A. Soler-Membrives, R. A. Lucena, J. B. Company, and G. Rotllant

Abstract We aim to study the sea spider specimens collected from a deep cruise in the economic exploitation zone of Uruguay from 250 to 3800 m with the Marianne fishing vessel. We gather the information from deep-sea Uruguayan pycnogonids from literature together with the ones reported in a recent checklist of sea spiders present in Uruguayan waters published by Scarabino et al. (2019) and including some biogeographic considerations. This new cruise includes two new species for Uruguayan waters and a great enlarging new bathymetric distribution of an already identified species in the area.

Keywords Pycnognida · Sea spiders · Deep-sea · Uruguay · Southwest Atlantic

2.1 Introduction

Pycnogonids or sea spiders are one of the most intriguing groups of arthropods. A resurgence in research interest in different aspects of pycnogonid biology has been seen in recent years. Pycnogonids are almost exclusively free-living marine invertebrates (Bamber et al. 2020). They are considered relatively rare, although they can be found from the poles to the tropics and from the littoral zone to the deepest sea (Arnaud and Bamber 1987). They range in size from littoral and interstitial species with leg spans of a few mm up to the large, deep-sea colossendeids, with leg spans

A. Soler-Membrives (✉)
Unitat de Zoologia, Universitat Autònoma de Barcelona, Bellaterra, Barcelona, Spain
e-mail: Anna.Soler@uab.cat

R. A. Lucena
Laboratório de Invertebrados Paulo Young, Departamento de Sistemática e Ecologia, Universidade Federal da Paraíba, João Pessoa, Paraíba, Brazil

J. B. Company · G. Rotllant
Institut de Ciències del Mar (CSIC), Barcelona, Spain

M. E. Hendrickx (ed.), *Deep-Sea Pycnogonids and Crustaceans of the Americas*, https://doi.org/10.1007/978-3-030-58410-8_2

up to 750 mm. A pycnogonid world species checklist is found in PycnoBase (www.marinespecies.org/pycnobase/) (Bamber et al. 2020), and their taxonomic classification was recently phylogenetically tested using multiple data classes (Ballesteros et al. 2020).

In general, species are epibenthic, although some are known to be bathypelagic being commensal or parasitic of other bathypelagic marine species (Arnaud and Bamber 1987). Pycnogonids are commonly considered as brooders with no planktonic dispersal stage, as fertilized eggs and sometimes post-larval stages are usually attached to the male (Arnaud and Bamber 1987). For this reason, a limited dispersal capability is assumed for this. In shallow continental where specific environmental constraints may play an important rate, the restricted dispersal may conduce to high speciation rates and the likely presence of cryptic species. However, deep-sea species seem to have broader distribution patterns thanks to deep currents, and many species are cosmopolitan but limited to deep-sea waters (Bamber and Thurston 1995).

Deep-sea pycnogonids are usually found dispersed in both rocky and soft-bottom habitats, although large aggregations of pycnogonids have been observed at some deep-sea habitats, such as hydrothermal vents (Brescia and Tunnicliffe 1998) or whale falls (Braby et al. 2009). The northeastern Atlantic Ocean has been well reported by severally deep-sea expeditions such as Stock (1978a, b, 1987), Bamber (1983), Bamber and Thurston (1993, 1995), Raiskii and Turpaeva (2006), and Turpaeva and Raiskii (2014).

Instead, the South Atlantic Ocean has been much neglected. The only studies on the Walvis Basin, offshore of Namibia, are those from Stock (1981) and Child (1982), and some deep-sea Pycnogonida from South Africa are recorded by Barnard (1954) and Stock (1963). The main pycnogonid study in the southeastern Pacific is the one by Araya (2016) along the Chilean coast.

The pycnogonid fauna from southwestern Atlantic has been sparsely studied, mainly by Child (1982, 1997) on the Argentina slope and basin and Uruguayan and Brazilian waters. Most of the knowledge of Uruguayan sea spiders available up to date is thanks to old literature studying specimens from the abovementioned foreign oceanographic expeditions in the coasts of Argentina, Uruguay, and Brazil (southwest Atlantic). National cruises in Uruguay were scarce until relatively recently (Scarabino 2006), but interest in Pycnogonida in these waters is growing up. Recently, Scarabino et al. (2019) published the basis of the Uruguayan sea spiders including a checklist of the species reported to date and some biogeographic considerations.

This is the first study on the deep continental margin of the economic exploitation zone (EEZ) of Uruguay, adding 10 new sampling stations to the 16 gathered up to date to the pycnogonid deep-sea community.

2.2 Materials and Methods

The goal of the cruise was to sample the deep megafauna community (between 250 and 3800 m) in the economic exploitation zone (EEZ) of Uruguay, from the mouth of the Río de la Plata to the border with Brazil, in three transects: one south (S) at the mouth of the river de la Plata, one central (C), and one north (N) near the Brazilian border. Trawling was done with a three-net 2.5 × 1.2 m Agassiz trawl (AGT) at 2 kn. The cruise was performed during April and May 2016 with a total of 26 trawls. From these, in ten trawls there were specimens belonging to Pycnogonida (Fig. 2.1, Table 2.1).

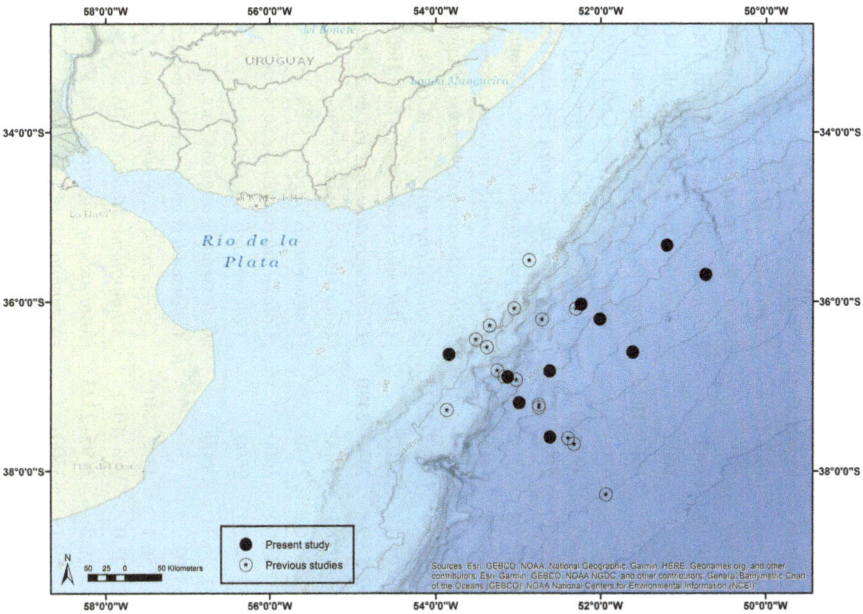

Fig. 2.1 Position of the stations sampled during the present cruise (black circles) in the exclusive economic zone (EEZ) of Uruguay and other stations were Pycnogonida have been recorded previously in literature (empty circles with star)

Table 2.1 Updated list of species of deep-sea Pycnogonida in Uruguayan waters, including localities, depth range, number of specimens, and sources

Date	Trawl number	Station	Lat (S)	Long (W)	Depth range	Species	N	References
3/11/1971	RVA-60	237	36°32.6	53°23.0	993–1011	*Ammothea longispina* Gordon, 1932	7	Child (1982)
7/13/1964	–	–	35°31.0	52°52.0	100	*Ammothea tetrapoda* Gordon, 1932	1	Lucena et al. (2019b)
5/3/2016	PS, P26	13	37°11.80	52°59.19	3300	*Ammothea spinosa* (Hodgson, 1907)	5	Present study
4/22/2016	PS, P7	56	36°36.00	51°36.71	3600	*Ammothea spinosa* (Hodgson, 1907)	4	Present study
3/11/1971	RVA-60	239A	36°49.0	53°15.4	1661–1679	*Ammothea spinosa* (Hodgson, 1907)	6	Child (1982)
2/4/1962	RVV-18	9	36°17	53°21	676	*Anoplodactylus vemae* Child, 1982	3	Child (1982)
3/11/1971	RVA-60	237	36°32.6	53°23.0	993–1011	*Ascorhynchus cuculus* Fry and Hedgpeth, 1969	1	Child (1982)
3/12/1971	RVA-60	240	36°53.4	53°10.2	2195–2323	*Ascorhynchus cuculus* Fry and Hedgpeth, 1969	1	Child (1982)
5/3/2016	PS, P26	13	37°11.80	52°59.19	3300	*Bathypallenopsis calcanea* (Stephensen, 1933)	1	Present study
4/21/2016	PS, P6	54	36°12.51	52°00.48	3100	*Bathypallenopsis calcanea* (Stephensen, 1933)	1	Present study
5/2/2016	PS, P24	78	35°20.34	51°11.72	2800	*Bathypallenopsis calcanea* (Stephensen, 1933)	1	Present study
3/12/1971	RVA-60	240	36°53.4	53°10.2	2195–2323	*Callipallene margarita* (Gordon, 1932)	1	Child (1982)
3/14/1971	RVA-60	245A	36°55.7	53°01.4	2707	*Cilunculus acanthus* Fry and Hedgpeth, 1969	11	Child (1997)
3/27/1971	RVA-60	262A	36°05.2	52°17.9	2440–2480	*Cilunculus acanthus* Fry and Hedgpeth, 1969	18	Child (1982)
4/23/2017	PS, P10	11	37°36.04	52°36.84	3800	*Colossendeis angusta* Sars, 1877	2	Present study
4/24/2016	PS, P12	14	36°53.57	53°07.62	2540	*Colossendeis angusta* Sars, 1877	21	Present study
5/2/2016	PS, P25	80	35°41.07	50°43.69	3600	*Colossendeis angusta* Sars, 1877	1	Present study
4/21/2016	PS, P5	53	36°01.99	52°14.20	2600	*Colossendeis angusta* Sars, 1877	1	Present study
4/22/2016	PS, P7	56	36°36.00	51°36.71	3600	*Colossendeis angusta* Sars, 1877	3	Present study
	–	–	36°05	53°03		*Colossendeis geoffroyi* Mañé-Garzón, 1944	2	Mañé-Garzón (1944)
3/13/1971	RVA-60	242	38°16.9	51°56.1	4382–4402	*Colossendeis scoresbii* Gordon, 1932	1	Child (1997)
3/11/1971	RVA-60	237	36°32.6	53°23.0	993–1011	*Mimipallene atlantis* Child, 1982	3	Child (1982)
3/12/1971	RVA-60	240	36°53.4	53°10.2	2195–2323	*Mimipallene atlantis* Child, 1982	13	Child (1982)
5/3/2016	PS, P26	13	37°11.80	52°59.19	3300	*Nymphon* sp. 1	1	Present study
4/21/2016	PS, P6	54	36°12.51	52°00.48	3100	*Nymphon* sp. 1	2	Present study

4/23/2016	PS, P8	31	36°49.37	52°37.13	3000	*Nymphon* sp. 2	3	Present study
3/14/1971	RVA-60	245A	36°55.7	53°01.4	2707	*Nymphon centrum* Child, 1997	31	Child (1997)
3/24/1971	RVA-60	256	37°40.9	52°19.3	3906–3917	*Nymphon dentiferum* Child, 1997	6	Child (1997)
3/11/1971	RVA-60	237	36°32.6	53°23.0	993–1011	*Nymphon hampsoni* Child, 1982	157	Child (1982)
3/12/1971	RVA-60	240	36°53.4	53°10.2	2195–2323	*Nymphon hampsoni* Child, 1982	77	Child (1982)
3/14/1971	RVA-60	243	37°36.8	52°23.6	3815–3822	*Nymphon inerme* Fage, 1956	9	Child (1982)
3/14/1971	RVA-60	245A	36°55.7	53°01.4	2707	*Nymphon inferum* Child, 1995	1	Child (1997)
3/15/1971	RVA-60	246	37°15.1	52°45.0	3343	*Nymphon laterospinum* Stock, 1963	36	Child (1982)
3/26/1971	RVA-60	259	37°13.3	52°45.0	3305–3317	*Nymphon laterospinum* Stock, 1963	25	Child (1997)
3/13/1971	RVA-60	242	38°16.9	51°56.1	4382–4402	*Nymphon longicollum* Hoek, 1881	7	Child (1997)
3/24/1971	RVA-60	256	37°40.9	52°19.3	3906–3917	*Nymphon longicollum* Hoek, 1881	2	Child (1997)
3/12/1971	RVA-60	240	36°53.4	53°10.2	2195–2323	*Nymphon longicoxa* Hoek, 1881	180	Child (1982)
3/14/1971	RVA-60	245A	36°55.7	53°01.4	2707	*Nymphon longicoxa* Hoek, 1881	2	Child (1997)
3/27/1971	RVA-60	262A	36°05.2	52°17.9	2440–2480	*Nymphon longicoxa* Hoek, 1881	2	Child (1982)
3/28/1971	RVA-60	264A	36°12.7	52°42.7	2041–2048	*Nymphon longicoxa* Hoek, 1881	1	Child (1982)
3/14/1971	RVA-60	243	37°36.8	52°23.6	3815–3822	*Nymphon sandersi* Child, 1982	2	Child (1982)
3/14/1971	RVA-60	245A	36°55.7	53°01.4	2707	*Nymphon scotiae* Stock, 1981	45	Child (1997)
3/14/1971	RVA-60	243	37°36.8	52°23.6	3815–3822	*Nymphon typhlops* (Hodgson, 1915)*	1	Child (1982)
3/14/1971	RVA-60	245A	36°55.7	53°01.4	2707	*Nymphon typhlops* (Hodgson, 1915)	6	Child (1997)
3/14/1971	RVA-60	245A	36°55.7	53°01.4	2707	*Nymphon vacans* Child, 1997	15	Child (1997)
3/11/1971	RVA-60	236	36°27	53°31	497–518	*Pallenopsis meinerti* Schimkewitsch, 1930	3	Child (1982)
3/11/1971	RVA-60	237	36°32.6	53°23.0	993–1011	*Pallenopsis meinerti* Schimkewitsch, 1930	3	Child (1982)
17/2/1876	HMS-C	320	37°17	53°52	1097	*Pallenopsis patagonica* (Hoek, 1881)**	1	Hoek (1881)
4/19/2016	PS, P1	17	36°37.56	53°50.35	250	*Pallenopsis* sp. 1	4	Present study
3/12/1971	RVA-60	240	36°53.4	53°10.2	2195–2323	*Pantopipetta longituberculata* (Turpaeva, 1955)	2	Child (1982)
3/13/1971	RVA-60	242	38°16.9	51°56.1	4382–4402	*Pantopipetta longituberculata* (Turpaeva, 1955)	27	Child (1997)
3/14/1971	RVA-60	243	37°36.8	52°23.6	3815–3822	*Pantopipetta longituberculata* (Turpaeva, 1955)	7	Child (1982)

(continued)

Table 2.1 (continued)

Date	Trawl number	Station	Lat (S)	Long (W)	Depth range	Species	N	References
3/14/1971	RVA-60	245A	36°55.7	53°01.4	2707	*Pantopipetta longituberculata* (Turpaeva, 1955)	30	Child (1997)
3/24/1971	RVA-60	256	37°40.9	52°19.3	3906–3917	*Pantopipetta longituberculata* (Turpaeva, 1955)	16	Child (1997)
3/27/1971	RVA-60	262A	36°05.2	52°17.9	2440–2480	*Pantopipetta longituberculata* (Turpaeva, 1955)	1	Child (1982)

PS present study, *RVA-60* RV Atlantis, cruise 60, *HMS-C* HMS Challenger, *RVV-18* RV Vema cruise 18

(*) *Nymphon typhlops* is referred as *N. spicatum* in Child (1982); (**) *Pallenopsis patagonica* is referred as *Phoxichilidium patagonicum* var. *elegans* in Hoek (1881)

2.3 Results

In this study, as part of ongoing studies documenting the invertebrates from the bycatch of commercial fisheries in the EEZ of Uruguay, a total of 50 specimens were analyzed to complete the sampling in this deep area (Table 2.1). Thanks to the present cruise, 10 new stations were added to the 16 previous stations from the literature of deep Uruguayan pycnogonids. Three nominal species were identified, *Ammothea spinosa* (Hodgson, 1907); *Colossendeis angusta* Sars, 1877; and *Bathypallenopsis calcanea* (Stephensen, 1933), belonging to three genera and three families. In addition, six specimens belonging two distinct species belonging to the genus *Nymphon* Fabricius, 1794, were identified. They did not match none of the 12 different species of *Nymphon* previously recorded in this area or any other *Nymphon* species of the areas nearby, including the Antarctic and sub-Antarctic waters. Further studies are needed to determine if these specimens may belong to other more distant *Nymphon* species or to a new species. Four other specimens could be only determined to the genus *Pallenopsis* Wilson, 1881, also discarding the *Pallenopsis* species recorded in the southwestern Atlantic.

In the present study, the deepest bathymetric record worldwide of *A. spinosa* is provided (3600 m depth) as well as the northernmost occurrence for this species. *Colossendeis angusta* and *B. calcanea* are recorded for the first time in Uruguay, adding two new species to the list of 27 species presently known (Scarabino et al 2019; Lucena et al. 2019a). Some geographical and bathymetrical discussion is provided for these species below.

The bathymetrical ranges for deep Uruguayan pycnogonids has been summarized (Table 2.2). Some families are typical from the deep slopes restricted to depths down to 2000 m, such as the Nymphonidae Wilson, 1878. This family and its main genus, *Nymphon*, is the most diverse in Uruguayan waters, with 12 species recorded and two more to be confirmed. Most of them are found at depths greater than 2000 m. The genus *Bathypallenopsis* Stock, 1975, has been discussed to be a subgenus of *Pallenopsis* (Cano-Sánchez and López-González 2019); however, nowadays it is considered a valid genus with 21 species (Bamber et al. 2020) due to some distinct morphological characteristics and the great depths at which it is generally found. Instead, the genus *Pallenopsis* seems to be restricted to the shallow slope, not found below the 2000 m. The genus *Colossendeis* Jarzynsky, 1870, the main genus of the family Colossendeidae Jarzynsky, 1870, is typical of deep waters and has many cosmopolitan species, being the second most abundant genus in Antarctic and sub-Antarctic waters (Child 1995; Gordon 1932; Fry and Hedgpeth 1969; Munilla and Soler-Membrives 2009), despite the few records for the Uruguayan region.

Table 2.2 Bathymetric range of deep-water Pycnogonida based on present study and records in literature (Scarabino et al. 2019; Lucena et al. 2019b)

Family	Species	100–1000 m	1000–2000 m	2000–3000 m	3000–4000 m	>4000 m
Ammotheidae Dorhn, 1881	*Ammothea longispina* Gordon, 1932		x			
	Ammothea spinosa (Hodgson, 1907)		x		x	
	Ammothea tetrapoda Gordon, 1932	x				
	Cilunculus acanthus Fry and Hedgpeth, 1969			x		
Ascorhynchidae Hoek, 1881	*Ascorhynchus cuculus* Fry and Hedgpeth, 1969		x	x		
Ascorhynchoidea *incertae sedis*	*Mimipallene atlantis* Child, 1982		x	x		
Austrodecidae Stock, 1954	*Pantopipetta longituberculata* (Turpaeva, 1955)			x	x	x
Callipallenidae Hilton, 1942	*Callipallene margarita* (Gordon, 1932)			x		
Colossendeidae Hoek, 1881	*Colossendeis angusta* Sars, 1877			x	x	
	Colossendeis geoffroyi Mañé-Garzón, 1944	x				
	Colossendeis scoresbii Gordon, 1932					x
Nymphonidae Wilson, 1878	*Nymphon centrum* Child, 1997			x		
	Nymphon dentiferum Child, 1997				x	
	Nymphon hampsoni Child, 1982		x	x		
	Nymphon inerme Fage, 1956				x	
	Nymphon inferum Child, 1995			x		
	Nymphon laterospinum Stock, 1963				x	
	Nymphon longicollum Hoek, 1881				x	x
	Nymphon longicoxa Hoek, 1881			x		
	Nymphon sandersi Child, 1982				x	
	Nymphon scotiae Stock, 1981			x		
	Nymphon sp. 1				x	
	Nymphon sp. 2				x	
	Nymphon typhlops (Hodgson, 1915)			x	x	
	Nymphon vacans Child, 1997			x		

Table 2.2 (continued)

Family	Species	100–1000 m	1000–2000 m	2000–3000 m	3000–4000 m	>4000 m
Pallenopsidae Fry, 1978	*Bathypallenopsis calcanea* (Stephensen, 1933)			x	x	
	Pallenopsis meinerti Schimkewitsch, 1930	x	x			
	Pallenopsis patagonica (Hoek, 1881)		x			
	Pallenopsis sp. 1	x				
Phoxichilidiidae Sars, 1891	*Anoplodactylus vemae* Child, 1982	x				

2.3.1 Colossendeis angusta *Sars, 1877 Fig. 2.2a*

Colossendeis angusta is a cosmopolitan species recorded in many deep oceans (Arctic, Atlantic, Pacific, and Indian oceans) and the Antarctica and sub-Antarctic waters, from shallow to deep waters (Raiskii and Turpaeva 2006). The species is known from the Argentina slope off Peninsula Valdés (Child 1982). However, this species has not been cited yet in Uruguayan waters and is recorded now for the first time, contributing a new area in its distribution range. This pan-oceanic species has a very wide bathymetric range (22–5480 m). Recent molecular studies (Dietz et al. 2013) including molecular and morphological data suggested that it is also plausible that *C. angusta* might be hiding more than a single species and that all *C. tenera*, *C. angusta*, and *C. megalonyx* complex species should be revised in detail.

2.3.2 Ammothea spinosa *(Hodgson, 1907) Fig. 2.2b*

Ammothea spinosa has been recorded yet in the Antarctica and sub-Antarctic waters (Ross Sea, Antarctic Peninsula Islands, Scotia Sea, and the Falkland Islands) as well as in the Argentine and Uruguayan basins. Therefore, this species is recorded for the second time in Uruguayan waters since Child (1982) recorded six specimens from the RV Atlantis II Cruise in 1971. This is a typical Antarctic species that could be transported by the cold bottom currents north into South Atlantic basins. Up to date, the present record is the northernmost occurrence of this species but also elucidates the deficient knowledge of the fauna of these basins. It should not be discarded that this species could be distributed further north at the deep shores off Brazil. Moreover, the bathymetric range known yet for this species was 76–1679 m, so the records here reported increases also significantly the worldwide bathymetric range for this species to 3600 m depth.

2.3.3 Bathypallenopsis calcanea *(Stephensen, 1933) Fig. 2.2c*

Bathypallenopsis calcanea is a cosmopolitan bathypelagic species at a depth range of 353–8400 m (Turpaeva and Raiskii 2014). It has been found in both the Atlantic and Pacific Ocean basins (Bamber and Thurston 1995, Bamber 2002, Raiskii and Turpaeva 2006, Gasca and Browne 2008). This is considered a bathypelagic form of associated with gelatinous organisms like Scyphomedusae and Doliolida (Hedgpeth 1962). This parasitic or commensal association with larger bathypelagic organisms may lead this species to move large distances attached to them and so could explain its cosmopolitan distribution.

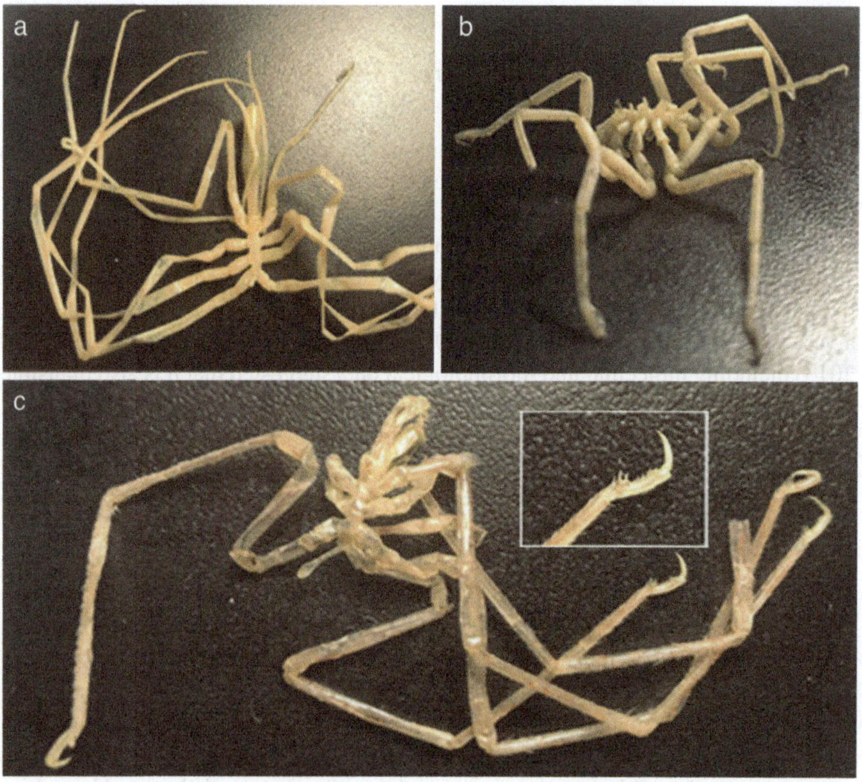

Fig. 2.2 Specimens of deep-water Pycnogonida captured during present study off Uruguay. (**a**) *Colossendeis angusta*. (**b**) *Ammothea spinosa*. (**c**) *Bathypallenopsis calcanea*

2.4 Conclusion

This study added two new species to the Uruguayan biodiversity, summing a total of 29 pycnogonid species recorded up to date. The presence of these species in Uruguay may give hints of a continuous distribution of these taxa along the deep-sea of the southwestern Atlantic Ocean (Bamber and Thurston 1995). Uruguay may represent part of a hidden biodiversity thriving in deep waters in the southwestern Atlantic. The Antarctic and sub-Antarctic fauna has been deeply studied. However, the connection between the Antarctica and the southwestern Atlantic has been merely studied, and northern areas such as the deep shore off Brazil have been even more neglected. Further deep cruises should be performed to understand to which extent can deep-sea species be distributed through long distances thanks to deep cold currents.

Acknowledgments We would like to thank to Fabrizio Scarabino for the good disposition he has had with this study and his advices. We would like to thank the crew of the F/V Marianne and to Alejandro V. Ariza, Claudio Barría, Ernesto Chiesa, Anabel Colmenero, Susana Díez, Alejandro Escánez, Valentina Leoni, and Carmen Morant for their participation in the CUCAGUAY cruise. Funding for cruising was supported by "Administración Nacional de Combustibles, Alcohol y Portland" (ANCAP) center in Uruguay.

References

Araya JF (2016) New records of deep-sea sea spiders (Chelicerata: Pycnogonida) in the southeastern Pacific. Mar Biodivers 46:725–729. https://doi.org/10.1007/s12526-015-0416-7

Arnaud F, Bamber RN (1987) The biology of Pycnogonida. Adv Mar Biol 24:1–96

Ballesteros JA, Setton EVW, Santibáñez López CE, Arango CP, Brenneis G, Brix S, Cano-Sánchez E, Dandouch M, Dilly GF, Eleaume MP, Gainett G, Gallut C, McAtee S, McIntyre L, Moran AL, Moran R, López-González PJ, Scholtz G, Williamson C, Woods HA, Wheeler WC, Sharma PP (2020) Phylogenomic resolution of sea spider diversification through integration of multiple data classes. bioRxiv. https://doi.org/10.1101/2020.01.31.929612

Bamber RN (1983) Some deep water Pycnogonids from the north-east Atlantic. Zool J Linn Soc-Lond 77(1):65–74. https://doi.org/10.1111/j.1096-3642.1983.tb01721.x

Bamber RN (2002) Bathypelagic pycnogonids (Arthropoda, Pycnogonida) from the discovery deep-sea cruises. J Nat Hist 36(6):715–727. https://doi.org/10.1080/00222930010025932

Bamber RN, Thurston MH (1993) Deep water pycnogonids of the Cape Verde slope. J Mar Biol Assoc UK 73:837–861. https://doi.org/10.1017/S0025315400034767

Bamber RN, Thurston MH (1995) The deep-water pycnogonids (Arthropoda: Pycnogonida) of the northeastern Atlantic Ocean. Zool J Linn Soc-Lond 115:117–162. https://doi.org/10.1006/zjls.1995.0035

Bamber RN, El Nagar A, Arango CP (2020) Pycnobase: world Pycnogonida database. Accessed at http://www.marinespecies.org/pycnobase on 1 Jan 2020. https://doi.org/10.14284/360

Barnard KH (1954) South African Pycnogonida. Ann S African Mus 41:81–158

Braby CE, Pearse VB, Bain BA, Vrijenhoek RC (2009) Pycnogonid-cnidarian trophic interactions in the deep Monterey Submarine Canyon. Invert Biol 128(4):359–363. https://doi.org/10.1111/j.1744-7410.2009.00176.x

Brescia LA, Tunnicliffe V (1998) Population biology of two pycnogonid species (Ammotheidae) at hydrothermal vents in the Northeast Pacific. Cah Biol Mar 39:233–236

Child CA (1982) Deep-Sea Pycnogonida from the North and South Atlantic Basins. Smithson Contr Zool 349:1–54. https://doi.org/10.5479/si.00810282.349

Child CA (1995) Antarctic and Subantarctic Pycnogonida IV. The families Colossendeidae and Rhynchotoraxidae. Antarct Res Ser 69:69–112. https://doi.org/10.1002/9781118668252.ch2

Child CA (1997) Some deep-sea Pycnogonida from the Argentine slope and basin. Proc Mar Biol Soc Wash 110(1):128–142

Dietz L, Krapp F, Hendrickx ME, Arango CP, Krabbe K, Spaak JM, Leese F (2013) Evidence from morphological and genetic data confirms that *Colossendeis tenera* Hilton, 1943 (Arthropoda: Pycnogonida), does not belong to the *Colossendeis megalonyx* Hoek, 1881 complex. Org Divers Evol 13:151–162

Fry WG, Hedgpeth JW (1969) Pycnogonida, 1. Colossendeidae, Pycnogonidae, Endeidae and Ammotheidae. Fauna of the Ross Sea, 7. Mem NZ Oceanogr Inst 49:1–149

Gasca R, Browne WE (2008) Symbiotic associations of crustaceans and a pycnogonid with gelatinous zooplankton in the Gulf of California. Mar Biodivers 48(4):1767–1775. https://doi.org/10.1007/s12526-017-0668-5

Gordon I (1932) Pycnogonida. Discover Rep 6:1–138

Hedgpeth JW (1962) A bathypelagic pycnogonid. Deep-Sea Res 9:487–491. https://doi.org/10.1016/0011-7471(62)90099-2

Hoek PPC (1881) Report on the Pycnogonida dredged by HMS Challenger during the years 1873–1876. Report of the Scientific Results of the Voyage of HMS Challenger during the years 1873–76. Zoology, 3(10), 1–167, 21 pls.

Lucena RA, Pereira de Araújo J, Christoffersen ML (2019a) Pycnogonida (Arthropoda) from coral reef environments along the Southwest Atlantic: new records and new species. Mar Biol Res 15(8–9):476–499. https://doi.org/10.1080/17451000.2019.1674876

Lucena RA, Scarabino F, Christoffersen ML (2019b) First record of *Ammothea tetrapora* Gordon, 1932 for Uruguayan waters (Ammotheidae, Pycnogonida). Spixiana 42(2):229–234

Mañé-Garzón F (1944) Notas sobre pantópodos, I : Colossendeis geoffroyi, nov.sp., de la plataforma continental en el Río de la Plata. p. 1–7

Munilla T, Soler-Membrives A (2009) Check-list of the pycnogonids from Antarctic and subantarctic waters: zoogeographic implications. Antarct Sci 21:99–111. https://doi.org/10.1017/S095410200800151X

Raiskii AK, Turpaeva EP (2006) Deep-sea pycnogonids from the North Atlantic and their distribution in the World Ocean. Oceanology 46(1):57–62. https://doi.org/10.1134/S0001437006010073

Scarabino F (2006) Faunística y taxonomía de invertebrados bentónicos marinos y estuarinos de la costa uruguaya. In: Menafra R, Rodríguez L, Scarabino F, Conde D (eds) Bases para la conservación y el manejo de la costa uruguaya. Vida Silvestre, Montevideo, pp 113–142

Scarabino F, Lucena RA, Munilla T, Soler-Membrives A, Ortega L, Schwindt E, López G, Orensanz JM, Christoffersen ML (2019) Pycnogonida (Arthropoda) from Uruguayan waters (Southwest Atlantic): annotated checklist and biogeographic considerations. Zootaxa 4550(2):185–200. https://doi.org/10.11646/zootaxa.4550.2.2

Stock JH (1963) South Africa deep-sea Pycnogonida, with descriptions of five new species. Ann S African Mus 46:321–340

Stock JH (1978a) Abyssal Pycnogonida from the Northeastern Atlantic Basin. Part I. Cah Biol Mar 19:189–219

Stock JH (1978b) Abyssal Pycnogonida from the Northeastern Atlantic Basin. Part II. Cah Biol Mar 19:397–413

Stock JH (1981) Abyssal Pycnogonida from the Walvis Basin, Southern Atlantic. Cah Biol Mar 22:453–471

Stock JH (1987) Faunistic transit between the Atlantic and the Mediterranean: the deep-water Pycnogonida. Cah Biol Mar 28:505–519

Turpaeva EP, Raiskii AK (2014) Deep-sea fauna of European seas: an annotated species check-list of benthic invertebrates living deeper than 2000 m in the seas bordering Europe. Pycnogonida. Invert Zool 11:240–247. https://doi.org/10.15298/invertzool.11.1.22

Chapter 3
The Deep-Water *Colossendeis tenera* Hilton, 1943 (Pycnogonida, Pantopoda, Colossendeidae) off Western Mexico

M. E. Hendrickx

Abstract The deep-water Pycnogonida *Colossendeis tenera* Hoek, 1881, is reported from 12 localities off western Mexico. In the eastern Pacific, it had been previously recorded from the Bering Sea to southern California and from one isolated record off western Mexico. A total of 43 specimens, ranging from 36.9 to 215.6 mm leg span, was collected in a depth range of 750–2054 m, all below the lower limit of the oxygen minimum zone. Ecological information is provided for his material which appears to be tolerant to mild and severe hypoxic conditions, i.e., from 0.11 to 0.78 ml/l.

Keywords Sea spiders · Pycnogonids · Eastern Pacific · Oxygen minimum zone

3.1 Introduction

Pycnogonida is an exclusively marine group of Arthropoda containing about 1300 species (Nakamura et al. 2007). Among the Pycnogonida, *Colossendeis* Jarzynsky, 1870, is a highly speciose genus composed of 72 species distributed worldwide, from shallow to deep waters (Araya 2016; Bamber et al. 2020). The genus is particularly diverse in deep, cold Antarctic and sub-Antarctic waters (Vinu et al. 2016). Some species are very large (over 70 cm leg span when expanded) (Vinu et al. 2016; Araya 2016), and their ecological role is therefore important in deep-water communities. As far as food webs are concerned, specimens of giant *Colossendeis* have been observed feeding on cnidarians in the North Atlantic and the North Central Pacific (Wicksten 2017), but members of this genus have also been reported feeding on detritus ("deposit feeders"), probably on sponges, hydroids, polychaetes,

M. E. Hendrickx (✉)
Laboratorio de Invertebrados Bentónicos, Unidad Académica Mazatlán, Instituto de Ciencias del Mar y Limnología, Universidad Nacional Autónoma de México,
Mazatlán, Sinaloa, Mexico
e-mail: michel@ola.icmyl.unam.mx

M. E. Hendrickx (ed.), *Deep-Sea Pycnogonids and Crustaceans of the Americas*, https://doi.org/10.1007/978-3-030-58410-8_3

nemertean, bryozoans, and limpets, and some scavenger behavior has also been observed (Vinu et al. 2016; Wicksten 2017; Dietz et al. 2018).

Munilla (2002) enlisted 18 species of pycnogonids for the Pacific coast of Mexico, most (50%) belonging to the Ammotheidae, and the rest distributed among the Callipallenidae (3 species), the Nymphonidae (4 species), and the Phoxichilididae (2 species). Hendrickx (2005) reports 15 species for the Gulf of California, essentially the same as those reported by Munilla (2002).

Deep-water pycnogonids have been little studied off western Mexico, and records are scarce (Dietz et al. 2013). While reviewing material of invertebrates collected in deep water off western Mexico, a small series of specimens of *C. tenera* Hilton, 1943, was recognized. One record from western Mexico was previously presented by Dietz et al. (2013). This additional material is reported herein with notes on this species ecology.

3.2 Material and Methods

The material on which this study is based was collected by the R/V "El Puma" of the Universidad Nacional Autónoma de México (UNAM), between 1991 and 2014 off western Mexico (TALUD project) (see Fig. 3.1). During this survey, 228 localities were sampled near the lower boundary of the OMZ core and deeper, from 377 to 2394 m depth. Positional coordinates were obtained using a GPS navigation system. Depth was measured with a digital recorder. Specimens were captured with a benthic sledge (BS) (2.35 m width, 0.9 m high) or a modified Agassiz dredge (AD) used during all cruises of this survey (see Hendrickx 2012). Specimens were deposited in the Regional Collection of Marine Invertebrates (ICML-EMU) at UNAM in Mazatlán, Mexico. Temperature, salinity, and oxygen were measured around 20 m above the sea bottom (20 mab) with a Seabird 19 CTD. Rosette-mounted 10 l Niskin bottles were also deployed at each station (St.), and dissolved oxygen concentration was estimated with the Winkler method (Strickland and Parsons 1972). Sediment was sampled at each station by means of a modified USNEL box core, and samples of the top 3 cm were stored at 4–8 °C. At the laboratory, sediments were dried to constant weight at 60 °C. The organic carbon (Corg) and organic matter (Morg) contents in sediments were determined following the technique of Loring and Rantala (1992). Leg span (LS) of specimens was measured between the tips of the first pair of walking leg (Lane et al. 2017).

3.3 Results

A total of 43 specimens of *C. tenera* was collected in 12 stations, between 750 and 2054 m depth.

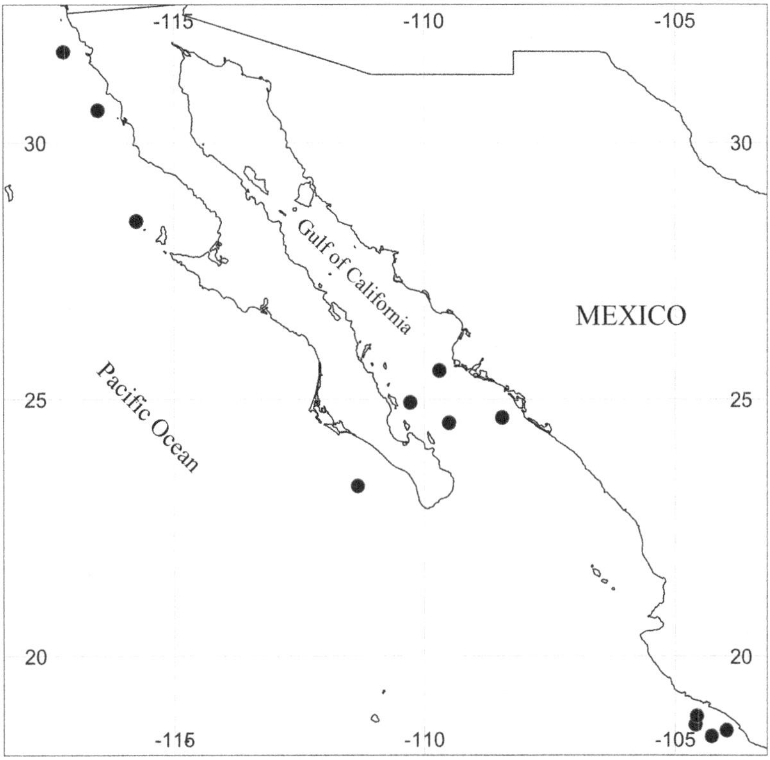

Fig. 3.1 Localities off western Mexico where *Colossendeis tenera* Hilton, 1943, was collected

Colossendeidae Jarzynsky, 1870

Colossendeis tenera Hoek, 1881

Material Examined. TALUD III, St. 14A (24°38′48″N; 108°26′54″W), August 19, 1991, AD, 1016–1020 m, 2 specimens (LS 123.2–150.3 mm) (ICML-EMU-8639); St. 24A (25°33′36″N; 109°42′01″W), August 21, 1991, 3 specimens (LS 130.4–143.0 mm), AD, 1043 m (ICML-EMU-8640).

TALUD VIII, St. 3 (24°32′36″N; 109°30′30″W), April, 16, 2005, 2 specimens (LS ca 98.9 and 103.1 mm), BS, 1600 m (ICML-EMU-8226).

TALUD IX, St. 10 (24°56′24″N; 110°16′42″W), November 12, 2005, 1 specimen (LS 102.7 mm), BS, 969–1225 m (ICML-EMU-12671).

TALUD XII, St. 23 (18°33′43″N; 103°57′45″W), April 1, 2008, BS, 1058–1088 m, 1 specimen (LS 115.3 mm) (ICML-EMU-8227) and 19 specimens (LS 78.9–140.1 mm) (ICML-EMU-8909); St. 25 (18°26′45″N; 104°16′10″W), April 1, 2008, 2 specimens (LS 112.4–297.0 mm), BS, 1858–1879 m (ICML-EMU-8910); St. 27 (18°40′28″N; 104°35′51″W), April 2, 2008, 1 specimen (LS 99.3 mm), BS,

1040–1095 m (ICML-EMU-12672); St. 28 (18°50′19″N; 104°34′14″W), April 2, 2008, 7 specimens (LS 90.4–102.8 mm), BS, 1101–1106 m (ICML-EMU-12673).

TALUD XV, St. 1 (23°18′40″N; 111°19′37″W), August 4, 2012, 2 specimens (LS 36.9–121.1 mm), BS, 750–850 m (ICML-EMU-12674).

TALUD XVI-B, St. 1 (28°28′18″N; 115°45′12″W), May 23, 2014, 1 specimen (LS ca 44.6 mm), BS, 2038–2054 m (ICML-EMU-12675); St. 19 (30°38′N; 116°31.67′W), May 25, 2014, 1 specimen (LS 215.6 mm), BS, 1385–1433 m (ICML-EMU-12676); St. 27 (31°46′18″N; 117°12′18″W), May 27, 2014, 1 specimen (LS 113.7 mm), BS, 1395–1397 m (ICML-EMU-12677).

Ecology. Depth records reported herein for *C. tenera* (750–2054 m) are within the depth range previously reported for this species (Child 1994, 1995). Values of environmental parameters associated with the presence of *C. tenera* were as follows: water temperature, 2.02–5.75 °C; dissolved oxygen, 0.11–0.78 ml/l; and salinity, 34.53–34.70. Sediments where the specimens were captured had an organic carbon content of 1.34–3.96%; sediments composition was predominantly silty (about 68–85%) with low, rather variable proportion of clay and fine sand (Table 3.1).

Geographic distribution. In the eastern Pacific, *Colossendeis tenera* is known to occur off western Mexico. Indeed, one specimen from station 28 of the TALUD XII cruise has been included in Dietz et al. (2013) contribution. The material examined herein indicates that *C. tenera* is widely distributed off western Mexico, including the southern Gulf of California (Fig. 3.1). The USNM collection contains material collected off Oregon, Washington, and Alaska (including the holotype from off Oregon: 45°30′N, 124°52′W); the species has been reported for these three western American states by Austin (1980) and Child (1994), who also included California in its distribution range. Child (1995) reported on material from the Bering Sea and off the Aleutians Islands. There is an additional record (as

Table 3.1 Environmental data measured at bottom level in localities where *C. tenera* Hilton, 1943, was collected. Sediments components in %

Locality	Oxygen (ml/l)	Temp. (°C)	Salinity	%Corg	Morg%	Clay	Lime	Sand
T III, St. 14A	0.15	ND	ND	ND	ND	ND	ND	ND
T III, St. 24A	ND	ND	ND	ND	ND	ND	ND	ND
T VIII, St. 3	0.70	3.00	34.70	ND	ND	ND	ND	ND
T IX, St. 10	0.18	4.22	34.55	ND	ND	ND	ND	ND
T XII, St. 23	0.22	4.39	34.54	ND	ND	ND	ND	ND
T XII, St. 24	0.95	3.10	34.59	1.34	ND	ND	ND	ND
T XII, St. 27	0.26	ND	34.53	1.96	ND	ND	ND	ND
T XII, St. 28	0.27	4.11	34.53	2.13	ND	ND	ND	ND
T XV, St 1.	0.11	5.75	34.54	3.93	6.77	13.37	67.94	18.69
T XVI, St. 1	ND	2.02	34.64	ND	ND	ND	ND	ND
T XVI-B, St. 19	0.90	3.16	34.55	3.96	6.81	11.07	84.62	4.31
T XVI-B, St. 27	0.78	3.32	34.54	3.68	6.32	12.58	84.99	2.44

Colossendeis megalonyx tenera) from off Peru at 18° 10′S, 71° 29′W. The material was collected by a Russian research ship and reported by Turpaeva (1975), a contribution which was apparently overlooked previously.

Bathymetry. The long series of records provided by Child (1994) are from 1000 to 4194 m depth, but in 1995 the same author reported a depth range of 882–3600 m (Child 1995). Turpaeva (1975) indicated a depth range of 600–610 m. The shallowest record observed during this study was 750–850 m.

3.4 Discussion

In addition to *C. tenera* (southern Alaska, Washington, and Oregon), Austin (1980) also reported two additional species of this genus for the west coast of America: *C. angusta* Sars, 1877, from the Arctic Sea to Oregon (Pacific and Atlantic), and *C. colossae* Wilson, 1881, from Oregon (West Pacific and Atlantic). Two other species, *C. gracilis* Hoek, 1881, and *C. minuta* Hoek, 1881, have been reported from off the west coast of Panama (Stock 1974). In his extensive review of Pycnogonida from the temperate West Coast of the United States, Child (1994) reported on material of several species of the genus, including two new to science: *C. cucurbita* Cole, 1909, from California and the Galapagos Islands (Atlantic and Pacific); *C. japonica* Hoek, 1898, from California and Washington (Japan and Kermadec Trench); *C. macerrima* Wilson, 1881, a cosmopolitan species recorded from off the west coast of Central America, California, Oregon, and Washington, USA (Bamber et al. 2020); *C. peloria* Child, 1994, from California; *C. spicula* Child, 1994, from Oregon; *C. colossae* from off California and Washington; and *C. tenera* from California, Oregon, and Alaska.

Environmental conditions other than depth ranges have been little documented for species of *Colossendeis*. Vinu et al. (2016) recorded the following data (sample taken near bottom; 1000 m depth) for *C. colossae*: temperature, 6.79 °C, salinity, 34.97; and dissolved oxygen, 0.86 ml/l. The sampling locality was near the SW tip of India, in an area where the oxygen minimum zone (OMZ) is intense (Helly and Levin 2004). All the specimens examined herein were also captured in an area strongly influenced by an intense OMZ (Hendrickx and Serrano 2010), below the lower limit of the OMZ core, in conditions similar to these reported by Vinu et al. (2016). The dissolved oxygen concentration range in this study, however, was much wider, i.e., 0.11–0.95 ml/l. *Colossendeis tenera* appears to be tolerant to mild (<0.5 ml/l) to severe (<0.2 ml/l) hypoxic conditions.

Acknowledgments Ship time aboard the R/V "El Puma" was provided by the Coordinación de la Investigación Científica, UNAM, and partly supported by CONACyT (project 179467 for the TALUD XV and XVI-B cruises). The TALUD project has also received laboratory and field work support from CONACyT (Project 31805-N for the TALUD IV to VII cruises; project 179467 for the TALUD XV and XVI–B cruises), Mexico. The authors thank all scientists, students, and crew members who took an active part in the TALUD cruises. We thank Mercedes Cordero Ruiz for preparing final version of Fig. 3.1 and Giselle A. Chavez Salgado for laboratory assistance (CONACyT, SNI Assistant).

References

Araya JF (2016) New records of deep-sea sea spiders (Chelicerata: Pycnogonida) in the southeastern Pacific. Mar Biodivers 46(3):725–729

Austin WC (1980) Pycnogonida. In: Austin WC (ed) An annotated checklist of marine invertebrates in the cold temperate Northeast Pacific. Khoyatan Marine Laboratory, Cowichan Bay, pp 450–455

Bamber RN, El Nagar A, Arango CP (2020) Pycnobase: World Pycnogonida Database. *Colossendeis* Jarzynsky, 1870. Accessed through: World Register of Marine Species at: https://www.marinespecies.org/aphia.php?p=taxdetails&id=134586 on 29 Jan 2020

Child CA (1994) Deep-sea Pycnogonida from the temperate west coast of the United States. Smithson Contrib Zool 556:1–23

Child CA (1995) Pycnogonida of the western Pacific Islands, XI: collections from the Aleutians and other Bering Sea islands, Alaska. Smithson Contrib Zool 569:1–29

Dietz L, Krapp F, Hendrickx ME, Arango CP, Krabbe K, Spaak JM, Leese F (2013) Evidence from morphological and genetic data confirms that *Colossendeis tenera* Hilton, 1943 does not belong to the *Colossendeis megalonyx* Hoek, 1881 complex. Org Divers Evol 13(2):151–162. https://doi.org/10.1007/s13127-012-0120-4

Dietz L, Dömel JS, Leese F, Lehmann T, Melzer RR (2018) Feeding ecology in sea spiders (Arthropoda: Pycnogonida): what do we know? Front Zool 15(1):7–16

Helly JJ, Levin LA (2004) Global distribution of naturally occurring marine hypoxia on continental margins. Deep Sea Res Part I: Oceanogr Res Papers 51(9):1159–1168

Hendrickx ME (2005) Chapter 8. Chelicerata: Pycnogonida. In: Hendrickx ME, Brusca RC, Findley LT (eds) A distributional checklist of the Macrofauna of the Gulf of California, Mexico. Part I. Invertebrates. [Listado y Distribución de la Macrofauna del Golfo de California, México, Parte I. Invertebrados]. Arizona-Sonora Desert Museum, Tucson, pp 117–119

Hendrickx ME (2012) Operaciones oceanográficas en aguas profundas: los retos del pasado, del presente y del proyecto TALUD en el Pacífico mexicano (1989–2009). In: Zamorano P, Hendrickx ME, Caso M (eds) Biodiversidad y comunidades del talud continental del Pacífico mexicano. Secretaría del Medio Ambiente y Recursos Naturales (SEMARNAT), Instituto Nacional de Ecología (INE), Mexico, pp 23–104. http://www2.ine.gob.mx/publicaciones/libros/690/operaciones.pdf

Hendrickx ME, Serrano D (2010) Impacto de la zona de mínimo de oxígeno sobre los corredores pesqueros en el Pacífico mexicano. Interciencia 35(1):12–18

Lane SJ, Shishido CM, Moran AL, Tobalske BW, Arango CP, Woods HA (2017) Upper limits to body size imposed by respiratory–structural trade-offs in Antarctic pycnogonids. Proc R Soc B Biol Sci 284(1865):1–9. https://doi.org/10.1098/rspb.2017.1779

Loring DH, Rantala RTT (1992) Manual for the geochemical analyses of marine sediments and suspended particulate matter. Earth-Sci Rev 32:235–283. https://doi.org/10.1016/0012-8252(92)90001-A

Munilla T (2002) Synopsis of the pycnogonids from Antarctic and Subantarctic waters. Polar Biol 24:941–945

Nakamura K, Kano Y, Suzuki N, Namatame T, Kosaku A (2007) 18S rRNA phylogeny of sea spiders with emphasis on the position of Rhynchothoracidae. Mar Biol 153(2):213–223

Stock JH (1974) Pycnogonida from the continental shelf, slope, and deep sea of the tropical Atlantic and East Pacific. Bull Mar Sci 24(4):957–1092

Strickland JDH, Parsons TR (1972) A practical handbook of seawater analysis, vol 167, 2nd edn. Bulletin of Fisheries Research Board of Canada, Ottawa, pp 1–293

Turpaeva EP (1975) On some deep-water pantopods (Pycnogonida) collected in north-western and South-Eastern Pacific. Trudy Inst Okeanol Akad Nauk SSSR 103:230–246

Vinu J, Kumar MR, Sumod KS, Deepa KP, Hashim M, Sanjeevan VN, Sudhakar M (2016) First record of deep-sea gigantic pycnogonid *Colossendeis colossea* Wilson, 1881 from the Western Indian Ocean. Mar Biodivers 46(2):495–498

Wicksten MK (2017) Feeding on cnidarians by giant pycnogonids (Pycnogonida: Colossendeidae Jarzinsky, 1870) in the north Central Pacific and North Atlantic oceans. J Crust Biol 37(3):359–360

Chapter 4
The Deep-Water Benthic Harpacticoida (Copepoda) of the Americas

S. Gómez and K. I. Rivera-Sánchez

Abstract Deep-sea harpacticoid copepods are poorly known worldwide. Due to sampling difficulties, these organisms are seldom captured. Recent investigation, however, has allowed for the discovery of many new species and genera in diverse regions of the Americas. The deep-water harpacticoid fauna from the Americas is reviewed, including their distribution and some aspects of their ecology. Information related to the occurrence of each species is provided and discussed in a regional context.

Keywords Benthic copepods · Diversity · Distribution

4.1 Introduction

The order Harpacticoida is one of the ten orders within the subclass Copepoda Milne Edwards, 1840 (Khodami et al. 2017). With about 5164 accepted species (Walter and Boxshall 2019), the ubiquitous order Harpacticoida is primitively benthic and can be found in a wide altitudinal range from well over 5000 m.a.s.l. (Kikuchi 1994; Ranga Reddy et al. 2014) to the deepest oceanic trenches (Becker 1974) and basins (Willen 2003; George et al. 2014; Mercado-Salas et al. 2019). Some harpacticoid species can be found in marine plankton samples at wide depth ranges (Boxshall 1979) but also in freshwater rivers (Beed 1962). Some others have successfully exploited hypersaline environments (Kolesnikova et al. 2017), and some can be found in damp moss (Menzel 1916; Chappuis 1931; Reid and Lesko 2003; Dahms and Qian 2004), leaf litter (Dumont and Maas 1988; Reid and Lesko 2003; Fiers and Jocque 2013), water-filled tree hollows (Reid and Lesko 2003), and phytotelmata (Por and Hadel 1986; Janetzky et al. 1996; Boxshall and Jaume 2000;

S. Gómez (✉)
Universidad Nacional Autónoma de México, Instituto de Ciencias del Mar y Limnología, Unidad Académica Mazatlán, Mazatlan, Mexico
e-mail: samuelgomez@ola.icmyl.unam.mx

K. I. Rivera-Sánchez
Posgrado en Ciencias del Mar y Limnología, Unidad Académica Mazatlán, Mazatlan, Mexico

M. E. Hendrickx (ed.), *Deep-Sea Pycnogonids and Crustaceans of the Americas*, https://doi.org/10.1007/978-3-030-58410-8_4

Kitching 2000; Reid and Lesko 2003). With about 500 hypogean species (Romero 2009), harpacticoids have successfully colonized wells (Bruno and Cottarelli 1999; Bruno and Perry 2004), caves (Cottarelli and Bruno 1993; Huys and Iliffe 1998; Suárez-Morales and Iliffe 2005; Apostolov 2007), cenotes (Wilson 1936; Gómez and Morales-Serna 2015), and the hyporheic zone of river beds (Corgosinho et al. 2007, 2010; Bruno and Cottarelli 2015). Harpacticoids can also be found as associates of a wide range of hosts from marine tetrapods and fishes to a variety of invertebrates, cyanobacteria, protozoans, macroalgae, and grasses (Huys 2016). Given their body size (<2.5 mm >0.2 mm; Giere 2009), benthic free-living harpacticoids form part of meiofaunal communities (also referred to as meiobenthos or interstitial communities), being the second most abundant taxon outnumbered only by nematodes both in the littoral (Giere 2009) and in the deep sea (Netto et al. 2005; Baguley et al. 2006a, b, 2008, 2015; Álvarez-Castillo et al. 2018; Neira et al. 2018).

Probably due to the constraints facing researchers for the study of deep-sea meiofauna (>200 m depth) but also due to the scarcity of specialized taxonomists, our knowledge of the harpacticoid deep-sea fauna from the Americas is relatively poor. At present, about 169 deep-sea nominotypical species of harpacticoid copepods distributed in 22 families have been reported from the Americas (4.1; see also Table 4.2 for references), being Miraciidae Dana, 1846 the best represented family (17 genera, 34 species), followed by Ancorabolidae Sars, 1909 (9 genera, 15 species), Argestidae Por, 1986 (9 genera, 31 species), Pseudotachidiidae Lang, 1936 (7 genera, 19 species), Aegisthidae Giesbrecht, 1893 (7 genera, 13 species), Cletodidae Scott T., 1904 (6 genera, 10 species), and Canthocamptidae Brady, 1880 (4 genera, 11 species). Ameiridae Boeck, 1865 (5 genera, 6 species); Dactylopusiidae Lang, 1936 (4 genera, 4 species); Ectinosomatidae Sars, 1903 (3 genera, 4 species); Parameiropsidae Corgosinho & Martínez Arbizu, 2010 and Zosimeidae Seifried, 2003 (one genus, 3 species each); Tisbidae Stebbing, 1910, Paramesochridae Lang, 1944, and Laophontidae Scott T., 1904 (2 genera, 2 species each); Neobradyidae Olofsson, 1917, Rhizothrichidae Por, 1986, and Tegastidae Sars, 1904 (one genus, 2 species each); and Cletopsyllidae Huys & Willems, 1989, Harpacticidae Dana, 1846, Idyanthidae Lang, 1948, and Normanellidae Lang, 1944 (one genus, one species each) are the less species-rich deep-sea families in the Americas. Of all the deep-sea harpacticoid species reported in published literature, 92 species have been described originally from the Americas (Tables 4.1 and 4.2, Figs. 4.1, 4.2, 4.3, and 4.4). On the other hand, the Eastern United States is the region with most records of deep-sea harpacticoids (20.74%), followed by Southern South America (15.05%), the Eastern Central Pacific (14.38%), the Gulf of Mexico (14.05%), the Northeastern Pacific (11.04%), the Southeast Pacific (10.37%), the Gulf of California (8.03%), the Beaufort Sea (4.01%), and the Campos Basin and Continental Slope of Sergipe, Brazil (2.34%) (Table 4.2).

Table 4.1 List of records of deep-sea harpacticoid nominotypical species (>200 m depth) known from the Americas. Fam., family

Taxon	Coordinates		Corresponding number in Figures
	Lat	Long	
Fam. Aegisthidae Giesbrecht, 1893			
Andromastax muricatus Conroy-Dalton and Huys, 1999*	**00°47.9′N**	**86°09.2′W**	1
Cervinia magna Smirnov, 1946	71°08.9′N	148°00.8′W	2
Cervinia unisetosa Montagna, 1980*	**71°45.1′N**	**150°35.0′W**	3
Cerviniella bodini Coull, 1973*	**34°20.5′N**	**75°45.8′W**	4
Cerviniella hamata Coull, 1973*	**34°6.5′N**	**75°46.8′W**	5
Cerviniella mirabilipes Smirnov, 1946	—	—	—
Cerviniella peruana Becker, 1974*	**12°03.0′S**	**78°45.0′W**	6
Cerviniopsis gorbunovi Smirnov, 1946	—	—	—
Cerviniopsis longicaudata Sars, 1903	—	—	—
Expansicervinia glacieria Montagna, 1980*	**72°42.0′N**	**143°40.0′W**	7
	72°58.0′N	**146°29.0′W**	8
Pontostratiotes peruanus Becker and Schriever, 1979*	**07°59.0′S**	**80°37.0′W**	9
Siphonis aureus Mercado-Salas, Khodami and Martínez Arbizu, 2019*	**11°4.3′N**	**119°36.3′W**	10
Siphonis ruehlemanni Mercado-Salas, Khodami and Martínez Arbizu, 2019*	**11°50.2′N**	**116°58.9′W**	11
	14°03.1′N	130°04.8′W	12
Fam. Ameiridae Boeck, 1865			
Ameira bathyalis Becker and Schriever, 1979*	**11°50.0′S**	**77°58.0′W**	13
Ameira longipes Boeck, 1865	27°06.4′N	91°09.9′W	14
Ameiropsis mixta Sars, 1907	27°06.4′N	91°09.9′W	15
	27°06.4′N	91°09.9′W	16
Argestigens celibis Gómez, 2018*	**27°42.0′N**	**111°37.9′W**	17
Malacopsyllus hades Becker, 1974*	**08°52.0′S**	**80°47.0′W**	18
Sarsameira knorri Reidenauer and Thistle, 1983*	**40°24.0′N**	**63°07.4′W**	19
	40°24.3′N	**63°09.6′W**	20
Fam. Ancorabolidae Sars, 1909			
Ancorabolus hendrickxi Gómez and Conroy-Dalton, 2002*	**25°54.0′N**	**110°11.0′W**	21
Arthuricornua anendopodia Conroy-Dalton, 2001*	**32°35.8′N**	**117°29.0′W**	22
Ceratonotus elongatus Gómez and Díaz, 2017*	**27°01.0′N**	**110°53.1′W**	23
Ceratonotus thistlei Conroy-Dalton, 2003*	**32°35.8N**	**117°29.0′W**	24
Dendropsyllus californiensis Gómez and Díaz, 2017*	**26°41.1′N**	**111°12.0′W**	25
Dendropsyllus magellanicus (George and Schminke, 1998)*	42°24.6′S	74°47.3′W	26
	53°21.2′S	**70°42.7′W**	27

(continued)

Table 4.1 (continued)

Taxon	Coordinates		Corresponding number in Figures
	Lat	Long	
Dendropsyllus thomasi Conroy-Dalton, 2003*	**32°35.8N**	**117°29.0′W**	28
Dorsiceratus triarticulatus Coull, 1973*	**34°14.4′N**	**75°49.5′W**	29
Echinopsyllus brasiliensis Wandeness, George and Santos, 2009*	**22°04.6′S**	**39°52.1′W**	30
Echinopsyllus grohmannae Wandeness, George and Santos, 2009*	**22°31.5′S**	**40°03.8′W**	31
Echinopsyllus nogueirae Wandeness, George and Santos, 2009*	**22°31.5′S**	**40°03.8′W**	32
Laophontodes sabinegeorgeae George and Gheerardyn, 2015*	**55°47.8′S**	**65°48.8′W**	33
Laophontodes gertraudae George, 2018*	**55°47.8′ S,**	**65°48.8′W**	34
Polyascophorus monoceratus George, Wandeness and Santos, 2013*	**21°54.6′S**	**39°54.6′W**	35
Uptionyx verenae Conroy-Dalton and Huys, 2000*	**48°27.5′N**	**128°42.5′W**	36
Fam. Argestidae Por, 1986			
Argestes analongises Gómez, 2018*	**22°01.0′N**	**106°40.0′W**	37
Argestes mollis Sars, 1910	—	—	—
Argestoides prehensilis Huys & Conroy-Dalton, 1997*	**00°47.9′N**	**86°09.2′W**	38
Eurycletodes (Eurycletodes) gorbunovi Smirnov, 1946	—	—	—
Eurycletodes (Oligocletodes) echinatus Lang, 1936	—	—	—
Eurycletodes (Oligocletodes) hoplurus Smirnov, 1946	—	—	—
Eurycletodes (Oligocletodes) major Sars, 1909	—	—	—
Eurycletodes (Oligocletodes) monardi Smirnov, 1946	54°58.7′S	69°01.9′W	39
Eurycletodes (Oligocletodes) oblongus Sars, 1920	54°58.7′S	69°01.9′W	40
	53°21.0′S	70°42.7W	41
	53°41.5′S	70°56.5′W	42
Eurycletodes (Oligocletodes) peruanus Becker, 1979*	**08°22.0′S**	**81°2.0′W**	43
Eurycletodes paraephippiger Gómez, 2018*	**27°09.1′N**	**111°39.9′W**	44
Fultonia bougisi Soyer, 1964	54°58.7′S	69°01.9′W	45
Fultonia sarsi (Smirnov, 1946)	54°58.7′S	69°01.9′W	46
	53°41.5′S	70°56.5′W	47

(continued)

Table 4.1 (continued)

Taxon	Coordinates		Corresponding number in Figures
	Lat	Long	
Mesocletodes abyssicola (Scott T. and Scott A., 1901)	53°33.0′S	70°39.2′W	48
	53°41.8′S	70°54.6′W	49
	53°43.4′S	70°56.0′W	50
Mesocletodes bathybia Por, 1964	—	—	—
Mesocletodes bicornis Menzel and George, 2009	—	—	—
Mesocletodes brevisetosus Gómez, 2018*	**25°54.7′N**	**110°11.0′W**	51
Mesocletodes comixtus Coull, 1973*	**34°14.4′N**	**75°49.5′W**	52
Mesocletodes dolichurus Smirnov, 1946	—	—	—
Mesocletodes elmari Menzel, 2011	09°50.5′N	104°17.5′W	53
Mesocletodes irrasus (Scott T. and Scott A., 1894)	—	—	—
Mesocletodes katharinae Soyer, 1964	—	—	—
Mesocletodes nudus Vakati, Thistle and Lee, 2017*	**32°35.8N**	**117°29.0′W**	54
Mesocletodes parirrasus Becker, 1979*	**12°03.0′S**	**78°45.0′W**	55
Mesocletodes robustus Por, 1965	—	—	—
Mesocletodes simplex Gómez, 2018*	**27°01.0′N**	**110°53.1′W**	56
Mesocletodes soyeri Bodin, 1968	54°58.1′S	68°49.9′W	57
	54°58.7′S	69°01.9′W	58
Mesocletodes tetrasetosus Vakati, Thistle and Lee, 2017*	**32°35.8N**	**117°29.0′W**	59
Mesocletodes unisetosus Gómez, 2018*	**25°54.7′N**	**110°11.0′W**	60
Neoargestes laevis Gheerardyn and George, 2019*	**14°42.4′N**	**125°26.5′W**	61
Odiliacletodes secundus Gómez, 2018*	**27°07.0′N**	**110°53.4′W**	62
Fam. Canthocamptidae Brady, 1880			
Bathycamptus eckmani Huys and Thistle, 1989*	**32°35.8N**	**117°29.0′W**	63
	32°51.0′N	117°46.9′W	64
Carolinicola trisetosa (Coull, 1973) *	**34°05.0′N**	**74°16.0′W**	65
Hemimesochra rapiens Becker, 1979*	**12°03.0′S**	**78°05.0′W**	66
Metahuntemannia crassa (Por, 1965)	—	—	—
Metahuntemannia drzycimskii Soyer, 1970	—	—	—
Metahuntemannia gorbunovi Smirnov, 1946	—	—	—
Metahuntemannia magniceps Becker, 1979*	**12°03.0′S**	**78°45.0′W**	67
Metahuntemannia pacifica Becker, 1979*	**11°50.0′S**	**77°58′W**	68
Metahuntemannia peruana Becker, 1979*	**12°03.0′S**	**78°05.0′W**	69
	12°04.0′S	**78°05.0′W**	70
Metahuntemannia spinosa (Klie, 1939)	—	—	—
Metahuntemannia talpa Becker, 1979*	**11°50.0′S**	**77°58.0′W**	71

(continued)

Table 4.1 (continued)

Taxon	Coordinates		Corresponding number in Figures
	Lat	Long	
Fam. Cletodidae Scott T., 1904			
Barbaracletodes barbara Becker, 1979*	**12°03.0′S**	**78°45.0′W**	72
Barbaracletodes carola Becker, 1979*	**11°50.0′S**	**77°58.0′W**	73
Cletodes latirostris Drzycimski, 1967	53°41.5′S	70°56.5′W	74
	53°41.8′S	70°54.6′W	75
	53°43.4′S	70°56.0′W	76
Cletodes longicaudatus (Boeck, 1872)	27°06.7′N	91°09.9′W	77
	27°06.7′N	91°09.9′W	78
Cletodes yotabis Por, 1967	—	—	—
Echinocletodes voightae George and Müller, 2013*	**41°00.0′N**	**127°29.7′W**	79
Enhydrosoma hopkinsi Lang, 1965	53°41.8′S	70°54.6′W	80
Pyrocletodes desuramus Coull, 1973*	**34°23.0′N**	**75°41.5′W**	81
Stylicletodes longicaudatus (Brady, 1880)	54°54.5′S	68°38.7′W	82
	54°58.7′S	69°01.9′W	83
	55°29.0′S	66°04.4′W	84
	53°21.0′S	70°42.7W	85
	53°41.5′S	70°56.5′W	86
	53°43.4′S	70°56.0′W	87
Stylicletodes oligochaeta Bodin, 1968	54°58.7′S	69°01.9′W	88
	53°21.0′S	70°42.7W	89
	53°41.5′S	70°56.5′W	90
	53°41.8′S	70°54.6′W	91
	53°43.4′S	70°56′W	92
Fam. Cletopsyllidae Huys and Willems, 1989			
Pentacletopsyllus montagni Bang, Baguley and Moon, 2014*	**28°43.2′N**	**88°20.7′W**	93
Fam. Dactylopusiidae Lang, 1936			
Dactylopodopsis dilatata Sars, 1911	—	—	—
Dactylopusia spinipes Brady, 1910	—	—	—
Diarthrodes aegidaeus (Brian, 1927)	—	—	—
Paradactylopodia brevicornis (Claus, 1866)	—	—	—
Fam. Ectinosomatidae Sars, 1903			
Bradya typica Boeck, 1873	—	—	—
Microsetella norvegica (Boeck, 1865)	09°50.0′N	—	—
Microsetella rosea (Dana, 1847)	24°15.2′N	108°10.2′W	94
	24°56.1′N	109°05.0′W	95
	24°56.4′N	109°05.6′W	96
	24°14.7′N	108°10.9′W	97
Pseudobradya confluens (Lang, 1936)	—	—	—

(continued)

Table 4.1 (continued)

Taxon	Coordinates		Corresponding number in Figures
	Lat	Long	
Fam. Harpacticidae Dana, 1846			
Harpacticus superflexus Willey, 1920	—	—	—
Fam. Idyanthidae Lang, 1948			
Dactylopia peruana Becker, 1974*	**11°50.0′S**	**77°58.0′W**	98
Fam. Laophontidae Scott T., 1904			
Archesola typhlops (Sars, 1908)	27°06.4′N	91°09.9′W	99
	27°06.7′N	91°09.9′W	100
	27°06.7′N	91°09.9′W	101
	27°06.4′N	91°09.9′W	102
	27°06.4′N	91°09.9′W	103
	27°06.4′N	91°09.9′W	104
	27°06.7′N	91°09.9′W	105
	27°06.7′N	91°09.9′W	106
	27°06.7′N	91°09.9′W	107
Bathylaophonte pacifica Lee and Huys, 1999*	11°24.9′N	103°47.2′W	108
	09°50.5′N	104°17.5′W	109
	09°50.7′N	104°17.4′W	110
	17°25.8′S	**113°12.3′W**	111
Fam. Miraciidae Dana, 1846			
Amphiascoides debilis (Giesbrecht, 1881)	—	—	—
Amphiascoides neglectus (Norman and Scott T., 1905)	27°06.4′N	91°09.9′W	112
	27°06.7′N	91°09.9′W	113
	27°06.4′N	91°09.9′W	114
	27°06.4′N	91°09.9′W	115
	27°06.7′N	91°09.9′W	116
Amphiascoides subdebilis (Willey, 1935)	55°29.0′S	66°04.4′W	117
Amphiascopsis cinctus (Claus, 1866)	—	—	—
Amphiascus giesbrechti Sars, 1906	—	—	—
Amphiascus parvulus (Claus, 1866)	27°06.4′N	91°09.9′W	118
	27°06.7′N	91°09.9′W	119
	27°06.7′N	91°09.9′W	120
	27°06.4′N	91°09.9′W	121
	27°06.4′N	91°09.9′W	122
	27°06.4′N	91°09.9′W	123
	27°06.7′N	91°09.9′W	124
	27°06.7′N	91°09.9′W	125
	27°06.7′N	91°09.9′W	126
Amphiascus soyeri (Lang, 1965)	—	—	—
Archaeohuysia huysi Gómez, 2020*	**30°55.5′N**	**116°49.6′W**	127

(continued)

Table 4.1 (continued)

Taxon	Coordinates		Corresponding number in Figures
	Lat	Long	
Beatricella calidafornax Gómez, 2020*	**27°42.0′N**	**111°37.9′W**	128
Delavalia asetosa Gómez, 2020*	**24°16.0′N**	**108°24.1′W**	129
Delavalia californiensis Gómez, 2020*	**26°32.6′N**	**113°56.3′W**	130
Delavalia diegensis (Thistle and Coull, 1979)*	**32°34.8′N**	**117°29.0′W**	131
Delavalia gundulae (Willen, 2003)	27°06.4′N	91°09.9′W	132
	27°06.7′N	91°09.9′W	133
Delavalia lima (Becker and Schriever, 1979)*	**11°50.0′S**	**77°58.0′W**	134
Delavalia profunda Gómez, 2020*	**27°07.1′N**	**114°36.1′W**	135
Delavalia reducta Gómez, 2020*	**31°48.3′N**	**116°55.5′W**	136
Diarthropodella secunda Gómez, 2020*	**27°09.1′N**	**111°39.9′W**	136
Diarthropodella prima Gómez, 2020*	24°16.0′N	108°24.1′W	138
	27°42.0′N	**111°37.9′W**	139
Haloschizopera abyssi Becker, 1974*	**12°04.0′S**	**78°05.0′W**	140
	53°41.8′S	70°54.6′W	141
	53°42.6′S	70°48.7′W	142
	53°43.4′S	70°56.0′W	143
Haloschizopera exigua (Sars, 1906)	54°54.5′S	68°38.7′W	144
	54°58.7′S	69°01.9′W	145
	55°27.8′S	66°09.1′W	146
	53°21.0′S	70°42.7W	147
	53°33.0′S	70°39.2′W	148
	53°41.5′S	70°56.5′W	149
	53°41.8′S	70°54.6′W	150
	53°42.6′S	70°48.7′W	151
	53°43.4′S	70°56.0′W	152
Haloschizopera lima Becker, 1974*	**11°50.0′S**	**77°58.0′W**	153
Haloschizopera pygmaea (Norman and Scott T., 1905)	—	—	—
Paramphiascella intermedia (Scott T., 1897)	—	—	—
Pseudostenhelia bathyalis Gómez, 2020*	**29°48.8′N**	**116°07.3′W**	154
Robertgurneya rostrata (Gurney, 1927)	—	—	—
Robertsonia knoxi (Thompson I.C. and Scott A., 1903)	—	—	—
Sarsamphiascus minutus (Claus, 1863)	55°27.8′S	66°09.1′W	155
Sarsamphiascus parvus (Sars M., 1906)	—	—	—
Sarsamphiascus profundus (Becker and Schriever, 1979)*	**11°50.0′S**	**77°58.0′W**	156
Sarsamphiascus varians (Norman and Scott T., 1905)	—	—	—

(continued)

Table 4.1 (continued)

Taxon	Coordinates Lat	Long	Corresponding number in Figures
Schizopera carolinensis Coull, 1971	—	—	—
Typhlamphiascus lamellifer lamellifer (Sars, 1911)	—	—	—
Wellstenhelia euterpoides Gómez and Cruz-Barraza, 2020*	**26°32.6′N**	**113°56.3′W**	157
	27°07.1′N	114°36.1′W	158
	27°01.0′N	110°53.7′W	159
Wellstenvalia wellsi Gómez and Cruz-Barraza, 2020*	23°17.7′N	110°21.9′W	160
	26°32.6′N	113°56.3′W	161
	27°07.1′N	**114°36.1′W**	162
	30°55.5′N	116°49.6′W	163
	27°42.0′N	111°37.9′W	164
	24°56.0′N	109°04.2′W	165
	27°09.9′N	111°47.1′W	166
Fam. Neobradyidae Olofsson, 1917			
Marsteinia bozici (Bodin, 1968)	—	—	—
Marsteinia similis Drzycimski, 1968	—	—	—
Fam. Normanellidae Lang, 1944			
Pseudocletodes vararensis Scott T. and Scott A., 1893	—	—	—
Fam. Parameiropsidae Corgosinho and Martínez Arbizu, 2010			
Parameiropsis kodosensis Cho, Wi and Suh, 2016*	**10°30.9′N**	**131°18.9′W**	167
Parameiropsis peruanus Becker, 1974*	**12°03.0′S**	**78°45.0′W**	168
Parameiropsis tetraspinosa Cho, Wi and Suh, 2016*	**10°30.1′N**	**131°19.9′W**	169
Fam. Paramesochridae Lang, 1944			
Emertonia minor (Vasconcelos, Veit-Köhler, Drewes and Parreira dos Santos, 2009)*	**11°29.7′S**	**37°09.7′W**	170
Wellsopsyllus (Wellsopsyllus) abyssalis (Becker, 1979)*	**12°03.0′S**	**78°45.0′W**	171
Fam. Pseudotachidiidae Lang, 1936			
Cylindronannopus primus Coull, 1973*	**34°14.0′N**	**75°20.0′W**	172
Keraia longiseta (Vasconcelos, George and Parreira dos Santos, 2008)*	**11°29.7′S**	**37°09.7′W**	173
Keraia tamara (Smirnov, 1946)	—	—	—
Parananopus echinipes Smirnov, 1946	—	—	—
Parananopus atlanticus Coull, 1973*	**34°09.0′N**	**75°54.0′W**	174
Parananopus longithorax Becker, 1979*	**12°03.0′S**	**78°45.0′W**	175
Parananopus truncatus Becker, 1979*	**08°56.0′S**	**80°47.0′W**	176

(continued)

Table 4.1 (continued)

Taxon	Coordinates		Corresponding number in Figures
	Lat	Long	
Psammis longipes Becker, 1974*	27°06.4′N	91°09.9′W	177
	12°03.0′S	**78°45.0′W**	178
Pseudomesochra gemina Coull, 1973*	**34°16.4′N**	**75°36.1′W**	179
Pseudomesochra longifurcata Scott T., 1902	54°58.7′S	69°01.9′W	180
Pseudomesochra media (Sars, 1911)	—	—	—
Pseudomesochra minor Becker, 1974*	**11°50.0′S**	**77°58.0′W**	181
Pseudomesochra similis Lang, 1935	—	—	—
Pseudotachidius abyssalis Becker, 1974*	**12°03.0′S**	**78°45.0′W**	182
Pseudotachidius biparttitus Montagna, 1980*	**70°42.8′N**	**141°39.5′W**	183
Pseudotachidius brevisetosus Montagna, 1980*	**70°42.8′N**	**147°39.5′W**	184
Pseudotachidius peruanus Becker, 1974*	**12°03.0′S**	**78°45.0′W**	185
Pseudotachidius vikingus Drzycimski, 1968	—	—	—
Xylora bathyalis Hicks, 1988	09°50.5′N	104°17.5′W	186
	09°50.7′N	104°17.4′W	187
	27°06.7′N	91°09.9′W	188
	27°06.7′N	91°09.9′W	189
	27°06.7′N	91°09.9′W	190
	27°06.4′N	91°09.9′W	191
	27°06.7′N	91°09.9′W	192
Fam. Rhizothrichidae Por, 1986			
Rhizothrix gracilis (Scott T., 1903)	—	—	—
Rhizothrix longiseta Gómez, 2018*	27°42.0′N	111°37.9′W	193
	27°09.9′N	**111°47.1′W**	194
Fam. Tegastidae Sars, 1904			
Smacigastes barti Gollner, Ivanenko and Martínez Arbizu, 2008*	**09°50.5′N**	**104°17.5′W**	195
Smacigastes methanophilus Plum and Martínez Arbizu, 2009*	**27°06.4′N**	**91°09.9′W**	196
Fam. Tisbidae Stebbing, 1910			
Cholidya polypi Farran, 1914	36°40.7—41′N	122°17.6—19.4′W	197
	29°43.0′N	80°10.0′W	198
	34°18.0′N	75°51.0′W	199
	35°04.0′N	75°06.0′W	200
	39°55.0′N	70°39.0′W	201
	48°27.4′N	128°42.5′W	202
	45°21.0′N	125°37.3′W	203
	45°57.6′N	125°46.2′W	204
	45°59.6′N	125°44.0′W	205

(continued)

Table 4.1 (continued)

Taxon	Coordinates		Corresponding number in Figures
	Lat	Long	
Genesis vulcanoctopusi López-González, Bresciani and Huys in López-González, Bresciani, Huys, González, Guerra and Pascual, 2000*	**12°48.4′N**	**103°56.4′W**	206
Fam. Zosimeidae Seifried, 2003			
Zosime atlantica Bodin, 1968	—	—	—
Zosime paramajor Bodin, 1968	—	—	—
Zosime typica Boeck, 1873	—	—	—

Asterisks (*) in the species authority indicate species originally described from the Americas
Coordinates in bold indicate the type locality
For references and depths see Table 4.2. Numbers of species correspond to figures

4.2 Material and Methods

The taxonomy and systematics of harpacticoid copepods is in a constant state of flux. A number of new taxa (families, genera, species, etc.) are described every year, and some others are reallocated to new or to already existing valid taxa in an effort to better understand their phylogenetic relationships. Some species names might have changed with time, and the reader is suggested to check the status of the species of interest in case an original reference is consulted. In this chapter, we followed the World Register of Marine Species (http://www.marinespecies.org/aphia.php?p=taxdetails&id=1102) as an authoritative source for the current classification and accepted names of harpacticoid copepods. Also, for the distribution of species described before 1948, we followed Lang (1948) as an authoritative source of information.

The description of new taxa and the report of new records are commonly accompanied by the coordinates where the taxon was found. These coordinates are presented in the original literature using different formats and degrees of precision. When available in the original publication, coordinates in this chapter follow the degrees decimal minutes (DDM) format, and the reader is suggested to check the original source if another format or if more precision is required.

This review is based largely on a revision of published descriptions and records of deep-sea harpacticoids from the Americas. Relevant information can also be found in gray literature (e.g., theses, dissertations, etc.). Comments and references of gray literature may appear in this chapter, but the species lists therein were not considered here. Finally, some papers on the ecology and distribution of harpacticoid copepods contain records of suprageneric and supraspecific taxa (George 2005; Thistle et al. 2007; Brooks et al. 2009; Gollner et al. 2015a, b; Plum et al. 2015;

Table 4.2 Published records (grey literature excluded) of deep-sea harpacticoids (>200 m depth) from each region in the Americas. Fam., family

Taxon	Basin	Station	Lat	Long	Depth (m)	Corresponding number in Figures	Associations	References
Fam. Aegisthidae Giesbrecht, 1893	**Eastern US**							
Cervinia magna Smirnov, 1946		Stn 13	—	—	—	—	Free-living	Coull (1973a)
Cerviniella bodini Coull, 1973		Stn **5**, 6	**34°20.5′N**	**75°45.8′W**	500	4	Free-living	Coull (1973a)*
Cerviniella hamata Coull, 1973		Stn **14**, 16	**34°6.5′N**	**75°46.8′W**	1000	5	Free-living	Coull (1973a)*
Cerviniella mirabilipes Smirnov, 1946		Stn 21	—	—	—	—	Free-living	Coull (1973a)
Cerviniopsis gorbunovi Smirnov, 1946		Stn 4, 16, 17	—	—	—	—	Free-living	Coull (1973a)
Cerviniopsis longicaudata Sars, 1903		Stn 10, 12	—	—	—	—	Free-living	Coull (1973a)
Eucanuella aff. reticulata Soyer, 1970		Stn 5, 17, 18	—	—	—	—	Free-living	Coull (1973a)
Fam. Ameiridae Boeck, 1865								
Sarsameira knorri Reidenauer and Thistle, 1983		—	**40°24.0′N**	**63°07.4′W**	4626	19	Free-living	Reidenauer and Thistle (1983)*
		—	**40°24.3′N**	**63°09.6′W**	4626	20	Free-living	Reidenauer and Thistle (1983)*

Fam. Ancorabolidae Sars, 1909								
Dorsiceratus triarticulatus Coull, 1973		Stn 11	**34°14.4′N**	**75°49.5′W**	500	29	Free-living	Coull (1973b)*
Fam. Argestidae Por, 1986								
Eurycletodes (Eurycletodes) gorbunovi Smirnov, 1946		Stn 13	—	—	—	—	Free-living	Coull (1973b)
Eurycletodes (Oligocletodes) echinatus Lang, 1936		Stn 6, 13	—	—	—	—	Free-living	Coull (1973b)
Eurycletodes (Oligocletodes) hoplurus Smirnov, 1946		Stn 13	—	—	—	—	Free-living	Coull (1973b)
Eurycletodes (Oligocletodes) major Sars, 1909		Stn 3	—	—	—	—	Free-living	Coull (1973b)
Eurycletodes (Oligocletodes) monardi Smirnov, 1946		Stn 4, 18, 29	—	—	—	—	Free-living	Coull (1973b)
Mesocletodes abyssicola (Scott T. and Scott A., 1901)		Stn 18	—	—	—	—	Free-living	Coull (1973b)
Mesocletodes comixtus Coull, 1973		Stn 7, **11**	**34°14.4′N**	**75°49.5′W**	500	53	Free-living	Coull (1973b)*
Mesocletodes dolichurus Smirnov, 1946		Stn 4	—	—	—	—	Free-living	Coull (1973b)

(continued)

Table 4.2 (continued)

Taxon	Basin	Station	Lat	Long	Depth (m)	Corresponding number in Figures	Associations	References
Mesocletodes irrasus (Scott T. and Scott A., 1894)		Stn 4, 6, 12, 13, 18, 21	—	—	—	—	Free-living	Coull (1973b)
Mesocletodes katharinae Soyer, 1964		Stn 4, 6, 27	—	—	—	—	Free-living	Coull (1973b)
Mesocletodes robustus Por, 1965		Stn 11	—	—	500	—	Free-living	Coull (1973b)
Mesocletodes soyeri Bodin, 1968		Stn 6	—	—	500	—	Free-living	Coull (1973b)
Fam. Canthocamptidae Brady, 1880								
Carolinicola trisetosa (Coull, 1973)		Stn **13**, 15, 18, 21	**34°05.0′N**	**74°16.0′W**	3940	66	Free-living	Coull (1973d)*
Metahuntemannia crassa (Por, 1965)		Stn 13, 24, 26	—	—	—	—	Free-living	Coull (1973b)
Metahuntemannia drzycimskii Soyer, 1970		Stn 7, 12	—	—	—	—	Free-living	Coull (1973b)
Metahuntemannia gorbunovi Smirnov, 1946		Stn 21	—	—	—	—	Free-living	Coull (1973b)
Metahuntemannia spinosa (Klie, 1939)		Stn 19	—	—	—	—	Free-living	Coull (1973b)
Fam. Cletodidae Scott T., 1904								
Cletodes yotabis Por, 1967		Stn 7	—	—	—	—	Free-living	Coull (1973b)

Pyrocletodes desuramus Coull, 1973		Stn 4	**34°23.0′N**	**75°41.5′W**	1050	82	Free-living	Coull (1973b)*
Fam. Dactylopusiidae Lang, 1936								
Dactylopodopsis dilatata Sars, 1911		Stn 4, 14	—	—	—	—	Free-living	Coull (1973c)
Dactylopusia spinipes Brady, 1910		Stn 3	—	—	—	—	Free-living	Coull (1973c)
Diarthrodes aegidaeus (Brian, 1927)		Stn 1, 3	—	—	—	—	Free-living	Coull (1973c)
Paradactylopodia brevicornis (Claus, 1866)		Stn 1	—	—	—	—	Free-living	Coull (1973c)
Fam. Miraciidae Dana, 1846								
Amphiascoides debilis (Giesbrecht, 1881)		Stn 9	—	—	—	—	Free-living	Coull (1973c)
Amphiascopsis cinctus (Claus, 1866)		Stn 1	—	—	—	—	Free-living	Coull (1973c)
Amphiascus soyeri (Lang, 1965)		Stn 3, 6, 9	—	—	—	—	Free-living	Coull (1973c)
Haloschizopera pygmaea (Norman and Scott T., 1905)		Stn 1, 3	—	—	—	—	Free-living	Coull (1973c)
Paramphiascella intermedia (Scott T., 1897)		Stn 9	—	—	—	—	Free-living	Coull (1973c)
Robertgurneya rostrata (Gurney, 1927)		Stn 3, 11	—	—	—	—	Free-living	Coull (1973c)

(continued)

Table 4.2 (continued)

Taxon	Basin	Station	Lat	Long	Depth (m)	Corresponding number in Figures	Associations	References
Robertsonia knoxi (Thompson I.C. and Scott A., 1903)		Stn 11	—	—	—	—	Free-living	Coull (1973c)
Sarsamphiascus parvus (Sars M., 1906)		Stn 11	—	—	—	—	Free-living	Coull (1973c)
Sarsamphiascus varians (Norman and Scott T., 1905)		Stn 1, 11	—	—	—	—	Free-living	Coull (1973c)
Schizopera carolinensis Coull, 1971		Stn 1	—	—	—	—	Free-living	Coull (1973c)
Fam. Neobradyidae Olofsson, 1917								
Marsteinia bozici (Bodin, 1968)		Stn 13	—	—	—	—	Free-living	Coull (1973c)
Marsteinia similis Drzycimski, 1968		Stn 4, 6, 13, 29	—	—	—	—	Free-living	Coull (1973c)
Fam. Normanellidae Lang, 1944								
Pseudocletodes vararensis Scott T. and Scott A., 1893		Stn 3, 6, 9, 10, 11	—	—	—	—	Free-living	Coull (1973b)
Fam. Pseudotachidiidae Lang, 1936								
Cylindronannopus primus Coull, 1973		Stn **13**, 18	**34°14.0′N**	**75°20.0′W**	3000	173	Free-living	Coull (1973d)*

Keraia tamara (Smirnov, 1946)		Stn 4, 12, 13, 18	—	—	—	—	Free-living	Coull (1973c)
Paranannopus atlanticus Coull, 1973		Stn **10**, 13, 15, 18	**34°09.0′N**	**75°54.0′W**	500	175	Free-living	Coull (1973d)*
Pseudomesochra gemina Coull, 1973		**Stn 12**	**34°16.4′N**	**75°36.1′W**	2300	180	Free-living	Coull (1973c)*
Pseudomesochra media (Sars, 1911)		Stn 3, 6, 7, 23, 29	—	—	—	—	Free-living	Coull (1973c)
Pseudomesochra similis Lang, 1935		Stn 4, 18, 27	—	—	—	—	Free-living	Coull (1973c)
Pseudotachidius vikingus Drzycimski, 1968		Stn 14, 18	—	—	—	—	Free-living	Coull (1973c)
Fam. Rhizothrichidae Por, 1986								
Rhizothrix gracilis (Scott T., 1903)		Stn 1	—	—	—	—	Free-living	Coull (1973b)
Fam. Tisbidae Stebbing, 1910								
Cholidya polypi Farran, 1914		—	29°43.0′N	80°10.0′W	357	199	Parasite of *Tetracheledone spinicirrhus* Voss, 1955 (Cephalopoda: Octopoda: Megaleledonidae)	Humes and Voight (1997)
		—	34°18.0′N	75°51.0′W	356	200	Parasite of *Tetracheledone spinicirrhus* Voss, 1955 (Cephalopoda: Octopoda: Megaleledonidae)	Humes and Voight (1997)

(continued)

Table 4.2 (continued)

Taxon	Basin	Station	Lat	Long	Depth (m)	Corresponding number in Figures	Associations	References
		—	35°04.0′N	75°06.0′W	356	201	Parasite of *Bathypolypus arcticus* (Prosch, 1849) (Cephalopoda: Octopoda: Bathypolypodidae)	Humes and Voight (1997)
		—	39°55.0′N	70°39.0′W	356—366	202	Parasite of *Bathypolypus arcticus* (Prosch, 1849) (Cephalopoda: Octopoda: Bathypolypodidae)	Humes and Voight (1997)
Fam. Zosimeidae Seifried, 2003								
Zosime atlantica Bodin, 1968		Stn 11	—	—	—	—	Free-living	Coull (1973c)
Zosime cf. incrassata incrassata Sars, 1910		Stn 14, 12, 13, 18	—	—	—	—	Free-living	Coull (1973c)
Zosime paramajor Bodin, 1968		Stn 4, 11, 13, 29	—	—	—	—	Free-living	Coull (1973c)
Zosime typica Boeck, 1873		Stn 13	—	—	—	—	Free-living	Coull (1973c)

Fam. Ancorabolidae Sars, 1909	**Southern South America (Magellan Straits, Beagle Channel and Patagonian Continental Slope)**							
Dendropsyllus magellanicus (George and Schminke, 1998)		—	**53°21.2′S**	**70°42.7′W**	200	27	Free-living	George and Schminke (1998)*, George (2005)
Laophontodes gertraudae George, 2018		#40/106	**55°47.8′ S,**	**65°48.8′W**	2550	34	Free-living	George (2018)*
Laophontodes sabinegeorgeae George and Gheerardyn, 2015		—	**55°47.8′S**	**65°48.8′W**	2250	33	Free-living	George and Gheerardyn (2015)*
Fam. Argestidae Por, 1986								
Eurycletodes (Oligocletodes) monardi Smirnov, 1946		—	54°58.7′S	69°01.9′W	219	40	Free-living	George (2005)
Eurycletodes (Oligocletodes) oblongus Sars, 1920		—	54°58.7′S	69°01.9′W	219	41	Free-living	George (2005)
		—	53°21.0′S	70°42.7W	200	42	Free-living	George (2005)
		—	53°41.5′S	70°56.5′W	227	43	Free-living	George (2005)
Fultonia bougisi Soyer, 1964		—	54°58.7′S	69°01.9′W	219	46	Free-living	George (2005)
Fultonia sarsi (Smirnov, 1946)		—	54°58.7′S	69°01.9′W	219	47	Free-living	George (2005)

(continued)

Table 4.2 (continued)

Taxon	Basin	Station	Lat	Long	Depth (m)	Corresponding number in Figures	Associations	References
		—	53°41.5′S	70°56.5′W	227	48	Free-living	George (2005)
Mesocletodes abyssicola (Scott T. and Scott A., 1901)		—	53°33.0′S	70°39.2′W	459	49	Free-living	George (2005)
		—	53°41.8′S	70°54.6′W	440	50	Free-living	George (2005)
		—	53°43.4′S	70°56.0′W	351	51	Free-living	George (2005)
Mesocletodes soyeri Bodin, 1968		—	54°58.1′S	68°49.9′W	257	58	Free-living	George (2005)
		—	54°58.7′S	69°01.9′W	219	59	Free-living	George (2005)
Fam. Cletodidae Scott T., 1904								
Cletodes latirostris Drzycimski, 1967		—	53°41.5′S	70°56.5′W	227	75	Free-living	George (2005)
		—	53°41.8′S	70°54.6′W	440	76	Free-living	George (2005)
		—	53°43.4′S	70°56.0′W	351	77	Free-living	George (2005)
Enhydrosoma hopkinsi Lang, 1965		—	53°41.8′S	70°54.6′W	440	81	Free-living	George (2005)
Stylicletodes longicaudatus (Brady, 1880)		—	54°54.5′S	68°38.7′W	320	83	Free-living	George (2005)
		—	54°58.7′S	69°01.9′W	219	84	Free-living	George (2005)
		—	55°29.0′S	66°04.4′W	1168	85	Free-living	George (2005)
		—	53°21.0′S	70°42.7W	200	86	Free-living	George (2005)
		—	53°41.5′S	70°56.5′W	227	87	Free-living	George (2005)

		—	53°43.4′S	70°56.0′W	351	88	Free-living	George (2005)
Stylicletodes oligochaeta Bodin, 1968		—	54°58.7′S	69°01.9′W	219	89	Free-living	George (2005)
		—	53°21.0′S	70°42.7W	200	90	Free-living	George (2005)
		—	53°41.5′S	70°56.5′W	227	91	Free-living	George (2005)
		—	53°41.8′S	70°54.6′W	440	92	Free-living	George (2005)
		—	53°43.4′S	70°56′W	351	93	Free-living	George (2005)
Fam. Miraciidae Dana, 1846								
Amphiascoides subdebilis (Willey, 1935)		—	55°29′S	66°04.4′W	1168	118	Free-living	George (2005)
Haloschizopera abyssi Becker, 1974		—	53°41.8′S	70°54.6′W	440	142	Free-living	George (2005)
		—	53°42.6′S	70°48.7′W	550	143	Free-living	George (2005)
		—	53°43.4′S	70°56.0′W	351	144	Free-living	George (2005)
Haloschizopera exigua (Sars, 1906)		—	54°54.5′S	68°38.7′W	320	145	Free-living	George (2005)
		—	54°58.7′S	69°01.9′W	219	146	Free-living	George (2005)
		—	55°27.8′S	66°09.1′W	336	147	Free-living	George (2005)
		—	53°21.0′S	70°42.7W	200	148	Free-living	George (2005)
		—	53°33.0′S	70°39.2′W	459	149	Free-living	George (2005)
		—	53°41.5′S	70°56.5′W	227	150	Free-living	George (2005)
		—	53°41.8′S	70°54.6′W	440	151	Free-living	George (2005)
		—	53°42.6′S	70°48.7′W	550	152	Free-living	George (2005)
		—	53°43.4′S	70°56.0′W	351	153	Free-living	George (2005)
Sarsamphiascus minutus (Claus, 1863)		—	55°27.8′S	66°09.1′W	336	154	Free-living	George (2005)

(continued)

Table 4.2 (continued)

Taxon	Basin	Station	Lat	Long	Depth (m)	Corresponding number in Figures	Associations	References
Fam. Pseudotachidiidae Lang, 1936								
Pseudomesochra longifurcata Scott T., 1902		—	54°58.7′S	69°01.9′W	219	181	Free-living	George (2005)
Fam. Ameiridae Boeck, 1865	**Gulf of Mexico**							
Ameira longipes Boeck, 1865		—	27°06.4′N	91°09.9′W	1408	14	Free-living	Plum et al. (2015)
Ameiropsis mixta Sars, 1907		—	27°06.4′N	91°09.9′W	1409	15	Free-living	Brooks et al. (2009)
		—	27°06.4′N	91°09.9′W	1409	16	Free-living	Plum et al. (2015)
Proameira dubia (Sars, 1920)/*simplex* (Norman and Scott T., 1905)		—	27°06.4′N	91°09.9′W	1408	208	Free-living	Plum et al. (2015)
Fam. Cletodidae Scott T., 1904								
Cletodes longicaudatus (Boeck, 1872)		—	27°06.7′N	91°09.9′W	1406	78	Free-living	Brooks et al. (2009)
		—	27°06.7′N	91°09.9′W	1406	79	Free-living	Plum et al. (2015)

Fam. Cletopsyllidae Huys and Willems, 1989								
Pentacletopsyllus montagni Bang, Baguley and Moon, 2014		—	**28°43.2′N**	**88°20.7′W**	1590	94	Free-living	Bang et al. (2014)*
Fam. Laophontidae Scott T., 1904								
Archesola typhlops (Sars, 1908)		—	27°06.4′N	91°09.9′W	1409	100	Free-living	Brooks et al. (2009)
		—	27°06.7′N	91°09.9′W	1406	101	Free-living	Brooks et al. (2009)
		—	27°06.7′N	91°09.9′W	1406	102	Free-living	Brooks et al. (2009)
		—	27°06.4′N	91°09.9′W	1409	103	Free-living	Brooks et al. (2009)
		—	27°06.4′N	91°09.9′W	1408	104	Free-living	Plum et al. (2015)
		—	27°06.4′N	91°09.9′W	1409	105	Free-living	Plum et al. (2015)
		—	27°06.7′N	91°09.9′W	1406	106	Free-living	Plum et al. (2015)
		—	27°06.7′N	91°09.9′W	1406	107	Free-living	Plum et al. (2015)
		—	27°06.7′N	91°09.9′W	1406	108	Free-living	Plum et al. (2015)
Fam. Miraciidae Dana, 1846								
Amphiascoides neglectus (Norman and Scott T., 1905)		—	27°06.4′N	91°09.9′W	1409	113	Free-living	Brooks et al. (2009)

(continued)

Table 4.2 (continued)

Taxon	Basin	Station	Lat	Long	Depth (m)	Corresponding number in Figures	Associations	References
		—	27°06.7′N	91°09.9′W	1406	114	Free-living	Brooks et al. (2009)
		—	27°06.4′N	91°09.9′W	1409	115	Free-living	Plum et al. (2015)
		—	27°06.4′N	91°09.9′W	1410	116	Free-living	Plum et al. (2015)
		—	27°06.7′N	91°09.9′W	1406	117	Free-living	Plum et al. (2015)
Amphiascus parvulus (Claus, 1866)		—	27°06.4′N	91°09.9′W	1409	119	Free-living	Brooks et al. (2009)
		—	27°06.7′N	91°09.9′W	1406	120	Free-living	Brooks et al. (2009)
		—	27°06.7′N	91°09.9′W	1406	121	Free-living	Brooks et al. (2009)
		—	27°06.4′N	91°09.9′W	1408	122	Free-living	Plum et al. (2015)
		—	27°06.4′N	91°09.9′W	1409	123	Free-living	Plum et al. (2015)
		—	27°06.4′N	91°09.9′W	1410	124	Free-living	Plum et al. (2015)
		—	27°06.7′N	91°09.9′W	1406	125	Free-living	Plum et al. (2015)
		—	27°06.7′N	91°09.9′W	1406	126	Free-living	Plum et al. (2015)
		—	27°06.7′N	91°09.9′W	1406	127	Free-living	Plum et al. (2015)

Delavalia gundulae (Willen, 2003)		—	27°06.4′N	91°09.9′W	1409	133	Free-living	Brooks et al. (2009)
		—	27°06.7′N	91°09.9′W	1406	134	Free-living	Brooks et al. (2009)
		—	27°06.4′N	91°09.9′W	1408	133	Free-living	Plum et al. (2015)
		—	27°06.4′N	91°09.9′W	1409	133	Free-living	Plum et al. (2015)
		—	27°06.7′N	91°09.9′W	1406	134	Free-living	Plum et al. (2015)
Fam. Pseudotachidiidae Lang, 1936								
Psammis longipes Becker, 1974		—	27°06.4′N	91°09.9′W	1408	178	Free-living	Plum et al. (2015)
Xylora bathyalis Hicks, 1988		—	27°06.7′N	91°09.9′W	1406	189	Free-living	Brooks et al. (2009)
		—	27°06.7′N	91°09.9′W	1406	190	Free-living	Brooks et al. (2009)
		—	27°06.7′N	91°09.9′W	1406	191	Free-living	Plum et al. (2015)
		—	27°06.4′N	91°09.9′W	1408	192	Free-living	Plum et al. (2015)
		—	27°06.7′N	91°09.9′W	1406	193	Free-living	Plum et al. (2015)

(continued)

Table 4.2 (continued)

Taxon	Basin	Station	Lat	Long	Depth (m)	Corresponding number in Figures	Associations	References
Fam. Tegastidae Sars, 1904								
Smacigastes methanophilus Plum and Martínez Arbizu, 2009		—	**27°06.4′N**	**91°09.9′W**	1409	197	Free-living	Plum and Martínez Arbizu (2009)* and Plum et al. (2015)
Fam. Aegisthidae Giesbrecht, 1893	**South East Pacific**							
Andromastax muricatus Conroy-Dalton and Huys, 1999		—	**00°47.9′N**	**86°09.2′W**	2494	1	Free-living	Conroy-Dalton and Huys (1999)*
Cerviniella peruana Becker, 1974		—	**12°03.0′S**	**78°45.0′W**	5000	6	Free-living	Becker (1974)*
Pontostratiotes peruanus Becker and Schriever, 1979		—	**07°59.0′S**	**80°37.0′W**	1000	9	Free-living	Becker and Schriever (1979)*
Fam. Ameiridae Boeck, 1865								
Ameira bathyalis Becker and Schriever, 1979		—	**11°50.0′S**	**77°58.0′W**	920	13	Free-living	Becker and Schriever (1979)*
Malacopsyllua hades Becker, 1974		—	**08°52.0′S**	**80°47.0′W**	6300	18	Free-living	Becker (1974)*

Fam. Ancorabolidae Sars, 1909								
Dendropsyllus magellanicus (George and Schminke, 1998)		—	42°24.6′S	74°47.3′W	296	26	Free-living	George (2006)
Fam. Argestidae Por, 1986								
Argestoides prehensilis Huys and Conroy-Dalton, 1997		—	**00°47.9′N**	**86°09.2′W**	2494	39	Free-living	Huys and Conroy-Dalton (1997)*
Eurycletodes (Oligocletodes) peruanus Becker, 1979		—	**08°22.0′S**	**81°2.0′W**	6260	44	Free-living	Becker (1979)*
Mesocletodes parirrasus Becker, 1979		—	**12°03.0′S**	**78°45.0′W**	2000	56	Free-living	Becker (1979)*
Fam. Canthocamptidae Brady, 1880								
Hemimesochra rapiens Becker, 1979		—	**12°03.0′S**	**78°05.0′W**	4100	67	Free-living	Becker (1979)*
Metahuntemannia magniceps Becker, 1979		—	**12°03.0′S**	**78°45.0′W**	5000	68	Free-living	Becker (1979)*
Metahuntemannia pacifica Becker, 1979		—	**11°50.0′S**	**77°58′W**	920	68	Free-living	Becker (1979)*
Metahuntemannia peruana Becker, 1979		—	**12°04.0′S**	**78°05.0′W**	4100	71	Free-living	Becker (1979)*
Metahuntemannia peruana Becker, 1979		—	**12°03.0′S**	**78°05.0′W**	5000	70	Free-living	Becker (1979)*
Metahuntemannia talpa Becker, 1979		—	**11°50.0′S**	**77°58.0′W**	920	72	Free-living	Becker (1979)*

(continued)

Table 4.2 (continued)

Taxon	Basin	Station	Lat	Long	Depth (m)	Corresponding number in Figures	Associations	References
Fam. Cletodidae Scott T., 1904								
Barbaracletodes barbara Becker, 1979		—	**12°03.0′S**	**78°45.0′W**	5000	73	Free-living	Becker (1979)*
Barbaracletodes carola Becker, 1979		—	**11°50.0′S**	**77°58.0′W**	920	74	Free-living	Becker (1979)*
Fam. Idyanthidae Lang, 1948								
Dactylopia peruana Becker, 1974		—	**11°50.0′S**	**77°58.0′W**	920	99	Free-living	Becker (1974)*
Fam. Laophontidae Scott T., 1904								
Bathylaophonte pacifica Lee and Huys, 1999		—	**17°25.8′S**	**113°12.3′W**	2572	112	Free-living	Lee and Huys (1999)*
Fam. Miraciidae Dana, 1846								
Delavalia lima (Becker and Schriever, 1979)		—	**11°50.0′S**	**77°58.0′W**	920	135	Free-living	Becker and Schriever (1979)*
Haloschizopera abyssi Becker, 1974		—	**12°04.0′S**	**78°05.0′W**	2000	141	Free-living	Becker (1974)*
Haloschizopera lima Becker, 1974		—	**11°50.0′S**	**77°58.0′W**	920	154	Free-living	Becker (1974)*
Sarsamphiascus profundus (Becker and Schriever, 1979)		—	**11°50.0′S**	**77°58.0′W**	920	157	Free-living	Becker and Schriever (1979)*

Fam. Parameiropsidae Corgosinho and Martínez Arbizu, 2010								
Parameiropsis peruanus Becker, 1974		—	**12°03.0′S**	**78°45.0′W**	5000	169	Free-living	Becker (1974)*
Fam. Paramesochridae Lang, 1944								
Wellsopsyllus (Wellsopsyllus) abyssalis (Becker, 1979)		—	**12°03.0′S**	**78°45.0′W**	5000	172	Free-living	Becker (1979)*
Fam. Pseudotachidiidae Lang, 1936								
Paranannopus longithorax Becker, 1979		—	**12°03.0′S**	**78°45.0′W**	4100	176	Free-living	Becker (1979)*
Paranannopus truncatus Becker, 1979		—	**08°56.0′S**	**80°47.0′W**	6100	177	Free-living	Becker (1979)*
Psammis longipes Becker, 1974		—	**12°03.0′S**	**78°45.0′W**	5000	179	Free-living	Becker (1974)*
Pseudomesochra minor Becker, 1974		—	**11°50.0′S**	**77°58.0′W**	920	182	Free-living	Becker (1974)*
Pseudotachidius abyssalis Becker, 1974		—	**12°03.0′S**	**78°45.0′W**	5000	183	Free-living	Becker (1974)*
Pseudotachidius peruanus Becker, 1974		—	**12°03.0′S**	**78°45.0′W**	5000	186	Free-living	Becker (1974)*
Fam. Aegisthidae Giesbrecht, 1893	**Eastern Central Pacific**							
Siphonis aureus Mercado-Salas, Khodami and Martínez Arbizu, 2019		—	**11°4.3′N**	**119°36.3′W**	4346.4	10	Probably associated to sponges or cnidarians	Mercado-Salas et al. (2019)*

(continued)

Table 4.2 (continued)

Taxon	Basin	Station	Lat	Long	Depth (m)	Corresponding number in Figures	Associations	References
Siphonis ruehlemanni Mercado-Salas, Khodami and Martínez Arbizu, 2019		—	**11°50.2′N**	**116°58.9′W**	4127.1	11	Probably associated to sponges or cnidarians	Mercado-Salas et al. (2019)*
		—	14°03.1′N	130°04.8′W	5024.9	12	Probably associated to sponges or cnidarians	Mercado-Salas et al. (2019)*
Fam. Ancorabolidae Sars, 1909								
Arthuricornua anendopodia Conroy-Dalton, 2001		—	**32°35.8′N**	**117°29.0′W**	1220	22	Free-living	Conroy-Dalton (2001)*
Ceratonotus thistlei Conroy-Dalton, 2003		—	**32°35.8N**	**117°29.0′W**	1220	24	Free-living	Conroy-Dalton (2003)*
Dendropsyllus thomasi Conroy-Dalton, 2003		—	**32°35.8N**	**117°29.0′W**	1220	28	Free-living	Conroy-Dalton (2003)*
Fam. Argestidae Por, 1986								
Argestes analongises Gómez, 2018		—	**22°01.0′N**	**106°40.0′W**	1540	38	Free-living	Gómez (2018a)*
Argestes angolaensis (?) George, 2008		—	09°50.4′N	104°18.11′W	2500	209	Free-living	Gollner et al. (2015b)

Mesocletodes elmari Menzel, 2011		—	—	—	4877—5042	—	Free-living	Menzel et al. (2011)
		—	09°50.5′N	104°17.5′W	2500	54	Free-living	Menzel et al. (2011)
Mesocletodes nudus Vakati, Thistle and Lee, 2017		—	**32°35.8N**	**117°29.0′W**	1218—1223	55	Free-living	Vakati et al. (2017)*
Mesocletodes tetrasetosus Vakati, Thistle and Lee, 2017		—	**32°35.8N**	**117°29.0′W**	1218—1223	60	Free-living	Vakati et al. (2017)*
Neoargestes laevis Gheerardyn and George, 2019		—	**14°42.4′N**	**125°26.5′W**	4501	62	Free-living	Gheerardyn and George (2019)*
Fam. Canthocamptidae Brady, 1880								
Bathycamptus eckmani Huys and Thistle, 1989		—	**32°35.8N**	**117°29.0′W**	1218—1223	64	Associated to *Kirkegaardia luticastella* (Jumars, 1975) (Polychaeta: Terebellida: Cirratulidae)	Huys and Thistle (1989)*
		—	32°51.0′N	117°46.9′W	1035	65	Associated to *Kirkegaardia luticastella* (Jumars, 1975) (Polychaeta: Terebellida: Cirratulidae)	Huys and Thistle (1989)*
Fam. Cletodidae Scott T., 1904								
Cletodes cf. latirostris sp. 2 Drzycimski, 1967		—	32°47.9′N	120°22.3′W	2698	210	Free-living	Easton and Thistle (2016)

(continued)

Table 4.2 (continued)

Taxon	Basin	Station	Lat	Long	Depth (m)	Corresponding number in Figures	Associations	References
Fam. Ectinosomatidae Sars, 1903								
Microsetella norvegica (Boeck, 1865)		—	09°50.0′N	—	2500	—	Free-living	Gollner et al. (2015b)
Fam. Laophontidae Scott T., 1904								
Bathylaophonte pacifica Lee and Huys, 1999		—	11°24.9′N	103°47.2′W	2500	109	Free-living	Gollner et al. (2006, 2007)
		—	09°50.5′N	104°17.5′W	2500	110	Free-living	Gollner et al. (2006, 2007)
		—	09°50.7′N	104°17.4′W	2500	111	Free-living	Gollner et al. (2006, 2007)
		—	11°24.9′N	103°47.2′W	2480	109	Free-living	Zekely et al. (2006)
Fam. Miraciidae Dana, 1846								
Archaeohuysia huysi Gómez, 2020		—	**30°55.5′N**	**116°49.6′W**	2037	128	Free-living	Gómez (2020b)*
Delavalia californiensis Gómez, 2020		—	**26°32.6′N**	**113°56.3′W**	479	131	Free-living	Gómez (2020b)*
Delavalia diegensis (Thistle and Coull, 1979)		—	**32°34.8′N**	**117°29.0′W**	1200	132	Free-living	Thistle and Coull (1979)*
Delavalia profunda Gómez, 2020		—	**27°07.1′N**	**114°36.1′W**	1039	136	Free-living	Gómez (2020b)*
Delavalia reducta Gómez, 2020		—	**31°48.3′N**	**116°55.5′W**	825	137	Free-living	Gómez (2020b)*

Pseudostenhelia bathyalis Gómez, 2020		—	**29°48.8′N**	**116°07.3′W**	1572	155	Free-living	Gómez (2020a)*
Sarsamphiascus sp. 1 (*aff. varians*) (Norman and Scott T., 1905)		—	09°50.0′N	—	2500	—	Free-living	Gollner et al. (2015a, b)
Wellstenhelia euterpoides Gómez and Cruz-Barraza, 2020		—	**26°32.6′N**	**113°56.3′W**	479	158	Free-living	Gómez and Cruz-Barraza (2020)*
		—	27°07.1′N	114°36.1′W	1039	159	Free-living	Gómez and Cruz-Barraza (2020)*
Wellstenvalia wellsi Gómez and Cruz-Barraza, 2020		—	23°17.7′N	110°21.9′W	665	161	Free-living	Gómez and Cruz-Barraza (2020)*
		—	26°32.6′N	113°56.3′W	479	162	Free-living	Gómez and Cruz-Barraza (2020)*
		—	**27°07.1′N**	**114°36.1′W**	1039	163	Free-living	Gómez and Cruz-Barraza (2020)*
		—	30°55.5′N	116°49.6′W	2037	164	Free-living	Gómez and Cruz-Barraza (2020)*
Fam. Neobradyidae Olofsson, 1917								
Marsteinia cf. ibericus (Becker, 1974)		—	32°47.9′N	120°22.3′W	2698	211	Free-living	Easton and Thistle (2016)
Marsteinia cf. similis sp. 2 Drzycimski, 1968		—	32°47.9′N	120°22.3′W	2698	212	Free-living	Easton and Thistle (2016)

(continued)

Table 4.2 (continued)

Taxon	Basin	Station	Lat	Long	Depth (m)	Corresponding number in Figures	Associations	References
Fam. Parameiropsidae Corgosinho and Martínez Arbizu, 2010								
Parameiropsis kodosensis Cho, Wi and Suh, 2016		—	**10°30.9′N**	**131°18.9′W**	4988	168	Free-living	Cho et al. (2016)*
Parameiropsis tetraspinosa Cho, Wi and Suh, 2016		—	**10°30.1′N**	**131°19.9′W**	5036	170	Free-living	Cho et al. (2016)*
Fam. Pseudotachidiidae Lang, 1936								
Xylora bathyalis Hicks, 1988		—	09°50.5′N	104°17.5′W	2500	187	Free-living	Gollner et al. (2006, 2007)
		—	09°50.7′N	104°17.4′W	2500	188	Free-living	Gollner et al. (2006, 2007)
Fam. Tegastidae Sars, 1904								
Smacigastes barti Gollner, Ivanenko and Martínez Arbizu, 2008		—	**09°50.5′N**	**104°17.5′W**	2500	196	Free-living	Gollner et al. (2008)*
Fam. Tisbidae Stebbing, 1910								

Genesis vulcanoctopusi López-González, Bresciani and Huys in López-González, Bresciani, Huys, González, Guerra and Pascual, 2000		—	**12°48.4′N**	**103°56.4′W**	2647	207	Parasite of *Muusoctopus hydrothermalis* (González and Guerra, 1998) (Cephalopoda: Octopoda: Enteroctopodidae)	López-González et al. (2000)*
Fam. Zosimeidae Seifried, 2003								
Zosime cf. atlantica Bodin, 1968		—	32°47.9′N	120°22.3′W	2698	213	Free-living	Easton and Thistle (2016)
Fam. Ameiridae Boeck, 1865	**Gulf of California**							
Argestigens celibis Gómez, 2018		—	**27°42.0′N**	**111°37.9′W**	1570	17	Free-living	Gómez (2018a)*
Fam. Ancorabolidae Sars, 1909								
Ancorabolus hendrickxi Gómez and Conroy-Dalton, 2002		—	**25°54.0′N**	**110°11.0′W**	1985	21	Free-living	Gómez and Conroy-Dalton (2002)*
Ceratonotus elongatus Gómez and Díaz, 2017		—	**27°01.0′N**	**110°53.1′W**	1642	23	Free-living	Gómez and Díaz (2017)*
Dendropsyllus californiensis Gómez and Díaz, 2017		—	**26°41.1′N**	**111°12.0′W**	1759	25	Free-living	Gómez and Díaz (2017)*
Fam. Argestidae Por, 1986								
Eurycletodes paraephippiger Gómez, 2018		—	**27°09.1′N**	**111°39.9′W**	1440	45	Free-living	Gómez (2018c)*

(continued)

Table 4.2 (continued)

Taxon	Basin	Station	Lat	Long	Depth (m)	Corresponding number in Figures	Associations	References
Mesocletodes brevisetosus Gómez, 2018		—	**25°54.7′N**	**110°11.0′W**	2018	52	Free-living	Gómez (2018b)*
Mesocletodes simplex Gómez, 2018		—	**27°01.0′N**	**110°53.1′W**	1642	57	Free-living	Gómez (2018b)*
Mesocletodes unisetosus Gómez, 2018		—	**25°54.7′N**	**110°11.0′W**	2018	61	Free-living	Gómez (2018b)*
Odiliacletodes secundus Gómez, 2018		—	**27°07.0′N**	**110°53.4′W**	1642	63	Free-living	Gómez (2018c)*
Fam. Ectinosomatidae Sars, 1903								
Microsetella rosea (Dana, 1847)		—	24°15.2′N	108°10.2′W	600	95	Free-living	Gómez and Morales-Serna (2012)
		—	24°56.1′N	109°05.0′W	1235	96	Free-living	Gómez and Morales-Serna (2012)
		—	24°56.4′N	109°05.6′W	1235	97	Free-living	Gómez and Morales-Serna (2012)
		—	24°14.7′N	108°10.9′W	600	98	Free-living	Gómez and Morales-Serna (2012)
Fam. Miraciidae Dana, 1846								
Beatricella calidafornax Gómez, 2020		—	**27°42.0′N**	**111°37.9′W**	1570	129	Free-living	Gómez (2020a)*

Delavalia asetosa Gómez, 2020		—	**24°16.0′N**	**108°24.1′W**	1240	130	Free-living	Gómez (2020b)*
Diarthopodella secunda Gómez, 2020		—	**27°09.1′N**	**111°39.9′W**	1440	138	Free-living	Gómez (2020b)*
Diarthropodella prima Gómez, 2020		—	24°16.0′N	108°24.1′W	1240	139	Free-living	Gómez (2020b)*
		—	**27°42.0′N**	**111°37.9′W**	1570	140	Free-living	Gómez (2020b)*
Wellstenhelia euterpoides Gómez and Cruz-Barraza, 2020		—	27°01.0′N	110°53.7′W	1642	160	Free-living	Gómez and Cruz-Barraza (2020)*
Wellstenvalia wellsi Gómez and Cruz-Barraza, 2020		—	27°42.0′N	111°37.9′W	1570	165	Free-living	Gómez and Cruz-Barraza (2020)*
		—	24°56.0′N	109°04.2′W	1235	166	Free-living	Gómez and Cruz-Barraza (2020)*
		—	27°09.9′N	111°47.1′W	1332	167	Free-living	Gómez and Cruz-Barraza (2020)*
Fam. Rhizothrichidae Por, 1986								
Rhizothrix longiseta Gómez, 2018		—	27°42.0′N	111°37.9′W	1570	194	Free-living	Gómez (2018d)*
		—	**27°09.9′N**	**111°47.1′W**	1332	195	Free-living	Gómez (2018d)*
Fam. Ancorabolidae Sars, 1909	**Northeastern Pacific**							
Uptionyx verenae Conroy-Dalton and Huys, 2000		—	**48°27.5′N**	**128°42.5′W**	2417	37	Free-living	Conroy-Dalton and Huys (2000)*

(continued)

Table 4.2 (continued)

Taxon	Basin	Station	Lat	Long	Depth (m)	Corresponding number in Figures	Associations	References
Fam. Argestidae Por, 1986								
Eurycletodes (Oligocletodes) cf. petiti Soyer, 1964		—	44°00.1′N	130°23.6′W	3247	214	Free-living	Easton and Thistle (2016)
		—	42°33.5′N	131°55.5′W	3593	215	Free-living	Easton and Thistle (2016)
Eurycletodes cf. ephippiger Por, 1964		—	42°33.5′N	131°55.5′W	3593	216	Free-living	Easton and Thistle (2016)
		—	39°59.6′N	125°52.6′W	3676	217	Free-living	Easton and Thistle (2016)
		—	36°47.8′N	123°41.9′W	3683	218	Free-living	Easton and Thistle (2016)
Mesocletodes cf. irrasus (Scott T. and Scott A., 1894)		—	36°41.9′N	123°00.1′W	3087	219	Free-living	Thistle et al. (2007)
Fam. Cletodidae Scott T., 1904								
Cletodes cf. latirostris sp. 2 Drzycimski, 1967		—	39°59.6′N	125°52.6′W	2733	220	Free-living	Easton and Thistle (2016)
		—	36°40.6′N	122°49.1′W	2717	221	Free-living	Easton and Thistle (2016)
		—	36°47.8′N	123°41.9′W	3683	222	Free-living	Easton and Thistle (2016)

Cletodes cf. longicaudatus (Boeck, 1872)		—	42°33.5′N	131°55.5′W	3593	223	Free-living	Easton and Thistle (2016)
		—	39°59.6′N	125°52.6′W	2733	224	Free-living	Easton and Thistle (2016)
Echinocletodes voightae George and Müller, 2013		—	**41°00.0′N**	**127°29.7′W**	3232	80	Free-living	George and Müller (2013)*
Fam. Idyanthidae Lang, 1948								
Nematovorax cf. gebkelinae Bröhldick, 2005		—	39°59.6′N	125°52.6′W	3676	225	Free-living	Easton and Thistle (2016)
		—	36°40.6′N	122°49.1′W	2717	226	Free-living	Easton and Thistle (2016)
Fam. Neobradyidae Olofsson, 1917								
Antarcticobradya cf. tenuis (Brady, 1910)		—	36°41.9′N	123°00.1′W	3087	227	Free-living	Thistle et al. (2007)
Marsteinia cf. ibericus (Becker, 1974)		—	42°33.5′N	131°55.5′W	3593	227	Free-living	Easton and Thistle (2016)
Marsteinia cf. similis sp. 2 Drzycimski, 1968		—	36°40.6′N	122°49.1′W	2717	229	Free-living	Easton and Thistle (2016)
Marsteinia cf. similis sp1. Drzycimski, 1968		—	36°47.8′N	123°41.9′W	3683	230	Free-living	Easton and Thistle (2016)
		—	42°33.5′N	131°55.5′W	3593	231	Free-living	Easton and Thistle (2016)

(continued)

Table 4.2 (continued)

Taxon	Basin	Station	Lat	Long	Depth (m)	Corresponding number in Figures	Associations	References
Fam. Pseudotachidiidae Lang, 1936								
Pseudotachidius cf. abyssalis Becker, 1974		—	44°00.1′N	130°23.6′W	3247	232	Free-living	Easton and Thistle (2016)
		—	42°33.5′N	131°55.5′W	3593	233	Free-living	Easton and Thistle (2016)
Fam. Tisbidae Stebbing, 1910								
Cholidya polypi Farran, 1914		—	36°40.7—41′N	122°17.6—19.4′W	1336—1347	198	Parasite of *Graneledone* Joubin, 1918, sp. B (Cephalopoda: Octopoda: Megaleledonidae)	Humes and Voight (1997)
		—	48°27.4′N	128°42.5′W	2500	203	Parasite of *Graneledone* Joubin, 1918, sp. A (Cephalopoda: Octopoda: Megaleledonidae)	Humes and Voight (1997)
		—	45°21.0′N	125°37.3′W	2500	204	Parasite of *Graneledone boreopacifica* Nesis, 1982 (Cephalopoda: Octopoda: Megaleledonidae)	Humes and Voight (1997)
		—	45°57.6′N	125°46.2′W	2500	205	Parasite of *Graneledone boreopacifica* Nesis, 1982 (Cephalopoda: Octopoda: Megaleledonidae)	Humes and Voight (1997)

		—	45°59.6′N	125°44.0′W	2265	206	Parasite of *Graneledone boreopacifica* Nesis, 1982 (Cephalopoda: Octopoda: Megaleledonidae)	Humes and Voight (1997)
Fam. Zosimeidae Seifried, 2003								
Zosime cf. atlantica Bodin, 1968		—	39°59.6′N	125°52.6′W	3676	234	Free-living	Easton and Thistle (2016)
		—	39°59.6′N	125°52.6′W	2733	235	Free-living	Easton and Thistle (2016)
		—	36°47.8′N	123°41.9′W	3683	236	Free-living	Easton and Thistle (2016)
		—	36°40.6′N	122°49.1′W	2717	237	Free-living	Easton and Thistle (2016)
Zosime cf. incrassata incrassata Sars, 1910		—	42°33.5′N	131°55.5′W	3593	238	Free-living	Easton and Thistle (2016)
		—	36°47.8′N	123°41.9′W	3683	239	Free-living	Easton and Thistle (2016)
Fam. Argestidae Por, 1986	**Campos Basin and Continental Slope of Sergipe (?), Brazil**							
Mesocletodes bathybia Por, 1964		—	—	—	4484	—	Free-living	Menzel et al. (2011)
Mesocletodes bicornis Menzel and George, 2009		—	—	—	5183	—	Free-living	Menzel et al. (2011)
Fam. Ancorabolidae Sars, 1909								

(continued)

Table 4.2 (continued)

Taxon	Basin	Station	Lat	Long	Depth (m)	Corresponding number in Figures	Associations	References
Echinopsyllus brasiliensis Wandeness, George and Santos, 2009		—	**22°04.6′S**	**39°52.1′W**	750	30	Free-living	Wandeness et al. (2009)*
Echinopsyllus nogueirae Wandeness, George and Santos, 2009*		—	**22°31.5′S**	**40°03.8′W**	1050	32	Free-living	Wandeness et al. (2009)*
Echinopsyllus grohmannae Wandeness, George and Santos, 2009*		—	**22°31.5′S**	**40°03.8′W**	1050	31		
Polyascophorus monoceratus George, Wandeness and Santos, 2013		—	**21°54.6′S**	**39°54.6′W**	1000	36	Free-living	George et al. (2013)*
Fam. Paramesochridae Lang, 1944								
Emertonia minor (Vasconcelos, Veit-Köhler, Drewes and Parreira dos Santos, 2009)		—	**11°29.7′S**	**37°09.7′W**	492	171	Free-living	Vasconcelos et al. (2009)*
Fam. Pseudotachidiidae Lang, 1936								
Keraia longiseta (Vasconcelos, George and Parreira dos Santos, 2008)		—	**11°29.7′S**	**37°09.7′W**	492	174	Free-living	Vasconcelos et al. (2008)*

Fam. Argestidae Por, 1986	**Beaufort Sea**							
Argestes mollis Sars, 1910		—	—	—	—	—	Free-living	Montagna and Carey (1978)
Fam. Aegisthidae Giesbrecht, 1893								
Cervinia magna Smirnov, 1946		—	71°08.9′N	148°00.8′W	25—355	2	Free-living	Montagna (1979, 1980a) and Montagna and Carey (1978)
Cervinia unisetosa Montagna, 1980		—	**71°45.1′N**	**150°35.0′W**	2325	3	Free-living	Montagna (1980a)*
Expansicervinia glacieria Montagna, 1980		—	**72°42.0′N**	**143°40.0′W**	3386	7	Free-living	Montagna (1980a)*
		—	**72°58.0′N**	**146°29.0′W**	3576	8	Free-living	Montagna (1980a)*
Fam. Ectinosomatidae Sars, 1903								
Bradya typica Boeck, 1873		—	—	—	—	—	Free-living	Montagna and Carey (1978)
Pseudobradya confluens (Lang, 1936)		—	—	—	—	—	Free-living	Montagna and Carey (1978)
Fam. Harpacticidae Dana, 1846								
Harpacticus superflexus Willey, 1920		—	—	—	—	—	Free-living	Montagna and Carey (1978)

(continued)

Table 4.2 (continued)

Taxon	Basin	Station	Lat	Long	Depth (m)	Corresponding number in Figures	Associations	References
Fam. Miraciidae Dana, 1846								
Amphiascus giesbrechti Sars, 1906		—	—	—	—	—	Free-living	Montagna and Carey (1978)
Typhlamphiascus lamellifer lamellifer (Sars, 1911)		—	—	—	—	—	Free-living	Montagna and Carey (1978)
Fam. Pseudotachidiidae Lang, 1936								
Paranannopus echinipes Smirnov, 1946		—	—	—	—	—	Free-living	Montagna and Carey (1978)
Pseudotachidius biparttitus Montagna, 1980		—	**70°42.8′N**	**141°39.5′W**	659	184	Free-living	Montagna (1980b)*
Pseudotachidius brevisetosus Montagna, 1980		—	**70°42.8′N**	**147°39.5′W**	659	185	Free-living	Montagna (1980b)*

Asterisks (*) in the references indicate original descriptions from the Americas
Coordinates in bold indicate the type locality
For approximate location of stations in bold (Eastern US) see Coull (1973d: 186, Fig. 4.1)
For approximate location of the species reported by Montagna and Carey (1978) from the Beaufort Sea, see Montagna and Carey (1978: 118, Fig. 4.1)

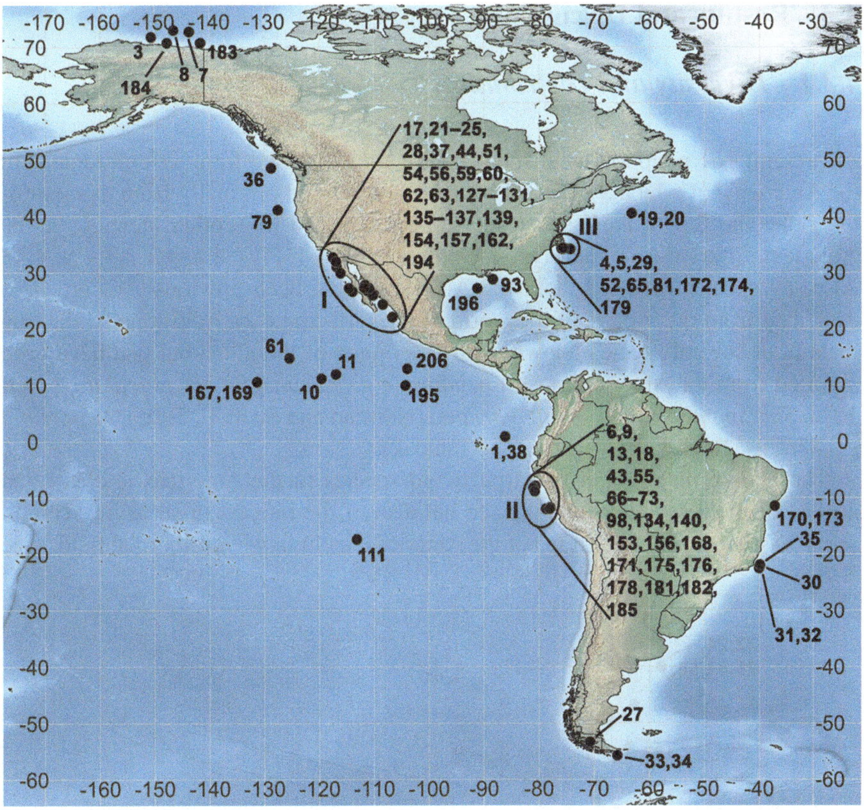

Fig. 4.1 Location of the type locality of the species originally described from the Americas. For correspondence of number of species, coordinates, depths, and references, see also Tables 4.1 and 4.2. For a detailed view of the west coast of the Baja California Peninsula and Gulf of California (I), the Peru Trench (II), and the region off North Carolina (III), see Figs. 4.2–4.4, respectively

Easton and Thistle 2016; Amon et al. 2017; Schmidt et al. 2018). Although some comments are given, these taxa were not included here, and a list is available upon request.

The oceanic limits of the continents are not always clear. Here we opted for a more inclusive approach and included the harpacticoid copepods from the Clarion-Clipperton Fracture Zone and the East Pacific Rise as part of the eastern Central Pacific. On the other hand, the regions presented below are purely arbitrary and have nothing to do with zoogeographic regions such as those proposed by Tunnicliffe et al. (1998), Watling et al. (2013), and Kiel (2016 and references cited therein).

4.3 Results and Discussion

4.3.1 Diversity and Distribution

At present, 161 nominotypical species have been reported from several regions in the Americas, of which 91 species have been described originally from this part of the globe and few of these species have been reported from other locations in the Americas different from their type localities (Table 4.1, Figs. 4.1, 4.2, 4.3, and 4.4). Also, 69 species reported from the Americas have been previously recorded or described from other parts of the world (Tables 4.1 and 4.2). Additionally, the true identity of several species known from other parts of the globe but that have been reported from the Americas is still pending (e.g., *Eucanuella aff. reticulata* Soyer, 1970, *Sarsamphiascus* sp. *1 (aff. varians)* (Norman and Scott T., 1905), *Marsteinia cf. ibericus* (Becker, 1974), etc.) (see Table 4.2).

The known distribution of harpacticoid species in the Americas is clearly an artifact of the researchers' workplace but also of the interest in surveying certain areas. Vast areas of the deep sea of the Americas, including hydrothermal fields (see

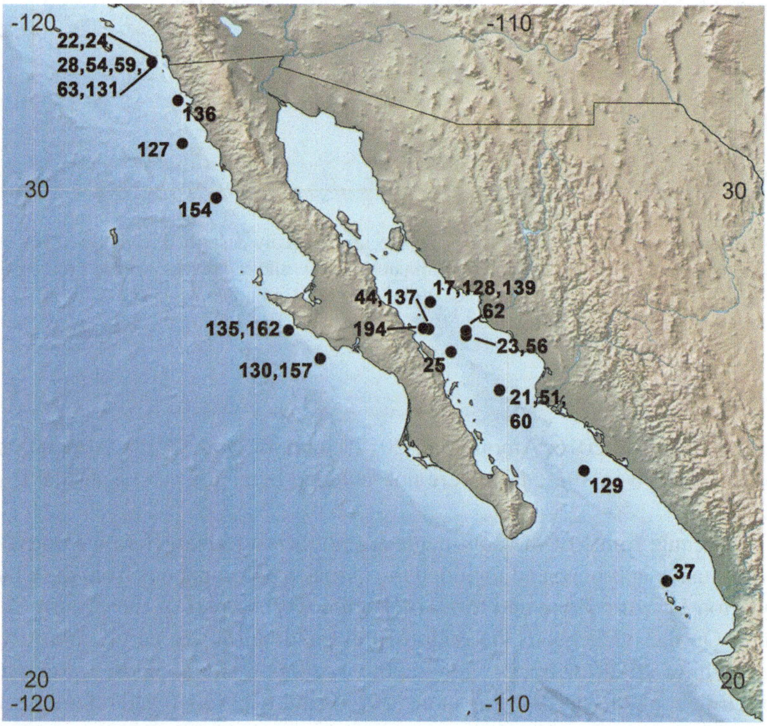

Fig. 4.2 Location of the type locality of the species originally described from the west coast of the Baja California Peninsula and the Gulf of California. For correspondence of number of species, coordinates, depths, and references, see also Tables 4.1 and 4.2

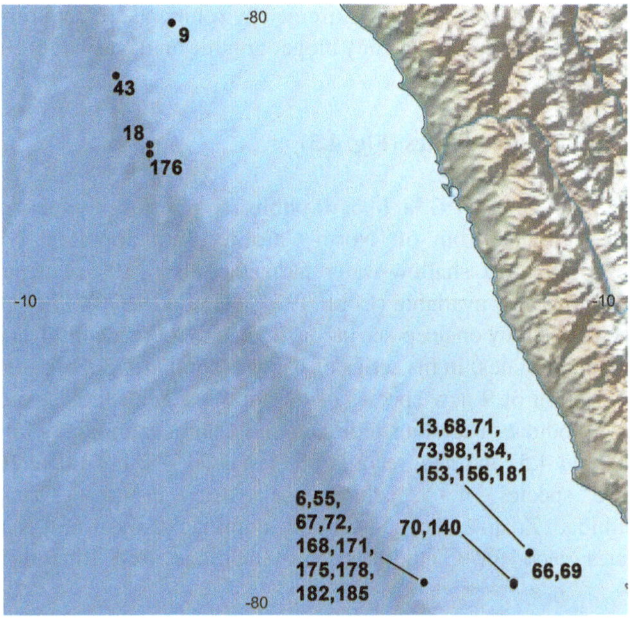

Fig. 4.3 Location of the type locality of the species originally described from the Peru Trench. For correspondence of number of species, coordinates, depths, and references, see also Tables 4.1 and 4.2

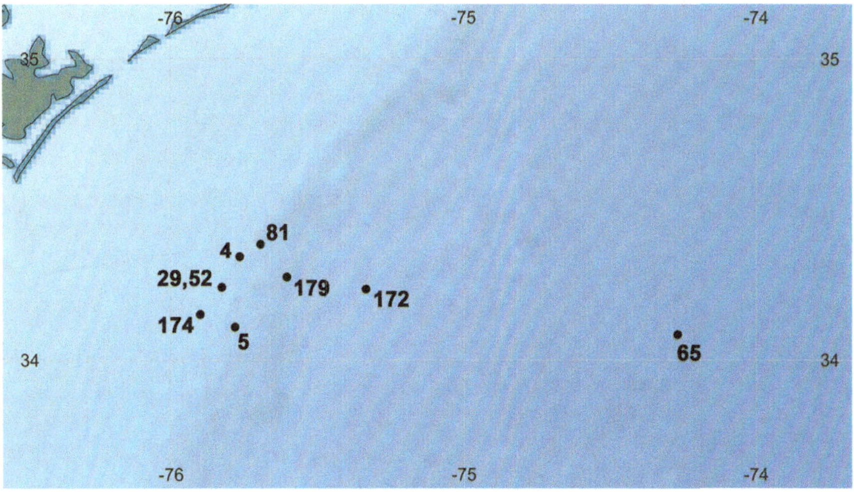

Fig. 4.4 Location of the type locality of the species originally described from off North Carolina. For correspondence of number of species, coordinates, depths, and references, see also Tables 4.1 and 4.2

also Bachraty et al., 2009) and other extreme environments, remain *terrae incognitae* for harpacticoid copepods, and any biogeographic analysis would be premature.

4.3.1.1 Eastern United States (Fig. 4.5)

In the early 1970s, Coull (1973a, b, c, d) published a series of papers on the deep-sea harpacticoid fauna from off North Carolina with important contributions. Previous descriptions of shallow-water harpacticoids (≤100 m depth) from off North Carolina are also available (Coull 1971; Hamond 1973), but Coull (1973d) presented the first survey on deep-sea harpacticoids (>200 m depth) from the eastern coast of the United States. In his series of papers, Coull (1973a, b, c, d) contributed with the description of 9 new species of the families Aegisthidae, Ancorabolidae, Argestidae, Cletodidae, Pseudotachidiidae, and Canthocamptidae (Tables 4.1 and 4.2, Figs. 4.4 and 4.5) and new records of 11 species of Argestidae; 10 species of Miraciidae; 5 species of Aegisthidae; 4 species each of Dactylopusiidae, Pseudotachidiidae, Zosimeidae, and Canthocamptidae; 2 species of Neobradyidae; and 1 species each for Cletodidae, Normanellidae, and Rhizothrichidae (no

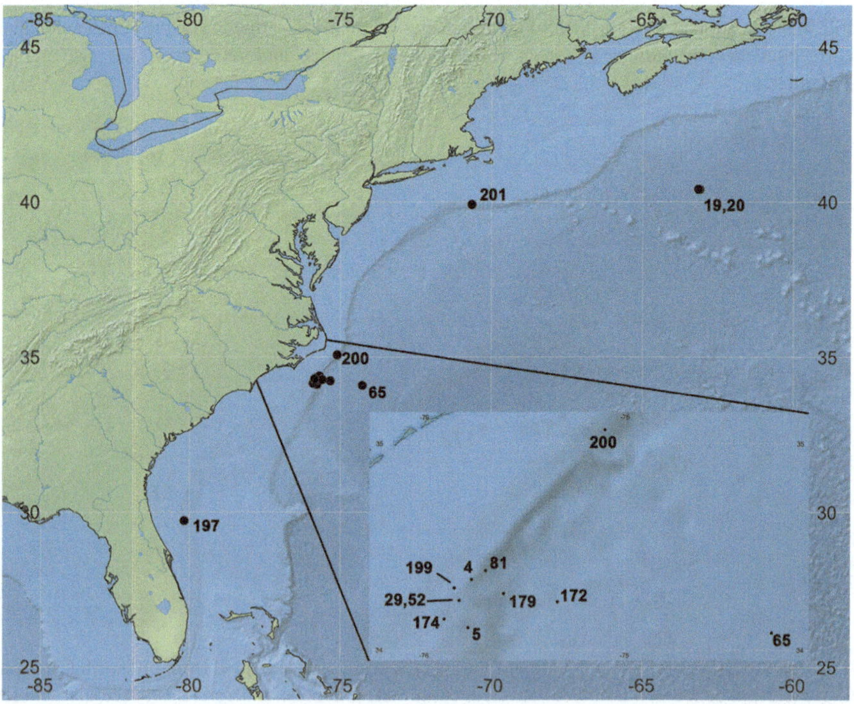

Fig. 4.5 Published records (gray literature excluded) of deep-sea harpacticoids from the eastern United States. For correspondence of number of species, coordinates, depths, and references, see also Tables 4.1 and 4.2

coordinates are available, Table 4.2). Additionally, Coull (1973b, c) reported on some unidentified species of the genera *Amphiascus* Sars, 1905 and *Schizopera* Sars, 1905 (Miraciidae), *Pseudomesochra* Scott T., 1902 (Pseudotachidiidae), *Zosime* Boeck, 1873 (Zosimeidae), *Echinopsyllus* Sars, 1909 (Ancorabolidae), and *Mesocletodes* Sars, 1909 (Argestidae). Coull (1973d) presented a map with a number of other sampling sites in the deep sea off North Carolina. One additional free-living deep-sea harpacticoid was originally described from the northwestern Atlantic, *Sarsameira knorri* Reidenauer and Thistle, 1983 (Ameiridae), from two sites at 4,626 m depth off New York (Reidenauer and Thistle 1983) (Tables 4.1 and 4.2, Figs. 4.1 and 4.5). Also, some of the few records of the parasitic copepod *Cholidya polypi* Farran, 1914 (Tisbidae) from the Americas that uses cephalopods as host are from this region (Humes and Voight 1997) (Table 4.2, Fig. 4.5; see below for some comments on *C. polypi*).

Intensive meiofauna samplings were done by Tietjen et al. (1989) at the Hatteras Abyssal Plain and Puerto Rico Trench; they did not mention harpacticoid species.

4.3.1.2 Southern South America (Fig. 4.6)

The Magellan Campaign to the Straits of Magellan and to the Beagle Channel on board RV "Victor Hensen" in 1994 (Arntz and Gorny 1994) and the expedition ANTARKTIS XIII/4 on board RV "Polarstern" in 1996 (Fahrbach and Gerdes 1997) provided George and Schminke (1998), George (2005, 2018), and George and Gheerardyn (2015) with enough sediment samples to document the deep-sea harpacticoid fauna of the southern tip of South America (Straits of Magellan, Beagle Channel, and the Patagonian continental shelf). *Dendropsyllus magellanicus*

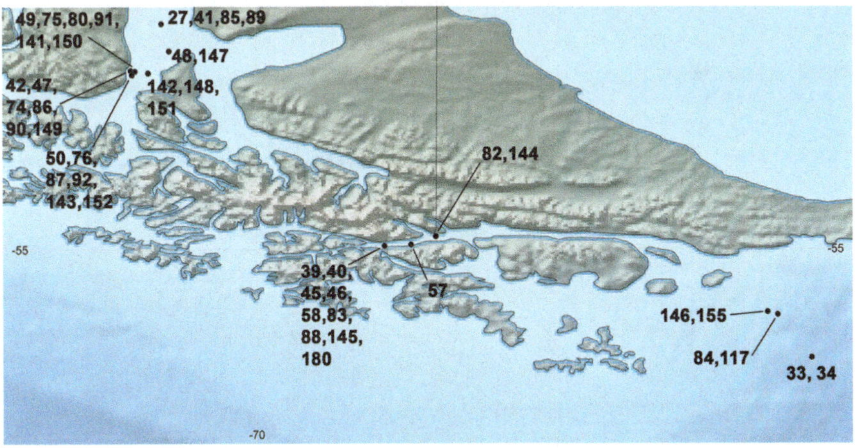

Fig. 4.6 Published records (gray literature excluded) of deep-sea harpacticoids from southern South America. For correspondence of number of species, coordinates, depths, and references, see also Tables 4.1 and 4.2

(George and Schminke, 1998), *Laophontodes sabinegeorgeae* George & Gheerardyn, 2015, and *Laophontodes gertraudae* George, 2018 (Ancorabolidae) are, to the best of our knowledge, the only species originally described from this region (Tables 4.1 and 4.2, Figs. 4.1 and 4.6). George (2001) found and described *Ancorabolus ilvae* George, 2001 from southern Brunswick Peninsula in Bahia Voces (Southwestern Paso Ancho, Straits of Magellan) at 440 m and 550 m depth. However, *A. ilvae* was described upon a female fifth copepodid, and Gómez and Conroy-Dalton (2002) relegated the species as *species inquirenda*. George (2005) reported on the presence of other 15 species (10 genera, 4 families) (Table 4.2, Fig. 4.6) and on several unidentified species and genera of Ancorabolidae, Miraciidae, Paramesochridae, Pseudotachidiidae, Argestidae, Canthocamptidae, and Cletodidae.

4.3.1.3 The Gulf of Mexico (Fig. 4.7)

The biota of the Gulf of Mexico (GM) has received some attention recently, and a very comprehensive review on its habitats and biota before the Deep-water Horizon (DWH) oil spill is available (Byrnes et al. 2017). Despite extensive studies in this important region (see Baguley et al. (2006a, b, 2015) and Montagna et al. (2013)), only 12 species of deep-sea harpacticoids (Ameiridae, 3 species; Cletodidae, 1 species; Cletopsyllidae, 1 species; Laophontidae, 1 species; Miraciidae, 3 species; Pseudotachidiidae, 2 species; Tegastidae, 1 species) have been reported from both the US and Mexican GM (Brooks et al. 2009; Plum and Martinez-Arbizu 2009; Bang et al. 2014; Plum et al. 2015) (Table 4.2, Fig. 4.7), and only 2 species (*Smacigastes methanophilus* Plum and Martínez Arbizu, 2009 (Tegastidae) and

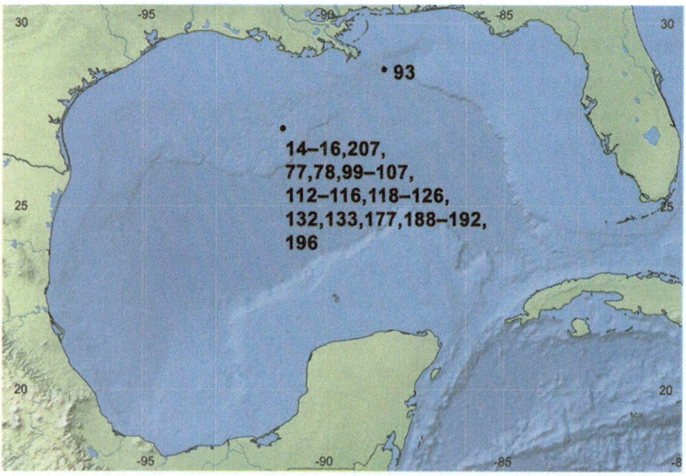

Fig. 4.7 Published records (gray literature excluded) of deep-sea harpacticoids from the Gulf of Mexico. For correspondence of number of species, coordinates, depths, and references, see also Tables 4.1 and 4.2

Pentacletopsyllus montagni Bang, Baguley & Moon, 2014 (Cletopsyllidae)) have been originally described from the deep sea of the GM (Plum and Martinez-Arbizu 2009; Bang et al. 2014) (Tables 4.1 and 4.2, Figs. 4.1 and 4.7). Additionally, Brooks et al. (2009) presented about 27 unidentified harpacticoid species of 21 genera and 10 families, and Plum et al. (2015) presented several unidentified species of about 21 genera and 9 families.

Baguley et al. (2015) analyzed the metazoan meiofauna community in 66 sampling stations affected by 4.9 million barrels of oil that were released into the GM during the DWH blowout and oil spill of 2010 (Lubchenco et al. 2010). They (Baguley et al. 2015) collected about 22,059 copepods in 66 sampling sites at water depths ranging from 76 to 2,767 m. Similar initiatives have been carried out recently in the Mexican GM during a number of intensive oceanographic campaigns by the CIGoM initiative (https://cigom.org/) resulting in an undetermined number of sediment samples waiting for inspection. Preliminary analyses (pers. obs.) of only six sediment samples from the Sigsbee Abyssal Plain revealed the presence of an undetermined number of harpacticoid species distributed in 66 genera and 17 families. Baguley et al. (2015) and the intensive oceanographic campaigns carried out by the CIGoM initiative are just two examples of the efforts displayed by research institutions and universities to reveal the diversity and function of meiofaunal communities in the GM.

Relevant information on the diversity of deep-sea harpacticoids from the northern GM can also be found in several dissertations. For example, Baguley (2004) presented a thorough analysis of the structure and function of the deep-sea meiofauna from 51 stations in the northern GM prior to the DWH oil spill. In his species list, he (Baguley 2004) presented 672 species, together with a list of unidentified adult specimens—of which a large proportion are deemed to be new taxa—and copepodids. Similarly, in her MSc dissertation, Degen (2010) listed an undetermined number of species of 43 genera associated with deep-sea tubeworms in cold seeps and hot vents from the upper Louisiana slope at 538–571 m depth.

Eventual analyses, description, and publication of the harpacticoid fauna of deep-sea sediment samples from the GM sheltered by different US and Mexican universities will undoubtedly unveil an extraordinary diversity of these microcrustaceans.

4.3.1.4 Southeast Pacific (Fig. 4.8)

In the southeast Pacific, the Peru Trench is, to the best of our knowledge, one of the two regions on the west coast of South America that has received some attention in the past and is by far the region with most descriptions of new species. Becker (1974, 1979) and Becker and Schriever (1979) gave the description of 27 new species of deep-sea harpacticoids of 19 genera in 10 families (Tables 4.1 and 4.2, Figs. 4.1, and 4.8). Additionally, Becker and Schriever (1979) reported on three unidentified species of *Amphiascoides* Nicholls, 1941 and *Bulbamphiascus* Lang,

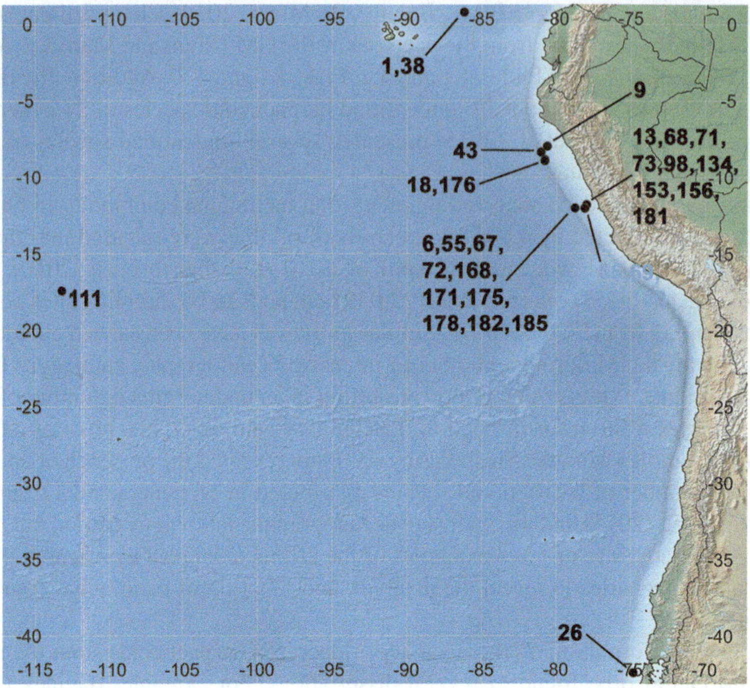

Fig. 4.8 Published records (gray literature excluded) of deep-sea harpacticoids from the Southeast Pacific. For correspondence of number of species, coordinates, depths, and references, see also Tables 4.1 and 4.2

1944 (Miraciidae) and *Paraschizopera* Wells, 1981 (Tetragonicipitidae). Also, *Bathylaophonte pacifica* Lee & Huys, 1999 (Laophontidae) was originally described from the north of Easter Island (Tables 4.1 and 4.2, Figs. 4.1 and 4.8).

Conroy-Dalton and Huys (1999) and Huys and Conroy-Dalton (1997) had the opportunity to inspect the hydrothermal vent harpacticoids found in samples taken at 2,494 m depth during a visit of the deep-sea submersible Alvin to the Galapagos Rift in 1979 (Huys and Conroy-Dalton 1997). Their efforts resulted in the description of two new species of two new genera, *Argestoides prehensilis* Huys & Conroy-Dalton, 1997 (Argestidae) and *Andromastax muricatus* Conroy-Dalton & Huys, 1999 (Aegisthidae) (Tables 4.1 and 4.2, Figs. 4.1 and 4.8).

George (2006) reported on the presence of *Dendropsyllus magellanicus* from the Chilean Pacific continental slope off Chiloé Island collected during Expedition SO 156C ("PUCK") on board RV "Sonne" (Tables 4.1 and 4.2, Fig. 4.8).

Danovaro et al. (2002) made a meiofaunal survey at the Atacama Trench to investigate the meiofaunal abundance, biomass, and community structure in relation to food availability. In their paper, they did not mention harpacticoid species.

4.3.1.5 Eastern Central Pacific (Fig. 4.9)

Twenty-eight free-living harpacticoids of 11 families, 2 aegisthids (*Siphonis aureus* Mercado-Salas, Khodami & Martínez Arbizu, 2019 and *Siphonis ruehlemanni* Mercado-Salas, Khodami & Martínez Arbizu, 2019) probably associated with sponges or cnidarians, 1 tisbid (*Genesis vulcanoctopusi* López-González, Bresciani & Huys, 2000) parasitizing octopodans, and 1 canthocamptid (*Bathycamptus eckmani* Huys and Thistle, 1989) associated with cirratulid polychaetes, have been reported from the eastern Central Pacific, and the type locality of 22 species lies within the limits of this region (Tables 4.1 and 4.2, Figs. 4.1 and 4.9). Most new species from the eastern Central Pacific have been found in samples taken during purely ecological/diversity surveys in the San Diego Trough like Quagmire Expedition (Thiel and Hessler 1974) from which Thistle (1978) gathered 3,935 individuals of 140 putative species (some were subsequently described as *Arthuricornua anendopodia* Conroy-Dalton, 2001; *Ceratonotus thistlei* Conroy-Dalton, 2003; *Dendropsyllus thomasi* Conroy-Dalton, 2003; *Bathycamptus eckmani*—associated with the cirratulid polychaete *Kirkegaardia luticastella* (Jumars, 1975); and *Delavalia diegensis* (Thistle and Coull, 1979), the TALUD project (Gómez and Morales-Serna 2012; Zamorano et al. 2014) in the west coast of the

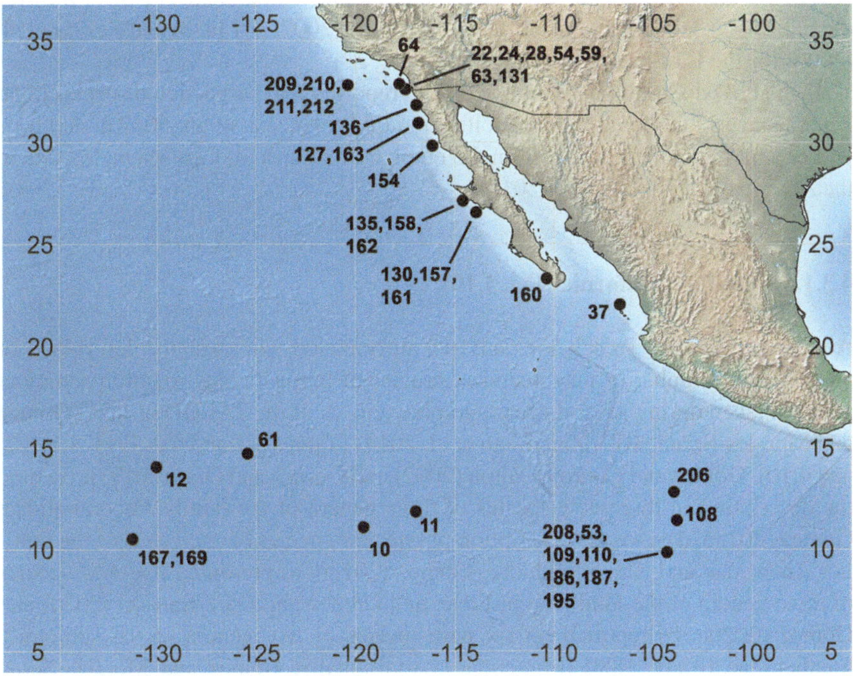

Fig. 4.9 Published records (gray literature excluded) of deep-sea harpacticoids from the Eastern Central Pacific. For correspondence of number of species, coordinates, depths, and references, see also Tables 4.1 and 4.2

Baja California Peninsula with the description of several new species (*Argestes analongises* Gómez, 2018, *Pseudostenhelia bathyalis* Gómez, 2020, *Archaeohuysia huysi* Gómez, 2020, *Delavalia californiensis* Gómez, 2020, *Delavalia profunda* Gómez, 2020, *Delavalia reducta* Gómez, 2020, *Wellstenhelia euterpoides* Gómez and Cruz-Barraza, 2020, *Wellstenvalia wellsi* Gómez and Cruz-Barraza, 2020) or in the East Pacific Rise with the description of *Smacigastes barti* Gollner et al., 2008 (Tables 4.1 and 4.2, Figs. 4.1, 4.2, and 4.9) and with records of other species in Gollner et al. (2006, 2007, 2015a, b). One species of the family Tisbidae, *Genesis vulcanoctopusi*, was found as parasite of the octopus *Muusoctopus hydrothermalis* (González and Guerra, 1998). Other species have been found in samples taken at sites of mining interest like the Clarion-Clipperton Fracture Zone (CCZ) which harbors the highest concentration of polymetallic nodules worldwide at depths ranging from 4500 to 5500 m (Mercado-Salas et al. 2019) (e.g., *Siphonis ruehlemanni* Mercado-Salas, Khodami and Martínez Arbizu, 2019, *Siphonis aureus* Mercado-Salas, Khodami and Martínez Arbizu, 2019, *Parameiropsis kodosensis* Cho, Wi and Suh, 2016, *Parameiropsis tetraspinosa* Cho, Wi and Suh, 2016, and *Neoargestes laevis* Gheerardyn and George, 2019 (Tables 4.1 and 4.2, Figs. 4.1 and 4.9)). An additional record of *Bathylaophonte pacifica* for the East Pacific Rise—but with its type locality in the north of Easter Island, southeast Pacific—can be found in Zekely et al. (2006). Additional unidentified species and genera can be found in Amon et al. (2017) for the abyssal Clarion-Clipperton Zone, in Eckman and Thistle (1988) for the San Diego Trough, and in Gollner et al. (2006, 2007, 2015a, b) and Zekely et al. (2006) for the East Pacific Rise.

In his PhD dissertation, Mahatma (2008) reported 340 harpacticoid species in 58 genera and 18 families from the Pacific Nodule Province at 4966–5027 m depth.

For a more complete account on the number of studies of deep-sea meiofauna in the Americas, see Rosli et al. (2018).

4.3.1.6 Gulf of California (Fig. 4.10)

Along with the Peru Trench, the Gulf of California has witnessed the description of an important number of new deep-sea species of harpacticoids found in sediment samples taken during several oceanographic cruises of the TALUD project (Gómez and Morales-Serna 2012; Zamorano et al. 2014) (Tables 4.1 and 4.2, Figs. 4.1, 4.2, and 4.10). Gómez and Conroy-Dalton (2002) and Gómez and Díaz (2017) presented the description of three new species of three genera of the family Ancorabolidae, Gómez (2018a, d) gave the description of one new species of one genus of the families Ameiridae and Rhizothrichidae, Gómez (2018b, c) presented five new species of three genera of the family Argestidae, and Gómez and Cruz-Barraza (2020) and Gómez (2020a, b) reported on six new species of five genera of the subfamily Stenheliinae Brady, 1880 (Miraciidae). Additionally, Gómez and Morales-Serna (2012) reported on the presence of an undetermined number of unidentified species of 23 genera in 12 families.

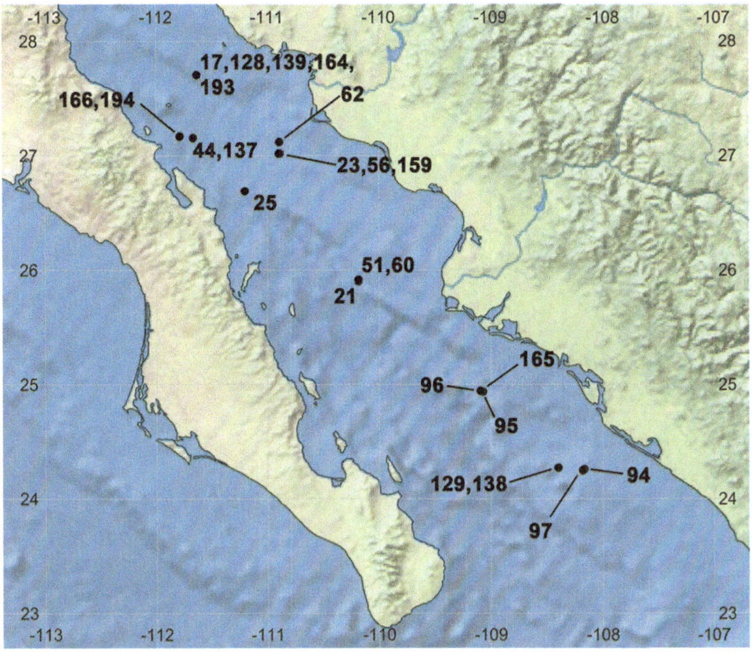

Fig. 4.10 Published records (gray literature excluded) of deep-sea harpacticoids from the Gulf of California. For correspondence of number of species, coordinates, depths, and references, see also Tables 4.1 and 4.2

4.3.1.7 Northeastern Pacific (Fig. 4.11)

Studies on the harpacticoid fauna from the northeastern Pacific by Thistle et al. (2007) and Easton and Thistle (2016) revealed the presence of 13 species of 8 genera and 6 families whose identity awaits confirmation (Table 4.2) along with a number of unidentified species of several genera of the families Aegisthidae, Ameiridae, Argestidae, Canthocamptidae, Zosimeidae, Ectinosomatidae, Miraciidae, Neobradyidae, and Pseudotachidiidae, as well as some material of uncertain affinities within Harpacticoida. Only two free-living species of harpacticoids have been described originally from the northeastern Pacific, *Uptionyx verenae* Conroy-Dalton and Huys, 2000 (Tables 4.1 and 4.2, Figs. 4.1 and 4.11) from hydrothermal vents (Conroy-Dalton and Huys 2000) and *Echinocletodes voightae* George and Müller, 2013 (Tables 1 and 2, Figs. 4.1 and 4.11) from the Gorda Ridge (George and Müller 2013). Additionally, Humes and Voight (1997) reported the parasitic *Cholidya polypi* (Table 4.2, Fig. 4.11) on the deep-sea octopus *Graneledone boreopacifica* Nesis, 1982 and on an unidentified species of the same genus—*Graneledone* sp. A in Humes and Voight (1997). Huys (2016) gave a complete account on the taxonomic history and some interesting notes on tisbid harpacticoids of the subfamily Cholidyinae Boxshall, 1979 that utilize deep-sea octopuses as hosts.

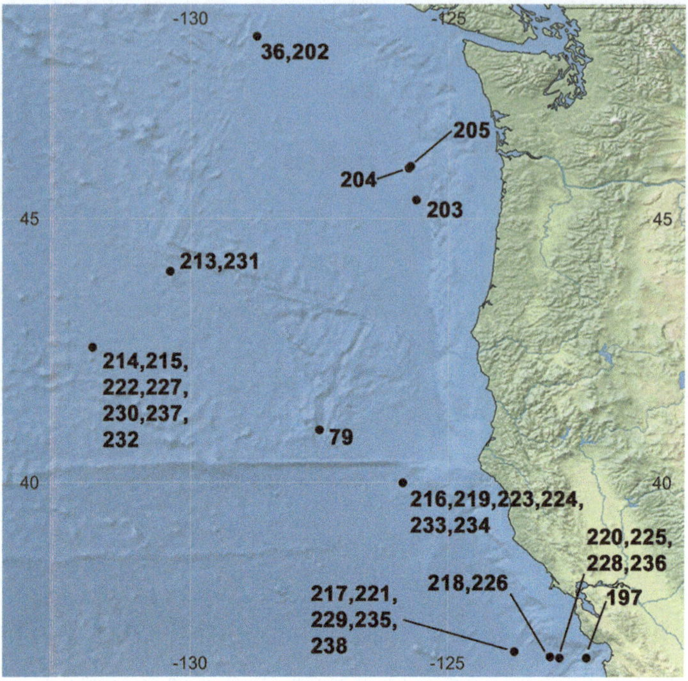

Fig. 4.11 Published records (gray literature excluded) of deep-sea harpacticoids from the Northeastern Pacific. For correspondence of number of species, coordinates, depths, and references, see also Tables 4.1 and 4.2

In her PhD dissertation, Sedlacek (2007) presented a species list from the continental slope off central California (3607 m depth) with 67 species distributed in 44 genera (some of which require confirmation) and 13 identified families and 1 unidentified family.

4.3.1.8 Campos Basin and Continental Slope off Sergipe, Brazil (Fig. 4.12)

Few records of deep-sea harpacticoids from off Brazil are available. In her dissertation on the distribution and ecology of deep-sea harpacticoids from off Sergipe, Vasconcelos (2008) recorded 42 genera and species, of which 2 of them were published some years later, *Keraia longiseta* (Vasconcelos, George and Santos, 2008) and *Emertonia minor* (Vasconcelos, Veit-Köhler, Drewes and Parreira dos Santos, 2009) (Tables 4.1 and 4.2, Figs. 4.1 and 4.12) found at 492 m depth (Vasconcelos et al. 2008, 2009). Also, as a result of the multidisciplinary environmental research project "Campos Basin Deep-Sea Environmental Program" financed by Petrobras, Wandeness et al. (2009) described three species of the genus *Echinopsyllus*, *E. brasiliensis* Wandeness, George and Santos, 2009 and *E. nogueirae* Wandeness,

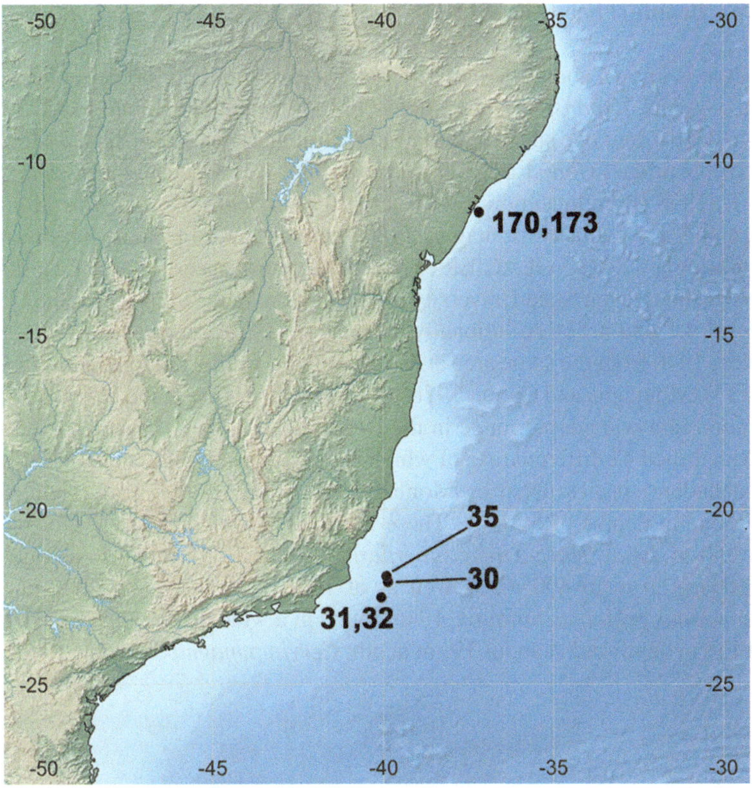

Fig. 4.12 Published records (gray literature excluded) of deep-sea harpacticoids from Campos Basin and Continental Slope of Sergipe, Brazil. For correspondence of number of species, coordinates, depths, and references, see also Tables 4.1 and 4.2

George and Santos, 2009 found at 750 m depth and *E. grohmannae* Wandeness, George and Santos, 2009 found at 1,050 m depth (Tables 4.1 and 4.2, Figs 4.1 and 4.12). Some years later, George et al. (2013) described *Polyascophorus monoceratus* George, Wandeness and Santos, 2013 found in sediment samples from the Grussaí Canyon (Campos Basin) at 1000 m depth (Tables 1 and 2, Figs. 4.1 and 4.12).

Ecological surveys of meiofauna with extensive collections of deep-sea harpacticoids from off Brazil have been carried out by Netto et al. (2005) without any comments on its taxonomic composition.

4.3.1.9 Beaufort Sea (Fig. 4.13)

Very little is known about the harpacticoid fauna of the Beaufort Sea. One species of a new genus, *Expansicervinia glacieria* Montagna, 1980 (found at 3575 m and 3386 m depth), and one species of *Cervinia* Norman, 1878, *C. unisetosa* Montagna, 1980 (Aegisthidae) (found at 2325 m depth), and two species of *Pseudotachidius* Scott T., 1898 (Pseudotachidiidae) found at 659 m depth, *P. bipartitus* Montagna, 1980, and *P. brevisetosus* Montagna, 1980, were described from sediment samples taken during a USCGC GLACIER cruise off the northeastern coast of Alaska conducted by the Oregon State University in 1977 (Montagna 1980a, b) (Tables 4.1 and 4.2, Figs. 4.1 and 4.13). Additionally, Montagna (1979) reported *Cervinia magna* Smirnov, 1946 from the same area at depths ranging from 25 m to 355 m (Table 4.2, Fig. 4.13). Montagna and Carey (1978: 119) presented a list of the harpacticoid species found in a wide depth range in the Beaufort Sea. They (Montagna and Carey 1978) identified 41 different taxa of which 28 are nominotypical species. Within the most abundant species listed by Montagna and Carey (1978), only eight species were found at depths of >200 m. These are *Bradya typica* Boeck, 1873, found at about 350 m depth; *Harpacticus superflexus* Willey, 1920, found in a wide depth range from 0.0 m to 400 m, as well as at 700 m, 1000 m, and 2000 m depth; *Paranannopus echinipes* Smirnov, 1946, found in a wide depth range from 0.0 m to about 130 m depth and at about 350 m depth; *Cervinia magna* Smirnov, 1946, found

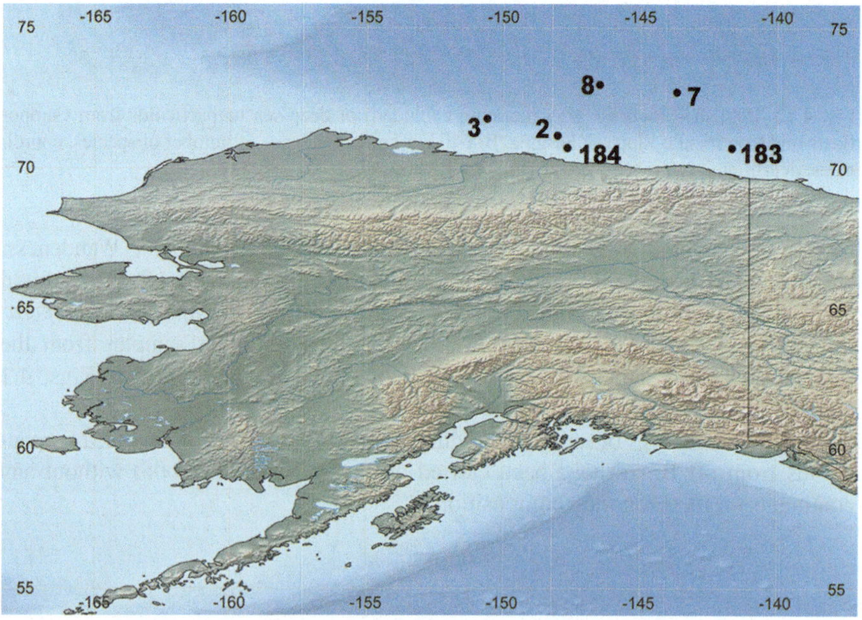

Fig. 4.13 Published records (gray literature excluded) of deep-sea harpacticoids from the Beaufort Sea. For correspondence of number of species, coordinates, depths, and references, see also Tables 4.1 and 4.2

in a wide depth range from about 20 m to about 120 m and at about 350 m depth; *Typhlamphiascus lamellifer lamellifer* (Sars, 1911), found in a wide depth range from about 30 m to 100 m and at 200 m and about 350 m depth; and *Amphiascus giesbrechti* Sars, 1906, *Parabradya confluens* (Lang, 1936) and *Argestes mollis* Sars, 1910, found at 700 m, 700 m, and 1000 m and 700 m, 1000 m, and 2000 m, respectively. Montagna and Carey (1978) did not give the coordinates of their sampling stations and were not included in Fig. 4.13. For an approximate position of the sampling stations, see Montagna and Carey (1978: 118, Fig. 1).

4.3.2 Ecology

4.3.2.1 Associations with other Metazoan Species

Harpacticoid copepods are mostly benthic free-living forms, and few genera (e.g., *Macrosetella* Scott. A., 1909, *Oculosetella* Dahl F., 1895, *Miracia* Dana, 1846, and *Distioculus* Huys and Böttger-Schnack, 1994 (Miraciidae: Miraciinae), *Clytemnestra* Dana, 1847 and *Goniopsyllus* Brady, 1883 (Peltidiidae Claus, 1860: Clytemnestrinae Scott A., 1909), *Microsetella* Brady and Robertson D., 1873 (Ectinosomatidae), *Euterpina* Norman, 1903 (Tachidiidae Sars, 1909)) occur in plankton samples. Also, about 172 species of 84 genera and 17 families—of which Laophontidae and Tisbidae possess most genera and species associated with the highest number of metazoan species among harpacticoids—have been reported to be associated with invertebrate and vertebrate hosts, with a preference for crustaceans (Huys 2016). However, the true nature of such associations is still uncertain, and most probably, only few harpacticoid species are genuine parasites (Huys 2016).

Huys and Thistle (1989) described the canthocamptid *Bathycamptus eckmani* from bathyal muds in the Sand Diego Trough. This species is seemingly associated with mud balls produced by the cirratulid polychaete *Kirkegaardia luticastella*.

Within Tisbidae, the subfamily Cholidyinae, with 9 genera and 13 species, is composed entirely of parasitic forms, and most species utilize octopodan cephalopods as hosts (Avdeev 2010; Huys 2016). The family Cholidyinae is unique among harpacticoids in displaying alternating endoparasitic and ectoparasitic phases in its life cycle (López-González et al. 2000). In their description of the parasitic *Genesis vulcanoctopusi*, López-González et al., (2000) described its effects on the tissue of its cephalopod host and gave an interesting description of its life cycle. From their observations, it seems that the naupliar phases are either benthic or planktonic free-living and that the first copepodids (or late nauplii) play a major role in dispersal and host location and subsequent infection (the female with well-developed swimming legs might also be involved in long-range host location). After infection, the different copepodid stages develop inside the host's tissue, and during the last molt, the later copepodid stages remain in contact with the exterior via a transverse slit in the host's integument from which the adults eventually emerge. López-González et al. (2000) suggested that mating may occur at the gills, arms, and mantle of the host. Copepods are absent from shallow-water octopods (Humes and Voight 1997), and *Genesis vulcanoctopusi* is the first record of the family Tisbidae infesting deep-sea octopods at tropical latitudes (Humes and Voight 1997; López-González et al. 2000). Humes and Voight (1997) presented previous records of Tisbidae associated with octopods in the North Atlantic, the North Pacific, and Antarctica. *Cholidya polypi* has been reported to be associated with several species of the megaleledonid octopod genus *Graneledone* in the Northeastern and eastern Central Pacific, with the bathypolypodid octopod genus *Bathypolypus* Grimpe, 1921 in the Northeastern Atlantic and in the eastern United States, and with the megaleledonid octopod genus *Tetracheledone* Voss, 1955 in the Eastern United States (Humes and Voight 1997). Avdeev (2010) described a new parasitic genus and species, *Amplipedicola pectinatus* Avdeev, 2010 found in the gills of *Muusoctopus profundorum* collected at 500 m depth and in the gills of *Enteroctopus dofleini* collected at 129 m depth from the Bering Sea. An extensive account on deep-sea parasitic tisbids that use octopods as hosts can be found in Huys (2016).

Siphonostomatoida Burmeister, 1835 is a well-defined order of copepods composed exclusively of parasites or associates of other invertebrate and vertebrate hosts (Huys and Boxshall 1991). This order is diagnosed by the presence of a stylet-like mandibular gnathobase with a two-segmented palp in some species of the families Asterocheridae Giesbrecht, 1899 and Entomolepididae Brady, 1899 or with a piston-like rod with teeth on one side of the apical blade and without palp as in the families Caligidae Burmeister, 1835 and Dichelesthiidae Milne Edwards, 1840 (Kabata 1979; Huys and Boxshall 1991; Dojiri and Ho 2013) inside the oral cone formed by the labrum and the medially fused paragnaths (labium). For a complete description of the mouth cone of caligids (Siphonostomatoida), see Dojiri and Ho (2013) and references cited therein. Some harpacticoids of the families Novocriniidae Huys and Iliffe, 1998, Superornatiremidae Huys, 1996, Rotundiclipeidae Huys, 1988, and Peltidiidae, Tegastidae, and Porcellidiidae Boeck, 1865 and the superfamily Tisboidea Stebbing, 1910, all belonging to the taxon Palinarthra (Seifried 2003; Mercado-Salas et al. 2019, and references cited therein), display an incipient oral cone similar to that of some siphonostomatoid families (see Mercado-Salas et al. 2019). Mercado-Salas et al. (2019) presented the description of a deep-sea genus and two new species of the family Aegisthidae, *Siphonis aureus* Mercado-Salas, Khodami and Martínez Arbizu, 2019 and *Siphonis ruehlemanni* Mercado-Salas, Khodami and Martínez Arbizu, 2019 from the Clarion-Clipperton Fracture Zone. The labrum and labium of these two species form an oral cone that evolved independently from that of siphonostomatoid and palinarthran species and are probably associated with cnidarians or sponges representing an independent colonization of deep-sea hosts (Mercado-Salas et al. 2019).

4.3.2.2 Regional/Long-Range Dispersal and Colonization of the Deep Sea

Meiofauna, which includes harpacticoid copepods among other groups, are intimately associated with the sediment, possess benthic larvae, and display very restricted dispersal capabilities (Higgins and Thiel 1988; Boeckner et al. 2009; Giere 2009), in such a way that their "development, morphology and biology all seem designed to assure one thing: that the organism never leaves its interstitial environment" (Sterrer 1973: 200). Yet, many species are widely distributed along continental shorelines, and many shows amphi-oceanic or cosmopolitan distribution (Foissner 2006; Giere 2009; Cerca et al. 2018). Two hypotheses have been formulated in the past to explain the wide and cosmopolitan distribution of micrometazoans in general and of meiofauna in particular. These are briefly mentioned below.

The EiE Hypothesis

That diverse types of bacteria could be cultured from almost any type of natural material was showed in the early twentieth century by Beijerinck (1913), who explained the ubiquitous distribution of particular species of bacteria through the "in micro-organisms, everything is everywhere, the environment selects" hypothesis (EiE hypothesis; but see also Foissner 2006). This hypothesis was further extended to protozoa and other microbial eukaryotes or protists (Finlay 2002; Fenchel and Finlay 2004). Finlay (2002) and Fenchel and Finlay (2004) explained the ubiquitous distribution of microbial eukaryotes based on their continuous large-scale dispersal as a result of their extremely high abundances—they do not have biogeographies, i.e., historical factors are irrelevant for the distribution patterns of microbial eukaryotes—with a subsequent low global species richness. Finlay (2002) challenged the existence of "endemic" microbial species and argued that the random spatial distribution of microbial eukaryotes is ruled by forces that are essentially random, so species maintain consistent patterns of abundance and rarity on a global scale. More recently, in his review on the diversity, dispersal, and geographic distribution of unicellular organisms, Foissner (2006) extended the EiE hypothesis to micrometazoans and also challenged the EiE hypothesis suggesting that vicariance events that separate groups of organisms by geographic barriers (e.g., continental drift, plate tectonics, etc.), intrinsic characteristics of the species (e.g., limited cyst viability), and time, are the major factors for dispersal and provinciality of microorganisms. For a more complete review of the EiE hypothesis, see Fontaneto (2011).

The Meiofaunal Paradox

The "meiofauna paradox" (Giere 2009) addresses the problem as to why distant areas share similar meiofauna species despite their very limited dispersal abilities and how meiofauna species may have bridged vast areas of open ocean between continents to colonize disjunct shores in absence of large populations and planktonic larval stages. Alternative dispersal hypotheses have been suggested to account for the inconsistency posed by the meiofauna paradox (e.g., dispersal models and vicariance-driven hypotheses) (see Cerca et al. 2018). There is, however, evidence suggesting that such paradox might not exist—wide distributions are rare and are also likely to be the result of stasis or recent speciation and could account for morphological similarity represented by the presence of cryptic species with restricted distribution ranges—and that one should focus on morphological similarity as a result of evolutionary factors and on marine connectivity (Cerca et al. 2018).

Dispersal and Distribution of Deep-Sea Harpacticoids

Most deep-sea harpacticoid species originally described from the Americas have been found only once and are known from their type localities only (Tables 4.1 and 4.2, Figs. 4.1, 4.2, 4.3, and 4.4). The find and occurrence of these species in their type localities only might have to do with high levels of endemism as, for example, in deep-sea hydrothermal vents and methane seeps (Tunnicliffe 1992; Van Dover et al. 2002; Wolff 2005; Ramírez-Llodra et al. 2007; Bachraty et al. 2009) and seamounts (McClain et al. 2009; Shank 2010; George 2013; Packmor et al. 2015; Packmor and Riedl 2016), but also with the lack of samplings in vast areas of Americas' deep-sea sediments. On the other hand, only few species originally described from a particular site in the Americas have been reported from distant localities in the same continent. *Dendropsyllus magellanicus* was originally described from the Strait of Magellan (Punta Arenas, Chile) at 200 m depth (George and Schminke 1998) but was found again more than 1000 km away in the Chilean Pacific continental slope off Chiloé Island at 296 m depth (George 2006). *Bathylaophonte pacifica*, originally described from deep-sea hydrothermal vents at 2572 m depth in north of Easter Island (Lee and Huys 1999), was found again more than 3000 km away at 2480 m–2500 m depth in the East Pacific Rise (Gollner et al. 2006, 2007; Zekely et al. 2006). Gómez (2020b) described *Diarthropodella prima* Gómez, 2020 from two sites in the Gulf of California at 1240 m and 1570 m depth, separated by 500 km. *Haloschizopera abyssi* Becker, 1974 described from the Peru Trench at 1900 m depth (Becker 1974) was reported by George (2005) from several sites at 351 m–550 m depth, 4000 km away in the Straits of Magellan. Gómez and Cruz-Barraza (2020) described *Wellstenvalia wellsi* from several sites at 479 m–2037 m depth in the west coast of the Baja California Peninsula (eastern Central Pacific) separated by a maximum distance of 1000 km and in the Gulf of California at depths from 1235 m to 1570 m. *Ceratonotus steiningeri* George, 2006 was found and described from the Angola Basin at 5389 m–5390 m depth (George 2006), and

Selenopsyllus antarcticus Moura and Pottek, 1998 was found and described from the Antarctic Weddell Sea at 3480 m depth (Moura and Pottek 1998), but in his PhD thesis, Mahatma (2008) reported these species in samples from the Pacific Nodule Province between the Clarion and the Clipperton Fracture Zone in the northeastern Pacific. Interestingly, the presence of *Bathylaophonte pacifica* in hydrothermal vents in north of Easter Island and in the East Pacific Rise might support the hypothesis of the role played by the East Pacific Rise as a center of dispersal for the hydrothermal faunas as previously suggested by Bachraty et al. (2009) and where seamounts (George and Schminke 2002; George 2013; Packmor et al. 2015; Packmor and Riedl 2016) and organic falls (Smith et al. 1989, 2015; Distel et al. 2000; Smith and Baco 2003; Bienhold et al. 2013; Cuvelier et al. 2014; but see also Kiel 2016) might play important roles as stepping stones. The role of seamounts in the dispersal of meiofauna in the Americas has not been assessed yet, but these formations could play a key role for the dispersal of harpacticoids as shown for the harpacticoid fauna of the Great Meteor Seamount (George and Schminke 2002) and Seine and Sedlo seamounts in the eastern Atlantic (Büntzow 2011; George 2013; Packmor et al. 2015; Packmor and Riedl 2016). As for regional dispersal, near-bottom flows may transport meiofauna and harpacticoid copepods (Fleeger et al. 1984; Thistle et al. 1999; Boeckner et al. 2009; Menzel et al. 2011) that were suspended as a result of sediment erosion or that entered the water column actively, suggesting higher dispersal potential (Cerca et al. 2018, and references cited therein). Near-bottom flows of a few centimeters per second that occur in much of the deep sea are too slow to erode the seabed and the meiofauna (Thistle and Levin 1998; Thistle et al. 2007), but Hagerman and Rieger (1981) found that suspended meiofauna may travel up to 10 km per day in subtidal currents that are strong enough. For example, the sites of major suspension under abyssal storms are those where surface eddy kinetic energy is high and coupled to a strong near-bottom mean flow and sufficient mud like the North American Basin under the Gulf Stream system (see also Gross and Williams 1991) and the Argentine Basin under the confluence of the Falkland and Brazil currents, among others, some of which are also regions of maximum turbidity in the ocean basins (Hollister and McCave 1984). On the other hand, the presence of, for example, *Diarthropodella prima* in the west coast of the Baja California Peninsula and in the Gulf of California might be due to a combination of dispersal during the marine incursions within the Proto-Gulf of California and its geological history (Ledesma-Vázquez 2002; Ledesma-Vázquez et al. 2009). However, the problem persists as to how genetic continuity is maintained between populations separated by 1000 km of open space. Easton and Thistle (2016) argue that if some deep-sea harpacticoids are suspended and stay in the water column for different amounts of time, the individuals could return to the sediment in different relative positions, creating the opportunity for genetic mixing, and depending on the strength of the current, harpacticoids could be transported tens of kilometers in days or weeks, and thus benthic storms could contribute to genetic continuity (Easton and Thistle 2016). Bottom currents may play an important role in the dispersal of some species of the genus *Mesocletodes* throughout the South and North Atlantic, southern Ocean, Southern Indian Ocean, Pacific Ocean, and eastern Mediterranean Sea (Menzel et al. 2011). However, Menzel et al. (2011) noted that many *Mesocletodes* species displayed high degrees of variability in setation of swimming legs, total body length and caudal rami, and number and shape of spinules on the body and appendages which may be indicative of the presence of different morphological groups or cryptic species.

Some deep-sea harpacticoid species from the Americas display wide distributions, and few could be considered cosmopolitan. The primary planktonic *Microsetella norvegica* (Boeck, 1865) and *Microsetella rosea* (Dana, 1847) reported from the Americas by Gollner et al. (2015b) and Gómez and Morales-Serna (2012) occur worldwide. Similarly, *Robertsonia knoxi* (Thompson I.C. and Scott A., 1903) known from Bermuda Islands, Algeria, Egypt, and Sri Lanka (see Lang 1948) has been found also in sediment samples from off North Carolina (Coull 1973c). On the other hand, some species recorded in the different regions of the Americas considered here (Table 4.2) have been previously described or reported from other distant localities.

According to our records, 13 nominotypical species of deep-sea harpacticoids are known from the Beaufort Sea (Montagna and Carey 1978; Montagna 1979, 1980a, b). Of these, five species have been originally described from the Beaufort Sea (Tables 4.1 and 4.2), but the harpacticoid fauna from this region seems to have strong affinities with the fauna from northern Europe (see below). Among the eight remaining species, *Cervinia magna* and *Paranannopus echinipes* Smirnov, 1946 were described from the Arctic Ocean (Smirnov 1946); *Bradya typica* Boeck, 1873 displays a wide distribution including Greenland, Franz Josef Land, Iceland, northern Europe, North America, and the Kara Sea (see Lang 1948: 224); *Harpacticus superflexus* Willey, 1920 is distributed in the Kara Sea, Franz Josef Land, Greenland, and Canada (see Lang 1948: 332); *Typhlamphiascus lamellifer* (Sars, 1911) is known from Iceland, northern Europe, and the Barents Sea (see Lang 1948: 712); *Amphiascus giesbrechti* Sars, 1906 is known to be present in Greenland and northern Europe (see Lang 1948: 689); *Parabradya confluens* (Lang, 1936) is known from Svalbard (see Lang 1948: 229); and *Argestes mollis* Sars, 1910 is known from northern Europe (see Lang 1948: 1317; George 2008).

The deep sea off Brazil seems to harbor two alien species (Menzel et al. 2011), *Mesocletodes bathybia* Por, 1964 originally described from off Rosh Hanikra (Israel), eastern Mediterranean, at about 468 m depth (66 fathoms) (Por 1964), and *Mesocletodes bicornis* Menzel and George, 2009 described from the Angola Basin (Menzel and George 2009).

The deep-sea harpacticoid fauna from the eastern Central Pacific seems to be composed of an amalgam of species known from several parts of the globe. Gollner et al. (2015b) reported a presumably closely related species of *Argestes angolaensis* George, 2008—originally described from the Angola Basin (George 2008)—as well as the cosmopolitan *Microsetella norvegica*, from the East Pacific Rise. *Sarsamphiascus varians* (Norman & Scott T., 1905) was originally described from the southern United Kingdom (Norman and Scott 1905) and was subsequently found in several sites in Northern Europe (see Lang (1948)); Gollner et al. (2015a, b) reported on a similar species from the East Pacific Rise. *Bathylaophonte pacifica* was described from the north of Easter Island at 2572 m depth (Lee and Huys 1999) and was found again in the East Pacific Rise by Zekely et al. (2006), Gollner et al. (2006), and Gollner et al. (2007). *Xylora bathyalis* Hicks, 1988 from New Zealand was found in samples from the East Pacific Rise (Gollner et al. 2006, 2007). Menzel (2011) and Menzel et al. (2011) commented on the wide distribution of *Mesocletodes elmari* Menzel, 2011 (see Menzel et al. (2011: 845, Table 2) and Menzel (2011: 41, Fig. 4.1, 43, Table 4.1)); the species was found again in the eastern Central Pacific (Menzel et al. 2011). The parasitic *Cholidya polypi*, originally described from the southwest coast of Ireland parasitizing the octopodan *Bathypolypus ergasticus* (Fischer P. and Fischer H., 1892) (Farran 1914), has been found to be attached to several octopodan host species from off Oregon, off North Carolina, off New Jersey, off Florida, and off central California (Humes and Voight 1997).

The eastern United States is home to 58 species of deep-sea harpacticoids. Of these, eight species have been described originally from off North Carolina (Coull 1973a, b, c, d), and one species was described from off New Jersey in the northwestern Atlantic. The remaining harpacticoid fauna is composed of, or have some affinities with:

1. Arctic forms (*Cervinia magna*, *Cerviniella mirabilipes* Smirnov, 1946, *Cerviniopsis gorbunovi* Smirnov, 1946, *Eurycletodes* (*Eurycletodes*) *gorbunovi* Smirnov, 1946, *Eurycletodes* (*Oligocletodes*) *hoplurus* Smirnov, 1946, *Eurycletodes* (*Oligocletodes*) *monardi* Smirnov, 1946, *Eurycletodes* (*Oligocletodes*) *echinatus* Lang, 1936, *Mesocletodes dolichurus* Smirnov, 1946, *Mesocletodes abyssicola* (Scott T. and Scott A., 1901), *Metahuntemannia gorbunovi* Smirnov, 1946, *Metahuntemannia spinosa* (Klie, 1939), *Keraia tamara* (Smirnov, 1946), *Dactylopodopsis dilatata* Sars, 1911, *Paramphiascella intermedia* (Scott T., 1897), with records from the Ellesmere Island in the Canadian Arctic Archipelago (see Lang 1948))
2. Northern European forms (*Cerviniopsis longicaudata* Sars, 1903; *Eurycletodes* (*Oligocletodes*) *major* Sars, 1909; *Mesocletodes irrasus* (Scott T. and Scott A., 1894); *Mesocletodes robustus* Por, 1965; *Mesocletodes soyeri* Bodin, 1968, *Mesocletodes crassa* (Por, 1965), *Paradactylopodia brevicornis* (Claus, 1866), also with records in the Mediterranean and North America (see Lang 1948); *Amphiascoides debilis* (Giesbrecht, 1881), with some records in eastern Canada (see Lang 1948); *Amphiascopsis cinctus* (Claus, 1866), with some records in the Mediterranean, eastern United States, Sri Lanka, Indonesia, and New Zealand (see Lang 1948); *Haloschizopera pygmaea* (Norman and Scott T., 1905); *Sarsamphiascus parvus* (Sars, 1906), with some records from eastern United States and Bermuda Islands (see Lang 1948); *Marsteinia bozici* (Bodin, 1968); *Marsteinia similis* (Drzycimski, 1968); *Pseudocletodes vararensis* Scott T. and Scott A., 1893; *Pseudomesochra media* (Sars, 1911); *Pseudomesochra similis* Lang, 1935; *Pseudotachidius vikingus* Drzycimski, 1968; *Rhizothrix gracilis* (Scott T., 1903); *Cholidya polypi* (see above); *Zosime atlantica* Bodin, 1968; *Zosime cf. incrassata incrassata* Sars, 1910 (Coull 1973c); *Zosime paramajor* Bodin, 1968; and *Zosime typica* Boeck, 1873)

3. Mediterranean forms (*Eucanuella aff. reticulata* Soyer, 1970 (Coull 1973a), *Mesocletodes katharinae* Soyer, 1964, *Metahuntemannia drzycimskii* Soyer, 1970, *Diarthrodes aegideus* (Brian, 1927), *Amphiascus soyeri* (Lang, 1965), *Robertgurneya rostrata* (Gurney, 1927), with some records from the eastern United States (see (Lang 1948))
4. The fauna from the Red Sea (*Cletodes yotabis* Por, 1967)
5. Forms of the southern Indian Ocean (*Dactylopusia spinipes* Brady, 1910)

Prior to 2002, no records of deep-sea harpacticoids were available from the Gulf of California. In addition to the records of the ubiquitous *Microsetella rosea* from the Gulf of California by Gómez and Morales-Serna (2012) (see also Lang (1948) for a more complete list of records worldwide), 16 species have been described from this region. The affinities of the deep-sea harpacticoid fauna from the Gulf of California is not clear yet, but some sort of affinities seem to exist with the fauna from northern Europe (Gómez and Conroy-Dalton 2002), the San Diego Trough (Gómez and Díaz 2017), the Angola Basin (Gómez 2018a, c), and probably Mozambique (Gómez 2018d) and the eastern Mediterranean and northern Europe (Gómez 2018b).

Only two species have been described from the Gulf of Mexico, *Pentacletopsyllus montagni* and *Smacigastes methanophilus*. Most records of deep-sea harpacticoids from the Gulf of Mexico appear in an ecological survey (Plum et al. 2015) and in a technical report (Brooks et al. 2009). The deep-sea harpacticoid fauna from the GM seems to bear some affinities with the fauna from (1) northern Europe (and probably with the Mediterranean fauna) (*Ameira longipes* Boeck, 1865, with records from eastern Canada (see Lang 1948); *Ameiropsis mixta* Sars, 1907, *Proameira dubia* (Sars, 1920)/*simplex* (Norman and Scott T., 1905), *Cletodes longicaudatus* (Boeck, 1872), with records from the Mediterranean and from eastern United States (see Lang 1948); *Archesola typhlops* (Sars, 1908), *Amphiascoides neglectus* (Norman and Scott T., 1905), with records from the Mediterranean (see Lang 1948); *Amphiascus parvulus* (Claus, 1866), with records from the Mediterranean, from the eastern United States and Bermuda Islands (see Lang 1948); *Delavalia gundulae* (Willen, 2003), with records from the southwestern Pacific, Papua New Guinea), (2) New Zealand (*X. bathyalis.*), and (3) the Peru Trench (*Psammis longipes* Becker, 1974).

The northeastern Pacific harbors 14 free-living and 1 parasitic species. Of these, only two species, *Uptionyx verenae* and *Echinocletodes voightae*, have been described from this region. Except for the parasitic *Cholidya polypi* reported in Humes and Voight (1997) attached to the octopodan genus *Graneledone*, the identity of all other free-living harpacticoids is uncertain. However, the deep-sea harpacticoid fauna of this region seems to have some affinities with the fauna (1) from Northern Europe (e.g., *Mesocletodes irrasus*, *Cletodes latirostris* Drzycimski, 1967, *Cletodes longicaudatus*, *Marsteinia similis*, *Zosime atlantica*, *Zosime incrassata incrassata*), (2) from the Mediterranean (e.g., *Eurycletodes (Eurycletodes) petiti* Soyer, 1964, *Eurycletodes ephippiger* Por, 1964), (3) from off Portugal (e.g., *Marsteinia ibericus* (Becker, 1974)), (4) from the Angola Basin (e.g., *Nematovorax gebkelinae* Bröhldick, 2005), (5) from the Kerguelen Islands (e.g., *Antarcticobradya tenuis* (Brady, 1910), and (6) from the Peru Trench (*Pseudotachidius abyssalis* Becker, 1974).

The Southeast Pacific has witnessed the description of 30 new species of 10 families (Tables 4.1, 4.2, Fig. 4.3). The only new record is that of *Dendropsyllus magellanicus* by George (2006); this species was originally described from the Straits of Magellan (George and Schminke 1998).

To the best of our knowledge, only three species have been described from southern South America, *Dendropsyllus magellanicus*, *Laophontodes sabinegeorgeae*, and *Laophontodes gertraudae*. The deep-sea harpacticoid fauna of this region is, as in other regions, a mixture of species already reported/described from distant localities: the Arctic (e.g., *Eurycletodes (Oligocletodes) monardi*, *Eurycletodes(Oligocletodes) oblongus* Sars, 1920, *Fultonia sarsi* (Smirnov, 1946), *Mesocletodes abyssicola*), the Mediterranean (*Fultonia bougisi* Soyer, 1964), Northern Europe (*Mesocletodes soyeri, Cletodes latirostris, Stylicletodes longicaudatus* (Brady, 1880), *Stylicletodes oligochaeta* Bodin, 1968, *Haloschizopera exigua* (Sars, 1906), *Sarsamphiascus minutus* (Claus, 1863), *Pseudomesochra longifurcata* Scott T., 1902), Central California (*Enhydrosoma hopkinsi* Lang, 1965), the eastern United States (*Amphiascoides subdebilis* (Willey, 1935)), and the Peru Trench (*Haloschizopera abyssi*).

The occurrence of wide, cosmopolitan, and eurybathic distributions has been addressed by Bik et al. (2012) and Cerca et al. (2018). Bik et al. (2012) showed that some deep-sea taxa display eurybathic ranges and cosmopolitan distributions, but the majority seem to be regionally restricted and seem to fit Foissner's (2006) view (see above). Similarly, Cerca et al. (2018) showed that wide distributions are rather rare and are likely the result of morphological stasis, parallelism, convergence, or recent speciation of cryptic species with a more restricted distribution, but also of the lack of accumulation of morphological differences between reproductively isolated species. Widely distributed, cosmopolitan, and eurybathic species are also present in Harpacticoida. For example, Montagna (1979) and Montagna and Carey (1978) found *Cervinia magna* and *Harpacticus superflexus* at depths ranging from 25 m to 355 m and from 0 m to 2000 m, respectively, in the Beaufort Sea, and George (2018) found *Laophontodes gertraudae* in samples taken from the Chilean Patagonian continental slope at depths of 105 m and 2550 m. On the other hand, the presence of a species similar to the shallow-water *Sarsamphiascus varians*, known from northern Europe, in deep-sea samples from the East Pacific Rise (Eastern Central Pacific) (Gollner et al. 2015a, b), and the presence of species similar to *Mesocletodes irrasus*, *Cletodes longicaudatus*, and *Zosime incrassata incrassata*, known from shallow-water situations in northern Europe, in the deep sea off central and northern California (Thistle et al. 2007; Easton and Thistle 2016) might be indicative of morphologically similar but reproductively isolated species.

The presence of widely or cosmopolitan species of harpacticoid copepods has always been controversial and "provide a roadmap for future research, suggesting a focus on morphological similarity and marine connectivity" (Cerca et al. 2018). Also, apparent wide distributions are likely the result of the presence of cryptic species and recent speciation (Cerca et al. 2018) but could also be indicative of speciation processes taking place at this very moment as a result of loss of genetic diversity driven by a variety of sources of chronic stress in the deep sea that could lower genetic diversity by eliminating less fit phenotypes and their associated genotypes (Street and Montagna 1996 and references cited therein). Street and Montagna (1996) found that the loss of mitochondrial haplotype diversity was linked to the presence of oil platforms in shallow areas of the Gulf of Mexico, and in their study on the effects of a xenobiotic on the genetic diversity of the ameirid *Nitokra lacustris lacustris* (Schmankevitsch, 1875), Street et al. (1998) observed that populations that encounter a novel environment can lose genetic diversity through intense directional selection leading to a gene pool dominated by relatively few genotypes.

The past decade has witnessed important advances in the study of the role of the interaction of genes in the mitochondrial and nuclear genomes in reproductive isolation and speciation. It is widely accepted that the successful coupling of the oxidative phosphorylation and energy production, among other biological functions (Rawson and Burton 2002; Wolff et al. 2014; Hill 2015), depends on the precise coordination of these genes, that mitonuclear allelic variation is under strong selection pressure, and that mitonuclear incompatibilities constitute intrinsic reproductive barriers and promote speciation through hybrid breakdown and postzygotic reproductive isolation and speciation (Ellison and Burton 2008; Barreto and Burton 2013; Wolff et al. 2014; Hill 2015; Hui 2018). Natural populations may undergo coadaptation in response to differences of their genetic backgrounds due to genetic drift or to selection for adaptation to local environments (Rawson and Burton 2002), but are most likely population-specific because each population is exposed to different spatial and temporal regimes of natural selection (Wolff et al. 2014; see also Hill 2015). "Adaptation to environmental conditions is a topic intimately connected to speciation" (Hill 2015: 1923) and is of particular interest for the study of speciation and diversity in extreme environments of the deep sea such as cold seeps, hydrothermal vents, and hydrothermal seeps (Joseph 2017) where organisms must be adapted to extreme pressure, limited light, and cold temperatures, among other factors. Cold seeps are characterized also by the release of methane and hydrogen sulfide. Hydrothermal vents are characterized by the absence of photosynthesis and by superheated water saturated with toxic chemicals through volcanic activity along spreading ridges. Hydrothermal seeps display a mix of characters typical for cold seeps and hydrothermal vents (Joseph 2017). Despite their extreme conditions, cold seeps and hydrothermal vents harbor a great variety of organisms adapted, for example, to hydrogen sulfide, methane, and other hydrocarbon-rich minerals typical for cold seeps and to overheated waters, presence of hydrogen sulfide, and heavy metals typical for hydrothermal vents (Minic et al. 2006; Joseph 2017) or to a mix of these extreme conditions in hydrothermal seeps (Joseph 2017). The mechanisms responsible for the loss of genetic diversity and mitonuclear incompatibilities that lead to hybrid breakdown and postzygotic reproductive isolation and further speciation are not well understood, but the directional selection driven by all these sources of severe stress in extreme environments of the deep sea is likely to trigger such mechanisms which are likely fueled also by time and intrinsic traits of species, e.g., short generation times in harpacticoid copepods and meiofauna in general, and could explain the high levels of endemism observed in such environments. On the other hand, this also raises the question about the true identity of similar forms of harpacticoid copepods found in different distant locations with slightly different extreme environmental conditions. In other words, do morphologically similar forms of harpacticoid copepods, found in different areas with slightly different extreme environmental conditions, belong to the same species, or are they complexes of cryptic species? How complex phenotypes originate and diversify? What is the role of phenotypic plasticity—the ability of an organism to alter its behavior, morphology, and/or physiology in response to changes in environmental conditions (Levis and Pfennig 2016; Levis et al. 2018; Perry et al. 2018)—in initiating a morphological novelty or derived trait? Are similar morphological novelties displayed by different harpacticoid taxa from remote locations with slightly different extreme environmental conditions the result of convergence? The plasticity-first evolution (PFE) hypothesis suggests that phenotypic plasticity often produces developmental variants that can enhance fitness under stressful conditions and that genetic accommodation—the refinement, driven by selection, of a trait from an initial suboptimal version through existing quantitative genetic changes—promotes adaptive phenotypes whose initial state has been modified in its morphological and physiological properties and in its environmental sensitivity (Levis and Pfennig 2016; Levis et al. 2018). Although the PFE might prove useful in trying to understand the forces and processes behind evolution and speciation processes in extreme environments of the deep sea, the four criteria for testing the PFE and the general framework to evaluate it (to know the background information of deep-sea extreme environments, the ancestral states of the taxa of interest to identify the most appropriate lineages that could serve as ancestral-proxies, and to perform experiments in which the ancestral-proxy and the derived lineages are cultured and subject to conditions similar to those of hydrothermal vents or cold seeps) as proposed by Levis and Pfennig (2016: 566–568)—and already performed by Levis et al. (2018) in spadefoot toads of the genus *Spea* Cope, 1866—are difficult to meet in studies on speciation processes and diversity of deep-sea fauna in general and of meiofauna and harpacticoid copepods in particular but undoubtedly open new venues of future research.

References

Álvarez-Castillo L, Hermoso-Salazar M, Estradas-Romero A (2018) Composition and spatial distribution of the meiofauna in the Wagner and Consag basins, Gulf of California, Mexico. Cah Biol Mar 59:245–256. https://doi.org/10.21411/CBM.A.ACD5ED1C

Amon DJ, Hilario A, Arbizu PM, Smith CR (2017) Observations of organic falls from the abyssal Clarion-Clipperton Zone in the tropical eastern Pacific Ocean. Mar Biodivers 47:311–321. https://doi.org/10.1007/s12526-016-0572-4

Apostolov A (2007) Notes sur les harpacticoïdes cavernicoles (Crustacea: Copepoda) de Vietnam du nord. Hist Nat Bulg 18:65–73

Arntz WE, Gorny M (1994) Cruise report of the joint Chilean-German-Italian Magellan "Victor Hensen" campaign in 1994

Avdeev GV (2010) *Amplipedicola pectinatus* gen. et sp. n. (Copepoda, Harpacticoida, Tisbidae), a parasite of octopuses in the Bering Sea. Crustaceana 83:1363–1370. https://doi.org/10.1163/001121610X535663

Bachraty C, Legendre P, Desbruyères D (2009) Biogeographic relationships among deep-sea hydrothermal vent faunas at global scale. Deep-Sea Res Part I Oceanogr Res Pap 56:1371–1378. https://doi.org/10.1016/j.dsr.2009.01.009

Baguley JG (2004) Meiofauna community structure and function in the northern Gulf of Mexico deep-sea. University of Texas at Austin

Baguley JG, Montagna PA, Cooksey C et al (2015) Community response of deep-sea soft-sediment metazoan meiofauna to the Deep-water Horizon blowout and oil spill. Mar Ecol Prog Ser 528:127–140. https://doi.org/10.3354/meps11290

Baguley JG, Montagna PA, Hyde LJ et al (2006a) Metazoan meiofauna abundance in relation to environmental variables in the northern Gulf of Mexico deep sea. Deep-Sea Res Part I Oceanogr Res Pap 53:1344–1362. https://doi.org/10.1016/j.dsr.2006.05.012

Baguley JG, Montagna PA, Woncheol L et al (2006b) Spatial and bathymetric trends in Harpacticoida (Copepoda) community structure in the Northern Gulf of Mexico deep-sea. J Exp Mar Biol Ecol 330:327–341. https://doi.org/10.1016/j.jembe.2005.12.037

Baguley JG, Montagna PA, Hyde LJ et al (2008) Metazoan meiofauna biomass, grazing, and weight-dependent respiration in the Northern Gulf of Mexico deep sea. Deep-Sea Res Part II Top Stud Oceanogr 55:2607–2616. https://doi.org/10.1016/j.dsr2.2008.07.010

Bang HW, Baguley JG, Moon H (2014) A new genus of *Cletopsyllidae* (Copepoda, Harpacticoida) from gulf of Mexico. Zookeys 391:37–53. https://doi.org/10.3897/zookeys.391.6903

Barreto FS, Burton RS (2013) Elevated oxidative damage is correlated with reduced fitness in interpopulation hybrids of a marine copepod. Proc R Soc B Biol Sci 280:7. https://doi.org/10.1098/rspb.2013.1521

Becker K-H (1974) Eidonomie und Taxonomie abyssaler Harpacticoidea (Crustacea, Copepoda) Teil I. Cerviniidae-Ameiridae. "Meteor" Forschungsergebnisse Reihe D 18:1–28

Becker K-H (1979) Eidonomie und taxonomie abyssaler Harpacticoidea (Crustacea, Copepoda) Teil II. Paramesochridae, Cylindropsyllidae und Cletodidae. "Meteor" Forschungsergebnisse Reihe D 31:1–37

Becker K-H, Schriever G (1979) Eidonomie und Taxonomie abyssaler Harpacticoidea (Crustacea, Copepoda) Teil III. 13 neue Tiefsee-Copepoda Harpacticoidea der Familien Canuellidae, Crviniidae, Tisbidae, Thalestridae, Diosaccidae und Ameiridae. "Meteor" Forschungsergebnisse Reihe D 31:38–62

Beed EB (1962) Freshwater plankton Crustacea of the Colville River area, Northern Alaska. Arctic 15:27–50

Beijerinck MW (1913) De infusies en de ontdekking der backteriën. In: Jaarboek van de Koninklijke Akademie v. Wetenschappen. Müller, Amsterdam

Bienhold C, Ristova PP, Wenzhöfer F et al (2013) How deep-sea wood falls sustain chemosynthetic life. PLoS One 8:e53590. https://doi.org/10.1371/journal.pone.0053590

Bik HM, Sung W, De Ley P et al (2012) Metagenetic community analysis of microbial eukaryotes illuminates biogeographic patterns in deep-sea and shallow water sediments. Mol Ecol 21:1048–1059. https://doi.org/10.1111/j.1365-294X.2011.05297.x

Boeckner MJ, Sharma J, Proctor HC (2009) Revisiting the meiofauna paradox: dispersal and colonization of nematodes and other meiofaunal organisms in low- and high-energy environments. Hydrobiologia 624:91–106. https://doi.org/10.1007/s10750-008-9669-5

Boxshall GA (1979) The planktonic copepods of the northeastern Atlantic Ocean: Harpacticoida, Siphonostomatoida and Mormonilloida. Bull Br Museum Nat Hist 35:201–264

Boxshall GA, Jaume D (2000) Making waves: the repeated colonization of fresh water by copepod crustaceans. Adv Ecol Res 31:61–79

Brooks JM, Fisher C, Roberts H, et al (2009) Investigations of chemosynthetic communities on the lower continental slope of the Gulf of Mexico: Interim Report 2. U. S. Dept. of Interior,

Minerals Management Service, Gulf of Mexico OCS Region, New Orleans, LA. OCS Study MMS 2009-046

Bruno MC, Cottarelli V (1999) Harpacticoids from groundwaters in the Philippines: *Parastenocaris mangyans*, new species, *Epactophanes philippinus*, new species, and redescription of *Phylloganthopus bassoti* (Copepoda). J Crustac Biol 19:510–529. https://doi.org/10.2307/1549260

Bruno MC, Cottarelli V (2015) First record of *Kinnecaris* (Copepoda: Harpacticoida: Parastenocarididae) from Turkey and Thailand; description of three new species and emended definition of the genus. Ital J Zool 82:69–94. https://doi.org/10.1080/11250003.2014.1002820

Bruno MC, Perry SA (2004) Exchanges of copepod fauna between surface- and ground-water in the Rocky Glades of Everglades National Park (Florida, U.S.A.). Arch für Hydrobiol 159:489–510. https://doi.org/10.1127/0003-9136/2004/0159-0489

Büntzow M (2011) Vergleichende gemeinschaftsanalytische und taxonomische Untersuchungen der Harpacticoidenfauna der Seeberge "Sedlo" und "Seine" (nördlicher Mittelatlantik). Carl von Ossietzky Universität Oldenburg

Byrnes MR, Davis RA, Kennicutt MC II et al (2017) Habitats and biota of the Gulf of Mexico: before the Deep-water Horizon oil spill. Vol. 1: Water quality, sediments, sediment contaminants, oil and gas seeps, coastal habitats, offshore plankton and benthos, and shellfish. Springer, New York

Cerca J, Purschke G, Struck TH (2018) Marine connectivity dynamics: clarifying cosmopolitan distributions of marine interstitial invertebrates and the meiofauna paradox. Mar Biol 165:1–21. https://doi.org/10.1007/s00227-018-3383-2

Chappuis PA (1931) Copepoda Harpacticoïda der Deutschen Limnologischen Sunda-Expedition. Arch für Hydrobiol Suppl 8. Trop Binnengewässer 1:512–584

Conroy-Dalton S, Huys R (1999) New genus of Aegisthidae (Copepoda: Harpacticoida) from hydrothermal vents on the Galapagos Rift. J Crustac Biol 19:408–431. https://doi.org/10.1163/193724099X00222

Conroy-Dalton S, Huys R (2000) Systematics and phylogeny of the Ancorabolidae (Copepoda: Harpacticoida). I. The *Ancorabolus*-lineage, with the description of three new genera. Cah Biol Mar 41:343–397. https://doi.org/10.21411/CBM.A.FF9630A7

Corgosinho PHC, Martínez Arbizu P, Santos-Silva EN (2007) Three new species of Remaneicaris Jakobi, 1972 (Copepoda, Harpacticoida, Parastenocarididae) from the Ribeirao do Ouro River, Minas Gerais, Brazil, with some remarks on the groundpattern of the Parastenocarididae. Zootaxa 1437:1–28. https://doi.org/10.5281/zenodo.175898

Corgosinho PHC, Martínez Arbizu P, Santos-Silva EN (2010) Three new interstitial species of *Remaneicaris* Jakobi, 1972 (Copepoda: Harpacticoida: Parastenocarididae) from Southern and Southeastern Brazil. Invertebr Zool 7:1–28. https://doi.org/10.15298/invertzool.07.1.01

Cottarelli V, Bruno MC (1993) Harpacticoida (Crustacea, Copepoda) from subterranean waters of Bue Marino cave, Sardinia, and St. Barthelemy cave, Corsica, and description of three new species. Int J Speleol 22:97–119

Coull BC (1971) Meiobenthic Harpacticoida (Crustacea, Copepoda) from the North Carolina continental shelf. Cah Biol Mar 12:195–237. https://doi.org/10.21411/CBM.A.116176E2

Coull BC (1973a) Meiobenthic Harpacticoida (Crustacea, Copepoda) from the deep sea off North Carolina II. The Family Cerviniidae Sars, Lang. Trans Am Microsc Soc 92:198–208. https://doi.org/10.2307/3224916

Coull BC (1973b) Meiobenthic Harpacticoida (Crustacea, Copepoda) from the deep sea off North Carolina IV. The families Cletodidae T. Scott and Ancorabolidae Sars. Trans Am Microsc Soc 92:604–620. https://doi.org/10.2307/3225271

Coull BC (1973c) Meiobenthic Harpacticoida (Crustacea, Copepoda) from the deep sea off North Carolina III. The families Tisbidae Stebbing emend. Lang, Thalestridae Sars emend. Lang, and Diosaccidae Sars. Trans Am Microsc Soc 92:592–603. https://doi.org/10.2307/3225270

Coull BC (1973d) Meiobenthic Harpacticoida (Crustacea, Copepoda) from the deep sea off North Carolina I. The genera *Hemimesochra* Sars, *Paranannopus* Lang, and *Cylindronannopus* n. g. Trans Am Microsc Soc 92:185–198. https://doi.org/10.2307/3224915

Cuvelier D, Beesau J, Ivanenko VN et al (2014) First insights into macro- and meiofaunal colonisation patterns on paired wood/slate substrata at Atlantic deep-sea hydrothermal vents. Deep-Sea Res Part I Oceanogr Res Pap 87:70–81. https://doi.org/10.1016/j.dsr.2014.02.008

Dahms HU, Qian PY (2004) Life histories of the Harpacticoida (Copepoda, Crustacea): a comparison with meiofauna and macrofauna. J Nat Hist 38:1725–1734. https://doi.org/10.1080/0022293031000156321

Danovaro R, Gambi C, Della Croce N (2002) Meiofauna hotspot in the Atacama Trench, eastern South Pacific Ocean. Deep-Sea Res I 49:843–857

Degen R (2010) Community study of tubeworm associated epizooic meiobenthos from deep sea cold seeps and hot vents. Universität Wien

Distel DL, Baco AR, Chuang E et al (2000) Do mussels take wooden steps to deep-sea vents? Nature 403:725–726

Dojiri M, Ho J-S (2013) Systematics of the Caligidae, copepods parasitic on marine fishes. Brill, Leiden

Dumont HJ, Maas S (1988) Five new species of leaf litter harpacticoids (Crustacea, Copepoda) from Nepal. Zool Scr 17:55–68. https://doi.org/10.1111/j.1463-6409.1988.tb00086.x

Easton EE, Thistle D (2016) Do some deep-sea, sediment-dwelling species of harpacticoid copepods have 1000-km-scale range sizes? Mol Ecol 25:4301–4318. https://doi.org/10.1111/mec.13744

Eckman JE, Thistle D (1988) Small-scale spatial pattern in meiobenthos in the San Diego Trough. Deep-Sea Res 35:1565–1578

Ellison CK, Burton RS (2008) Interpopulation hybrid breakdown maps to the mitochondrial genome. Evolution (N Y) 62:631–638. https://doi.org/10.1111/j.1558-5646.2007.00305.x

Fahrbach E, Gerdes D (1997) Die Expedition ANTARKTIS XIII/4-5 des Forschungsschiffs "Polarstern" 1996. Berichte zur Polarforsch 239:1–126

Farran GP (1914) Description of a harpacticoid copepod parasitic on an octopus. Ann Mag Nat Hist 8:472–475

Fenchel T, Finlay BJ (2004) The ubiquity of small species: patterns of local and global diversity. Bioscience 54:777–784. https://doi.org/10.1641/0006-3568(2004)054[0777:tuossp]2.0.co;2

Fiers F, Jocque M (2013) Leaf litter copepods from a cloud forest mountain top in Honduras (Copepoda: Cyclopidae, Canthocamptidae). Zootaxa 3630:270–290. https://doi.org/10.11646/zootaxa.3630.2.4

Finlay BJ (2002) Global dispersal of free-living microbial eukaryote species. Science (80-) 296:1061–1063. https://doi.org/10.1126/science.1070710

Fleeger JW, Chandler GT, Fitzhugh GR, Phillips FE (1984) Effects of tidal currents on meiofauna densities in vegetated salt marsh sediments. Mar Ecol Prog Ser 19:49–53. https://doi.org/10.3354/meps019049

Foissner W (2006) Biogeography and dispersal of micro-organisms: a review emphasizing protists. Acta Protozool 45:111–136

Fontaneto D (ed) (2011) Biogeography of microscopic organisms: is everything small everywhere? Cambridge University Press, New York

George KH (2001) First record of the 'genus' *Ancorabolus* Norman 1903 from the Southern Hemisphere, including analyses of copepodid development (Crustacea, Copepoda, Harpacticoida, Ancorabolidae). Senck Biol 81(1–2):23–3

George KH (2005) Sublittoral and bathyal Harpacticoida (Crustacea: Copepoda) of the Magellan region. Composition, distribution and species diversity of selected major taxa. Sci Mar 69:147–158. https://doi.org/10.3989/scimar.2005.69s2147

George KH (2006) New Ancorabolinae Sars, 1909 (Copepoda: Harpacticoida: Ancorabolidae) of the Atlantic and the Pacific Ocean. The taxa *Ceratonotus* Sars, and *Dendropsyllus* Conroy-Dalton. Meiofauna Mar 15:87–122

George KH (2008) *Argestes angolaensis* sp. nov. (Copepoda: Harpacticoida: Argestidae) from the Angola Basin (Southeast Atlantic), and the phylogenetic characterization of the taxon *Argestes* Sars, including the redescription of *A. mollis* Sars, 1910, and *A. reductus* (Itô, 1983). Zootaxa 262:223–262. https://doi.org/10.5281/zenodo.183720

George KH (2013) Faunistic research on metazoan meiofauna from seamounts - a review. Meiofauna Mar 20:1–32

George KH (2018) *Laophontodes typicus* T. Scott (Copepoda, Harpacticoida, Laophontodinae) — discovering a species complex by morphology. Eur J Taxon 439:1–43. https://doi.org/10.5852/ejt.2018.439

George KH, Gheerardyn H (2015) Remarks on the genus *Laophontodes* T. Scott, 1894 (Copepoda, Harpacticoida, Ancorabolidae, Laophontodinae), including the (re-)description of four species. Zool Anz 259:61–96. https://doi.org/10.1016/j.jcz.2015.02.009

George KH, Müller F (2013) Characterization of a monophylum *Echinocletodes*, its exclusion from Ancorabolinae (Copepoda, Harpacticoida), and displacement of *E. bodini* and *E. walvisi* to *Cletodes*, including the description of two new species. Zootaxa 3666:101–136. https://doi.org/10.11646/zootaxa.3666.2.1

George KH, Schminke HK (1998) First records of the genus *Ceratonotus* G. O. Sars, 1909 (Copepoda, Ancorabolidae) from the Southern Hemisphere, with the description of two new species. Crustaceana 71:801–817. https://doi.org/10.1163/156854098X00059

George KH, Schminke HK (2002) Harpacticoida (Crustacea, Copepoda) of the Great Meteor Seamount, with first conclusions as to the origin of the plateau fauna. Mar Biol 144:887–895. https://doi.org/10.1007/s00227-002-0878-6

George KH, Veit-Köhler G, Arbizu PM et al (2014) Community structure and species diversity of Harpacticoida (Crustacea: Copepoda) at two sites in the deep sea of the Angola Basin (Southeast Atlantic). Org Divers Evol 14:57–73. https://doi.org/10.1007/s13127-013-0154-2

George KH, Wandeness AP, Santos PJP (2013) A new species of Ancorabolidae (Crustacea: Copepoda: Harpacticoida) from the Campos Basin, Brazil, including a phylogenetic review of the taxon *Polyascophorus* George, 1998, and a key to its species. Zool Anz 252:20–33. https://doi.org/10.1016/j.jcz.2012.02.002

Giere O (2009) Meiobenthology. The microscopic motile fauna of aquatic sediments, 2nd edn. Springer, Berlin Heidelberg

Gollner S, Govenar B, Arbizu PM et al (2015a) Differences in recovery between deep-sea hydrothermal vent and vent-proximate communities after a volcanic eruption. Deep-Sea Res I 106:167–182. https://doi.org/10.1016/j.dsr.2015.10.008

Gollner S, Govenar B, Fisher CR, Bright M (2015b) Size matters at deep-sea hydrothermal vents: different diversity and habitat fidelity patterns of meio- and macrofauna. Mar Ecol Prog Ser 520:57–66. https://doi.org/10.3354/meps11078

Gollner S, Ivanenko VN, Martínez Arbizu P (2008) A new species of deep-sea Tegastidae (Crustacea: Copepoda: Harpacticoida) from 9°50′N on the East Pacific Rise, with remarks on its ecology. Zootaxa 1866:323–326. https://doi.org/10.5281/zenodo.183742

Gollner S, Zekely J, Govenar B et al (2007) Tubeworm-associated permanent meiobenthic communities from two chemically different hydrothermal vent sites on the East Pacific Rise. Mar Ecol Prog Ser 337:39–49. https://doi.org/10.3354/meps337039

Gollner S, Zekely J, Van Dover CL et al (2006) Benthic copepod communities associated with tubeworm and mussel aggregations on the East Pacific Rise. Cah Biol Mar 47:397–402. https://doi.org/10.21411/CBM.A.936358EA

Gómez S (2018a) Two new deep-sea species of Argestidae and Ameiridae (Copepoda: Harpacticoida) from the Eastern Mexican Pacific and Gulf of California, with proposal of a new genus of the family Argestidae. J Nat Hist 52:2613–2638. https://doi.org/10.1080/00222933.2018.1546915

Gómez S (2018b) New species of *Eurycletodes* Sars, 1909 and *Odiliacletodes* Soyer, 1964 from the deep Gulf of California (Copepoda, Harpacticoida, Argestidae). Zookeys 764:1–25. https://doi.org/10.3897/zookeys.764.24511

Gómez S (2018c) New species of the genus *Mesocletodes* Sars, 1909 from the deep Gulf of California (Copepoda, Harpacticoida). Zookeys 751:75–112. https://doi.org/10.3897/zookeys.751.20387

Gómez S (2018d) A new record and a new species of the genus Rhizothrix (Copepoda: Harpacticoida: Rhizothrichidae) from the deep sea of the Gulf of California. Proc Biol Soc Washingt 131:7–18. https://doi.org/10.2988/17-00021

Gómez S (2020a) On some deep-sea Stenheliinae from the Gulf of California and the west coast of the Baja California Peninsula (Mexico): *Pseudostenhelia bathyalis* sp. nov., and *Beatricella calidafornax* sp. nov. (Copepoda: Harpacticoida: Miraciidae). Zootaxa

Gómez S (2020b) On some deep-sea Stenheliinae from the Gulf of California and the west coast of the Baja California Peninsula (Mexico): the genus *Delavalia* and proposal of *Archaeohuysia* gen. nov. and *Diarthropodella* gen. nov. (Copepoda: Harpacticoida: Miraciidae). Zootaxa

Gómez S, Conroy-Dalton S (2002) Description of *Ancorabolus hendrickxi* sp. nov. (Copepoda: Harpacticoida: Ancorabolidae) from the neotropics and notes on caudal ramus development within oligoarthran harpacticoids. Cah Biol Mar 43:111–129. https://doi.org/10.21411/CBM.A.5B7EF6D7

Gómez S, Cruz-Barraza JA (2020) On some deep-sea Stenheliinae from the Gulf of California and the west coast of the Baja California Peninsula (Mexico): *Wellstenhelia euterpoides* sp. nov., and *Wellstenvalia wellsi* gen. et sp. nov. (Copepoda: Harpacticoida: Miraciidae). Zootaxa

Gómez S, Díaz K (2017) On some new species of Ancorabolidae Sars, 1909 from the Gulf of California: the genera *Ceratonotus* Sars, 1909, and *Dendropsyllus* Conroy-Dalton, 2003 (Crustacea, Copepoda, Harpacticoida). Zookeys 657:43–65. https://doi.org/10.3897/zookeys.657.10725

Gómez S, Morales-Serna FN (2012) Meiofauna de mar profundo del golfo de California: algunos aspectos acerca de la distribución y abundancia de Copepoda. In: Zamorano P, Hendrickx ME, Caso M (eds) Biodiversidad y comunidades del talud continental del Pacífico Mexicano. Secretaria de Medio Ambiente y Recursos Naturales (SEMARNAT); Instituto Nacional de Ecología (INE), Mexico City, p 466

Gómez S, Morales-Serna FN (2015) On a small collection of Tetragonicipitidae Lang, 1944 (Copepoda: Harpacticoida) from Mexico: new records and new species. J Nat Hist 49:1–44. https://doi.org/10.1080/00222933.2012.757657

Gross TF, Williams AJ (1991) Characterization of deep-sea storms. Mar Geol 99:281–301. https://doi.org/10.1016/0025-3227(91)90045-6

Hagerman GM, Rieger RM (1981) Dispersal of benthic meiofauna by wave and current action in Bogue Sound, North Carolina, USA. Mar Ecol 2:245–270. https://doi.org/10.1111/j.1439-0485.1981.tb00099.x

Hamond R (1973) Some Laophontidae (Crustacea: Harpacticoida) from off North Carolina. Trans Am Microsc Soc 92:44–59. https://doi.org/10.2307/3225170

Higgins RP, Thiel H (1988) Introduction to the study of meiofauna. Smithsonian Institution Press, Washington

Hill GE (2015) Mitonuclear ecology. Mol Biol Evol 32:1917–1927. https://doi.org/10.1093/molbev/msv104

Hollister CD, McCave IN (1984) Sedimentation under deep-sea storms. Nature 309:220–225. https://doi.org/10.1038/309220a0

Hui JHL (2018) Copepod incompatibilities. Nat Ecol Evol 2:1203–1204. https://doi.org/10.1038/s41559-018-0615-2

Humes AG, Voight JR (1997) *Cholidya polypi* (Copepoda: Harpacticoida: Tisbidae), a parasite of deep-sea octopuses in the North Atlantic and northeastern Pacific. Ophelia 46:65–81. https://doi.org/10.1080/00785326.1997.10432478

Huys R (2016) Harpacticoid copepods—their symbiotic associations and biogenic substrata: a review. Zootaxa 4174:448–729. https://doi.org/10.11646/zootaxa.4174.1.28

Huys R, Boxshall GA (1991) Copepod evolution. The Ray Society, London

Huys R, Conroy-Dalton S (1997) Discovery of hydrothermal vent Tantulocarida on a new genus of Argestidae (Copepoda: Harpacticoida). Cah Biol Mar 38:235–249. https://doi.org/10.21411/CBM.A.2E122F4E

Huys R, Iliffe TM (1998) Novocriniidae, a new family of harpacticoid copepods from anchihaline caves in Belize. Zool Scr 27:1–15. https://doi.org/10.1111/j.1463-6409.1998.tb00425.x

Huys R, Thistle D (1989) *Bathycamptus eckmani* gen. et spec. nov. (Copepoda, Harpacticoida) with a review of the taxonomic status of certain other Deep-water harpacticoids. Hydrobiologia 185:101–126. https://doi.org/10.1007/BF00010809

Janetzky W, Martinez Arbizu P, Reid JW (1996) *Attheyella (Canthosella) mervini* sp. n. (Canthocamptidae, Harpacticoida) from Jamaican bromeliads. Hydrobiologia 339:123–135

Joseph A (2017) Seafloor hot chimneys and cold seeps: mysterious life around them. In: Joseph A (ed) Investigating seafloors and oceans. From mud volcanoes to giant squid. Elsevier, Amsterdam, p 612. https://doi.org/10.1016/C2015-0-05842-7

Kabata Z (1979) Parasitic Copepoda of British fishes. Ray Society, London

Khodami S, McArthur JV, Blanco-Bercial L, Martinez Arbizu P (2017) Molecular phylogeny and revision of copepod orders (Crustacea: Copepoda). Sci Rep 7:9164. https://doi.org/10.1038/s41598-017-06656-4

Kiel S (2016) A biogeographic network reveals evolutionary links between deep-sea hydrothermal vent and methane seep faunas. Proc R Soc B Biol Sci 283:20162337. https://doi.org/10.1098/rspb.2016.2337

Kikuchi Y (1994) *Glaciella*, a new genus of freshwater Canthocamptidae (Copepoda, Harpacticoida) from a glacier in Nepal, Himalayas. Hydrobiologia 292–293:59–66. https://doi.org/10.1007/BF00229923

Kitching RL (2000) Food webs and container habitats: the natural history and ecology of phytotelmata. Cambridge University Press, Cambridge

Kolesnikova EA, Anufriieva EV, Latushkin AA, Shadrin NV (2017) *Mesochra rostrata* Gurney, 1927 (Copepoda, Harpacticoida) in Sivash Bay (Sea of Azov): Is it a new alien species or a relict of Tethys? Russ J Biol Invasions 8:244–250. https://doi.org/10.1134/S2075111717030079

Lang K (1948) Monographie der Harpacticiden Vols I & II. Nordiska Bokhandeln, Stockholm

Ledesma-Vázquez J (2002) A gap in the Pliocene invasion of seawater, Gulf of California. Rev Mex Ciencias Geológicas 19:145–151

Ledesma-Vázquez J, Johnson ME, González-Yajimovich O, Santamaría-del-Angel E (2009) Gulf of California geography, geological origins, oceanography, and sedimentation patterns. In: Atlas of Coastal Ecosystems in the Gulf of California: past and present, pp 1–10

Lee W, Huys R (1999) *Bathylaophonte* gen.nov. from deep-sea hydrothermal vents and the polyphyly of *Paronychocamptus* (Copepoda: Harpacticoida). Cah Biol Mar 40:293–328. https://doi.org/10.21411/CBM.A.36920EDF

Levis NA, Isdaner AJ, Pfennig DW (2018) Morphological novelty emerges from pre-existing phenotypic plasticity. Nat Ecol Evol 2:1289–1297. https://doi.org/10.1038/s41559-018-0601-8

Levis NA, Pfennig DW (2016) Evaluating "Plasticity-First" evolution in nature: Key criteria and empirical approaches. Trends Ecol Evol 31:563–574. https://doi.org/10.1016/j.tree.2016.03.012

López-González P, Bresciani J, Huys R et al (2000) Description of *Genesis vulcanoctopusi* gen. et sp. nov. (Copepoda: Tisbidae) parasitic on a hydrothermal vent octopod and a reinterpretation of the life cycle of cholidynid harpacticoids. Cah Biol Mar 41:241–253. https://doi.org/10.21411/CBM.A.94121E5B

Lubchenco J, McNutt M, Lehr B, et al (2010) BP Deep-water Horizon oil budget: What happened to the oil? NOAA/USGS Fed Sci Rep

Mahatma R (2008) Meiofauna communities of the Pacific Nodule Province: abundance, diversity and community structure. Carl von Ossietzky Universität Oldenburg.

McClain CR, Lundsten L, Ream M et al (2009) Endemicity, biogeography, composition, and community structure on a Northeast Pacific seamount. PLoS One 4:19–26. https://doi.org/10.1371/journal.pone.0004141

Menzel L (2011) First descriptions of copepodid stages, sexual dimorphism and intraspecific variability of *Mesocletodes* Sars, 1909 (Copepoda, Harpacticoida, Argestidae), including the description of a new species with broad abyssal distribution. Zookeys 96:39–80. https://doi.org/10.3897/zookeys.96.1496

Menzel L, George KH (2009) Description of four new species of *Mesocletodes* Sars, 1909 (Copepoda, Harpacticoida, Argestidae) and redescription of *Mesocletodes robustus* Por, 1965 from the South Atlantic, including remarks on the *Mesocletodes abyssicola*-group. Zootaxa 2096:214–256. https://doi.org/10.11646/zootaxa.2096.1.14

Menzel L, George KH, Martínez Arbizu P (2011) Submarine ridges do not prevent large-scale dispersal of abyssal fauna: a case study of *Mesocletodes* (Crustacea, Copepoda, Harpacticoida). Deep-Sea Res I 58:839–864. https://doi.org/10.1016/j.dsr.2011.05.008

Menzel R (1916) Über das Auftreten der Harpacticidengattungen Epactophanes Mrázek aud Parastenocaris Keβler in Surinam. Zool Anz 47:145–152

Mercado-Salas NF, Khodami S, Martínez Arbizu P (2019) Convergent evolution of mouthparts morphology between Siphonostomatoida and a new genus of deep-sea Aegisthidae Giesbrecht, 1893 (Copepoda: Harpacticoida). Mar Biodivers:1635–1655. https://doi.org/10.1007/s12526-018-0932-3

Minic Z, Serre V, Hervé G (2006) Adaptation des organismes aux conditions extrêmes des sources hydrothermales marines profondes. Comptes Rendus Biol 329:527–540. https://doi.org/10.1016/j.crvi.2006.02.001

Montagna PA (1979) *Cervinia langi* n. sp. and *Pseudocervinia magna* (Copepoda: Harpacticoida) from the Beaufort Sea (Alaska, USA). Trans Am Microsc Soc 98:77–88. https://doi.org/10.2307/3225941

Montagna PA (1980a) A new species and a new genus of Cerviniidae (Copepoda: Harpacticoida) from the Beaufort Sea, with a revision of the family. Proc Biol Soc Washingt 93:1204–1219

Montagna PA (1980b) Two new bathyal species of *Pseudotachidius* (Copepoda: Harpacticoida) from the Beaufort Sea (Alaska, U.S.A.). J Nat Hist 14:567–578. https://doi.org/10.1080/00222938000770461

Montagna PA, Baguley JG, Cooksey C et al (2013) Deep-sea benthic footprint of the Deep-water Horizon blowout. PLoS One 8:e70540. https://doi.org/10.1371/journal.pone.0070540

Montagna PA, Carey AG Jr (1978) Distributional notes on Harpacticoida (Crustacea: Copepoda) collected from the Beaufort Sea (Arctic Ocean). Astarte 11:117–122

Moura G, Pottek M (1998) *Selenopsyllus*, a new genus of Cylindropsyllinae (Copepoda, Harpacticoida) from Atlantic and Antarctic deep waters. Senckenbergiana 28:185–209. https://doi.org/10.1007/BF03043149

Neira C, Ingels J, Mendoza G et al (2018) Distribution of meiofauna in bathyal sediments influenced by the oxygen minimum zone off Costa Rica. Front Mar Sci 5:1–17. https://doi.org/10.3389/fmars.2018.00448

Netto SA, Gallucci F, Fonseca GFC (2005) Meiofauna communities of continental slope and deep-sea sites off SE Brazil. Deep-Sea Res I 52:845–859. https://doi.org/10.1016/j.dsr.2004.11.009

Norman AM, Scott T (1905) Crustacea Copepoda new to science from Devon and Cornwall. Ann Mag Nat Hist 15:284–300

Packmor J, Müller F, George KH (2015) Oceanic islands and seamounts as staging posts for Copepoda Harpacticoida (Crustacea) - Shallow-water Paramesochridae Lang, 1944 from the North-East Atlantic Ocean, including the (re-)description of three species and one subspecies from the Madeiran Archi. Prog Oceanogr 131:59–81. https://doi.org/10.1016/j.pocean.2014.11.012

Packmor J, Riedl T (2016) Records of Normanellidae Lang, 1944 (Copepoda, Harpacticoida) from Madeira island support the hypothetical role of seamounts and oceanic islands as "stepping stones" in the dispersal of marine meiofauna. Mar Biodivers:1–17. https://doi.org/10.1007/s12526-016-0448-7

Perry BW, Schield DR, Castoe TA (2018) Evolution: Plasticity versus selection, or plasticity and selection? Curr Biol 28:R1104–R1106. https://doi.org/10.1016/j.cub.2018.07.050

Plum C, Gollner S, Martínez-Arbizu P, Bright M (2015) Diversity and composition of the copepod communities associated with megafauna around a cold seep in the Gulf of Mexico with remarks on species biogeography. Mar Biodivers 45:419–432. https://doi.org/10.1007/s12526-014-0310-8
Plum C, Martínez-Arbizu P (2009) Discovery of *Smacigastes* Ivanenko & Defaye, 2004 (Copepoda: Harpacticoida: Tegastidae) in a deep-sea cold seep, with the description of a new species from the Gulf of Mexico. Zootaxa 2096:338–355. https://doi.org/10.11646/zootaxa.2096.1.20
Por FD (1964) A study of the Levantine and Pontic Harpacticoida (Crustacea, Copepoda). Zool Verh 64:1–128
Por FD, Hadel VF (1986) Two new species of *Attheyella* (Copepoda: Harpacticoidea: Canthocamptidae) from bromeliads of the Serra da Juréia (Sao Paulo, Brazil). J Crustac Biol 6:777–788. https://doi.org/10.1163/193724086X00578
Ramírez-Llodra E, Shank T, German C (2007) Biodiversity and Biogeography of Hydrothermal Vent Species: Thirty Years of Discovery and Investigations. Oceanography 20:30–41
Ranga Reddy Y, Totakura VR, Corgosinho PHC (2014) *Himalayacaris alaknanda* n. gen., n. sp. (Copepoda: Harpacticoida: Parastenocarididae) from the hyporeic zone of a Himalayan River, northern India. J Crustac Biol 34:801–819. https://doi.org/10.1163/1937240X-00002281
Rawson PD, Burton RS (2002) Functional coadaptation between cytochrome c and cytochrome c oxidase within allopatric populations of a marine copepod. Proc Natl Acad Sci U S A 99:12955–12958. https://doi.org/10.1073/pnas.202335899
Reid JW, Lesko LT (2003) A new species of *Moraria* (Crustacea: Copepoda: Harpacticoida) from the Laurentian Great Lakes. Zootaxa 19:1–19. https://doi.org/10.11646/zootaxa.205.1.1
Reidenauer JA, Thistle D (1983) *Sarsameira* (Copepoda: Harpacticoida): an update and a new species from the deep sea. Trans Am Microsc Soc 102:105–112. https://doi.org/10.2307/3225879
Romero A (2009) Cave biology Life in Darkness, 1st edn. Cambridge University Press, Cambridge
Rosli N, Leduc D, Rowden AA, Probert PK (2018) Review of recent trends in ecological studies of deep-sea meiofauna, with focus on patterns and processes at small to regional spatial scales. Mar Biodivers 48:13–34. https://doi.org/10.1007/s12526-017-0801-5
Schmidt C, Lins L, Brandt A (2018) Harpacticoida (Crustacea, Copepoda) across a longitudinal transect of the Vema Fracture Zone and along a depth gradient in the Puerto Rico trench. Deep-Sea Res Part II 148:236–250. https://doi.org/10.1016/j.dsr2.2017.12.024
Sedlacek L (2007) Impacts of deep-sea carbon sequestration and of ciliate epibionts on harpacticoid copepods. The Florida State University
Seifried S (2003) Phylogeny of Harpacticoida (Copepoda): revision of "Maxillipedasphalea" and Exanechentera. Cuvillier Verlag, Göttingen
Shank TM (2010) Seamounts: deep-ocean laboratories of faunal connectivity, evolution, and endemism. Oceanography 23:108–122. https://doi.org/10.5670/oceanog.2010.65
Smirnov SS (1946) New species of Copepoda Harpacticoida from the Artic Ocean. Tr Dreifuyushchei Ekspeditsyai Glausemov Ledokol Por "Sedov" 3:231–263
Smith CR, Baco AR (2003) Ecology of whale falls at the deep-sea floor. Oceanogr Mar Biol 41:311–354
Smith CR, Glover AG, Treude T et al (2015) Whale-fall ecosystems: recent insights into ecology, paleoecology, and evolution. Ann Rev Mar Sci 7:571–596. https://doi.org/10.1146/annurev-marine-010213-135144
Smith CR, Kukert H, Wheatcroft RA et al (1989) Vent fauna on whale remains. Nature 341:27–28. https://doi.org/10.1038/341027a0
Sterrer W (1973) Plate tectonics as mechanism for dispersal and speciation in interstitial sand fauna. Netherlands J Sea Res 7:200–222
Street GT, Lotufo GR, Montagna PA, Fleeger JW (1998) Reduced genetic diversity in a meiobenthic copepod exposed to a xenobiotic. J Exp Mar Biol Ecol 222:93–111. https://doi.org/10.1016/S0022-0981(97)00140-8
Street GT, Montagna PA (1996) Loss of genetic diversity in Harpacticoida near off shore platforms. Mar Biol 126:271–282. https://doi.org/10.1007/BF00347452

Suárez-Morales E, Iliffe TM (2005) A new *Stygonitocrella* Petkovski (Copepoda: Harpacticoida) from a cave in Northern Mexico with comments on the taxonomy of the genus. Hydrobiologia 544:215–228. https://doi.org/10.1007/s10750-005-0621-7

Thiel H, Hessler RR (1974) Ferngesteuertes Unterwasserfahrzeug erforscht Tiefseeboden. Umsch Wiss Tech 74:451–453

Thistle D (1978) Harpacticoid dispersion patterns: Implications for deep-sea diversity maintenance. J Mar Res 36:377–397

Thistle D, Levin LA (1998) The effect of experimentally increased near-bottom flow on metazoan meiofauna at a deep-sea site, with comparison data on macrofauna. Deep-Sea Res I 45:625–638

Thistle D, Levin LA, Gooday AJ et al (1999) Physical reworking by near-bottom flow alters the metazoan meiofauna of Fieberling Guyot (northeast Pacific). Deep-Sea Res Part I Oceanogr Res Pap 46:2041–2052. https://doi.org/10.1016/S0967-0637(99)00040-0

Thistle D, Sedlacek L, Carman KR et al (2007) Emergence in the deep sea: evidence from harpacticoid copepods. Deep-Sea Res I 54:1008–1014. https://doi.org/10.1016/j.dsr.2007.03.002

Tietjen JH, Deming JW, Rowe GT et al (1989) Meiobenthos of the Hatteras Abyssal Plain and Puerto Rico Trench: abundance, biomass and associations with bacteria and particulate fluxes. Deep-Sea Res 36:1567–1577. https://doi.org/10.1016/0198-0149(89)90058-7

Tunnicliffe V (1992) The nature and origin of the modern hydrothermal vent fauna. Palaios 7:338–350. https://doi.org/10.2307/3514820

Tunnicliffe V, McArthur AG, McHugh D (1998) A biogeographical perspective of the deep-sea hydrothermal vent fauna

Van Dover CL, German CR, Speer KG, et al (2002) Evolution and biogeography of deep-sea vent and seep invertebrates. Science (80-) 295:1253–1257. https://doi.org/10.1126/science.1067361

Vasconcelos DM (2008) Distribuição dos Copepoda Harpacticoida da meiofauna em área de talude no litoral de Sergipe. Universidade Federal de Pernambuco, Brasil

Vasconcelos DM, George KH, Parreira Dos Santos PJ (2008) First record of the genus *Pseudomesochra* T. Scott (Harpacticoida: Pseudotachidiidae) in the South Atlantic with description of a deep-sea species: *Pseudomesochra longiseta* sp. nov. Zootaxa 1866:337–348. https://doi.org/10.11646/zootaxa.1866.1.17

Vasconcelos DM, Veit-Köhler G, Drewes J, Parreira dos Santos PJ (2009) First record of the genus *Kliopsyllus* Kunz, 1962 (Copepoda Harpacticoida, Paramesochridae) from Northeastern Brazil with description of the deep-sea species of *Kliopsyllus minor* sp. nov. Zootaxa 2096:327–337. https://doi.org/10.11646/zootaxa.2096.1.19

Walter TC, Boxshall G (2019) World of Copepods database. Harpacticoida. In: World Regist. Mar. Species. http://www.marinespecies.org/aphia.php?p=taxdetails&id=1102. Accessed 2 Oct 2019

Wandeness AP, George KH, Santos PJP (2009) First record of the taxon *Echinopsyllus* (Copepoda, Harpacticoida, Ancorabolidae) from the deep sea of Campos Basin, Brazil, with the description of three new species and their contribution to phylogenetic analysis. Zool J Linn Soc 156:52–78. https://doi.org/10.1111/j.1096-3642.2008.00479.x

Watling L, Guinotte J, Clark MR, Smith CR (2013) A proposed biogeography of the deep ocean floor. Prog Oceanogr 111:91–112. https://doi.org/10.1016/j.pocean.2012.11.003

Willen E (2003) A new species of Stenhelia (Copepoda, Harpacticoida) from a hydrothermal, active, submarine volcano in the New Ireland Fore-Arc system (Papua New Guinea) with notes on deep sea colonization within the Stenheliinae. J Nat Hist 37:1691–1711. https://doi.org/10.1080/00222930110114437

Wilson CB (1936) Copepods from the cenotes and caves of the Yucatan peninsula, with notes on Cladocerans. Carnegie Inst Washingt 457:77–88

Wolff JN, Ladoukakis ED, Enríquez JA, Dowling DK (2014) Mitonuclear interactions: Evolutionary consequences over multiple biological scales. Philos Trans R Soc B Biol Sci 369:20130443. https://doi.org/10.1098/rstb.2013.0443

Wolff T (2005) Composition and endemism of the deep-sea hydrothermal vent fauna. Cah Biol Mar 46:97–104. https://doi.org/10.21411/CBM.A.28CE1239

Zamorano P, Hendrickx ME, Méndez N, et al (2014) La exploración de las aguas profundas del Pacífico Mexicano: El proyecto TALUD. In: Low Pfeng A, Peters Recagno EM (eds) La frontera final: El oceano profundo, 1st editio. Secretaría de Medio Ambiente y Recursos Naturales (SEMARNAT), Instituto Nacional de Ecología y Cambio Climático (INECC), p 304
Zekely J, Van Dover CL, Nemeschkal HL, Bright M (2006) Hydrothermal vent meiobenthos associated with mytilid mussel aggregations from the Mid-Atlantic Ridge and the East Pacific Rise. Deep Sea Res I 53:1363–1378. https://doi.org/10.1016/j.dsr.2006.05.010

Chapter 5
Updated Checklist of Deep-Sea Amphipods (Amphilochidea and Senticaudata) from Western Mexico, NE Pacific Ocean

I. Winfield, M. E. Hendrickx, and M. Ortiz

Abstract An updated checklist of deep-sea Amphilochidea and Senticaudata amphipods (excluding the caprelloids) documented for Western Mexico, in the NE Pacific, from bathyal to hadal zones (210–6000 m) is presented. Depth range, geographic distribution, and references are provided. A total of 118 deep-sea species have been reported from the area, grouped in 66 genera, 30 families, and 2 suborders. The suborder Amphilochidea is highly diverse with 110 species, 62 genera, and 27 families; in contrast, suborder Senticaudata includes 8 species, 4 genera, and 3 families. The most diversified amphipod families are Phoxocephalidae (12 genera, 18 species), Synopiidae (7 and 16), Oedicerotidae (4 and 14), Ampeliscidae (2 and 12), and Tryphosidae (6 and 10). The Western Mexico borderland includes 94 species, the Gulf of California 3, and the SW Mexican Pacific 5; in addition, *Ampelisca pacifica*, *Epimeria morronei*, *Leucothoe spinicarpa*, *Aruga oculata*, *Heterophoxus oculatus*, *Metaphoxus frequens*, and *Hippomedon propinquus* inhabit both the borderland and the Gulf of California. *Stephonyx californiensis* occurs in both the Gulf of California and the SW Mexican Pacific. Five species, *Eyakia calcarata*, *Synopia scheeleana*, *Stephonyx laqueus*, *Bonnierella linearis*, and *Bonnierella californica*, have been documented in the borderland and in the SW Mexican Pacific. *Ampelisca romigi*, *Byblis veleronis*, and *Trischizostoma unam* inhabit all three areas. Species numbers are very dissimilar according to depth, with 94% (111 species) in the bathyal zone vs. only 3 species in the abyssal zone and one species in the hadal zone.

Keywords Deep water · Amphipods · Mexican Pacific · Diversity · Distribution

---

I. Winfield (✉) · M. Ortiz
Laboratorio de Crustáceos, Facultad de Estudios Superiores Iztacala, Universidad Nacional Autónoma de México, Tlalnepantla, Estado de México, Mexico
e-mail: ignacioc@unam.mx

M. E. Hendrickx
Laboratorio de Invertebrados Bentónicos, Unidad Académica Mazatlán, Instituto de Ciencias del Mar y Limnología, Universidad Nacional Autónoma de México,
Mazatlán, Sinaloa, Mexico
e-mail: michel@ola.icmyl.unam.mx

M. E. Hendrickx (ed.), *Deep-Sea Pycnogonids and Crustaceans of the Americas*, https://doi.org/10.1007/978-3-030-58410-8_5

5.1 Introduction

The order Amphipoda represents a group of peracarid crustaceans that mostly inhabits the marine environment worldwide. They have successfully occupied sedimentary ecosystems of the continental margin and the abyssal plain as members of the benthic and suprabenthic communities (Barnard and Karaman 1991; Mees and Jones 1997). These environments included deltaic systems, calcareous and soft bottoms of the continental shelf and slope, submarine canyons, and abyssal plain (Bellan-Santini 1999; Cartes et al. 2001), where the amphipods constitute one of the most diverse, numerous, and dominant groups of the macrofauna (Bachelet et al. 2003).

Their morphological and physiological adaptations have led to species diversification in both the pelagic and benthic realms. The amphipods exhibit a wide range of lifestyles, from inquilines and tube-dwellers, to infaunal sediment burrowers, nestlers, suprabenthic swimmers, and free-ranging plankters (Thurston 1990; Thomas 1992). According to Lowry and Myers (2017), most marine amphipods are benthic detrital, deposit, and suspension feeders. Other species, however, are predators and scavengers, and fewer species are parasitoids. Thurston (1990) has shown that necrophagous lysianassoids occur in large numbers on the abyssal seafloor and feed there exclusively, but can be found 1000 m or more up into the water column.

Until 2017, the amphipods included almost 10,000 species, grouped in 1618 genera, 223 families, and 5 suborders: Pseudingolfiellidea, Hyperiidea, Colomastigidea, Hyperiopsidea, Senticaudata, and Amphilochidea (Lowry and Myers 2017). Families recognized as exclusively found in the very deep sea (>3000 m depth) included Thurstonellidae, Cyphocarididae, Vitjazianidae, Cebocaridae, Cyclocaridae, Thoriellidae, Alicellidae, and Valettiopsidae (Horton 2004; Horton and Thurston 2009, 2011; Horton et al. 2013). Thurston (2000) characterized other ten families as common in the deep sea (>200 m depth) worldwide: Amathillopsidae, Ampeliscidae, Aristiidae, Lepechinellidae, Calliopiidae, Epimeriidae, Eusiridae, Iphimediidae, Ischyroceridae, Lysianassoidea, Leucothoidae, and Phoxocephalidae. These families represent about 1% of species of amphipod worldwide.

Several authors have described new species and made major additions to the list of deep-sea amphipod species and families in the world oceans; E. Dahl, E.L. Mills, J.L. Barnard, G.O. Sars, E. Chevreux, T.R.R. Stebbing, K.H. Barnard, K. Stephensen, A. Schellenberg, J.M. Pirlot, E.F. Gurjanova, M. Ledoyer, M. Thurston, and T. Horton (see contributions in Thurston 2000, 2001). Others have contributed specifically to our knowledge of the deep-sea amphipod fauna of the Northeastern Pacific Ocean, from the United States of America to Baja California Sur, Northern Mexico. These American and Canadian amphipod specialists have increasing the number of new species and new records since 1920s: C.R. Shoemaker, J.L. Barnard, E.L. Bousfield, D.R. Laubitz, M.E. Vinogradov, and K.E. Conlan (see supplementary authors and publications on amphipods of this area in García-Madrigal 2007).

Although deep-sea amphipods have been recognized as forming a key element in the deep-sea communities worldwide, there is no checklist available for Western Mexico, in the NE Pacific Ocean. This chapter contributes with an updated checklist of deep-sea amphipods (Amphilochidea and Senticaudata) documented for this geographic area, based on previous records and recent descriptions of species.

5.2 Material and Methods

A complete list of Amphilochidea and Senticaudata (with the exception of the caprelloids) was compiled from published references dealing with the eastern Pacific Ocean. The western Mexican Pacific was divided in three geographic zones (Fig. 5.1): (1) the western Mexico borderland (the west coast of the Baja California Peninsula); (2) the Gulf of California, and (3) the SW Mexican Pacific, from southern Sinaloa to Chiapas. The classification scheme of Lowry and Myers (2017) for Amphipoda was followed. The presence of each species in these geographic zones has been compiled, together with the depth range on record. For further analysis of the bathymetric occurrence of species, the bathyal zone (201–3000 m), the abyssal zone (3001–6000 m; also referred to as the abyssal plain), and the hadal zone (>6000 m) were considered. The bathyal zone (also known as the continental slope, CS) was further divided into the upper (201–1000 m; UCS), the middle (1001–2000 m; MCS), and the lower (2001–3000 m; LCS) zones.

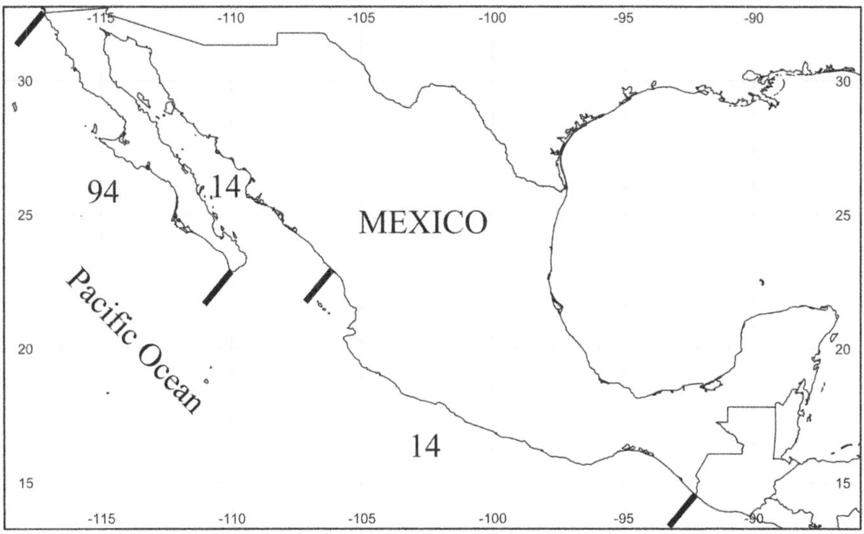

Fig. 5.1 Areas considered in this study and number of species of Amphipoda recorded in each area. Western Mexico borderland (west coast of the Baja California Peninsula), 94 species; the Gulf of California (to the southern limit of the State of Sinaloa), 14 species; and the SW Mexican Pacific (from southern Sinaloa to the southern limit of the State of Chiapas), 14 species

5.3 Results

A total of 118 species of deep-sea Amphilochidea and Senticaudata amphipods, grouped in 66 genera, 30 families, and 2 suborders, has been documented in western Mexico (Table 5.1). The suborder Amphilochidea was highly diverse with 110 species, 62 genera, and 27 families; in contrast, the suborder Senticaudata included only 8 species, 4 genera, and 3 families (Table 5.2). At family level, number of

Table 5.1 Numbers of genera and species of deep-sea amphipod in western Mexico

Family	Genera	Species	Depth zone
Suborder Amphilochidea			
1. Acidostomatidae	1	1	Bathyal-LCS
2. Alicellidae	1	1	Bathyal-LCS
3. Amathillopsidae	1	1	Abyssal
4. Ampeliscidae	2	12	Bathyal-ALL-CS
5. Argissidae	1	1	Bathyal-UCS
6. Aristiidae	1	1	Bathyal-LCS
7. Cyphocarididae	1	2	Bathyal-ALL-CS
8. Epimeriidae	1	1	Bathyal-MCS
9. Eusiridae	1	4	Bathyal-UCP-MCS
10. Lepechinellidae	1	2	Bathyal-MCS-LCP
11. Lepidepecreellidae	1	1	Bathyal-MCS
12. Leucothoidae	1	1	Bathyal-UCP-MCS
13. Liljeborgiidae	2	3	Bathyal-UCP-MCS
14. Lysianassidae	2	2	Bathyal-UCP
15. Oedicerotidae	4	14	Bathyal-ALL-CS
16. Pakinidae	1	1	Bathyal-UCP
17. Pardaliscidae	4	5	Bathyal-ALL-CS
18. Phoxocephalidae	12	18	Bathyal-abyssal
19. Pleustidae	1	1	Bathyal-abyssal
20. Stegocephalidae	4	5	Bathyal-abyssal
21. Stenothoidae	2	2	Bathyal-UCP-MCS
22. Synopiidae	7	16	Bathyal-ALL-CS
23. Trischizostomatidae	1	1	Bathyal-MCS
24. Tryphosidae	6	10	Bathyal-abyssal
25. Uristidae	1	2	Bathyal-UCP-MCS
26. Urothoidae	1	1	Bathyal-LCS
27. Vemanidae	1	1	Abyssal
Suborder Senticaudata			
28. Aoridae	1	1	Bathyal-UCP
29. Ischyroceridae	1	3	Bathyal-MCS and hadal
30. Photidae	2	4	Bathyal-UCP-LCS

LCS lower continental slope, *MCS* middle continental slope, *UCS* upper continental slope, *ALL-CS* throughout the continental slope

Table 5.2 Bathymetric distribution of species of amphipods from bathyal and hadal zones off Western Mexico

Depth zone	Species	Genera	Families
Bathyal zone	115	63	27
ALL-CS	49	18	5
UCS	28	24	10
MCS	32	24	10
LCS	13	12	6
Abyssal zone	2	2	2
Hadal zone	1	1	1

genera and species varied considerably, from 1 to 12 genera and from 1 to 18 species. In both cases, the highest values corresponded to the Phoxocephalidae. Most families (63%) were represented by 1–2 species only (Table 5.2).

Overall, species of western Mexico with records in deep water (>200 m) have been collected between 1 and 6324 m depth (Table 5.1). Depth ranges at which these 118 species have been reported in the literature are often surprising, with several species found from very shallow to very deep water: e.g., *Ampelisca macrocephala* Liljeborg, 1852, 21–1248 m; *A. unsocalae* Barnard, 1960, 50–1720 m; *Rhachotropis ludificor* Barnard, 1967, 17–1748 m; *Leucothoe spinicarpa* s.l. (Abildgaard, 1789), 1–1505 m; and *Heterophoxus oculatus* (Homes, 1908), 2–1941 m (Table 5.1).

When considering major habitat of the deep-water Amphilochidea and Senticaudata amphipods of western Mexico, 105 (88%) were benthic (3 were commensal); 3 were either pelagic-free, pelagic-commensal, or pelagic-demersal (Table 5.1); and 4 were scavenger (Table 5.1).

Based on our present knowledge, with 94 recorded species, biodiversity of amphipods is markedly higher in the western Mexico borderland compared with only 14 species in the Gulf of California and 14 in SW Mexico (Table 5.1). All species considered in this study, only *Ampelisca romigi*, *Byblis veleronis*, and *Trischizostoma unam* have been found on all three geographic zones (Table 5.1). Seven of the 14 Gulf of California species, *Ampelisca pacifica*, *Epimeria morronei*, *Leucothoe spinicarpa*, *Aruga oculata*, *Heterophoxus oculatus*, *Metaphoxus frequens*, and *Hippomedon propinquus*, have also been recorded in the borderland area. *Stephonyx californiensis* is the only species that has been reported only in both the Gulf of California and in SW Mexico, while 5 species, *Eyakia calcarata*, *Synopia scheeleana*, *Stephonyx laqueus*, *Bonnierella linearis*, and *Bonnierella californica*, are represented in western Mexico borderland and in SW Mexico (Table 5.1).

Distribution of families, genera, and species of Amphilochidea and Senticaudata according to depth was analyzed based on the division in bathymetric zones proposed herein (Table 5.3). In western Mexico, a total of 113 species were documented for the entire bathyal zone (>201 m to 3000 m). A total of 61 species are limited to the upper continental slope, of which a sizable number, 27 (44%), are

Table 5.3 List of species of amphipods recorded in western Mexico borderland (WM), Gulf of California (GC), and SW Mexico (SW) and depth range on record

Family and species	WM	GC	SW	Depth (m)	Sources
Acidostomatidae					
Acidostoma ortum Barnard, 1967	X			2398–2475/B	Barnard (1967)
Alicellidae					
Tectovalopsis fusilus Barnard and Ingram, 1990			X	2884/B	Barnard and Ingram (1990)
Amathillopsidae					
Amathillopsis pacifica margo Barnard, 1967	X			3481–3518/B	Barnard (1967)
Ampeliscidae					
Ampelisca amblyopsoides Barnard, 1960	X			1123–1481/B	Barnard (1967)
Ampelisca anversensis Karaman, 1975		X		500–2000/B	Brusca and Hendrickx (2005)
Ampelisca eoa Gurjanova, 1951	X			421–3718/B	Barnard (1967)
Ampelisca furcigera Bulycheva, 1936	X			21–1248/B	Barnard (1967)
Ampelisca macrocephala Liljeborg, 1852	X			403–1720/B	Barnard (1967)
Ampelisca pacifica Holmes, 1908	X	X		20–550/B	Barnard (1954) and Brusca and Hendrickx (2005)
Ampelisca plumosa Holmes, 1908	X			813–2667/B	Barnard (1967)
Ampelisca pugetica Stimpson, 1864	X			1248–1292/B	Barnard (1967)
Ampelisca romigi Barnard, 1954	X	X	X	3–504/B	Barnard (1954, 1964a) and Brusca and Hendrickx (2005)
Ampelisca unsocalae Barnard, 1960	X			50–1720/B	Barnard (1967)
Byblis teres Barnard, 1967	X			791–842/B	Barnard (1967)
Byblis veleronis Barnard, 1954	X	X	X	5–300/B	Barnard (1963), Blake et al. (1997), and Brusca and Hendrickx (2005)
Argissidae					
Argissa hamatipes (Norman, 1869)	X			4–1096/B	Barnard (1967) and Brusca and Hendrickx (2005)
Aristiidae					
Aristias expers Barnard, 1967	X			2398–2475/B	Barnard (1967)
Cyphocarididae					
Cyphocaris anonyx Boeck, 1871	X			600–1800/P	Barnard (1967)

(continued)

Table 5.3 (continued)

Family and species	WM	GC	SW	Depth (m)	Sources
Cyphocaris faurei K.H. Barnard, 1916	X			175–2800/P,C	Schellenberg (1929) and Hurley (1963)
Epimeriidae					
Epimeria morronei Winfield, Ortiz and Hendrickx, 2013	X	X		1526–2093/SC	Winfield et al. (2013)
Eusiridae					
Rhachotropis cervus Barnard, 1957	X			1000/B	Barnard (1967)
Rhachotropis clemens Barnard, 1967	X			92–842/B	Barnard (1967) and Blake et al. (1997)
Rhachotropis ludificor Barnard, 1967	X			17–1748/B	Barnard (1967)
Rhachotropis multesimus Barnard, 1967	X			1720–1748/B	Barnard (1967)
Lepechinellidae					
Lepechinella arctica Schellenberg, 1926	X			2667–2706/B	Barnard (1967)
Lepechinella turpis Barnard, 1967	X			1205–2667/B	Barnard (1973)
Lepidepecreellidae					
Lepidepecreella charno Barnard, 1966	X			1895/B	Barnard (1966)
Leucothoidae					
Leucothoe spinicarpa (Abildgaard, 1789)	X	X		1–1505/C,B	Barnard (1963, 1969, 1979) and Blake et al. (1997)
Liljeborgiidae					
Liljeborgia cota Barnard, 1962	X			366–1821/B	Barnard (1967)
Idunella albina (Barnard, 1959)	X			16–721/B	Barnard (1963) and Blake et al. (1997)
Idunella goleta (Barnard, 1959)	X			12–459/B	Barnard (1963) and Blake et al. (1997)
Lysianassidae					
Aruga oculata Holmes, 1908	X	X		20–340/C	Hurley (1963)
Socarnes vahlii (Kroyer, 1838)	X			8–300/P,D	Barnard (1964b)
Oedicerotidae					
Aceroides callida Barnard, 1967	X			1095–1205/B	Barnard (1967)
Aceroides edax Barnard, 1967	X			1095–1205/B	Barnard (1967)
Bathymedon caino Barnard, 1967	X			1095–1205/B	Barnard (1967)
Bathymedon candidus Barnard, 1961	X			2000–2398/B	Barnard (1961, 1967)

(continued)

Table 5.3 (continued)

Family and species	WM	GC	SW	Depth (m)	Sources
Bathymedon covilhani Barnard, 1961	X			549–1720/B	Barnard (1967)
Bathymedon flebilis Barnard, 1967	X			1748–2398/B	Barnard (1967)
Bathymedon nepos Barnard, 1967	X			1748–2398/B	Barnard (1967)
Monoculodes diversisexus Barnard, 1967	X			1720–1748/B	Barnard (1967)
Monoculodes latissimanus Stephensen, 1931	X			344–2398/B	Barnard (1967) and Blake et al. (1997)
Monoculodes necopinus Barnard, 1967	X			1720–1748/B	Barnard (1967)
Monoculodes recandesco Barnard, 1967	X			2398–2475/B	Barnard (1967)
Monoculodes sudor Barnard, 1967	X			791–842/B	Barnard (1967)
Oediceroides morosa (Barnard, 1966)	X			1095–1205/B	Barnard (1966, 1967)
Oediceroides trepadora (Barnard, 1961)			X	875/B	Barnard (1961)
Pakinidae					
Prachynella lodo Barnard, 1964	X			791–842/B	Barnard (1963, 1967)
Pardaliscidae					
Caleidoscopsis copal (Barnard, 1967)	X			1475–2398/B	Barnard (1967)
Caleidoscopsis tikal (Barnard, 1967)	X			1720–1748/B	Barnard (1967)
Halicoides synopiae Barnard, 1962	X			52–1748/B	Barnard (1963, 1967) and Blake et al. (1997)
Nicippe tumida Bruzelius, 1859	X			34–1367/B	Barnard (1964a)
Tosilus arroyo Barnard, 1966	X			976–1095/B	Barnard (1967)
Phoxocephalidae					
Cephalophoxoides homilis (Barnard, 1960)	X			62–2059/B	Blake et al. (1997)
Cephalophoxoides kergueleni (Stebbing, 1888)	X			200–2300/B	Blake et al. (1997)
Eyakia calcarata (Gurjanova, 1938)	X		X	18–695/B	Barnard (1960) and Barnard and Barnard (1981)
Foxiphalus cognatus (Barnard, 1960)		X		1–324/B	Blake et al. (1997)
Harpiniopsis emeryi Barnard, 1960	X			344–2800/B	Barnard (1967) and Blake et al. (1997)
Harpiniopsis epistomata Barnard, 1960	X			344–2800/B	Barnard (1967)

(continued)

Table 5.3 (continued)

Family and species	WM	GC	SW	Depth (m)	Sources
Harpiniopsis fulgens Barnard, 1960	X			128–2667/B	Barnard (1967) and Blake et al. (1997)
Harpiniopsis naiadis Barnard, 1960	X			338–2800/B	Barnard (1967) and Blake et al. (1997)
Harpiniopsis petulans Barnard, 1966	X			1265–1720/B	Barnard (1967)
Harpiniopsis profundis Barnard, 1960	X			385–2398/B	Barnard (1967)
Heterophoxus oculatus (Holmes, 1908)	X	X		2–1941/B	Barnard (1960) and Brusca and Hendrickx (2005)
Hopiphoxus simillimus (Barnard, 1967)	X			2667–2706/B	Barnard (1967)
Leptophoxus falcatus (Sars, 1882)	X			248–1120/B	Barnard (1967)
Metaphoxus frequens Barnard, 1960	X	X		41–458/B	Barnard (1960) and Brusca and Hendrickx (2005)
Paraphoxus oculatus (Sars, 1879)			X	27–2800/B	Barnard (1960)
Phoxocephalus kergueleni Stebbing, 1888	X			220–2398/B	Barnard (1967)
Pseudharpinia excavata (Chevreux, 1887)	X			425–5110/B	Barnard (1967)
Rhepoxynius bicuspidatus (Barnard, 1960)	X			11–475/B	Barnard (1963, 1970) and Blake et al. (1997)
Pleustidae					
Mesopleustes abyssorum (Stebbing, 1888)	X			694–3479/B	Barnard (1967)
Stegocephalidae					
Austrocephaloides camoti (Barnard, 1967)	X			791–842/B	Barnard (1967)
Parandaniexis mirabilis Schellenberg, 1929	X			3479–3515/B	Barnard (1967)
Pseudo viscaina (Barnard, 1967)	X			791–842/B	Barnard (1967)
Stegocephalus pajarella (Barnard, 1967)	X			1720/B	Barnard (1967)
Stegocephalus viscaina (Barnard, 1967)	X			791/B	Barnard (1967)
Stenothoidae					
Metopa samsiluna Barnard, 1966	X			1096–1620/B	Barnard (1967)
Metopella aporpis Barnard, 1962	X			24–591/B	Barnard (1963) and Blake et al. (1997)
Synopiidae					
Austrosyrrhoe rinconis Barnard, 1967	X			1095–1205/B	Barnard (1967, 1972)

(continued)

Table 5.3 (continued)

Family and species	WM	GC	SW	Depth (m)	Sources
Bruzelia ascua Barnard, 1966	X			1687–1720/B	Barnard (1967)
Bruzelia inlex Barnard, 1967	X			1720–2398/B	Barnard (1967, 1972)
Ileraustroe ilergetes (Barnard, 1967)	X			1720–1748/B	Barnard (1967)
Priscosyrrhoe priscis (Barnard, 1972)	X			791–1720/B	Barnard (1967, 1972)
Pseudotiron coas Barnard, 1967	X			2667–2706/B	Barnard (1967, 1972)
Pseudotiron longicaudatus Pirlot, 1934	X			1720–1748/B	Barnard (1972)
Pseudotiron pervicax Barnard, 1967	X			1095–1205/B	Barnard (1967, 1972)
Synopia angustifrons Dana, 1853			X	1095–1720/B	Barnard (1972) and Barnard and Karaman (1991)
Synopia scheeleana Bovallius, 1886	X		X	1095–1720/B	Barnard (1972) and Barnard and Karaman (1991)
Syrrhoites cohasseta Barnard, 1967	X			1095–1720/B	Barnard (1972)
Syrrhoites dulcis Barnard, 1967	X			1095–1205/B	Barnard (1967)
Syrrhoites redox Barnard, 1967	X			1720–1748/B	Barnard (1967, 1972)
Syrrhoites silex Barnard, 1967	X			1095–1205/B	Barnard (1967, 1972)
Syrrhoites terceris Barnard, 1964			X	1609–1746/B	Barnard (1972) and Barnard and Karaman (1991)
Syrrhoites trux Barnard, 1967	X			842–1205/B	Barnard (1967, 1972)
Trischizostomatidae					
Trischizostoma unam Winfield, Hendrickx and Ortiz, 2016	X	X	X	1392–1420/SC	Winfield et al. (2016)
Tryphosidae					
Cedrosella fomes (Barnard, 1967)	X			3705–3745/B	Barnard (1967)
Cedrosella perspinis (Barnard, 1967)	X			1720–1774/B	Barnard (1967)
Hippomedon denticulatus (Bate, 1857)	X			924 B–C	Barnard (1964a)
Hippomedon propinquus Sars, 1895	X	X		5–1095/B,C	Barnard (1963, 1970) and Hurley (1963)
Orchomenella tabasco (Barnard, 1967)	X			1720–1748/B	Barnard (1967)

(continued)

Table 5.3 (continued)

Family and species	WM	GC	SW	Depth (m)	Sources
Pseudonesimus abyssi Chevreux, 1926	X			2667–2706/B	Barnard (1967)
Pseudonesimus robustus (Barnard, 1967)	X			1720–1748/B	Barnard (1967)
Pseudonesimus tacita (Barnard, 1961)	X			791–842/B	Barnard (1967)
Thrombasia grabensis (Barnard, 1967)	X			1720–1748/B	Barnard (1967)
Tryphosella metacaecula (Barnard, 1967)	X			791–842/B	Barnard (1967)
Uristidae					
Stephonyx californiensis Winfield, Hendrickx, and Ortiz, 2017		X	X	1150/SC	Winfield et al. (2017)
Stephonyx laqueus (Barnard, 1967)	X		X	330–1850/SC	Barnard (1967)
Urothoidae					
Urothoides inops Barnard, 1967	X			2667–2706/B	Barnard (1967)
Vemanidae					
Vemana lemuresa Barnard, 1967	X			3705–3745/B	Barnard (1967)
Aoridae					
Bemlos edentulus (Barnard, 1967)	X			791–842/B	Barnard (1967)
Ischyroceridae					
Bonnierella linearis Barnard, 1964	X		X	6324/B	Barnard (1964c) and Barnard and Karaman (1991)
Bonnierella californica Barnard, 1966	X		X	1292/B	Barnard (1967) and Barnard and Karaman (1991)
Bonnierella palenquia Barnard, 1967	X			1095–1205/B	Barnard (1967) and Barnard and Karaman (1991)
Photidae					
Gammaropsis grasslei Soto and Corona, 2007		X		2000–2100/C	Soto and Corona (2007)
Photis brevipes Shoemaker, 1942	X			1–492/B	Shoemaker (1942), Barnard (1963, 1969), and Blake et al. (1997)
Photis malinalco Barnard, 1967	X			2398–2475/B	Barnard (1967) and Blake et al. (1997)
Photis viuda Barnard, 1962	X			37–400/B	Barnard (1964a)

General habitat indicated after depth range

B benthic, *C* commensal, *D* demersal, *P* pelagic, *SC* scavenger, *WP* Western Pacific, *GC* Gulf of California, *ETP* eastern tropical Pacific found exclusively in that depth range. The highest number of continental slope species occur in the middle slope (76), of which 34 (45%) are exclusive of that depth range. In the lower continental slope, 33 species have been recorded, 10 exclusively so (30%). All in all, only 15 species feature a wide distribution range and inhabit throughout the continental slope. Some other species exhibit a relatively wide distribution range, occurring in two of the three continental slope zones: 16 species occur in both the upper and middle continental slopes, and 6 species occur in the middle and lower continental slopes (Table 5.1). Seven species are found at abyssal depths of which four (*Amathillopsis pacifica margo*, *Parandaniexis mirabilis*, *Cedrosella fomes*, and *Vemana lemuresa*) are confined to depths >3000 m. Three species, *Ampelisca eoa*, *Pseudharpinia excavata*, and *Mesopleustes abyssorum*, inhabit both bathyal and abyssal zones, and *Bonnierella linearis* is the only species reported exclusively in the hadal zone, although the genus also occur in the middle continental slope (Table 5.1).

5.4 Discussion

Thurston (2000) characterized 19 families as common (>200 m depth) in the deep-sea worldwide; of these, 15 were documented in the present study in the bathyal zone: Amathillopsidae, Ampeliscidae, Aristiidae, Epimeriidae, Eusiridae, Lepechinellidae, Leucothoidae, Oedicerotidae, Pardaliscidae, Phoxocephalidae, Stegocephalidae, Stenothoidae, Synopiidae, Tryphosidae, and Uristidae (Table 5.2). Three families, Cyphocarididae, Vemanidae, and Ischyroceridae, have been recognized as occurring mostly in very deep water (>3000 m depth) according to Horton (2004), Horton and Thurston (2009, 2011), and Horton et al. (2013). In western Mexico, however, Cyphocarididae and Ischyroceridae have been reported for continental slope depths (Table 5.1).

Bousfield (2000, 2001) proposed that speciose families groups containing "primitive" swimmers, clingers, and scavengers forms which typically mate in the water column are dominant among the North American marine amphipod, while "advantage" forms mating in benthic substratum (burrowers, nestlers, tube-dwellers) are less frequent. In the present study, primitive forms (i.e., Phoxocephalidae, Synopiidae, Oedicerotidae, Tryphosidae, and Ampeliscidae) were the most diversified amphipod families in terms of numbers of species and genera in western Mexico. Benthic forms, i.e., Amathillopsidae, Vemanidae, and Ischyroceridae, dominated in both the abyssal and the hadal zones (Table 5.1).

Number of species and density pattern of benthic amphipods decrease with the increase in depth and distance from the shoreline (Bellan-Santini 1990; Thurston 2000). In the middle continental slope, however, an increase in species richness has been documented (Bachelet et al. 2003; Cartes et al. 2003) and is linked to the increase of organic matter exported from the euphotic zone and the continental shelf (Sorbe 1999). In this study, the Amphilochidea and Senticaudata amphipod fauna was richer in the upper (61 species) and middle (76) continental slope, with a sharp decrease of richness in the lower continental slope (33) (Table 5.3). Abyssal and hadal depths have not been adequately explored in the region, and the very low number of species recorded there (seven and one, respectively) is certainly a reflection of this lack of sampling operations.

Several dominant families and genera considered as primitive groups among amphipods (i.e., plesiomorphic calceoli, strongly developed rostrum, dactylus-shaped pereopods, loss or reduction of antennal 2 flagellum, setae on pereopods, extended coxae) have experienced a successful radiation in shallow waters in the high latitudes in the NE Pacific Ocean, colonizing bathyal and abyssal depths, previously to continental drift (Bousfield 1982; Barnard and Karaman 1991; Thurston 2000). Subsequently, and based on vicariance events (Barnard and Drummond 1978; Barnard and Karaman 1991; Jarrett and Bousfield 1994a, b), dispersion and speciation towards the rest of the eastern Pacific and the western Atlantic took place. The endemism, including the tendency to lose eyes and morphological variations in several deep-sea species of the genera *Byblis*, *Bathymedon*, *Monoculodes*, *Syrrhoites*, *Synopia*, *Pseudharpinia*, *Harpiniopsis*, and *Hippomedon*, has been documented and constitutes some evidence of colonization in different ocean basins and reinforces the hypothesis that Pacific amphipods fauna is dominated by primitive forms (Bousfield 2000).

The difference in the species richness of deep-sea Amphilochidea and Senticaudata amphipods observed among the three zones can certainly be attributed mainly to the strong variation of the sampling intensity reported by authors during the last 50 years (mostly in contributions by J.L. Barnard, E.L. Bousfield, and C.R. Shoemaker). The amphipods that inhabit the deep sea of the Gulf of California and SW Mexico remain essentially unknown due to a lack of intensive survey of the continental slope and the abyssal plain in these two areas. Without any doubts, further sampling in these areas using proper gear will considerably increase our knowledge on the amphipod biodiversity in western Mexico.

References

Bachelet G, Dauvin JC, Sorbe JC (2003) An updated checklist of marine and brackish water Amphipoda (Crustacea: Peracarida) of the southern Bay of Biscay. Cah Biol Mar 44:121–151

Barnard JL (1954) Amphipoda of the family Ampeliscidae collected in the Eastern Pacific Ocean by the Velero III and Velero IV. Allan Hancock Pacific Exp 18:1–137

Barnard JL (1960) The amphipod family Phoxocephalidae in the Eastern Pacific Ocean, with analyses of other species and notes for a revision of the family. Allan Hancock Pacific Exp 18(3):170–375

Barnard JL (1961) Gammaridean Amphipoda from depths of 400 to 6000 meters. Galathea Rep 5:23–128

Barnard JL (1963) Los anfípodos bentónicos marinos de la costa occidental de Baja California. Rev Soc Mex Hist Nat 24:205–273

Barnard JL (1964a) Los anfípodos bentonicos marinos de la costa occidental de Baja California. Rev Soc Mex Hist Nat 24:205–273

Barnard JL (1964b) Marine Amphipoda of san Quintín Bay, Baja California. Pacific Naturalist 4:55–139

Barnard JL (1964c) Deep-Sea Amphipoda (Crustacea) collected by the R/V Vema in the eastern Pacific and the Caribbean and Mediterranean seas. Bull Am Mus Nat Hist 127:3–46

Barnard JL (1966) Submarine canyons of Southern California part 5, systematics: Amphipoda. Allan Hancock Pacific Exp 27(5):1–166

Barnard JL (1967) Bathyal and abyssal gammaridean Amphipoda of Cedros trench, Baja California. US Nat Mus Bull 260:1–205

Barnard JL (1969) A biological survey of Bahía de Los Angeles Gulf of California, México. IV. Benthic Amphipoda (Crustacea). Trans San Diego Soc Nat Hist 15(13):175–228

Barnard JL (1970) Benthic ecology of Bahia de San Quintin, Baja California. Smithson Contrib Zool 44:1–60

Barnard JL (1972) A review of the family Synopiidae (= Tironidae), mainly distributed in the deep sea (Crustacea: Amphipoda). Smithson Contrib Zool 124:1–94

Barnard JL (1973) Deep-sea Amphipoda of the genus *Lepechinella* (Crustacea). Smithson Contrib Zool 133:1–31

Barnard JL (1979) Littoral Gammaridean Amphipoda from the Gulf of California and the Galapagos Islands. Smithson Contrib Zool 271:1–149

Barnard JL, Barnard CM (1981) The amphipod genera *Eobrolgus* and *Eyakia* (Crustacea: Phoxocephalidae) in the Pacific Ocean. Proc Biol Soc Wash 94(1):295–313

Barnard JL, Drummond MM (1978) Gammaridean Amphipoda of Australia, part III: the Phoxocephalidae. Smithson Contrib Zool 245:1–551

Barnard JL, Ingram C (1990) Lysianassoid Amphipoda (Crustacea) from Deep-Sea thermal vents. Smithson Contrib Zool 499:1–80

Barnard JL, Karaman G (1991) The families and genera of marine gammaridean Amphipoda (except marine gammaroidea). Rec Austral Mus, suppl 13:1–866

Bellan-Santini D (1990) Mediterranean deep-sea Amphipoda: composition, structure and affinities of the fauna. Prog Oceanogr 24:275–387

Bellan-Santini D (1999) Ordre des Amphipodes (Amphipoda Latreille, 1816). In: Grassé P-P (ed) Traité de Zoologie. Anatomie, Systématique, Biologie. Tome VII. Fascicule, IIIA. Crustacés Péracarides. Masson, Paris/Milan/Barcelone, pp 95–168

Blake JA, Watling L, Valentich Scott P (1997) Taxonomic atlas of the benthic Fauna of the Santa Maria Basin and the western Santa Barbara Channel. Santa Barbara Mus Nat Hist 12(3):1–251

Bousfield EL (1982) Amphipoda. Gammaridea. In: Parker SP (ed) Synopsis and classification of living organisms, vol 2. McGraw-Hill, New York, pp 254–285

Bousfield EL (2000) Biogeographical analysis of gammaridean amphipod faunas based on their phyletic classification. Polskie Arch Hydrobiol 47(3–4):335–351

Bousfield EL (2001) An updated commentary on phyletic classification of the amphipod Crustacea and its application to the North American fauna. Amphipacifica 3(1):49–119

Brusca RC, Hendrickx ME (2005) Crustacea 4: Peracarida: Lophogastrida, Mysida, Amphipoda, Tanaidacea and Cumacea. pp 139–154 in Listado y Distribución de la macrofauna del golfo de California, México. Parte I. Invertebrados. A distributional checklist of the macrofauna of the Gulf of California, Mexico. I. Invertebrates

Cartes JE, Elizalde M, Sorbe JC (2001) Contrasting life-histories, secondary production, and trophic structure of Peracarid assemblages of the bathyal suprabenthos from the Bay of Biscay (NE Atlantic) and the Catalan Sea (NW Mediterranean). Deep-Sea Res I 48:2209–2232

Cartes JE, Jaume D, Madurell T (2003) Local changes in the composition and community structure of suprabenthic peracarid crustaceans on the bathyal Mediterranean: influence of environmental factors. Mar Biol 143:745–758. https://doi.org/10.1007/s00227-003-1090-z

García Madrigal MS (2007) Annotated checklist of the amphipods (Peracarida: Amphipoda) from the tropical eastern Pacific. Contrib Estudio Crustáceos Pacífico Este 4(2):63–195

Horton T (2004) Revision of the amphipod genus *Valettiopsis* Holmes, 1908 (Crustacea: Lysianassoidea), with the addition of three new species. J Nat History 38:1735–1755

Horton T, Thurston M (2009) *Hirondellea sindhusagar* (Crustacea: Amphipoda: Lysianassoidea), a new deep-water scavenger species from the Indian Ocean, with a key to the genus. Zootaxa 2096:433–441

Horton T, Thurston M (2011) *Centromedon zoe* (Crustacea: Amphipoda: Lysianassoidea: Uristidae), a new deep-water scavenger from the North Atlantic, with a key to the genus *Centromedon*. Zootaxa 2869:54–62

Horton T, Thurston MH, Duffy G (2013) Community composition of scavenging amphipods at bathyal depths on the Mid-Atlantic Ridge. Deep Sea Res II 98:352–359. https://doi.org/10.1016/j.dsr2.2013.01.032

Hurley DE (1963 Amphipoda of the family Lysianassidae from the West Coast of North and Central America. Allan Hancock Found Publ, Occas Paper 25, pp 1–160

Jarrett NE, Bousfield EL (1994a) The amphipod superfamily Phoxocephaloidea on the Pacific Coast of North America. Family Phoxocephalidae. Part 1. Metharpiniinae, new subfamily. Amphipacifica 1(1):58–140

Jarrett NE, Bousfield EL (1994b) The amphipod superfamily Phoxocephaloidea on the Pacific Coast of North America. Family Phoxocephalidae. Part II. Subfamilies Pontharpiniinae, Parharpiniinae, Brolginae, Phoxocephalinae, and Haspiniinae. Systematics and distributional ecology. Amphipacifica 1(2):71–149

Lowry JK, Myers AA (2017) A phylogeny and classification of the Amphipoda. (Crustacea: Peracarida). Zootaxa 4265:1):1–1)89

Mees J, Jones M (1997) The Hyperbenthos. Oceanogr Mar Biol: An Ann Rev 35:221–255

Schellenberg A (1929) Die abyssale und pelagische Gammariden. Reports on the Scientific Results of the Expedition to the Tropical Eastern Pacific "Albatross"...1904...1905.... Bull Mus Compar Zool 69:191–201

Shoemaker CR (1942) Amphipod crustaceans collected on the Presidential Cruise of 1938. Smithson Misc Collect 101:11):1–11)52

Sorbe JC (1999) Deep-Sea macrofaunal assemblages within the Benthic Boundary Layer of the Cap-Ferret Canyon (Bay of Biscay, NE Atlantic). Deep-Sea Res II 46:2309–2329

Soto AL, Corona A (2007) *Gammaropsis* (*Podoceropsis*) *grasslei* (Amphipoda: Photidae) a new species of commensal amphipod of the deep-water lithodid *Neolithodes diomedeae* from the Guaymas Basin, Gulf of California. Zootaxa 1406:33–39

Thurston M (1990) Abyssal necrophagous amphipods (Crustacea: Amphipoda) in the northeast and tropical Atlantic Ocean. Prog Oceanogr 24:257–274

Thurston M (2000) Benthic Gammaridea (Crustacea: Amphipoda) in the deep sea. Paper presented at 10th International Colloquium on Amphipoda, Alpha Taxonomy Workshop, Heraklion, Crete, 16–21 April 2000

Thurston M (2001) Benthic Gammaridea (Crustacea Amphipoda) in the deep sea. Polskie Arch Hydrobiol 7(3–4):353–377

Thomas JD (1992) Identification manual for marine Amphipoda (Gammaridea): I common coral reef and rocks, bottom amphipods of South Florida. Final report DEP contract number SP290. Smithsonian Institution, Washington, DC

Winfield I, Ortiz M, Hendrickx ME (2013) A new deep-water species of *Epimeria* (Amphipoda: Gammaridea: Epimeriidae) from the continental slope of western Mexico. J Mar Biol Assoc UK 93:991–997

Winfield I, Hendrickx ME, Ortiz M (2016) A new deep-water species of *Trischizostoma* (Crustacea: Amphipoda: Gammaridea: Trischizostomatidae) from western Mexico. J Mar Biol Assoc UK 97(1):141–149

Winfield I, Hendrickx ME, Ortiz M (2017) *Stephonyx californiensis* sp. nov. (Amphipoda: Lysianassoidea: Uristidae), a new bathyal scavenger species from the Central Gulf of California, Mexico, and comments on the bathymetric and geographic distribution of the *Stephonyx* species group. J Nat Hist 51(47–48):2793–2807. https://doi.org/10.1080/00222933.2017.1384076

Chapter 6
Isopoda Epicaridea from Deep Water Around North and Central America

J. C. Markham

Abstract A total of 39 species of epicarid isopods, four of these unidentified, are recorded from deep water around North and Central America. Most species (27) belong to the Bopyridae with 17 genera, including five species of *Munidion* and at least four species of *Pseudione*. Other families are the Colypuridae (1 sp.), the Cryptoniscidae (1 sp.), the Dajidae (9 spp.), and the Entophilidae (1 sp.). Of the 35 identified species, about half (17) were described in 1930 or earlier, and the rest after 1972.

Keywords Deep water · Epicarids · West Atlantic · East Pacific · Diversity Distribution

6.1 Introduction

Although there is uncertainty about the proper status of the Epicaridea (see Brandt and Poore 2003; Boyko et al. 2013), it is here handled as an infraorder of the suborder Cymothoida of the order Isopoda, in accordance with the presentation of the online catalog WoRMS. All epicarideans are parasites of other crustaceans, their larvae usually infesting copepods and adults found on cirripedes and malacostracans. By far the most specious and widespread family is the Bopyridae, with just over 600 currently described species worldwide out of a total of about 700 epicaridean species. All bopyrids are ectoparasites, of calanoid copepods as larvae and of decapods as adults. Species in other families parasitize a great diversity of crustaceans, most externally, but several also internally. With very few exceptions, epicarideans are marine, and they are rarely found deeper than 200 m in any part of the oceans. An exception is the dajid genus *Holophryxus*, all of species of which, including two covered herein, are known only from depths greater than 200 m. For reasons not determined, the number of species of bopyrids in the eastern Pacific from Alaska

J. C. Markham (✉)
Arch Cape Marine Laboratory, Arch Cape, OR, USA
e-mail: jmarkham@seasurf.net

M. E. Hendrickx (ed.), *Deep-Sea Pycnogonids and Crustaceans of the Americas*, https://doi.org/10.1007/978-3-030-58410-8_6

to Chile is far smaller than the number of potential host species would suggest (see Markham 1992), with whole subfamilies rare or absent. Nonetheless, two monotypic subfamilies, Phyllodurinae in shallow water (Markham 1977b) and Bathygyginae in deep water (Markham 2016), are known primarily from the eastern Pacific.

In this tabulation of epicaridean species from depths of 200 m or more in the western north Atlantic (including Bermuda and the Gulf of Mexico) and the eastern north Pacific (including Hawaii), the total number of identified species is 36; the totals by family are Bopyridae 24, Colypuridae 1, Cryptoniscidae 1, Dajidae 9, and Entophilidae 1. Half of all species considered, 18, are in the bopyrid subfamily Pseudioninae, whose hosts are primarily anomurans, in this case mostly munidids, but also some carideans found only in deeper waters.

Many of the species listed are known from only their type-specimens and type-localities. This is the case with nine bopyrid species and several of the species described by Harriet Richardson in various cited references. Furthermore, the descriptions H. Richardson presented are far less complete than desired, though all of her type-specimens evidently are still extant in the collections of the Smithsonian Institution, so redescriptions would be possible. For example, Butler (1964) redescribed the types of *Holophryxus alaskensis* Richardson, 1905, in fine detail, identified its host for the first time, and entered two other species into its synonymy. For species described in H. Richardson's papers, I have converted the recorded depths from fathoms to meters and, as necessary, supplemented collection data from other sources, such as station-logs of the cruises and ships cited in the descriptions, which are accessible online.

6.2 Material and Methods

Names of the epicaridean species have been updated as needed, with all published synonyms indicated. Names of the host species have also been updated as needed. The authority consulted for the current validity of names of both the epicarideans and their hosts is the online catalog World Register of Marine Species, WoRMS (WoRMS Editorial Board 2019). Genera and species are ordered alphabetically within their respective higher taxa, while families and subfamilies are presented in order from those considered to be closest to the ancestral forms to those considered more derived.

6.3 Results

6.3.1 Systematic Section

Infraorder Epicaridea Latreille, 1825
Superfamily Bopyroidea Rafinesque, 1815

Family Bopyridae Rafinesque, 1815
Subfamily Pseudioninae Codreanu, 1967

1. *Anuropodione carolinensis* Markham, 1974

Type-locality. Off coast of North Carolina, USA, West Atlantic, 35°05′N, 75°11.5′W, 220–260 m. Host: *Munida iris* A. Milne-Edwards, 1880. References: Williams and Brown (1972), Markham (1974a). **Other localities**. Norfolk Canyon, off coast of North Carolina, USA, West Atlantic, 36°56′ to 37°09′N, 75°09′ to 74°33′W and 36°36′ to 37°09′N, 75°06′ to 74°33′W, 83–662 m. Host: *Munida iris*. Reference: Wenner and Windsor (1979). **Range outside of region**. Potiguar Basin, Brazil. Host: *Munida iris*. Reference: Alves-Júnior et al. (2019).

2. *Aporobopyrina anomala* Markham, 1974

Type-locality. Off coast of Colombia, Caribbean, 12°40′N, 72°00′W, 620- *M. erinaceus* 660 m. Host: *Munida valida* Smith, 1883. Reference: Markham (1974a). **Other localities**. Northwest Gulf of Mexico, 27°38′N, 95°22′N, 490–640 m, and 27°35′N, 95°23′W, 730 m. Host: *M. valida*. South of Marquesas Keys, Florida, USA, Gulf of Mexico, 24°07′N, 82°00′W, 686–715 m. Host: *M. valida*. South of Dry Tortugas, Florida USA, Gulf of Mexico, 24°12′N, 82°50′W, 622 m. Host: *M. valida*. Reference: Markham (1974a). Northern Gulf of Mexico, 27°22′ to 29°17′N, 87°47′ to 91°16′W, 185–770 m. Host: *M. valida*. Reference: Adkison and Collard (1990). **Range outside of region**. None known.

3. *Bathione magnafolia* Román-Contreras and Boyko, 2007

Type-locality. Southern Gulf of California, Mexico, 24°53′N, 108°59′W, 835–870 m. Host: *Munidopsis depressa* Faxon, 1893. Reference: Román-Contreras and Boyko (2007). **Range outside of region**. None known.

4. *Galathocrypta acaudata* Román-Contreras and Soto, 2002

Type-locality. Southwest Gulf of Mexico, 18°54′N, 94°18′W, 502 m. Host: *Munidopsis erinaceus* (A. Milne-Edwards, 1880). Reference: Román-Contreras and Soto (2002). **Possible other records**. [As "*Pseudione* sp."] Southern Caribbean. Host: *M. erinaceus*. Reference: Mayo (1974). [As "Bopyrid"]. Southern Gulf of Mexico. Host: *M. erinaceus*. Reference: Vázquez-Bader and Gracia (2016). **Range outside of region**. None known.

5. *Goleathopseudione bilobata* Román-Contreras, 2008

Type-locality. Off coast of California, USA, Pacific, 34°45′N, 123°04′W, 4100 m. Host: *Munidopsis beringanus* Benedict, 1902. Reference: Román-Contreras (2008). **Range outside of region**. None known.

6. *Munidion cubense* Bourdon, 1972

Type-locality. Bahía de Cochinos, Cuba, 22°13′N, 81°10′W, 230–390 m. Host: *Munida stimpsoni* A. Milne-Edwards, 1880. Reference: Bourdon (1972). **Other locality**. Near Península de Paraguaná, Venezuela, Caribbean, 11°40′N, 69°22′W,

134–161 m. Host: *Munida flinti* Benedict, 1902. Reference: Markham (1975). **Range outside of region**. None known. [Described as *Munidion parvum cubensis* Bourdon, 1972].

7. *Munidion irritans* Boone, 1927

Type-locality. North of Glover Reef, Belize, Caribbean, 16°N, 88°W, 670 m. Host: *Munida irrasa* A. Milne Edwards, 1880. Reference: Boone (1927). **Other localities**. Along southeast Florida, USA, Atlantic, 24°12′ to 24°48′N, 80°26′ to 80°46′W, 90–620 m. Host: *M. irrasa*. Reference: Markham (1975). Hudson Canyon to Norfolk Canyon, off Virginia, USA, Atlantic, 37–40°N, 72–74°W, >200 m. Host: *Munida iris*. Reference: Bursey (1978). **Range outside of region**. None known.

8. *Munidion longipedis* Markham, 1975

Type-locality. East coast of Florida, Atlantic, 25°45′ to 29°59′N, 79°49′to 80°11′W, 70–330 m. Host: *Agononida longipes* (A. Milne-Edwards, 1880). Reference: Markham (1975). **Other localities**. Isla de Providencia to Dry Tortugas, Florida USA, Southeast Gulf of Mexico, 13°37′N, 81°51′W to 24°18′N, 82°54′W, 330–460 m. Host: *M. longipes*. Reference: Markham (1975). Norfolk Canyon, off Virginia, USA, Atlantic, 37°N, 75°W, ca 200 m. Host: *M. longipes*. Reference: Wenner and Windsor (1979). **Range outside of region**. None known.

9. *Munidion parvum* Richardson, 1904

Type-locality. Straits of Juan de Fuca, Washington, USA, Pacific, 48°20′N, 123°30′W, 278 m. Host: *Munida quadrispina* Benedict, 1902. References: Richardson (1904), Markham (1975). **Possible other record**. [As "*Munidon* (sic) *parvum/Pseudione galacanthae*"]. Sechelt Peninsula, British Columbia, Canada, Pacific, 50°N, 124°W, depth unspecified. Host: unidentified munidid. Reference: Lamb and Hanby (2005). **Range outside of region**. None known.

10. *Munidion princeps* Hansen, 1897

Type-locality. Near Malpelo Island, Pacific. 03°58′N, 81°36′W, 205 m. Host: *Munida refulgens* Faxon, 1893. References: Hansen (1897), Markham (1975). **Range outside of region**. None known.

11. *Paragigantione americana* (Markham, 1974)

Type-locality. Off coast of Guyana, Atlantic, 08°40′N, 57°38′W, 1220–1440 m. Host: *Munida microphthalma* A. Milne-Edwards, 1880. Reference: Markham (1974a). **Other locality**. Norfolk Canyon, off Virginia, USA, Atlantic, 36°56′ to 37°09′N, 75°06′ to 74°33′W, 1408–1698 m. Host: *M. microphthalma*. Reference: Wenner and Windsor (1979). **Range outside of region**. None known. [Described as *Bonnieria americana* Markham, 1974].

12. *Parapleurocryptella elasmonoti* Bourdon, 1972

Type-locality. Martinique, Caribbean, 14°29′N, 61°06′W, 349 m. Host: *Munidopsis squamosus* (A. Milne-Edwards, 1880). Reference: Bourdon (1972). **Range outside of region**. None known.

13. *Pleurocryptella fimbriata* Markham, 1974

Type-locality. Southwest of Jamaica, Caribbean, 17°27′N, 78°10′W, 605–665 m. Host: *Munida constricta* A. Milne Edwards, 1880. Reference: Markham (1974a). **Other localities**. South of Jamaica, Caribbean, 17°16′N, 77°48′W, 720 m; Bahía de Cochinos, Cuba, Caribbean, 22°07′N, 81°08′W, 400–500 m. Host: *M. constricta*. Reference: Markham (1974a). **Range outside of region**. None known.

14. *Pleurocryptella wolffi* Bourdon, 1972

Type locality. Gulf of Panamá, Pacific, 05°44′N, 79°20′W, 2950–3190 m. Host: *Munidopsis antonii* (Filhol, 1884). Reference: Bourdon (1972). **Range outside of region**. None known.

15. *Pontobopyrus abyssorum* Markham, 1979

Type-locality. Northwestern Atlantic, 38°04′ to 38°09′N, 70°26′ to 70°22′W, 3850 m. Host: *Parapontophilus abyssi* (Smith, 1884). Reference: Markham (1979). **Range outside of region**. None known.

16. *Pseudione ampla* Markham, 1988

Type-locality. Off Florida, USA, Gulf of Mexico, 24°33′N, 83°22′W, 366 m. Host: *Heterocarpus ensifer* A. Milne Edwards, 1881. Reference: Markham (1988). **Range outside of region**. None known.

17. *Pseudione confusa maxillipedis* Bourdon, 1972

Type-locality. Bahía de Cochinos, Cuba, Caribbean, 22°13′N, 81°10′W, 229–393 m. Host: *Munida stimpsoni* A. Milne Edwards, 1880. Reference: Bourdon (1972). **Other locality**. Punta Alegre, Cuba, Caribbean, 22°50′N, 78°55′N, 366 m. Host: *M. stimpsoni*. Reference: Bourdon (1972). St. Croix, US Virgin Islands, Caribbean, ca 17°44′N, 64°50′W, depth uncertain. Host: *Munida* sp. [Reported as *Pseudione minimocrenulata* Nierstrasz and Brender à Brandis (1931) pro parte, separated from that species by Kazmi and Boyko (2005)]. Reference: Nierstrasz and Brender à Brandis (1931). **Range outside of region**. None known.

18. *Pseudione galacanthae* Hansen, 1897

Type-locality. Off east side of Baja California, Mexico, Golfo de California, Mexico, 26°48′N, 110°45′W, 1571 m. Host: *Galacantha diomedeae* Faxon, 1893. Reference: Hansen (1897). **Other localities**. Nanaimo, British Columbia, Canada, Pacific, ca 49°10′N, 123°57′W, depth unrecorded, and San Juan Islands, Washington, USA, Pacific, ca 48°30′N, 123°W, depth unrecorded. Host: *Munida quadrispina* J. E. Benedict, 1902. Reference: Fraser (1932). [Misidentified as *Pseudione giardi* Calman, 1898]. Strait of Georgia, British Columbia, Canada, Pacific, ca 49°18′N, 123°48′W, 157–448 m. Host: *Munida* sp. Reference: Nierstrasz and Brender à Brandis (1931). **Range outside of region**. Off east coast of Patagonia, Argentina, South Atlantic. Host: *Munida rugosa* (Fabricius, 1793). Reference: Richardson (1904). Falkland Islands, South Atlantic. Host: *Munida gregaria* (Fabricius, 1793). Reference: Rayner (1935). Golfo de Penas and Magallanes, Chile, South Pacific.

Host: *Munida gregaria*. Reference: Stuardo et al. (1986). Off Arica, Chile, South Pacific. Host: *Galacantha diomedeae*. Reference: Pardo et al. (2009). Golfo de San Jorge, Argentina, South Atlantic. Host: *Munida gregaria*. Reference: Varisco and Vinuesa (2011).

19. *Pseudione* sp. or spp., western Atlantic

Known localities. South of Jamaica, 17°21′N, 77°35′W, 805–1089 m. Host: *Munidopsis abbreviata* (A. Milne Edwards, 1880). Reference: Mayo (1974). Off Guadeloupe, Caribbean, 16°06′N, 61°22–24′W, 531–733 m. Host: *M. alaminos* Pequegnat and Pequegnat, 1970. Reference: Mayo (1974). Florida Straits, Atlantic, 24°28–51′N, 79°52′–80°16′W, 802–805 m. Host: *M. longimanus* (A. Milne Edwards, 1880). Reference: Mayo (1974). Off Campeche, Campeche, Mexico, Gulf of Mexico, ca 20°N, 71°W, depth unspecified. Host: *M. nitida* (A. Milne Edwards, 1880). Reference: Salazar-Vallejo and Leija-Tristán (1989). Off Caribbean coast of Colombia, and off Guadeloupe, Caribbean, 11°30–34′N, 73°23–26′W, 507–531 m. Host: *M. riveroi* Chace, 1939. Reference: Mayo (1974), Florida Straits, Gulf of Mexico, 24°24′N, 80°52′W, 512 m. Host: *M. robusta* (A. Milne Edwards, 1880). Reference: Mayo (1974). Straits of Florida, 25°17–36′N, 79°40–45′W, 570–824 m; and north coast of Cuba, Gulf of Mexico 23°24′N, 80°11′W, 677–1107 m; and south of Jamaica, Caribbean. Host: *M. serricornis* (Lovén, 1852). Reference: Mayo (1974).

20. Unidentified Pseudioninae, western Atlantic.

Known localities. Norfolk Canyon, off Virginia, USA, Atlantic, 36°56′ to 37°09′N, 75°06′ to 74°33′W, 1876–2427 m. Host: *Galacantha rostrata* A. Milne-Edwards, 1880. Reference: Wenner and Windsor (1979). Venezuela Basin, Caribbean, 13°34′N, 64°47′W, 3517–3549 m. Hosts: *Munidopsis crassa* Smith, 1885, and *M. geyeri* Pequegnat and Pequegnat (1970). Reference: Gore (1983). Off Cuba, ca 25°N, 80°W, ca 800 m. Host: *M. longimanus*. Reference: Chace (1942). Southern Gulf of Mexico. Hosts: *M. alaminos* and *M. robusta*. Reference: Vázquez-Bader and Gracia (2016).

21. Unidentified Pseudioninae, eastern Pacific

Known locality. Guaymas Basin, Mexico, 27°29′N, 111°24′W, depth uncertain. Host: *Galacantha diomedeae*. References: Luke (1977), Wicksten (2012).

Subfamily Argeiinae Markham, 1977

22. *Argeia atlantica* Markham, 1977

Type-locality. Tongue of the Ocean, Bahamas, Atlantic, 23°54′N, 77°12′W, 1298–1335 m. Host: *Metacrangon jacqueti* (A. Milne-Edwards, 1881). Reference: Markham (1977a). **Other localities**. Southeast of Newfoundland, Canada, Atlantic, 42°59′N, 51°15′W, 1100 m. Host: *M. jacqueti*. References: Sivertsen and Holthuis (1956); Markham (1977a). Middle Atlantic Bight, off Virginia, USA. Atlantic, ca 37°N, 75°W, 452–2679 m. Host: *M. jacqueti*. Reference: Wenner (1978). **Range**

outside of region. Coast of China, West Pacific. Host: *Crangon affinis* de Haan, 1849. Reference: An et al. (2008).

23. *Parargeia ornata* Hansen, 1897

Type-locality. Off Acapulco, Mexico, Pacific, 16°33′N, 99°52′W, 1460–1650 m. Host: *Metacrangon procax* (Faxon, 1893). Reference: Hansen (1897). **Range outside of region**. None known.

Subfamily Bathygyginae Markham, 2016

24. *Bathygyge grandis* Hansen, 1897

Type-locality. Off Acapulco, Mexico, Pacific, 16°33′N, 99°52′W, 1460–1650 m. Host: *Glyphocrangon spinulosa* Faxon, 1896. Reference: Hansen (1897). **Other localities**. Middle Atlantic Bight, off Virginia USA, Atlantic, ca 37°N, 75°W, 452–2679 m. Hosts: *Glyphocrangon sculpta* (Smith, 1882) and *G. longirostris* (Smith, 1882). Reference: Wenner (1978). Magellan Rise, Northeast Pacific, 07°05′N, 176°55′W, 3100 m. Host unspecified. Reference: Kaufmann et al. (1989). San Clemente Basin, off California, USA, Pacific, 32°28′N, 118°08′W, 1782 m. Host: *Glyphocrangon vicaria* Faxon, 1897. Reference: Markham (2016). **Range outside of region**. Off Cape Point, South Africa, South Atlantic Host: *Glyphocrangon sculpta* [possibly actually an undescribed species of *Bathygyge*]. Reference: Stebbing (1906). Off coast of China, western Pacific. Host: *Glyphocrangon* sp. [possibly actually an undescribed species of *Bathygyge*]. Reference: An et al. (2007).

Subfamily Athelginae Codreanu & Codreanu, 1956

25. *Anathelges hyphalus* (Markham, 1974)

Type-locality. Channel Islands, California, Pacific, 33°46′ to 33°18′N, 118°22′ to 118°33′W, 95319 m. Host: *Parapagurodes laurentae* McLaughlin and Haig, 1973. Reference: Markham (1974b). **Other localities**. Off Carmel, California USA, Pacific, 36°27′N, 122°00′W, 141 m. Host: *P. makarovi* McLaughlin & Haig, 1973. Near Punta de Abreoja, Baja California Sur, Mexico, Pacific, 26°17′N, 113°41′W, 99 m. Host: *P. laurentae*. Islas de San Benito, Baja California, Mexico, Pacific, 28°13′N, 115°34′W, 115–179 m. Host: *P. makarovi*. Reference: Markham (1974b). **Range outside of region**. None known. [Described as *Stegophryxus hyphalus* Markham, 1974].

26. *Parathelges tumidipes* Markham, 1972

Type-locality. Southwest of Jamaica, Caribbean, 17°09′N, 78°57′W, 26 m. Host: *Dardanus fucosus* Biffar and Provenzano. Reference: Markham (1972). **Other localities**. South of Bermuda, Atlantic, 32°17′N, 64°43′W, ca 200 m. Host: *Allodardanus bredini* Haig and Provenzano, 1965. Reference: Markham (1978). Belize, Caribbean, ca 17°N, 88°W, shallow water. Hosts: *Calcinus tibicen* (Herbst, 1791), *Clibanarius tricolor* Gibbes, 1850, and *C. vittatus* Bosc, 1802. Reference: Boyko and Williams (2016). **Range outside of region**. None known.

Subfamily Hemiarthrinae Markham, 1972

27. *Hemiarthrus nematocarcini* Stebbing, 1914

Type-locality (outside of region). Off South Africa, southern Atlantic. Host: *Nematocarcinus longirostris* Bate, 1888. Reference: Stebbing (1914). **Other locality** (within the region). Norfolk Canyon, off Virginia USA, Atlantic, 36°56′ to 37°09′N, ca 75°W, 1430–3083 m. Host: *Nematocarcinus ensifer* (S.I. Smith, 1882). Reference: Wenner (1979). **Range outside of region**. Chatham Rise, southeast of New Zealand, southern Pacific. Host: *Campylonotus rathbunae* Schmitt, 1926. Reference: Page (1985).

Family Colypuridae Richardson, 1905

28. *Colypurus agassizi* Richardson, 1905

Type-locality. Off Punta Mariato, Panamá, Pacific, 06°36′N, 81°44′W, 1063 m. Host unknown. Reference: Richardson (1905b). **Range outside of region**. None known.

Superfamily Cryptoniscoidea Kossmann, 1880
Family Cryptoniscidae Kossmann, 1880

29. *Zeuxokoma setosa* (Nierstrasz and Brender à Brandis, 1930)

Type-locality. Central California USA, Pacific, 38°17′N, 123°29′W, 305 m. Host: *Spirontocaris holmesi* Holthuis, 1947. References: Nierstrasz and Brender à Brandis (1930), Boyko (2015). **Range outside of region**. None known. [Described as *Faba setosa* Nierstrasz and Brender à Brandis, 1930].

Family Dajidae Giard and Bonnier, 1887

30. *Arthrophryxus beringanus* Richardson, 1908

Type-locality. Coast of British Columbia, Canada, Pacific. 53°53′N, 144°53′W, 4023 m. Host: unknown. Reference: Richardson (1908a). **Range outside of region**. None known.

31. *Colophryxus novangliae* Richardson, 1908

Type-locality. Off Long Island, New York, USA, Atlantic. 39°00′N, 72°04′W, 1293 m. Host: unknown. Reference: Richardson (1908b). **Range outside of region**. None known.

32. *Holophryxus acanthephyrae* Stephensen, 1912

Type-locality. Davis Strait, between Canada and Greenland, North Atlantic, 60°07′N, 48°26′W, depth unrecorded, probably ca 1000 m. Host: *Acanthephyra pelagica* (Risso, 1816). Reference: Stephensen (1912). **Range outside of region**. South of England, northeast Atlantic. Host: *A. pelagica*. Reference. Jones and Smaldon (1986). East of New Zealand, southern Pacific. Host: *Acanthephyra pelagica*. Reference: Wasmer (1988). Rocas Atoll, Brazil, south Atlantic. Host: *A. acanthitelsonis* Spence Bate, 1888. Reference: Alves-Júnior et al. (2018). [As

Isophryxus concavus Schultz, 1977]. Antarctic Ocean. Host: probably *A. pelagica*. Reference: Schultz (1977).

33. *Holophryxus alaskensis* Richardson, 1905

Type-locality. Revillagigedo Island, Alexander Archipelago, Alaska, USA, Pacific, 56°N, 131°W, 274–512 m. Host: *Pasiphaea pacifica* Rathbun, 1902. References: Richardson (1905a), Butler (1964). **Other localities**. Granby Bay, Strait of Georgia and west coast of Vancouver Island, British Columbia, Canada, Pacific, ca 49°24′N to 55°24′N,123°54′N to 129°49′W, 77–198 m. Host: *P. pacifica*, where known. Reference: Butler (1964). [As *Holophryxus californiensis* Richardson, 1908]. Santa Barbara Channel, California, USA, Pacific, ca 34°15′N, 119°50′W, 512 m. Host: unknown. References: Richardson (1908a); Butler (1964). Near Seattle, Washington, USA, Pacific, ca 47°30′N, 122°20′W, 183–512 m. Host: *P. pacifica*. References: Richardson (1908a), Butler (1964). [As *Hypodajus georgiensis* Nierstrasz and Brender à Brandis, 1931]. Strait of Georgia, British Columbia, Canada, Pacific, ca. 49°N, 124°W, ca 200 m. Host: unidentified pasiphaeid. References: Nierstrasz and Brender à Brandis (1931), Butler (1964). **Range outside of region**. None known.

34. *Oculophryxus bicaulis* Shields and Gómez-Gutiérrez, 1996

Type-locality. Off Baja California, Mexico, Pacific, 20°13′ to 28°52′N, 112°23′ to 117°48′W, to 210 m. Host: *Stylocheiron affine* Hansen, 1910. Reference: Shields and Gómez-Gutiérrez (1996). **Range outside of region**. South China Sea, western Pacific. Host: *S. affine*. Reference: Gómez-Gutiérrez and Shields (1998).

35. *Prophryxus alascensis* Richardson, 1909

Type-locality. Goletus Channel, British Columbia, Canada, Pacific, 53°05′N, 138°31′W, 3658 m. Host unknown. Reference: Richardson (1909). **Range outside of region**. None known.

36. *Zonophryxus retrodens* Richardson, 1903

Type-locality. South of Oahu, Hawaii, Pacific, ca 21°20′N, 158°W, ca 300 m. Host unknown. Reference: Richardson (1903). **Range outside of region**. None known.

37. *Zonophryxus similis* Richardson, 1914

Type-locality. Off Punta Mariato, Panamá, Pacific, 06°36′N, 81°44′W, 1063 m. Host unknown. Reference: Richardson (1914). **Range outside of region**. None known.

38. *Zonophryxus* sp.

Known localities. California Bight, off California, USA, Pacific, 32°35′N, 117°55′W to 33°59′N, 118°43′W, 98–305 m. Hosts: *Pantomus affinis* Chace, 1973, and *Plesionika trispinus* Squires and Barragán, 1976. Reference: Montagne and Cadien (2001). **Range outside of region**. None known.

Family Entophilidae Richardson, 1903

39. *Entophilus omnitectus* Richardson, 1903

Type-locality. Between Maui and Molokai, Hawaii, Pacific, 20°57′N, 156°41′W, ca 300 m. Host: *Agononida normani* (Henderson, 1885). Reference: Richardson (1903). **Other localities**. Palolo Channel, Hawaii, Pacific, 21°04′N, 156°42′W, to 258 m. Host: *A. normani*. Reference: Richardson (1903). Northern Gulf of Mexico, 27°22′ to 29°17′N, 85°52′ to 93°31′W, 185–850 m. Hosts. *Munida valida* S. I. Smith, 1883, and *M. microphthalma*. Reference: Adkison and Collard (1990). **Range outside of region**. Madagascar, Indian Ocean. Host: *Agononida incerta* (Henderson, 1888). Reference: Bourdon (1976). Chesterfield Islands, southwestern Pacific. Host: *A.* "*incerta*". Reference: Markham (1994).

6.4 Conclusions

A thorough review of previous records of epicarid isopods from deep water around North and Central America allows for the listing of 39 species, 4 of these unidentified. Most species (27) belong to the Bopyridae with 17 genera, including five species of *Munidion* and at least four species of *Pseudione*. Other families are the Colypuridae (1 sp.), the Cryptoniscidae (1 sp.), the Dajidae (9 spp.), and the Entophilidae (1 sp.).

Markham (1986) reported on the evolution and zoogeography of Bopyridae stating that 469 species had been described worldwide at that time. In a more recent review, Williams and Boyko (2012) estimated the number of epicarideans isopods (Bopyroidea and Cryptoniscoidea) at 794 species: 605 Bopyridae and 99 Cryptoniscoidea. Bopyrids were more diverse in the NW Pacific, while cryptoniscoids were more diverse in the NE Atlantic (Williams and Boyko 2012).

Of the 35 identified species, about half (17) were described in 1930 or earlier and the rest after 1972. Epicarid isopods of the area infest a wide variety of hosts, including euphausiids of the genus *Stylocheiron*; caridean shrimps of the genera *Acanthephyra*, *Campylonotus*, *Crangon*, *Glyphocrangon*, *Heterocarpus*, *Parapontophilus*, *Pasiphaea*, *Metacrangon*, *Nematocarcinus*, and *Spirontocaris*; hermit crabs of the genera *Clibanarius*, *Dardanus*, and *Parapagurodes*; and squat lobsters of the genera *Agononida*, *Munida*, *Munidopsis*, and *Galacantha*.

References

Adkison DL, Collard SB (1990) Description of the cryptoniscium larva of *Entophilus omnitectus* Richardson. 1903 (Crustacea: Isopoda: Epicaridea) and records from the Gulf of Mexico. Proc Biol Soc Washington 103:649–654

Alves-Júnior F de A, Leitão BA, Araújo C de, Carvalho RJ de P, Souza-Filho JF de (2018) First report of the ectoparasitic isopod, *Holophryxus acanthephyrae* Stephensen 1912 (Cymothoida: Dajidae) in the South Atlantic: recovered from a new Host: the deep-sea shrimp, *Acanthephyra acanthitelsonis* Spence Bate, 1888. Thalassas: Intern J Mar Sci. https://doi.org/10.1007/s41208-018-0072-3

Alves-Júnior F de A, Boyko CB, Carvalho Paiva RJ de, Negromonte AO, Souza-Filho JF de (2019) First record of the genus *Anuropodione* Bourdon, 1967 (Crustacea: Isopoda: Bopyridae) from the South Atlantic. Nauplius. J Brazilian Crust Soc 27:1–7

An JM, Yu HY, Li XZ (2008) Subfamily Argeiinae (Crustacea, Isopoda, Epicaridea) from China. Acta Zootaxo Sinica 33(1):225–226. [In Chinese]

An JH, Yu HY, Williams JD (2007) Four new records and a new species of *Dactylokepon* Stebbing, 1910 (Epicaridea: Bopyridae: Ioninae) from Chinese waters. J Nat Hist 41(33-36):2063–2079

Boone PL (1927) Crustacea from tropical east American seas. Bull Bingham Oceanogr Coll 1(2):1–147

Bourdon R (1972) Sur quelques Bopyridae (Crustacea, Isopoda) parasites de Galatheidae. Bull Mus Natl Hist Nat, Paris (3) no. 66 Zool 52:817–838

Bourdon R (1976) Épicarides de Madagascar. 1. Bull Mus Natl Hist Nat, Paris (3) no. 371, Zool 259: 353–392

Boyko CB (2015) A Revision of *Danalia* Giard, 1887, *Faba* Nierstrasz & Brender à Brandis, 1930 and *Zeuxokoma* Grygier, 1993 (Crustacea: Isopoda: Epicaridea: Cryptoniscoidea: Cryptoniscidae) with description of a new genus and four new species. Bishop Mus Bull Zool 9:65–92

Boyko CB, Williams JD (2016) Parasitic isopods (Isopoda: Bopyridae) of hermit crabs from Belize and other localities of the Caribbean. Abstracts – Poster Session. The Crustacean Society Mid-year Meeting 2016, 11–13 July 2016. Natl Univ Singapore, Singapore. Abstract PO-23, p 54

Bursey CR (1978) Histopathology of the parasitization of *Munida iris* (Decapoda: Galatheidae) by *Munidion irritans* (Isopoda: Bopyridae). Bull Mar Sci 28:566–570

Butler TH (1964) Redescription of the parasitic Isopod *Holophryxus alaskensis* Richardson, and a note on its synonymy. J Fish Res Bd Canada 21(5):971–976

Chace FA Jr (1942) The anomuran Crustacea, 1. Galatheidea (families Chirostylidae, Galatheidae and Porcellanidae). Reports on the scientific results of the Atlantis expeditions to the West Indies in 1938 and 1939, under the joint auspices of the University of Havana and Harvard University. Torreia 11:1–106

Fraser C McL (1932) A comparison of the marine fauna of the Nanaimo region with that of the San Juan Archipelago. Trans Roy Soc Canada (3) 26 (Section V – Biol Sci):49–70

Gómez-Gutiérrez J, Shields JD (1998) Range extension for *Oculophryxus bicaulis* Shields & Gómez (Isopoda, Dajidae) in the South China Sea. Crustaceana 71(2):167–170

Gore RH (1983) Notes on rare species of *Munidopsis* (Anomura: Galathidae) and *Ethusina* (Brachyura: Dorippidae) collected by the USNS Bartlett in the Venezuelan Basin, Caribbean Sea. Proc Acad Nat Sci Philadelphia 135:200–217

Hansen HJ (1897) Reports on the dredging operations off the west coast of Central America to the Galápagos Islands, to the west coast of Mexico, and in the Gulf of California, in charge of Alexander Agassiz, carried on by the U.S. Fish Commission Steamer "Albatross" during 1891, Lieut. Commander Z. L. Tanner, U.S. Navy, commanding. XXII. The Isopods. Bull Mus Comp Zool, Harvard College 31(5):96–129

Jones MB, Smaldon G (1986) On the genus *Holophryxus* (Isopoda:Epicaridea), with description of the male and redescription of the female of *Holophryxus acanthephyrae*. J Mar Biol Assoc UK 66:303–314

Kaufmann RS, Wakefield WW, Genin A (1989) Distribution of epibenthic megafauna and lebensspuren on two central North Pacific seamounts. Deep-Sea Res 36(12):1863–1896

Kazmi QB, Boyko CB (2005) A new locality and host for *Pseudione minimocrenulata* Nierstrasz & Brender à Brandis, 1931 (Crustacea: Isopoda: Bopyridae) in the Indian Ocean, with comments on the identity of the type specimens. Zootaxa 925(1):10

Lamb A, Hanby BP (2005) Marine life of the Pacific Northwest. A photographic encyclopedia of invertebrates, seaweeds and selected fishes. Harbour Publishing, Madeira Park

Luke SR (1977) Catalog of the benthic invertebrate collections. I. Decapod Crustacea and Stomatopoda. Scripps Inst Oceanogr Ref no. 77–9:1–72

Markham JC (1972) Four new species of *Parathelges* Bonnier, 1900 (Isopoda, Bopyridae), the first record of the genus from the western Atlantic. Crustaceana Suppl 3:57–78

Markham JC (1974a) Six new species of bopyrid isopods parasitic on galatheid crabs of the genus *Munida* in the western Atlantic. Bull Mar Sci 23:613–648

Markham JC (1974b) Parasitic bopyrid isopods of the amphi-American genus *Stegophryxus* Thompson with the description of a new species from California. Bull South Calif Acad Sci 73:3–41

Markham JC (1975) A review of the bopyrid isopod genus *Munidion* Hansen, 1897, parasitic on galatheid crabs in the Atlantic and Pacific Oceans. Bull Mar Sci 25:422–441

Markham JC (1977a) Description of a new western Atlantic species of *Argeia* Dana with a proposed new subfamily for this and related genera (Crustacea Isopoda, Bopyridae). Zool Meded 52(9):107–123

Markham JC (1977b) The status and systematic position of the species of the bopyrid isopod genus *Phyllodurus* Stimpson, 1857. Proc Biol Soc Washington 90:813–818

Markham JC (1978) Bopyrid isopods parasitizing hermit crabs in the northwestern Atlantic Ocean. Bull Mar Sci 28:102–117

Markham JC (1979) A new genus and species of bopyrid isopod infesting the crangonid shrimp *Pontophilus abyssi* Smith in deep water of the northwestern Atlantic Ocean. Proc Biol Soc Washington 92:769–772

Markham JC (1986) Evolution and zoogeography of the Isopoda Bopyridae, parasites of Crustacea Decapoda. In: Gore RH, Heck KL (eds) Crustacean biogeography. AA Balkema, Rotterdam/Boston, pp 143–164

Markham JC (1988) Descriptions and revisions of some species of Isopoda Bopyridae of the north western Atlantic Ocean. Zool Verhand 246:1–63

Markham JC (1992) The Isopoda Bopyridae of the eastern Pacific – missing or just hiding? Proc San Diego Soc Nat Hist 17:1–4

Markham JC (1994) Crustacea Isopoda: Bopyridae in the MUSORSTOM collections from the tropical Indo-Pacific I. Subfamilies Pseudioninae (in part), Argeiinae, Orbioninae, Athelginae and Entophilinae. In: Crosnier A (ed) Résultats des Campagnes MUSORSTOM Vol. 10. Mém Mus Natl Hist Nat, Paris 161:225–253

Markham JC (2016) Redescription of *Bathygyge grandis* Hansen, 1897 (Crustacea, Isopoda, Bopyridae) from southern California with erection of a new subfamily for it. Bull South Calif Acad Sci 115(1):72–80

Mayo BS (1974) The systematics and distribution of the deep-sea genus *Munidopsis* (Crustacea, Galatheidae) in the western Atlantic Ocean. Dissertation, University of Miami

Montagne DE, Cadien DB (2001) Northern range extension into the southern California Bight of ten decapod Crustacea related to the 1991/92 and 1997/98 El Niño events. Bull South Calif Acad Sci 100(3):199–211

Nierstrasz HF, Brender à Brandis GA (1930) Three new genera and five new species of parasitic Crustacea. Proc U S Natl Mus 77(2832):1–9

Nierstrasz HF, Brender à Brandis GA (1931) Papers from Dr. Th. Mortensen's Pacific Expedition 1914–16. 57. Epicaridea 2. Vidensk Meddel Dansk Naturhist Kjøbenhavn 91:147–225

Page RDM (1985) Review of the New Zealand Bopyridae (Crustacea: Isopoda: Epicaridea). New Zeal J Zool 12:185–212

Pardo LM., Boyko CB and. Matelatto FL (2009) Description of a new species of *Asymmetrione* (Isopoda: Bopyridae: Pseudioninae) infesting the hermit crab *Paguristes tomentosus* (Anomura: Diogenidae) from Peru, with a key to species and a review of southeastern Pacific bopyrids. J Nat Hist 43 (33-34): 2041–2055

Rayner GW (1935) The Falkland species of the crustacean genus *Munida*. Discovery Rep 10:209–245

Richardson H (1903) Isopods collected at the Hawaiian Islands by the U. S. Fish Commission Steamer Albatross. Bull U S Fish Comm 23(3):819–826

Richardson H (1904) Contributions to the natural history of the Isopoda. V. Isopod crustaceans of the Northwest Coast of North America. Proc U S Natl Mus 27(1369):657–681

Richardson H (1905a) Isopods of the Alaska Salmon Investigation. Bull U S Bur Fish Comm 24:209–221

Richardson H (1905b) A monograph on the isopods of North America. Bull U S Natl Mus 54:i-liii, 1–727

Richardson H (1908a) On some isopods of the family Dajidae from the northwest Pacific Ocean, with descriptions of a new genus and two new species. Proc U S Natl Mus 33(1586):689–696

Richardson H (1908b) Description of a new isopod genus of the family Dajidae. Proc U S Natl Mus 34:391–392

Richardson H (1909) Isopods collected in the northwest Pacific by the U.S. Bureau of Fisheries steamer "Albatross" in 1906. Proc U S Natl Mus 37(1701):75–129

Richardson H (1914) Isopoda. Reports on the scientific results of the expedition to the eastern tropical Pacific in charge of Alexander Agassiz by the U. S. Fish Commission steamer "Albatross", from October, 1904, to March, 1905, Lieut. Commander L. M. Garnett, U. S. N., commanding, XVIII. Bull Mus Comp Zool, Harvard College 58:356–372

Román-Contreras R (2008) A new genus of bopyrid isopod infesting *Munidopsis beringana* (Decapoda: Galatheidae) from deep waters. J Crust Biol 28(2):378–384

Román-Contreras R, Boyko CB (2007) A new genus and species of bopyrid isopod infesting the crab *Munidopsis depressa* (Anomura: Galatheidae) from the Gulf of California, with notes on its ecology. J Crust Biol 27(2):370–379

Román-Contreras R, Soto LA (2002) A new deep water genus and species of branchial bopyrid infesting the galatheid crab *Munidopsis erinaceus* from the southwestern Gulf of Mexico. J Crust Biol 22(2):279–286

Salazar-Vallejo SI, Leija-Tristan A (1989) *Progebiophilus bruscai* n. sp., a new bopyrid isopod parasitic on the mud shrimp, *Upogebia dawsoni* Williams (Thalassinidea), from the Gulf of California. Cah Biol Mar 30(4):423–432

Schultz GA (1977) Bathypelagic isopod Crustacea from the Antarctic and southern seas. Antarct Res Ser Biol Antarct Seas 23(3):69–128

Shields JD and Gómez-Gutiérez J (1996) Oculophryxus bicaulis, a new genus and species of dajid isopod parasitic on the euphausiid Stylocheiron affine Hansen. Internatl J Parasitol 26(3:):261–268

Sivertsen E, Holthuis LB (1956) Crustacea Decapoda (the Penaeidea and Stenopodidea excepted). Rep Sci Res "Michael Sars" North Atlantic Deep-Sea Exped 1910 5(12):1–56

Stebbing,TRR (1906) Marine crustaceans. XII. Isopoda, with description of a new genus. pp. 699-721 *in*; Gardiner, J. S. The fauna and geography of the Maldive and Laccadive Archipelagoes, being the account of the work carried on and of the collections made by an expedition during the years 1899 and 1900. Volume 2. Cambridge: Cambridge University Press

Stebbing TRR (1914) South African Crustacea. (part VII of S. A. Crustacea, for the Marine Investigations in South Africa). Ann South Afr Mus 15(1):1–55

Stephensen K (1912) Report on the Malacostraca collected by the "Tjalfe"-Expedition, under the direction of cand. mag. Ad. S. Jensen, especially at W. Greenland. Vidensk Meddel Dansk Naturh Kjøbenhavn 64:57–134

Stuardo J, Vega R and Céspedes I (1986). New bopyrid isopod parasitic on *Callianassa uncinata* H. Milne Edwards: with functional and ecological remarks. Gayana, Zool 50 (1-4): 3-15

Varisco M, Vinuesa J (2011) Infestation of the squat lobster *Munida gregaria* (Anomura: Galatheidae) by *Pseudione galacanthae* (Isopoda: Bopyridae) in San Jorge Gulf, Argentina. J Mar Biol Assoc UK 91(8):1681–1688

Vázquez-Bader AR, Gracia A (2016) Diversity and distribution of Chirostyloidea and Galatheoidea (Decapoda, Anomura) in the southern Gulf of Mexico. Zookeys 612:1–30

Wasmer RA (1988) The parasitic isopod *Holophryxus acanthephyrae* Stephensen (Epicaridea: Dajidae) from the Subantarctic South Pacific, with notes on its synonymy and Host. Proc Biol Soc Wash 101(1):20–30

Wenner EL (1978) Comparative biology of four species of glyphocrangonid and crangonid shrimp from the continental slope of the Middle Atlantic Bight. Canadian J Zool 56:1052–1065
Wenner EL (1979) Distribution and reproduction of nematocarcinid shrimp (Decapoda: Caridea) from the northwestern North Atlantic. Bull Mar Sci 29:380–393
Wenner EL, Windsor NT (1979) Parasitism of galatheid crustaceans from the Middle Atlantic Bight by bopyrid isopods. Crustaceana 37(3):293–303
Wicksten MK (2012) Decapod Crustacea of the Californian and Oregonian zoogeographic provinces. Zootaxa 3371:1–307
Williams AB, Brown WS (1972) Notes on structure and parasitism of *Munida iris* A. Milne Edwards (Decapoda, Galatheidae) from North Carolina, U. S. A. Crustaceana 22(3):303–308
Williams JD, Boyko CB (2012) The global diversity of parasitic isopods associated with crustacean hosts (Isopoda: Bopyroidea and Cryptoniscoidea). PLoS One 7(4):e35350
WoRMS Editorial Board (2019). World Register of Marine Species. Available from http://www.marinespecies.org at VLIZ. Accessed Jan–Dec 2019)

Chapter 7
Biodiversity of the Deep-Sea Isopods, Cumaceans, and Amphipods (Crustacea: Peracarida) Recorded off the Argentine Coast

B. L. Doti, I. L. Chiesa, and D. Roccatagliata

Abstract Our knowledge on the deep-sea peracarids at a global scale is limited, and this gap in knowledge is still larger when we refer to the peracarids from off the coast of Argentina. With the aim of improving this situation, a complete and accurate inventory of the deep-sea isopods, cumaceans, and amphipods from off the coast of Argentina is presented. This inventory is based mainly on data taken from the literature, but some records provided in the GBIF database were included as well. A total of 126 stations taken during 24 oceanographic surveys carried out by seven countries were compiled. Isopoda showed the highest number of species (107 spp.) followed by Cumacea (50 spp.) and Amphipoda (47 spp.). A large amount of specimens, including many new species, has been collected in recent years and wait to be described. Thus, it is expected that the number of species recorded from the area substantially increases in the near future.

Keywords Checklist · Crustaceans · Deep water · Peracarids · Southwestern Atlantic

B. L. Doti · D. Roccatagliata (✉)
Universidad de Buenos Aires, Facultad de Ciencias Exactas y Naturales, Departamento de Biodiversidad y Biología Experimental, Buenos Aires, Argentina

Universidad de Buenos Aires, Instituto de Biodiversidad y Biología Experimental y Aplicada (IBBEA, UBA-CONICET), Buenos Aires, Argentina
e-mail: bdoti@bg.fcen.uba.ar; rocca@bg.fcen.uba.ar

I. L. Chiesa
Centro Austral de Investigaciones Científicas (CADIC-CONICET), Ushuaia, Tierra del Fuego, Argentina

M. E. Hendrickx (ed.), *Deep-Sea Pycnogonids and Crustaceans of the Americas*, https://doi.org/10.1007/978-3-030-58410-8_7

7.1 Introduction

From the low number of species and individuals yielded by earlier deep-sea collections, there has arisen the generally accepted concept that the fauna of deep-sea was depauperated compared to that of shallow waters (see Hessler and Sanders 1967). However, in the 1960s, a high faunal diversity was obtained from a transect of the ocean floor between southern New England and Bermuda isles (the Gay Head-Bermuda transect). Such diversity was far in excess of anything reported before, and the crustacean peracarids (isopod, cumacean, amphipod, and tanaid species) were the second most common group recovered (Sanders and Hessler 1969). Samples were taken with the epibenthic sled (EBS) that move over the surface of the sediment for distances of a kilometer or more, and washed through a small-size-aperture screen, in order to retain the most abundant smaller animals. Certainly, the relatively few specimens obtained in earlier surveys were due to the technical limitations in sampling the deep-sea benthos. It is worth noticing that after the Gay Head-Bermuda transect, a remarkably high diversity of peracarids was also reported from many other deep-sea localities, i.e., the Tropical Atlantic, the Angola Basin, the Southern Ocean, the Kurile-Kamchatka Trench, the basin of the Sea of Japan, and the Greenland-Iceland-Faeroe Ridge, among others (Jones and Sanders 1972; Hessler et al. 1979; Brandt et al. 2005, 2007, 2018; Golovan et al. 2013, 2019; Brix et al. 2018a, b).

Regarding the isopods, cumaceans, and amphipods obtained off the coast of Argentina, the earliest records from this fauna were obtained by the HMS Challenger (Beddard 1884, 1886; Stebbing 1888). After that, 23 surveys, most of them foreign expeditions, reported on additional records. Despite the sampling efforts and taxonomic works already done, vast areas remain unexplored. Many species have just single records, and others are awaiting a formal description. Furthermore, there are just a few comprehensive studies dealing with the peracarids from this area. For amphipods, there are two mandatory references, i.e., López Gappa et al. (2006) and De Broyer et al. (2007). For isopods, only the Asellota from the continental shelf and slope were compiled (see Doti et al. 2014). No global revisions or catalogs have been published until now for the cumaceans and the remaining isopods.

With the aim of improving the knowledge of the deep-sea isopods, cumaceans, and amphipods off the Argentine coast, the opportunity has been taken to compile all the records reported from this area. Although remarks on the bathymetric and geographic distributions of some taxa are presented, it should be kept in mind that a biogeographic analysis of this fauna is beyond the scope of the current study. We hope the information herein presented may prove to be useful to conserve this fragile deep-sea fauna and their habitats and to promote further taxonomic works. We also hope that this contribution promotes further sampling programs in the area as well as biogeographic analyses of this fauna, both of which are much needed.

7.2 Material and Methods

Previous records of isopods, cumaceans, and amphipods (excluding hyperiids) from the study area were compiled from the literature. In addition, the GBIF (Global Biodiversity Information Facility) was checked, and the species not published but identified by a trained taxonomist were also included. Those species identified with a preliminary "interim" name in the GBIF (2019) were not taken into account. The boundary of the area under study is between 35°S and 56°S and from 200 m isobath to 46°W (sky-blue area in Fig. 7.1). The stations considered within this area are listed in Table 7.1 and plotted on a map (Fig. 7.1a). The whole area was divided into a one-degree square grid. Squares containing published records were numbered from the 200 m isobath to the west and from north to south (Fig. 7.1b). For the sake of clarity, the coordinates (Table 7.1) are expressed as they appear in the original publications, i.e., in degrees, minutes, seconds (DMS) or in decimal degrees (DD). The records for each one-degree square were counted. In Table 7.2, the column "<200 m" includes those species that occur at the study area and also in the adjacent continental shelf (most of the white area in Fig. 7.1); the column "A/SA" refers to those species that occur at the study area and also in the Antarctic/Sub-Antarctic region; and the column "Others" to those species that occur at the study area and elsewhere (any region other than the A/SA). The northern limit of the Sub-Antarctic (SA) region was defined after De Broyer and Danis (2011). It should be kept in mind that the SA region overlaps in part with the study area. Species are listed according to current taxonomic classification (WoRMS, 2020), or as appeared in the latest taxonomical revision in which they were mentioned.

7.3 Results

7.3.1 General Comments

The study area has been poorly explored. Only 73 (24.6%) out of the 296 one-degree squares plotted on the maps encompass records from the taxa analyzed (Fig. 7.1). Moreover, 25 of these 73 one-degree squares contain just a single record (Fig. 7.2).

The study of the peracarids from off the coast of Argentina began with the Challenger expedition in 1873–1876, which recorded a few species of isopods and amphipods. Half a century had to elapse before additional records of these two taxa were reported from the area. Furthermore, cumaceans were recorded for the first time much later, as late as 1976 (Fig. 7.3). Of the 428 records reported from the area, 170 were obtained between 200 and 1000 m depth, and the number of records dropped abruptly below 1000 m depth. The same trend was observed for the stations (Fig. 7.4).

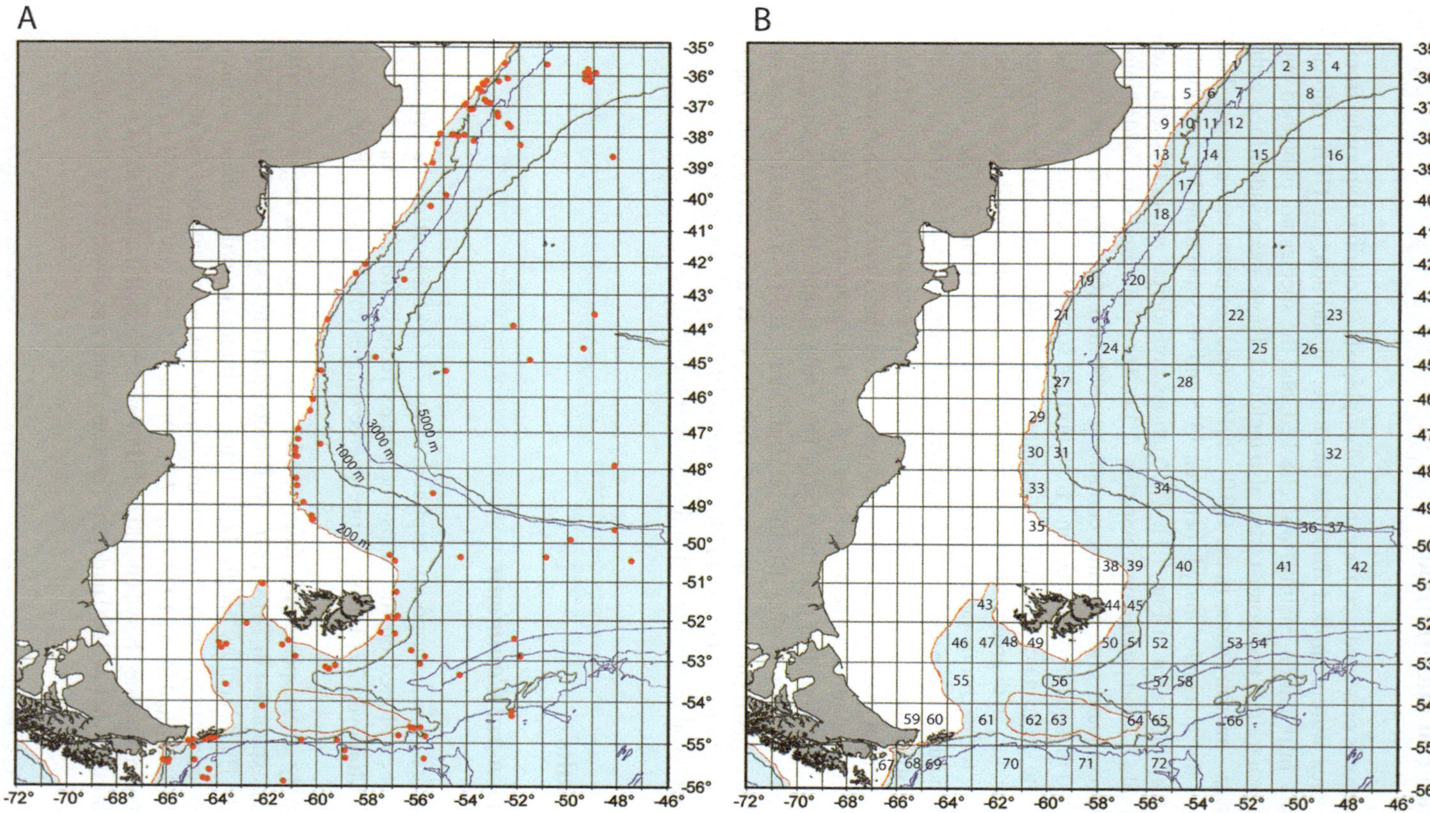

Fig. 7.1 Map of the study area divided into a one-degree square grid. (**a**) Distribution of the stations listed in Table 7.1. (**b**) One-degree squares containing records, sequentially numbered from west to east and from north to south

Table 7.1 Surveys and station data of all the records compiled in this study

Survey/ship	Station	Latitude (S)	Longitude (W)	Depth (m)	Sq. no.	No. of species		
						I	C	A
HMS Challenger Expedition (1873–1876)	317	48°37′	55°17′	1893	34	0	0	1
	318	42°32′	56°27′	3730	20	1	0	0
	320	37°17′	53°52′	1097	11	4	0	2
	323	35°39′	50°47′	3475	2	0	0	1
Swedish Antarctic Expedition (1901–1903)	34b	44°49′	57°34′	700–500	24	4	0	0
	Shag quadrant	50°19′	50°50′	2675	41	0	0	1
	–	49°56′	49°56′	0–2700	36	0	0	1
Terra Nova Expedition (1910)	38	West of Malvinas Islands		229	48	0	0	2
RRS *William Scoresby* (1927–1931)	WS 76	51°00′	62°02′30″	207–205	43	1	0	1
	WS 212	49°22′	60°10′	242–249	35	3	0	0
	WS 213	49°22′	60°10′	249–239	35	1	0	0
	WS 214	48°25′	60°40′	208–219	33	3	0	0
	WS 215	47°37′	60°50′	219–146	30	3	0	0
	WS 216	47°37′	60°50′	219–133	30	2	0	0
	WS 227	51°08′	56°50′	295	45	1	0	0
	WS 228	50°50′	56°58′	229–236	39	1	0	0
	WS 229	50°35′	57°20′	210–271	38	2	0	0
	WS 234	48°52′	60°25′	195–207	33	2	0	0
	WS 236	46°55′	60°40′	273–300	29	3	0	0
	WS 237	46°00′	60°05′	150–256	29	3	0	0
	WS 244	52°00′	62°40′	253–248	47	3	0	0
	WS 245	52°36′	63°40′	304–290	46	3	0	0
	WS 246	52°25′	61°00′	267–208	48	2	0	0
	WS 766	45°13′	59°56′30″	545	27	2	0	0
	WS 772	47°28′	60°51′	309–162	30	2	0	0
	WS 773	47°28′	60°51′	291–296	30	2	0	0
	WS 818	52°31°15″	63°28′	272–278	46	4	0	0
	WS 820	52°53′15″	51°51′	351–367	54	1	0	0
	WS 821	52°55′45″	60°55′	461–468	49	2	0	0
	WS 839	53°30′15″	63°29′	403–434	55	2	0	0
	WS 871	53°16′	54°12′	336–341	58	2	0	0
Soviet Antarctic Expedition, RV *Ob*, Cruise III (1958)	477	48°14′	60°49′	280–303	33	1	0	0
	479	45°16′	54°54′	680	28	5	0	0
	480	43°40′	59°34′	399–500	21	14	0	0

(continued)

Table 7.1 (continued)

Survey/ship	Station	Latitude (S)	Longitude (W)	Depth (m)	Sq. no.	No. of species I	C	A
Lamont Geological Observatory (LGO), RV *Vema*, Cruises 14, 15, 17 and 18 (1958–1962)	14-12	52°32′	61°15′	361	48	0	3	0
	15-114	55°02.6′	64°17′	1737	69	1	0	0
	15-116	55°42.9′	64°21.6′	3813	69	3	0	0
	15-117	55°31.2′	64°07.5′	3839–3845	69	7	1	1
	15-118	55°44.2′	64°11.5′	3776–3777	69	1	0	1
	15-123	50°23.2′	47°25′	2681	42	1	0	0
	15-124	49°35′	48°04.6′	2738	37	1	0	0
	15-126	47°57.5′	48°03′	6079	32	1	0	0
	15-128	44°53.3′	51°26.5′	5843	25	6	0	0
	15-131	40°14.6′	55°24.7′	1475–1479	18	1	4	6
	15-132	39°57.5′	54°49.5′	1911	17	1	0	0
	17-51	55°17.5′	66°00′	205–207	67	0	0	1
	17-53	55°20′	65°50′	1185–1240	68	0	3	0
	17-54	55°19.5′	65°49′	1274–1362	68	0	1	0
	17-59	54°53.5′	60°26.5′	426	62	0	2	0
	17-65	50°18′	54°11′	1498–1501	40	0	0	1
	17-80	43°58′	52°07′	5781	22	0	1	0
	17-81	44°33′	49°19′	5332	26	0	3	0
	17-101	38°13′	55°19′	450	13	1	5	0
	18-8	36°06′	53°18′	278–282	6	1	0	0
	18-9	36°17′	53°21′	547–676	6	0	0	1
	18-12	47°09′	60°38′	424–428	30	0	1	1
RV *Eltanin,* Cruises 4, 6, 7, 11, 22 (1962–1964)	4-107	55°49′12″	61°10′12″	?	70	1	0	0
	6-339	53°05′	59°31′	512–586	56	1	1	1
	6-340	53°07.6′	59°23.2′	578–567	56	4	0	1
	6-348	54°52′58.8″	59°1′30″	644	63	0	0	1
	6-350	55°02.7′	58°57.4′	2450–2452	71	1	0	4
	6-353	55°16′30″	58°56′31″	3578	71	1	0	0
	6-375	53°00′	55°50′	712–933	57	2	0	0
	6-381	55°15′29″	55°36′	1867	72	0	0	1
	7-557	51°55.5′	56°38.5′	864–854	45	1	0	0
	7-558	51°58′	56°38′	646–845	45	0	0	2
	11-969	54°55.7′	65°03.2′	229–265	59	2	0	2
	11-970	54°59′	64°53.2	586–641	60	1	0	0

(continued)

Table 7.1 (continued)

Survey/ship	Station	Latitude (S)	Longitude (W)	Depth (m)	Sq. no.	No. of species		
						I	C	A
	11-973	55°18.2′	64°42′	1920–2210	69	2	0	0
	11-977	52°32′	63°53′	229	46	0	0	5
	22-1521	54°08′	52°08′	419–483	66	1	0	0
	22-1525	52°22′	52°02′	1351–1618	53	1	0	0
	22-1592	54°43′58.8″	55°33′28.80″	1845	65	0	0	1
	22-1593	54°43′	56°37′	339–357	64	0	0	2
Knipovich Survey (1965)	249	35°34.5′	52°40.3′	310	1	0	0	2
Walther Herwig 15 (1966)	245	36°49′	54°02′	600	5	0	0	1
Woods Hole Oceanographic Institution (WHOI), RV *Atlantis II*, Cruise 60 (1970)	236	36°27′	53°31.0′	497–518	6	4	1	0
	237	36°32.6′	53°23.0′	993–1011	6	7	6	0
	239	36°49.0′	53°15.4′	1661–1679	6	2	13	0
	240	36°53.4′	53°10.2′	2195–2323	6	0	11	0
	242	38°16.9′	51°56.1′	4382–4402	15	6	7	0
	243	37°36.8′	52°23.6′	3815–3822	12	0	5	0
	245	36°55.7′	53°1.4′	2707	6	12	15	0
	246	37°15.1′	52°45.0′	3343	12	0	6	0
	247	43°33′	48°58.1′	5208–5223	23	6	5	0
	256	37°40.9′	52°19.3′	3906–3917	12	7	5	0
	259	37°13,3′	52°45′	3305–3317	12	4	5	0
	262	36°05.2′	52°17.9′	2440–2480	7	4	10	0
	264	36°12.7′	52°42.7′	2041–2048	7	3	8	0
	280	36°18′	53°23.9′	256–293	6	0	1	0
RV *Hero*, Cruise 715 (1971)	880	54°50′	64°00′	205–208	60	1	0	0
	893	54°54.8′	64°19′	303–358	60	0	0	3
	894	54°54.8′	64°18′	263–285	60	0	0	4
	895	54°59.9′	64°50′	438	60	1	0	0

(continued)

Table 7.1 (continued)

Survey/ship	Station	Latitude (S)	Longitude (W)	Depth (m)	Sq. no.	No. of species		
						I	C	A
RV *Akademik Kurchatov,* Cruise 11 (1971), Cruise 43 (1985–1986)	11-926	52°56′0″	55°36′0″	1966–2016	52	4	0	0
	11-927	52°40′0″	56°08′4″	1660–1664	51	2	0	0
	11-928	52°15′	56°51′	1105	51	1	0	0
	11-929	52°08′5″	57°16′11″	720–724	50	3	0	0
	11-930	51°57′	57°36′	401	44	3	0	0
	43-4893	36°12′	49°09′	4630	8	3	0	0
FV *Sund*, Cruise 4 (1974)	136	46°18′	60°12′	?	29	1	0	0
	-	47°17′	59′54′	750	31	2	0	0
RV *Dmitriy Mendeleev,* Cruise 43 (1989)	4109	38°40′	48°10′	5225	16	4	0	0
RV *Eduardo Holmberg* (2001)	EH 04/01	42°01′	58°02′	227	19	0	1	0
RV *Polarstern* ANT XIX/5 (LAMPOS) (2002)	145-1	54°1.36′	62°1.3′	272	61	0	0	1
	147-1	54°32′	55°55′	413	65	0	0	1
	150-1	54°30.22′	56°8.2′	290	64	0	0	1
	153-1	54°32.75′	56°09.84′	296–299	64	0	0	1
"ICEFISH" cruise (2004)	14	53°02′32″	59°08′03	470	56	0	0	1
Red Crab Spring Survey (DINARA) (2006)	-	37°00′	53°59′	500	11	1	0	0
	-	37°00′	54°08′	500–800	10	1	0	0
RV *Meteor*, DIVA III (2009)	527	35°59.25′	49°00.96′	4568	3	0	1	0
	531	35°56.50′	48°53.90′	4605	4	0	0	1
	532	35°59.16′	49°00.75′	4605	3	4	5	0
	533	36°00.20′	49°01.96′	4602	8	5	5	0
	534	36°00.61′	49°01.54′	4608	8	5	4	0
	535	35°59.22′	49°00.95′	4589	3	0	1	0
RV *Puerto Deseado*, COPLA II (2010)	3	38°41.05′	55°56.93′	208	13	1	1	0
	11	42°19.99′	58°24.96′	284	19	1	1	0
RV *Polarstern,* Cruise PS77 ANT-XXVII/3 (CAMBIO) (2011)	208-2	54°32.27′	56°9.78′	282–283	64	0	0	1
RV *Puerto Deseado*, Talud I (2012)	6	37°58.337′	55°8.915′	530	9	0	1	0
	12	37°57.9′	54°31.92′	1140	10	3	0	0
RV *Puerto Deseado*, Talud II (2013)	36	37°57′31″	54°24′00″	1289	10	1	0	0
RV *Puerto Deseado*, Talud III (2013)	46	38°5.310′	53°39.988′	3282	14	1	0	0
	39	37°58′01″	54°31′43″	1144	10	1	0	0
	59	37°47′42″	54°05′15″	1398	10	1	0	0

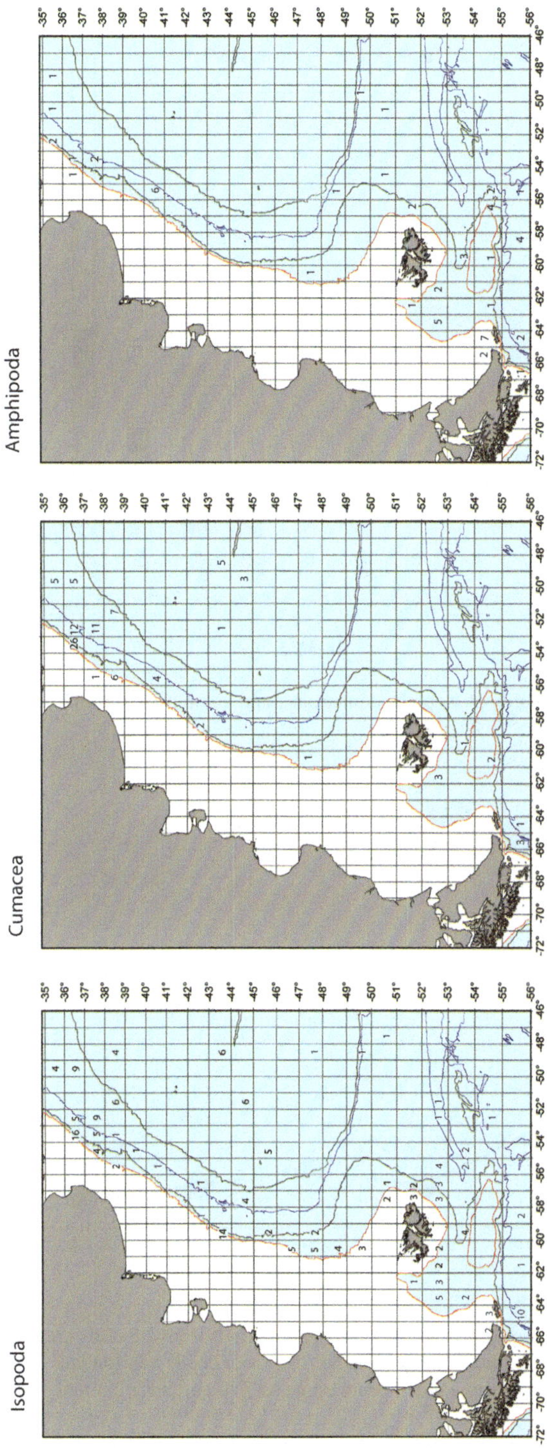

Fig. 7.2 Maps showing the number of species of isopods, cumaceans, and amphipods, recorded in each one-degree square of the study area

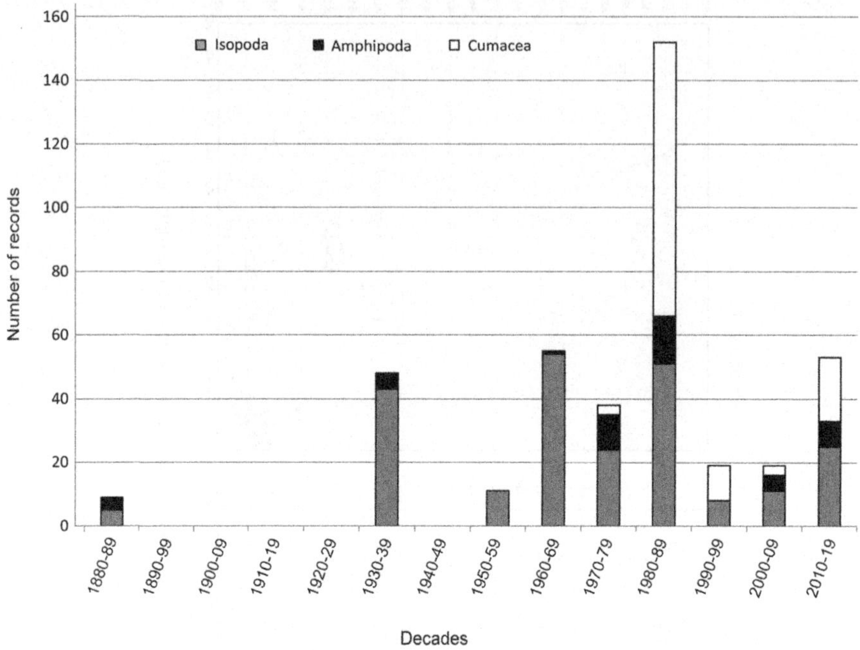

Fig. 7.3 Number of records of isopods, cumaceans, and amphipods from the study area, sorted by decades

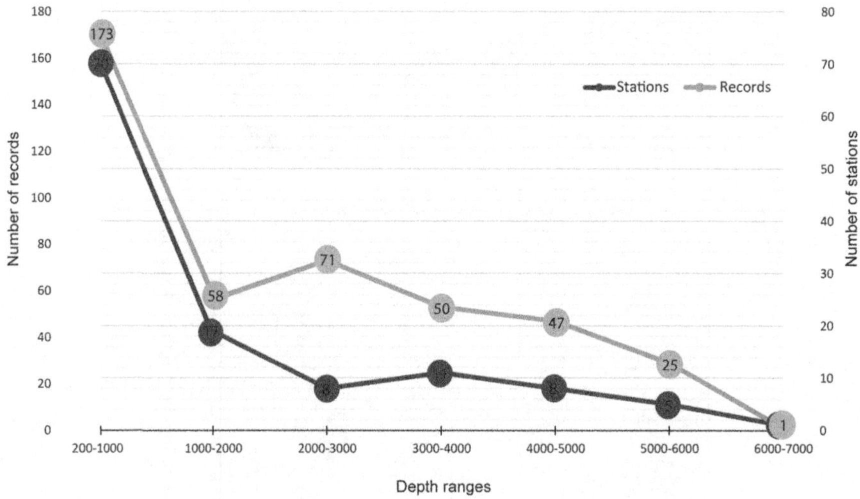

Fig. 7.4 Total number of stations and total number of records from the study area, sorted by depth ranges. For stations falling into two depth ranges, the lower depth was applied. Two stations with missing depth data and one with a broad depth range (0–2700 m) were omitted

Data for the isopods, cumaceans, and amphipods from the study area came from 24 expeditions/vessels (Table 7.1). Although the highest number of stations (23) corresponded to the expedition of the RRS William Scoresby, the highest number of species (58) was registered by the RV Atlantis II 60 (Table 7.2). Remarkably, no amphipods have been reported from the latter expedition. A total of 204 species of Isopoda, Cumacea, and Amphipoda were compiled from the study area. Of these, 47 (23.0%) species were found exclusively in this area; however, since the deep-sea bottoms are under-sampled, these species are not necessarily endemic ones.

Many species extend both in shallow and deep waters. Of the 204 species recorded, 51 (25.0%) were also found on the continental shelf adjacent to the study area (Table 7.2). In addition, remarkably wide distribution ranges have been reported for some species (Table 7.2). In particular, 48 (23.5%) of the 204 species recorded from the study area are also known from the North Atlantic. About half of the study area overlaps with the Sub-Antarctic region. Most precisely, localities south of 43°S belong to the Sub-Antarctic region (De Broyer and Danis 2011). Considering this overlap, it is not surprising that many species reported from the study area were also known from other localities of the Antarctic/Sub-Antarctic region (see column A/SA in Table 7.2). However, it is noteworthy that 41 (20.1%) species reported from the study area north of 43°S were also found in the A/SA region (the SA sector of the study area was omitted in this calculation). Brief comments on each taxon are provided below.

7.3.2 *Isopoda*

Isopods were recorded from 86 stations (Table 7.1), beginning in the late nineteenth century with the Challenger Expedition (Beddard 1884, 1886). In terms of the number of species, the RV Vema and the RV Atlantis II 60 were the most successful expeditions (22 spp. recorded in each expedition), followed by the Soviet Antarctic Expedition (17 spp.) (Table 7.2). The highest number of stations (23) was reported by the RRS William Scoresby; however, only a few serolids (4 spp.) and valviferans (5 spp.) were obtained by this expedition (Tables 7.1 and 7.2). A total of 107 species was recorded from the study area (Table 7.2); Asellota being the most diverse suborder (63 spp., 13 families), followed by Valvifera (22 spp., 7 families), Sphaeromatidea (15 spp., 2 families), and Cymothoida (7 spp., 5 families). Among the Asellota, the largest number of species was found in Munnopsidae (24), but since this family is highly speciose in deep waters, this result is not unexpected.

Among the 107 species recorded, only 10 were found in five or more stations. In contrast, 64 (59.8%) species were found only at single stations (Table 7.2). Of the 86 stations analyzed, 47 corresponded to depth between 200 and 1000 m (48 spp. recorded), 21 between 1000 and 3000 m (33 spp.), and 18 below 3000 m depth (44 spp.). *Acutiserolis neaera* presented the broader bathymetric distribution, i.e., from 239 m to 3730 m depth. The families Sphaeromatidae, Chaetiliidae, Rectarcturidae, Pseudidotheidae, Xenarcturidae, Janiridae, and Joeropsididae were reported

Table 7.2 Species of isopods, cumaceans, and amphipods recorded from the study area

	Survey/ship (stations)	1° square numbers	References	<200 m	A / SA	Others
Isopoda						
Asellota						
Acanthaspidiidae						
Iolanthe longiramosa (Kussakin & Vasina, 1982)	Ak. Kurchatov 11 (926, 927, 929)	52, 51, 50	Vasina and Kussakin (1982) and Kussakin and Vasina (1984)	0	G	0
Iolanthe mucronata (Menzies & Schultz, 1968)	Eltanin (6-340, 7-557)	56, 45	Menzies and Schultz (1968)	0	0	0
Echinothambemathidae						
Vemathambema argentinensis (Malyutina , Wägele & Brenke, 2001)	Ak. Kurchatov (43-4893)	8	Malyutina et al. (2001)	0	0	0
Haplomunnidae						
Thylakogaster majusculus (Wilson & Hessler, 1974)	Atlantis II 60 (247, 259)	23, 12	Wilson and Hessler (1974)	0	0	0
Haploniscidae						
Antennuloniscus dimeroceras (Barnard, 1920)	Vema (15-128)	25	Menzies (1962)	0	0	1
Antennuloniscus ornatus (Menzies, 1962)	Vema (15-117)	69	Menzies (1962)	0	1	1
Haploniscus tridens (Menzies, 1962)	Vema (15-128)	25	Menzies (1962)	0	0	1
Ischnomesidae						
Gracilimesus hanseni (Kavanagh, Wilson & Power, 2006)	Atlantis II 60 (245)	6	Kavanagh et al. (2006)	0	0	0
Stylomesus elegans (Menzies, 1962)	Vema (15-124)	37	Menzies (1962)	0	0	1
Stylomesus inermis (Vanhöffen, 1914)	Vema (15-126)	32	Menzies (1962)	0	1	1
Stylomesus simplex Menzies, 1962	Vema (15-128)	25	Menzies (1962)	0	0	1

Janiridae						
Iathrippa longicauda (Chiltom, 1884)	Ob (480)	21	Kussakin (1967)	1	1	1
Iathrippa sarsi (Pfeffer, 1887)	Swedish Ant. Exp. (34b)	24	Nordenstam (1933)	1	1	1
Neojaera antarctica (Pfeffer, 1887)	Swedish Ant. Exp. (34b)	24	Nordenstam (1933)	1	1	1
Joeropsididae						
Joeropsis curvicornis (Nicolet, 1849)	Eltanin (6-340, 11-969)	56, 59	Menzies and Schultz (1968)	1	1	1
Joeropsis intermedius (Nordenstam, 1933)	Eltanin (11-969, 11-970)	59, 60	Menzies and Schultz (1968)	1	1	1
Munnidae						
Astrurus ornatus (Vanhöffen, 1914)	Ak. Kurchatov (11-927)	51	Kussakin and Vasina (1984)	0	1	G
Munna argentinae (Menzies, 1962)	Vema (15-117)	69	Menzies (1962)	0	0	0
Munnopsidae						
Acanthocope argentinae (Menzies, 1962)	Ak. Kurchatov (43-4893)	8	Malyutina (1999a)	0	0	1
Bellibos (Bellibos) buzwilsoni (Haugsness & Hessler, 1979)	Atlantis II 60 (245)	6	Haugsness and Hessler (1979)	0	0	1
Betamorpha fusiformis (Barnard, 1920)	Atlantis II 60 (242, 245, 247, 256, 259), Vema (15-116)	15, 6, 23,12, 69	Thistle and Hessler (1977) and Menzies (1962)	0	1	1
Betamorpha megalocephalis (Thistle & Hessler, 1977)	Atlantis II 60 (245, 256)	6, 12	Thistle and Hessler (1977)	0	1	1
Coperonus comptus (Wison, 1989)	Atlantis II 60 (236, 237, 239, 245)	6	Wilson (1989)	0	0	0
Disconectes antarcticus (Vanhöffen, 1914)	Vema (15-117)	69	Menzies (1962)	0	1	0
Dubinectes infirmus (Malyutina & Brandt, 2011)	Diva III (532, 534)	3, 8	Malyutina and Brandt (2011)	0	0	0
Dubinectes intermedius (Malyutina & Brandt, 2006)	Diva III (532, 533, 534)	3, 8	Malyutina and Brandt (2011)	0	1	1

(continued)

Table 7.2 (continued)

Dubinectes nodosus (Menzies, 1962)	Diva III (533)	8	Malyutina and Brandt (2011)	0	1	1
Eurycope vicarius (Vanhöffen, 1914)	Vema (15-117)	69	Menzies (1962)	0	1	0
Ilyarachna antarctica (Vanhöffen, 1914)	Atlantis II 60 (237, 239, 242, 245, 264)	6, 15, 7	Thistle (1980)	0	1	1
Ilyarachna argentina (Thistle, 1980)	Atlantis II 60 (245, 256, 259, 262)	6, 12, 7	Thistle (1980)	0	0	1
Ilyarachna triangulata (Menzies, 1962)	Atlantis II 60 (242)	15	Thistle (1980)	0	1	1
Munneurycope antarctica (Schultz, 1977)	Eltanin (6-375, 22-1525)	57, 53	Schultz (1978)	0	1	0
Storthyngurella digitata (Menzies, 1962)	Ak. Kurchatov (43-4893), D. Mendeleev (4109), Vema (15-128)	8, 16, 25	Malyutina (1999b) and Menzies (1962)	0	0	0
Storthyngurella menziesi (Malyutina, 1999)	D. Medeleev (4109)	16	Malyutina (1999b)	0	1	0
Storthyngurella triplispinosa (Menzies, 1962)	Vema (15-128)	25	Menzies (1962)	0	1	1
Sursumura angulata (Malyutina, 2003)	Ak. Kurchatov (11-926)	52	Malyutina and Brandt (2004)	0	1	0
Syneurycope heezeni (Menzies, 1962)	Vema (15-116, 15-117)	69	Menzies (1962) and Haugsness and Hessler (1979)	0	1	1
Syneurycope parallela (Hansen, 1916)	Atlantis II 60 (242, 245, 256)	15, 6, 12	Haugsness and Hessler (1979)	0	1	1
Tytthocope divae (Malyutina & Brandt, 2014)	Diva III (532, 533, 534)	3, 8	Malyutina and Brandt (2014)	0	0	0
Vanhoeffenura birsteini (Menzies, 1962)	Eltanin (11-973), Vema (15-114, 15-118)	69	George and Menzies (1968) and Menzies (1962)	0	1	0
Vanhoeffenura georgei (Malyutina, 2003)	D. Medeleev (4109)	16	Malyutina (2003)	0	0	1

Vanhoeffenura scotia (George & Menzies, 1968)	Eltanin (6-350)	71	George and Menzies (1968)	0	1	0
Nannoniscidae						
Hebefustis hexadentium (Siebenaller & Hessler, 1977)	Atlantis II 60 (247)	23	Siebenaller and Hessler (1977)	0	0	0
Hebefustis par (Siebenaller & Hessler, 1977)	Atlantis II 60 (256)	12	Siebenaller and Hessler (1977)	0	0	1
Nannoniscoides gigas (Siebenaller & Hessler, 1977)	Atlantis II 60 (256)	12	Siebenaller and Hessler (1977)	0	0	0
Nannoniscus oblongus (Sars, G.O., 1890)	Vema (15-128)	25	Menzies (1962)	0	0	1
Paramunnidae						
Abyssianira acutilobi (Doti & Roccatagliata, 2006)	Atlantis II 60 (245)	6	Doti and Roccatagliata (2006)	0	0	0
Abyssianira argentenensis (Menzies, 1962)	Vema (15-123)	42	Menzies (1962)	1	0	0
Abyssianira dentifrons (Menzies, 1956)	Atlantis II 60 (247)	23	Doti and Roccatagliata (2006)	0	0	1
Neasellus argentinensis (Doti, 2016)	DINARA (w/n), Talud I (12)	10, 11	Doti (2016)	0	0	0
Neasellus bicarinatus (Doti, 2016)	COPLA II (3, 11)	13, 19	Doti (2016)	1	0	0
Neasellus kerguelensis (Beddard, 1885)	Challenger (320)	11	Beddard (1886)	0	1	0
Stenetriidae						
Tenupedunculus pulchrum (Schultz, 1982)	Vema (15-131, 15-132)	18, 17	Schultz (1982)	0	0	0
Tenupedunculus virginale (Schultz, 1982)	Eltanin (6-340)	56	Schultz (1982)	0	1	0
Stenetriidae *incertae sedis* (after Song et al., 2018)						
Tenupedunculus acutum (Vanhöffen, 1914)	Vema (17-101)	13	Schultz (1982)	1	1	1
Tenupedunculus beddardi (Kussakin, 1967)	Ob (479, 480)	28, 21	Kussakin (1967)	0	1	0
Tenupedunculus dentimanum (Kussakin, 1967)	Ak. Kurchatov 11 (929, 930), Ob (479, 480)	50, 44, 28, 21	Kussakin and Vasina (1984), Kussakin (1967) and Schultz (1982)	0	0	0
Tenupedunculus drakensis (Schultz, 1982)	Hero (895)	60	Schultz (1982)	0	0	0

(continued)

Table 7.2 (continued)

Tenupedunculus haswelli (Beddard, 1886)	Challenger (320)	11	Beddard (1886) and Schultz (1982)	0	0	0
Tenupedunculus inflectofrons (Schultz, 1982)	Eltanin (6-339, 6-340), Vema (18-8)	56, 6	Schultz (1982)	1	0	0
Tenupedunculus serraticaudum (Kussakin & Vasina, 1984)	Ak. Kurchatov (11-926)	52	Kussakin and Vasina (1984)	0	0	0
Tenupedunculus smirnovi (Vasina, 1982)	Sund (136)	29	Vasina (1982)	0	0	0
Urstylidae						
Urstylis zapiola (Riehl, Wägele & Brenke, 2014)	Atlantis II 60 (247)	23	Riehl et al. (2014)	0	0	0
Cymothoidea						
Anthuridae						
Cyathura profunda (Kensley, 1982)	Atlantis II 60 (236, 237, 245, 262, 264)	6, 7	Kensley (1982)	0	0	0
Hyssuridae						
Hyssura vimsae (Kensley, 1978)	Atlantis II 60 (237)	6	Kensley (1982)	0	0	1
Leptanthuridae						
Leptanthura argentinae (Kensley, 1982)	Atlantis II 60 (237, 242, 245, 259, 262)	6, 15, 12, 7	Kensley (1982)	0	1	0
Leptanthura glacialis (Hodgson, 1910)	Atlantis II 60 (237, 242, 247, 256)	6, 15, 23, 12	Kensley (1982)	0	1	0
Paranthuridae						
Paranthura argentinae (Kussakin, 1967)	Atlantis II 60 (236), Ob (480)	6, 21	Kensley (1982) and Kussakin (1967)	0	0	0
Paranthura possessia (Kensley, 1980)	Atlantis II 60 (236, 237, 245, 262, 264)	6, 7	Kensley (1982)	0	1	0
Dajidae						
Notophryxus sp.	Eltanin (6-375)	57	Schultz (1978)			

Sphaeromatidea						
Serolidae						
Acanthoserolis schythei (Lütken, 1858)	W. Scoresby (76, 214, 215, 216, 229, 234, 236, 237, 244, 245, 772, 818)	43, 33, 30, 38, 29, 47, 46	Sheppard (1933)	1	1	1
Acutiserolis margaretae (Menzies, 1962)	Vema (15-116)	69	Menzies (1962)	0	0	0
Acutiserolis neaera (Beddard, 1884)	Challenger (318, 320); W. Scoresby (212, 213, 236, 244, 773, 820, 821, 839); Ob (477)	20, 11, 35, 29, 47, 30, 54, 49, 55, 33	Beddard (1886), Menzies (1962), Sheppard (1933) and Kussakin (1967)	0	0	0
Atlantoserolis vemae (Menzies, 1962)	Diva III (532, 533, 534)	3, 8	Brandt et al. (2014)	0	0	1
Brucerolis bromleyana (Suhm, 1876)	Eltanin (4-107, 6-353, 11-973), Vema (15-117)	70, 71, 69	GBIF (Menzies R. and Moreira P.S.) and Menzies (1962)	0	G	1
Brucerolis maryannae (Menzies, 1962)	Vema (15-117)	69	Menzies (1962)	0	0	0
Glabroserolis occidentalis (Brandt & Brix, 2014)	Diva III (533, 534)	8	Brandt et al. (2014)	0	0	0
Neoserolis exigua (Nordenstam, 1933)	W. Scoresby (212, 214, 215, 216, 227, 228, 229, 234, 236, 237, 246, 766, 773, 818, 821)	35, 33, 30, 45, 39, 38, 29, 48, 27, 46, 49	Sheppard (1933)	1	G	1
Paraserolis polita (Pfeffer, 1887)	Swedish Ant. Exp. (34b)	24	Nordenstam (1933)	1	1	0
Septemserolis septemcarinata (Miers, 1875)	Swedish Ant. Exp. (34b)	24	Nordenstam (1933)	0	1	0
Serolis kempi (Sheppard, 1933)	W. Scoresby (244, 245, 246, 818); Ob (480)	47, 46, 48, 21	Sheppard (1933) and Kussakin (1967)	1	0	0
Sphaeromatidae						
Caecocassidias patagonica (Kussakin, 1967)	Ob (479, 480)	28, 21	Kussakin (1967)	0	0	0

(continued)

Table 7.2 (continued)

Cassidinopsis tuberculata (Schultz, 19780	Eltanin (22-1521)	66	Schultz (1978)	0	0	0
Moruloidea darwini (Cunningham, 1871)	Ob (480)	21	Kussakin (1967)	1	1	0
Waiteolana tuberculata (Kussakin, 1967)	Ob (479)	28	Kussakin (1967)	0	0	0
Valvifera						
Antarcturidae						
Acantharcturus brevipleon (Kussakin & Vasina, 1998)	Ak. Kurchatov (11-928)	51	Kussakin and Vasina (1998)	0	0	0
Chaetarcturus aculeatus (Kussakin, 1967)	Ob (480)	21	Kussakin (1967)	0	1	0
Chaetarcturus franklini (Hodgson, 1902)	Ob (480)	21	Kussakin (1967)	0	1	0
Chaetarcturus tenuispinatus (Kussakin & Vasina, 1998)	Ak. Kurchatov (11-930)	44	Kussakin and Vasina (1998)	0	0	0
Fissarcturus granulosus (Nordenstam, 1933)	Ob (480)	21	Kussakin (1967)	0	1	0
Fissarcturus paxillaris (Kussakin & Vasina, 1998)	D. Medeleev (4109)	16	Kussakin and Vasina (1998)	0	0	0
Glaberarcturus stellae (Kussakin & Vasina, 1998)	Ak. Kurchatov (11-926)	52	Kussakin and Vasina (1998)	0	0	0
Litarcturus americanus (Beddard, 1886)	Ob (479)	28	Kussakin (1967)	1	0	0
Mixarcturus abnormis (Kussakin, 1967)	Ob (480)	21	Kussakin (1967)	0	1	0
Oxyarcturus dubius (Kussakin, 1967)	Ob (480)	21	Kussakin (1967)	0	0	0
Xiphoarcturus carinatus (Pereira, Roccatagliata & Doti, 2019)	Talud I (12), Talud III (39)	10	Pereira et al. (2019)	0	0	0
Xiphoarcturus kussakini (Pereira, Roccatagliata & Doti, 2019)	Talud I (12), Talud II (36), Talud III (59)	10	Pereira et al. (2019)	0	0	0
Austrarcturellidae						
Dolichiscus anna (Beddard, 1886)	Challenger (320), Ob (480)	11, 21	Beddard (1886) and Kussakin (1967)	0	0	0
Dolichiscus georgei (Kussakin & Vasina, 1980	Sund (w/n); Ak Kurchatov 11(929, 930)	31, 50, 44	Kussakin and Vasina (2001)	0	0	0

Dolichiscus marinae (Kussakin & Vasina, 2001)	Sund (w/n)	31	Kussakin and Vasina (2001)	0	0	0
Chaetiliidae						
Macrochiridothea stebbingi (Ohlin, 1901)	W. Scoresby (772)	30	Sheppard (1957)	1	0	0
Idoteidae						
Edotia abyssalis (Pereira & Doti, 2017)	Talud III (46)	14	Pereira and Doti (2017)	0	0	0
Edotia bilobata (Nordenstam, 1933)	W. Scoresby (245, 871)	46, 58	Sheppard (1957)	1	0	0
Edotia oculata (Ohlin, 1901)	W. Scoresby (215)	30	Sheppard (1957)	1	0	0
Pseudidotheidae						
Pseudidothea miersii (Studer, 1884)	DINARA (w/n), W. Scoresby (212, 214, 766, 818, 839, 871), Ob (480)	10, 35, 33, 27, 46, 55, 58, 21	Doti et al. (2008), Sheppard (1957) and Kussakin (1967)	1	1	0
Rectarcturidae						
Rectarcturus kophameli (Ohlin, 1901)	Hero (880)	60	Schultz (1981)	1	0	0
Xenarcturidae						
Xenarcturus spinulosus (Sheppard, 1957)	W. Scoresby (237)	29	Sheppard (1957)	1	0	0
Cumacea						
Bodotriidae						
Apocuma brasiliense (Jones, 1973)	Atlantis II 60 (239)	6	Roccatagliata et al. (2012)	0	0	1
Bathycuma elongatum (Hansen, 1895)	Atlantis II 60 (240, 242, 243, 245, 246, 247, 256, 259, 262)	7, 6, 12, 15, 23	GBIF (Jones N.S.)	0	0	1
Cyclaspis longicaudata (Sars, 1865)	Atlantis II 60 (239, 240, 242, 245, 262, 264)	7, 6, 15,	Roccatagliata (1989) and GBIF (Jones N.S.)	0	0	1
Ceratocumatidae						
Cimmerius reticulatus (Jones, 1973)	Atlantis II 60 (245, 262)	7, 6	GBIF (Jones N.S.)	0	0	G
Diastylidae						

(continued)

Table 7.2 (continued)

Austrostylis paradivacumata (Mühlenhardt-Siegel, 2005)	Diva III (532, 533, 534)	3, 8	Mühlenhardt-Siegel (2015)	0	0	1
Diastylis hammoniae (Zimmer, 1902)	EH 04/01	19	Alberico and Roccatagliata (2011)	1	1	0
Holostylis helleri (Zimmer, 1907)	Atlantis II 60 (246), Eltanin (6-339)	12, 56	GBIF (Jones N.S., Blazewicz-Paszkowycz M., Heard R.W.)	0	1	0
Holostylis uniramosa (Roccatagliata & Alberico, 2016)	COPLA II (3, 11), Talud I (6)	9, 13, 19	Roccatagliata and Alberico (2016)	1	0	0
Leptostylis dentantipus (Mühlenhardt-Siegel, 2016)	Diva III (532)	3	Mühlenhardt-Siegel (2018)	0	1	0
Leptostylis lacunatus species complex (Mühlenhardt-Siegel, 2016)	Diva III (527, 532, 533, 534, 535)	3, 8	Mühlenhardt-Siegel (2018)	0	1	1
Leptostyloides secundus (Mühlenhardt-Siegel, 2015)	Diva III (532, 533, 534)	3, 8	Mühlenhardt-Siegel (2015)	0	0	0
Makrokylindrus tubulicauda (Calman, 1905)?	Atlantis II 60 (243)	12	GBIF (Jones N.S.)	0	0	1
Pseudoleptostyloides mediosetosus (Mühlenhardt-Siegel, 2015)	Diva III (532, 533)	3, 8	Mühlenhardt-Siegel (2015)	0	1	1
Lampropidae						
Chalarostylis brenkei (Mühlenhardt-Siegel, 2005)	Diva III (533, 534)	8	Alberico et al. (2014)	0	1	1
Leuconidae						
Bytholeucon hiscens (Bishop, 1981)	Atlantis II 60 (242, 247, 264)	7, 15, 23	Bishop (1981b)	0	0	1
Eudorella hispida (Sars, 1871)	Vema (17-101)	13	Petrescu (1991)	0	0	1
Leucon (Crymoleucon) sagitta (Zimmer, 1907)	Vema (15-131, 17-80, 17-81)	18, 22, 26	Petrescu (1991)	1	1	0
Leucon (Crymoleucon) septemdentatus (Zimmer, 1902)	Vema (17-53, 17-101)	13, 68	Petrescu (1994)	1	1	0

Leucon (Epileucon) cf. *tenuirostris* Sars, 1887, form B Bishop (1981)	Atlantis II 60 (245, 262, 264)	7, 6	Bishop (1981a)	0	0	0
Leucon (Leucon) assimilis (Sars, 1887)	Vema (14-12, 17-59)	48, 62	Petrescu (1994)	1	1	0
Leucon (Leucon) homorhynchus (Bishop, 1981)	Atlantis II 60 (242, 256)	12, 15	Bishop (1981b)	0	0	1
Leucon (Leucon) serratus (Norman, 1879)	Atlantis II 60 (242, 256)	12, 15	Shalla and Bishop (2004)	0	0	1
Nannastacidae						
Atlantocuma benguelae (Băcescu & Muradian, 1974)	Atlantis II 60 (247), Vema (17-81)	23, 26	Jones (1984) and Petrescu (1995)	0	0	1
Atlantocuma tenue (Jones, 1984)	Atlantis II 60 (239, 242, 245)	6, 15	Jones (1984)	0	0	1
Campylaspis bacescui (Muradian, 1976)	Atlantis II 60 (237, 239) Vema (14-12)	6, 48	Jones (1984) and Muradian (1976)	1	1	0
Campylaspis bicarinata (Jones, 1974)	Atlantis II 60 (237, 239, 243, 245, 246, 256)	6, 12	Jones (1984)	0	0	1
Campylaspis exarata (Jones, 1974)	Atlantis II 60 (245)	6	Jones (1984)	0	0	1
Campylaspis frigida (Hansen, 1908)	Atlantis II 60 (237, 239, 240, 245); Vema (14-12, 15-117, 15-131,17-53, 17-54, 17-101)	6, 13, 18, 48, 68, 69	Jones (1984) and Muradian-Ciamician (1980)	1	1	0
Campylaspis guttata (Jones, 1974)	Atlantis II 60 (237)	6	GBIF (Jones N.S.)	0	0	1
Campylaspis johnstoni (Hale, 1937)	Atlantis II 60 (237, 239, 240, 242, 245, 262), Vema (18-12)	7, 6, 15, 30	Jones (1984) and Muradian (1976)	0	1	1
Campylaspis laevigata (Jones, 1974)	Atlantis II 60 (236)	6	Jones (1984)	0	0	1
Campylaspis multinodosa (Jones, 1974)	Atlantis II 60 (259)	12	Jones (1984)	0	0	1
Campylaspis nodulosa (Sars, 1887)	Atlantis II 60 (240, 245, 246, 247) Vema (15-131, 17-101)	6, 12, 13, 18, 23	Jones (1984) and Muradian-Ciamician (1980)	1	1	0
Campylaspis nuda (Jones, 1974)	Atlantis II 60 (240, 264)	7, 6	Jones (1984)	0	0	1

(continued)

Table 7.2 (continued)

Campylaspis pilosa (Jones, 1974)	Atlantis II 60 (239)	6	Jones (1984)	0	0	1
Campylaspis quadriplicata (Lomakina, 1968)	Atlantis II 60 (240, 245) Vema (15-131)	6, 18	Jones (1984) and Muradian (1976)	0	1	0
Campylaspis sticta (Jones, 1974)	Atlantis II 60 (240)	6	Jones (1984)	0	0	1
Campylaspis valleculata (Jones, 1974)	Atlantis II 60 (243, 246, 247, 256, 259, 262, 264)	7, 12, 23	Jones (1984) and GBIF (Jones N.S.)	0	0	1
Cumella argentinae (Jones, 1984)	Atlantis II 60 (280), Vema (17-59, 17-101)	6, 13, 62	Jones (1984) and Petrescu (1995)	1	1	0
Cumella decipiens (Jones, 1984)	Atlantis II 60 (259)	12	Jones (1984)	0	0	1
Cumella jonesi (Petrescu, 2005)	Vema (17-81)	26	Petrescu (1995)	0	0	0
Cumella meridionalis (Jones, 1984)	Atlantis II 60 (237, 239, 240, 243, 245, 246, 259, 262, 264)	7, 6, 12	Jones (1984)	0	1	1
Cumellopsis bicostata (Jones, 1984)	Atlantis II 60 (239)	6	Jones (1984)	0	0	1
Platycuma holti (Calman, 1905)	Atlantis II 60 (239, 240, 262, 264)	7, 6	Jones (1984)	0	0	1
Procampylaspis procurrens (Jones, 1984)	Atlantis II 60 (239, 240, 245, 262, 264)	7, 6	Jones (1984)	0	0	1
Schizocuma spinosum (Jones, 1984)	Atlantis II 60 (262)	7	Jones (1984)	0	0	1
Styloptocuma bishopi (Jones, 1984)	Vema (17-53)	68	Petrescu (2000a, b)	0	0	1
Styloptocuma dayae (Jones, 1984)	Atlantis II 60 (245)	6	Jones (1984) and Petrescu (2000b)	0	0	1
Styloptocuma formosum (Jones, 1984)	Atlantis II 60 (245)	6	Jones (1984) and Petrescu (2000b)	0	0	0
Styloptocuma gracillimum (Calman, 1905)	Atlantis II 60 (239)	6	Jones (1984)	0	1	1
Amphipoda						
Amphilochoidea						
Sebidae						
Seba saundersii (Stebbing, 1875)	Hero (893), W. Herwig (245)	5, 60	Holman and Watling (1983a) and Alonso (2012)	1	1	1

Seba cf. *subantarctica* (Schellenberg, 1931)	Hero (893)	60	Holman and Watling (1983a)	1	1	0
Seba typica (Chilton, 1884)	Hero (893)	60	Holman and Watling (1983a)	1	0	1
Calliopioidea						
Pontogeneiidae						
Eusiroides crassi (Stebbing, 1888)	Challenger (320)	11	Stebbing (1888)	0	1	1
Eusiroides monoculoides (Haswell, 1879)	Eltanin (11-969)	59	GBIF (Gray S.)	1	1	1
Caprelloidea						
Caprellidae						
Aeginoides gaussi (Schellenberg, 1926)	Vema (17-65), Eltanin (22-1592)	40, 65	McCain and Gray (1971); GBIF (Guerra-García J.)	0	1	0
Caprella penantis (Leach, 1814)	Eltanin (11-969)	59	McCain and Gray (1971)	1	1	1
Caprellinoides mayeri (Pfeffer, 1888)	Vema (17-51)	67	McCain and Gray (1971)	1	1	0
Dodecas elongata (Stebbing, 1883)	Vema (18-12)	30	McCain and Gray (1971)	1	1	0
Dodecasella elegans (K.H. Barnard, 1931)	Eltanin (6-339)	56	GBIF (Guerra-García J.)	G	1	0
Dodecasella georgiana (Schellenberg, 1931)	Vema (15-131)	18	McCain and Gray (1971)	0	1	0
Mayerella magellanica (McCain & Gray, 1971)	Vema (18-9)	6	McCain and Gray (1971)	1	1	0
Protella trilobata (McCain and Gray, 1971)	Eltanin (7-558)	45	McCain and Gray (1971)	0	0	0
Dexaminoidea						
Lepechinellidae						
Lepechinella cachi (J.L. Barnard, 1973)	Vema (15-118)	69	Barnard (1973)	0	1	0
Lepechinella huaco (J.L. Barnard, 1973)	Eltanin (6-350)	71	Barnard (1973)	0	1	0
Eusiroidea						
Eusiridae						
Harcledo curvidactyla (Pirlot, 1929)	Eltanin (6-350)	71	GBIF (Gray S.)	0	1	1

(continued)

Table 7.2 (continued)

Hadzioidea						
Maeridae						
Elasmopus delaplata (Stebbing, 1888)	Challenger (320)	11	Stebbing (1888)	0	0	0
Haustorioidea						
Urothoidae						
Urothoe falcata (Schellenberg, 1931)	W. Scoresby (76)	43	Barnard (1932)	1	1	0
Iphimedioidea						
Amathillopsidae						
Parepimeria irregularis (Schellenberg, 1931)	Vema (15-131)	18	GBIF (Gray S.)	1	1	0
Dikwidae						
Dikwa andresi (Lörz & Coleman, 2003)	LAMPOS (145-1, 150-1), CAMBIO (208-2)	61, 64	Lörz and Coleman (2003) and d'Udekem d'Acoz and Verheye (2017)	0	0	0
Epimeriidae						
Epimeria (Metepimeria) acanthurus (Schellenberg, 1931)	Terra Nova Exp. (38), Hero (894), LAMPOS (153-1)	48, 60, 64	Barnard (1930); Watling and Holman (1981) and d'Udekem d'Acoz and Verheye (2017)	1	1	0
Iphimediidae						
Iphimedia imparilabia (Watling & Holman, 1980)	Eltanin (6-340)	56	Watling and Holman (1980)	0	1	0
Iphimedia magellanica (Watling & Holman, 1980)	Eltanin (11-977)	46	Watling and Holman (1980)	1	0	0
Iphimedia multidentata (Schellenberg, 1931)	Hero (894)	60	Watling and Holman (1981)	1	0	0
Pseudiphimediella glabra (Schellenberg, 1931)	Eltanin (11-977)	46	Watling and Holman (1980)	1	0	0

Stilipedidae						
Alexandrella polarsterni (Berge & Vader, 2005)	Vema (15-131)	18	GBIF (Berge J. and Vader W.)	0	1	0
Alexandrella schellenbergi (Holman & Watling, 1983)	Swedish Ant. Exp. (Shag quadrant)	41	Schellenberg (1931) and Holman and Watling (1983b)	0	0	0
Astyra antarctica (Andres, 1997)	Vema (15-131)	18	GBIF (Berge J.)	0	1	0
Liljeborgioidea						
Liljeborgiidae						
Liljeborgia georgiana (Schellenberg, 1931)	Vema (15-117)	69	GBIF (d'Udekem d'Acoz C.)	1	1	0
Liljeborgia homospora (d'Udekem d'Acoz, 2008)	Vema (15-131)	18	GBIF (d'Udekem d'Acoz C.)	0	1	0
Liljeborgia macrodon (Schellenberg, 1931)	Hero (894)	60	Holman and Watling (1983a)	1	0	0
Liljeborgia octodentata (Schellenberg, 1931)	Eltanin (11-977)	46	Holman and Watling (1983a)	1	G	0
Lysianassoidea						
Cyphocarididae						
Cyphocaris richardi (Chevreux, 1905)	Eltanin (6-381)	72	GBIF (Hurley D.E.)	0	1	1
Eurytheneidae						
Eurythenes maldoror (d'Udekem d' Acoz & Havermans, 2015)	Diva III (531)	4	d'Udekem d'Acoz and Havermans (2015)	0	1	1
Tryphosidae						
Pseudorchomene debroyeri (d'Udekem d'Acoz & Havermans, 2012)	ICEFISH (14), LAMPOS (147-1)	56, 65	d'Udekem d'Acoz and Havermans (2012)	0	1	0
Tryphosites chevreuxi (Stebbing, 1914)	Knipovich (249)	1	Alonso (2012)	1	1	1
Uristidae						

(continued)

Table 7.2 (continued)

Abyssorchomene abyssorum (Stebbing, 1888)	Challenger (323)	2	Stebbing (1888) and Thurston and Allen (1969)	0	G	G
Uristes schellenbergi (Lowry & Bullock, 1976)	Knipovich (249)	1	Alonso (2012)	1	1	1
Oedicerotoidea						
Oedicerotidae						
Oediceroides cinderella (Stebbing, 1888)	Challenger (317)	34	Stebbing (1888)	0	1	1
Pagetinoidea						
Pagetinidae						
Pagetina reducta (Holman & Watling, 1981)	Eltanin (11-977)	46	Holman and Watling (1981)	0	0	0
Stegocephaloidea						
Stegocephalidae						
Andaniotes linearis (K.H. Barnard, 1932)	Eltanin (7-558, 11-977, 22-1593), Hero (894)	45, 46, 60, 64	Watling and Holman (1981) and Berge et al. (2000)	1	1	0
Metandania tordi (Berge & Vader, 2003)	Eltanin (6-350)	71	Berge and Vader (2003)	0	1	0
Parandania boecki (Stebbing, 1888)	Swedish Ant. Exp. (-), Eltanin (6-348)	36, 63	Schellenberg (1931) and GBIF (Berge J.)	0	1	1
Tetradeion crassum (Chilton, 1883)	Eltanin (22-1593)	64	Berge et al. (2000)	0	1	1
Synopioidea						
Synopiidae						
Bruzelia poton (J.L. Barnard, 1972)	Eltanin (6-350)	71	Barnard (1972)	0	0	0
Syrrhoe serrima (J.L. Barnard, 1972)	Vema (15-131)	18	Barnard (1972)	0	0	G
Amphipoda *incertae sedis*						
Sanchoidae						
Chosroes incisus (Stebbing, 1888)	Terra Nova Exp. (38)	48	Barnard (1930)	1	1	0

exclusively in less than 1000 m depth, whereas the families Echinothambemathidae, Haplomunnidae, Urstylidae, and most of the Munnopsidae were reported exclusively in waters deeper than 3000 m depth.

Of the 107 species, 39 were found exclusively in the study area, whereas 20 were also recorded in the continental shelf adjacent to the study area. In addition, 43 species were recorded from Antarctic and Sub-Antarctic waters. Finally, many species show a wide geographic distribution, e.g., 15 species recorded from the study area were also found in the North Atlantic (Table 7.2).

7.3.3 Cumacea

The cumaceans from off the Argentine coast had not received any attention until the second half of the last century. As in isopods, the highest number of species was collected by the RV Atlantis II 60 (36 spp.) and the RV Vema (13 spp.). In addition, the DIVA III expedition reported six species, and the RV Eltanin 6, RV Holmberg, COPLA II, and Talud I expeditions one species each (Table 7.2).

Cumaceans were retrieved from 34 stations, and the largest number of stations were visited by the RV Atlantis II 60 and the RV Vema (14 and 10 stations, respectively) (Table 7.1). A total of 50 species was recorded from the study area. The family Nannastacidae includes 28 species (56.0% of the species reported; see Table 7.2). Regarding the families Diastylidae and Leuconidae, 9 and 8 species were reported, respectively. In turn, for the family Bodotriidae, 3 species were reported, and for families Lampropidae and Ceratocumatidae, one species each. The family Nannastacidae has benefited from the comprehensive taxonomic revision published by the late Dr. Norman S. Jones (Port Erin, Isle of Man), who reported most of the cumaceans collected by the RV Atlantis II 60 (see Jones 1984). Concerning the Bodotriidae, as most of its members occurred in warm/temperate waters at depths of less than 200 m (Day 1975, 1978), the low number of species reported from the study area is not surprising. In contrast, being the Lampropidae a typical cold/deep-water family, the single species registered might be the result of the little attention paid to the Cumacea from the study area. However, Mühlenhardt-Siegel (2011) reported that Lampropidae was one of the least speciose families in Antarctic deep-sea basins. Therefore, lampropids are not expected to be very diverse in the study area.

Of the 50 species recorded, 19 (38.0%) were found only at a single station (Table 7.2). Of the 34 stations analyzed, 12 were located between 200 and 1000 m depth (16 spp. recorded), 8 between 1000 and 3000 m (29 spp.), and 14 below 3000 m depth (26 spp.). Thus, the depth range with the lowest number of stations has the highest number of species. Nine species reported from the study area were also found on the adjacent continental shelf. Many of the species listed have a large range of distribution, i.e., 27 species were found on both the South and the North Atlantic. In contrast, just four species have not been found outside the study area. However, as vast extensions of the Atlantic remain unexplored, these four species cannot be referred to as endemic. Finally, 17 species found off the Argentine coast were also reported from Antarctic and Sub-Antarctic waters.

7.3.4 Amphipoda

A total of 47 amphipod species were recorded from the study area, which were collected at 35 stations carried out by 13 oceanographic expeditions (Tables 7.1 and 7.2). Almost half of the 35 stations reported were sampled by the RV Eltanin and the RV Vema expeditions (10 and 7 stations, respectively). These research vessels also collected the highest number of species, 18 and 12, respectively (Tables 7.1 and 7.2).

The 47 amphipods recorded belong to 22 families and 35 genera. The family with the highest number of species was Caprellidae (8 spp.), followed by Liljeborgiidae and Iphimediidae (4 spp. each). As Caprellidae is not a highly diverse family in the deep sea (De Broyer et al. 2004), the relatively high number of species reported from the study area seems to be the result of the exhaustive taxonomic revision prepared by McCain and Gray (1971) (Table 7.2). Of the 35 stations reported, 22 were located between 200 and 1000 m depth (27 spp. recorded), 9 between 1000 and 3000 m depth (17 spp.), and 4 below 3000 m depth (4 spp.). Most of the species collected came from the shelf break zone, i.e., of the 47 species recorded, as many as 20 were collected exclusively between 200 and 500 m depth and just 9 below 1500 m depth.

Most of the amphipods reported have single records, i.e., 40 of the 47 species recorded were found only once (Table 7.2). This result reflects the little attention paid to the amphipods from off the Argentine coast. Certainly, the amphipod diversity is underestimated. Regarding the geographical distribution, of the 47 amphipod recorded, just 5 species were reported exclusively from the study area, whereas 22 were also recorded from the adjacent continental shelf. In addition, 35 (74.5%) species are also known from the Antarctic/Sub-Antarctic waters, and 15 species show a wide geographic distribution (Table 7.2).

7.4 Discussion

Our knowledge on the deep-sea isopod, cumacean, and amphipod species from off the Argentine coast is quite incomplete. The scarcity in taxonomic information became evident when the number of species reported from this area is compared with nearby deep-sea localities, such as the Atlantic sector of the Southern Ocean and the Angola Basin (see Brandt et al. 2005, 2007). The lack of knowledge on regional deep-sea peracarids is linked to several issues. One problem is the wide areas that still remain under-sampled. Another impediment is the scarce number of trained taxonomists working on this fauna (see also Coleman 2015). The fact that most of the species reported from the study area are deposited in museums mainly in the USA and Europe also represents a significant inconvenience at the time of identifying the local fauna.

The Atlantic II 60 has been the most successful survey carried out in the Argentine deep-sea floor during the last century. A high number of isopod and cumacean species were reported by this survey, e.g., as many as 27 species were retrieved from a single station (Sta. 245). The isopods from this survey were studied by Joseph F. Siebeneller and Robert R. Hessler (nannoniscids), David Thistle (ilyarachnids), Brian Kensley (anthuriids), and George D. F. Wilson (haplomunnids and some munnopsids), among others. The cumaceans were studied mainly by Norman S. Jones (1984), who focused on the nannastacids. In contrast, nobody has studied the amphipods from this survey.

Many of the species recorded from the study area have a wide distribution range; i.e., 48 of the species recorded from this area are also known from the North Atlantic. Molecular studies on peracarids, however, have recently documented that some widely distributed species are indeed assemblages of cryptic species (e.g., Brandt et al. 2012; Raupach and Wägele 2006; Raupach et al. 2007; Brökeland and Raupach 2008; d'Udekem d'Acoz and Havermans 2015). Some of the 48 species recorded from both the South and North Atlantic may be complexes of cryptic species as well. Hence, the status of these widespread species should be confirmed based on morphological and molecular evidence.

The finding of as many as 122 singletons (species recorded only once) in the study area may be an artifact of under-sampling (only 24.6% of the one-degree squares shown on the maps have been sampled; see Fig. 7.1). Another possibility is that singletons reflect a patchy distribution of these species. It should be noted that a high number of singletons were also reported from other deep-sea areas (see Brandt et al. 2007; Brix et al. 2018a).

Recent studies have identified submarine canyons as potential hotspots of biodiversity (Tyler et al. 2009; De Leo et al. 2010; De Leo and Puig 2018, among others). Along the Argentine Continental Margin, four main submarine canyon systems have been reported (Bozzano et al. 2017); however, their biodiversity remains almost unknown. To investigate this fauna, the surveys TALUD I-III were carried out to the Mar del Plata submarine canyon by the RV Puerto Deseado in 2012 and 2013. A large number of isopods, cumaceans, and amphipods have been collected during these surveys, and at least 89 species, many of them new to science, have been gathered from a single station (unpubl. data); thus, many new species and new records from this submarine canyon will be published in the near future.

Finally, it should be noticed that the scattered records and the dissimilar sampling effort performed in the study area prevent investigating any potential latitudinal diversity gradients (LSDGs). Furthermore, due to the scarce knowledge on South Atlantic deep-sea isopods, LSDGs analyses are restricted to the species inhabiting the North Atlantic (see Rex et al. 1993; Gage et al. 2004). Besides, Watling (2009), in an attempt to define biogeographic provinces in the deep Atlantic based on the deep-sea cumacean distribution, failed to separate South-Atlantic provinces, once again because of the relatively few species recorded in the southern Atlantic. Further sampling programs and exhaustive taxonomic works are greatly needed to overcome this situation.

Acknowledgments We are grateful to Michel E. Hendrickx (UNAM) for his invitation to participate with the present chapter of this book. We are also thankful to Ute Mühlenhardt-Siegel (CeNak), Saskia Brix (DZMB), and Anna Jażdżewska (University of Lodz), whose comments and suggestions allowed us to improve the final version of this chapter. This work was partially founded by the Universidad de Buenos Aires (UBACyT 2014-2017 20020130100711BA) and the Consejo Nacional de Investigaciones Científicas y Técnicas (PIP 11220130100434CO and PIP 11220170100643CO).

References

Alberico NA, Roccatagliata D (2011) On two species of *Diastylis* (Cumacea) from the Southwestern Atlantic: *D. sexpectinata* n. sp. and *D. hammoniae* Zimmer, 1902. J Crust Biol 31(1):192–208

Alberico NA, Roccatagliata D, Mühlenhardt-Siegel U (2014) Remarks on the deep-sea genus *Chalarostylis* (Cumacea: Lampropidae). Zootaxa 3753(4):323–334

Alonso GM (2012) Amphipod crustaceans (Corophiidea and Gammaridea) associated with holdfasts of *Macrocystis pyrifera* from the Beagle Channel (Argentina) and additional records from the Southwestern Atlantic. J Nat Hist 29–30:1799–1899

Barnard KH (1930) Crustacea. Part XI. Amphipoda. British Antarctic "Terra Nova" expedition 1910. Nat Hist Rep Zool 8:307–454

Barnard KH (1932) Amphipoda. Discovery Rep 5:1–326

Barnard JL (1972) A review of the family Synopiidae (=Tironidae), mainly distributed in the deep sea (Crustacea: Amphipoda). Smithson Contrib Zool 124:1–94

Barnard JL (1973) Deep-sea Amphipoda of the genus *Lepechinella* (Crustacea). Smithson Contrib Zool 133:1–31

Beddard FE (1884) Report on the Isopoda collected by H.M.S. Challenger during the years 1873–76. Part 1. The genus *Serolis*. Challenger Rep Zool 11:1–85

Beddard FE (1886) Report on the Isopoda collected by H.M.S. Challenger during the years 1873–76. Part 2. Challenger Rep Zool 17:1–178

Berge J, Vader W (2003) *Metandania tordi*, a new stegocephalid (Crustacea: Peracarida: Amphipoda) species from the Southern Ocean. Proc Biol Soc Wash 116:986–999

Berge J, De Broyer C, Vader W (2000) Revision of the Antarctic and sub-Antarctic species of the family Stegocephalidae (Crustacea: Amphipoda) with description of two new species. Bull Inst R Sci Nat Belg 70:217–233

Bishop JDD (1981a) A revised definition of the genus *Epileucon* Jones (Crustacea, Cumacea), with descriptions of species from the deep Atlantic. Philos Trans R Soc Lond Ser B Biol Sci 291(1052):353–409

Bishop JDD (1981b) Two new leuconids (Peracarida, Cumacea) of widespread occurrence in the deep Atlantic. Crustaceana 40(2):144–159

Bozzano G, Martín J, Spoltore DV, Violante RA (2017) Los cañones submarinos del Margen Continental Argentino: una síntesis sobre su génesis y dinámica sedimentaria. Lat Am J Sedimentol Basin Anal 24(1):85–101

Brandt A, Brenke N, Andres HG, Brix S, Guerrero-Kommritz J, Mühlenhardt-Siegel U, Wägele JW (2005) Diversity of peracarid crustaceans (Malacostraca) from the abyssal plain of the Angola Basin. Org Divers Evol 5(1):105–112

Brandt A, Brix S, Brökeland W, Choudhury M, Kaiser S, Malyutina M (2007) Deep-sea isopod biodiversity, abundance, and endemism in the Atlantic sector of the Southern Ocean-results from the ANDEEP I-III expeditions. Deep-Sea Res II 54(16–17):1760–1775

Brandt A, Błażewicz-Paszkowycz M, Bamber RN, Mühlenhardt-Siegel U, Malyutina MV, Kaiser S, De Broyer C, Havermans C (2012) Are there widespread peracarid species in the deep sea (Crustacea: Malacostraca)? Pol Polar Res 33(2):139–162

Brandt A, Brix S, Held C, Kihara TC (2014) Molecular differentiation in sympatry despite morphological stasis: deep-sea *Atlantoserolis* Wägele, 1994 and *Glabroserolis* Menzies, 1962 from the south-west Atlantic (Crustacea: Isopoda: Serolidae). Zool J Linnean Soc 172(2):318–359

Brandt A, Frutos I, Bober S, Brix S, Brenke N, Guggolz T, Heitland N, Malyutina M, Minzlaff U, Riehl T, Schwabe E, Zinkann A-C, Linse K (2018) Composition of abyssal macrofauna along the Vema fracture zone and the hadal Puerto Rico trench, northern tropical Atlantic. Deep-Sea Res II 148:35–44

Brix S, Stransky B, Malyutina M, Pabis K, Svavarsson J, Riehl T (2018a) Distributional patterns of isopods (Crustacea) in Icelandic and adjacent waters. Mar Biodivers 48:783–811

Brix S, Lörz A-N, Jazdzewski A, Hughes L, Tandberg AHS, Pabis K, Stransky B, Krapp-Schickel T, Sorbe JC, Hendrycks E, Vader W, Frutos I, Horton T, Jazdzewski K, Peart R, Beermann J, Coleman CO, Buhl-Mortensen L, Corbari L, Havermans C, Tato R, Jimenez Campean A (2018b) Amphipod family distribution around Iceland. ZooKeys 731:1–53

Brökeland W, Raupach MJ (2008) A species complex within the isopod genus *Haploniscus* (Crustacea: Malacostraca: Peracarida) from the Southern Ocean deep sea: a morphological and molecular approach. Zool J Linnean Soc 152:655–706

Coleman CO (2015) Taxonomy in times of the taxonomic impediment – examples from the community of experts on amphipod crustaceans. J Crust Biol 36(6):729–740

d'Udekem d'Acoz C, Havermans C (2012) Two new *Pseudorchomene* species from the Southern Ocean, with phylogenetic remarks on the genus and related species (Crustacea: Amphipoda: Lysianassoidea: Lysianassidae: Tryphosinae). Zootaxa 3310:1–50

d'Udekem d'Acoz C, Havermans C (2015) Contribution to the systematics of the genus *Eurythenes* S.I. Smith in Scudder, 1882 (Crustacea: Amphipoda: Lysianassoidea: Eurytheneidae). Zootaxa 3971:1):1–1)80

d'Udekem d'Acoz C, Verheye ML (2017) *Epimeria* of the Southern Ocean with notes on their relatives (Crustacea, Amphipoda, Eusiroidea). Eur J Taxon 359:1–553

Day J (1975) South African Cumacea. Part 1. Family Bodotriidae, subfamily Vaunthompsoniinae. Ann S Afr Mus 66(9):177–220

Day J (1978) Southern African Cumacea. Part 2. Family Bodotriidae, subfamily Bodotriinae. Ann S Afr Mus 75(7):159–290

De Broyer C, Danis B (2011) How many species in the Southern Ocean? Towards a dynamic inventory of the Antarctic marine species. Deep-Sea Res II 58:5–17

De Broyer C, Guerra-García J, Takeuchi I, Robert H, Meerhaeghe A (2004) Biodiversity of the Southern Ocean: a catalogue of the Antarctic and subAntarctic Caprellidae and Cyamidae (Crustacea: Amphipoda) with distribution and ecological data. Bull Inst R Sci Nat Belg, Biol 74:61–99

De Broyer C, Lowry JK, Jażdżewski K, Robert H (2007) Catalogue of the Gammaridean and Corophiidean Amphipoda (Crustacea) of the Southern Ocean with distribution and ecological data. In: De Broyer C (ed) Census of Antarctic marine life: synopsis of the Amphipoda of the Southern Ocean. Bull Inst R Sci Nat Belg, Biol 77(Suppl 1):1–35

De Leo FC, Puig P (2018) Bridging the gap between the shallow and deep oceans: the key role of submarine canyons. Prog Oceanogr 169:1–5

De Leo FC, Smith CR, Rowden AA, Bowden DA, Clark MR (2010) Submarine canyons: hotspots of benthic biomass and productivity in the deep sea. Proc R Soc B 277:2783–2792

Doti BL (2016) Three new paramunnids (Isopoda: Asellota: Paramunnidae) from the Argentine Sea, Southwest Atlantic. J Mar Biol Assoc UK 97(8):695–1709

Doti BL, Roccatagliata D (2006) On the Atlantic species of the genus *Abyssianira* Menzies, 1956 (Isopoda: Asellota: Paramunnidae). Zootaxa 1252:1–35

Doti BL, Roccatagliata D, Scarabino F (2008) Range extension of *Pseudidothea miersii* (Studer, 1884) (Isopoda, Valvifera) and new evidence of its likely commensal relationship with a sea urchin. Crustaceana 81(7):883–888

Doti BL, Roccatagliata D, López Gappa J (2014) An inverse latitudinal biodiversity pattern in asellote isopods (Crustacea, Peracarida) from the Southwest Atlantic between 35° and 56°S. Mar Biodivers 44(1):115–125

Gage J, Lambshead PJD, Bishop JDD, Stuart CT, Jones NS (2004) Large-scale biodiversity pattern of Cumacea (Percarida: Crustacean) in the deep Atlantic. Mar Ecol Prog Ser 277:181–196
GBIF.org (2019) The Global Biodiversity Information Facility. Available from: https://www.gbif.org [13 December 2019]
George RY, Menzies RJ (1968) Species of *Storthyngura* (Isopoda) from the Antarctic with descriptions of six new species. Crustaceana 14(3):275–301
Golovan OA, Blażewicz-Paszkowycz M, Brandt A, Budnikova LL, Elsner NO, Ivin VV, Lavrenteva AV, Malyutina MV, Petryashov VV, Tzareva LA (2013) Diversity and distribution of peracarid crustaceans (Malacostraca) from the continental slope and the deep-sea basin of the sea of Japan. Deep-Sea Res II 86–87:66–78
Golovan OA, Błażewicz M, Brandt A, Jażdżewska AM, Jóźwiak P, Lavrenteva AV, Malyutina MV, Petryashov VV, Riehl T, Sattarova VV (2019) Diversity and distribution of peracarid crustaceans (Malacostraca) from the abyss adjacent to the Kuril-Kamchatka trench. Mar Biodivers 49:1343
Haugsness JA, Hessler RR (1979) A revision of the subfamily Syneurycopinae (Isopoda: Asellota: Eucopidae) with a new genus and species (*Bellibos buzwilsoni*). Trans San Diego Soc Nat Hist 19:121–151
Hessler RR, Sanders HL (1967) Faunal diversity in the deep-sea. Deep-Sea Res 14:65–78
Hessler RR, Wilson GDF, Thistle D (1979) The deep-sea isopods: a biogeographic and phylogenetic overview. Sarsia 64:67–75
Holman H, Watling L (1981) *Pagetina reducta* sp. n. (Crustacea: Amphipoda) with a review of the family Pagetinidae. Sarsia 66:213–215
Holman H, Watling L (1983a) Amphipoda from the Southern Ocean: families Colomastigidae, Dexaminidae, Leucothoidae, Liljeborgiidae, and Sebidae. Antarct Res Ser 38:215–262
Holman H, Watling L (1983b) A revision of the Stilipedidae (Amphipoda). Crustaceana 44(1):29–53
Jones NS (1984) The family Nannastacidae (Crustacea: Cumacea) from the deep Atlantic. Bull Br Mus Nat Hist Zool 46(3):207–289
Jones NS, Sanders HL (1972) Distribution of Cumacea in the deep Atlantic. Deep-Sea Res 19:737–745
Kavanagh FA, Wilson GDF, Power AM (2006) Heterochrony in *Haplomesus* (Crustacea: Isopoda: Ischnomesidae): revision of two species and description of two new species. Zootaxa 1120:1–33
Kensley B (1982) Deep-water Atlantic Anthuridea (Crustacea: Isopoda). Smithson Contrib Zool 346(346):1–60
Kussakin OG (1967) Fauna of Isopoda and Tanaidacea in the coastal zones of the Antarctic and Subantarctic waters. Biol Rep Soviet Antarct Exped (1955–1958) 3:220–389
Kussakin OG, Vasina GS (1984) Deep-sea lower asellotes from the Scotia Sea and South Sandwich Trench. Biol Morya 6:9–17
Kussakin OG, Vasina GS (1998) New bathyal and abyssal arcturids from the western Antarctic and Subantarctic (Crustacea: Isopoda: Arcturidae). Zoosyst Ross 7(1):55–75
Kussakin OG, Vasina GS (2001) Supplements for the fauna of Arcturid Isopods of the genus *Dolichiscus* (Crustacea, Isopoda, Valvifera) of the Antarctic and Subantarctic. Russ J Mar Biol 24(4):208–217
López Gappa J, Alonso GM, Landoni NA (2006) Biodiversity of benthic Amphipoda (Crustacea: Peracarida) in the Southwest Atlantic between 35°S and 56°S. Zootaxa 1342:1–66
Lörz AN, Coleman CO (2003) *Dikwa andresi*, a new amphipod crustacean (Dikwidae) from the Scotia Arc. J Mar Biol Assoc UK 83:785–788
Malyutina MV (1999a) New species of *Acanthocope* (Isopoda, Munnopsidae). Russ J Mar Biol 25(4):320–329
Malyutina MV (1999b) *Storthyngurella*, new genus of Munnopsidae (Crustacea: Isopoda), with descriptions of three new species from deep-sea basins of the Southern Hemisphere. Mem Mus Vic 57(2):267–285
Malyutina MV (2003) Revision of *Storthyngura* Vanhöffen, 1914 (Crustacea: Isopoda: Munnopsididae) with descriptions of three new genera and four new species from the deep South Atlantic. Org Divers Evol 3(13):1–101

Malyutina M, Brandt A (2004) Storthyngurinae (Isopoda, Asellota, Munnopsidea) from the Antarctic deep sea with the descriptions of three new species. Beaufortia 54:1):1–1)38

Malyutina M, Brandt A (2011) *Dubinectes infirmus*, a new species of deep-water Munnopsidae (Crustacea, Isopoda, Asellota) from the Argentine Basin, South Atlantic Ocean. ZooKeys 144:1–19

Malyutina M, Brandt A (2014) New species of the deep-sea munnopsid genus *Tytthocope* (Crustacea, Isopoda, Asellota) from the South Atlantic and the Southern Ocean. Zootaxa 3786:1):1–1)43

Malyutina MV, Wägele JW, Brenke N (2001) New records of little known deep-sea Echinothambematidae (Crustacea: Isopoda: Asellota) with redescription of *Vemathambema elongata* Menzies, 1962 and description of a new species from the Argentina Basin. Org Divers Evol 1(6):1–28

McCain JC, Gray WS Jr (1971) Antarctic and Subantarctic Caprellidae (Crustacea: Amphipoda). Antarct Res Ser 17:111–139

Menzies RJ (1962) The isopods of abyssal depths in the Atlantic Ocean. Abyssal Crustacea Vema Res Ser 1:79–206

Menzies RJ, Schultz GA (1968) Antarctic isopod Crustacea. II. Families Haploniscidae, Acanthaspidiidae, and Jaeropsidae, with diagnoses of new genera and species. Biol Antarct Seas III Antarct Res Ser 11:141–184

Mühlenhardt-Siegel U (2011) Cumacean (Peracarida, Crustacea) endemism and faunal overlap in Antarctic deep-sea basins. Deep-Sea Res II 58(1–2):68–73

Mühlenhardt-Siegel U (2015) Some remarks on selected diastylid genera. Part I: *Leptostyloides*, *Divacuma*, *Austrostylis* n. gen. and *Pseudoleptostyloides* n. gen. (Crustacea, Cumacea, Diastylidae) from the deep South Atlantic. Mar Biodivers 45(2):287–310

Mühlenhardt-Siegel U (2018) Some remarks on selected diastylid genera. Part II: *Leptostylis* (Crustacea, Cumacea) from the deep southern Atlantic and Pacific, with the description of six new species. Mar Biodivers 48(1):465–516

Muradian Z (1976) Species belonging to the genus *Campylaspis* (Cumacea, Nannastacidae) collected from the western Atlantic. Trav Mus Hist Nat "Gr Antipa" 17:65–83

Muradian-Ciamician Z (1980) On some species belonging to the genus *Campylaspis* (Cumacea, Nannastacidae) from the collections of the Natural History Museum "Grigore Antipa". Trav Mus Hist Nat "Gr Antipa" 21:73–88

Nordenstam Å (1933) Marine Isopoda of the families Serolidae, Idotheidae, Pseudidotheidae, Arcturidae, Parasellidae and Stenetriidae mainly from South Atlantic. Further Zool Res Swed Antarct Exped 1901–1903 3:1–284

Pereira E, Doti BL (2017) *Edotia abyssalis* n. sp. from the Southwest Atlantic Ocean, first record of the genus (Isopoda, Valvifera, Idoteidae) in the deep sea. Zool Anz 268:19–31

Pereira E, Roccatagliata D, Doti BL (2019) *Xiphoarcturus* – a new genus and two new species of the family Antarcturidae (Isopoda: Valvifera) from the Mar del Plata submarine canyon and its phylogenetic relationships. Arthropod Syst Phylo 77(2):303–323

Petrescu I (1991) Contribution to the knowledge of genus *Eudorella* Norman, 1867 (Crustacea, Cumacea, Leuconidae) with description of two new species: *Eudorella bacescui* n.sp. and *Eudorella menziesi* n.sp. Trav Mus Hist Nat "Gr Antipa" 31:375–385

Petrescu I (1994) Contributions to the knowledge of *Leucon* genus (Crustacea: Cumacea) from the waters of South America. Trav Mus Hist Nat "Gr Antipa" 34:325–345

Petrescu I (1995) Cumaceans (Crustacea: Peracarida) from the south American coasts collected by the R/V "Vema". Trav Mus Hist Nat "Gr Antipa" 35:49–86

Petrescu I (2000a) Additional data on some deep-sea Nannastacidae (Crustacea: Cumacea) collected by R/V "Vema". Trav Mus Hist Nat "Gr Antipa" 42:55–74

Petrescu I (2000b) Remarks on genera *Schizocuma* Băcescu, 1972 and *Styloptocuma* Băcescu & Muradian, 1974 (Crustacea, Cumacea). Bull Nat Hist Mus, London (Zool) 66(2):147–166

Raupach M, Wägele JW (2006) Distinguishing cryptic species in Antarctic Asellota (Crustacea: Isopoda) – a preliminary study of mitochondrial DNA in *Acanthaspidia drygalskii*. Antarct Sci 18:191–198

Raupach MJ, Malyutina M, Brandt A, Wägele JW (2007) Molecular data reveal a highly diverse species flock within the munnopsoid deep-sea isopod *Betamorpha fusiformis* (Barnard, 1920) (Crustacea: Isopoda: Asellota) in the Southern Ocean. Deep-Sea Res II 54:1820–1830

Rex MA, Stuart CT, Hessler RR, Allen JA, Sanders HL, Wilson GDF (1993) Global-scale latitudinal patterns of species diversity in the deep-sea benthos. Nature 365(6447):636–639

Riehl T, Wilson GDF, Malyutina MV (2014) Urstylidae – a new family of abyssal isopods (Crustacea: Asellota) and its phylogenetic implications. Zool J Linnean Soc 170:245–296

Roccatagliata D (1989) The genus *Cyclaspis* (Cumacea): a new species and additional records of distribution. Crustaceana 56(1):39–46

Roccatagliata D, Alberico N (2016) Two new cumaceans (Crustacea: Peracarida) from the south-West Atlantic with remarks on the problematic genus *Holostylis* Stebbing, 1912. Mar Biodivers 46(1):163–181

Roccatagliata D, Alberico NA, Heard RW (2012) *Apocuma* (Cumacea: Bodotriidae): two new species from the West-Atlantic and a significant extension of the known distribution of this genus in the Atlantic Ocean. Zootaxa 3436:24–40

Sanders HL, Hessler RR (1969) Ecology of the deep-sea benthos. Science 163(3874):1419–1424

Schellenberg A (1931) Gammariden und Caprelliden des Magellangebietes, Südgeorgiens und der Westantarktis. Further Zool Res Swed Antarct Exped 1901–1903 2:1–290

Schultz GA (1978) More planktonic isopod crustaceans from Subantarctic and Antarctic seas. Biol Antarct Seas VII Antarct Res Ser 27:69–89

Schultz GA (1981) Arcturidae from the Antarctic and Subantarctic seas (Isopoda, Valvifera) part 1. Biol Antarct seas X. Antarct Res Ser 32:63–94

Schultz GA (1982) Species of Protallocoxoidea and Stenetrioidea (Isopoda, Asellota) from the Antarctic and southern seas. Biol Antarct Seas X Antarct Res Ser 32:17–62

Shalla SH, Bishop JDD (2004) Four new species of the genus *Leucon* (Crustacea: Cumacea) from the Atlantic frontier margin. J Mar Biol Assoc UK 84(1):139–153

Sheppard EM (1933) Isopod Crustacea, part I. the family Serolidae. Discovery Rep 7:253–362

Sheppard EM (1957) Isopod Crustacea, part II. The sub-order Valvifera. Families: Idoteidae, Pseudidotheidae and Xenarcturidae fam. n. with a supplement to isopod Crustacea, part 1. The family Serolidae. Discovery Rep 29:141–198

Siebenaller JF, Hessler RR (1977) The Nannoniscidae (Isopoda, Asellota): *Hebefustis* n. gen. and *Nannoniscoides* Hansen. Trans San Diego Soc Nat Hist 19(2):17–44

Song JH, Bruce NL, Min GS (2018) The first records of *Stenobermuda* Schultz, 1982 and *Tenupedunculus* Schultz, 1979 from Australia, with description of two new species from the Great Barrier Reef (Isopoda, Asellota, Stenetriidae). ZooKeys 733:1–24

Stebbing TRR (1888) Report on the Amphipoda collected by H.M.S. Challenger during the years 1873–1876. Challenger Rep Zool 29:1–1737

Thistle D (1980) A revision of *Ilyarachna* (Crustacea, Isopoda) in the Atlantic with four new species. J Nat Hist 14:111–143

Thistle D, Hessler RR (1977) A revision of *Betamorpha* (Isopoda: Asellota) in the world ocean with three new species. Zool J Linnean Soc 60:275–295

Thurston MH, Allen E (1969) Type material of the families Lysianassidae, Stegocephalidae, Ampeliscidae and Haustoriidae (Crustacea: Amphipoda) in the collections of the British Museum (Natural History). Bull Br Mus Nat Hist Zool 17:349–388

Tyler P, Amaro T, Arzola R, Cunha MR, De Stigter H, Gooday A, Huvenne V, Ingels J, Kiriakoulakis K, Lastras G, Masson D, Oliveira A, Pattenden A, Vanreusel A, Van Weering T, Vitorino J, Witte U, Wolff G (2009) Europe's grand canyon Nazaré submarine canyon. Oceanography 22(1):46–57

Vasina GS (1982) New species of the genus *Stenetrium* (Crustacea, Isopoda, Stenetriidae) from macrophytes of the Patagonian shelf. In: Kafanov AI, Kussakin OG (eds) Fauna and distribution of Crustacea in Southern and Antarctic waters. Akademiya Nauk CCCP, Dal'nevostochnyi Nauchyi Tsenter (Far East Science Centre), Vladivostok, pp 106–109 [In Russian]

Vasina GS, Kussakin OG (1982) Four new species of the isopod crustacean genus *Acanthaspidia* (Crustacea: Isopoda: Janiridae) from Antarctica. In: Kafanov AI, Kussakin OG (eds) Fauna and distribution of Crustacea in Southern and Antarctic waters. Akademiya Nauk CCCP, Dal'nevostochnyi Nauchyi Tsenter (Far East Science Centre), Vladivostok, pp 62–72 [In Russian]

Watling L (2009) Biogeographic provinces in the Atlantic deep sea determined from cumacean distribution patterns. Deep-Sea Res II 56:1747–1753

Watling L, Holman H (1980) New Amphipoda from the Southern Ocean, with partial revisions of the Acanthonotozomatidae and Paramphithoidae. Proc Biol Soc Wash 93:609–654

Watling L, Holman H (1981) Additional acanthonotozomatid, paramphithoid and stegocephalid Amphipoda from the Southern Ocean. Proc Biol Soc Wash 94:181–227

Wilson GDF (1989) A systematic revision of the deep-sea subfamily Lipomerinae of the isopod crustacean family Munnopsidae. Bull Scripps Inst Oceanogr 27:1–138

Wilson GD, Hessler RR (1974) Some unusual Paraselloidea (Isopoda, Asellota) from the deep benthos of the Atlantic. Crustaceana 27(1):47–67

WoRMS Editorial Board (2020) World Register of Marine Species. Available from https://www.marinespecies.org at VLIZ. Accessed 27 Feb 2020. https://doi.org/10.14284/170

Chapter 8
Benthic Invertebrate Communities in the Continental Margin Sediments of the Monterey Bay Area

J. S. Oliver, K. K. Hammerstrom, L. A. Kuhnz, P. N. Slattery, J. M. Oakden, and S. L. Kim

Abstract We observed high diversity (species density) of infaunal invertebrates from the mid-shelf (50 m) to the upper slope (325 m) with high abundance and low dominance along 4 depth transects ranging from 10 to 2000 m. The highest shallow water diversity recorded worldwide was at the shelf-slope break (109–150 m) with a peak of 185 species 0.1 m^{-2} (449 m^{-2}, this work and Oliver JS, Hammerstrom K, McPhee-Shaw E, Slattery P, Oakden J, Kim S, Hartwell SI, Mar Ecol 32:278–288, 2011). The peak included a large number of species and individuals of small nestling amphipods. The most abundant genera were *Photis*, *Aoroides*, and *Gammaropsis*. The percentages of crustacean species (40%) and individuals (60%) were extremely high as well. A large caprellid amphipod, *Tritella tenuissima*, was the most abundant animal found in our survey (387 individuals 0.1 m^{-2}), and characterized the mixed gravel bottoms on the upper slope. We discovered a dense tube mat of relatively large ampeliscid amphipods at the upper margin of the oxygen minimum zone (700 m). This was the most distinct community cluster and was dominated by *Ampelisca unsocalae* and *Byblis barbarensis*. The percentage of crustacean species (40%) and individuals (67%) were also highest here. This is the only ampeliscid tube mat known from deep water and is ecologically similar to extensive shallow-water ampeliscid mats in the Bering and Chukchi Seas, the primary feeding grounds of gray whales. The diverse and abundant continental margin communities occurred in a dynamic, current-swept upwelling center with complex topography. Diversity decreased below the upper slope and on the wave-swept inner shelf, where there was another dramatic crustacean pattern. These shifting sands were dominated by burrowing amphipods (phoxocephalids and haustoriids) and small ostracods in the 1970s that were rare in the present survey. In total, we collected 938 invertebrate species, including 431 polychaetes, 270 crustaceans, 171 mollusks, and 38 echinoderms in 123 samples. More than any other taxa, the crustaceans characterized the

J. S. Oliver · K. K. Hammerstrom (✉) · L. A. Kuhnz · P. N. Slattery · J. M. Oakden
S. L. Kim
Moss Landing Marine Laboratories, Moss Landing, CA, USA
e-mail: khammerstrom@mlml.calstate.edu

M. E. Hendrickx (ed.), *Deep-Sea Pycnogonids and Crustaceans of the Americas*, https://doi.org/10.1007/978-3-030-58410-8_8

most unique and interesting community patterns along the continental margin of the Monterey Bay area.

Keywords Benthic community patterns · California · Continental margin infauna · High diversity · High species density · Monterey Bay · USA

8.1 Introduction

Thorson (1957) was the first to emphasize the zonation of benthic invertebrates on the continental shelf. He called these parallel bottom communities, where often there is a faunal band or zone in wave swept sands that gradually grades to other faunal zones in finer sediments with increasing water depth. Along the west coast of North America, similar parallel bottoms or faunal zones were first described in Southern California (Barnard and Ziesenhenne 1961), where there has been extensive benthic sampling related to sewage discharges and, later, regional monitoring programs (Stull et al. 1986; Swartz et al. 1986; Ranasinghe et al. 2007). The taxonomic work and standardization that accompanied these surveys benefited benthic sampling to the south and far to the north. The work we present here would have been much more difficult and perhaps impossible without the taxonomy fueled by pollution surveys in the Southern California Bight.

Here we present the first extensive sampling of infaunal invertebrates along the continental margin of central and northern California. The continental margin includes the shelf, slope, and rise that fringe most ocean basins. Only the shallow shelf communities were previously described in Monterey Bay (Hodgson and Nybakken 1973; Oliver et al. 1980, 2008; Slattery 1985). Sampling around local sewer discharges was conducted in shallow water, was unpublished, and was recently terminated in favor of a regional sampling program focused in the center of the mud belt at 80 m (Oliver et al. 2011). The original quantitative descriptions provide the first baselines to compare with future ecological conditions. Of course this is a sliding baseline, but any baseline is better than none.

Telecommunication companies funded the present survey as part of a failed effort to bring fiber optic cables across the Pacific Ocean and into the Monterey Bay National Marine Sanctuary. This survey produced the first extensive ecological baselines for seafloor communities along our continental margin. Benthic invertebrate communities were described from 10 m to 2000 m, highlighted by the surprising contributions from the crustaceans, mostly amphipods, other peracarids, and a few species of ostracods.

8.2 Methods

Benthic invertebrate communities living in sedimentary environments (infauna) were sampled along four depth transects extending from the inner continental shelf to the deep slope (Figs. 8.1 and 8.2, Appendix 8.5.1). Each transect was a potential path for a submarine fiber optic cable. Hard substrates and areas with massive sediment movements like the axes of the Monterey Canyon were avoided because the cable needed to be buried into relatively stable sediment at a minimum depth of 50 cm. We selected sampling stations at regular depth intervals for comparisons among transects.

Benthic samples were taken with a Smith-McIntyre grab between June and August 1999. Five replicate grabs (each 0.1 m^2) were taken from each station for benthic invertebrate communities, and one grab was taken for sediment sampling. A box corer (0.1 m^2) was used to sample the deepest stations (1000–2000 m) in October 1999. Community samples were washed through 0.5 mm mesh Nitex

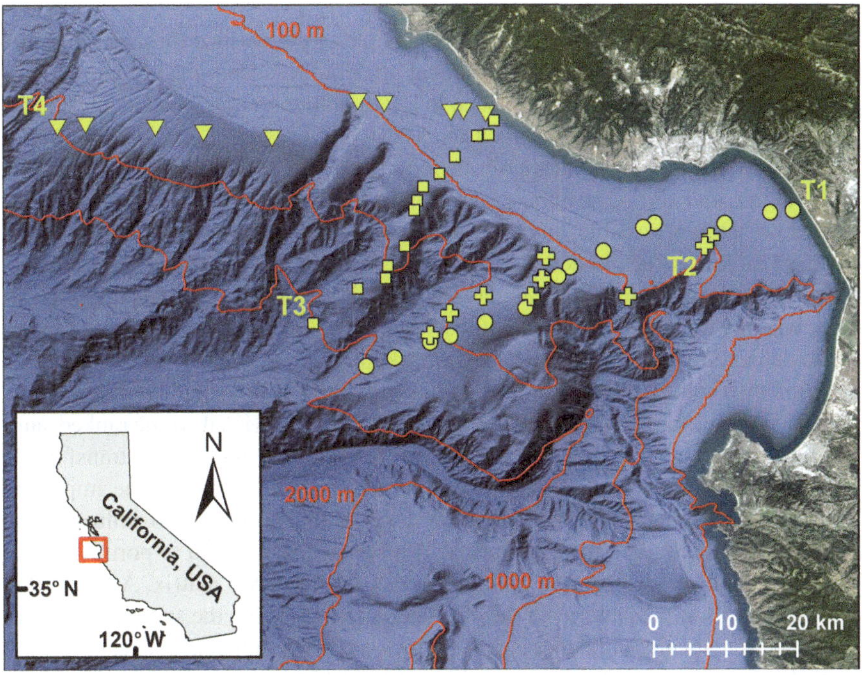

Fig. 8.1 Location of benthic sampling stations along four depth transects in the Monterey Bay area (Basemap: Google Earth 2019). Transect 1 (T1) = circles, Transect 2 (T2) = plus signs, Transect 3 (T3) = squares, and Transect 4 (T4) = triangles. 100, 1000, and 2000 m contour lines shown in red. Inset shows location of Monterey Bay in California

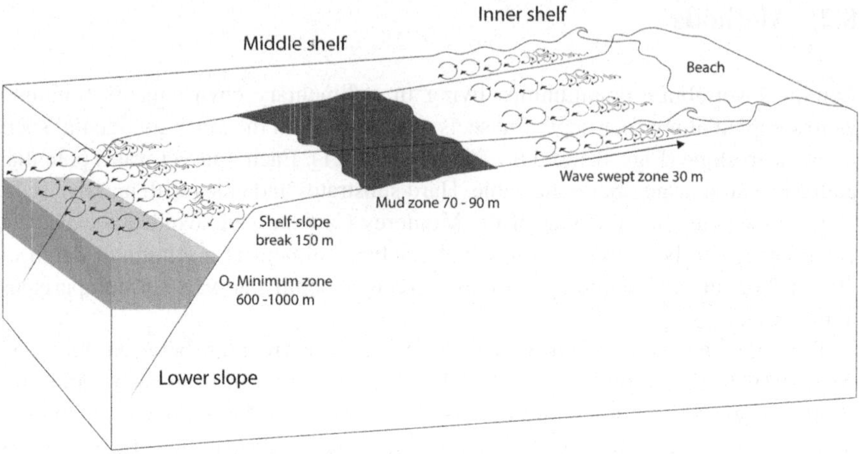

Fig. 8.2 Schematic of major sedimentary habitats along the continental margin of the Monterey Bay Area showing the inner shelf, mid-shelf and mud zone, shelf-slope break, upper slope, oxygen minimum zone, and lower slope. Areas of increased water motion are indicated by arrows

(nylon) sieves. Approximately 10% of the samples were washed through a 0.3 mm screen after the 0.5 mm screen to document potential loss from the 0.5 mm mesh.

Sieved infaunal samples were placed in a 7% solution of magnesium chloride for 2 h to relax the animals prior to fixation. They were preserved in a 4% solution of formaldehyde for 2 days and then transferred to 70% isopropyl alcohol for long-term storage and sorting. Animals were sorted from the screen residues, identified to the lowest taxon possible, and counted. Biomass of the main animal groups (crustaceans, mollusks, echinoderms, and other) was measured as total blotted wet weight on an electronic scale. Biomass of polychaete worms was measured by volume, which was converted to weight, because polychaetes are particularly vulnerable to damage from desiccation during the weighing process.

Primer v 6 was used to examine community patterns (Clarke and Warwick 2001). Non-metric multidimensional scaling (MDS) was used to visualize the ranked sample similarities using a Bray-Curtis similarity matrix and fourth-root transformed abundance data. Fourth root transformation was selected to lessen the impact of numerically dominant species and allow less abundant species to influence the resulting MDS plot patterns. Similarly, cluster analysis was used to portray groupings of samples as determined by the Bray-Curtis similarity matrix. A SIMPROF test identified sample patterns in the cluster analysis in which the species and abundance are statistically indistinguishable at the $\alpha = 0.05$ significance level.

Sediment grain size was measured by wet sieving and weighing surface sediments in general size categories (clay, silt, sand, gravel). These sediment samples were the top 2 cm of a 3 cm diameter core taken from the grab or box core sample designated for sediment sampling. Microscopic analysis of sediments larger than 63 microns was conducted to obtain results for glauconite pellet analysis.

Two ROVs were used to examine each transect as a potential path for the submarine cable, providing photographs of the general habitat as well as estimates of physical and biogenic structures and the abundance of megafaunal invertebrates. We took 10 replicate samples from ROV videos around each infaunal station along transects 1 and 2 from 30 to 2000 m (Fig. 8.1 and Appendix 8.5.1). Each replicate sample was 1 × 25 m (25 m^2). We counted megafaunal invertebrates and large pits and mounds (<15 cm in diameter) in each video sample. A Phantom DS4 ROV with a Benthos Model 387 35 mm camera and a Simrad Color Zoom Video camera (model OE1366) was towed along each transect from 25 to 500 m. A Remora ROV was used deeper than 500 m and was equipped with a Photosea 1000 35 mm camera, a Simrad 3500 YC video camera, a Simrad Color Zoom camera (model OE1366), a NTSC wide-angle black and white video camera, five spot lights, and 2 MW C-map lasers.

8.3 Results and Discussion

A total of 123 0.1 m^{-2} grab samples and 47 grain size samples were taken and processed from the four transects (Appendix 8.5.1). Not all depths were sampled on all transects because of logistical problems, storms, or firm bottoms that the grab could not adequately penetrate. The most complete sampling and sample processing was done on transect 1, because our main interest was central Monterey Bay (Appendix 8.5.1). We were unable to process completely all five replicate samples at each station, but we processed at least two replicates from each station. The number of grab samples processed from each station is shown in Appendix 8.5.1.

Samples sieved through a 0.3 mm sieve in addition to the 0.5 mm sieve demonstrated that we captured the vast majority of the infauna on the larger screen. Only a few small individuals were found on the 0.3 mm screen, which is typically used in deeper water where animal size is often smaller than along the shelf. Even the animals from the deepest water were relatively large along this high production, upwelling continental margin. In addition, the screen residues often contained large volumes of glauconite and woody debris, which made it difficult for smaller infauna to pass through the 0.5 mm mesh. All the data reported here are from the 0.5 mm screen.

8.3.1 Sediments

Sand dominated the inner shelf sediments (10–50 m) (Figs. 8.2 and 8.3). The mud belt was common from 60 to 90 m, often centered in 80 m. Silt and clay (mud) characterized the mud zone on the shelf and the slope sediments beyond the oxygen minimum zone (Figs. 8.2 and 8.3). Gravel was common mostly at the 325 m and 450 m depths (Fig. 8.3), where the grab was often unable to obtain good samples, requiring additional grab deployments until an adequate sample depth was obtained.

Soft, coarse, sand-sized pellets covered in a thin veneer of dark glauconite (Appendix 8.5.2.1) were a major component of the sediment from 450 to 1000 m

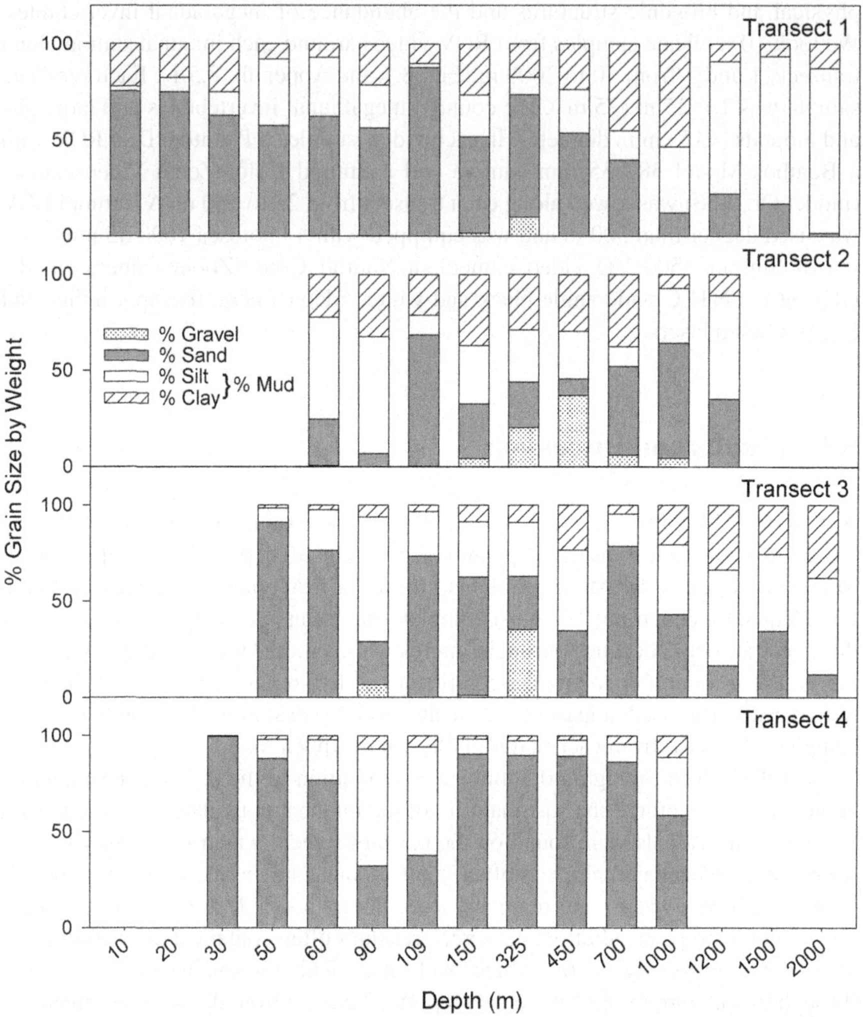

Fig. 8.3 Sediment grain size along the four depth transects by % wet weight and size fraction

along our transects (Table 8.1). Since glauconite pellets were sand-sized, they are included in that fraction (Fig. 8.3). Like our results, previous studies at Point Sur (south of Monterey Bay and our study area) documented abundant glauconite pellet deposits at the shallow edge and throughout the oxygen minimum zone (Mullins et al. 1985). In both locations, glauconite was rare along most of the continental shelf (10–90 m).

Table 8.1 Percent glauconite found in microscopic analysis of sediments. Shown are values for selected depths and transects. nd = no data

	Transects					
Depth	1	2	3	4	Mean	SD
90	5	0	0	0	1	2.5
109	25	15	0	0	10	12.3
150	65	8	90	0	41	43.8
325	20	10	90	50	43	35.9
450	90	70	90	70	80	11.6
700	80	90	nd	90	87	5.8
1000	90	nd	65	80	78	12.6

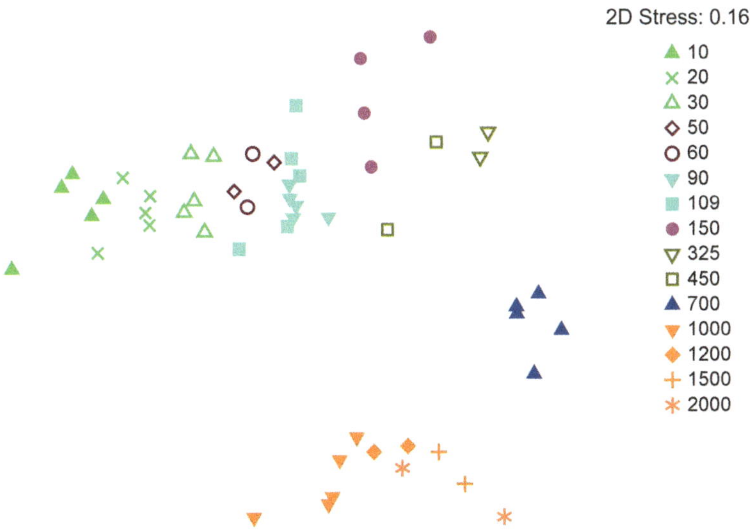

Fig. 8.4 Multidimensional scaling diagram (MDS plot) based on ranked sample similarities for transect one at the Monterey Submarine Canyon. Symbols indicate sample depths

8.3.2 Benthic Communities

Cluster analysis showed a distinct similarity in species composition and secondarily relative abundance with changes in water depth (Figs. 8.4, 8.5, and 8.6). There were two ways to visualize this pattern: MDS plots and dendrograms. We created an MDS plot for transect 1, which was the most intensively sampled (Fig. 8.4), and for all four transects combined (Fig. 8.5). The depth patterns were easier to see along the single transect in the MDS plot (Fig. 8.4). The most distinct cluster was the samples from 700 m (blue triangles). The deepest samples from 1000 to 2000 m clustered together and were distinct from all other depths (orange symbols). The shelf and upper slope samples form a gradient of clusters at the top of the figure with

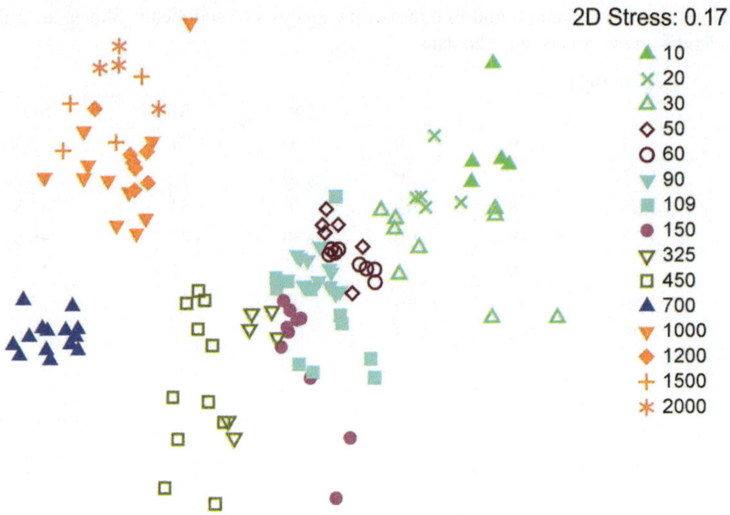

Fig. 8.5 MDS plot based on ranked sample similarities for all four transects in the Monterey Bay area. Symbols indicate sample depths with transects pooled by depth

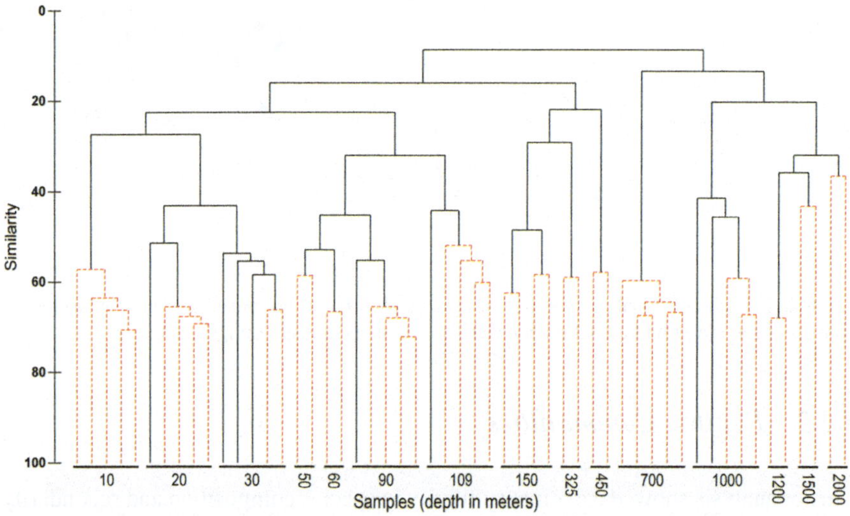

Fig. 8.6 Dendogram based on cluster analysis for transect one at the Monterey Submarine Canyon. Clusters of water depths connected by dashed orange lines (e.g., all 10 m samples) are not significantly different from each other at P-value = 0.05 but are significantly different from other clusters of connected orange lines

the shallowest stations (10 m) on the left end and the deepest (325, 450 m) on the right end of the MDS plot (Fig. 8.4). In this broad cluster cloud, there were distinct sample clusters for 10, 20, 30, 50–60, 90–109, 150, and 325–450 m (Fig. 8.4).

The same general depth patterns emerged when all transects were combined (Fig. 8.5). Samples from 700 m were most distinct, followed by those from 1000 to 2000 m, with a broad cluster cloud of samples from the shelf and upper slope, showing a progression of clusters from 10 to 450 m (Fig. 8.5).

The dendogram showed the clustering of samples by depth very clearly along the bottom axis of the diagram (Fig. 8.6). These samples were only from transect 1. The dendogram also showed the different levels of similarity within and between depths, groups of samples that were statistically distinct within and between depths (clusters connected by orange dotted lines), and outliers (solid black lines) that clustered with the right water depth but were statistically different from the other samples at that depth (SIMPROF test, Fig. 8.6). The dendogram for all the transects combined showed the same distinct clustering with water depth, but the details were more difficult to see here without a magnifying glass (Appendix 8.5.3). Since the samples clustered so strongly with water depth, we present the data by depth in the following sections.

8.3.3 *Diversity and Abundance*

The number of species was extremely high from the mid-shelf (50–90 m) to the shelf-slope break (109–150 m) to the upper slope (325 m) (Figs. 8.7 and 8.8). The number of species per unit area is the best measure of community diversity at our sampling scale (Oliver et al. 2011). Therefore, in this presentation, diversity always refers to species density, species per unit area of seafloor. Diversity increased along the inner shelf to the mid-shelf high and decreased from the upper to the lower slope (Figs. 8.7 and 8.8). The inner shelf (10–30 m) and the deeper slope (700–2000 m) were the least diverse locations. The general depth patterns were similar when all four transects were combined (Fig. 8.7) and presented separately (Fig. 8.8).

The number of individuals showed a similar pattern (Fig. 8.9), increasing from the inner to mid-shelf, peaking around the shelf-slope break and upper slope, and decreasing from the upper to the lower slope with a much smaller secondary peak in the oxygen minimum zone (700 m). These patterns were similar when the four transects were presented separately (Fig. 8.9). Simpson's dominance showed the opposite pattern, peaking at the shallowest depth (10 m), and from the oxygen minimum zone into deeper water (Figs. 8.7 and 8.10). Dominance was very low from 20 to 450 m. Even the highest peak in dominance (1500 m) was relatively low compared to other geographic locations with high dominance (above 0.5). Evenness is the reciprocal of dominance. So in general, the number of individuals was evenly dispersed among the species along the entire continental margin; dominance was low (Figs. 8.7 and 8.10).

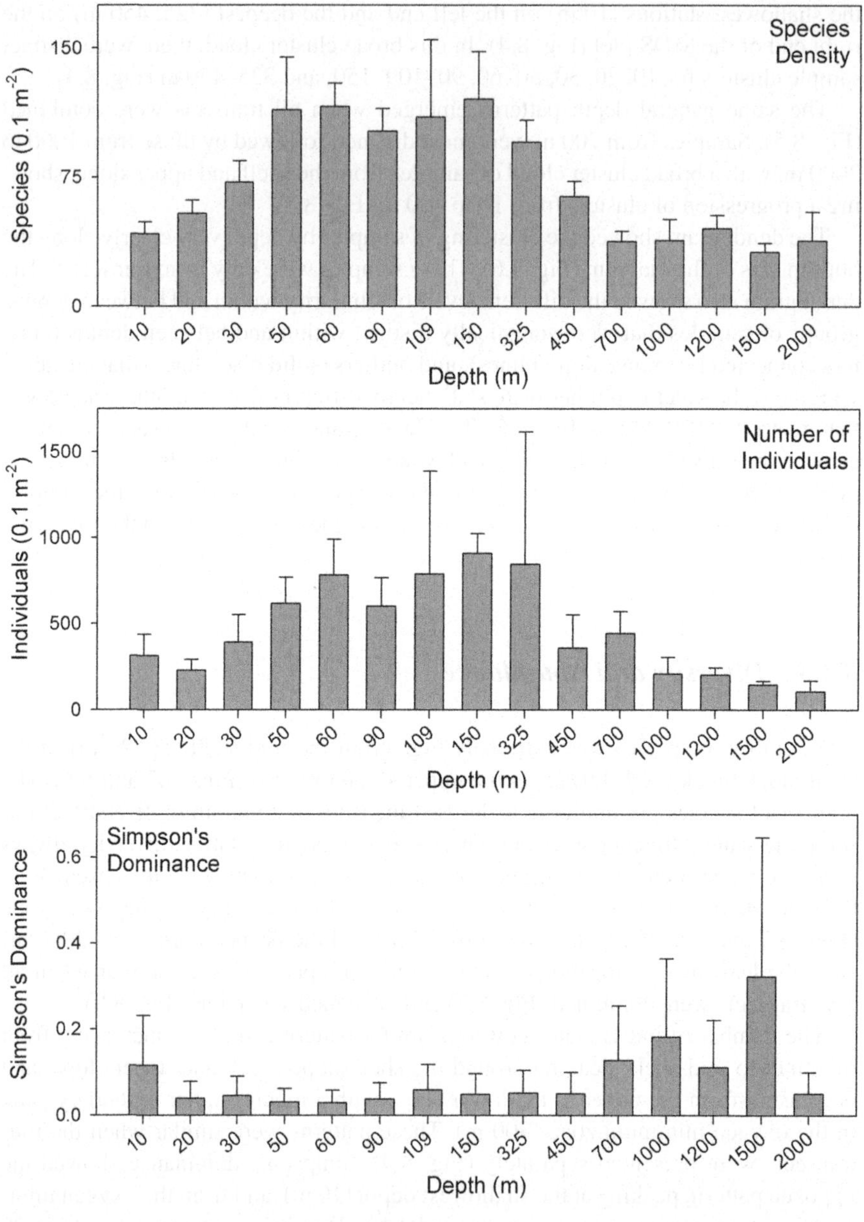

Fig. 8.7 Total number of species (top), individuals (middle), and dominance (bottom) with water depth when all transects are pooled (means and standard deviations based on grab numbers shown in Tables 8.2, 8.3, and 8.4)

The benthic communities did not show a shift in motility with water depth (Table 8.2). Each species was assigned to a motility group after Fauchald and Jumars (1979) and Macdonald et al. (2010). The three groups are completely sessile, discretely motile (able to move but movement is not necessary for feeding), and motile

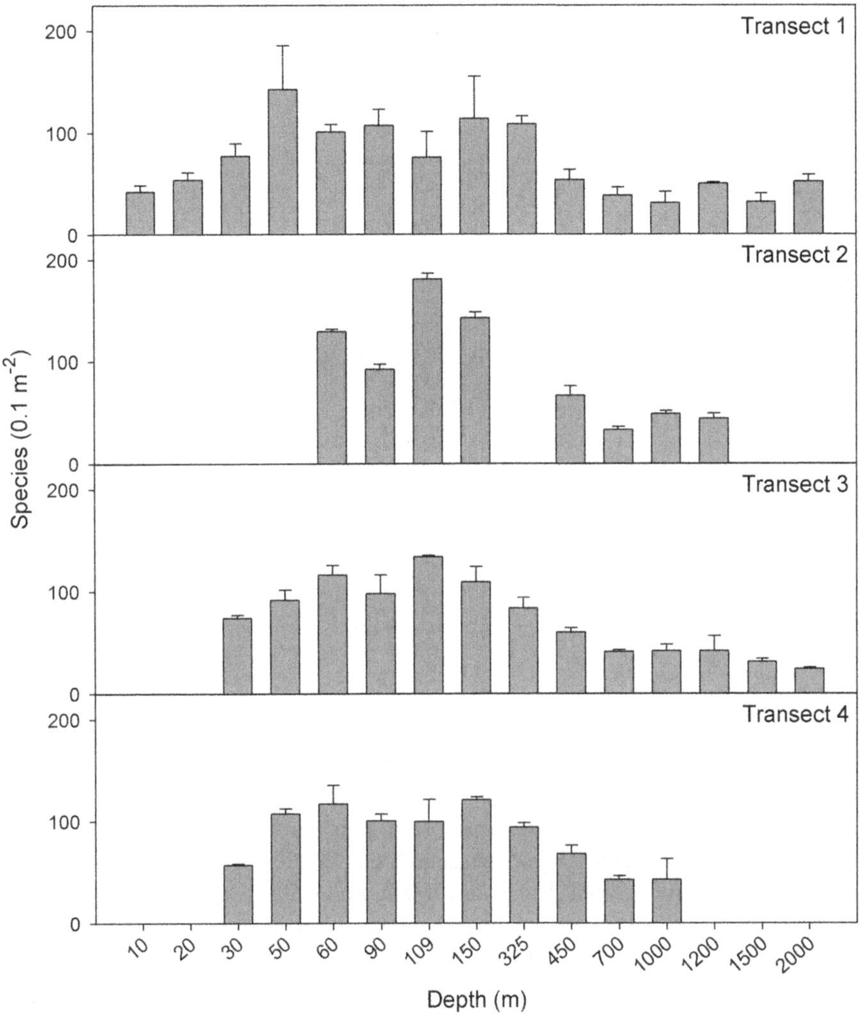

Fig. 8.8 Total number of species along the four depth transects (means and standard deviations based on grab numbers shown in Tables 8.5 and 8.6)

(movement is required for feeding). As in Fig. 8.7, all transects were combined. Sessile infaunal animals were rare along the entire continental margin, for both the number of species and individuals (Table 8.2). Even if the sessile and discretely motile groups were combined into a relatively sessile group, there was little difference between this relatively sessile group and the motile group from 10 to 2000 m (Table 8.2). We expected an increase in the percentage of motile groups in deeper water, but this did not occur along this highly productive, upwelling continental margin (Table 8.2).

Table 8.2 Percentage of three mobility groups with depth when all transects are pooled. % Sp = percent of total number of species, % Ind = percent of total number of individuals

	Sessile		Discretely motile		Motile				
Depth (m)	% Sp	% Ind	% Sp	% Ind	% Sp	% Ind	Total species	Total individuals	Number of grabs
10	3	<1	45	51	52	48	89	1570	5
20	2	9	52	56	46	35	104	1143	5
30	2	<1	43	38	55	62	262	3527	9
50	4	1	50	58	46	41	276	3696	6
60	4	1	52	62	44	37	284	6398	8
90	5	2	48	61	47	37	300	6612	11
109	4	2	45	69	51	28	401	8690	11
150	4	2	43	51	53	48	367	9094	10
325	3	1	44	38	53	61	266	5077	6
450	2	1	46	45	52	53	264	3955	11
700	3	6	45	79	52	15	143	6219	14
1000	3	1	50	65	47	34	149	2290	11
1200	5	1	53	63	42	35	123	1429	7
1500	6	2	45	78	49	20	82	586	4
2000	3	3	53	62	44	35	104	427	4

The biomass of benthic invertebrates was strongly influenced by a few large animals, usually ophiuroids or other echinoderms (Fig. 8.11). Large moon snails (*Euspira pallida*) accounted for the biomass peak at 30 m. Excluding these few large animals, most of the biomass was from polychaete worms, which was usually about 10 times greater than crustacean biomass. However, the crustacean biomass had two major peaks at 150 m and 700 m. There also was a peak in crustacean species and individuals at 150 m and a secondary peak in crustacean individuals at 700 m, where there were many large amphipods (see Oxygen Minimum Zone below).

The megafaunal invertebrates observed with the ROVs were mostly epifaunal species with much lower numbers of species and individuals than the macrofauna (animals collected on a 0.5 mm screen) discussed thus far. Unless otherwise indicated, all the megafauna numbers were means of the ten replicates at each depth, pooling replicates from transects 1 and 2. There was a distinct depth pattern. The shallowest stations sampled (30–90 m) had fewer species and individuals per 25 m^2 than all of the deeper stations (109–2000 m). There was an average of 2 species and 8 individuals at the shallow stations, compared to 5 species and 36 individuals per 25 m^2 from 109 to 2000 m. There was no overlap in the numbers from the two depth zones, so the differences would likely be statistically significant.

Table 8.3 Percent number of species and individuals in major taxonomic groups with depth when all transects are combined. % Sp = percent of total number of species; % Ind = percent of total number of individuals. Means are per grab

Depth (m)	Crustacea		Polychaeta		Mollusca		Echinodermata		Mean number		
	% Sp	% Ind	% Sp	% Ind	% Sp	% Ind	% Sp	% Ind	Sp	Ind	N grabs
10	28	15	52	77	15	6	1	<1	42	314	5
20	17	20	60	64	17	11	<1	<1	54	229	5
30	21	20	49	41	25	31	3	2	72	392	9
50	13	10	60	58	18	18	3	4	114	616	6
60	12	7	60	59	19	19	5	8	116	784	8
90	18	10	61	73	14	10	4	5	102	601	11
109	34	35	46	52	11	7	4	3	110	790	11
150	41	59	42	33	11	5	3	1	120	909	10
325	29	37	50	45	11	9	6	4	95	846	6
450	26	20	55	58	12	11	4	5	64	359	11
700	40	67	46	30	9	2	1	<1	38	444	14
1000	15	4	61	72	10	11	4	2	38	208	11
1200	10	4	69	81	12	6	3	1	45	204	7
1500	7	2	69	90	12	4	4	1	31	147	4
2000	10	5	77	81	7	10	<1	<1	38	107	4

8.3.4 Major Taxonomic Groups

Polychaete worms and crustaceans dominated the number of species and individuals in marine sediments (Table 8.3). Combining all transects and water depths, these two taxonomic groups account for an average of 78% of the species (ranging from 69 to 87%), and 82% of the individuals (ranging from 62 to 95%) (Table 8.3). We collected a total of 938 species of benthic invertebrates, including 431 polychaete species, 270 crustacean species, 171 molluscan species, and 38 species of echinoderms.

The percentage of crustacean species was highest at the shelf-slope break (34% at 109 m, 41% at 150 m) and the oxygen minimum zone (40% at 700 m). They were also high at the shallowest, most wave-swept station (28% at 10 m) and the upper slope (29% at 325 m, 26% at 450 m). The percentage of crustacean individuals was high from the shelf-slope break to the oxygen minimum zone, where they peaked at 67% of the number of individuals sampled at 700 m (Table 8.3). The peracarids accounted for 92% of the crustacean species, and 64% of these were species of amphipods. We collected a total of 159 species of amphipods, 38 species of cumaceans, 29 species of isopods, 19 species of tanaids, and 3 species of mysids. The percentage of amphipods was high at all depths, peaking in the oxygen minimum zone for both species and individuals. Cumaceans, isopods, and mysids were more common on the shelf, and tanaids along the deeper slope (Table 8.4). Amphipods were clearly the dominant crustacean group (Table 8.4). So the two

Table 8.4 Percent number of species and individuals of five orders of peracarid crustaceans with depth when all transects are combined. % Sp = percent of total number of peracarid species; % Ind = percent of total number of peracarid individuals. Means are per grab

	Amphipoda		Cumacea		Isopoda		Mysida		Tanaidacea		Mean Peracarida		
Depth (m)	% Sp	% Ind	% Sp	% Ind	% Sp	% Ind	% Sp	% Ind	% Sp	% Ind	Sp	Ind	N grabs
10	41	48	31	31	26	21	2	<1	0	0	11	45	5
20	46	31	32	49	20	19	2	<1	0	0	8	39	5
30	57	30	25	59	17	11	0	0	1	<1	13	67	9
50	57	43	24	37	16	19	2	<1	1	1	14	52	6
60	58	53	19	26	21	20	2	1	0	0	13	59	8
90	62	63	18	20	17	16	0	0	2	1	16	50	11
109	63	72	20	12	12	14	0	0	5	2	31	298	11
150	67	86	14	4	12	7	<1	0	7	3	40	388	10
325	66	82	6	1	12	3	0	0	17	14	25	382	6
450	75	84	5	3	6	4	1	<1	13	9	15	57	11
700	75	96	13	2	2	<1	0	0	10	1	15	299	14
1000	70	70	20	21	2	1	0	0	9	8	6	9	11
1200	37	39	14	12	23	20	0	0	26	28	4	8	7
1500	31	31	29	25	0	0	0	0	40	44	2	3	4
2000	45	51	20	18	11	7	0	0	24	23	4	6	4

major groups of benthic invertebrates along the continental margin were amphipod crustaceans and polychaete worms.

The percentage of polychaete species was high at all water depths, ranging from 42 to 69% (Table 8.3). The percentage of polychaete individuals ranged from 41 to 90%, excluding the 150 m (33%) and 700 m (30%) stations, where the percentage of crustacean individuals was highest (Table 8.3). Unlike the crustaceans, the polychaetes showed no pattern of change with water depth, in both the number of species and individuals (Tables 8.3, 8.5, and 8.6). There were many families of polychaetes, but although some families appeared to be more characteristic of the deeper slope (paranoids, cossurids, cirratulids, pilargids), even this pattern was weak.

Mollusks and echinoderms were the only other major taxonomic groups that were relatively common or abundant in our survey. They showed little pattern with depth (Table 8.3). Mollusks averaged 13% of the species and 11% of the individuals. Echinoderms were 3% of the species and 2% of the individuals. The percentage of mollusk species (25%) and individuals (31%) was only high at 30 m (Table 8.3), because of many abundant species of gastropods on transect 3 (Tables 8.5 and 8.6; Appendix 8.5.4). Most of the changes in major taxonomic groups discussed so far were derived from the percentages in Table 8.3. If the actual number of species and individuals per sample are examined (Tables 8.5 and 8.6), the patterns among the taxonomic groups and with water depth were similar. Below we consider community patterns in relation to the major sedimentary habitats shown in Fig. 8.2.

Table 8.5 Mean number of species and the mean number in each major taxonomic group along the four depth transects (means and standard deviations based on N grab samples)

	Total Species		Crustacea		Polychaeta		Mollusca		Echinodermata		
Depth	Mean	SD	Mean	SD	Mean	SD	Mean	SD	Mean	SD	N
Transect 1											
10	42	7	12	1	22	4	7	3	1	1	5
20	54	8	9	2	32	4	9	4	<1	<1	5
30	77	12	14	7	41	6	18	3	1	1	5
50	142	43	27	21	75	16	27	4	8	2	2
60	101	7	12	4	57	3	23	0	5	1	2
90	107	16	18	2	67	9	16	4	3	3	5
109	76	25	28	3	33	21	9	4	2	3	5
150	114	41	51	2	45	34	11	3	3	2	4
325	109	8	39	1	51	5	6	2	8	1	2
450	54	10	20	1	22	8	7	1	2	1	2
700	38	8	14	5	18	3	3	<1	<1	<1	5
1000	31	11	5	3	16	9	4	2	2	1	5
1200	50	1	5	2	32	2	8	2	1	0	2
1500	32	8	1	0	22	3	4	3	1	1	2
2000	52	7	7	1	39	8	5	3	1	1	2
Depth	Mean	SD	Mean	SD	Mean	SD	Mean	SD	Mean	SD	N
Transect 2											
60	130	2	16	2	77	1	28	3	6	1	2
90	92	5	13	4	56	6	15	2	5	1	2
109	181	6	54	1	80	14	29	21	9	1	2
150	143	6	49	0	72	2	14	1	4	0	2
450	67	9	13	3	43	5	5	3	3	1	5
700	33	3	13	2	16	2	3	2	<1	<1	5
1000	49	3	8	4	31	3	4	1	2	1	2
1200	44	5	4	2	31	1	5	3	2	1	3
Depth	Mean	SD	Mean	SD	Mean	SD	Mean	SD	Mean	SD	N
Transect 3											
30	74	3	17	1	31	4	23	1	3	1	2
50	92	10	11	2	61	9	11	1	3	0	2
60	116	9	15	1	69	6	23	2	6	0	2
90	98	18	23	7	60	13	8	4	6	1	2
109	134	1	36	2	75	0	10	1	9	2	2
150	109	15	33	6	56	8	14	1	3	0	2
325	84	11	22	0	41	8	11	4	4	1	2
450	60	4	13	0	34	0	10	3	2	1	2
700	41	1	22	0	12	2	5	0	1	1	2
1000	42	6	5	3	29	3	3	1	2	0	2
1200	41	15	4	2	30	8	4	4	<1	1	2
1500	31	3	3	1	21	1	4	4	2	1	2
2000	24	1	2	0	19	1	1	0	0	0	2

(continued)

Table 8.5 (continued)

	Total Species		Crustacea		Polychaeta		Mollusca		Echinodermata		
Depth	Mean	SD	Mean	SD	Mean	SD	Mean	SD	Mean	SD	N
Transect 4											
30	57	1	16	1	25	2	13	1	3	1	2
50	108	5	10	1	65	4	25	1	3	1	2
60	117	18	15	2	76	1	15	14	7	1	2
90	101	6	18	1	59	2	14	4	5	3	2
109	99	22	29	2	53	24	8	2	7	1	2
150	121	3	47	3	49	2	18	1	4	0	2
325	94	4	21	3	51	3	13	1	5	4	2
450	68	9	24	3	30	1	10	1	3	2	2
700	43	3	16	3	25	6	0	0	0	0	2
1000	43	21	5	3	30	11	5	4	1	1	2

8.3.5 *Depth Patterns*

8.3.5.1 Inner Shelf (10–30 m)

The inner shelf was exposed to extensive wave disturbance at 10 m, decreasing in intensity as depth increased to 30 m (Fig. 8.2). The sandy sediments (Fig. 8.3) were transported into distinct parallel ripple marks. The diversity and abundance were the lowest for the inner shelf compared to the mid-shelf, shelf break, and upper slope (Figs. 8.7, 8.8, and 8.9). But the diversity at the inner shelf was still higher than any stations from the oxygen minimum zone to 2000 m, and abundance was higher than any of the deep slope stations (1000–2000 m) (Figs. 8.7, 8.8, and 8.9). The 30 m stations were a transition from strong wave disturbance (at 10–20 m) to little wave disturbance on the mid-shelf (Fig. 8.2).

The 30 m station at transect 3 had a uniquely high number of species and individuals of gastropod mollusks. The station had a typical level of diversity (Table 8.5), but nearly twice as many total individuals as the other 30 m stations (Table 8.6). Half of the ten most abundant species were small gastropods: *Alvania rosana*, *Cingula* sp., *Odostomia angularis*, *Odostomia* sp., and *Turbonilla* sp. (Appendix 8.5.4). In total, there were 25 species of gastropods from the two replicate grabs taken here with a mean abundance of 18/grab. In comparison, the 30 m station on transect 1 had 13 species in five grabs with a mean abundance of less than 1/grab. Transect 4 had the only other 30 m station, where there were 8 species of gastropods in two grabs with a mean abundance of less than 1/grab. No other station at any depth had the large number of gastropods (species and individuals) found at 30 m on transect 3. Perhaps they were associated with a mixed rocky bottom, since grab samples were difficult to obtain here and there were fringing kelp forests on shale reefs along this section of coast. We did not obtain a sediment sample from this station because of low grab penetration.

Table 8.6 Total number of individuals and number in each major taxonomic group along the four depth transects (means and standard deviations based on N grab samples)

	Individuals		Crustacea		Polychaeta		Mollusca		Echinodermata		
Depth	Mean	SD	Mean	SD	Mean	SD	Mean	SD	Mean	SD	N
Transect 1											
10	314	124	48	21	241	94	19	13	1	1	5
20	229	63	46	10	147	42	27	14	<1	<1	5
30	313	49	71	42	129	15	80	15	1	2	5
50	659	125	118	90	337	19	140	13	32	0	2
60	573	18	32	1	284	19	145	13	93	30	2
90	527	104	51	8	394	84	62	11	10	16	5
109	340	127	103	23	192	101	29	12	5	8	5
150	886	148	657	83	185	163	20	4	10	14	4
325	1752	692	1072	806	473	149	18	13	70	16	2
450	183	32	56	16	79	23	26	2	2	1	2
700	362	116	221	83	126	48	8	2	<1	1	5
1000	151	81	7	5	103	95	25	24	2	2	5
1200	227	9	7	0	177	8	18	1	4	3	2
1500	145	34	1	0	129	23	8	8	2	3	2
2000	161	11	10	4	123	7	24	20	1	1	2
Depth	Mean	SD	Mean	SD	Mean	SD	Mean	SD	Mean	SD	N
Transect 2											
60	1035	148	98	8	531	89	268	29	104	33.9	2
90	460	5	68	37	299	34	53	11	23	17.0	2
109	1647	516	1032	430	475	134	66	52	35	22.6	2
150	917	146	459	22	379	81	49	21	7	2.8	2
450	526	164	69	27	355	120	20	13	55	72.2	5
700	427	64	303	57	113	27	4	4	<1	<1	5
1000	257	58	14	6	192	52	9	6	2	0	2
1200	205	40	10	3	168	39	14	9	2	1	3
Depth	Mean	SD	Mean	SD	Mean	SD	Mean	SD	Mean	SD	N
Transect 3											
30	658	103	111	21	214	123	311	36	20	8	2
50	464	143	34	0	274	112	64	16	12	6	2
60	704	208	50	16	466	129	95	3	37	11	2
90	804	100	85	15	583	134	50	21	84	30	2
109	1282	129	309	7	910	102	14	1	30	15	2
150	913	160	453	25	361	157	73	27	6	4	2
325	342	118	106	7	179	82	28	21	14	1	2
450	221	5	36	1	135	5	35	19	3	3	2
700	66	26	464	2	184	16	14	2	1	1	2
1000	213	30	9	7	172	9	5	3	4	2	2
1200	180	68	7	3	147	50	7	4	1	1	2
1500	148	13	4	0	135	19	6	6	2	1	2
2000	52	9	2	1	45	9	2	1	0	0	2

(continued)

Table 8.6 (continued)

	Individuals		Crustacea		Polychaeta		Mollusca		Echinodermata		
Depth	Mean	SD	Mean	SD	Mean	SD	Mean	SD	Mean	SD	N
Transect 4											
30	325	23	70	8	162	7	81	23	8	1	2
50	724	98	27	1	457	64	131	9	25	18	2
60	886	100	60	1	619	37	111	47	20	0	2
90	725	211	57	14	564	215	60	17	36	4	2
109	564	369	92	20	363	395	55	1	38	7	2
150	944	11	414	25	437	22	64	3	13	9	2
325	445	16	102	40	237	45	77	10	19	13	2
450	259	25	84	2	117	18	49	3	6	1	2
700	468	1	328	29	137	28	0	0	0	0	2
1000	299	168	10	4	269	140	8	9	2	3	2

The two most common megafaunal invertebrates were a large tube-dwelling polychaete, *Diopatra ornata*, and sea pens, probably *Stylatula* sp. Each species averaged one individual per 25 m^2 at 30 m.

8.3.5.2 Mid-shelf and Mud Zone (50–90 m)

The 50 m and 60 m stations were beyond the influence of strong wave action, were usually sandy bottoms (Appendix 8.5.2.2), and bordered the mud zone, which was centered at about 80 m (Figs. 8.2 and 8.3). The 50–60 m stations were very similar as indicated by the cluster analysis: only the 700 m communities clustered as well (Figs. 8.4, 8.5, and 8.6). The 90 m stations were on the seaward edge of the mud zone and usually had high proportions of silt and clay: mud (Figs. 8.2 and 8.3, Appendix 8.5.2.3). Diversity was high at all of the mid-shelf depths (Table 8.5). The highest was at 50 m on transect 1 with an average of 143 species per grab (0.1m^2). The average number of species at the other 50–90 m stations had a range of 92–130 species per grab (Table 8.5): 100 species in a bottom area of 0.1m^2 is a very high diversity (see Conclusions). In addition, animal abundance was high at the mid-shelf depths (Figs. 8.7, 8.8, and 8.9). We also captured the largest numbers of ophiuroids from the mid-shelf (often at 60 m), including *Amphiodia urtica* and arms of larger species (Appendix 8.5.2.2) that were deeper in the sediment (Appendix 8.5.4). Although ophiuroids were not abundant in grabs from other depths, we did see larger species at high densities with the ROV images from the upper slope to 2000 m.

The two most common megafaunal invertebrates were the same as in the wave-swept zone. *Diopatra* averaged 3, and sea pens 6 per 25 m^2. The mud bottoms at the 90 m stations had the highest numbers of large (<15 cm) pits and mounds, probably made by burrowing shrimp and crabs. In transect 1, there was an average of 12 mounds per 25 m^2 (range 8–20), and 8 pits (2–15). Transect 2 had an average of 4 mounds (range 0–17), and 4 pits (0–18). We also encountered patches of other abundant epifauna (Appendix 8.5.2.4).

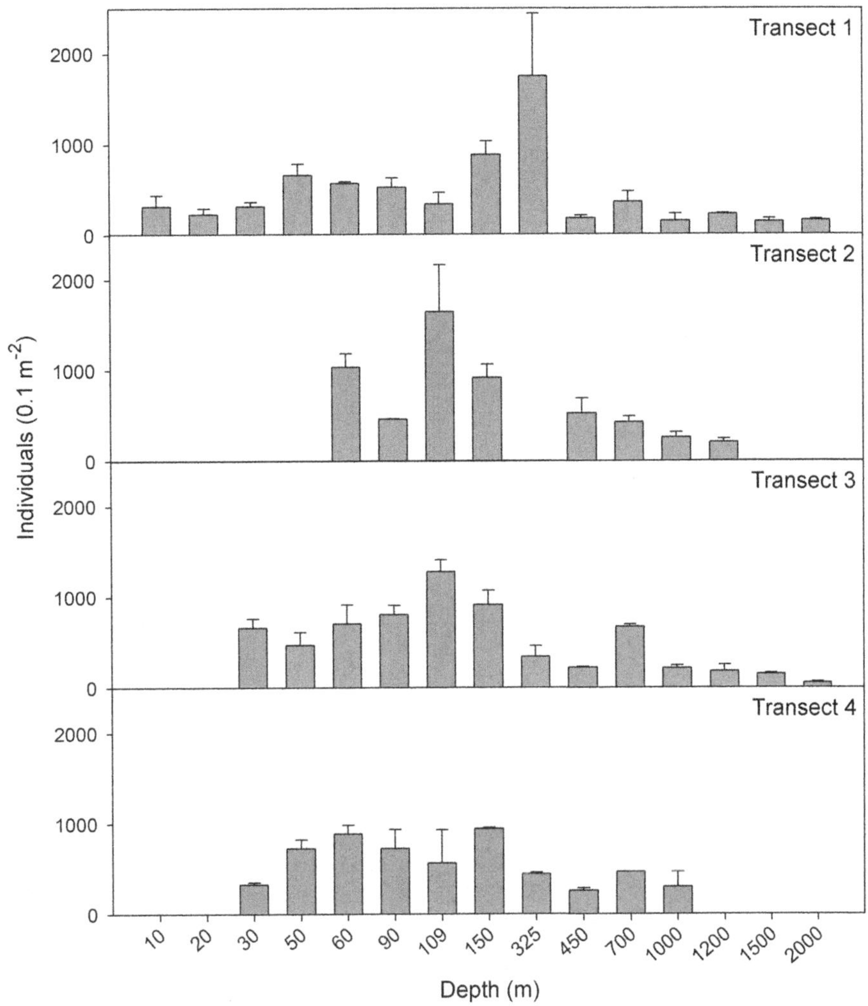

Fig. 8.9 Total number of individuals along the four depth transects (means and standard deviations based on grab numbers shown in Tables 8.5 and 8.6)

8.3.5.3 Shelf-Slope Break (109–150 m)

The highest diversities were at the shelf-slope break (Figs. 8.7 and 8.8). The highest was on transect 2 at 109 m with a mean of 181 species/grab (Fig. 8.8, Table 8.5): the two grabs here had 177 and 185 species/grab. However, diversity was also very high at some mid-shelf and upper slope depths (Figs. 8.7 and 8.8; Table 8.5). The diversity of crustaceans peaked at 109–150 m along all four transects, except along transect 4 where 150 m was the highest (52/grab), 325 m was next (40/grab), and 109 m was the third highest (29/grab) (Table 8.5). So crustaceans accounted for the peak in diversity at the shelf-slope break. Number of individuals was often highest here as

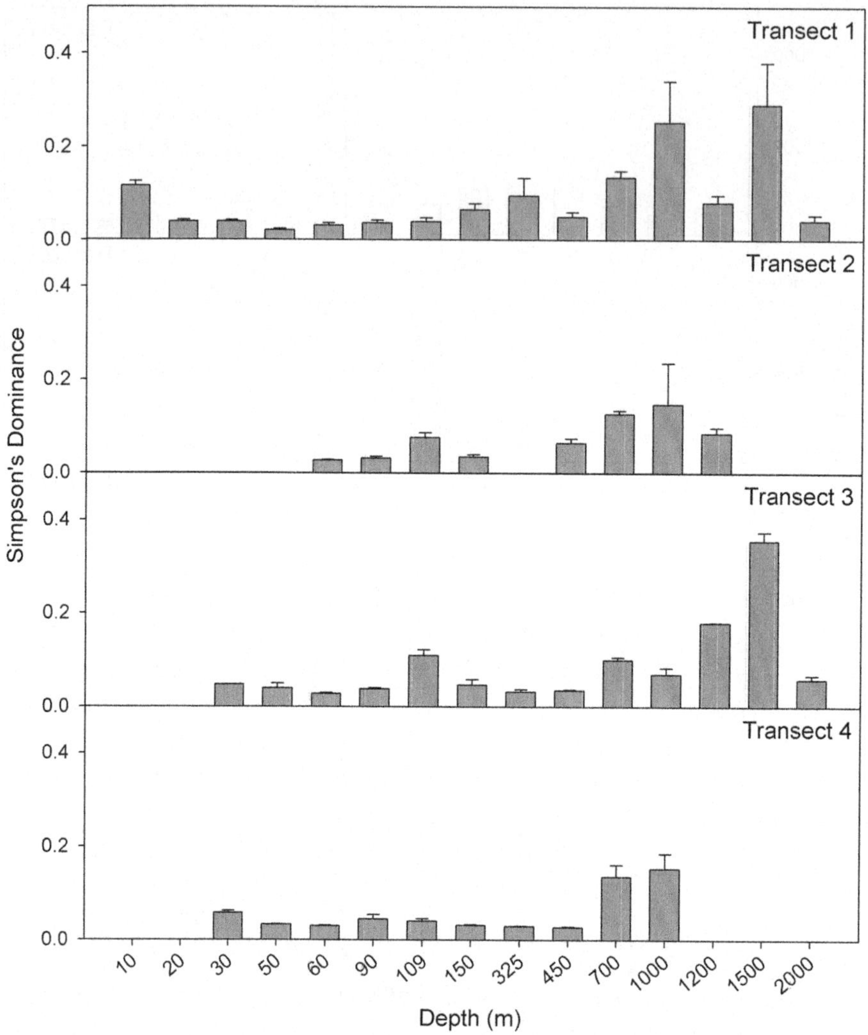

Fig. 8.10 Changes in dominance along the four depth transects (higher numbers = greater dominance) (means and standard deviations based on grab numbers shown in Tables 8.5 and 8.6)

well or was at least among the greatest abundances (Figs. 8.7 and 8.9; Table 8.6). Species of *Photis* (Fig. 8.12A), *Aoroides*, and *Gammaropsis* were among the most abundant amphipods, along with the ostracod *Euphilomedes* (Appendix 8.5.4), but the list of numerous crustacean species was much longer as noted above. The percentage of crustacean species (41%) and individuals (59%) was extremely high (Table 8.3). The highest total (1648/grab) and crustacean abundance (1032/grab) was at 109 m on transect 2 (Table 8.6). Only one other station had higher total and crustacean abundances (325 m on transect 1; see next section).

Several species had mean abundances over 100 individuals/grab at the shelf-slope stations (Appendix 8.5.4). They were crustaceans and one polychaete worm.

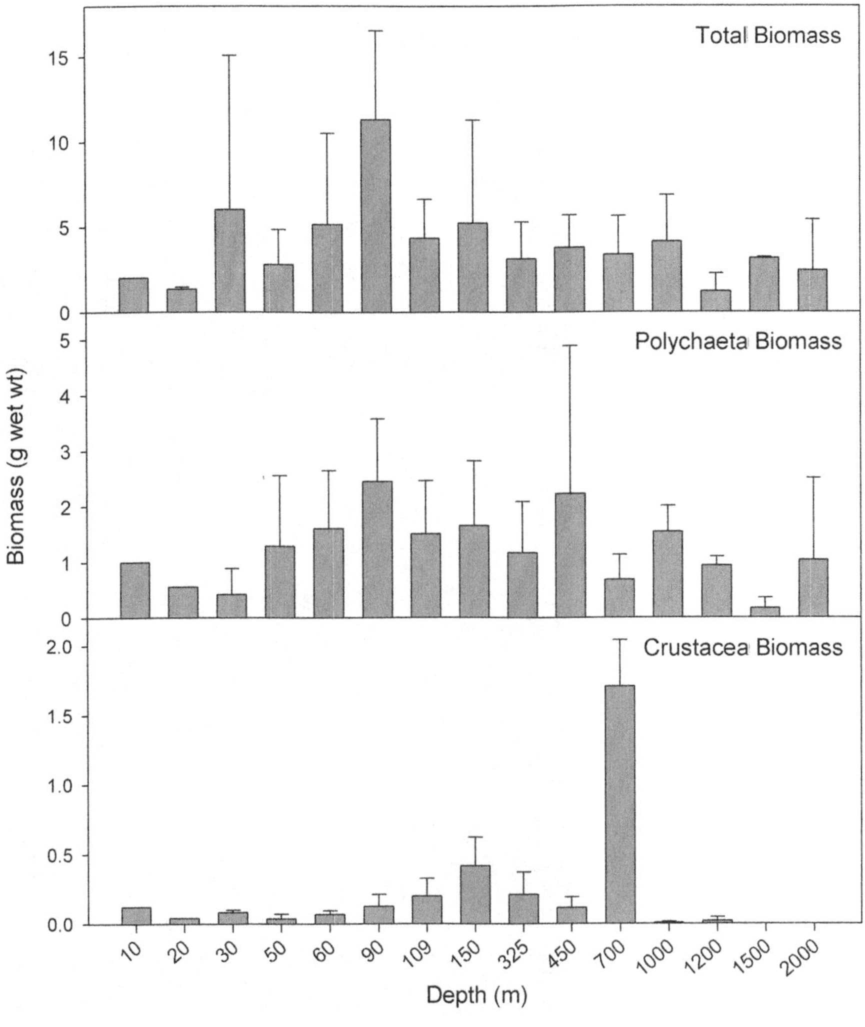

Fig. 8.11 Biomass of the entire fauna (top), the polychaete worms (middle), and crustaceans (bottom) when all transects are combined (means and standard deviations based on grab numbers shown in Tables 8.2, 8.3, and 8.4)

Two abundant amphipods were at 109 m on transect 2: *Photis* sp. 371/grab and *Aoroides exilis* 230/grab. *Photis* sp. was 126/grab at 150 m on transect 1 and 118/grab at 150 m on transect 2. *Euphilomes carcharodonta* (an ostracod) was 137/grab at 150 m on transect 3. And the most abundant was *Myriochele* sp. (owneniid polychaete) at 150 m on transect 3 (395/grab). *Myriochele* is capable of asexual reproduction (Oliver 1984), which may account for the large numbers of small worms. Even with these very abundant species, dominance was low at 109–150 m (Figs. 8.7 and 8.10), because other species were abundant as well (Appendix 8.5.4).

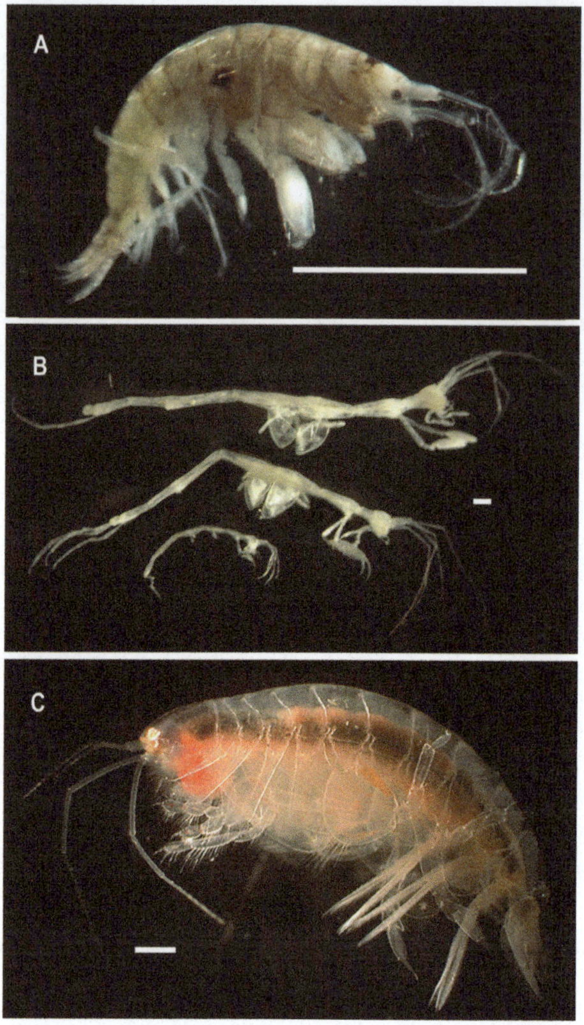

Fig. 8.12 Three of the most common crustacean species observed in our infaunal samples. Shown are *Photis californica* (subadult) (**A**), *Tritella tenuissima* (**B**), and *Ampelisca unsocalae* (**C**). All scale bars represent 1 mm. Photo credits: Martinez Lara, Ricardo. "*Photis californica*." 2010. City of San Diego, MBOO (**A**); Miller, John. "*Tritella tenuissima*." 2003. Los Angeles County Sanitation Districts, Marine Biology Laboratory (**B**); Kuhnz, Linda. "*Ampelisca unsocalae*." 2007 (C)

We also found remarkable beds of brachiopods (*Laqueus californianus*) at the shelf-slope break with the ROVs. Sometimes there were large patches of live and dead shells covering the sea floor (Appendix 8.5.2.5, 8.5.2.6). We only found these dense populations near the canyon on transect 2. The other most abundant megafaunal invertebrates were sea pens, averaging 10 per 25 m^2. The only other station with high numbers of large pits and mounds was 109 m on transect 1. There were 12 mounds per 25 m^2 (range 0–22); and 10 pits (0–24). The larger-scale bottom topography was flat with distinct erosion features and sometimes with dense patches of brittle stars (Appendix 8.5.2.7, 8.5.2.8, 8.5.2.9, 8.5.2.10).

8.3.5.4 Upper Slope (325–450 m)

The upper slope contained rocky outcrops with patch reefs, regions with mixed hard and soft bottoms, and large invertebrates that live on hard or firm substrates, including sponges, hydroids, soft corals, bryozoans, sea stars, large anemones (*Metridium giganteum*), tube-dwelling anemones (*Ceriantharia*), and dense patches of urchins (*Allocentrotus fragilis*). These were observed with the ROVs (Appendix 8.5.2.11, 8.5.2.12). Gravel was a significant component of the sediment (Fig. 8.3), and glauconite was abundant, especially closer to the oxygen minimum zone (Table 8.1). About half of the grab samples were rejected here because of poor penetration. The hard bottoms were most common on transect 1, which was indicated by the large number of caprellid amphipods at 325 m (Appendix 8.5.4).

Tritella tenuissima is a large caprellid and like most caprellids lives on hard substrates often perched on sessile invertebrates (Fig. 8.12B). It was the most abundant animal at 325 m (387/grab), and from any station we sampled (Appendix 8.5.4). It co-occurred with three other species with very high abundances: a small polychaete (*Sphaerosyllis ranunculus*, 219/grab) and two small gammarid amphipods (*Mesometopa neglecta*, 138/grab; and *Photis* sp., 136/grab). No other single station had four species with mean abundances greater than 100/grab. This 325 m station also had the highest abundance of crustaceans (1072/grab) and total fauna (1752/grab) (Table 8.6), which accounted for the large variation in abundance between transects at 325 m (Figs. 8.7 and 8.9). Despite the high abundances, dominance was low at 325 m because a number of animals were abundant, not just one (Figs. 8.7 and 8.10; Appendix 8.5.4). Like abundance, diversity was high at 325 m and was lower at 450 m (Figs. 8.7 and 8.8).

8.3.5.5 Oxygen Minimum Zone (700 m)

The 700 m station was in the upper margin of the oxygen minimum zone (Fig. 8.2), which was in water depths of 600–1000 m. The community here indicated this margin. Diversity was relatively low with station means on the four transects ranging from 33 to 43 species/grab (Figs. 8.7 and 8.8, Table 8.5). However, diversity of crustaceans was the second highest of all stations as a percentage of species (40%) from 700 m (Table 8.3). The mean number of individuals on each transect varied from 362 to 668/grab (Figs. 8.7 and 8.9; Table 8.6). This was also relatively low over all depths but was a small peak for stations below 450 m (Figs. 8.7 and 8.8). Similar to diversity, the abundance of crustaceans was the highest of all stations as a percentage of individuals (67%) from the 700 m stations (Table 8.3).

Two species of ampeliscid amphipods, *Ampelisca unsocalae* (Fig. 8.12C) and *Byblis barbarensis*, were the two most abundant species at all four 700 m stations (Appendix 8.5.4). These are large, tube-dwelling amphipods, which formed a conspicuous dense tube mat at all 700 m stations (Appendix 8.5.2.13). Both species had a mean abundance of over 100 individuals/grab at two of the 700 m stations, ranging from 102 to 134/grab (Appendix 8.5.4). Half or more of the ten most abundant species at each 700 m station were amphipod species (Appendix 8.5.4). Surprisingly, this ampeliscid tube mat was ecologically similar to dense ampeliscid beds from the shallow water Bering and Chukchi Seas (see Conclusions).

The most common megafauna living with the tube mat were sea pens (22 per 25 m^2), *Ceriantharia* (7 per 25 m^2), and *Caridina* shrimp (7 per 25 m^2).

8.3.5.6 Deeper Slope (1000–2000 m)

The deeper slope stations had the lowest diversities and abundances, and the highest dominance among all depths (Figs. 8.7, 8.8, 8.9, and 8.10). The percentages of crustacean species and individuals were the lowest (Table 8.3) for any of the major habitat types (Fig. 8.2). Crustacean diversity ranged from 1 to 8 species/grab (Table 8.5). Crustacean abundance ranged from 1 to 14 individuals/grab, the lowest from any other station (Table 8.6). The few tanaids that we found were mostly from the deeper slope (Table 8.4). Polychaetes were the dominant taxonomic group for both species and individuals (Tables 8.3, 8.5, and 8.6). The two most abundant species were polychaetes, *Aricidea (Acmira) simplex* and a cirratulid. Their highest mean abundances were 103 and 88 individuals/grab, respectively (Appendix 8.5.4). The higher abundances of these two species compared to other species at the deep-slope stations accounted for the peak in dominance here (Figs. 8.7 and 8.10), although even these peaks were relatively low for dominance.

We encountered large brittle stars in very dense patches (as in Appendix 8.5.2.9, 8.5.2.10) at 1000 m (averaging 27 per 25 m^2) and 2000 m (mean of 6 per 25 m^2, with 0 per 25 m^2 on transect 1 and 12 per 25 m^2 on transect 2). Sea cucumbers were common at 1500 m (18 per 25 m^2) and 2000 m (14) and so were sea pens (1500 m = 8 per 25 m^2, 2000 m = 17 per 25 m^2). Sea stars and soft corals were the most abundant megafauna at 1200 m. However, the bottoms were often large muddy areas (Appendix 8.5.2.14, 8.5.2.15).

8.4 Conclusions

The benthic infaunal communities in Monterey Bay were the most diverse observed from any continental shelf or other shallow-water ecosystem in the world (Oliver et al. 2011). Oliver et al. 2011 found 449 species in a square meter. Crustaceans caused the peak numbers of species and individuals at the shelf-slope break and along the upper slope (Figs. 8.2, 8.7, and 8.8; Table 8.5; also in Oliver et al. 2011). The next most diverse shelf community was from Bass Straights in Australia (Coleman et al. 1997; Gray et al. 1997), which likely will be as high or higher than Monterey Bay with further sampling (Oliver et al. 2011). There are a number of slope communities with high diversity, but only one of these has the extremely high number of species found in Monterey Bay (Blake and Grassle 1994; Oliver et al. 2011). Two high diversity slope regions were sampled south (Hyland et al. 1991) and north of our study region (Blake et al. 2009), but none along the shelf (Oliver et al. 2011). We were unable to find any high diversity regions on the continental shelf in other geographic regions (Jayaraj et al. 2008; Carroll and Ambrose 2012; Christopher et al. 2012).

There was a dramatic decrease in the number of crustaceans at the inner shelf over time. In the 1970s, there was intensive sampling of this zone on both sides of the canyon (Hodgson and Nybakken 1973; Oliver et al. 1980, 2008). The communities at 10–20 m were numerically dominated by phoxocephalid and haustoriid amphipods and ostracods that were rare in the present sampling. Hodgson and Nybakken (1973) documented the inner and mid-shelf communities and the increase in brittle stars in the mid-shelf. They collected many species of small crustaceans at 10–20 m despite the use of a 1 mm sieve, which loses 16–25% of the species and 50–85% of the individuals of macrofaunal invertebrates captured on a 0.5 mm screen (Hodgson and Nybakken 1973). Even though the crustaceans were undersampled, they characterized the inner shelf in the 1970s and were rare by the 1990s. Oliver et al. 2008 documented the same decrease on the south side of the canyon using a 0.5 mm screen.

Sandy deposits cover the Washington continental shelf to at least 100 m with muddy bottoms beyond (Lie and Kisker 1970). Species density gradually increased with depth, but was generally low at all depths. There was no high diversity community at the shelf break or anywhere sampled from 13 to 329 m (Lie and Kisker 1970). Unfortunately, the samples were washed through a 1 mm screen, making comparisons with other studies difficult. The number of species and individuals of macrofaunal invertebrates passing through a 1 mm mesh is highly variable and thus unpredictable (Reish 1959; Hodgson and Nybakken 1973; Bachelet 1990; James et al. 1995; Hammerstrom et al. 2010). However, the shallow-water sand community (13–48 m) in Washington also contained a number of relatively abundant amphipods and cumaceans. These data are adequate for assessing a potential loss of crustaceans from the Washington shallows, as we observed in Monterey Bay.

The extensively sampled shelf communities in Southern California (Stull et al. 1986; Swartz et al. 1986; Ranasinghe et al. 2007) have much lower species density than we found in the Monterey Bay area. The highest number (100 species 0.1 m^{-2} grab) was observed from Catalina Island (Ranasinghe et al. 2007), one of several islands forming a unique biogeographic region offshore of Southern California (Henkel and Nelson 2018). The peak species density in the present study was 185 species 0.1 m^{-2} (Table 8.5, Fig. 8.7) and 449 m^{-2} (Oliver et al. 2011). There were similarities in the major faunal zones in Southern California, including an inner wave-disturbed zone with many crustaceans (Barnard 1963, Ranasinghe et al. 2007) and brittle stars becoming common along the middle shelf (Barnard and Ziesenhenne 1961; Ranasinghe et al. 2007). However, no high diversity benthic communities were documented from the shelf-slope break or upper slope (Ranasinghe et al. 2007). The Southern California Bight is a complex, broad continental margin with deep basins and outer islands. The shelf break and adjacent slopes fringe large anoxic basins. This is quite different from the narrow shelf, dynamic currents, and high production of the Monterey Bay area.

Monterey Bay is a very dynamic environment in the middle of a highly productive upwelling system (Bruland et al. 2001; Ryan et al. 2005). The narrow shelf is often swept by offshore water during upwelling and from the California current

(Breaker and Mooers 1986; Breaker and Broenkow 1994; Ramp et al. 1997). The shelf is cut by one of the largest canyons in the world, doubling the length of shelf-slope break in Monterey Bay (Fig. 8.1). Up canyon currents are common (Breaker and Broenkow 1994) with significant movements of nutrients available for production (Shea and Broenkow 1982; Ryan et al. 2005). Just north of our study region, there is a well-defined upwelling plume at Año Nuevo. It spreads food-rich water over the entire study area, slope, and shelf (Ramp et al. 2005). Upwelling shadow fronts help move water and particles shoreward across the shelf (Woodson et al. 2009). Internal waves move water and sediment into the bay and shallows (Storlazzi et al. 2003). Topography and internal tides interact most dramatically at the shelf-slope break, where currents erode sediments and transport food across the sea floor (Cacchione et al. 2002; Ryan et al. 2005; McPhee-Shaw 2006). Resuspended sediment from the outer shelf can be moved inshore and deposited in the mud belt (Cheriton et al. 2014). Intermediate nepheloid layers (suspended sediment) move sediment from the shelf offshore (McPhee-Shaw et al. 2004). Long-period swell and internal tides combine to resuspend outer shelf sediments, moving them both offshore and onshore (Rosenberger et al. 2016). Mud belt fines are resuspended and moved offshore. Slope bottoms are swept with strong currents (Noble and Ramp 2000). Variations in plankton production on the shelf are linked to internal tides (Ryan et al. 2005, 2010; Sevadjian et al. 2014). Groundwater nutrients can increase phytoplankton growth in the bay (Lecher et al. 2015). There are few if any regions in the world with such a complex topography and dynamic currents moving highly productive waters over the self and slope benthos.

The bottom sediments and communities indicate sediment erosion is common along the inner shelf, at the shelf-slope break, and along the upper slope (Fig. 8.3). The inner shelf sediments are moved by waves, which have a strong control over community structure (Oliver et al. 1980). However, along the outer shelf and upper slope, only fine sediments are eroded, and the dynamic currents described above bathe these areas with highly productive waters. Strong currents move more food over the bottom, and the result appears to be extremely high ecological diversity, high abundances, and low community dominance (Oliver et al. 2011, Fig. 8.7).

The upper margin of the oxygen minimum zone was dominated and indicated by tube-building ampeliscid amphipods, forming the most distinct community cluster we documented from the mud belt to 2000 m (Figs. 8.4, 8.5, and 8.6). The conspicuous tube mat was not observed south of Monterey Bay around the well-defined oxygen minimum zone at Point Sur (Mullins et al. 1985; Thompson et al. 1985, personal observations). In contrast, the tube mat at 700 m was the most distinct zone of animals in our study area (Figs. 8.4, 8.5, and 8.6, Appendix 8.5.4) and extended from the north side of the canyon to north of Santa Cruz (Figs. 8.1 and 8.2).

This crustacean-dominated margin could be a sensitive indicator of the expansion of the regional oxygen minimum zone. The tube mat could disappear, shrink, or move up slope. Crustaceans are generally rare in low oxygen environments (Levin 2003; Gooday et al. 2009; Menot et al. 2010). However, distinct margin communities occur where oxygen levels are less stressful and food is apparently abundant (Mullins et al. 1985; Levin 2003; Gooday et al. 2009; Menot et al. 2010). Since some oxygen minimum zones have expanded during the recent global warming (Stramma et al. 2010), a margin full of suspension-feeding amphipods should be a sensitive indicator of increasing stress from lower oxygen.

The 700 m ampelicid community was ecologically unique in several ways. First was the remarkable similarity of the assemblages among replicates on the same transect and among transects covering a large geographic region. This high similarity was indicated both by the strong clustering (Figs. 8.4, 8.5, and 8.6) and the presence of the same numerically dominant species along each transect (Appendix 8.5.4). Second was the high percentage of species (40%) and individuals (67%) of crustaceans, despite the fact that crustaceans are generally intolerant of low oxygen. A community dominated by crustaceans is unexpected and unique among oxygen minimum margin communities. Finally, this is the only dense ampeliscid tube mat known from deep water. All others are found in shallow water. The most extensive ampeliscid tube mats occur in the Bering and Chukchi Seas, where they cover much of the continental shelf and are the major feeding grounds of gray whales (Nerini and Oliver 1983; Oliver et al. 1983, 1984; Oliver and Slattery 1985). Perhaps this deep-water crustacean community is maintained by strong bottom currents sweeping down the slope and bathing the margin with oxygenated water and food.

Only one other dense amphipod tube mat was reported from deep water. It was dominated by one small species, *Photis typhlops*, peaking at 9500 individuals m^{-2} at 1770 to 1990 m offshore of San Francisco (Conlan 1995). This species of *Photis* and others were common in our survey as well (Appendix 8.5.4). In our survey, the abundances of all crustaceans were over 10,000 per m^2 at 109 m and 325 m, where the total abundance of the entire fauna was 16,480 m^2 and 17,520 m^2, respectively (Table 8.6). However, these high-abundance communities also had high diversity and low dominance (Figs. 8.7, 8.8, 8.9, and 8.10, Tables 8.5 and 8.6).

Crustaceans characterized bottom communities with the highest diversity of species. At the shelf-slope break, they were mainly small amphipods in tubes or nestling in surface sediments. Along the gravel and mixed hard and soft bottoms of the upper slope, the community was characterized by a large caprellid amphipod, which was the most abundant animal sampled from 10 to 2000 m. Since it co-occurred with a number of other abundant species, community dominance was not high. The ampeliscid amphipod community at the upper margin of the oxygen minimum zone was not diverse, but this site had some of the highest percentages of both crustacean species and individuals. In contrast, in shallow water, there was a dramatic loss of amphipods and ostracods, when comparing the 1970s to our sampling in 1999. Here the crustaceans indicated ecological stress, correlated with decades of global warming. Whether positive or negative ecological indicators, the crustacean patterns along this continental margin highlighted the most impressive changes in benthic community structure.

Acknowledgments This work is impossible without good taxonomy. The late Jim Nybakken was instrumental in fostering taxonomic expertise in many students. We depended especially on Paul Valentich-Scott (bivalves), Hank Chaney (gastropods), Megan Lily (ophiuroids), Gene Ruff (polychaetes), and Peter Slattery (crustaceans). We dedicate this effort to the late Gene Ruff, who, like Olga Hartman, loved polychaete worms. We are always indebted for the support from the staff, faculty, and students at the Moss Landing Marine Laboratories, in particular Gary Greene. Jessica Soriano created the schematic of the California coast used in Fig. 8.2. This project was funded by a contract from MCI Inc. (via Dallas Meggitt) to ABA Consultants, allowing us to establish a unique baseline for the Monterey Bay benthos.

Appendices

Appendix 8.5.1: Station locations and the number (N) of replicate grab samples (0.1 m²) taken and processed at each station for community analyses along the four depth transects in the Monterey Bay area

Depth	Transect 1			Transect 2			Transect 3			Transect 4		
(m)	Lat	Long	N	Lat	Long	N	Lat	Long	N	Lat	Long	N
10	36.9076	121.8573	5	–	–	–	–	–	–	–	–	–
20	36.9050	121.8854	5	–	–	–	–	–	–	–	–	–
30	36.8996	121.9465	5	–	–	–	37.0166	122.2298	2	37.0243	122.2398	2
50	36.8938	122.0295	2	–	–	–	36.9992	122.2364	2	37.0275	122.2675	2
60	36.8867	122.0447	2	36.8749	121.9603	2	36.9980	122.2508	2	37.0260	122.2830	2
90	36.8580	122.0932	5	36.8648	121.9679	2	36.9727	122.2783	2	37.0366	122.3662	2
109	36.8386	122.1342	5	36.8034	122.0634	2	36.9523	122.2980	2	37.0393	122.3997	2
150	36.8275	122.1493	4	36.8524	122.1651	2	36.9362	122.3176	2	36.9950	122.5061	2
325	36.8249	122.1680	2	36.8245	122.1703	1	36.9191	122.3250	2	37.0015	122.5918	2
450	36.7963	122.1938	2	36.8038	122.1836	5	36.9078	122.3287	2	37.0077	122.6516	2
700	36.7717	122.2410	5	36.8042	122.2433	5	36.8642	122.3410	2	37.0115	122.7389	2
1000	36.7520	122.2830	5	36.7830	122.2850	2	36.8405	122.3618	2	37.0089	122.7739	2
1200	36.7481	122.3077	2	36.7566	122.3094	3	36.8256	122.3649	2	–	–	–
1500	36.7280	122.3548	2	–	–	–	36.8133	122.4002	2	–	–	–
2000	36.7180	122.3895	2	–	–	–	36.7708	122.4554	2	–	–	–

Appendix 8.5.2: Images of the seafloor and seafloor elements in Monterey Bay

Appendix 8.5.2.1: Glauconite from a sediment sample from 877 m depth off Monterey, California. © Linda Kuhnz, 2006. Scale bar = 0.25 mm

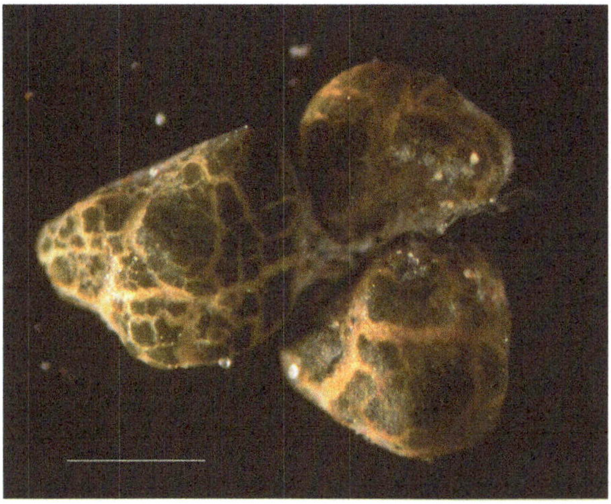

Appendix 8.5.2.2: Sand ripple bottom at 45 m. ABA © 1999 (ROV Phantom/Remora)

Appendix 8.5.2.3: Mud bottom at 90 m. ABA © 1999 (ROV Phantom/Remora)

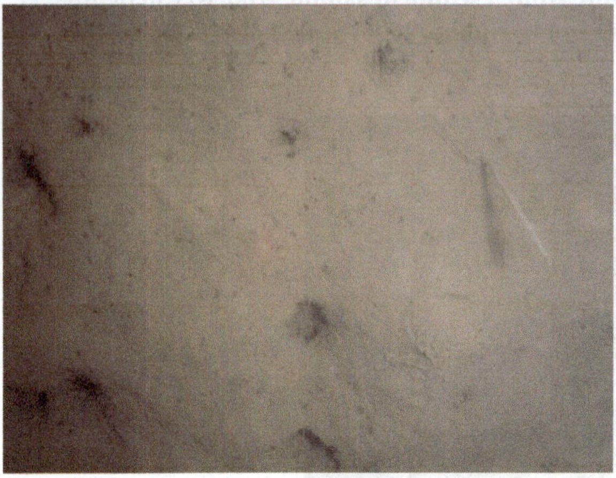

Appendix 8.5.2.4: Large group of sea urchins, *Strongylocentrotus fragilis*, at 91 m. ABA © 1999 (ROV Phantom/Remora)

Appendix 8.5.2.5 Live and dead brachiopods, *Laqueus californianus*, at 150 m. ABA © 1999 (ROV Phantom/Remora)

Appendix 8.5.2.6: High density of brachiopods, *Laqueus californianus*, at 112 m. ABA © 1999 (ROV Phantom/Remora)

Appendix 8.5.2.7: Mud bottom at 112 m. ABA © 1999 (ROV Phantom/Remora)

Appendix 8.5.2.8 Mixed bottom at 122 m. ABA © 1999 (ROV Phantom/Remora)

Appendix 8.5.2.9: Ophiuroids and a sea star, *Rathbunaster californicus*, at 191 m. ABA © 1999 (ROV Phantom/Remora)

Appendix 8.5.2.10: Ophiuroids at 193 m. ABA © 1999 (ROV Phantom/Remora)

Appendix 8.5.2.11: Mixed bottom at 324 m. ABA © 1999 (ROV Phantom/Remora)

Appendix 8.5.2.12: Mixed bottom at 434 m. ABA © 1999 (ROV Phantom/Remora)

Appendix 8.5.2.13: Ampeliscid tube mat found at 700 m. ABA © 1999 (ROV Phantom/Remora)

Appendix 8.5.2.14: Mud bottom at 1000 m. ABA © 1999 (ROV Phantom/Remora)

Appendix 8.5.2.15: Mud bottom at 1200 m. ABA © 1999 (ROV Phantom/Remora)

8.4.1 Appendix 8.5.3: Dendrogram displaying the results of a cluster analysis on samples from all four transects combined. Clusters of water depths connected by dashed orange lines are not significantly different from each other, but are significantly different from other clusters of connected orange lines

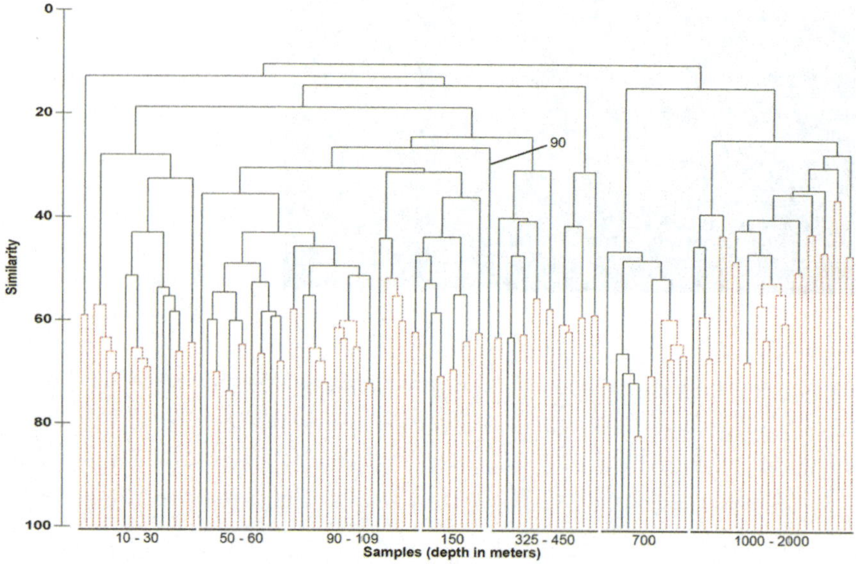

Appendix 8.5.4: Top ten species at each depth and transect (means and standard deviations for N grab samples shown in Tables 4-5 and Appendix 8.5.1). Blank sections have no data to report for that depth and transect combination

10 m

Transect 1				Transect 2				Transect 3				Transect 4			
Species	Group	Mean	SD	Species	Group	Mean	SD	Species	Group	Mean	SD	Species	Group	Mean	SD
Mediomastus acutus	Polychaeta	69.0	20.6												
Scoletoma tetraura Cmplx	Polychaeta	46.2	20.0												
Prionospio pygmaeus	Polychaeta	41.8	37.9												
Glycera macrobranchia	Polychaeta	30.6	6.5												
Chaetozone lunula	Polychaeta	9.6	2.3												
Americhelidium shoemakeri	Crustacea	7.6	4.2												
Pacifoculodes barnardi	Crustacea	7.4	4.4												
Glycinde armigera	Polychaeta	6.0	4.8												
Alamprops sp.	Crustacea	6.0	6.0												
Bathycopea daltonae	Crustacea	5.8	5.4												

20 m

Transect 1				Transect 2				Transect 3				Transect 4			
Species	Group	Mean	SD	Species	Group	Mean	SD	Species	Group	Mean	SD	Species	Group	Mean	SD
Mediomastus sp.	Polychaeta	21.2	7.6												
Euchone sp. A	Polychaeta	17.8	10.5												
Prionospio pygmaeus	Polychaeta	17.4	7.3												
Mediomastus acutus	Polychaeta	13.8	9.2												
Hemilamprops californicus	Crustacea	12.0	7.5												
Mediomastus californiensis	Polychaeta	8.0	2.3												
Glycinde armigera	Polychaeta	7.8	3.1												
Pleurogonium californiense	Crustacea	6.2	2.9												
Rhepoxynius fatigans	Crustacea	6.2	3.9												
Zeugophilomedes oblonga	Crustacea	6.2	4.0												

30 m

Transect 1				Transect 2				Transect 3				Transect 4			
Species	Group	Mean	SD	Species	Group	Mean	SD	Species	Group	Mean	SD	Species	Group	Mean	SD
Hemilamprops californicus	Crustacea	31.2	7.0					*Alvania rosana*	Mollusca	73.5	7.8	*Prionospio pygmaeus*	Polychaeta	42.5	3.5
Actiniaria	Cnidaria	24.0	8.5					*Hemilamprops californicus*	Crustacea	50.0	11.3	*Gadila* sp.	Mollusca	34.5	4.9
Gadila sp.	Mollusca	18.4	7.9					*Hesionura coineaui difficilis*	Polychaeta	40.5	34.6	*Zeugophilomedes oblonga*	Crustacea	31.5	10.6
Mediomastus sp.	Polychaeta	18.0	10.1					*Ophelia limacina*	Polychaeta	34.0	25.5	*Glycinde armigera*	Polychaeta	23.5	4.9
Sigalion spinosus	Polychaeta	16.0	6.3					*Cingula* sp.	Mollusca	32.0	45.3	*Nephtys caecoides*	Polychaeta	17.0	4.2
Aricidea (Acmira) catherinae	Polychaeta	11.0	8.0					*Odostomia angularis*	Mollusca	31.0	15.6	*Scoloplos armiger armiger*	Polychaeta	16.5	2.1
Macoma yoldiformis	Mollusca	8.4	10.9					*Turbonilla* sp.	Mollusca	27.5	3.5	*Siliqua* sp.	Mollusca	16.0	21.2
Photis sp.	Crustacea	8.2	11.4					Odostomia sp.	Mollusca	26.5	37.5	*Onuphis* sp.	Polychaeta	10.5	2.1
Magelona hartmanae	Polychaeta	7.8	6.2					*Spiophanes norrisi*	Polychaeta	22.0	5.7	*Sigalion spinosus*	Polychaeta	9.5	0.7
Macoma sp.	Mollusca	7.2	8.8					*Dendraster excentricus*	Echinodermata	17.0	5.7	*Bathycopea daltonae*	Crustacea	8.5	7.8

50 m

Transect 1				Transect 2				Transect 3				Transect 4			
Species	Group	Mean	SD	Species	Group	Mean	SD	Species	Group	Mean	SD	Species	Group	Mean	SD
Sternaspis affinis	Polychaeta	37.0	8.5					Actiniaria	Cnidaria	55.5	10.6	Actiniaria	Cnidaria	68.0	2.8
Photis sp.	Crustacea	30.5	43.1					*Aricidea* sp.	Polychaeta	29.0	1.4	*Prionospio pygmaeus*	Polychaeta	44.5	12.0
Aricidea (Strelzovia) sp. A	Polychaeta	26.5	2.1					*Gadila* sp.	Mollusca	27.0	4.2	*Bipalponephtys cornuta*	Polychaeta	40.5	10.6
Nemertea	Nemertea	23.0	0.0					*Spiophanes berkeleyorum*	Polychaeta	21.0	8.5	*Aricidea* sp.	Polychaeta	40.0	14.1
Aricidea (Acmira) catherinae	Polychaeta	22.5	3.5					*Mediomastus* sp.	Polychaeta	16.0	18.4	*Mediomastus* sp.	Polychaeta	38.5	14.8
Mediomastus sp.	Polychaeta	17.5	2.1					*Bipalponephtys cornuta*	Polychaeta	15.5	9.2	*Spiophanes berkeleyorum*	Polychaeta	37.0	2.8
Tellina sp. A	Mollusca	16.0	22.6					*Glycinde armigera*	Polychaeta	15.0	1.4	*Gadila* sp.	Mollusca	35.0	5.7
Hemilamprops californicus	Crustacea	15.5	21.9					*Hemilamprops californicus*	Crustacea	14.5	9.2	*Pholoe glabra*	Polychaeta	20.0	11.3
Tellinidae	Mollusca	14.5	20.5					Bivalvia	Mollusca	11.0	14.1	*Amphioplus strongyloplax*	Echinodermata	19.0	17.0
Astyris gausapata	Mollusca	14.0	2.8					*Pholoe glabra*	Polychaeta	10.5	4.9	Cirratulidae	Polychaeta	16.5	2.1

60 m

Transect 1				Transect 2				Transect 3				Transect 4			
Species	Group	Mean	SD	Species	Group	Mean	SD	Species	Group	Mean	SD	Species	Group	Mean	SD
Ophiuroidea	Echinodermata	46.5	38.9	Ophiuroidea	Echinodermata	76.0	25.5	*Bipalponephtys cornuta*	Polychaeta	59.0	11.3	*Bipalponephtys cornuta*	Polychaeta	64.5	0.7
Axinopsida serricata	Mollusca	39.0	5.7	*Sternaspis affinis*	Polychaeta	56.5	3.5	*Aricidea (Acmira) catherinae*	Polychaeta	43.0	19.8	*Aricidea (Acmira) catherinae*	Polychaeta	62.0	14.1
Sternaspis affinis	Polychaeta	32.0	1.4	*Gadila* sp.	Mollusca	45.5	12.0	Actiniaria	Cnidaria	37.5	37.5	*Spiophanes berkeleyorum*	Polychaeta	51.0	9.9
Gadila sp.	Mollusca	30.5	4.9	*Tellina* sp. A	Mollusca	42.5	60.1	*Spiophanes berkeleyorum*	Polychaeta	31.5	21.9	Nemertea	Nemertea	45.0	22.6
Amphiodia urtica	Echinodermata	26.0	1.4	*Protomedeia articulata* Cmplx	Crustacea	38.0	22.6	*Mediomastus* sp.	Polychaeta	26.0	15.6	*Mediomastus* sp.	Polychaeta	43.0	15.6
Mediomastus sp.	Polychaeta	22.0	7.1	*Lumbrineris cruzensis*	Polychaeta	35.0	8.5	*Cossura pygodactylata*	Polychaeta	25.0	11.3	*Aricidea (Strelzovia)* sp. A	Polychaeta	35.0	9.9
Aricidea (Strelzovia) sp. A	Polychaeta	19.0	4.2	*Cossura pygodactylata*	Polychaeta	32.5	26.2	*Pholoe glabra*	Polychaeta	22.5	2.1	*Gadila* sp.	Mollusca	32.0	7.1
Pholoe glabra	Polychaeta	15.0	9.9	*Bipalponephtys cornuta*	Polychaeta	32.5	29.0	*Glycinde armigera*	Polychaeta	21.0	0.0	Actiniaria	Cnidaria	28.5	6.4
Aricidea (Acmira) catherinae	Polychaeta	14.5	2.1	Tellinidae	Mollusca	32.0	45.3	Cirratulidae	Polychaeta	20.5	3.5	*Gadila aberrans*	Mollusca	28.0	0.0
Scoloplos armiger alaskensis	Polychaeta	13.5	2.1	*Mediomastus* sp.	Polychaeta	26.5	10.6	*Hemilamprops californicus*	Crustacea	18.5	7.8	*Cossura candida*	Polychaeta	23.0	11.3

90 m

Transect 1				Transect 2				Transect 3				Transect 4			
Species	Group	Mean	SD	Species	Group	Mean	SD	Species	Group	Mean	SD	Species	Group	Mean	SD
Mediomastus sp.	Polychaeta	55.6	16.3	*Protomedeia articulata* Cmplx	Crustacea	35.0	17.0	*Mediomastus* sp.	Polychaeta	67.0	18.4	*Mediomastus* sp.	Polychaeta	84.0	63.6
Spiophanes berkeleyorum	Polychaeta	50.6	37.3	*Levinsenia gracilis*	Polychaeta	34.0	0.0	*Spiophanes berkeleyorum*	Polychaeta	67.0	5.7	*Cossura pygodactylata*	Polychaeta	65.0	22.6
Cossura pygodactylata	Polychaeta	17.2	9.0	*Prionospio lighti*	Polychaeta	32.0	9.9	*Cossura pygodactylata*	Polychaeta	63.5	10.6	*Bipalponephtys cornuta*	Polychaeta	46.0	8.5
Pholoe glabra	Polychaeta	17.0	5.3	*Cossura pygodactylata*	Polychaeta	21.5	13.4	*Bipalponephtys cornuta*	Polychaeta	56.0	24.0	*Mediomastus californiensis*	Polychaeta	39.0	52.3
Axinopsida serricata	Mollusca	15.8	7.8	*Mediomastus* sp.	Polychaeta	17.0	8.5	*Levinsenia gracilis*	Polychaeta	39.5	12.0	*Spiophanes berkeleyorum*	Polychaeta	39.0	7.1
Euclymeninae	Polychaeta	15.8	4.1	*Gadila* sp.	Mollusca	16.0	1.4	*Aricidea (Strelzovia)* sp. A	Polychaeta	29.0	9.9	*Levinsenia gracilis*	Polychaeta	37.0	25.5
Bipalponephtys cornuta	Polychaeta	15.8	9.9	Lumbrineridae	Polychaeta	15.5	16.3	*Amphiodia urtica*	Echinodermata	28.0	5.7	*Pholoe glabra*	Polychaeta	34.0	0.0
Lumbrineris cruzensis	Polychaeta	14.2	3.1	*Maldane sarsi*	Polychaeta	13.0	1.4	*Prionospio lighti*	Polychaeta	22.5	4.9	*Prionospio lighti*	Polychaeta	25.0	9.9
Spiophanes duplex	Polychaeta	12.2	6.4	*Chloeia pinnata*	Polychaeta	12.5	0.7	*Pholoe glabra*	Polychaeta	20.5	0.7	*Axinopsida serricata*	Mollusca	18.0	8.5
Astyris gausapata	Mollusca	11.2	8.4	*Sternaspis affinis*	Polychaeta	12.5	6.4	Amphiuridae	Echinodermata	20.0	5.7	*Rutiderma lomae*	Crustacea	18.0	11.3

109 m

Transect 1				Transect 2				Transect 3				Transect 4			
Species	Group	Mean	SD	Species	Group	Mean	SD	Species	Group	Mean	SD	Species	Group	Mean	SD
Mediomastus sp.	Polychaeta	26.6	13.0	*Photis* sp.	Crustacea	371.0	118.8	*Myriochele* sp.	Polychaeta	395.0	80.6	*Mediomastus* sp.	Polychaeta	66.0	93.3
Spiophanes berkeleyorum	Polychaeta	21.8	12.5	*Aoroides exilis*	Crustacea	230.0	152.7	*Aoroides intermedia*	Crustacea	75.0	19.8	*Spiophanes berkeleyorum*	Polychaeta	25.5	36.1
Scoloplos armiger armiger	Polychaeta	16.8	23.0	*Aoroides inermis*	Crustacea	60.5	47.4	*Spiophanes berkeleyorum*	Polychaeta	67.0	1.4	*Gadila* sp.	Mollusca	24.5	10.6
Photis californica	Gammaridea	14.2	8.8	*Gammaropsis ociosa*	Crustacea	52.0	8.5	*Mediomastus* sp.	Polychaeta	54.5	26.2	*Cossura pygodactylata*	Polychaeta	23.5	33.2
Pholoe glabra	Polychaeta	11.4	3.9	*Exogone lourei*	Polychaeta	44.0	17.0	*Diastylis crenellata*	Crustacea	39.5	7.8	*Pholoe glabra*	Polychaeta	22.0	31.1
Tellina carpenteri	Bivalvia	10.2	9.3	*Gammaropsis ocellata*	Crustacea	29.0	41.0	*Photis californica*	Crustacea	38.5	6.4	*Levinsenia gracilis*	Polychaeta	19.0	26.9
Pleurogonium californiense	Isopoda	9.6	5.3	Tanaid sp. W	Crustacea	29.0	25.5	*Decamastus gracilis*	Polychaeta	34.0	9.9	*Bipalponephtys cornuta*	Polychaeta	15.0	21.2
Prionospio jubata	Polychaeta	9.2	8.3	Terebellidae	Polychaeta	27.5	10.6	*Capitella capitata* Cmplx	Polychaeta	30.0	38.2	*Protomedeia articulata* Cmplx	Crustacea	14.5	7.8
Rhepoxynius variatus	Gammaridea	8.8	4.5	Ampharetidae	Polychaeta	27.0	11.3	*Photis* sp.	Crustacea	25.5	12.0	*Mediomastus californiensis*	Polychaeta	14.0	19.8
Sigalion spinosus	Polychaeta	8.8	2.3	*Chondrochelia dubia* Cmplx	Crustacea	27.0	22.6	*Munnogonium tillerae*	Crustacea	21.0	19.8	Nemertea	Nemertea	13.5	2.1

150 m

Transect 1				Transect 2				Transect 3				Transect 4			
Species	Group	Mean	SD	Species	Group	Mean	SD	Species	Group	Mean	SD	Species	Group	Mean	SD
Photis sp.	Gammaridea	126.2	48.5	*Photis* sp.	Crustacea	118.0	12.7	*Euphilomedes carcharodonta*	Crustacea	137.0	18.4	*Photis* sp.	Crustacea	83.5	30.4
Orchomene sp. L	Gammaridea	85.0	53.3	*Aoroides intermedia*	Crustacea	57.0	8.5	*Mediomastus* sp.	Polychaeta	44.0	14.1	*Euphilomedes producta*	Crustacea	62.5	19.1
Gammaropsis ociosa	Gammaridea	66.5	78.6	*Orchomene* sp. L	Crustacea	43.0	8.5	*Photis* sp.	Crustacea	43.0	25.5	*Mediomastus* sp.	Polychaeta	62.0	4.2
Aoroides exilis	Gammaridea	47.7	32.5	*Spiophanes berkeleyorum*	Polychaeta	38.0	2.8	*Euphilomedes producta*	Crustacea	40.5	10.6	*Scoloplos armiger alaskensis*	Polychaeta	47.5	3.5
Euphilomedes carcharodonta	Ostracoda	47.5	33.8	*Euphilomedes carcharodonta*	Crustacea	37.0	11.3	*Photis californica*	Crustacea	37.0	25.5	Terebellidae	Polychaeta	38.5	0.7
Euphilomedes producta	Ostracoda	25.7	25.6	*Euphilomedes producta*	Crustacea	30.5	2.1	*Rutiderma lomae*	Crustacea	32.5	17.7	*Decamastus gracilis*	Polychaeta	33.5	6.4
Philomedes dentatus	Ostracoda	22.2	14.6	Cirratulidae	Polychaeta	21.5	12.0	*Tellina* sp.	Mollusca	31.0	38.2	Cirratulidae	Polychaeta	32.0	4.2
Bemlos macromanus	Gammaridea	19.0	22.5	*Mediomastus* sp.	Polychaeta	21.5	6.4	*Orchomenella decipiens*	Crustacea	28.0	26.9	*Aricidea* sp.	Polychaeta	31.0	5.7
Guernea reduncans	Ostracoda	16.7	6.5	Onuphidae	Polychaeta	20.5	14.8	*Guernea reduncans*	Crustacea	22.0	9.9	*Tellina carpenteri*	Mollusca	28.0	8.5
Orchomenella decipiens	Gammaridea	15.7	7.4	*Aphelochaeta* sp.	Polychaeta	17.5	6.4	*Armandia brevis*	Polychaeta	21.0	19.8	*Rutiderma lomae*	Crustacea	25.5	9.2

325 m

Transect 1				Transect 2				Transect 3				Transect 4			
Species	Group	Mean	SD	Species	Group	Mean	SD	Species	Group	Mean	SD	Species	Group	Mean	SD
Tritella tenuissima	Caprellida	386.5	475.9					*Chloeia pinnata*	Polychaeta	38.0	8.5	*Tellina carpenteri*	Mollusca	35.5	3.5
Sphaerosyllis ranunculus	Polychaeta	219.0	29.7					*Orchomenella decipiens*	Crustacea	21.5	0.7	*Decamastus gracilis*	Polychaeta	30.0	18.4
Mesometopa neglecta	Gammaridea	138.0	21.2					*Protomedeia articulata* Cmplx	Crustacea	18.5	4.9	*Rhepoxynius daboius*	Crustacea	24.5	7.8
Photis sp.	Gammaridea	136.0	132.9					*Decamastus gracilis*	Polychaeta	15.0	4.2	*Aricidea (Acmira) catherinae*	Polychaeta	21.0	7.1
Philomedes dentatus	Ostracoda	73.5	48.8					*Tanaopsis cadieni*	Crustacea	11.0	5.7	*Protomedeia articulata* Cmplx	Crustacea	19.0	7.1
Oligochaeta	Oligochaeta	54.0	42.4					*Guernea reduncans*	Crustacea	10.0	5.7	*Astyris gausapata*	Mollusca	18.5	16.3
Maera similis	Gammaridea	41.5	21.9					Onuphidae	Polychaeta	9.0	2.8	*Chloeia pinnata*	Polychaeta	15.5	6.4
Actiniaria	Cnidaria	40.0	18.4					*Photis* sp.	Crustacea	9.0	2.8	*Prionospio jubata*	Polychaeta	14.0	2.8
Chondrochelia dubia Cmplx	Tanaidacea	35.0	4.2					Cirratulidae	Polychaeta	8.5	9.2	*Mediomastus* sp.	Polychaeta	13.0	1.4
Amphipholis pugetana	Ophiuroidea	27.5	23.3					*Akanthophoreus phillipsi*	Crustacea	7.5	3.5	*Monoculodes emarginatus*	Crustacea	11.5	6.4

450 m

Transect 1				Transect 2				Transect 3				Transect 4			
Species	Group	Mean	SD	Species	Group	Mean	SD	Species	Group	Mean	SD	Species	Group	Mean	SD
Chloeia pinnata	Polychaeta	23.0	17.0	*Protodorvillea gracilis*	Polychaeta	56.4	15.7	Onuphidae	Polychaeta	16.5	2.1	*Decamastus gracilis*	Polychaeta	17.0	1.4
Oligochaeta	Oligochaeta	16.5	7.8	*Amphipholis pugetana*	Echinodermata	49.8	74.6	*Bipalponephtys cornuta*	Polychaeta	16.0	8.5	*Tellina carpenteri*	Mollusca	14.0	9.9
Hardametopa nasuta	Crustacea	14.0	12.7	*Prosphaerosyllis bilineata*	Polychaeta	46.6	17.2	*Protomedeia articulata* Cmplx	Crustacea	12.5	2.1	*Bipalponephtys cornuta*	Polychaeta	11.5	7.8
Tellina carpenteri	Mollusca	12.0	1.4	Onuphidae	Polychaeta	39.2	34.1	Cirratulidae	Polychaeta	12.0	0.0	Onuphidae	Polychaeta	11.0	5.7
Photis californica	Crustacea	9.0	0.0	*Chloeia pinnata*	Polychaeta	17.4	12.2	*Chloeia pinnata*	Polychaeta	9.5	10.6	*Protomedeia articulata* Cmplx	Crustacea	10.0	1.4
Onuphidae	Polychaeta	8.5	10.6	Oligochaeta	Oligochaeta	17.0	14.6	*Amphissa undata*	Mollusca	8.0	11.3	Euclymeninae	Polychaeta	9.0	7.1
Laonice nuchala	Polychaeta	6.5	3.5	*Tritella tenuissima*	Crustacea	17.0	6.9	*Polycirrus* sp.	Polychaeta	8.0	1.4	*Araphura breviaria*	Crustacea	7.5	0.7
Decamastus gracilis	Polychaeta	5.5	0.7	*Mooreonuphis segmentispadix*	Polychaeta	14.4	19.3	*Amphissa bicolor*	Mollusca	7.5	10.6	*Chauliopleona dentata*	Crustacea	7.5	0.7
Mediomastus sp.	Polychaeta	5.5	3.5	*Dipolydora* sp.	Polychaeta	12.8	25.3	*Tellina carpenteri*	Mollusca	7.5	9.2	*Harbansus mayeri*	Crustacea	7.5	7.8
Ampelisca unsocalae	Crustacea	5.0	2.8	*Syllis heterochaeta*	Polychaeta	12.8	24.4	*Sphaerosyllis ranunculus*	Polychaeta	6.5	2.1	*Amphissa bicolor*	Mollusca	7.0	9.9

700 m

Transect 1				Transect 2				Transect 3				Transect 4			
Species	Group	Mean	SD	Species	Group	Mean	SD	Species	Group	Mean	SD	Species	Group	Mean	SD
Byblis barbarensis	Gammaridea	82.2	45.5	*Byblis barbarensis*	Crustacea	105.4	28.0	*Ampelisca unsocalae*	Crustacea	105.5	2.1	*Ampelisca unsocalae*	Crustacea	134.0	32.5
Ampelisca unsocalae	Gammaridea	74.8	31.2	*Ampelisca unsocalae*	Crustacea	54.0	24.1	*Byblis barbarensis*	Crustacea	102.0	0.0	*Byblis barbarensis*	Crustacea	72.5	13.4
Myriochele gracilis	Polychaeta	57.2	30.4	*Gammaropsis ocellata*	Crustacea	50.6	18.2	*Myriochele gracilis*	Polychaeta	99.5	27.6	*Phyllochaetopterus limicolus*	Polychaeta	47.5	6.4
Phyllochaetopterus limicolus	Polychaeta	24.8	15.9	*Byblis thyablis*	Crustacea	45.4	14.0	*Harpiniopsis epistomata*	Crustacea	68.0	22.6	*Gammaropsis ocellata*	Crustacea	40.0	32.5
Gammaropsis ocellata	Gammaridea	19.0	14.2	*Myriochele gracilis*	Polychaeta	43.4	21.8	*Lepidepecreum kasatka*	Crustacea	54.5	17.7	*Photis typhlops*	Crustacea	21.0	7.1
Euclymeninae sp. A	Polychaeta	13.0	2.8	*Lepidepecreum kasatka*	Crustacea	19.8	14.2	*Gammaropsis ocellata*	Crustacea	45.0	35.4	Euclymeninae sp. A	Polychaeta	18.5	2.1
Harpiniopsis epistomata	Gammaridea	9.8	5.1	*Phyllochaetopterus limicolus*	Polychaeta	16.8	9.1	*Photis typhlops*	Crustacea	29.0	7.1	*Lepidepecreum kasatka*	Crustacea	15.5	17.7
Lepidepecreum kasatka	Gammaridea	9.0	8.9	*Chaetozone* sp. SD3	Polychaeta	15.0	6.0	*Phyllochaetopterus limicolus*	Polychaeta	28.0	1.4	*Harpiniopsis epistomata*	Crustacea	10.5	2.1
Eudorella pacifica	Cumacea	5.6	4.2	*Laonice nuchala*	Polychaeta	12.0	4.8	*Maldane sarsi*	Polychaeta	21.0	5.7	*Byblis bathyalis*	Crustacea	8.5	9.2
Prionospio marsupialia	Polychaeta	5.6	4.0	*Photis typhlops*	Crustacea	10.6	4.8	Euclymeninae	Polychaeta	13.0	2.8	*Chaetozone* sp. SD3	Polychaeta	8.0	0.0

1000 m

Transect 1				Transect 2				Transect 3				Transect 4			
Species	Group	Mean	SD	Species	Group	Mean	SD	Species	Group	Mean	SD	Species	Group	Mean	SD
Aricidea (Acmira) simplex	Polychaeta	57.8	76.3	*Aricidea (Acmira) simplex*	Polychaeta	75.0	94.8	Oligochaeta	Oligochaeta	20.0	12.7	*Aricidea (Acmira) simplex*	Polychaeta	102.5	41.7
Rochefortia coani	Mollusca	19.0	23.3	Cirratulidae	Polychaeta	20.0	19.8	Cirratulidae	Polychaeta	19.5	17.7	Euclymeninae	Polychaeta	23.0	11.3
Bipalponephtys cornuta	Polychaeta	10.2	5.2	Actiniaria	Cnidaria	15.0	4.2	*Cossura* sp.	Polychaeta	19.0	24.0	Ampharetidae	Polychaeta	21.0	8.5
Oligochaeta	Oligochaeta	4.8	4.2	Ampharetidae	Polychaeta	15.0	15.6	*Aricidea (Acmira) simplex*	Polychaeta	18.0	4.2	*Cossura modica*	Polychaeta	15.5	10.6
Travisia brevis	Polychaeta	4.8	2.2	Oligochaeta	Oligochaeta	13.5	4.9	*Chaetozone* sp.	Polychaeta	13.5	19.1	Cirratulidae	Polychaeta	15.0	5.7
Neoheteromastus lineus	Polychaeta	4.6	3.0	Nemertea	Nemertea	6.5	4.9	*Bipalponephtys cornuta*	Polychaeta	12.5	0.7	Lumbrineridae	Polychaeta	13.0	8.5
Actiniaria	Cnidaria	4.4	2.4	*Ampelisca unsocalae*	Crustacea	6.0	0.0	*Ancistrosyllis groenlandica*	Polychaeta	12.0	9.9	*Cossura candida*	Polychaeta	10.0	9.9
Levinsenia gracilis	Polychaeta	3.0	1.9	*Cossura modica*	Polychaeta	6.0	8.5	Lumbrineridae	Polychaeta	9.5	4.9	*Cossura rostrata*	Polychaeta	9.5	7.8
Leiochrides hemipodus	Polychaeta	2.4	1.5	*Neomediomastus glabrus*	Polychaeta	6.0	2.8	Euclymeninae	Polychaeta	8.0	2.8	*Bipalponephtys cornuta*	Polychaeta	9.0	7.1
Nemertea	Nemertea	2.4	2.3	*Cossura candida*	Polychaeta	5.5	7.8	*Prionospio* sp.	Polychaeta	7.0	2.8	Nemertea	Nemertea	7.0	9.9

1200 m

Transect 1				Transect 2				Transect 3				Transect 4			
Species	Group	Mean	SD	Species	Group	Mean	SD	Species	Group	Mean	SD	Species	Group	Mean	SD
Cirratulidae	Polychaeta	52.0	14.1	Cirratulidae	Polychaeta	51.3	19.4	*Aricidea (Acmira) simplex*	Polychaeta	72.5	29.0				
Ampharetidae	Polychaeta	15.5	0.7	Ampharetidae	Polychaeta	16.7	4.0	Oligochaeta	Oligochaeta	15.0	5.7				
Actiniaria	Cnidaria	15.0	4.2	*Bipalponephtys comuta*	Polychaeta	11.3	3.5	Cirratulidae	Polychaeta	10.5	6.4				
Cossura modica	Polychaeta	15.0	11.3	Euclymeninae	Polychaeta	9.7	2.1	Ampharetidae	Polychaeta	9.5	4.9				
Bipalponephtys cornuta	Polychaeta	11.0	1.4	*Cossura candida*	Polychaeta	9.3	2.5	Lumbrineridae	Polychaeta	6.5	0.7				
Neomediomastus glabrus	Polychaeta	9.0	4.2	Nemertea	Nemertea	9.0	3.6	*Cossura rostrata*	Polychaeta	5.0	2.8				
Falcidens longus	Mollusca	8.5	3.5	*Cossura modica*	Polychaeta	7.3	2.1	*Cossura candida*	Polychaeta	4.5	0.7				
Prionospio sp.	Polychaeta	8.0	2.8	*Neomediomastus glabrus*	Polychaeta	7.3	4.0	*Bipalponephtys comuta*	Polychaeta	4.5	4.9				
Capitellidae sp. 2	Polychaeta	7.0	2.8	Aplacophora	Mollusca	5.7	3.5	*Synidotea calcarea*	Crustacea	3.0	1.4				
Lumbrineridae	Polychaeta	7.0	1.4	*Pholoe courtneyae*	Polychaeta	5.3	2.9	Nephtyidae	Polychaeta	2.5	0.7				

1500 m

Transect 1				Transect 2				Transect 3				Transect 4			
Species	Group	Mean	SD	Species	Group	Mean	SD	Species	Group	Mean	SD	Species	Group	Mean	SD
Cirratulidae	Polychaeta	73.0	1.4					Cirratulidae	Polychaeta	87.5	10.6				
Ampharetidae	Polychaeta	14.5	7.8					Ampharetidae	Polychaeta	8.0	0.0				
Cossura modica	Polychaeta	5.5	3.5					*Cossura rostrata*	Polychaeta	8.0	2.8				
Euclymeninae	Polychaeta	5.5	0.7					*Neomediomastus glabrus*	Polychaeta	5.0	2.8				
Cossura candida	Polychaeta	4.0	5.7					Euclymeninae	Polychaeta	4.5	2.1				
Aricidea sp.	Polychaeta	3.0	2.8					*Aphrodita japonica*	Polychaeta	2.5	0.7				
Gadila tolmiei	Mollusca	2.5	2.1					*Cossura modica*	Polychaeta	2.0	0.0				
Actiniaria	Cnidaria	2.0	0.0					Aplacophora	Mollusca	1.5	2.1				
Aplacophora	Mollusca	2.0	2.8					*Cossura candida*	Polychaeta	1.5	0.7				
Capitellidae sp. 2	Polychaeta	1.5	2.1					Lumbrineridae	Polychaeta	1.5	0.7				

2000 m

Transect 1				Transect 2				Transect 3				Transect 4			
Species	Group	Mean	SD	Species	Group	Mean	SD	Species	Group	Mean	SD	Species	Group	Mean	SD
Cirratulidae	Polychaeta	14.0	5.7					Lumbrineridae	Polychaeta	8.0	4.2				
Ampharetidae	Polychaeta	13.0	5.7					Aricidea sp.	Polychaeta	7.0	2.8				
Aricidea sp.	Polychaeta	11.0	11.3					Ampharetidae	Polychaeta	5.5	2.1				
Mendicula ferruginosa	Mollusca	11.0	15.6					*Aricidea (Strelzovia)* sp. A	Polychaeta	3.0	0.0				
Lumbrineridae	Polychaeta	9.0	5.7					Aplacophora	Mollusca	2.5	0.7				
Cossura pygodactylata	Polychaeta	6.5	2.1					*Chaetozone* cf. *spinosa*	Polychaeta	2.0	1.4				
Aplacophora	Mollusca	5.0	1.4					Cirratulidae	Polychaeta	2.0	0.0				
Aricidea (Strelzovia) sp. A	Polychaeta	4.0	4.2					*Levinsenia gracilis*	Polychaeta	1.5	0.7				
Ninoe longibranchia	Polychaeta	4.0	5.7					*Pettiboneia* sp. nov. 1	Polychaeta	1.5	2.1				
Terebellides sp.	Polychaeta	4.0	1.4					*Prionospio delta*	Polychaeta	1.5	2.1				

References

Bachelet G (1990) The choice of a sieving mesh size in the quantitative assessment of marine macrobenthos: a necessary compromise between aims and constraints. Mar Environ Res 30:21–35

Barnard JL (1963) Relationship of benthic Amphipoda to invertebrate communities of inshore sublittorial sands of Southern California. Pacif Nat 3:439–467

Barnard JL, Ziesenhenne FC (1961) Ophiuroid communities of Southern Californian coastal bottoms. Pacif Nat 2:131–152

Blake JA, Grassle JF (1994) Benthic community structure on the US South Atlantic slope off the Carolinas: spatial heterogeneity in a current-dominated system. Deep-Sea Res II 41:835–874

Blake JA, Maciolek NJ, Ota AY, Williams IP (2009) Long- term benthic infaunal monitoring at a deep-ocean dredged material disposal site off Northern California. Deep-Sea Res II 56:1775–1803

Breaker LC, Broenkow WW (1994) The circulation of Monterey Bay and related processes. Oceanogr Mar Biol 32:1–64

Breaker LC, Mooers CNK (1986) Oceanic variability off the Central California coast. Prog Oceanogr 17:61–135

Bruland KW, Rue EL, Smith GJ (2001) The influence of iron and macronutrients in coastal upwelling regimes off central California: implications for extensive blooms of large diatoms. Limnol Oceanogr 46:1661–1674

Cacchione DA, Pratson LF, Ogston AS (2002) The shaping of continental slopes by internal tides. Science 296:724–727

Carroll ML, Ambrose WG Jr (2012) Benthic infaunal community variability on the northern Svalbard shelf. Polar Biol 35:1259–1272

Cheriton OM, McPhee-Shaw EE, Shaw WJ, Stanton TP, Bellingham JG, Storlazzi CD (2014) Suspended particulate layers and internal waves over the southern Monterey Bay continental shelf: an important control on shelf mud belts? J Geophys Res 119(1):428–444

Christopher RSBF, Bolam SG, Eggleton JD, Mason C (2012) Large-scale faunal characterisation of marine benthic sedimentary habitats around the UK. J Sea Res 69:53–65

Clarke KR, Warwick RM (2001) Change in marine communities: an approach to statistical analysis and interpretation, 2nd edn. PRIMER-E, Plymouth

Coleman NC, Gason A, Poore GCB (1997) High species richness in the shallow marine waters of south-east Australia. Mar Ecol Prog Ser 154:17–26

Conlan K (1995) 4. Superfamily Corophoidea. In: Blake J, Watling L, Scott P (eds) Taxonomic atlas of the benthic fauna of the Santa Maria Basin and Western Santa Barbara Channel. Vol 12 – the Crustacea part 3 the Amphipoda. Santa Barbara Museum of Natural History, Santa Barbara

Fauchald K, Jumars PA (1979) The diet of worms: a study of polychaete feeding guilds. Oceanogr Mar Biol 17:193–284

Gooday AJ, Levin LA, Aranda da Silva A, Bett B, Cowie G, Dissard D, Gage J, Hughes D, Jeffreys R, Larkin K, Murty SJ, Shumaker S, Whitcraft C, Woulds C (2009) Faunal responses to oxygen gradients on the Pakistan margin: a comparison of foraminifera, macrofauna and megafauna. Deep-Sea Res II 56:488–502

Google Earth Pro 7.3.2.5776 (2019) Monterey Bay 36.8037°N, 121.9439°W, elevation to −2000 m. 3D map, Oceans data layer, viewed 18 November 2019. http://www.google.com/earth/index.html

Gray JS, Poore GCB, Ugland KI, Wilson RS, Olsgard F, Johannessen Ø (1997) Coastal and deep-sea benthic diversities compared. Mar Ecol Prog Ser 159:97–103

Hammerstrom KK, Ranasinghe JA, Weisberg SB, Oliver JS, Fairey WR, Slattery PN, Oakden JM (2010) Effect of sample area and sieve size on benthic macrofaunal community condition assessments in California enclosed bays and estuaries. Integr Environ Asses Manag 8(4):649–658

Henkel SK, Nelson WG (2018) Assessment of spatial patterns in benthic macrofauna of the U.S. west coast continental shelf. J Biogeogr 45:2701–2717

Hodgson AT, Nybakken JW (1973) A quantitative survey of the benthic infauna of northern Monterey Bay, California. Moss Landing Marine Laboratories Technical Publication, 73–78, p 245

Hyland J, Baptiste E, Campbell J, Kennedy J, Kropp R, Williams S (1991) Macroinfaunal communities of the Santa Maria Basin on the California outer continental shelf and slope. Mar Ecol Prog Ser 78:147–161

James RJ, Smith MPL, Fairweather PG (1995) Sieve mesh-size and taxonomic resolution needed to describe natural spatial variation of marine macrofauna. Mar Ecol Prog Ser 118:187–198

Jayaraj KA, Josia J, Dinesh Kumar PK (2008) Infaunal macrobenthic community of soft bottom sediment in a tropical shelf. J Coast Res 24:708–718

Lecher AL, Mackey K, Kudela R, Ryan J, Fisher A, Murray J, Paytan A (2015) Nutrient loading through submarine groundwater discharge and phytoplankton growth in Monterey Bay, CA. Environ Sci Technol 49(11):6665–6673

Levin LA (2003) Oxygen minimum zone benthos: adaptation and community response to hypoxia. Oceanogr Mar Biol 41:1–45

Lie U, Kisker DS (1970) Species composition and structure of benthic infaunal communities off the coast of Washington. J Fish Res Board Can 27:2273–2285

Macdonald TA, Burd BJ, Macdonald VI, van Roodselaar A (2010) Taxonomic and feeding guild classification for the marine benthic macroinvertebrates of the Strait of Georgia, British Columbia. Can Tech Rep Fish Aquatic Sci 2874:1–63

McPhee-Shaw E (2006) Boundary–interior exchange: reviewing the idea that internal-wave mixing enhances lateral dispersal near continental margins. Deep-Sea Res II 53(1–2):42–59

McPhee-Shaw EE, Sternberg RW, Mullenbach B, Ogston AS (2004) Observations of intermediate nepheloid layers on the northern California margin. Cont Shelf Res 24:693–720

Menot L, Sibuet M, Carney RS, Levin LA, Rowe GT, Billett DSM, Poore G, Kitazato H, Vanreusel A, Galeron J, Lavrado HP, Sellanes J, Ingole B, Krylova E (2010) New perceptions of continental margin biodiversity. In: AD MI (ed) Life in the world's oceans: diversity, distribution, and abundance. Wiley-Blackwell, Oxford, pp 79–101

Mullins HT, Thompson JB, McDougall K, Vercoutere TL (1985) Oxygen-minimum zone edge effects: evidence from the central California coastal upwelling system. Geology 13:491–494

Nerini MK, Oliver JS (1983) Gray whales and the structure of the Bering Sea benthos. Oecologica 59:224–225

Noble MA, Ramp SR (2000) Subtidal currents over the central California slope: evidence for offshore veering of the undercurrent and for direct, wind-driven slope currents. Deep-Sea Res II 47:871–906

Oliver JS (1984) Selection for asexual reproduction in an Antarctic polychaete worm. Mar Ecol Prog Ser 19:33–38

Oliver JS, Slattery PN (1985) Destruction and opportunity on the sea floor: effects of gray whale feeding. Ecology 66:1965–1975

Oliver JS, Slattery PN, Hulberg LW, Nybakken JW (1980) Relationships between wave disturbance and zonation of benthic invertebrate communities along a high-energy subtidal beach in Monterey Bay, California. Fish Bull 78:437–454

Oliver JS, Slattery PN, Silberstein MA, O'Connor EF (1983) A comparison of gray whale, *Eschrichtius robustus*, feeding in the Bering Sea and Baja California. Fish Bull 81:513–522

Oliver JS, Slattery PN, Silberstein MA, O'Connor EF (1984) Gray whale feeding on dense ampeliscid amphipod communities near Bamfield, British Columbia. Can J Zool 62:41–49

Oliver JS, Slattery PN, Hammerstrom K, Kim S, Oakden JM, Barnes EM (2008) Sandy bottom communities at the end of a cold (1971–1975) and warm (1997–1998) regime in the California Current: impacts of high and low plankton production. Nat Precedings. https://doi.org/10.1038/npre.2008.2103.1

Oliver JS, Hammerstrom K, McPhee-Shaw E, Slattery P, Oakden J, Kim S, Hartwell SI (2011) High species density patterns in macrofaunal invertebrate communities in the marine benthos. Mar Ecol 32:278–288

Ramp SR, Rosenfeld LK, Tisch TD, Hicks MR (1997) Moored observations of the current and temperature structure over the continental slope off central California: 1. A basic description of the variability. J Geophys Res 102:22,877–22,902

Ramp SR, Paduan JD, Shulman I, Kindle J, Bahr FL, Chavez FP (2005) Observations of upwelling and relaxation events in the northern Monterey Bay during August 2000. J Geophys Res 110:C07013. https://doi.org/10.1029/2004JC002538

Ranasinghe JA, Barnett AM, Schiff K, Montagne DE, Brantley C, Beegan C, Weisberg S B (2007) Southern California Bight 2003 regional monitoring program: III. Benthic Macro- fauna, Costa Mesa. Southern California Coastal Water Research Project

Reish DJ (1959) A discussion of the importance of the screen size in washing quantitative marine bottom samples. Ecology 40:307–309

Rosenberger KJ, Storlazzi CD, Cheriton OM (2016) Variability of the internal tide on the southern Monterey Bay continental shelf and associated bottom boundary layer sediment transport. Cont Shelf Res 120:68–81

Ryan JP, Chavez FP, Bellingham JG (2005) Physical-biological coupling in Monterey Bay, California: topographic influences on phytoplankton ecology. Mar Ecol Prog Ser 287:23–32
Ryan JP, McManus MA, Sullivan JM (2010) Interacting physical, chemical and biological forcing of phytoplankton thin-layer variability in Monterey Bay, California. Cont Shelf Res 30(1):7–16
Sevadjian JC, McManus MA, Ryan J, Greer AT, Cowen RK, Woodson CB (2014) Across-shore variability in plankton layering and abundance associated with physical forcing in Monterey Bay, California. Cont Shelf Res 72:138–151
Shea RE, Broenkow WW (1982) The role of internal tides in the nutrient enrichment of Monterey Bay, California. Estuar Coast Shelf Sci 15:57–66
Slattery PN (1985) Life histories of infaunal amphipods from subtidal sands of Monterey Bay, California. J Crustac Biol 5:635–649
Storlazzi CD, McManus MA, Figurski JD (2003) Long-term, high-frequency current and temperature measurements along central California: insights into upwelling/relaxation and internal waves on the inner shelf. Cont Shelf Res 23:901–918
Stramma L, Schmidt S, Levin LA, Johnson GC (2010) Ocean oxygen minima expansions and their biological impacts. Deep-Sea Res I 210:587–595
Stull JK, Haydock CI, Smith RW, Montagne DE (1986) Long-term changes in the benthic community on the coastal shelf off Palos Verdes, southern California. Mar Biol 91:539–551
Swartz RC, Cole FA, Shults DW, DeBen WA (1986) Ecological changes in the Southern California Bight near a large sewage outfall: benthic conditions in 1980 and 1983. Mar Ecol Prog Ser 31:1–13
Thompson JB, Mullins HT, Newton CR, Vercoutere TL (1985) Alternate biofacies model for dysaerobic communities. Lethaia 18:169–179
Thorson G (1957) Bottom communities (sublittoral or shallow shelf). In: Hedgpeth J (ed) Treatise on marine ecology and paleoecology. Geological Society of America Memoirs, Boulder, pp 461–534
Woodson CB, Washburn AL, Barth JA, Hoover DJ, Kirincich AR, McManus MA, Ryan JP, Tyburczy J (2009) The northern Monterey Bay upwelling shadow front: observations of a coastally- and surface-trapped buoyant plume. J Geophys Res 114:C12013. https://doi.org/10.1029/2009JC005623

Chapter 9
Deep-Water Penaeoid Shrimp of the Southern Gulf of Mexico Upper Slope: Distribution, Abundance, and Fishery Potential

A. Gracia and A. R. Vázquez-Bader

Abstract A systematic study to investigate the epibenthic megafauna biodiversity and potential fishing resources in the upper slope (290–1200 m depth) was carried along the Mexican Gulf of Mexico (off Tamaulipas-Yucatán). Samples were collected with a commercial shrimp trawl net (18 m mouth aperture, 4.5 cm stretched mesh, 1.5 cm stretched mesh cod-end). Fourteen species of the Aristaeidae, Penaeidae, Solenoceridae, and Benthesicymidae families were caught: *Aristaeomorpha foliacea*, *Aristaeopsis edwardsiana*, *Aristeus antillensis*, *Hepomadus tener*, *Parapenaeus americanus*, *Penaeopsis serrata*, *Parapenaeus politus*, *Funchalia villosa*, *Pleoticus robustus*, *Hymenopenaeus debilis*, *Solenocera vioscai*, *S. necopina*, *S. atlantidis*, and *Benthoecetes bartletti*. Two species were first records and two species extended its distribution to the southern Gulf of Mexico. Depth range of ten species was extended in its deeper limit. Shrimp of the Aristaeidae family were the most abundant with 6173 individuals (41%) followed by Penaeidae with 4772 organisms (31%), Solenoceridae 2914 with individuals (19%), and Benthesicymidae with 1352 specimens (9%). Seven species represented 99% of the total penaeoid catch in numbers and biomass (*A. foliacea*, *P. serrata*, *P. robustus*, *A. antillensis*, *P. americanus*, *B. bartletti*, and *A. edwardsiana*). Largest sizes were recorded in *A. foliacea*, *A. edwardsiana*, and *P. robustus* deep-water shrimp (45.57–48.09 mm mean CL). Maximum CL were also registered in these species (86.4–97.85 mm CL). Estimated penaeoid deep-water catch per unit effort (kg/h) varied from zero to18.62 kg/h. High mean CPUE values (>1.0 kg/h) were estimated at the 300–799 m depth range. Four areas of high deep-water shrimp abundance were identified in an area estimated of 60,000 km^2 in the upper slope at 300–1000 m depth. The penaeoid fishery potential and possible utilization is discussed.

Keywords Penaeoid · Deep-water shrimp · Distribution · Gulf of Mexico
Potential fishery resources · CPUE · Fishing grounds

A. Gracia (✉) · A. R. Vázquez-Bader
Instituto de Ciencias del Mar y Limnología, Unidad Académica de Biodiversidad y Ecología Marina, Universidad Nacional Autónoma de México, CDMX, México
e-mail: gracia@unam.mx

M. E. Hendrickx (ed.), *Deep-Sea Pycnogonids and Crustaceans of the Americas*, https://doi.org/10.1007/978-3-030-58410-8_9

9.1 Introduction

The Gulf of Mexico is a semi-enclosed sea bordered by three nations (Mexico, USA, and Cuba). It is the ninth largest body of the world that has an extension of 1,540,000 km^2 (Ward and Tunnell 2017) and has been classified as one of the Large Marine Ecosystems of the planet (Kumpf et al. 1999). Several ecosystems are found around the Gulf, like salt marshes, oyster reefs, mangrove swamps, and seagrasses. About 55% of its surface area (0.9 million km^2) belongs to Mexico Economic Exclusive Zone. The Gulf of Mexico basin has an average depth of 1485 m with a maximum depth near to 4000 m in the central area and the Sigsbee Canyon (Darnell 2015). According to Ward and Tunnell (2017) approximately 32% of the Gulf of Mexico is continental shelf (up to 200 m), 41% is continental slope (200–3000 m), and 24% is abyssal plain (>3000 m). Continental shelves have different sediment types; western shelves in the North and South are mainly composed of fine-grained mud and clay sediments of terrigenous origin, whereas the broad shelves adjacent to Florida and Yucatán Peninsulas are sandy carbonated areas.

General circulation pattern is influenced by the Loop Current that originates from the Caribbean Sea and enters to the Gulf between the Yucatán Peninsula and Cuba and leaves through Florida Straits (Monreal-Gómez et al. 2004). The current generates a net current West-North-East movement around the Gulf from Campeche Bank to Florida with the presence of several cyclonic-anticyclonic gyres of different scales due to wind direction and pressure effects (Monreal-Gómez and Salas de León 1997). Occasionally, large eddies spin off the Loop Current and move westward across the Gulf to Tamaulipas coasts (Sturges and Lugo Fernández 2005). The Gulf of Mexico receives freshwater load of several rivers being the most important the Mississippi River in the north and the Grijalva-Usumacinta River system in the South.

The Gulf of Mexico also supports largely the fishery production of the three bordering countries. Fisheries landings are based on an array of fishes and shellfishes exploited in the diverse ecosystems around the Gulf of Mexico (Gracia et al. 2020). Fisheries rely on several species inhabiting inshore, coastal, benthic, demersal, and oceanic-pelagic realms. Although many fish species like snappers, groupers, croakers, menhaden, mackerel, dolphinfish, billfish, and tunas are important, crustaceans represent the most valuable resource in the Gulf of Mexico. Among crustaceans shallow-water shrimp of the superfamily Penaeoidea are the most important fishing resource (Gracia et al. 2010). According to Vázquez-Bader and Gracia (1994) and Gracia and Hernández-Aguilera (2005), at least 15 species of this Superfamily occur in shallow waters of the continental shelf of the Mexican Gulf of Mexico; however, mainly six species (brown, pink, white, spotted pink, rock, and seabob shrimps) support the bulk of shrimp fishery due to its abundance and commercial size. Brown, pink, and white shrimp have been extensively exploited along the Mexican Gulf of Mexico since the early 1950s through artisanal and industrial fisheries which conducted the stocks to fully and overexploited states (Gracia 1995, 1996, 2004; Gracia and Vázquez-Bader 1998, 1999). Also due to the intense fishery activities in the continental platform of the Gulf of Mexico, it is not possible that shrimp production could increase based on the traditional fishing grounds. One alternative is to find potential fishery resources in deeper waters to fulfill the increasing demand of food. This demand has caused that fisheries activities have gone deeper around the world. In Latin America, there are 17 species of commercial interest that support deep-water shrimp fisheries mainly in Chile, Colombia, Brazil, and some activities in Costa Rica and Guyana (Arana et al. 2009; Dallagnolo et al. 2009; Wehrtmann et al. 2012; Pérez et al. 2019).

In the Gulf of Mexico, 1007 decapod crustacean species have been reported in the planktonic pelagic and benthic environments of coastal, inshore, and oceanic realms (Felder et al. 2009a). According to Wicksten and Packard (2005), 130 decapod species occur on the continental slope to the abyssal plain (200–3840 m). Felder et al. (2009a), based mainly on data of the northern Gulf of Mexico, compiled 56 shrimplike species of the Aristeidae, Benthesicymidae, Penaeidae, Sicyoniidae, and Solenoceridae families and stand out the scarce knowledge for the southern Gulf. Of these species, only Royal Red Shrimp *Pleoticus robustus* is subjected to commercial exploitation by a small-scale fishery in the northern Gulf of Mexico. In the southern Gulf of Mexico, there are no deep-water fisheries except the one carried on pelagic fish species like yellowfin tuna and other tuna, marlin, billfish, as well as incidental caught species. Deep-water benthic potential fishing resources and megafauna biodiversity have been poorly studied in the southern Gulf of Mexico. Recently, the National Autonomous University of Mexico (UNAM) through the Institute of Marine Sciences and Limnology has conducted a systematic survey to investigate the epibenthic megafauna biodiversity and to identify potential fishing resources in the upper slope of the Mexican Gulf of Mexico.

Several contributions have increased remarkably the knowledge about the species composition and abundance of crustaceans (Lozano-Álvarez et al. 2007; Briones-Fourzán et al. 2010; Vázquez-Bader and Gracia 2013, 2016; Vázquez-Bader et al. 2014; Lemaitre et al. 2014), echinoderms (Vázquez-Bader et al. 2008; Solís-Marín et al. 2014), and fishes (Ramírez et al. 2019), including the presence of deep-water shrimp of potential fisheries interest (Gracia et al. 2010) in the upper slope of Mexican Gulf. The knowledge of deep-water biodiversity and living resources of the Gulf of Mexico also is important because the Gulf of Mexico has been subjected to different stressors from diverse sources derived from land and marine environments including industrial, agriculture, urban activities, oil industry activities, and two mega oil spills in the southern (Ixtoc 1, 1979–1980) and northern Gulf (deep-water horizon, 2010). The impact of oil spills on living resources and in general on the Gulf ecosystem is of outstanding concern and recently deserved great scientific research effort (Murawski et al. 2020a). The increasing trend of oil exploration and production in ultra-deep-water fields of the Gulf of Mexico with potential possibility of another accident (Murawski et al. 2020b) stand out the need for having a better knowledge of ecological communities of the Gulf of Mexico deep waters.

9.2 Material and Methods

9.2.1 Sampling Procedure

The study developed by the National Autonomous University of Mexico (UNAM) to investigate the epibenthic megafauna biodiversity and to identify potential fishing resources in the upper slope of the Mexican Gulf of Mexico started in 1999 and then was resumed in 2007 with a yearly basis up to now. It was supported partly with funds of the project "Biodiversity and potential fishing resources in deep waters of Gulf of Mexico" (PAPIIT IN223109, DGAPA-UNAM) and then through the Institute of Marine Sciences and Limnology and ship time provided by Scientific Research Coordination of UNAM.

This unique exploration by its coverage, scope, and frequency was conducted onboard the R/V Justo Sierra of UNAM in the upper slope of the Mexican Gulf of Mexico in a depth range of 290–1200 m. Surveys were done along the entire Mexican Gulf from the Mexican-USA border (offshore Tamaulipas state) to the Yucatan Channel in the Mexican Caribbean Sea (Fig. 9.1). This study encompass 17 cruises: BATO (May 1999), BIOREPES (August 2005), BIOREPES 2 (May–June 2007), BIOREPES 3 (November 2008), COBERPES (August 2009), COBERPES 2011 (April 2011), COBERPES 3 (November 2011), COBERPES 4 (August 2012), COBERPES 5 (May 2012), COBERPES 6 (August 2014), COBERPES 7 (April 2016), COBERPES 8 (October 2016), COBERPES 9 (July–August 2017), SOGOM 1 (June 2015), SOGOM 2 (September 2016), SOGOM 3 (May 2017), and SOGOM 4 (September 2018).

Samples were obtained with a commercial shrimp trawl net (18 m mouth aperture, 4.5 cm stretched mesh, 1.5 cm stretched mesh cod-end). Since there was not enough data about sea bottom characteristics, a seabed exploration was carried with a multibeam echosounder EM 300 and a sub-bottom profiler Topas PS 18 before casting the trawl net. Procedure consisted in detecting soft bottoms large enough for operating safely the trawl net. This required a distance that could allow launching, recovering, and a 30 min trawling operation. Total distance varied because time for launching and recovering increased with depth. This means that in deeper locations (>900 m) a linear distance up to 4 miles was required for safe trawling operation. Due to ship time availability, bottom surveys were limited to 2 h and then moved to other potential areas if exploration was not successful.

Once a suitable soft bottom was found, a 30-min trawl was performed at a speed of 2.5–3 knots along the explored area. The initial and final position of each tow were registered during each cast. The trawl operation was constantly surveilled with the EM 300 echosounder and the ship GPS system whose high precision (1.2 m) practically allows to follow back the same path explored.

The catch from each haul was sorted by species, quantified, weighed, and preserved in ethanol 70%. Organisms were deposited in the Crustacean Reference Collection of the Laboratorio de Ecología Pesquera de Crustáceos (LEPC-ICML-UNAM). Shrimp were measured to the nearest mm 0.01 mm with vernier calipers

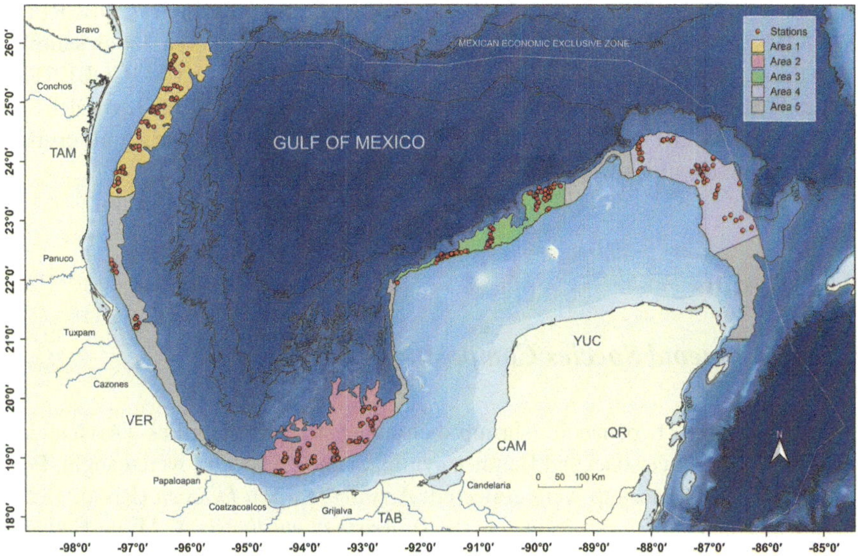

Fig. 9.1 Sampling locations in the upper slope of the Mexican Gulf of Mexico. Main river and its names are shown. Coastal states names are abbreviated. *TAM* Tamaulipas, *VER* Veracruz, *TAB* Tabasco, *CAM* Campeche, *YUC* Yucatán, *QR* Quintana Roo

from the posterior orbital margin to posterolateral margin of the carapace length (CL). Catch per unit effort (CPUE) was standardized in kilograms per hour per single net. Mean CPUE data was estimated in a 100-m-depth strata basis.

9.2.2 *Sampling Locations*

During the 17 cruises, 460 hauls were done along the Mexican Gulf. Sampling strategy tried to cover the whole gulf; however, bottom surveys did not show suitable bottoms for trawling in all the area. Trawlable bottoms (Fig. 9.1) were found mainly off Tamaulipas State (Area 1), off Tabasco and Campeche (Area 2), in the east Yucatán upper slope (Area 3) and northwest of Yucatán Peninsula near to the Mexican Caribbean Sea (Area 4). In the area denominated 5 (Fig. 9.1) trawlable bottoms were not frequently found so trawls done were a few. Bottom in this area was mainly steep, rugged, or rocky, unsuitable for trawl sampling. Soft bottoms when found in Area 5 were scattered among irregular topography and/or did not have enough distance for safe sampling. However, it is important to stand out that these soft bottoms with a patchy distribution could represent a habitat for penaeoid shrimp.

The research cruises conducted in Area 1 were BIOREPES 3, COBERPES 4, and COBERPES 9. Cruises developed in the Area 2 were comparatively more numerous: COBERPES, COBERPES 3, COBERPES 5, SOGOM 1, COBERPES, SOGOM 2, SOGOM 3, COBERPES 8, and SOGOM 4. The area 3 was sampled with cruises BATO, BIOREPES 1, BIOREPES 2, COBERPES 2011, COBERPES 3, and COBERPES 6. Cruises that covered Area 4 were BIOREPES 1, BIOREPES 2, COBERPES 2011, and COBERPES 6. Area 5 was surveyed during all cruises, but sampling locations were detected in cruise COBERPES.

9.3 Results

9.3.1 Penaeoid Species Composition

A total of 15,221 penaeoid shrimp belonging to four families (Aristaeidae, Penaeidae, Solenoceridae, and Benthesicymidae) and 14 species were caught. Four species of aristeid shrimp were collected: *Aristaeomorpha foliacea* (Risso, 1827), *Aristaeopsis edwardsiana* (Johnson, 1868), *Aristeus antillensis* A. Milne-Edwards & Bouvier, 1909, and *Hepomadus tener* Smith, 1884. Penaeoid shrimp found were *Parapenaeus americanus* Rathbun, 1901, *Penaeopsis serrata* Bate, 1881, *Parapenaeus politus* (Smith, 1881), and *Funchalia villosa* (Bouvier, 1905); the family Solenoceridae was represented by five species: *Pleoticus robustus* (Smith, 1885), *Hymenopenaeus debilis* Smith, 1882, *Solenocera vioscai* Burkenroad, 1934, *S. necopina* Burkenroad, 1939, and *S. atlantidis* Burkenroad, 1939. In the family Benthesicymidae only one species was collected: *Benthoecetes bartletti* Smith, 1882.

Shrimp of the Aristaeidae family were the most abundant with 6173 individuals (41%) followed by Penaeidae family with 4772 (31%), Solenoceridae 2914 individuals (19%), and Benthesicymidae with 1352 organisms (9%) (Fig. 9.2). In the Aristeidae family (Fig. 9.3), *A. foliacea* was the most abundant species with 3584 (58%) specimens, followed by *A. antillensis* with 1317 (21%) organisms, *A. edwardsiana* with 1255 (20%), and a low presence of *H. tener* (<1%). *Aristaeopsis edwardsiana* and *A. antillensis* were found along the whole Mexican Gulf of Mexico in a depth range of 300–1011 and 300–1011 m, respectively (Table 9.1). *Aristaeomorpha foliacea* was registered in almost all the study area except near the Caribbean Sea in the eastern-southeast sector of the Gulf following Felder et al. (2009b) division (see Briones-Fourzán et al., this volume). *A foliacea* distributed in a depth range of 226–1144 m. *H. tener* was collected only in the southwestern Gulf of Mexico in a narrow depth range (863–1144 m) (Table 9.1).

Penaeoid shrimp amounted a total catch of 4771 organisms (Fig. 9.4). Within the Penaeidae family, *P. serrata* was largely the most representative with 3400 (71%) shrimp caught followed by *P. americanus* with 1311 (28%) individuals, *P. politus* 59 (1%) organisms, and *F. villosa* with a scarce occurrence of two organisms (<1%). *Parapenaeus americanus*, *P. politus*, and *P. serrata* were collected along the entire southern Gulf of Mexico. Bathymetric distribution of *P. americanus* was registered in a shallow range of the upper slope from 244 to 462 and 206–385 m depth,

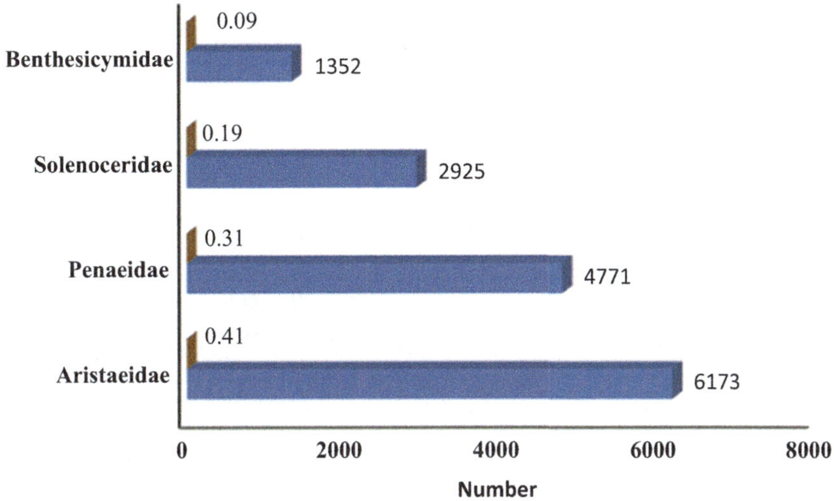

Fig. 9.2 Comparative abundance of the four deep-water penaeoid shrimp families registered in the upper slope of the Mexican Gulf of Mexico

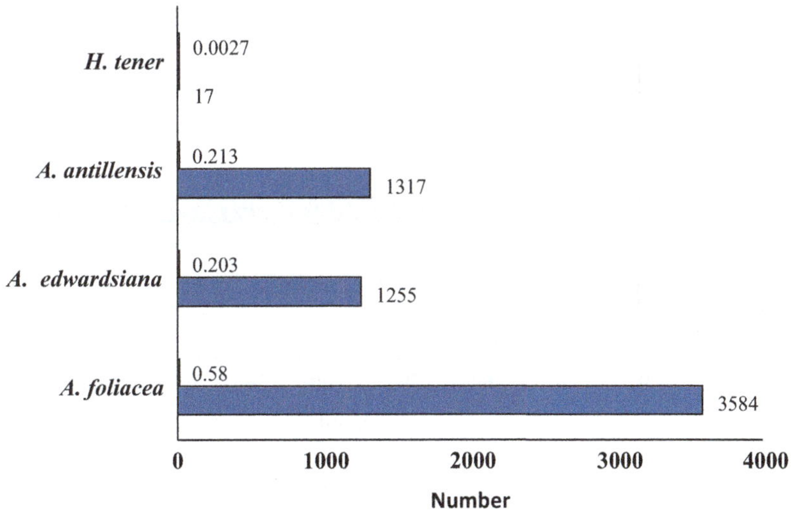

Fig. 9.3 Abundance and percentage of deep-water shrimp of the Aristaeidae family

respectively, whereas *P. politus* was found in almost all the depth range studied (309–904 m) (Table 9.1).

The Solenoceridae family was represented by five species with a total of 2925 shrimp caught (Fig. 9.5). *Pleoticus robustus* was the most abundant with 2772 individuals representing 95% of the solenocerid shrimp total catch. *Hymenopenaeus debilis* (104 organisms, 4%), *S. necopina* (36 specimens, 1.2%), *S. vioscai* (12

Table 9.1 Bathymetric distribution of penaeoid deep-water shrimp in the Mexican Gulf of Mexico

Species	Depth range (m)
A. foliacea	226–1144
A. edwardsiana	300–1011
A. antillensis	300–1108
H. tener	863–1147
P. americanus	244–462
P. serrata	309–904
P. politus	200–385
F. villosa	560.00
H. debilis	251–1104
P. robustus	296–735
S. vioscai	257–433
S. necopina	306–791
S. atlantidis	298.00
B. bartletti	546–1044

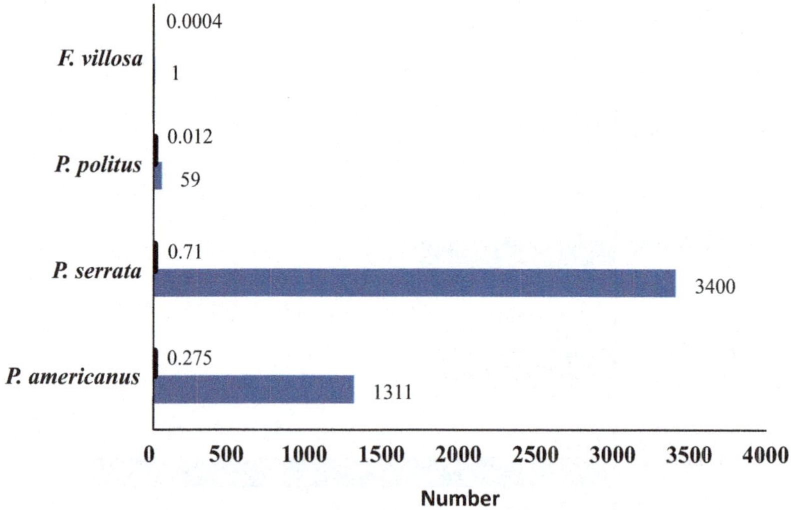

Fig. 9.4 Abundance and percentage of deep-water shrimp of the Penaeidae family

shrimp, <1%), and a unique presence of *S. atlantidis* (≪1%) constituted the rest of shrimps of this family. *Pleoticus robustus*, *H. debilis*, and *S. vioscai* were registered along the whole upper slope of the Mexican Gulf. Bathymetric distribution of *P. robustus* was recorded in a depth range of 296–735 m. *H. debilis* presented an extended depth range (251–1104 m) along the upper slope, whereas *S. vioscai* bathymetric distribution was limited to shallow upper slope depths (257–433 m). *S. necopina* was found in almost all the study area except in the northwest part of

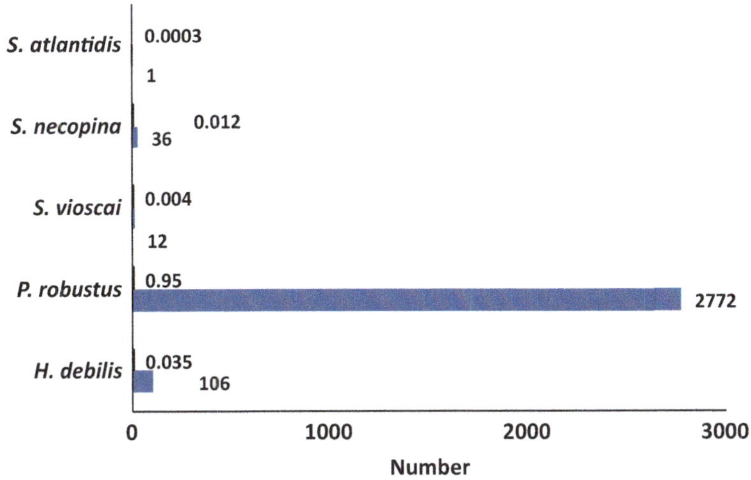

Fig. 9.5 Abundance and percentage of Deep-water shrimp of the Solenoceridae family

the Mexican waters of the Gulf of Mexico. This shrimp was registered in a depth range of 206–798 m. The only individual of *S. atlantidis* was found in the southwestern Gulf of Mexico at 298 m depth (Table 9.1).

In this study the Benthesicymidae family was only represented by *B. bartletti* with 1352 organisms. It was present in all the upper slope of the southern Gulf of Mexico. Its bathymetric distribution was registered in a 546–1094 m depth (Table 9.1).

9.3.2 Biomass and Catch per Unit Effort (CPUE)

The overall comparison of penaeoid shrimp catch shows that (99%) of the catch in numbers was composed mainly by seven species: *A. foliacea* (24%), *P. Serrata* (22%), *P. robustus* (18%), *A. antillensis* (9%), *P. americanus* (9%), *B. bartletti* (9%), and *A. edwardsiana* (8%). The other eight species only represented 1% of the total penaeoid shrimp catch varying from 0.007 to 0.68%. In terms of biomass these species represented more than 99% of total penaeoid shrimp weight (293 kg). *A. edwardsiana* registered the highest total biomass with 97.2 kg which represented 33% of the seven species total weight (290 kg). *P. robustus* was the second one with a total biomass of 72.7 kg (25%), followed by *A. foliacea* with 62.4 kg (21%). *P. serrata*, *A. antillensis*, and *P. americanus* presented lower biomass values with 26 kg (9%), 16.1 (6%), and 11.1 (4%), respectively. The biomass of the tiny shrimp *B. bartletti* amounted 4.9 kg and only represented 2% of the total biomass, so it was not considered for further analysis (Fig. 9.6).

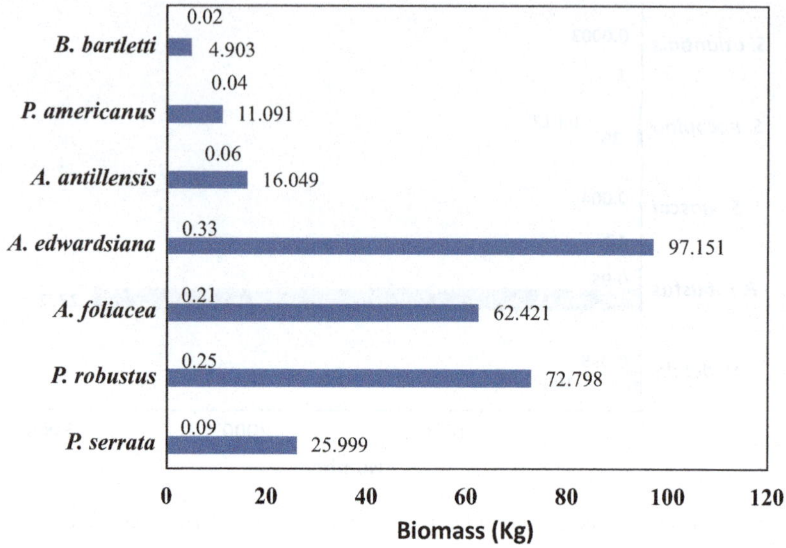

Fig. 9.6 Biomass of the seven penaeoid deep-water shrimp species amounting 99% of the total catch

Table 9.2 Size range, average size, and standard deviation (Carapace Length, mm) of the six most abundant penaeoid deep-water shrimp in the southern Gulf of Mexico

Species	Minimum size	Average ± SD	Maximum size
A. edwardsiana	18.79	48.09 ± 15.01	88.25
A. foliacea	31.56	49.80 ± 7.00	86.4
P. robustus	18.04	45.57 ± 16.73	97.85
A. antillensis	18.32	28.31 ± 7.25	76.63
P. serrata	11.37	23.00 ± 3.49	46.31
P. americanus	10.51	21.5 ± 4.0	30.4

The penaeoid shrimp that constituted the 98% of the catch biomass showed sizes that ranged from small to large ones and can be divided in two groups. The first one comprised *A. foliacea*, *A. edwardsiana*, and *P. robustus* which presented the largest sizes with mean cephalothoracic length (CL) varying from 45.57 to 48.09 mm. Maximum CL were also registered in these species from 86.40 to 97.85 mm CL. The second group was composed of *A. antillensis*, *P. serrata*, and *P. americanus* with a CL mean size range of 21.50–28.31 mm. *P. americanus* was the smallest one with a mean size of 21.5 mm CL and minimum CL of 10.51 mm (Table 9.2).

The total catch per unit effort (kg/h) of the six penaeoid species registered values that ranged from 0 to18.62 kg/h. Average CPUE ordered by 100 m strata showed higher values at shallows depths with a maximum in the 400–499 m stratum (3.1 kg/h). CPUE record of this stratum was 50 and 60% higher than the adjacent depth strata. After the maximum value, the CPUE presented a decreasing trend with

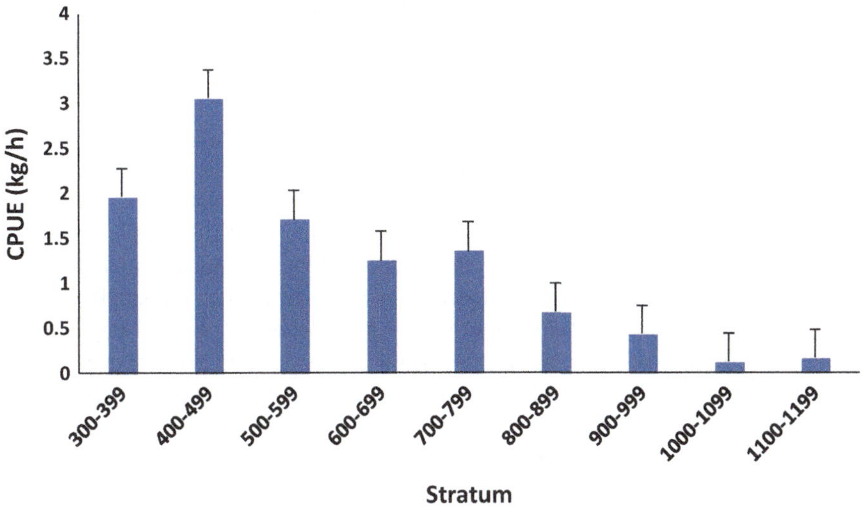

Fig. 9.7 Catch per unit effort (kg/h) of deep-water penaeoid shrimp per100 m stratum in the upper slope of the Mexican Gulf of Mexico

a slight increase in the 700–799-m-depth stratum. High mean CPUE values (> 1.0 kg/h) were estimated at the 300–799 m depth range (Fig. 9.7).

9.4 Discussion

9.4.1 Distribution and Depth Range

Most of the 14 penaeoid shrimp were widely distributed along the southern Gulf of Mexico. The aristeid *A. foliacea* and *A. edwardsiana* have a cosmopolitan distribution including the Gulf of Mexico (Pérez-Farfante and Kensley 1997; Tavares 2002; Gracia et al. 2010; Wehrtmann et al. 2012). Felder et al. (2009a) reported a narrower depth range distribution for *A. foliacea* and *A. edwardsiana* in the Gulf of Mexico of 400–800 and 680–990, respectively. In this study, the bathymetric distribution of these two species is extended in the shallow and deep limits which is according to the reported general bathymetric distribution for *A. foliacea* (250–1300 m) and *A. edwardsiana* (274–1850 m) (Tavares 2002). *A. antillensis* distribution is restricted to the Western Atlantic from Florida to Brazil (Holthuis 1980; Pérez-Farfante and Kensley 1997; Tavares 2002). Previous reports in the Gulf of Mexico were limited to the northeast (Roberts and Pequegnat 1970; Pérez-Farfante and Kensley 1997). In this study, and also according to Gracia et al. (2010) and Wehrtmann et al. (2012), *A. antillensis* occurred along the whole Mexican Gulf of Mexico, so its distribution range is extended. Depth range distribution is also extended in the deeper limit to 1108 m. *H. tener* geographical distribution was reported for the western Atlantic including the Gulf of Mexico (Felder et al. 2009a). Inside the Gulf, it was only reported for the northern area in a deeper range of 1386–3780 m (Roberts and Pequegnat 1970; Pérez-Farfante and Kensley 1997), but in this study its bathymetric range is extended to lower depths (Table 9.1).

The penaeoid shrimp have a broad distribution in the Atlantic. *Parapenaeus americanus* is distributed from New Jersey, the Caribbean Sea to Uruguay (Pérez-Farfante 1977; Pérez-Farfante & Kensley 1997). It was previously reported for the northern (Felder et al. 2009a) and southern (Gracia et al. 2010) Gulf of Mexico. This shrimp was frequently found in the catches during this study with a wide distribution along the Gulf. Its bathymetric range was limited to the shallow upper slope (241–462). This depth distribution is almost like the reported depth range (190–412) for this species (Felder et al. 2009a), but extending 50 m deeper. *P. serrata* was the most abundant of the penaeoid family distributing along the whole study area. It has an amphi-Atlantic distribution and was also reported for the northern (Roberts and Pequegnat 1970; Pérez-Farfante and Kensley 1997) and southern Gulf of Mexico (Gracia et al. 2010). The bathymetric range is extended to 904 m compared with the reported deeper limit (Pérez-Farfante and Kensley 1997; Felder et al. 2009a). *Parapenaeus politus* has a wide distribution from Massachusetts to the Caribbean Sea including the entire Gulf of Mexico (Roberts and Pequegnat 1970; Williams 1984; Pérez-Farfante and Kensley 1997; Gracia and Hernández-Aguilera 2005). Although it was not very abundant in the catches, it occurred along the whole study area. Depth data records allowed to extend *P. politus* distribution in its deeper limit to 385 m compared with the 330 m previous register (Felder et al., 2009a). The only specimen of *F. villosa* was found in the south-southwestern sector of the Gulf of Mexico. This species is a pelagic-benthopelagic shrimp that has a worldwide distribution in the North and South Atlantic, Mediterranean Sea, and South Pacific (Crosnier and Forest 1973; Abele and Kim 1986; Hopkins et al. 1994). Felder et al. (2009a) reported the distribution of *F. villosa* for the northeastern Gulf of Mexico in a depth range of 50–1430 m. The presence of *F. villosa* in the southwestern Gulf of Mexico is the first record for this area which is within the reported depth range of this species.

Four of the solenocerid shrimp registered were found along the whole study area. *Pleoticus robustus* was the most abundant one representing 95% of total shrimp catch. *Pleoticus robustus* distribution range is restricted to the western Atlantic from Massachusetts and the Caribbean Sea to French Guiana (Pérez-Farfante 1977; Holthuis 1980; Pérez-Farfante and Kensley 1997; Tavares 2002). According to Felder et al. (2009a), it is found in the entire Gulf of Mexico in a depth range of 200–1000 m. In this study, *P. robustus* was collected along the whole Mexican Gulf, but the deeper register was limited to 735 m depth. *H. debilis* geographical range was reported for the Atlantic from New Jersey to Guyana, Azores, and the eastern Atlantic (Roberts and Pequegnat 1970; Pérez-Farfante 1977; Pérez-Farfante and Kensley 1997). Inside the Gulf of Mexico, its presence has been registered mainly in northern Gulf of Mexico (Felder et al. 2009a). *Hymenopenaeus debilis* specimens represent the first records in the southern Gulf of Mexico. Although this species was not very abundant, it had a constant presence along the area, confirming that its distribution comprises the entire Gulf. The depth range observed in the southern Gulf fits within the larger bathymetric distribution (300–2163 m) reported for *F. villosa* (Felder et al. 2009a). *S. vioscai* distribution was reported from North Carolina to the Gulf of Mexico (Roberts and Pequegnat 1970; Pérez-Farfante and Kensley 1997; Williams 1984; Gracia and Hernández-Aguilera 2005; Vázquez-Bader and Gracia 1994). *Solenocera vioscai* was found along all the study area, but in a small number. According to Vázquez-Bader and Gracia (1994), this solenocerid species was the most abundant in the continental platform of the southern Gulf of Mexico compared with *S. necopina* and *S. atlantidis*. The low occurrence of *S. vioscai* in the upper slope can be understood because its main distribution is in shallow waters of the continental shelf. However, it should stand out that the depth range recorded (257–433 m) extends largely the *S. vioscai* bathymetric distribution compared with the reported one (37–239 m, Felder et al. 2009a). *Solenocera necopina* and *S. atlantidis* have been reported to distribute in the entire Gulf of Mexico. Besides, its geographical range is almost similar, from North Carolina to Uruguay and Brazil, respectively (Roberts and Pequegnat 1970; Williams 1984; Pérez-Farfante and Kensley 1997), but the reported bathymetric distribution range is larger for *S. necopina*. Depth ranges of both species were extended in its deeper limit with data registered in the Mexican Gulf of Mexico. The only specimen of *S. atlantidis* was found at 298 m depth higher than the maximum reported depth of 232 m (Felder et al., 2009a, b), whereas *S. necopina* extended to 791 m compared with the previous 550 m report (Roberts and Pequegnat 1970; Williams 1984; Pérez-Farfante and Kensley 1997; Gracia and Hernández-Aguilera 2005).

Benthoecetes bartletti was the only species caught of the Benthesicymidae family in spite that 11 species are reported for the Gulf of Mexico, mainly in the north area (Felder et al. 2009a). *Benthoecetes bartletti* has a cosmopolitan distribution in a depth range of 509–5777 m depth, including the entire Gulf of Mexico (Roberts and Pequegnat 1970; Pérez-Farfante and Kensley 1997). The depth range recorded in the southern Gulf of Mexico fall within this wide bathymetric distribution.

9.4.2 Potential Fishery Resource

The six most abundant penaeoid that represented 98% of the biomass and 90% of shrimp catch number (*A. foliacea*, *A. edwardsiana*, *P. robustus*, *A. antillensis*, and *P. americanus*) were reported by Gracia et al. (2010) as a potentially important deep-water shrimp fishery resource in the Gulf of Mexico. Most of these species have a worldwide distribution or along the Atlantic and many of them are the target of important economic fisheries in some world deep-water oceans. *Aristaeomorpha foliacea* (giant red shrimp) sustains a valuable deep-water shrimp fishery in the Mediterranean Sea and the eastern Atlantic off Portugal that is one of the most important fishery resources in the area (D'Onghia et al. 1998; Figueiredo et al. 2001; Ragonese et al. 2001; Belcari et al. 2003). In Latin America the giant red shrimp fishery was initiated in Brazil since 2003 (Pezutto et al. 2006; Dallagnolo et al. 2009; Wehrtmann 2012). *Aristaeopsis edwardsiana* (Scarlet shrimp) is commercially exploited in the eastern Atlantic from Africa to Portugal and Spain (Holthuis 1980). In Latin America, it was caught in French Guiana (Guéguen 2001) and constituted an important fishery in Brazil deep waters (Pezutto et al. 2006; Dallagnolo et al. 2009; Wehrtmann et al. 2012). *Aristeus antillensis* (purple shrimp) was also fished in Brazil with the giant red and scarlet shrimp and seasonally in French Guiana (Guéguen 1998, 2001; Pezutto et al. 2006). *Pleoticus robustus* (royal red shrimp) had been exploited in the northern Gulf for a long time since 1960. Royal red shrimp is the target of a small shrimp trawl fleet fitted to fish in deep-waters off Alabama and Florida States. Average annual royal red shrimp catch is comparatively lower than the shallow water penaeoid shrimp catch (Jones et al. 1994; Stiles et al. 2007; Wehrtmann et al. 2012). *Penaeopsis serrata* (Speckled Shrimp) and *P. americanus* (Rose Shrimp) are not directly subjected to a fishery exploitation elsewhere probably due to its small size, but they are frequently present in deep-water shrimp catch.

Gracia et al. (2010) and Wehrtmann et al. (2012) reported that *A. foliacea*, *A. edwardsiana*, and *P. robustus* were the most important deep-water shrimp in terms of biomass representing 90% of the total weight catch. The results obtained in this study, with more sampling effort, show that these three species amount 79% of total catch, which still represent the largest part of the total deep-water shrimp catch. As stated above, these three species presented larger sizes than *A. antillensis*, *P. serrata*, and *P. americanus* which only amounted for 21% of the total biomass. Large sizes of giant red, scarlet, and red royal shrimp make them commercially more attractive than the purple, speckled, and rose shrimp.

The general penaeoid CPUE pattern (kg/h) with respect to depth is consistent with data reported by Gracia et al. (2010) and Wehrtmann et al. (2012) who coincided that high CPUE values are mainly found in the 300–700 depth range, although in this study a high mean value was also registered in the 700–799 m depth stratum. On the other hand, Gracia et al. (2010) registered higher CPUE values in the 600–699 m stratum in the Yucatán upper slope (Area 3), whereas Wehrtmann et al. (2012) reported higher values in the 500–699 m depth range. These variations could be expected because cruises were carried on different years, seasons, and geographical areas and also can be influenced by the typical patchy distribution of penaeoid shrimp (D'Onghia et al. 1998; Belcari et al. 2003; Gracia et al. 2010). Nonetheless, the mean deep-water shrimp CPUE pattern estimated in this study could be robust to describe the penaeoid shrimp abundance related to depth as it encompasses many of these variability sources.

According to Gracia et al. (2010) the mean CPUE values estimated in the Gulf of Mexico could be comparable with CPUE registered in deep-water shrimp fisheries of the world. The mean values are in the range reported for several deep-water shrimp fisheries; however they seem to be lower. For example, mean annual CPUE records for *A. edwardsiana* in Brazil fisheries were in a 4.7–14 kg/h range, whereas *A. foliacea* and *A. antillensis* were about 0.76–6.3 and 0.005–2.4 kg/h, respectively (Dallagnolo et al. 2009). CPUE values recorded for *A. foliacea* and *A. antennatus* in the Mediterranean Sea and Eastern Atlantic off Portugal (0–12 kg/h) (D'Onghia et al. 1998; Figueiredo et al. 2001; Carbonell and Azevedo 2003; Can and Atkas 2005) are in a range similar to the one registered in the Mexican Gulf of Mexico. It must be pointed out that CPUE registers for the Gulf of Mexico were estimated from an exploratory scientific survey which may not be as accurate as the ones derived of a commercial fleet specialized on deep-water trawl fishing. However, it may serve as an index of the deep-water shrimp abundance in the Gulf of Mexico (Gracia et al. 2010; Wehrtmann et al. 2012).

Penaeoid deep-water shrimp have a potential distribution along the Gulf of Mexico in an area estimated of 60,000 km^2 of the upper slope between 300 and 1000 m depth (Fig. 9.8). This area has a complex topography with submarine canyons, escarpments, rugged bottoms, and basins. Al least four potential fishing grounds were located in this study (Fig. 9.8) with plain soft bottoms suitable for trawling. CPUE values were very variable, but some registers were up to 18.62 kg/h. The highest mean CPUE estimated in this area is like the average CPUE estimated (~2–2.4 kg/h by net) in the shallow water penaeoid overexploited white (*L. setiferus*) and pink shrimp (*F. duorarum*) fisheries in the southwestern Gulf of Mexico (Gracia et al. 2010; Wehrtmann 2012; INAPESCA 2014). This is not the case for the *F. aztecus* (brown shrimp) fishery whose stock is in good condition and currently sustains most of the Mexican shrimp fishery in the Gulf of Mexico (Gracia 2004). Estimated shallow water brown shrimp CPUE based on data reported by INAPESCA (2010) varied between 20 and 65 kg/h by net. This CPUE value is remarkably higher than the one recorded for deep-water shrimp in this scientific survey, but usually deep-water shrimp yield is lower that the shallow one. Annual shrimp deep-water

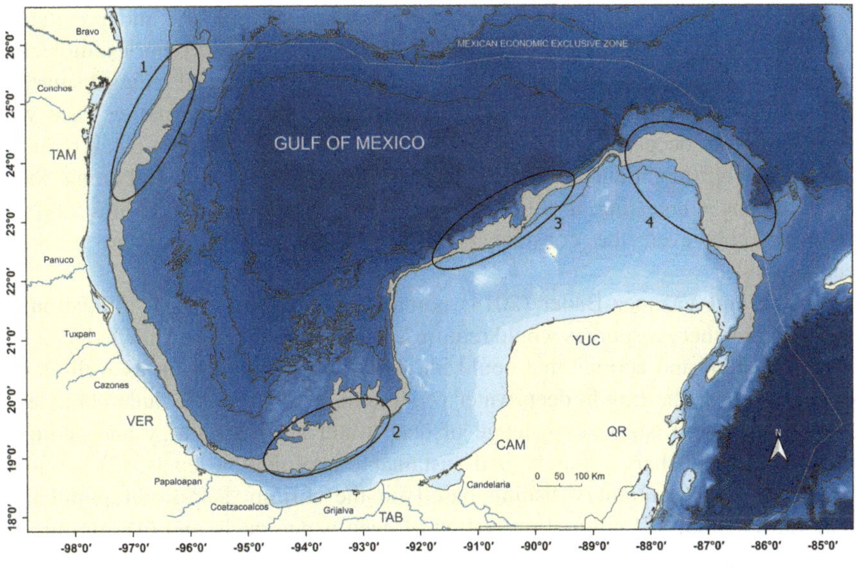

Fig. 9.8 Deep-water penaeoid shrimp distribution area and potential fishing grounds in the southern Gulf of Mexico

fishery registers vary between 100 and 200 metric tons (Stiles et al. 2007; Dallagnolo et al. 2009), although some like *A. foliacea* in the Mediterranean Sea can reach around 1000 metric tons. According to Gracia and Vázquez-Bader (2014), a moderate annual yield could be expected for a potential deep-water fishery in the southern Gulf of Mexico, although it could be important due to deep-water shrimp commercial value. However, it is still necessary to carry on further studies to assess the deep-water shrimp stocks potential.

9.5 Conclusions

The increasing demand for seafood propitiated that fishing activities moved farther and deeper around the world. In the Gulf of Mexico, most of the fishing resources reached its maximum sustainable level or are in an overexploited stage (DOF 2018), particularly shallow penaeoid shrimp (Gracia 2004). Finding and assessing potential fishery resources is contemplated in the Mexican National Plan of Scientific and Technological Research for Fisheries and Aquaculture to increase fishery production. deep-water penaeoid shrimp represent a valuable potential fishery resource as well as other species found in the catch like lobsters (see Briones-Fourzán et al. this volume), fishes (Ramírez et al. 2019), and other crustaceans that could be caught as by-catch. Utilization of deep-water penaeoid shrimp would require adapting ships for operating deeper, but more important is to assess its potential and to evaluate possible strategies for its utilization and conservation of the deep-water ecosystem.

Deep-water ecosystems and fishery resources are very vulnerable to overexploitation. Deep-water fishery resources could be depleted quickly, and population rebuilding could be slow (Koslow et al. 2000; Large et al., 2002). For example, the rapid fishing effort growth in the *A. edwardsiana* fishery in Brazil reduced substantially the deep-water shrimp stock (Dallagnolo et al. 2009). Another case was reported in the deep-water shrimp fishery in Costa Rica where the catch reduced strongly (Wehrtmann and Nielsen Muñoz 2009). At the international arena, there are opinions suggesting a total protection of deep-water ecosystems and its resources; however, the constant protein demand poses a pressure in these ecosystems.

Gracia and Vázquez-Bader (2014) pointed out that an eventual utilization of deep-water fishery resources would require a strategy based on the knowledge of the stock potential and actions that could allow its integral utilization as well as the conservation of the fragile deep-water ecosystem. This should include (1) assessment of the stock sizes, knowledge of deep-water shrimp biology and ecology, delimitation of fishery grounds, and sustainable exploitation levels, (2) adopting strategies of precautionary fishing based on the shrimp deep-water population renewal potential, (3) optimal utilization of deep-water catch, and (4) minimizing the impact on deep-water ecosystem, especially on cold water coral reefs.

Acknowledgments Officers and crew of the R/V Justo Sierra are greatly appreciated for their support during research cruises. We thank many graduate and undergraduate students that participated along the different research cruises. We are grateful to the invaluable technical support onboard and in the laboratory of Magaly Galván Palmerín, Hermelinda Trejo Rosas, Sandra Antonio Bueno, Brenda Barbosa Nieto, Ingrid Antillón Zaragoza, and León F. González Morales, who also elaborated the maps for this contribution. This study was developed mainly with funds and research vessel time provided by the Universidad Nacional Autónoma de Mexico and partly supported through the research project PAPIIT IN223109 of the Dirección General del Personal Académico, UNAM. SOGOM cruises were funded by the Mexican National Council for Science and Technology – Mexican Ministry of Energy – Hydrocarbon Fund, project 201441 as part of the Gulf of Mexico Research Consortium (CIGoM) due to PEMEX's specific request to the Hydrocarbon Fund to address the environmental effects of oil spills in the Gulf of Mexico.

References

Abele IG, Kim W (1986) An illustrated guide to the marine decapod crustaceans of Florida. State Fla Dept Env Reg, Tech Ser 8((1), Parts 1 & 2):1–760

Arana P, Álvarez Pérez JA, Pezutto PA (2009) Deep-sea fisheries off Latin America: an introduction. In: Arana P, Pérez JAA, Pezutto PR (eds) Deep-sea fisheries off Latin America. Lat Am J Aquat Res 37(3):281–284

Belcari P, Viva C, Mori M, de Ranieri S (2003) Fishery and biology of *Aristaeomorpha foliacea* (Risso, 1827) (Crustacea, Decapoda) in the northern Tyrrhenian Sea (western Mediterranean). J Northwest Atl Fish Sci 31:195–204

Briones-Fourzán P, Barradas-Ortiz C, Negrete-Soto F, Lozano-Álvarez E (2010) Reproductive traits of tropical deep-water pandalid shrimps (*Heterocarpus ensifer*) from the SW Gulf of Mexico. Deep Sea Res I Top Stud Oceanogr 57:978–987

Briones-Fourzán P, Lozano-Álvarez E, Vázquez-Bader AR, Gracia A (this volume) Deep-sea lobsters from the continental slope of the southern Gulf of Mexico: distribution and morphometric relationships. In: Hendrickx ME (ed) Deep-sea pycnogonids and crustaceans of the Americas. Springer, Cham

Can MF, Atkas M (2005) A preliminary study on population structure and abundance of *Aristaeomorpha foliacea* (Risso, 1827) in the deep water of the northeastern Mediterranean. Crustaceana 78:941–946

Carbonell A, Azevedo M (2003) Application of non-equilibrium production models to the red shrimp (*Aristeus antennatus*, Risso 1816) fishery in the northwestern Mediterranean. Fish Res 65:323–334

Crosnier A, Forest J (1973) Les crevettes profondes de l'Atlantique Oriental tropical. ORSTOM, Faune Trop 19:4–410

D'Onghia G, Tursi A, Maiorano P, Panza M (1998) Distribution, biology and population dynamics of *Aristaeomorpha foliacea* (Risso, 1827) (Decapoda, Natantia, Aristeidae) in the northwestern Ionian Sea (Mediterranean Sea). Crustaceana 71:518–544

Dallagnolo R, Álvarez Pérez JA, Pezutto PR, Wahrlich R (2009) The deep-sea shrimp fishery off Brazil (Decapoda: Aristeidae) development and present status. In: Arana P, Pérez JAA, Pezutto PR (eds) Deep-sea fisheries off Latin America. Lat Am J Aquat Res 37(3):327–346

Darnell R (2015) The American Sea: a natural history of the Gulf of Mexico. Texas A&M University Press, College Station, p 554

Diario Oficial de la Federación (DOF) (2018) Actualización de la Carta Nacional Pesquera. Secretaría de Agricultura, Ganadería, Desarrollo Rural, Pesca y Alimentación. Ciudad de México, México

Felder DL, Álvarez F, Goy JW, Lemaitre R (2009a) Decapoda (Crustacea) of the Gulf of Mexico with comment on the Amphionidacea. In: Felder DL, Camp DK (eds) Gulf of Mexico origins and biota. Volume I. Texas A&M University Press, College Station, pp 1019–1104

Felder DL, Kamp DK, Tunnell JW Jr (2009b) An introduction to Gulf of Mexico biodiversity assessment. In: Felder DL, Camp DK (eds) Gulf of Mexico origins and biota. Volume I. Texas A&M University Press, College Station, pp 3–13

Figueiredo MJ, Figueiredo I, Machado PB (2001) Deep-water penaeoid shrimps (Crustacea: Decapoda) from the Portuguese continental slope: an alternative fishery resource? Fish Res 51:321–326

Gracia A (1995) Impacto de la pesca artesanal sobre la producción de camarón rosado *Penaeus Farfantepenaeus duorarum* Burkenroad, 1939. Cienc Mar 21:343–359

Gracia A (1996) White shrimp *Penaeus setiferus* recruitment overfishing. Mar Freshw Res 47:59–63

Gracia A (2004) Aprovechamiento y Conservación del Recurso Camarón. In: Caso M, Pisanty I, Ezcurra E (eds) Diagnóstico Ambiental del Golfo de México. Secretaría de Medio Ambiente y Recursos Naturales, Instituto Nacional de Ecología, A. C. Harte Res Inst Gulf Mex Stud 2:713–725

Gracia A, Hernández-Aguilera JL (2005) Camarones Penaeoideos. Camarones, Langostas y Cangrejos de la Costa Este de México. Vol. 1. (Shrimps, Lobsters and Crabs, Vol. 1). In: Hernández-Aguilera JL, Ruiz Nuño JA, Toral Almazán RE, Arenas Fuentes, VA (eds) Estudio y Conservación de la Naturaleza A.C. Comisión Nacional para el Conocimiento y Uso de la Biodiversidad, Mexico DF, pp 23–65

Gracia A, Vázquez-Bader AR (1998) The effects of artisanal fisheries on penaeoid shrimp stocks in the Gulf of México. In: Funk F, Quinn TJ II, Heifetz J, Ianelli JN, Powers JE, Schweigert JF, Sullivan PJ, Zhang CI (eds) Proceedings of the international symposium fishery stock assessment models for the 21st century. University of Alaska, Sea Grant College, pp 977–998

Gracia A, Vázquez-Bader AR (1999) Shrimp fisheries in the south of the Gulf of México. Present state and future management alternatives. In: Kumpf H, Steidinger K, Sherman K (eds) The Gulf of Mexico large marine ecosystem. Assessment, sustainability, and management. Blackwell Science, Malden, pp 205–234

Gracia A, Vázquez-Bader AR (2014) Recursos Pesqueros de Mar Profundo. In: Low Pfeng A, Peters Recagno EM (eds) La Frontera Final: El Océano Profundo. Secretaría de Medio Ambiente y Recursos Naturales (SEMARNAT). Instituto Nacional de Ecología (INE), Mexico DF, pp 255–272

Gracia A, Vázquez-Bader AR, Lozano-Alvarez E, Briones-Fourzán P (2010) Deep-water shrimps (Crustacea: Penaeoidea) off the Yucatan Peninsula (Southern Gulf of Mexico): a potential fishing resource? J Shellfish Res 29(1):37–43

Gracia A, Murawski SA, Vázquez-Bader AR (2020) Impacts of deep spills on fish and fisheries. In: Murawski SA, Ainsworth C, Gilbert S, Hollander D, Paris CB, Schluter M, Wetzel D (eds) Deep oil spills: facts, fate, effects. Springer, Cham, pp 414–430

Guéguen F (1998) Biologie de la crevette profonde *Plesiopenaeus edwardsiana* en Guyane Française. CR Acad Sci Paris 321:757–770

Guéguen F (2001) Notes sur la biologie de la crevette profonde *Aristaeus antillensis* en Guyane Française. CR Acad Sci Paris 324:689–700

Holthuis LB (1980) Shrimps and prawns of the world. An annotated catalogue of species of interest to fisheries. FAO Fish Synop 125:1–271

Hopkins TL, Flock ME, Gartner JV Jr, Torres JL (1994) Structure and trophic ecology of a low latitude midwater decapod and mysid assemblage. Mar Ecol Progr Ser 109:143–156

INAPESCA (2010) Programa Nacional de Investigación Científica y Tecnológica en Pesca y Acuacultura. Instituto Nacional de la Pesca. Secretaría de Agricultura, Ganadería, Desarrollo Rural, Pesca y Alimentación. www.inapesca.com

INAPESCA (2014) Fundamento Técnico para el establecimiento de vedas para la pesca de camarón en el Golfo de México y Mar Caribe (2014). Dirección General Adjunta de Investigación

Pesquera en el Atlántico Instituto Nacional de Pesca. Secretaría de Agricultura, Ganadería, Desarrollo Rural, Pesca y Alimentación, Mexico DF

Jones CJ, Nance JM, Antozzi WO Jr (1994) A review of the royal red shrimp fishery in the Gulf of Mexico. Gulf of Mexico Management Council, Tampa

Koslow JA, Boehlert GW, Gordon JDM, Haedrich RL, Lorance P, Parin N (2000) Continental slope and deep-sea fisheries: implications for a fragile ecosystem. ICES J Mar Sci 57:548–557

Kumpf H, Steindinger K, Sherman K (eds) (1999) The Gulf of Mexico large marine ecosystem. Assessment, sustainability, and management. Blackwell Science, Malden

Large PA, Hammer C, Bergstad OA, Gordon JDM, Lorance P (2002) Fisheries of the Northeast Atlantic: II assessment and management approaches. J Northwest Atl Fish Sci 31:151–163

Lemaitre R, Vázquez-Bader AR, Gracia A (2014) An unusual new species of paguroid (Crustacea, Anomura, Paguridae) from deep waters of the Gulf of Mexico. ZooKeys 449:57–67

Lozano-Álvarez E, Briones-Fourzán P, Gracia A, Vázquez-Bader AR (2007) Relative growth and size at first maturity of the deep water shrimp *Heterocarpus ensifer* (Decapoda, Pandalidae) from the southern Gulf of Mexico. Crustaceana 80:555–568

Monreal-Gómez MA, Salas de León DA (1997) Circulación y estructura termohalina del Golfo de México. In: Lavín MF (ed) Contribuciones a la Oceanografía Física en México, Monografía no. 3. Unión Geofísica Mexicana, México DF, pp 183–189

Monreal-Gómez MA, Salas de León DA, Velasco-Mendoza H (2004) La hidrodinámica del Golfo de México. In: Caso M, Pisanty I, Ezcurra E (eds) Diagnóstico Ambiental del Golfo de México. SEMARNAT, Mexico DF, pp 47–68

Murawski SA, Ainsworth C, Gilbert S, Hollander D, Paris CB, Schluter M, Wetzel D (eds) (2020a) Deep oil spills: facts, fate, effects. Springer, Cham

Murawski SA, Hollander D, Gilbert S, Gracia A (2020b) Deep-water oil and gas production in the Gulf of Mexico, and related global trends. In: Murawski SA, Ainsworth C, Gilbert S, Hollander D, Paris CB, Schluter M, Wetzel D (eds) Scenarios and responses to future deep oil spills: fighting the next war. Springer, Cham, pp 16–32

Pérez D, Paramo J, Wolff M (2019) Distribution, abundance and fishing potential of mega-invertebrates in the sub-euphotic zone (150–535m) in the Colombian Caribbean. Reg Stud Mar Sci 32:100868

Pérez-Farfante I (1977) American Solenocerid shrimps of the genera *Hymenopenaeus*, *Haliporoides, Pleoticus*, *Hadropenaeus* new genus, and *Mesopenaeus* new genus. Fish Bull 75(2):261–346

Pérez-Farfante I, Kensley B (1997) Penaeoid and sergestoid shrimps and prawns of the world: keys and diagnoses for the families and genera. Mém Mus Hist Nat, París 175:1–233

Pezutto PR, Pérez JAA, Wahrlich R (2006) Deep sea shrimps (Decapoda: Aristeidae): new targets of the deep water trawling fishery in Brazil. Braz J Ocean 54(2/3):123–134

Ragonese S, Zagra M, Di Stefano L, Bianchii ML (2001) Effect of codend mesh size on the performance of the deep-water bottom trawl used in the red shrimp fishery Strait of Sicily (Mediterranean Sea). Hydrobiologia 229:279–291

Ramírez JM, Vázquez-Bader AR, Gracia A (2019) Icthyofaunal list of the continental slope of the southern Gulf of Mexico. Zookeys 846:117–132

Roberts WT, Pequegnat WE (1970) Deep water decapod shrimp of the family Penaeidae. In: Pequegnat WE, Chace FA Jr (eds) Contributions on the biology of the Gulf of Mexico, Texas A&M oceanographic studies. Gulf Publishing Company, Houston, pp 21–57

Solís-Marín FA, Laguarda-Figueras A, Durán-González A, Vázquez-Bader AR, Gracia A (2014) Biodiversidad de los equinodermos (Echinodermata) del mar profundo mexicano. In: Low Pfeng A, Peters Recargno EM (eds) La Frontera Final: El Océano Profundo. Secretaría de Medio Ambiente y Recursos Naturales (SEMARNAT). Instituto Nacional de Ecología (INE), Mexico DF, pp 215–272

Stiles ML, Harrould-Kolieb E, Faure P, Yitalo-Ward H, Hirshfield M (2007) Deep sea trawl fisheries of the southeast US and the Gulf of Mexico: rock shrimp, royal red shrimp, Callico scallops. Oceana, Washington, DC

Sturges W, Lugo Fernández A (2005) Circulation of the Gulf of Mexico: observations and models, volume 161, Geophisycal monograph series. American Geophysycal Union, Washington, DC
Tavares M (2002) Shrimps. In: Carpenter KE (ed) The living marine resources of the Western Central Atlantic, FAO species identification guide for fisheries purpose, vol I. FAO, Rome, pp 251–291
Vázquez-Bader AR, Gracia A (1994) Macroinvertebrados bénticos de la plataforma continental del Suroeste del Golfo de México. Pub Esp Inst Biol UNAM 12:1–113
Vázquez-Bader AR, Gracia A (2013) Crangonidae and Glyphocrangonidae (Decapoda; Caridea) of the Southern Gulf of Mexico. Zootaxa 3669(3):367–383
Vázquez-Bader AR, Gracia A (2016) Diversity and distribution of Chyrostiloidea and Galatheoidea (Decapoda, Anomura) in the Southern Gulf of Mexico. Zookeys 612:1–30
Vázquez-Bader AR, Laguarda-Figueras A, Gracia A, Solís-Marín FA, Celaya-Hernández EV, Durán-González A (2008) Seasonal changes in the density and species composition of the epifaunal echinoderms recorded from the southwestern Gulf of Mexico. Rev Biol Trop 56(3):297–310
Vázquez-Bader AR, Gracia A, Lemaitre R (2014) A new species of *Munidopsis* Whiteaves, 1874 (Crustacea: Anomura: Galatheoidea: Munidopsidae) from the Gulf of Mexico and Caribbean Sea. Zootaxa 3821(3):354–362
Ward CH, Tunnell JW Jr (2017) Habitats and biota of the Gulf of Mexico: an overview. In: Ward CH (ed) Habitats and biota of the Gulf of Mexico: before the Deep-water horizon oil spill, vol I. Springer Nature, New York, p 868
Wehrtmann IS, Nielsen Muñoz V (2009) The deep-water fishery along the Pacific coast of Costa Rica, Central America. Lat Am J Aquat Res 37(3):543–554
Wehrtmann IS, Arana P, Barriga E, Gracia A, Pezzuto PR (2012) Deep-water shrimp fisheries in Latin America: a review. Lat Am J Aquat Res 40(1):497–535
Wicksten M, Packard JM (2005) A qualitative zoogeographic analysis of decapod crustaceans of the continental slope and abyssal plain of the Gulf of Mexico. Deep Sea Res I Oceanogr Res Papers 52:1745–1765
Williams AB (1984) Shrimps, lobsters and crabs of the Atlantic Coast of the Eastern United States, Maine to Florida. Smithsonian Institution Press, Washington, DC

Chapter 10
Sex Distribution and Reproductive Trends in the Deep-Water Species of *Nematocarcinus* (Crustacea: Decapoda: Nematocarcinidae) from Western Mexico

J. C. Hernández-Payán and M. E. Hendrickx

Abstract Two species of the deep-water Caridea genus *Nematocarcinus* occurring in western Mexico were studied: *N. faxoni* (1463 specimens) and *N. agassizii* (169 specimens). All samples considered, *N. faxoni*, were represented by 35% of males, 59% of females, and only 6% of ovigerous females, the latter occurring throughout the seasons; in the case of *N. agassizii*, ovigerous females represented 29%. In *N. faxoni*, size at first maturity was 16.95 mm carapace length. The number of eggs in *N. faxoni* varied from 1269 to 6882 (average, 3700), and the relationship between ovigerous female size and number of eggs was moderately good. Egg size in *N. faxoni* (0.50 mm to 0.80 mm long axis; 0.40 mm to 0.69 mm short axis) did not vary with female size.

Keywords Caridean shrimps · Deep sea · East Pacific · Eggs size Nematocarcinids · Reproduction · Sex rate

10.1 Introduction

Because of technical issues and operation cost when collecting organisms, information related to reproduction trends and sex ratio of deep-water species are often scanty (Gage and Tyler 1992; Young 2003). This information, however, is of paramount importance for an eventual management of deep-sea resources in order to avoid depletion of adults and reproducing individuals. For obvious reasons (i.e., accessibility of samples, possibilities of in vitro experiments, abundant material available from commercial fishery), reproduction patterns of shallow-water species

J. C. Hernández-Payán (✉) · M. E. Hendrickx
Laboratorio de Invertebrados Bentónicos, Unidad Académica Mazatlán Instituto de Ciencias del Mar y Limnología, Universidad Nacional Autónoma de México, Mazatlán, Sinaloa, Mexico

M. E. Hendrickx (ed.), *Deep-Sea Pycnogonids and Crustaceans of the Americas*, https://doi.org/10.1007/978-3-030-58410-8_10

have received more attention than deep-water species (see Corey and Reid 1991; Wenner and Kuris 1991; Bauer 2004; Mente 2008).

Although less frequently, reproduction trends in deep-water decapod crustaceans have been investigated in the Mediterranean Sea (e.g., Company and Sardà 1997; Maiorano et al. 2004; Chilari et al. 2005; Kapiris and Thessalou Legaki 2006; Arculeo et al. 2014), the northern and eastern Atlantic (e.g., Wenner 1979a, 1980; Melville-Smith 1987; Tuset et al. 2009; Burukovsky 2017), the central and western Pacific Ocean (e.g., Dailey and Ralston 1986; Erdman et al. 1991; Ohtomi 1997; Ahamed and Ohtomi 2011), the Indian Ocean (e.g., Suseelan 1974; Benjamin et al. 2017), and to some extent among species associated with thermal vents (e.g., Van Dover et al. 1985; Ramirez-Llodra et al. 2000; Ramirez-Llodra and Segonzac 2006; Nakamura et al. 2015).

In natural marine ecosystems of the Americas, there also is a very large amount of data available on the fertility and reproduction trends of shallow-water species of decapod crustaceans. Many concerned species inhabiting temperate waters of the northern hemisphere (e.g., Butler 1964; Paul and Adams 1984; Clarke et al. 1991; Dugan et al. 1991; Hines 1991; Hartnoll et al. 1992; Moriyasu and Lanteigne 1998; McDermott 1999; Lawton et al. 2002). There is, however, a sizable amount of information related to reproduction trends of shallow-water species in other areas of the Americas (e.g., Lardies and Wehrtmann 1996; Navarrete et al. 1999; Wehrtmann and Lardies 1999; Lardies and Castilla 2001; García-Guerrero and Hendrickx 2004; Brante et al. 2004; Castilho et al. 2007; Terossi et al. 2010).

For the same reasons as in other parts of the world, contributions dealing with reproductive aspects of deep-water decapods inhabiting off the coasts of the American continent are scarce. Some good examples are the contributions by Hines (1988), Erdman et al. (1991), Lozano-Álvarez et al. (2007), and Briones-Fourzan et al. (2010) in the western Atlantic and Quiroga and Soto (1997), Wehrtmann and Echeverría-Sáenz (2007), Echeverría-Sáenz and Wehrtmann (2011), and Hendrickx and Papiol (2015, 2019) in the eastern Pacific.

Species of *Nematocarcinus* have been reported in many deep-water surveys (e.g., Wenner and Boesch 1979; Cartes and Sardà 1993; Hanamura and Evans 1996; Company et al. 2004; Raupach et al. 2010; Cardoso and Burukovsky 2014; Hernández-Payán and Hendrickx 2016; Hendrickx and Hernández-Payán 2018), yet information related to their ecology is limited and mostly refers to the depth range and habitat they occupy (e.g., Company et al. 2004; Komai and Segonzac 2005; Burukovsky 2012) and to their diet (e.g., Cartes 1993; Burukovsky 2009). Some other aspects have been addressed related to their internal anatomy (e.g., Storch et al. 2002), body composition (e.g., Allen et al. 2000; Graeve and Wehrtmann 2003), larval development (e.g., Thatje et al. 2005), and molecular markers (e.g., Dambach et al. 2013). Very few contributions are fully related to their reproduction patterns or to some aspects dealing with mature females (e.g., Wenner 1979b; Clarke 1993; Gorny and George 1997; Lovrich et al. 2005; Burukovsky and Sudnik 2014).

Six species of *Nematocarcinus* have been reported in the eastern Pacific (see Hendrickx and Hernández-Payán 2016, 2018), but only two have so far been recorded off western Mexico: *N. agassizii* Faxon, 1893 and *N. faxoni* Burukovsky, 2001. These two species inhabit at the edge and below the lower boundary of the oxygen minimum zone (OMZ) where they are commonly captured in benthic sledges and represent a sizable portion of the deep-water decapod crustaceans community (Hendrickx and Hernández-Payán 2018). Their taxonomy, geographic and depth distributions, and density were recently addressed (Hendrickx and Hernández-Payán 2016, 2018). Their relationship with environmental parameters has been addressed in this volume (see Papiol et al., this volume). Present contribution contributes to basic knowledge of their reproductive pattern.

10.2 Material and Methods

During sampling operations off the west coast of Mexico, a large series of specimens of two species of *Nematocarcinus* (*N. agassizii* and *N. faxoni*) were collected during bottom trawling operations (see Hernández-Payán and Hendrickx 2016). This series of specimens, the largest on record available for study, was collected by the R/V "El Puma" of the Universidad Nacional Autónoma de México (UNAM), between 1991 and 2014 (see Hernández-Payán and Hendrickx 2016). Specimens of *Nematocarcinus* were captured during sampling operations in three areas off western Mexico (Fig. 10.1) during ten cruises (see Hernández-Payán and Hendrickx 2016) in the lower boundary of the OMZ core and deeper, from 377 to 2394 m depth. All the specimens were captured with benthic gear, including an Agassiz dredge (2.5 m width, 1 m high) and a benthic sledge (2.35 m width, 0.9 m high). Both were equipped with a modified shrimp net (ca. 5.5 cm stretched mesh size) with a ca. 2.0 cm internal lining net. All organisms in the catch were identified, and specimens of *Nematocarcinus* were sexed, counted, measured (carapace length, CL in mm ± 0.1 mm), and weighted (total weight, W in g ± 0.1 g) at the laboratory. Eggs attached to ovigerous females were separated with a mild chlorine solution and counted. A selected batch of 20 eggs of each size range of the ovigerous females was measured to the nearest 0.01 mm (small and large diameters). Statistical analysis was applied to data (regression curves, χ^2, Kruskal-Wallis, and Mann-Whitney tests) following Zar (1998). In the case of the test of Mann-Whitney, the value of nM and nF is the number of male and female for each parameter, respectively.

10.3 Results

A total of 1463 specimens of *Nematocarcinus faxoni* and 169 specimens of *N. agassizii* were available for this study. All samples considered, proportions by sex varied considerably between these two species: *N. faxoni*, 35% males, 59% females, and 6% ovigerous females; and *N. agassizii*, 39% males, 32% females, and 29% ovigerous females (Fig. 10.2). Overall male/female ratio (M:F) was 1:1.87 for *N. faxoni* (χ^2 = 129.69,

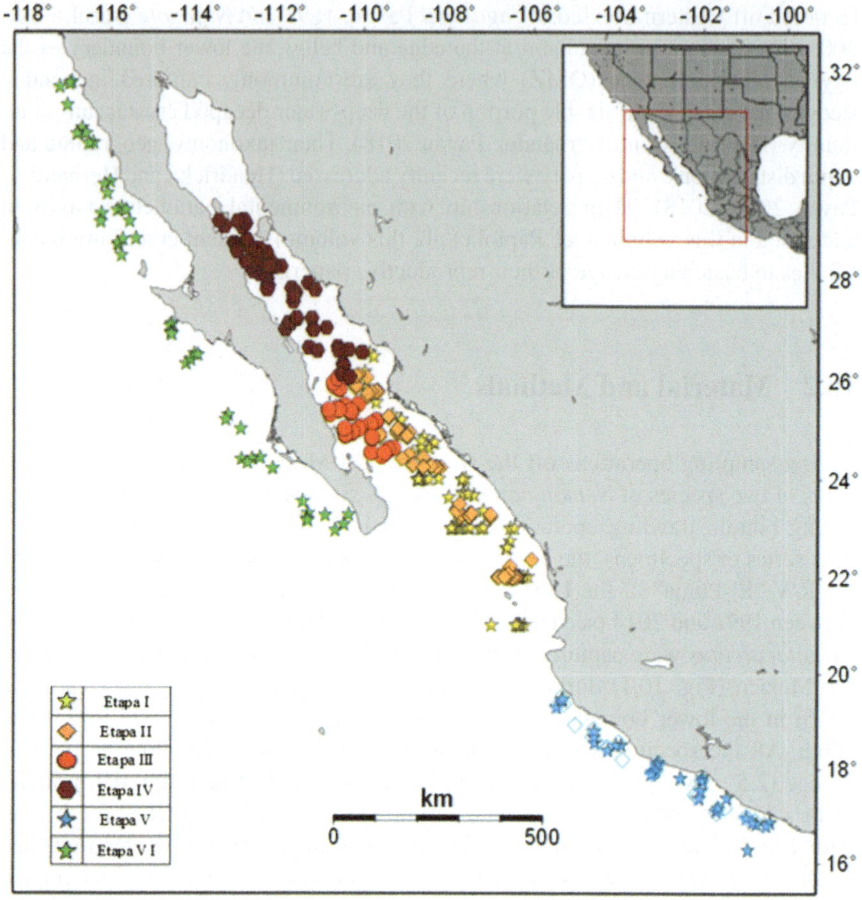

Fig. 10.1 Localities sampled during the TALUD survey in western Mexico (1989–2014)

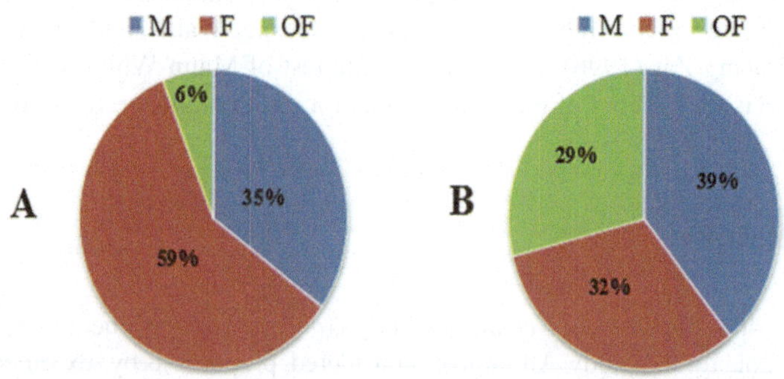

Fig. 10.2 Proportion of males (M), females (F), and ovigerous females (OF) of *Nematocarcinus faxoni* (**a**) and *Nematocarcinus agassizii* (**b**) considering all specimens collected during the TALUD survey

$p<0.05$) and 1:1.53 for *N. agassizii* ($\chi^2 = 8.14$, $p<0.05$). In the case of *N. faxoni*, the presence of these three groups did not vary considerably from one period of the year to another (Fig. 10.3), and ovigerous females were collected all year long except in July and December (Fig. 10.4). No samples were available for January, September, and October. In the case of *N. agassizii*, data are few to draw conclusions; ovigerous females were found only in March and August (i.e., in two out of nine cruises).

In *N. faxoni*, males ranged from 9.04 to 26.00 mm CL and females from 13.05 to 28.00 mm CL. The smallest ovigerous female was 17.96 mm CL and the largest 28.33 mm CL. The Kruskal-Wallis test showed a significant difference in size distribution among sexes (Fig. 10.5) (χ^2, (1463) = 213.97; $p < 0.001$), with females being significantly larger than males (Mann-Whitney test; $nM = 513$, $nF = 948$, $p < 0.05$). In the case of *N. agassizii* (Fig. 10.6), males ranged from 12.03 to 27.04 mm CL and females from 12.03 to 27.04 mm CL. The smallest ovigerous female was 23.32 mm CL and the largest 31.90 mm CL. The Kruskal-Wallis test

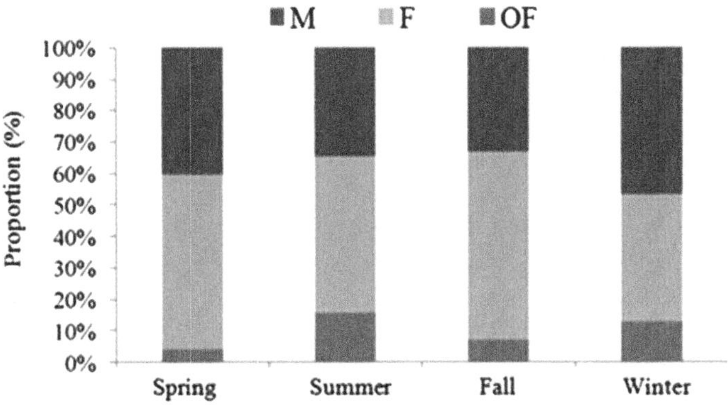

Fig. 10.3 Proportion of males (M), females (F), and ovigerous females (OF) of *Nematocarcinus faxoni* collected in the Gulf of California during four periods of the year

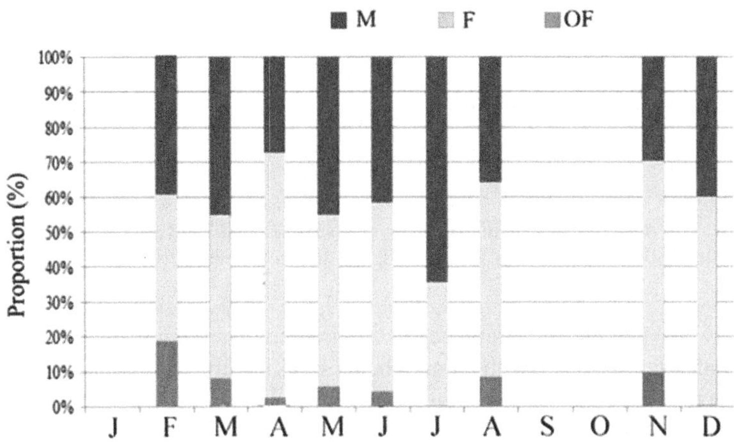

Fig. 10.4 Proportion of males (M), females (F), and ovigerous females (OF) of *Nematocarcinus faxoni* collected in the Gulf of California for each month over the period of the survey

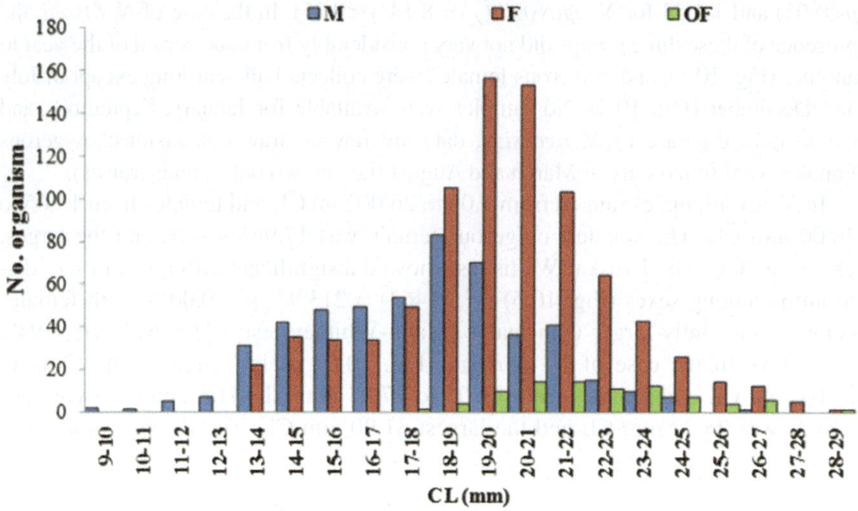

Fig. 10.5 Size (CL) distribution of specimens of *Nematocarcinus faxoni* collected during the TALUD survey by sex. Males (M), females (F), and ovigerous females (OF)

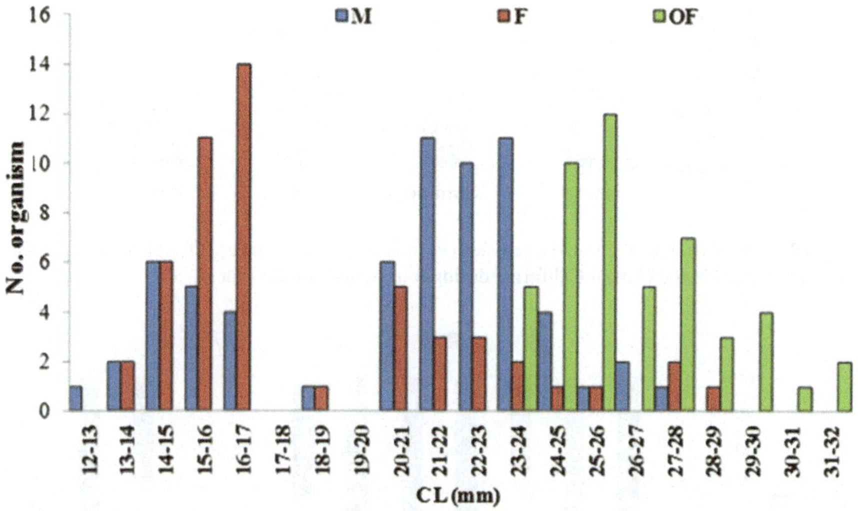

Fig. 10.6 Size (CL) distribution of specimens of *Nematocarcinus agassizii* collected during the TALUD survey by sex. Males (M), females (F), and ovigerous females (OF)

also showed a significant difference in size distribution among sexes (Fig. 10.6) (χ^2, (169) = 85.65; $p < 0.001$), with females being significantly larger than males (Mann-Whitney test; $nM = 65$, $nF = 104$, $p < 0.05$).

In the case of *N. faxoni*, distribution of specimens by sex and by bathymetric range indicates that a larger proportion of males generally occurred in deep water

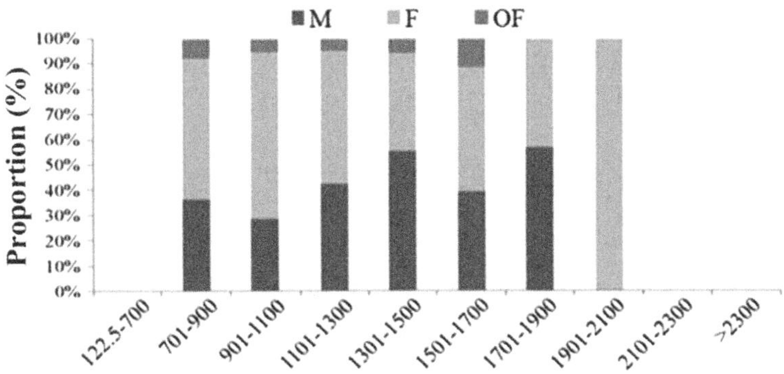

Fig. 10.7 Proportion of males (M), females (F), and ovigerous females (OF) of *Nematocarcinus faxoni* collected at different depth (m) intervals during the TALUD survey

(i.e., 1101–1900 m) than in shallow water (i.e., 701–1100 m depth) (Fig. 10.7). Ovigerous females of *N. faxoni* were not found in the 1701–2100 depth range, although at these depths females were abundant. Again, data for *N. agassizii* were too few to draw conclusions.

The number of eggs per female of *N. faxoni* in the material examined was high, from 1269 to 6882 (average, 3700), with ca 2.5 times more eggs in the largest ovigerous females than in the smallest. The relationship between female size (CL) and number of eggs in *N. faxoni* was found to be loosely related ($R^2 = 0.762$) (Fig. 10.8). All measurements considered, egg size in ovigerous females of *N. faxoni* with CL from 18.09 to 26.21 mm varied from 0.50 to 0.80 mm (long axis) and from 0.40 to 0.69 mm (short axis). The relationship between length of long and short axis showed a loosely correlated increase of length with increase of width (Fig. 10.9). Average egg size (20 eggs per female) was 0.59 to 0.72 mm (long axis) and 0.44 to 0.58 mm (short axis). There was no clear variation of egg size with female size. The lack of sufficient ovigerous females in *N. agassizii* did not allow for fecundity analysis.

10.4 Discussion

The proportion of individuals per sex (males/females) was similar in *N. faxoni* and in *N. agassizii* but with a much higher proportion of ovigerous females in the case of *N. agassizii* (29% vs. 6% in *N. faxoni*). Vélez et al. (1992) reported an astonishing dissimilarity in sex proportion for *N. agassizii* off Peru, with about 90% of the specimens being females. Quite the opposite, in *N. africanus*, Burukovsky and Sudnik (2014) found an average incidence of mature females of 9.5% in the 75–86 mm size class. The same authors assumed that the lower percentage of mature females in populations of *Nematocarcinus* inhabiting in tropical and subtropical areas is linked to their more frequent spawning vs. the single spawning of a lower latitude species such as *N. lanceopes* which experiences a single spawning over a

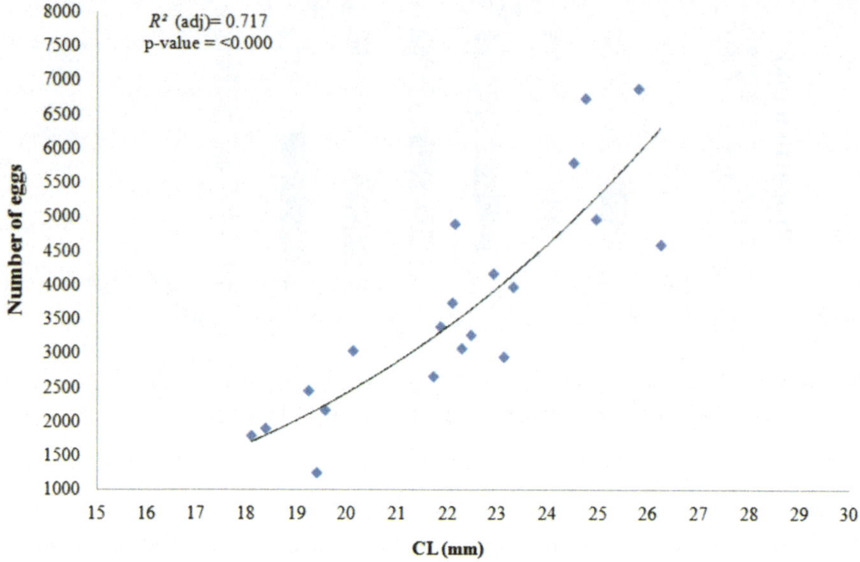

Fig. 10.8 Relationship between egg numbers and size (CL) of ovigerous females in *Nematocarcinus faxoni*

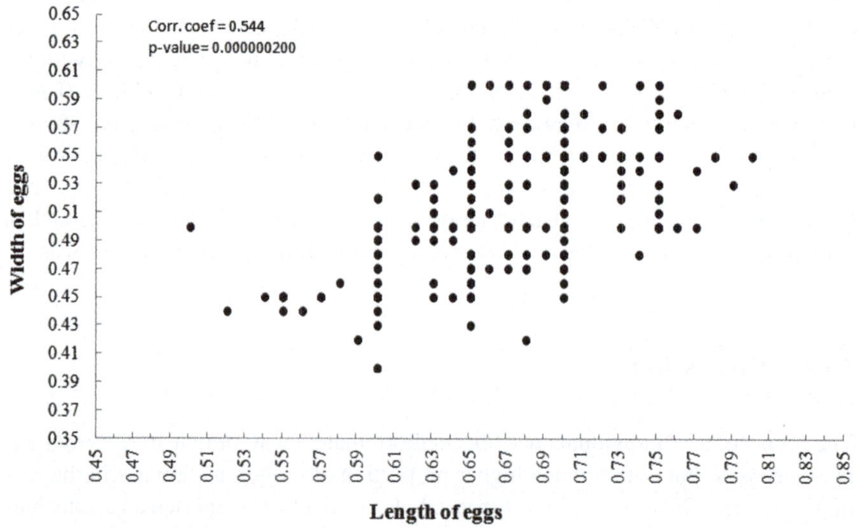

Fig. 10.9 Relationship between long and short axis of eggs in ovigerous females of *Nematocarcinus faxoni*

2-year period, with 14–18% incidence of mature females. Ovigerous females in *N. lanceopes* have been observed in a mean CL of 26.3 mm (Graeve and Wehrtmann 2003).

Sizes of *N. faxoni* and *N. agassizii* specimens examined were rather similar, with maximum of 26–27 mm CL in males and 28–32 mm CL in females. In the case of *N. agassizii*, this was considerably smaller than the maximum size reported by Vélez et al. (1992) for Peru (maximum of 34 mm for both sexes). Cornejo (2010) reported ovigerous females of *N. agassizii* of 31.5 mm CL. In both species, mature females were considered as those carrying eggs; the development of ovaries was not evaluated, thus sexual maturity (i.e., oocytes development) is probably attained earlier in both species (see Gorny and George 1997).

Brood size in caridean shrimps is extraordinary variable, from less than 10 eggs in *Fennera chacei* Holthuis, 1951, to over 35,000 in *Heterocarpus laevigatus* Bate, 1888 (Bauer 2004). Today it is generally recognized that many deep-water carideans feature a reduced brood size with large lecitotrophic eggs, as in several species of Crangonidae and Glyphocrangonidae (e.g., Hendrickx and Papiol 2015, 2019). There are, however, many exceptions to this rule particularly among the deep-water Pandalidae (e.g., *Plesionika* and *Heterocarpus*) (e.g., Wehrtmann and Andrade 1998, Triay-Portella et al. 2017) and within the genus *Nematocarcinus* (e.g., Suseelan 1974; Echeverría-Sáenz and Wehrtmann 2011). In *N. faxoni*, the number of eggs per female was high, up to 6882 per female, thus confirming the trend observed by previous authors. Gorny and George (1997) reported maximum brood size of 1522 eggs for *N. lanceopes* Bate, 1888 of 13,847 eggs for *N. rotundus* Crosnier and Forest, 1973, and 15,573 for *N. ensifer* (Smith, 1882), equivalent to a tenfold variation among the genus. Graeve and Wehrtmann (2003) reported an average brood pouch size of 1219 eggs in *N. lanceopes*. Burukovsky and Sudnik (2014) reported minimum-maximum value of 7252 (female total length, 68 mm) and 17,700 (female total length, 85 mm) eggs per female (average, 8580) in *N. africanus* Crosnier and Forest, 1973, notably more than in *N. gracilipes* Filhol, 1884 (average, 1362 eggs per female of 69–70 mm of length) (Burukovsky and Sudnik 2014).

In caridean shrimps, the number of eggs carried by females is usually correlated to the size of the individuals (Bilgin and Samsun 2006; Ahamed and Ohtomi 2011). In *N. faxoni* the relationship between female size (CL) and brood pouch size was moderately good. King and Butler (1985) reported on the relationship between brood size and the female body wet weight in four species of deep-water *Heterocarpus*, with results varying from poor ($R^2 = 0.47$) to very good ($R^2 = 0.92$). Wehrtmann and Andrade (1998) reported a good relationship ($R^2 = 0.85$) for the commercial *H. reedi* Bahamonde, 1955, in Chile.

Nematocarcinus faxoni seems to experience reproduction activity all year long, contrary to the Antarctic *N. lanceopes* in which a clear seasonally limited oogenesis has been reported, apparently associated with the primary productivity cycle in the euphotic zone (Gorny and George 1997).

Gorny and George (1997) studied the oocyte development and gonad production in *N. lanceopes*. In addition to providing a large amount of information obtained from histological observations, they estimated the completely matured oocytes diameter (females CL 27–34 mm) at about 1.4 mm. Burukovsky and Sudnik (2014) reported on the maximum size of mature oocytes in both *N. africanus* (up to 0.28 mm × 0.36 mm) and in *N. gracilipes* (up to 0.6 mm × 0.7 mm). Comparatively, egg size in *N. faxoni* was on average 0.58 mm × 0.72 mm. This average size is very similar to what has been observed in other groups of deep-water carideans, like the genera *Plesionika* and *Heterocarpus*. King and Butler (1985) reported egg length varying from 0.68 mm to 0.87 mm in four different deep-water species of *Heterocarpus*, while Burukovsky (2017) and Tray-Portella et al. (2017) reported egg length from 0.5 mm to 0.65 mm and from 0.60 mm to 0.82 mm, respectively, in two species of deep-water *Plesionika*.

Based on preliminary information obtained from early deep-water exploration, Thorson (1950) predicted that deep-water species would probably have low fecundity and reduced or no pelagic larvae. One argument in favor of this prediction is the long migrating journey to surface the pelagic larvae would have to endure in order to feed in the rich, primary productive epipelagic zone. In addition, production of a large amount of larvae would be necessary to guarantee survival of some during this migrating journey (Gage and Tyler 1992). The presence of numerous, small-sized eggs with typical planktotrophic larvae (i.e., with an extended larval life) in *N. lanceopes* has been documented and considered as an outstanding feature for this deep-water species (Thatje et al. 2005). In *N. faxoni* as in some other species of *Nematocarcinus*, a similar situation seems to occur, with numerous small eggs. Little is known, however, on the larval morphology and larval behavior within the genus, especially for the very deep-water species, and further investigation is needed in order to fully understand the dynamics of the migration and feeding processes of these larvae. In western Mexico, where a wide and strong oxygen minimum zone occurs (Hendrickx and Serrano 2010), vertical migration of planktotrophic larvae from deep water to the euphotic zone might be limited for lack of physiological adaptations to hypoxic conditions (see Ekau et al. 2010; Hendrickx 2015).

Acknowledgments Ship time aboard the R/V "El Puma" was provided by the Instituto de Ciencias del Mar y Limnología and by the Coordinación de la Investigación Científica, UNAM, and partly supported by CONACyT (project 31805-N for the TALUD IV-VII cruises; project 179467 for the TALUD XV and XVI-B cruises). The TALUD project has received laboratory support from the DGAPA (PAPIIT IN-217306-3 and PAPIIT IN-203013-2) and from CONACyT (Project 31805-N for the TALUD IV to VII cruises; project 179467 for the TALUD XV and XVI-B cruises), Mexico. We thank Manuel Ayón-Parente for providing tips and opinions on this manuscript.

References

Ahamed F, Ohtomi J (2011) Reproductive biology of the pandalid shrimp *Plesionika izumiae* (Decapoda: Caridea). J Crustac Biol 31(3):441–449

Allen CE, Tyler PA, Varney MS (2000) Lipid profiles of *Nematocarcinus gracilis* a deep-sea shrimp from below the Arabian Sea oxygen minimum zone. In: Jones MB, Azevedo JMN,

Neto AI, Costa AC, Frias Martins AM (eds) Island, ocean and deep-sea biology. Springer, Dordrecht, pp 273–279

Arculeo M, Brutto SL, Cannizzaro L, Vitale S (2014) Growth and reproduction of the deep-water rose shrimp, *Parapenaeus longirostris* (Lucas, 1846) (Decapoda, Penaeidae), in the Southern Tyrrhenian Sea. Crustaceana 87(10):1168–1184

Bauer RT (2004) Remarkable shrimps: adaptations and natural history of the carideans, vol 7. University of Oklahoma Press, Norman

Benjamin D, Harikrishnan M, Rozario JV, Jose D, Kurup BM, Sreedhar U, Cubelio SS (2017) Reproductive traits of deep-sea armoured shrimp *Glyphocrangon investigatoris* from Bay of Bengal, Indian Ocean. J Mar Biol Assoc UK 99(1):93–100

Bilgin S, Samsun O (2006) Fecundity and egg size of three shrimp species, *Crangon crangon, Palaemon adspersus,* and *Palaemon elegans* (Crustacea: Decapoda: Caridea), off Sinop Peninsula (Turkey) in the Black Sea. Turkish J Zool 30(4):413–421

Brante A, Cifuentes S, Pörtner HO, Arntz W, Fernández M (2004) Latitudinal comparisons of reproductive traits in five brachyuran species along the Chilean coast. Rev Chil Hist Nat 77:15–27

Briones-Fourzan P, Barradas-Ortíz C, Negrete-Soto F, Lozano-Alvarez E (2010) Reproductive traits of tropical deep-water pandalid shrimps (*Heterocarpus ensifer*) from the SW Gulf of Mexico. Deep Sea Res I Oceanogr Res Pap 57(8):978–987

Burukovsky RN (2009) Morphology and biology of shrimp *Nematocarcinus gracilipes* Filhol, 1884 (Decapoda: Nematocarcinidae). Inverteb Zool 6(2):81–88

Burukovsky RN (2012) Deep sea shrimps of the family Nematocarcinidae (history of study, systematic, distribution, and biology). Prospekt Nauki, St. Petersburg

Burukovsky RN (2017) Distribution, reproduction, and feeding of West African shrimp *Plesionika carinata* Holthuis, 1951 (Decapoda, Pandalidae). Oceanology 57(5):668–674

Burukovsky RN, Sudnik SA (2014) Oogenesis and maturation of ovarian as the basis of reproductive strategies of female shrimps *Nematocarcinus africanus* Crosnier et Forest 1973 (Crustacea, Caridea, Nematocarcinidae). Sci World 12(16), II:16–24 [In Russian]

Butler TH (1964) Growth, reproduction, and distribution of pandalid shrimps in British Columbia. J Fish Board Canada 21(6):1403–1452

Cardoso IA, Burukovsky RN (2014) *Nematocarcinus* Milne Edwards, 1881 (Crustacea, Decapoda) from southwestern Atlantic, including the southern mid-Atlantic ridge area. Zootaxa 3887(3):437–458

Cartes JE (1993) Day-night feeding by decapod crustaceans in a deep-water bottom community in the western Mediterranean. J Mar Biol Assoc UK 73(4):795–811

Cartes JE, Sardà F (1993) Zonation of deep-sea decapod fauna in the Catalan Sea (Western Mediterranean). Mar Ecol Prog Ser 94:27–27

Castilho AL, Gavio MA, Costa RC, Boschi EE, Bauer RT, Fransozo A (2007) Latitudinal variation in population structure and reproductive pattern of the endemic South American shrimp *Artemesia longinaris* (Decapoda: Penaeoidea). J Crustac Biol 27(4):548–552

Chilari A, Thessalou-Legaki M, Petrakis G (2005) Population structure and reproduction of the deep-water shrimp *Plesionika martia* (Decapoda: Pandalidae) from the eastern Ionian Sea. J Crust Biol 25(2):233–241

Clarke A (1993) Reproductive trade-offs in caridean shrimps. Funct Ecol 7:411–419

Clarke A, Hopkins CCE, Nilssen EM (1991) Egg size and reproductive output in the deep-water prawn *Pandalus borealis* Kroyer, 1838. Funct Ecol 5:724–730

Company JB, Sardà F (1997) Reproductive patterns and population characteristics in five deep-water pandalid shrimps in the Western Mediterranean along a depth gradient (150–1100 m). Mar Ecol Prog Ser 148:49–58

Company JB, Maiorano P, Tselepides A, Politou CY, Plaity W, Rotllant G, Sardá F (2004) Deep-sea decapod crustaceans in the western and Central Mediterranean Sea: preliminary aspects of species distribution, biomass and population structure. Sci Mar 68(S3):73–86

Corey S, Reid DM (1991) Comparative fecundity of decapod crustaceans: I. The fecundity of thirty-three species of nine families of caridean shrimp. Crustaceana 60(3):270–294

Cornejo Antepara M (2010) Los Crustáceos Decápodos de Aguas Profundas del Mar Ecuatoriano (Master Thesis), Universidad de Guayaquil, Facultad de Ciencias Naturales, Maestría en Ciencias Manejo Sustentable de Recursos Bioacuáticos y Medio Ambiente

Dailey MD, Ralston S (1986) Aspects of the reproductive biology, spatial distribution, growth and mortality of the deep water caridean shrimp *Heterocarpus laevigatus* in Hawaii. Fish Bull 84:915–925

Dambach J, Raupach MJ, Mayer C, Schwarzer J, Leese F (2013) Isolation and characterization of nine polymorphic microsatellite markers for the deep-sea shrimp *Nematocarcinus lanceopes* (Crustacea: Decapoda: Caridea). BMC Res Notes 6(1):75. https://doi.org/10.1186/1756-0500-6-75

Dugan JE, Wenner AM, Hubbard DM (1991) Geographic variation in the reproductive biology of the sand crab *Emerita analoga* (Stimpson) on the California coast. J Exp Mar Biol Ecol 150(1):63–81

Echeverría-Sáenz S, Wehrtmann IS (2011) Egg production of the commercially exploited Deepwater shrimp, *Heterocarpus vicarius* (Decapoda: Pandalidae), Pacific Costa Rica. J Crustac Biol 31(3):434–440

Ekau W, Auel H, Pörtner HO, Gilbert D (2010) Impacts of hypoxia on the structure and processes in pelagic communities (zooplankton macro-invertebrates and fish). Biogeosciences 7:1669–1699. https://doi.org/10.5194/bg-7-1669-2010

Erdman RB, Blake NJ, Lockhart FD, Lindberg WJ, Perry HM, Waller RS (1991) Comparative reproduction of the deep-sea crabs *Chaceon fenneri* and *C. quinquedens* (Brachyura: Geryonidae) from the northeast Gulf of Mexico. Invertebr Reprod Dev 19(3):175–184

Gage JD, Tyler PA (1992) Deep-sea biology: a natural history of organisms at the deep-sea floor. Cambridge University Press, Cambridge

García-Guerrero M, Hendrickx ME (2004) Fecundity traits of seven species of brachyuran crabs (Crustacea: Brachyura) from the Pacific coast of Mexico. In: Hendrickx ME (ed) Contributions to the study of East Pacific Crustaceans 3. [Contribuciones al Estudio de los Crustáceos del Pacífico Este 3]. Instituto de Ciencias del Mar y Limnología, UNAM, Mexico, pp 79–87

Gorny M, George MR (1997) Oocyte development and gonad production of *Nematocarcinus lanceopes* (Decapoda: Caridea) in the eastern Weddell Sea, Antarctica. Polar Biol 17(3):191–198

Graeve M, Wehrtmann I (2003) Lipid and fatty acid composition of Antarctic shrimp eggs (Decapoda: Caridea). Polar Biol 26(1):55–61

Hanamura Y, Evans DR (1996) Deep-water caridean shrimps of the families Nematocarcinidae, Stylodactylidae, Pandalidae and Crangonidae (Crustacea: Decapoda) from western Australia. Bull Nansei Nat Fish Res Inst 29:1–18

Hartnoll RG, Rice AL, Attrill YM (1992) Aspects of the biology of the galatheid genus *Munida* (Crustacea, Decapoda) from the Porcupine Seabight, Northeast Atlantic. Sarsia 76(4):231–246

Hendrickx ME (2015) Further records of species of *Gennadas* (Crustacea, Decapoda, Dendrobranchiata, Benthesicymidae) in the Mexican Pacific. Zootaxa 3980(3):417–426

Hendrickx ME, Hernández-Payán JC (2018) Distribution and abundance of species of *Nematocarcinus* (Crustacea: Decapoda: Caridea: Nematocarcinidae) off western Mexico, eastern Pacific. Regional Studies Mar Sc 2018 23:47-52. https://doi.org/10.1016/j.rsma.2017.10.012

Hendrickx ME, Papiol V (2015) Insights on the biology and ecology of the deep-water shrimp *Parapontophilus occidentalis* (Faxon, 1893) (Crustacea: Caridea: Crangonidae) in the eastern Pacific with notes on its morphology. Zootaxa 4007(3):370–388

Hendrickx ME, Papiol V (2019) Distribution, relative growth, ecology and fecundity of the deep-water squat lobster *Galacantha diomedeae* (Crustacea: Decapoda: Galatheoidea: Munidopsidae) in the Mexican Pacific. Inverteb Biol 138(2):1–12. https://doi.org/10.1111/ivb.12248

Hendrickx ME, Serrano D (2010) Impacto de la zona de mínimo de oxígeno sobre los corredores pesqueros en el Pacífico mexicano. Interciencia 35(1):12–18
Hernández-Payán JC, Hendrickx ME (2016) Two species of the deep-water genus *Nematocarcinus* A. Milne–Edwards, 1881 (Crustacea, Caridea, Nematocarcinidae) from the Mexican Pacific. Zootaxa 4126(4):587–599
Hines AH (1988) Fecundity and reproductive output in two species of deep-sea crabs, *Geryon fenneri* and *G. quinquedens* (Decapoda: Brachyura). J Crustac Biol 8(4):557–562
Hines AH (1991) Fecundity and reproductive output in nine species of *Cancer* crabs (Crustacea, Brachyura, Cancridae). Can J Fish Aquat Sci 48(2):267–275
Kapiris K, Thessalou Legaki M (2006) Comparative fecundity and oocyte size of *Aristaeomorpha foliacea* and *Aristeus antennatus* in the Greek Ionian Sea (E. Mediterranean) (Decapoda: Aristeidae). Acta Zool 87(4):239–245
King MG, Butler AJ (1985) Relationship of life-history patterns to depth in deep-water caridean shrimps (Crustacea: Natantia). Mar Biol 86(2):129–138
Komai T, Segonzac M (2005) Two new species of *Nematocarcinus* A. Milne-Edwards, 1881 (Crustacea: Decapoda: Caridea: Nematocarcinidae) from hydrothermal vents on the North and South East Pacific rise. Zoosystema 27(2):343–364
Lardies M, Castilla J (2001) Latitudinal variation in the reproductive biology of the commensal crab *Pinnaxodes chilensis* (Decapoda: Pinnotheridae) along the Chilean coast. Mar Biol 139(6):1125–1133
Lardies MA, Wehrtmann IS (1996) Aspects of the reproductive biology of *Petrolisthes laevigatus* (Guérin, 1835) (Decapoda, Anomura, Porcellanidae). Part I: reproductive output and chemical composition of eggs during embryonic development. Arch Fish Mar Res 43:121–135
Lawton P, Duggan D, Moriyasu M, Benhalima K, Robichaud D (2002) Reproductive biology of male Jonah crab, *Cancer borealis* Stimpson, 1859 (Decapoda, Cancridae) on the Scotian shelf, northwestern Atlantic. Crustaceana 75(7):891–913
Lovrich GA, Romero MC, Tapella F, Thatje S (2005) Distribution, reproductive and energetic conditions of decapod crustaceans along the scotia arc (Southern Ocean). Sci Mar 69(2):183–193
Lozano-Álvarez E, Briones-Fourzán P, Gracia A, Vázquez-Bader AR (2007) Relative growth and size at first maturity of the deep water shrimp, *Heterocarpus ensifer* (Decapoda, Pandalidae) from the southern Gulf of Mexico. Crustaceana 80:555–568. https://doi.org/10.1163/156854007780765588
Maiorano P, Tselepides A, Politou CY, Plaity W, Rotllant G, Sardá F (2004) Deep-sea decapod crustaceans in the western and central Mediterranean Sea: preliminary aspects of species distribution, biomass and population structure. Sci Mar 68(S3):73–86
McDermott JJ (1999) Reproduction in the hermit crab *Pagurus longicarpus* (Decapoda: Anomura) from the coast of New Jersey. J Crustac Biol 19(3):612–621
Melville-Smith R (1987) The reproductive biology of *Geryon maritae* (Decapoda, Brachyura) off south west Africa/Namibia. Crustaceana 53:259–275
Mente E (2008) Reproductive biology of crustaceans: case studies of decapod crustaceans. CRC Press
Moriyasu M, Lanteigne C (1998) Embryo development and reproductive cycle in the snow crab, *Chionoecetes opilio* (Crustacea: Majidae), in the southern gulf of St. Lawrence, Canada. Canadian J Zool 76(11):2040–2204
Nakamura M, Chen C, Mitarai S (2015) Insights into life-history traits of *Munidopsis* spp. (Anomura: Munidopsidae) from hydrothermal vent fields in the Okinawa trough, in comparison with the existing data. Deep Sea Res I Oceanogr Res Papers 100:48–53
Navarrete N, Soto R, Quiroga E, Vargas M, Wehrtmann IS (1999) Egg production in *Eurypodius latreillei* Guérin, 1828 (Decapoda: Majidae) in the Strait of Magellan, southern Chile. Sci Mar 63(S1):333–337
Ohtomi J (1997) Reproductive biology and growth of the deep-water pandalid shrimp *Plesionika semilaevis* (Decapoda: Caridea). J Crust Biol 17(1):81–89

Papiol V, Hernández-Payán JC, Hendrickx ME (this volume) Influence of environmental factors on the abundance and distribution of the deep-water shrimps *Nematocarcinus faxoni* Burukovsky, 2001 and *N. agassizii* Faxon, 1893 (Crustacea: Decapoda: Nematocarcinidae) off western Mexico. In: Hendrickx ME (ed) Deep-water pycnogonids and crustaceans from the Americas. Springer Latin-America, p xxx-www

Paul AJ, Adams AE (1984) Breeding and fertile period for female *Chionoecetes bairdi* (Decapoda, Majidae). J Crust Biol 4(4):589–594

Quiroga EJ, Soto RM (1997) Relaciones biométricas y fecundidad de *Glyphocrangon alata* (Faxon, 1893) en la zona norte de Chile (21819′S; 70820′W). Invest Mar 25:281–285

Ramírez-Llodra E, Segonzac M (2006) Reproductive biology of *Alvinocaris muricola* (Decapoda: Caridea: Alvinocarididae) from cold seeps in the Congo Basin. J Mar Biol Assoc UK 86(6):1347–1356

Ramírez-Llodra E, Tyler PA, Copley JT (2000) Reproductive biology of three caridean shrimp, *Rimicaris exoculata*, *Chorocaris chacei* and *Mirocaris fortunata* (Caridea: Decapoda), from hydrothermal vents. J Mar Biol Assoc UK 80(3):473–484

Raupach MJ, Thatje S, Dambach J, Rehm P, Misof B, Leese F (2010) Genetic homogeneity and circum-Antarctic distribution of two benthic shrimp species of the Southern Ocean, *Chorismus antarcticus* and *Nematocarcinus lanceopes*. Mar Biol 157(8):1783–1797

Storch V, Bluhm BA, Arntz WE (2002) Microscopic anatomy and ultrastructure of the digestive system of three Antarctic shrimps (Crustacea: Decapoda: Caridea). In: Arntz WE, Clarke A (eds) Ecological studies in the Antarctic Sea ice zone. Springer, Berlin/Heidelberg, pp 66–76

Suseelan C (1974) Observations on the deep-sea prawn fishery off the south-west coast of India with special reference to pandalids. J Mar Biol Assoc India 16(2):491–511

Terossi M, Wehrtmann IS, Mantelatto FL (2010) Interpopulation comparison of reproduction of the Atlantic shrimp *Hippolyte obliquimanus* (Caridea: Hippolytidae). J Crustac Biol 30(4):571–579

Thatje S, Bacardit R, Arntz WE (2005) Larvae of the deep-sea Nematocarcinidae (Crustacea: Decapoda: Caridea) from the southern ocean. Polar Biol 28(4):290–302

Thorson G (1950) Reproduction and larval ecology of marine bottom invertebrates. Biol Rev 25:1–45

Triay-Portella R, Ruiz-Díaz R, Pajuelo JG, González JA (2017) Ovarian maturity, egg development, and offspring generation of the deep-water shrimp *Plesionika edwardsii* (Decapoda, Pandalidae) from three isolated populations in the eastern North Atlantic. Mar Biol Res 13(2):174–187

Tuset VM, Pérez-Peñalvo J, Delgado J, Pinho MR, Santana J, Biscoito M et al (2009) Biology of the deep-water shrimp *Heterocarpus ensifer* (Caridea: Pandalidae) off the canary, Madeira and the Azores islands (Northeastern Atlantic). J Crustac Biol 29(4):507–515

Van Dover CL, Factor JR, Williams AB, Berg CJ Jr (1985) Reproductive patterns of decapod crustaceans from hydrothermal vents. Bull Biol Soc Washington 6:223–227

Velez J, Kameya A, Yamashiro C, Lostaunau N, Valiente O (1992) Investigación del recurso potencial langostino rojo de profundidad a bordo del BIC "Fridtjof Nansen" (25 de abril–25 de mayo, 1990). Informe. Instituto del Mar del Perú, C.E.E 104:3–24

Wehrtmann IS, Andrade G (1998) Egg production in *Heterocarpus reedi* from northern Chile, with a comparison between iced and living females (Decapoda: Caridea: Pandalidae). Ophelia 49(1):71–82

Wehrtmann IS, Echeverría-Sáenz S (2007) Crustacean fauna (Stomatopoda: Decapoda) associated with the deepwater fishery of *Heterocarpus vicarius* (Decapoda: Pandalidae) along the Pacific coast of Costa Rica. Rev Biol Trop 55(Suppl 1):121–130

Wehrtmann IS, Lardies MA (1999) Egg production of *Austropandalus grayi* (Decapoda, Caridea, Pandalidae) from the Magellan region, South America. Sci Mar 63(S1):325–331

Wenner EL (1979a) Some aspects of the biology of deep-sea lobsters of the family Polychelidae (Crustacea, Decapoda) from the western North-Atlantic. Fish Bull 77(2):435–444

Wenner EL (1979b) Distribution and reproduction of nematocarcinid shrimp (Decapoda: Caridea) from the northwestern North Atlantic. Bull Mar Sci 29(3):380–393

Wenner EL (1980) Notes on the biology of a deep-sea penaeid, *Benthesicymus bartletti* Smith. Crustaceana 38(3):290–294

Wenner EL, Boesch DF (1979) Distribution patterns of epibenthic decapod Crustacea along the shelf-slope coenocline, Middle Atlantic Bight, USA. Bull Biol Soc Washington 3:106–133

Wenner A, Kuris A (1991) Crustacean eggs production, Crustacean Issues 7. Balkema AA, Roterdam

Young CM (2003) Chapter 12. Reproduction, development and life-history traits. In: Tyler PA (ed) Ecosystems of the world, 28, ecosystems of the deep oceans. Elsevier Science BV, Amsterdam, pp 381–426

Zar JH (1998) Biostatistical Analysis (4th ed) Prentice Hall, New Jersey, USA, 907 pp

Chapter 11
Influence of Environmental Variables on the Abundance and Distribution of the Deep-Water Shrimps *Nematocarcinus faxoni* Burukovsky, 2001 and *N. agassizii* Faxon, 1893 (Crustacea, Decapoda, Nematocarcinidae) off Western Mexico

V. Papiol, J. C. Hernández-Payán, and M. E. Hendrickx

Abstract Two species of the deep-water caridean shrimp genus *Nematocarcinus* were collected off western Mexico: *N. agassizii* (6 samples) and *N. faxoni* (56 samples). The specimens were collected using benthic gear (i.e., benthic sledge and Agassiz dredge) during a series of 12 cruises (228 samples) in the Mexican Pacific and inside the Gulf of California. At each locality, near-bottom temperature, salinity, and dissolved oxygen were measured, and the organic carbon content in the sediments was analyzed. *Nematocarcinus faxoni* was distributed along all the Mexican Pacific coasts, and *N. agassizii* was only collected off the western Baja California Peninsula. The density of the two species peaked at different depths, *N. agassizii* at 701–1000 m and *N. faxoni* at 1001–1300 m, bathymetric changes in density being overall significant for both species. The females of both species were significantly larger than males, and ovigerous females were larger than the rest of the females. Ovigerous females of *N. faxoni* measured 17.96–28.33 mm (CL) and those of *N. agassizii* measured 23.32–31.90 mm. The size of *N. faxoni* changed with depth, smaller organisms were not collected deeper than 1600 m, and greater proportions of large organisms were found at greater depths. Overall sex ratio was M/F = 1:2, except at 1301–1600 (M/F = 1:1). High densities of *N. faxoni* were recorded at temperature ranging 3.5–6.5 °C, DO ranging 0–0.5 ml/l, and salinity ranging 34.5–34.6 and 34.7–34.8, and in sediments with 1–2% organic carbon content. Specimens of *N. agassizii* were mainly found at temperature between 5 and

V. Papiol (✉)
Escuela Nacional de Estudios Superiores, Unidad Mérida, Universidad Nacional Autónoma de México, Ucú, Yucatán, México

J. C. Hernández-Payán · M. E. Hendrickx
Laboratorio de Invertebrados Bentónicos, Unidad Académica Mazatlán Instituto de Ciencias del Mar y Limnología, Universidad Nacional Autónoma de México, Mazatlán, Sinaloa, Mexico

M. E. Hendrickx (ed.), *Deep-Sea Pycnogonids and Crustaceans of the Americas*, https://doi.org/10.1007/978-3-030-58410-8_11

6 °C and aggregated mostly at DO of 0–0.5 ml/l. No particular trend was found regarding salinity and organic carbon in the sediments. Generalized additive models revealed that the distribution of *N. faxoni* was associated with environments with intermediate salinity, low organic carbon content in the sediments, and high temperature. *Nematocarcinus agassizii* distribution was associated with lower salinity and higher organic carbon content than *N. faxoni*.

Keywords Deep-sea · Caridean shrimps · Ecology · East Pacific · Oxygen minimum zone

11.1 Introduction

The Nematocarcinidae is a large family of caridean shrimps including five genera: *Lenzicarcinus* Burukovsky, 2005 (1 species), *Macphersonus* Burukovsky, 2012 (1 species), *Nematocarcinus* A. Milne-Edwards, 1881 (48 species), *Nigmatullinus* Burukovsky, 1991 (1 species), and *Segonzackomaius* Burukovsky, 2011 (2 species) (WoRMS editorial board 2020). The genus *Nematocarcinus* is the one comprising a largest number of species, about half of which (24) have been described by Rudolph Burukovsky, who also provided exhaustive reviews of the genus (see Burukovsky 2001, 2004, 2012).

Nematocarcinus spp. are distributed worldwide in deep-water (Chace 1986; De Grave and Fransen 2011), including hydrothermal vents (e.g., Komai and Segonzac 2005), and can be dominant in decapod crustacean communities: e.g., *Nematocarcinus exilis* (Bate, 1888) in the NW Mediterranean Sea (Cartes and Sardà 1992), *N. rotundus* Crosnier and Forest, 1973 in the SW Gulf of Mexico (Escobar-Briones et al. 2008), *N. africanus* Crosnier and Forest, 1973 in CE Atlantic (Muñoz et al. 2012), and *N. ensifer* (Smith, 1882) in Atlantic seamounts (Cartes et al. 2014). In the eastern Pacific, six species of *Nematocarcinus* have been reported: *N. agassizii* Faxon, 1893 (Mexico to Peru, including localities off Coco, Malpelo, and Galapagos Islands), *N. faxoni* Burukovsky, 2001 (Mexico to Galapagos Islands, southwestern Atlantic), *N. lanceopes* Bate, 1888 (off Chile), *N. proximatus* Bate, 1888 (off Chile), *N. longirostris* Bate, 1888 (off Chile), and *N. serratus* Bate, 1888 (off Chile) (Guzmán and Quiroga 2005; Cardoso and Burukovsky 2014; Hernández-Payán and Hendrickx 2016). *Nematocarcinus faxoni* is dominant in the slope of the northeastern Mexican Pacific (Papiol et al. 2017) and inside the Gulf of California (Papiol and Hendrickx 2016a). Despite the large distribution and abundance of *Nematocarcinus* spp., information of their ecology is scarce and mostly refers to their depth distribution range and habitat (e.g., Türkay 1998, Komai and Segonzac 2005, Burukovsky 2012) and to their diet (e.g., Wenner 1979; Cartes 1993a). Additionally, some knowledge is available on their internal anatomy (e.g., Storch et al. 2001), reproduction patterns (e.g., Wenner 1979; Gorny and George 1997; Burukovsky and Sudnik 2014), body composition (e.g., Allen et al. 2000), larval development (e.g., Thatje et al. 2005), and molecular markers (e.g., Dambach et al. 2013). Yet, no integrative ecological study addressing the environmental drivers of species' distributions has been performed for this genus.

In the northeastern Pacific, the distribution of decapod crustaceans (and of macrofauna and megafauna in general) is strongly defined by the presence of a permanent oxygen minimum zone (OMZ) in the water column. OMZs are large water layers where dissolved oxygen (DO) values fall persistently below 0.5 ml/l (22 μM) (Levin 2003). They have a hypoxic core (DO<0.15 ml/l or 6.6 μM) where macrofauna and megafauna are virtually absent (Diaz and Rosenberg 1995; Levin 2003; Murty et al. 2009; Hunter et al. 2011), and therefore the consumption of particulate organic carbon (OC) in the water column is low, and sediments are organically enriched (Devol and Hartnett 2001; Cowie 2005; Roullier et al. 2013). Below the hypoxic core, DO increases gradually with depth as temperature decreases. Right under this core, in the lower OMZ boundary (0.15 ml/l < DO<0.5 ml/l), increased DO and the high availability of labile OC in the sediments favor aggregations of megafauna (Murty et al. 2009; Papiol and Hendrickx 2016a), and the strong environmental (DO, OC, and temperature) gradients found are linked to rapid bathymetric successions of species or large taxonomic groups (Zamorano et al. 2007; Méndez 2007; Murty et al. 2009). Below the OMZ boundary, OC in the sediments and megafauna abundance decline with depth. In the eastern Pacific off western Mexico, recent comprehensive studies of the deep-sea decapod community ecology described similar patterns of fauna distribution as described above and located the lower OMZ boundary in the region at ca. 700–1300 m (Papiol and Hendrickx 2016a; Papiol et al. 2017). In this region, the horizontal convergence of water masses (i.e., the North Pacific Intermediate Water and the Equatorial Subsurface Water) at 26°N, off the southern Baja California Peninsula, was recognized as an additional factor defining the distribution of decapod crustaceans together with the OMZ, and strong changes in community composition have been described at this latitude, not only for decapod crustaceans (Wicksten 1989; Papiol et al. 2017) but also for fish (Cruz-Acevedo et al. 2018).

During sampling operations off the west coast of Mexico, large series of specimens of two species of *Nematocarcinus* (*N. agassizii* and *N. faxoni*) were collected during bottom trawling operations (see Hernández-Payán and Hendrickx 2016; Hendrickx and Hernández-Payán 2018). These two species inhabit the lower boundary of the OMZ and below. Hendrickx and Hernández-Payán (2018) provided some insight of the occurrence and density of these two species along western Mexico, mostly of their depth distribution and the ranges of environmental variables they occupied. In this contribution, we provided a more complete analysis of the bathymetric patterns of distribution of the species, including sex and size patterns, and we addressed the influence of the environmental variables available with the aim to understand which factors drive the abundance and distribution of these species.

11.2 Material and Methods

11.2.1 Biological Sampling

The material on which this study is based was collected by the R/V “El Puma” of the Universidad Nacional Autónoma de México (UNAM), between 1991 and 2014 (see Hernández-Payán and Hendrickx 2016). Specimens of *Nematocarcinus* (*N. faxoni* and *N. agassizii*) were captured during sampling operations performed during 12 cruises in three areas off western Mexico (Fig. 11.1a): off the west coast of the Baja California Peninsula (three cruises: TALUD XV, July–August 2012; TALUD XVI, July 2013; TALUD XVI-B, May–June 2014), in the Gulf of California (eight cruises: TALUD III, September 1991; TALUD IV, August 2000; TALUD V, December 2000; TALUD VI, March 2001; TALUD VII, June 2001; TALUD VIII, April 2005; TALUD IX, November 2005; TALUD X, February 2007), and off the SW coast of Mexico, from Jalisco to Guerrero (one cruise: TALUD XII, March–April 2009). During these cruises, a total of 228 localities were sampled from 377 to 2394 m depth. Positional coordinates for each sampling station were obtained using a GPS navigation system. Depth was measured with an Edo Western analog recorder (TALUD III–VIII) or a digital recorder (TALUD IX–XVI-B).

All the specimens were captured with benthic gear, including an Agassiz dredge (2.5 m width, 1 m high) used in the 1991 cruise (TALUD III, 17 hauls) and a benthic sledge (2.35 m width, 0.9 m high) used in the other surveys (211 hauls; see Hendrickx 2012; Papiol et al. 2017). Both were equipped with a modified shrimp net (ca. 5.5 cm stretched mesh size) with a ca. 2.0 cm internal lining net. All samplings were performed at an average speed of 1.75 knots (3.2 km/h), and the time on the bottom of each haul was 30 min in most cases. All organisms in the catch were identified, and specimens of *Nematocarcinus* were sexed (M, males; F, females; OF, ovigerous females), counted, measured (carapace length (CL) in mm ± 0.1 mm), and weighted (total weight (W) in g ± 0.1 g) at the laboratory. Most specimens were

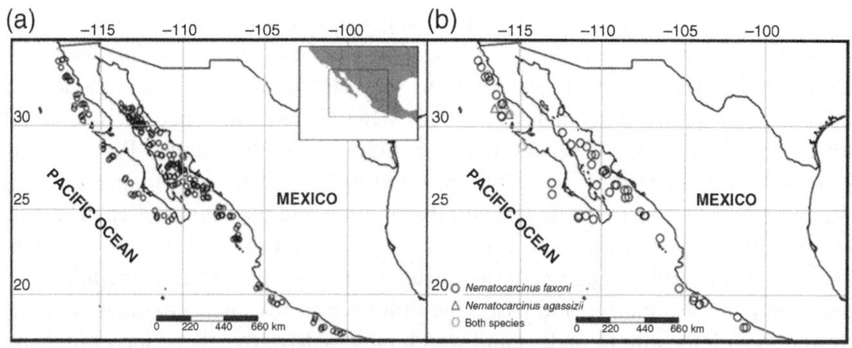

Fig. 11.1 (**a**) Position of the samplings performed during the TALUD cruises. (**b**) Localities in the Mexican Pacific where *Nematocarcinus faxoni* Burukovsky 2001 (open circles), *Nematocarcinus agassizii* Faxon, 1893 (open triangles), or both species (open hexagons) were collected

deposited in the Regional Collection of Marine Invertebrates (ICML-EMU) at the UNAM in Mazatlán, Mexico (see Hernández-Payán and Hendrickx 2016).

11.2.2 Environmental Data

Temperature (T), salinity (S), and dissolved oxygen (DO) were measured about 20 m above the sea bottom (20 mab) with a Seabird 19 CTD. Rosette-mounted 10 l Niskin bottles were also deployed, and dissolved oxygen concentration was estimated with the Winkler method (Strickland and Parsons 1972) during all samplings and compared with CTD results. Sediment was sampled at each sledge station by means of a modified USNEL box core, and samples of the top 3 cm were stored at 4–8 °C. At the laboratory, sediments were dried to constant weight at 60 °C. The organic carbon content in sediments (OC_{sed}) was determined following the technique of Loring and Rantala (1992) and used as an indicator of food availability for benthos (see Beaulieu 2002).

11.2.3 Data Analysis

Abundance of *N. faxoni* and *N. agassizii* was standardized to a common swept area of 1 hectare (inds/ha). Density patterns with depth were analyzed for the Mexican Pacific as a whole grouping samples every 300 m in order to reflect the main bathymetric gradients in dissolved oxygen and fauna composition previously detected in this OMZ (see Papiol and Hendrickx 2016a; Papiol et al. 2017). Statistical differences in density with depth were tested. Sexual and bathymetric patterns in carapace length (CL) distribution were analyzed, and differences were also tested. All data were tested for normality using the Kolmogorov-Smirnov test. For data that were normally distributed, ANOVA was used. For data that did not satisfy the assumptions of normality, even after transformation, a nonparametric Kruskal-Wallis (K-W) or Mann-Whitney *U* test was used. Sex ratio (males/females (M/F)) per depth stratum was looked at, and variations with depth were analyzed.

Considering the similar characteristics of the two sampling gear used in this study (both benthic, similar sizes and similar mesh size) and the importance of increasing sample size for the analyses of population patterns, the data obtained by both sampling gear were analyzed jointly. Yet, bearing in mind the potential effects of differences in catchability between the gear on population size (Flannery and Przeslawski 2015), the main patterns in abundance/density for each gear were compared qualitatively. Besides, the samples obtained with Agassiz dredge were not considered for joint analyses of fauna and environmental data.

Environmental maximum and minimum thresholds of distribution of the species were recorded for each variable. Patterns in density with respect to each environmental variable were analyzed. In order to identify which variables explained the patterns of near-bottom distribution of both species of *Nematocarcinus*, Spearman rank correlations were calculated between values of density and the available environmental variables using two different approaches: using all samples, aiming to detect the main variables determining the presence of the species, and using samples containing specimens of each of the species, with the purpose to establish which variables define abundance patterns of *Nematocarcinus* specimens. Generalized additive models (GAMs; Hastie and Tibshirani 1986) were also performed on density data using the same two approaches. First, correlation matrices were calculated on the environmental data set in order to identify correlations among the variables. Correlated variables (Pearson's correlation coefficient, $\rho > 0.70$) were not fitted in the same model. GAMs were used to evaluate the potential contribution of selected environmental variables in explaining the variation in density of individuals of *Nematocarcinus*. GAMs were computed using the statistical programming environment R (version 3.5.0) and the package mgcv (Wood 2006; Wood and Augustin 2002). GAMs are flexible statistical predictive models which allow for nonlinear, nonparametric relationships between a set of predictors and a dependent variable. We used smoothing splines to represent the nonlinear effect of predictors. The dependent variable was modeled using the Tweedie distribution function with logarithmic link. The Tweedie set of distributions include a range of distributions including the normal, Poisson, gamma, and inverse Gaussian distributions. In this study, a power function (p) between 1 and 2 was used. In this special case, Tweedie distributions are also known as "compound Poisson" (Smyth and Jørgensen 2002), because they can be obtained as Poisson mixtures of gamma distributions. We computed GAMs using density of individuals of the *Nematocarcinus* species included in this chapter in each haul as a response variable and the available set of environmental variables as predictor variables. Model selection was based on minimizing Akaike information criterion (AIC) values and including covariates with $p < 0.05$. All statistical analyses were carried out with STATISTICA 10 (StatSoft, Inc) and R (version 3.5.1: https://www.r-project.org/) software. Map was obtained using SimpleMappr (https://www.simplemappr.net/).

11.3 Results

11.3.1 Population Structure

A total of 1632 specimens were collected during the study, including 1463 of *N. faxoni* and 169 of *N. agassizii* (see Hendrickx and Hernández-Payán 2018). Considering all cruises, a total of 58 (25%) samples contained specimens of *Nematocarcinus* (one or two species). *Nematocarcinus faxoni* was collected in 56 samples distributed along the entire western Mexico coast (northeastern Pacific and inside the Gulf of California) (Fig.11.1b); *N. agassizii* was collected in only six samples distributed off the west coast of the Baja California Peninsula (Fig. 11.1b).

No species of *Nematocarcinus* were found in water shallower than 701 m or deeper than 2100 m, corresponding to the continental slope. Considering their bathymetric and geographic distributions, further analyses were performed only considering samples obtained deeper than 700 m, and for *N. agassizii*, samples were restricted to the west coast of the Baja California Peninsula.

Nematocarcinus faxoni was caught in a global depth range of 779–2055 m and was more frequently sampled between 1301 and 1500 m, appearing in 48% of the samples (Table 11.1). The highest density was at 1001–1300 m (57.6 ± 198.4 inds/ha), where the species was collected in 45% of the samples. Lowest density was found at depths greater than 1600 m (3 ± 14.6 inds/ha). Frequency of appearance of *N. faxoni* with depth in Agassiz dredge was slightly greater at 1001–1300 m (75%) and lower at 701–1000 m (0%), but the overall bathymetric patterns in frequency of appearance and density previously described were not altered by the inclusion of such data, likely owing to the low number of samples.

Nematocarcinus agassizii was collected in a narrower depth range (730–1318 m) (Table 11.1, Hendrickx and Hernández-Payán 2018) than *N. faxoni*. Virtually all specimens (99%) were collected at the 701–1000 m depth range, where density was 50.5 (±69.34) inds/ha, and the species was collected in 10% of the samples. No specimens of this species were collected with the Agassiz dredge, which was only used inside the Gulf of California.

Large standard deviations in density values estimated in each depth stratum were much greater than mean values of density, as a result of the large number of samples

Table 11.1 The number of hauls performed below 700 m during this study (only west of the Baja California Peninsula for *N. agassizii*) is grouped every 300 m. The frequency of appearance, number of individuals, mean density per hectare (±standard deviation), and the numbers of individuals per each sex are indicated for the two species

Depth range	701–1000	1001–1300	1301–1600	>1600
	Nematocarcinus faxoni			
Number of hauls	41	49	42	32
Frequency of appearance (%)	12	45	48	16
Number of individuals	303	964	211	42
Mean density (±standard deviation)	23.3 (±95.0)	57.6 (±198.4)	12.5 (±28.9)	3.0 (±14.6)
Number of males	105	299	110	16
Number of females	167	594	87	22
Number ovigerous females	31	37	14	4
	Nematocarcinus agassizii			
Depth range	701–1000	1001–1300	1301–1600	>1600
Number of hauls	10	8	14	8
Frequency of appearance (%)	50	0	7	0
Number of individuals	167	0	2	0
Mean density (±standard deviation)	50.5 (±69.3)	0	0.4 (±1.4)	0
Number of males	64	0	1	0
Number of females	52	0	1	0
Number ovigerous females	51	0	0	0

not containing individuals of the targeted species and of the heterogeneity in density distribution. Bathymetric changes in density of *N. faxoni* were overall significant (K-W test, $H_{3,179} = 20.161$, $p < 0.001$). Yet, related with the large standard deviations, bathymetric changes in density were only significant between 701–1000 and 1001–1300 m ($p = 0.04$). Changes in *N. agassizii* density with depth were also overall significant (K-W test, $H_{3,40} = 1.505$, $p < 0.01$), but pairwise comparisons were not.

Analysis of size distribution of specimens of *N. faxoni* shows that females (F) (including OF) were larger (CL range 9.05–28.33 mm) than males (M) (size range 9.04–26.00 mm CL) (Mann-Whitney U test, $n_{F+OF} = 956$, $n_M = 530$, $p < 0.001$) (Fig. 11.2). The smallest OF measured 17.96 mm CL and the largest 28.33 mm. Significant differences in size distribution among sexual categories were also observed when F and OF were considered separately (K-W test, $H_{2,1461} = 169.24$; $p < 0.001$) and all pairwise comparisons were significant (OF > F > M; $p < 0.001$).

A similar trend was observed in *N. agassizii* (Fig. 11.3), and F + OF were larger than M (Mann-Whitney U test, $n_{F+OF} = 104$; $n_M = 65$; $p < 0.05$). OF (23.32–31.90 mm) were significantly larger than F (13.75–28.96 mm), and the latter were larger than M (12.03–27.04 mm) (K-W test, $H_{2,169} = 85.64$; $p < 0.001$).

The size of *N. faxoni* changed with depth (K-W test, $H_{3,1486} = 33.22$; $p < 0.001$), and the main patterns observed were the restriction of organisms smaller than 11 cm to 1001–1600 m and the greater proportions of organisms larger than 24 cm with increasing depth (Fig. 11.4). Consistently, size distribution at 701–1000 m significantly differed from those at the other depth strata (all $p > 0.01$). Overall sex ratio (M/F) was 1:2, which was constant in the depth gradient except at 1301–1600 where sex ratio was 1:1 (Table 11.1).

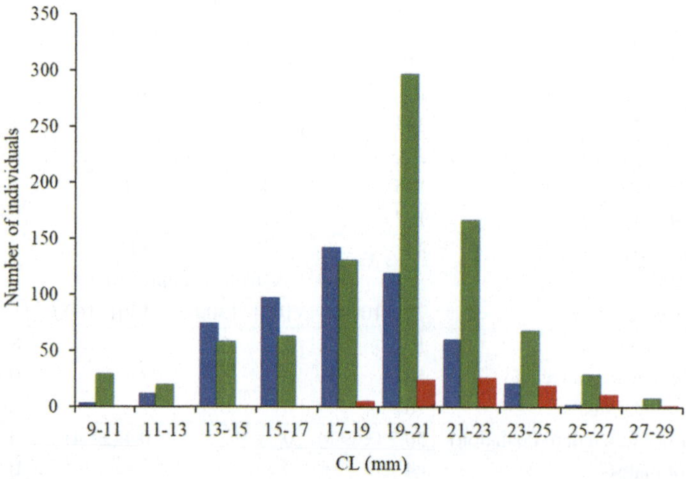

Fig. 11.2 Size distribution of males (blue), females (green), and ovigerous females (red) of *Nematocarcinus faxoni* collected during this survey

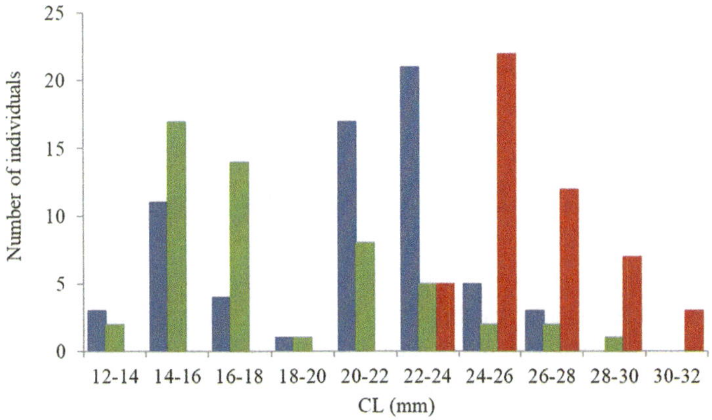

Fig. 11.3 Size distribution of males (blue), females (green), and ovigerous females (red) of *Nematocarcinus agassizii* collected during this survey

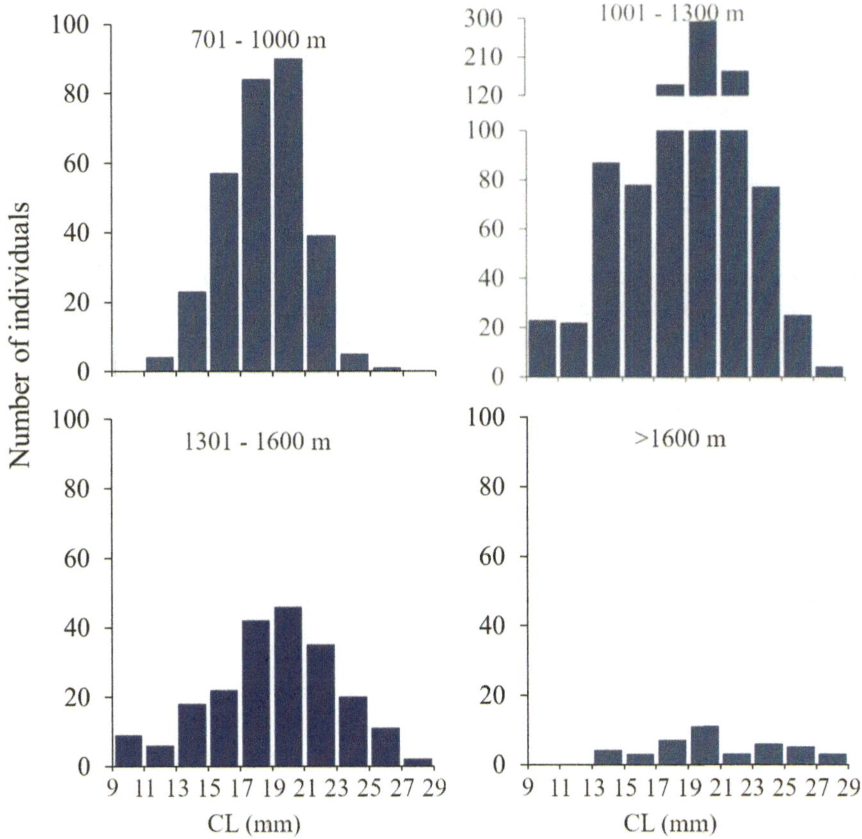

Fig. 11.4 Size distribution of *Nematocarcinus faxoni* by depth range

The size and sex bathymetric distribution of *N. agassizii* could not be analyzed because of the narrow depth range occupied by the species and the low number of individuals obtained.

11.3.2 Relationship Between Density and Environmental Variables

Hendrickx and Hernández-Payán (2018) presented the range of each environmental variable where *Nematocarcinus* species were found. In summary, *N. faxoni* was observed at near-bottom temperature (T_{20mab}) between 2.35 and 5.75 °C, salinity (S_{20mab}) between 34.40 and 34.70, dissolved oxygen (DO_{20mab}) between 0.11 and 1.60 ml/l, and organic carbon in the sediments (OC_{sed}) between 1.2 and 5.9%. *Nematocarcinus agassizii* was at T_{20mab} ranging 3.15–5.81 °C, S_{20mab} between 34.40 and 34.63, DO_{20mab} ranging 0.11–0.93 ml/l, and OC_{sed} ranging 1.3–5.2%.

High densities of *N. faxoni* (mean values >60 inds/ha) were recorded at T_{20mab} between 3.5 and 6.5 °C. Below 3.5 °C, recorded mean density was lower than 15 inds/ha (Fig. 11.5a). Mean density of *N. faxoni* was high (57.9 inds/ha) at DO_{20mab} of 0–0.5 ml/l and decreased with increasing DO_{20mab} (Fig. 11.5b). Density values above 55 inds/ha were recorded at S_{20mab} ranges 34.5–34.6 and 34.7–34.8, and values lower than 15 inds/ha were found at S_{20mab} ranges 34.4–34.5 and 34.6–34.7 (Fig. 11.5c). High density of *N. faxoni* (78.8 inds/ha) was recorded at sediments with OC_{sed} between 1 and 2%. At higher OC_{sed} values, density was always below 30 inds/ha (Fig. 11.5d).

Nematocarcinus agassizii specimens were mainly found at T_{20mab} between 5 and 6 °C at mean density of 75.6 inds/ha, with few specimens observed at 3–4 °C (0.3 inds/ha) (Fig. 11.6a). Organisms aggregated (79.2 inds/ha) mostly at DO_{20mab} ranging 0–0.5 ml/l and low density (1.2 inds/ha) were observed at DO_{20mab} 0.5–1 ml/l (Fig. 11.6b). Mean density was similar at S_{20mab} 34.4–34.5 and 34.5–34.6, about 22 and 25 inds/ha (Fig. 11.6c). Patterns of density were irregular in relation with OC_{sed}. Higher density was observed at areas were OC_{sed} was 3–4% (36.9 inds/ha) or 5–6% (20.7 inds/ha) (Fig. 11.6d). Low density was observed at sediments with 1–2% (7.5 inds/ha) or 4–5% (3.4 inds/ha) of organic carbon.

Spearman rank correlations calculated between *N. faxoni* density data and the available environmental variables revealed that when using data of all samples, *N. faxoni* density was positively correlated with T_{20mab} ($n = 45$, $\rho = 0.300$, $p < 0.05$) and negatively correlated with DO_{20mab} ($n = 54$, $\rho = -0.315$, $p < 0.05$). When only considering samples containing organisms, *N. faxoni* density was still positively correlated with T_{20mab} ($n = 43$, $\rho = 0.474$, $p < 0.01$) and negatively correlated with DO_{20mab} ($n = 52$, $\rho = -0.413$, $p < 0.01$), correlations being stronger than when all samples were considered. Almost significant negative correlations were detected between density and S_{20mab} ($n = 44$, $\rho = -0.278$, $p = 0.07$) and OC_{sed} ($n = 34$, $\rho = -330$, $p = 0.06$).

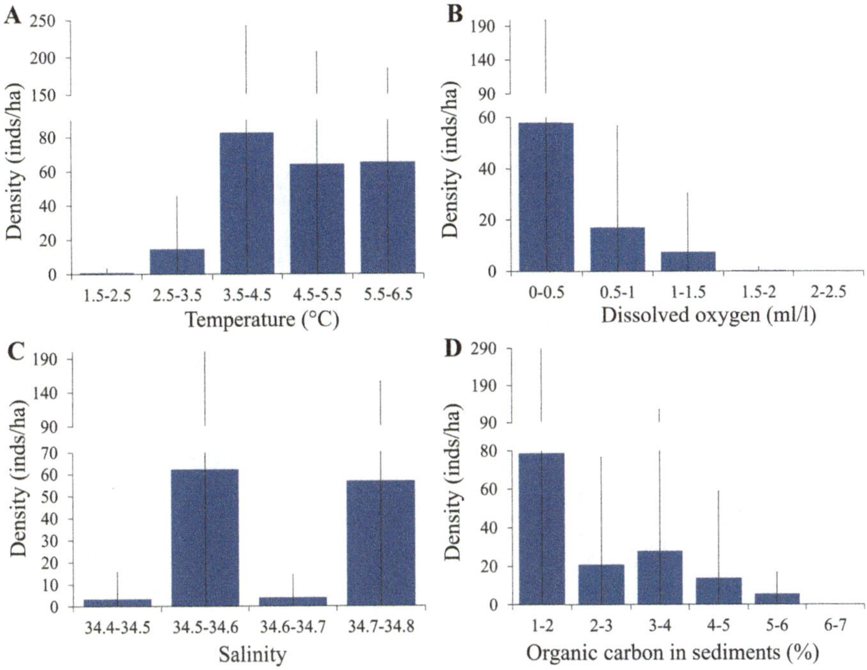

Fig. 11.5 Density of *Nematocarcinus faxoni* per (**a**) near-bottom temperature, (**b**) oxygen, (**c**) salinity, and (**d**) organic carbon in sediments intervals

For *N. agassizii*, when all samples were considered, significant positive Spearman rank correlations were found between density and T_{20mab} ($n = 40$, $\rho = 0.489$, $p < 0.01$). Density was negatively correlated with DO_{20mab} ($n = 40$, $\rho = -0.512$, $p < 0.001$) and S_{20mab} ($n = 40$, $\rho = -0.447$, $p < 0.01$). Only considering samples containing *N. agassizii*, density was negatively correlated with DO_{20mab} ($n = 6$, $\rho = -0.942$, $p < 0.01$).

T_{20mab} and DO_{20mab} were negatively correlated ($\rho = -0.86$), and thus, when computing GAMs, they were never included in the same model. The best GAM model for samples with and without individuals of *N. faxoni* was density of individuals = T_{20mab} + s(S_{20mab}, $k = 20$) + s(OC_{sed}, $k = 20$), where s() denotes the use of a spline smoother. This model explained 48.3% of the total deviance, and AIC was 356.5. Density of individuals of *N. faxoni* was higher at higher T_{20mab}, peaked at intermediate (34.5–34.6) S_{20mab} values, and was greater in areas with low OC_{sed} (Fig. 11.7a). When considering only samples containing individuals of *N. faxoni*, the best GAM model was density = s(T_{20mab}) + s(OC_{sed}, $k = 12$). This model explained 90.8% of the total variance, and AIC was 256.0. Density increased steadily with T_{20mab} from 2.5 to 4.5 °C and remained constant at temperature higher than 4.5 °C. Patterns of density related with OC_{sed} were irregular, with higher abundance at low OC_{sed} and secondary peaks at 2% and 4% OC_{sed} (Figure 11.7b).

For *N. agassizii*, best GAM explaining density patterns when samples both containing and not containing specimens were considered was density of individu-

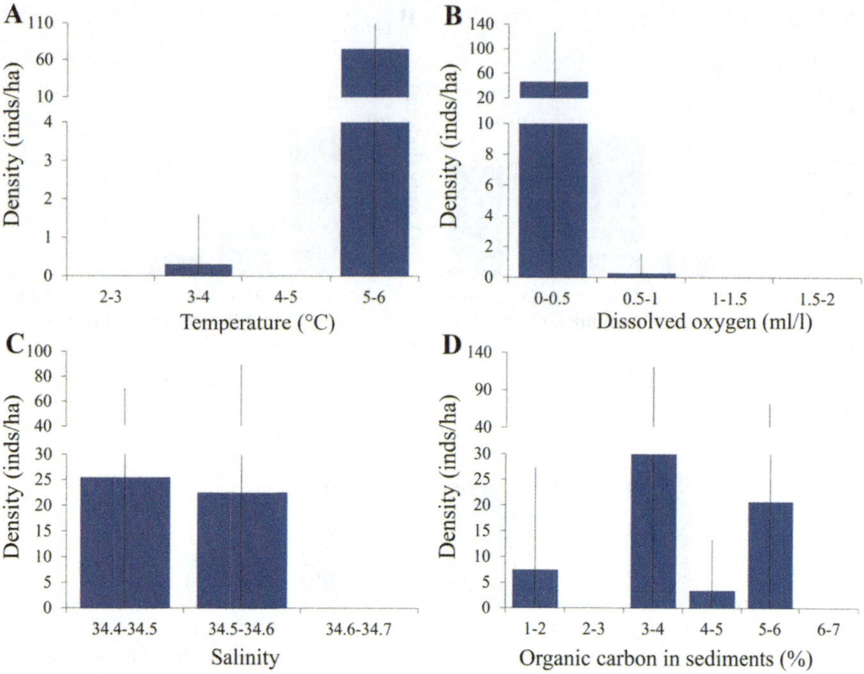

Fig. 11.6 Density of *Nematocarcinus agassizii* per (**a**) near-bottom temperature, (**b**) oxygen, (**c**) salinity, and (**d**) organic carbon in sediments intervals

als = DO_{20mab} + s(S_{20mab}, $k = 20$), explaining 86.6% of the total deviance with an AIC of 102.36. Density remained high at low S_{20mab} (34.4–34.5) and decreased at higher S_{20mab} values (Fig. 11.8). Density decreased steadily as DO_{20mab} increased. GAMs only considering samples containing organisms were not calculated because of the low sample size (only 6).

11.4 Discussion

11.4.1 Geographic and Bathymetric Distributions

The two *Nematocarcinus* species studied in this contribution are distributed over tropical and temperate areas of the eastern Pacific, coinciding in their latitudinal range from Mexico to Perú (Kameya et al. 1997; Martínez-Guerrero and López-Pérez 2018). *Nematocarcinus faxoni* also occurs in the SW Atlantic Ocean (Cardoso and Burukovsky 2014). Off western Mexico (northeastern tropical Pacific), the geographic distribution of both species overlapped only partially, and while *N. faxoni* was distributed along all the Mexican Pacific coasts with the only exception of the northern Gulf of California, *N. agassizii* was only found off the west coast of

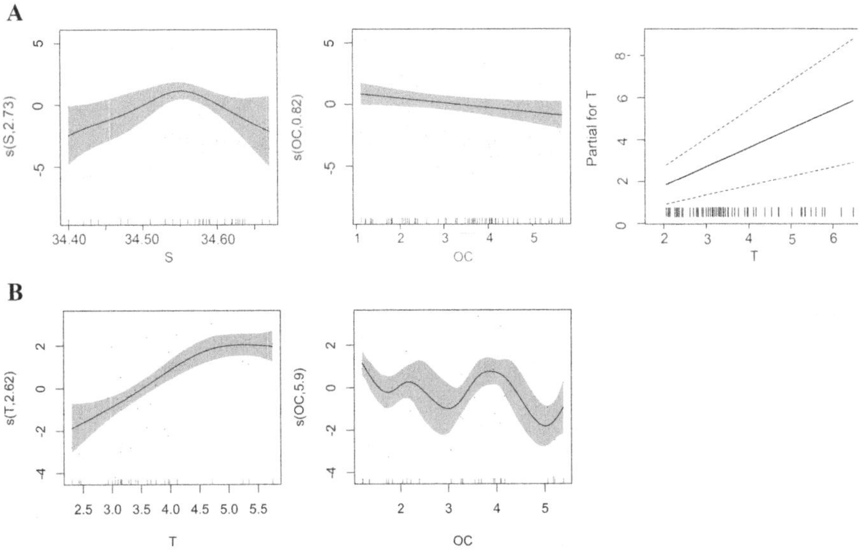

Fig. 11.7 Generalized additive model estimates of partial effects of environmental variables on density of *Nematocarcinus faxoni* in the Mexican Pacific. (**a**) Partial effects of near-bottom salinity and temperature and of organic carbon content in the sediments (samples with and without individuals of *N. faxoni* included). (**b**) Partial effects of near-bottom temperature and organic carbon in the sediments estimated on samples that contained individuals of *N. faxoni*. Shaded areas on each variable show the 95% confidence intervals, and tick marks on the *x*-axis indicate sampling data points

the Baja California Peninsula. The two species mainly inhabited different depth ranges: *N. agassizii* was distributed above 900 m, and most *N. faxoni* organisms occupied depths below 1000 m (this study, Hendrickx and Hernández-Payán 2018). Bathymetric replacement among species with similar ecological requirements is a common niche segregation mechanism in the marine environment (e.g., Wenner 1978, Cartes 1998) and has already been reported in other species of *Nematocarcinus* (*N. ensifer* and *N. rotundus*: Wenner 1979). This is thought to be an effective mechanism for reducing interspecific predation or competition, e.g., for food resources, and may act restricting *N. agassizii* from dwelling greater depths. Still, some bathymetric overlap between both species was observed, and they even were collected simultaneously in a few samples. Such coexistence may be supported by a high availability of food at the lower OMZ boundary, located at depths ranging ca. 700–1300 m in the study area (Papiol and Hendrickx 2016a; Papiol et al. 2017). *Nematocarcinus* spp. are described as deposit and low-mobile benthos feeders with an opportunistic scavenging behavior on fish remains (Wenner 1979; Cartes 1993a; Allen et al. 2000). Accumulations of labile organic material in the sediments (Levin 2003; Cowie 2005) and proliferation of benthic macrofauna (Wishner et al. 1990; Levin 2003; Levin et al. 2009) at the lower OMZ boundary may provide enough food to support *Nematocarcinus* species and other benthic and deposit feeders that aggregate at these depth ranges (Papiol and Hendrickx 2016a). Concomitant high

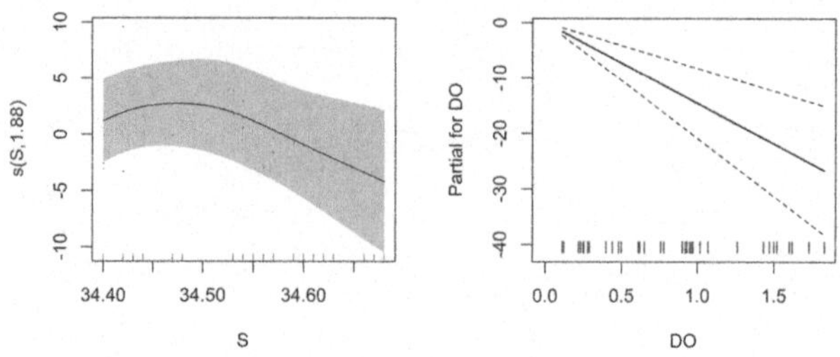

Fig. 11.8 Generalized additive model estimates of partial effects of environmental variables on density of *Nematocarcinus agassizii* in the Mexican Pacific. Partial effects of near-bottom salinity and dissolved oxygen (samples with and without individuals of *N. agassizii* included). Shaded areas on each variable show the 95% confidence intervals, and tick marks on the *x*-axis indicate sampling data points

abundances of benthic and benthopelagic megafauna may result in an additional food source for *Nematocarcinus* through the presence of dead animal remains.

Below OMZs, species usually occupy narrow bathymetric ranges of 200–400 m (Levin et al. 2009; Murty et al. 2009; Quiroga et al. 2009) determined by the steep bathymetric gradients in oxygen, temperature, and food availability. Decapod crustaceans found at depth intervals of 600 m or more, such as *N. faxoni*, are rare (Quiroga et al. 2009; Papiol et al. 2017). Species-specific combinations of temperature and oxygen restrict metazoan distributions to bathymetric ranges by regulating their aerobic capacity (Ekau et al. 2010; Seibel 2011). Below OMZs, such regulation is especially significant at defining the shallowest distributions of the species in association with their lower oxygen demand thresholds. Both species studied in the present contribution were well distributed in hypoxic conditions (DO $\leq$0.18 ml/l: Kamykowski and Zentara 1990), which is considered a minimum threshold affecting the distribution of pelagic organisms in OMZs of the eastern tropical Pacific and the Indian Ocean (see Seibel 2011 and references cited therein). It is expected that *Nematocarcinus* have lower metabolic demands associated with their benthic (rather than pelagic) habits and their deposit (rather than predatory) feeding activity (Childress et al. 1990; Company and Sardà 1998; Maynou and Cartes 1998), which may allow them to proliferate in such oxygen-depleted context. Although it has not been specifically documented for *Nematocarcinus* species, adaptations that enhance their capacity for oxygen extraction and transport (Childress and Seibel 1998; Seibel et al. 1999; Drazen and Seibel 2007; Jeffreys et al. 2012) likely also contribute to their capacity to inhabit severely hypoxic environments (Drazen and Seibel 2007). It is important to note that in this survey, samplings were not performed at depths less than 710 m off the Baja California Peninsula and therefore a shallower distribution of *N. agassizii* cannot be discarded. Yet, distribution at significantly lower oxygen concentration is unlikely because it should be typically associated with temporary vertical migrations to feed or diapause (Escribano et al. 2009; Gooday et al. 2009), and in the Arabian Sea OMZ, the absence of anaerobic bacteria markers in *Nematocarcinus gracilis* proved that this species does not feed inside the OMZ core (Allen et al. 2000).

The strong quantitative and qualitative gradients in the food available below OMZ cores (Cowie 2005) may also contribute to the confinement of species in narrow bathymetric bands. Yet, the opportunistic feeding of *Nematocarcinus* characterized by a marked scavenging behavior can help them overcome food variations, and it may support the organisms of *N. faxoni* dwelling great depths, where food is notably scarce. Anyway, the observed decrease in density with depth is consistent with the well-known concurrent food decline.

11.4.2 Size and Sexual Maturity

The bathymetric variations in the size structure of *N. faxoni* population suggested a somehow bigger-deeper trend that is common in the deep-sea (Morales-Nin et al. 2003). Bathymetric size segregation within a species likely results from the combination of depth variations in food availability (Wishner et al. 1995; Cowie 2005; Levin et al. 2009) and ontogenic trophic niche partitioning, and therefore contributes to the reduction of intraspecific competition (Rowe 1971; Carey 1981). Besides, the spatial segregation between large and small specimens might contribute to minimization of potential cannibalism on smaller specimens by large conspecifics, as already described for fish (e.g., Holt et al. 2013 and references cited therein). Sexual vertical partitioning of the space was not observed, and the presence of ovigerous females and large females in all depth strata concurs with the ability of the species to find the required food (e.g., to cover the high energetic demands for reproduction) in the entire depth range, likely through their opportunistic scavenging behavior.

Sex ratios in favor of females, common in many deep-water caridean shrimps (e.g., Wenner 1979, Vafidis et al. 2008 and references cited therein, Paramo and Núñez 2015), are likely related with the presence of spermatophores in this family. Each male can fertilize more than one female by the transfer of spermatophores, which they can synthesize in a relatively short time compared to female egg production (Mauchline 1972). Spermatophores can be stored by females who will fertilize the eggs as they are laid (Nagaraju 2011). Larger size of ovigerous females compared to males will likely allow for the production of a larger number of eggs (Bauer 2004), which is definitely important in *r*-strategists such as *Nematocarcinus* spp. (Wenner 1979).

11.4.3 Environmental Drivers

The distribution of both species of *Nematocarcinus* in the Mexican Pacific was partially explained by salinity (S), which is one of the variables defining water masses. Water masses have already been identified as drivers of fauna shifts (e.g., Muñoz et al. 2012; Cartes et al. 2014). In the Mexican Pacific, two different water masses (i.e., the North Pacific Intermediate Water (NPIW) and the Equatorial Subsurface Water (ESsW)) converge horizontally at ca. 26 °N, resulting in the presence of a relatively marked salinity gradient to depths of ca. 1000 m and in important fauna changes (Wicksten 1989; Papiol et al. 2017). The presence of *N. agassizii* off the western coast of the Baja California Peninsula appeared to be linked to the influence of the NPIW, characterized by low salinity. Alternatively, *N. faxoni* was associated with higher S from the ESsW, with greater influence south of 26°N.

Temperature (T) or dissolved oxygen (DO) also regulated the distribution of both *Nematocarcinus* species. These variables are interdependent, and under OMZs they covary with depth: T typically decreases, and DO increases with depth (Levin 2003). Coherently, in the study area, strong negative correlations between depth and T and positive correlations between depth and DO have been documented (see Papiol and Hendrickx 2016a; Cruz-Acevedo et al. 2018). Species-specific combinations of T and DO regulate the aerobic capacity of most metazoans limiting their distribution (Ekau et al. 2010; Seibel 2011) to specific bathymetric ranges in this case. The broader geographic and bathymetric distributions of *N. faxoni* is probably linked to its capacity to proliferate at wider T and DO ranges. The influence of T and DO on the bathymetric distribution of other benthic species similar to *Nematocarcinus* spp. has already been detected (Hendrickx and Papiol 2015, 2019; Papiol et al. 2016) in the Mexican Pacific OMZ, but such influence was not so strong in the pelagic vertically migrating shrimp *Benthesycimus tanneri* (Papiol and Hendrickx 2016b) who presumably performs temporary migrations into the OMZ.

The different OC_{sed} ranges mainly inhabited by the two species agree with their bathymetric replacement with depth and likely also contributed to the more stenobathic distribution of *N. agassizii*. This species seemed to require larger food amounts (as reflected by OC_{sed}) than its congener and was mainly distributed in areas with $OC_{sed} \geq 3\%$. Such large amounts of organic carbon in the sediments can be found at the lower OMZ core transition zone, where *N. agassizii* was distributed, and are typically linked with aggregations of benthic macrofauna (Wishner et al. 1990; Levin 2003; Levin et al. 2009), which are also potential food items for *Nematocarcinus* (Wenner 1979; Cartes 1993a). *Nematocarcinus faxoni* was able to maintain high abundance in deeper, less productive areas, which are found below the lower OMZ boundary, suggesting lower feeding intensity by this species. In the southern region of the Mexican Pacific, *N. faxoni* is further replaced by the oplophorid shrimp *Acanthephyra brevicarinata* along the bathymetric gradient (Papiol et al. 2017), arguably associated with the low feeding intensity characteristic of some species of this genus (Cartes 1993b). Thus, the significant effect of the steep bathymetric gradients in food availability on species composition under OMZs is highlighted.

Acknowledgments Ship time aboard the R/V "El Puma" was provided by the Instituto de Ciencias del Mar y Limnología and by the Coordinación de la Investigación Científica, UNAM, and partly supported by CONACyT (project 31805-N for the TALUD IV–VII cruises, project 179467 for the TALUD XV and XVI-B cruises). The TALUD project has received laboratory support from the DGAPA (PAPIIT IN-217306-3 and PAPIIT IN-203013-2) and from CONACyT (project 31805-N for the TALUD IV to VII cruises, project 179467 for the TALUD XV and XVIB cruises), Mexico.

References

Allen CE, Tyler PA, Varney MS (2000) Lipid profiles of *Nematocarcinus gracilis* a deep-sea shrimp from below the Arabian Sea oxygen minimum zone. In: Jones MB, JMN A, Neto AI, Costa AC, Frias Martins AM (eds) Island, ocean and deep-sea, Developments in hydrobiology 152. Kluwer Academic, Dordrecht/Boston/London, pp 273–279

Bauer RT (2004) Remarkable shrimps: adaptations and natural history of the carideans, vol 7. University of Oklahoma Press, Norman

Beaulieu SE (2002) Accumulation and fate of phytodetritus on the sea floor. Oceanogr Mar Biol 40:171–232

Burukovsky RN (2001) Taxonomy of *Nematocarcinus* (Decapoda, Nematocarcinidae). Description of *Nematocarcinus* from waters of the American continent. Zoolog Z 80(11):1429–1443. [In Russian with English abstract]

Burukovsky RN (2004) Taxonomy of shrimps of the genus *Nematocarcinus* (Decapoda, Nematocarcinidae). A review of taxonomic characteristics and a key to identifying species of the genus. Zoolog Z 83(5):549–561. [In Russian with English abstract]

Burukovsky RN (2012) Deep sea shrimps of the family Nematocarcinidae (history of study, systematic, distribution, and biology). Prospekt Nauki, St. Petersburg. [In Russian with English abstract]

Burukovsky RN, Sudnik SA (2014) Oogenesis and maturation of ovarian as the basis of reproductive strategies of female shrimps *Nematocarcinus africanus* Crosnier et Forest 1973 (Crustacea, Caridea, Nematocarcinidae). Sci World 12(16), II:16–24 [In Russian]

Cardoso IA, Burukovsky RN (2014) *Nematocarcinus* Milne Edwards, 1881 (Crustacea, Decapoda) from southwestern Atlantic, including the southern mid-Atlantic ridge area. Zootaxa 3887(3):437–458

Carey AG (1981) A comparison of benthic infaunal abundance on two abyssal plains in the Northeast Pacific Ocean. Deep-Sea Res 28:467–179

Cartes JE (1993a) Diets of two deep-sea decapods: *Nematocarcinus exilis* (Caridea: Nematocarcinidae) and *Munida tenuimana* (Anomura: Galatheidae) on the western Mediterranean slope. Ophelia 37(3):213–229

Cartes JE (1993b) Feeding habits of oplophorid shrimps in the deep Western Mediterranean. J Mar Biol Assoc UK 73:193–206

Cartes JE (1998) Feeding strategies and partition of food resources in deep-water decapod crustaceans (400–2300 m). J Mar Biol Assoc UK 78:509–524

Cartes JE, Sardà F (1992) Abundance and diversity of decapod crustaceans in the deep Catalan Sea (Western Mediterranean). J Nat Hist 26:1305–1323

Cartes JE, Papiol V, Frutos I, Macpherson E, González-Pola C, Punzón A, Valeiras X, Serrano A (2014) Distribution and biogeographic trends of decapod assemblages from Galicia Bank (NE Atlantic) with connections to regional water masses. Deep-Sea Res II Top Stud Oceanogr 106(6):165–178

Chace FA (1986) The caridean shrimps (Crustacea: Decapoda) of the "Albatross" Philippine Expedition 1907–1910, Part 4: Oplophoridae and Nematocarcinidae. Smithson Contrib Zool 432:1–82

Childress JJ, Cowles DL, Favuzzi JA, Mickel TJ (1990) Metabolic rates of benthic deep-sea decapod crustaceans decline with increasing depth primarily due to the decline in temperature. Deep-Sea Research 37(6):929–949

Childress JJ, Seibel BA (1998) Life at stable low oxygen: adaptations of animals to oceanic oxygen minimum layers. J Exp Biol 201:1223–1232

Company JB, Sardà F (1998) Metabolic rates and energy content of deep-sea benthic decapod crustaceans in the western Mediterranean Sea. Deep-Sea Res I Oceanogr Res Pap 45:1861–1880

Cowie G (2005) The biogeochemistry of Arabian Sea surficial sediments: a review of recent studies. Prog Oceanogr 65:260–289

Cruz-Acevedo E, Tolimieri N, Aguirre-Villaseñor H (2018) Deep-sea fish assemblages (300–2100 m) in the eastern Pacific off northern Mexico. Mar Ecol Progr Ser 592:225–242

Dambach J, Raupach MJ, Mayer C, Schwarzer J, Leese F (2013) Isolation and characterization of nine polymorphic microsatellite markers for the deep-sea shrimp *Nematocarcinus lanceopes* (Crustacea: Decapoda: Caridea). BMC Res Notes 6(1):75

De Grave S, Fransen CHJM (2011) Carideorum catalogus: the recent species of the dendrobranchiate, stenopodidean, procarididean and caridean shrimps. Zool Meded, Leiden 85:195–588

Devol AH, Hartnett HE (2001) Role of the oxygen-deficient zone in transfer of organic carbon to the deep ocean. Limnol Oceanogr 46:1684–1690

Diaz RJ, Rosenberg R (1995) Marine benthic hypoxia: a review of its ecological effects and the behavioural responses of benthic macrofauna. Oceanogr Mar Biol 33:245–303

Drazen JC, Seibel BA (2007) Depth-related trends in metabolism of benthic and benthopelagic deep-sea fishes. Limnol Oceanogr 52:2306–2316

Ekau W, Auel H, Pörtner HO, Gilbert D (2010) Impacts of hypoxia on the structure and processes in pelagic communities (zooplankton, macro-invertebrates and fish). Biogeosciences 7:1669–1699

Escobar-Briones EG, Gaytán-Caballero A, Legendre P (2008) Epibenthic megacrustaceans from the continental margin, slope and abyssal plain of the Southwestern Gulf of Mexico: factors responsible for variability in species composition and diversity. Deep-Sea Res II Top Stud Oceanogr 55:2667–2678

Escribano R, Hidalgo P, Krautz C (2009) Zooplankton associated with the oxygen minimum zone system in the northern upwelling region of Chile during March 2000. Deep-Sea Res II Top Stud Oceanogr 56:1083–1094

Flannery E, Przeslawski R (2015) Comparison of sampling methods to assess benthic marine biodiversity: are spatial and ecological relationships consistent among sampling gear? Record 2015/07. Geoscience Australia, Canberra

Gooday AJ, Levin LA, Aranda da Silva A, Bett BJ, Cowie GL, Dissard D, Gage JD, Hughes DJ, Jeffreys R, Lamont PA, Larkin KE, Murty SJ, Schumacher S, Whitcraft C, Woulds C (2009) Faunal responses to oxygen gradients on the Pakistan margin: a comparison of foraminiferans, macrofauna and megafauna. Deep-Sea Res II Top Stud Oceanogr 56:488–502

Gorny M, George MR (1997) Oocyte development and gonad production of *Nematocarcinus lanceopes* (Decapoda: Caridea) in the eastern Weddell Sea, Antarctica. Polar Biol 17(3):191–198

Guzmán G, Quiroga E (2005) Nuevos registros de camarones (Decapoda: Dendrobranchiata y Caridea) en aguas profundas de Chile. Gayana (Concepción) 69(2):285–290

Hastie T, Tibshirani R (1986) Generalized additive models. Stat Sci 1:297–318

Hendrickx ME (2012) Crustáceos decápodos (Arthropoda: Crustacea: Decapoda) de aguas profundas del Pacífico mexicano: Lista de especies y material recolectado durante el proyecto TALUD. In: Zamorano P, Hendrickx ME, Caso M (eds) Biodiversidad y comunidades del talud continental del Pacífico mexicano. Secretaría del Medio Ambiente y Recursos Naturales (SEMARNAT), Instituto Nacional de Ecología (INE), Mexico City, pp 283–317

Hendrickx ME, Hernández-Payán JC (2018) Distribution and abundance of *Nematocarcinus* spp. (Crustacea: Decapoda: Caridea: Nematocarcinidae) off western Mexico, eastern Pacific. Reg Stud Mar Sci 23:47–52

Hendrickx ME, Papiol V (2015) Insights on the biology and ecology of the deep-water shrimp *Parapontophilus occidentalis* (Faxon, 1893) (Crustacea, Caridea, Crangonidae) in the eastern Pacific with notes on its morphology. Zootaxa 4007(3):370–388

Hendrickx ME, Papiol V (2019) Distribution, relative growth, ecology and fecundity of the deep-water squat lobster *Galacantha diomedeae* (Crustacea: Decapoda: Galatheoidea: Munidopsidae) in the Mexican Pacific. Inverteb Biol 2019:1–12. (Online). https://doi.org/10.1111/ivb.12248

Hendrickx ME, Serrano D (2010) Impacto de la zona de mínimo de oxígeno sobre los corredores pesqueros en el Pacífico mexicano. Interciencia 35(1):12–18

Hernández-Payán JC, Hendrickx ME (2016) Two species of the deep-water shrimp genus *Nematocarcinus* A. Milne–Edwards, 1881 (Crustacea, Decapoda, Caridea, Nematocarcinidae) from the Mexican Pacific. Zootaxa 4126(4):587–599

Holt R, Foggo A, Neat FC, Howell KL (2013) Distribution patterns and sexual segregation in chimaeras: implications for conservation and management. ICES J Mar Sci 70(6):1198–1205
Hunter WR, Oguri K, Kitazato H, Ansari ZA, Witte U (2011) Epi-benthic megafaunal zonation across an oxygen minimum zone at the Indian continental margin. Deep-Sea Res I Oceanogr Res Pap 58:699–710
Jeffreys RM, Levin LA, Lamont PA, Woulds C, Whitcraft CR, Mendoza GF, Wolff GA, Cowie GL (2012) Living on the edge: single-species dominance at the Pakistan oxygen minimum zone boundary. Mar Ecol Prog Ser 470:79–99
Kameya A, Castillo R, Escudero L, Tello E, Blaskovi'c V, Córdova J, Hooker Y, Gutiérrez M, Mayor S (1997) Localizaci6n, distribución y concentraci6n de langostinos rojos de profundidad, Crucero BIC Humboldt 9607-08 18 de julio al 6 de agosto de 1996. Publ Espec Inst Mar Perú 7:1–47
Kamykowski D, Zentara SJ (1990) Hypoxia in the world ocean as recorded in the historical dataset. Deep-Sea Res I Oceanogr Res Pap 37:1861–1874
Komai T, Segonzac M (2005) Two new species of *Nematocarcinus* A. Milne-Edwards, 1881 (Crustacea: Decapoda: Caridea: Nematocarcinidae) from hydrothermal vents on the North and South East Pacific Rise. Zoosystema 27(2):343–364
Levin LA (2003) Oxygen minimum zone benthos: adaptation and community response to hypoxia. Oceanogr Mar Biol 41:1–45
Levin LA, Whitcraft CR, Mendoza GF, Gonzalez JP, Cowie G (2009) Oxygen and organic matter thresholds for benthic faunal activity on the Pakistan margin oxygen minimum zone (700–1100 m). Deep-Sea Res II Top Stud Oceanogr 56:449–471
Loring DH, Rantala RTT (1992) Manual for the geochemical analyses of marine sediments and suspended particulate matter. Earth-Sci Rev 32:235–283
Martínez-Guerrero B, López-Pérez A (2018) Richness and large-scale distribution of marine benthic caridean shrimps (Decapoda: Caridea) from the eastern tropical Pacific. Nauplius 26:1–16. https://doi.org/10.1590/2358-2936e2018035
Mauchline J (1972) The biology of bathypelagic organisms, especially Crustacea. Deep-Sea Res 19:753–780
Maynou F, Cartes JE (1998) Daily ration estimates and comparative study of food consumption in nine species of deep-water decapod crustaceans of the NW Mediterranean. Mar Ecol Progr Ser 171:221–231
Méndez N (2007) Relationships between deep-water polychaete fauna and environmental factors in the southeastern Gulf of California, Mexico. Sci Mar 71:605–622
Morales-Nin B, Maynou F, Cartes JE, Moranta J, Massutí E, Company B, Rotllant G, Bozzano A, Stefanescu C (2003) Size influence in zonation patterns in fishes and crustaceans from deep-water communities of the western Mediterranean. J Northwest Atl Fish Sci 31:413–430
Muñoz I, García-Isarch E, Sobrino I, Burgos C, Funny R, González-Porto M (2012) Distribution, abundance and assemblages of decapod crustaceans in waters off Guinea-Bissau (north-west Africa). J Mar Biol Assoc UK 92(3):475–494
Murty SJ, Bett BJ, Gooday AJ (2009) Megafaunal responses to strong oxygen gradients on the Pakistan margin of the Arabian Sea. Deep-Sea Res II Top Stud Oceanogr 56:472–487
Nagaraju GP (2011) Reproductive regulators in decapod crustaceans: an overview. J Exp Biol 214:3–16
Papiol V, Hendrickx ME (2016a) Community structure and ecology of deep-water decapod crustaceans below the oxygen minimum zone in the SE Gulf of California, Mexico. Mar Freshw Res 67(12):1862–1879
Papiol V, Hendrickx ME (2016b) Ecology of *Benthesicymus tanneri* Faxon, 1893 off Mexican Pacific slopes. J Crustac Biol 36(1):50–60
Papiol V, Hendrickx ME, Serrano D (2016) Distribution and ecology of the Pacific lobsterette *Nephropsis occidentalis* Faxon, 1893 (Crustacea, Decapoda) in the continental slope off western Mexico. In: Riosmena-Rodriguez R (ed) Marine benthos: biology, ecosystem functions and environmental impact. Nova Science, New York, pp 197–223

Papiol V, Hendrickx ME, Serrano D (2017) Effects of latitudinal changes in the oxygen minimum zone of the northeast Pacific on the distribution of bathyal benthic decapod crustaceans. Deep-Sea Res II Top Stud Oceanogr 137:113–130

Paramo J, Núñez S (2015) Estructura de tallas, talla media de madurez sexual y razón sexual de camarones de aguas profundas de importancia comercial en el Caribe colombiano. Rev Acad Colomb Cienc Exact Fis Nat 39(152):408–415

Quiroga E, Sellanes J, Arntz WE, Gerdes D, Gallardo VA, Hebbeln D (2009) Benthic megafaunal and demersal fish assemblages on the Chilean continental margin: the influence of the oxygen minimum zone on bathymetric distribution. Deep-Sea Res II Top Stud Oceanogr 56:1112–1123

Roullier F, Berline L, Guidi L, Sciandra A, Durrieu De Madron X, Picheral M, Pesant S, Stemmann L (2013) Particles size distribution and carbon flux across the Arabian Sea oxygen minimum zone. Biogeosci Discuss 10:19271–19309

Rowe GT (1971) Benthic biomass and surface productivity. In: Costlow JD Jr (ed) Fertility of the sea, vol 2. Gordon and Breach, New York, pp 441–454

Seibel BA (2011) Critical oxygen levels and metabolic suppression in oceanic oxygen minimum zones. J Exp Biol 214:326–336

Seibel BA, Chausson FC, Lallier FH, Zal F, Childress JJ (1999) Vampire blood: respiratory physiology of the vampire squid (Cephalopoda: Vampyromorpha) in relation to the oxygen minimum layer. Exp Biol Online 4:1–10

Smyth GK, Jørgensen B (2002) Fitting Tweedie's compound Poisson model t insurance claims data: dispersion modelling. ASTIN Bull 32:143–157

Storch V, Bluhm BA, Arntz WE (2001) Microscopic anatomy and ultrastructure of the digestive system of three Antarctic shrimps (Crustacea: Decapoda: Caridea). Polar Biology 24:604–614

Strickland JDH, Parsons TR (1972) A practical handbook of seawater analysis. Fish Res Board Can Bull 167:1–310

Thatje S, Bacardit R, Arntz WE (2005) Larvae of the deep-sea Nematocarcinidae (Crustacea: Decapoda: Caridea) from the southern ocean. Polar Biol 28(4):290–302

Türkay M (1998) Notes on the Mediterranean nematocarcinid shrimps (Crustacea: Decapoda: Caridea). J Nat Hist 32(10–11):1787–1794

Vafidis D, Leontarakis PK, Dailianis T, Kallianiotis A (2008) Population characteristics of four deep-water pandalid shrimps (Decapoda: Caridea) in the northern Aegean Sea (NE Mediterranean). J Nat Hist 42(31–32):2079–2093

Wenner EL (1978) Comparative biology of four species of glyphocrangonid and crangonid shrimp from the continental slope of the middle Atlantic Bight. Can J Zool 56(5):1052–1065

Wenner EL (1979) Distribution and reproduction of nematocarcinid shrimp (Decapoda: Caridea) from the northwestern North Atlantic. Bull Mar Sci 29(3):380–393

Wicksten MK (1989) Ranges of offshore decapod crustaceans in the eastern Pacific Ocean. Trans San Diego Soc Nat Hist 21(19):291–316

Wishner KF, Levin LA, Gowing MM, Mullineaux L (1990) Involvement of the oxygen minimum in the benthic zonation on a deep seamount. Nature 346:57–59

Wishner KF, Ashjian CJ, Gelfman C, Gowing MM, Levin LA, Mullineaux S, Saltzman J (1995) Pelagic and benthic ecology of the lower interface of the eastern tropical Pacific oxygen minimum zone. Deep-Sea Res I Oceanogr Res Pap 42(1):93–115

Wood SN (2006) Generalized additive models, an introduction with R. Chapman and Hall/CRC, Boca Raton

Wood SN, Augustin NH (2002) GAMs with integrated model selection using penalized regression splines and applications to environmental modelling. Ecol Model 157:157–177

WoRMS Editorial Board (2020) http://www.marinespecies.org/index.php. Accessed January–May 2020

Zamorano P, Hendrickx ME, Toledano-Granados A (2007) Distribution and ecology of deep-water mollusks from the continental slope, southeastern Gulf of California, Mexico. Mar Biol 150:883–892

Chapter 12
Pelagic Shrimps (Crustacea, Decapoda, Dendrobranchiata, and Caridea) in the Southeast Pacific

G. L. Guzman and C. Nicole Olguin

Abstract Updated information on pelagic shrimps collected in the southeast Pacific is provided. The study is based on recent expeditions during the last decade off the Chilean coast and on the material of the 1962 Eltanin expedition to Peru and Chile (cruises 3–5). This material was deposited in the Museum Support Center of the Smithsonian Institution but have so far not been studied. The study area was from 07°S to 62°S, from Punta Chicama, Peru, to Drake Passage, and from the continental slope of southeast of South America to Chilean Island (Desventuradas Islands, Easter Island and Juan Fernandez Archipelago). Samples were taken from the epipelagic to the bathypelagic zones. The information obtained is compared with a previous review published by G. Guzman in 2008. A total of 78 species were registered, and the differences with the 79 species reported earlier are discussed. It is concluded that some of the inconsistencies between both values correspond to species that were collected only once due to their low abundance and characterized as rare species, probably due to sampling methods. This is the case of *Glyphus marsupialis*, collected only once off Peru. Biogeographical connections between the shrimp fauna of the study area and other zones of the Pacific Ocean are discussed. Two distributional patterns were observed, one related with the circum-Antarctic waters and another related with the West Pacific. Despite these recent efforts, the pelagic fauna occurring in large zones of the Southeast Pacific remains unknown. Surveys in the area have been episodic, thus impeding the possibility to establish accurate biogeographical patterns and relationships with other regions and to increase our knowledge on the biology of the shrimp pelagic fauna.

Keywords Decapoda · Pelagic shrimps · Biogeography · Southeastern Pacific · Chile

G. L. Guzman (✉) · C. Nicole Olguin
Facultad de Recursos Naturales Renovables, Universidad Arturo Prat, Iquique, Chile
e-mail: gguzman@unap.cl

M. E. Hendrickx (ed.), *Deep-Sea Pycnogonids and Crustaceans of the Americas*, https://doi.org/10.1007/978-3-030-58410-8_12

12.1 Introduction

The pelagic shrimps considered here correspond to a fraction of the holopelagic decapods in the Nielsen's sense (Nielsen 2013). This group is composed of a few families of Decapoda, all included in the Dendrobranchiata and Caridea. Within the Dendrobranchiata, the families Sergestidae and Benthesicymidae are mainly pelagic, and the Penaeidae, Solenoceridae, and Aristeidae have some species that live in this environment. Within the Caridea, pelagic species belong to the Pasiphaeidae and Oplophoridae s.l. (Acanthephyridae and Oplophoridae).

In the Southeast Pacific (SEP), the pelagic shrimp fauna is relatively well-known, and some reviews are available for Peru (Mendez 1981; Moscoso 2012), the vicinity of the Seamounts of the Salas y Gomez ridge and Nazca Ridge (Vereshchaka 1990), for the circum-Antarctic waters (von Tiefenbacher 1991, 1994), and for Chile (Guzmán 2008). A total of 39 species of pelagic shrimps has been recorded from off Peru, 65 species from off Chile (Guzmán 2008), and 39 around the Seamounts of the Nazca Plate (Vereshchaka 1990). In total, 92 species of pelagic shrimps have been reported in the SEP.

New expeditions were organized recently bringing additional material. Material available in collections was also examined, finding species that had not been previously reported for the area. These new records are reported herein, and a new, updated analysis related to the pelagic shrimps in the SEP is provided.

12.2 Materials and Methods

We analyze the material deposited in the Museum Support Center of the Smithsonian Institutions collected during the R/V "Eltanin" expeditions to South America between 1961 and 1964. The material examined is reported. We also included the known depth range and references of interest for each species.

12.3 Results

Dendrobranchiata Bate, 1888
Family Aristeidae Wood-Mason, 1891

1. *Aristaeomorpha foliacea* (Risso, 1827)

Distribution in SEP. Dome Seamount in the Salas y Gomez ridge (25°04′S, 97°26′W), 218 to 800 m deep (Burukovsky 1990).

Family Benthesicymidae Wood-Mason in Wood-Mason and Alcok, 1891

2. *Gennadas pasithea* de Man, 1907

   Distribution in SEP. Soldier Seamount of Salas y Gomez Ridge (21°41′S, 81°46′W), 900–2000 m (Vereshchaka 1990); off Iquique (20°20′S, 71°11′W), 540 m (Guzmán and Wicksten 2000).

3. *Gennadas barbari* Vereshchaka, 1990

   Distribution in SEP. Eclipse Seamount, Nazca Ridge (22°06′S, 81°19′W), between 230 and 2000 m; in the vicinity of the Seamounts Big (25°40′S, 85°27′W), 160 and 2000 m, Yarala (25°40′S, 86°34′W), 370 and 2280 m, Amber (24°58′S, 88°31′W), 510 and 1500 m, Pearl (25°33′S, 89°12′W), 530 and 1120 m, Cliff (25°58′S, 100°41′W), 330 and 1800 m, Ichthyologist (25°07′S, 99°35′W), 328 and 790 m, Salas y Gomez Ridge (Vereshchaka 1990); from 300 miles off Caldera Harbour (27°00′12″S, 77°34′53″W) to vicinity of Easter Island (27°00′12″S, 107°35′00″W), CIMAR 5 cruise (Guzmán 2004a).

4. *Gennadas brevirostris* Bouvier, 1905

   Distribution in SEP. Arauco Gulf (Retamal 2000), Valparaiso to Juan Fernandez Archipelago (Guzmán 2004a), Gulf of Penas (46°59′34″S, 75°40′58″W), CIMAR 14 cruise.

   Remarks. New record and southernmost record for this species.

5. *Gennadas gilchristi* Calman, 1925

   Distribution in SEP. 90 miles off Valparaiso (32°59′S, 73°31′W) to vicinity of Juan Fernandez Archipelago (33°00′S, 78°4′W), off Desventuradas Islands (26°18′S, 80°06′W), and 300 miles off Caldera (26°00′S, 77°34′W) (Guzmán 2004a).

6. *Gennadas incertus* (Balss, 1927)

   Distribution in SEP. Off Arica (18°25′S 71°22′W), 603 m (Guzmán and Wicksten 2000); 300 miles off Valparaiso (33°34′S, 78°52′W), close to San Ambrosio (26°22′S, 79°54′W) and San Felix (26°20′S, 80°00′W) Islands. Caught during the CIMAR 6 cruise (Guzmán 2004a).

7. *Gennadas kempi* Stebbing, 1914

   Distribution in SEP. Drake Passage (58°29′S, 65°30′W), west of Penas Gulf (47°14′S, 76°28′W), Eltanin USAP expedition 1962.

   Remarks. First record in the Southeast Pacific. In the same expedition this species was caught on the Atlantic side of the Drake Passage (57°11′S, 63°51′W); new record.

8. *Gennadas propinquus* Rathbun, 1906

   Distribution in SEP. Off Arica (18°25′S, 71°22′W) 603 m (Guzmán and Wicksten 2000). Vereshchaka (1990) report this species in the vicinity of Seamounts Soldier (21°41′S, 81°46′W), 960–2000 m, and Pearl (25°33′S, 89°12′W), 530–1120 m.

9. *Gennadas scutatus* Bouvier, 1906

   Distribution in SEP. Off Peru (Mendez 1981); north of Chile from Arica (18°25′S, 71°43′W) to Punta Lobos (21°04′S, 70°51′W) (Guzmán and Wicksten 2000); close to Desventuradas Islands (26°16′S, 80°04′W), 577 m (Guzmán 2004a); in the vicinity of Seamounts Soldier (21°41′S, 81°46′W), 960–2000 m, and Eclipse (22°06′S, 81°19′W) 230 to 2000 m, and at Salas y Gomez Ridge

(Vereshchaka 1990); in Peruvian waters off Lobos de Tierra Island (06°35′S, 81°03′W; 06°25′S, 81°03′W), 900–910 m deep (Moscoso 2012).

10. *Gennadas sordidus* Kemp, 1910

    Distribution in SEP. Probably in Peruvian waters (Moscoso 2012); North of Chile from Arica (18°25′S, 71°01′W), 643 m, to Iquique (20°20′S, 70°58′W), 450 m (Guzmán and Wicksten 2000); surroundings of San Ambrosio Island (26°16′S, 80°04′W), 577 m (Guzmán, 2004a).

11. *Gennadas tinayrei* Bouvier, 1906

    Distribution in SEP. South of Peru (16°30′S, 74°34′W), 1162 m, Eltanin expedition USAP 1962; Seamounts Soldier (21°41′S, 81°46′W), 960–2000 m, and Professor Mesyatsev (21°25′S, 81°38′W), 320–1200 m, Nazca Ridge (Vereshchaka 1990); NE off Juan Fernandez Archipelago (33°00′S, 77°57′W; 33°34′S, 78°52′W) and close to San Ambrosio Island (26°22′S, 79°54′W), CIMAR 6 cruise (Guzmán 2004a).

    Remarks. First record of this species in Peruvian water, although it has been previously recorded from both north and south of Peru.

Family Penaeidae Rafinesque, 1815

12. *Funchalia woodwardi* Johnson, 1868

    Distribution in SEP. Close to "May-Day" Seamount, Salas y Gomez ridge (27°00′08″S, 94°05′05″W), CIMAR 5 cruise; 62 miles SE of Desventuradas Islands (27°00′08″S, 79°05′49″W), CIMAR 21 cruise (Guzmán and Escribano 2019).

Family Solenoceridae Wood-Mason and Alcock, 1891

13. *Hadropenaeus lucasii* Bate, 1881

    Distribution in SEP. Salas y Gomez Ridge; Seamounts Big (25°40′S, 85°27′W), Yarala (25°40′S, 86°34′W), Pearl (25°33′S, 89°12′W), May-Day (25°11′S, 94°29′W), Dome (25°04′S, 97°26′W), and Cliff (25°58′S, 100°41′W) (Burukovsky 1990).

14. *Haliporoides diomedeae* (Faxon, 1893)

    Distribution in SEP. From Mancora to 18°19′S, off Peru (Mendez 1981); Iquique (Retamal 1993), central to southern Chile (33–42°S) (Noziglia and Arana 1976; Arana et al. 2003).

15. *Hymenopenaeus halli* Bruce, 1966

    Distribution in SEP. Close to Ichthyologist Seamount (25°07′S, 99°35′W), 330–800 m (Burukovsky 1990).

    Remarks. Another species of *Hymenopenaeus* has been reported from Peru, but it is a benthonic species (Moscoso 2012).

Family Sergestidae Dana, 1852

16. *Allosergestes pectinatus* (Sund, 1920)

Distribution in SEP. 840 miles off Caldera, between 86°33′W and 93°97′W; vicinity of Easter Island (109°15′W) (Guzmán 2004a).

17. *Allosergestes pestafer* (Burkenroad, 1937)

Distribution in SEP. Caldera, Chile (27°00′07″S, 71°80′05″–79°05′05″W), from surface to 1000 m, CIMAR 5 cruise (Guzmán 2004a); from 10 miles off Caldera to 190 miles SW of Desventuradas Islands, CIMAR 21 cruise; Seamounts Soldier and Eclipse, Nazca Ridge (21°41′S, 81°46′W; 22°06′S, 81°19′W); Salas y Gomez Ridge, Seamounts Yarala, Amber, Pearl, Cliff, Dome, and Ichthyologist (Vereshchaka 1990).

18. *Deosergestes corniculus* (Kroyer, 1855)

Distribution in SEP. Vicinity of Seamount Big (25°40′S, 85°27′W), Salas y Gomez Ridges (Vereshchaka 1990).

19. *Eusergestes arcticus* (Kroyer, 1859)

Distribution in SEP. Southern Chile (41°43′18″S, 72°38′15″W); W of Rio Puelo (Holthuis 1852); 50–100 miles NW of Juan Fernandez Archipelago (32°58′S, 81°13′W; 31°45′S, 79°59′W), CIMAR 6 cruise (Guzmán 2003).

20. *Eusergestes similis* (Hansen, 1903)

Distribution in SEP. Off Constitucion (35°10′S, 74°08′W) (Guzmán 2004a); off Papudo (32°32′56″S, 72°35′45″W).
Remarks. The material from off Papudo represents a new record.

21. *Gardinerosergia bigemmea* (Burkenrad, 1940)

Distribution in SEP. Off Caldera Harbour (27°00′07″S, 79°05′05″W), CIMAR 5 cruise (Guzmán 2004a).

22. *Neosergestes brevispinatus* (Judkins, 1978)

Distribution in SEP. Off Peru (11°02′S and 14°48′S) (Judkins 1978; Mendez 1981); off Arica to Iquique (Guzmán 1999a); Caldera Harbour (27°00′0″S, 70°52′48″W), and 50 miles SW of Desventuradas Islands (26°59′24″S, 80°34′48″W), CIMAR 22 cruise.
Remarks. The CIMAR 22 cruise records are new for this species.

23. *Neosergestes consobrinus* (Milne, 1968)

Distribution in SEP. Easter Island (27°11′S, 109°15′W) (Guzmán 2004a); Chañaral (29°00′S, 72°00′W).
Remarks. The Chañaral material is a new record.

24. *Neosergestes orientalis* Hansen, 1919

Distribution in SEP. Off Peru (6°N to 20°S), 230–2280 m (Mendez 1981); vicinity of Soldier and Eclipse Seamounts, Nazca Ridge (21°41′S, 81°46′W;

22°06′S, 81°19′W), 200–2000 m (Vereshchaka 1990); Amber Seamount, Salas y Gomez Ridge (24°58′S, 88°31′W), 500–1500 m (Vereshchaka 1990).

Remarks. Previously recorded in the SEP as *Neosergestes geminus* (Judkins, 1978) and in the Nazca and Salas y Gomez Seamounts as *N. gibbilobatus* (Judkins, 1978) (Judkins 1978), the latter now considered a junior synonym of *N. orientalis* (Hansen, 1919).

25. *Neosergestes tantillus* (Burkenroad, 1940)

Distribution in SEP. 50 miles off Iquique and 65 miles SW of Loa River mouth (Guzmán 1999a); 10 and 270 miles off Caldera Harbour; 190 miles west of Desventuradas and Easter Islands during the CIMAR 21 cruise.

26. *Parasergestes extensus* (Hanamura, 1983)

Distribution in SEP. Off Loa River mouth (21°04′S, 70°51′W), 378 m (Guzmán 1999a).

27. *Parasergestes halia* (Faxon, 1893)

Distribution in SEP. In the neighborhood of Soldier (21°41′S, 81°46′W) and Eclipse (22°06′S, 81°19′W) Seamounts of Nazca Ridge, and Seamounts of Salas y Gomez Ridge, Yarala (25°40′S, 86°34′S), Amber (24°58′S, 88°31′W), Pearl (25°33′S, 89°12′W) Cliff (25°58′S, 100°41′W), and Ichthyologist (25°07′S, 99°35′W), between 230 and 2200 m (Vereshchaka 1990).

28. *Parasergestes vigilax* (Stimpson, 1860)

Distribution in SEP. Seamounts of Salas y Gomez Ridge, Big (25°40′S, 85°27′W), "Yarala" (25°40′S, 86°34′S), Amber (24°58′S, 88°31′W) Pearl (25°33′S, 89°12′W), Cliff (25°58′S, 100°41′W), Ichthyologist (25°07′S, 99°35′W), and Dome (25°04′S, 97°26′W), between 160 and 2200 m deep (Vereshchaka 1990).

29. *Petalidium foliaceum* Bate, 1881

Distribution in SEP. Sub Antarctic waters (Gorny 1999); Drake Passage (57°11′S, 63°51′W), expedition USAP on board of RV "Eltanin," July 1962.

Remarks. This is a new record for the southern tip of South America.

30. *Phorcosergia maxima* (Burkenroad, 1940)

Distribution in SEP. 100 to 20 miles W of Camarones River (19°09′S, 72°00′W; 19°09′S, 70°36′W), to 30 to SW of Loa River (21°41′S, 70°31′W), and NW of Loa River (21°04′S, 71°11′W) (Guzmán 1999a).

31. *Phorcosergia phorca* (Faxon, 1893)

Distribution in SEP. NW of Lobos de Tierra Island (06°13′S, 81°17′W) and off Supe (10°54′S, 81°14′W), Peru (Mendez 1981); 100 miles SW of Arica (19°09′S, 72°00′W) to 30 miles SW mouth of Loa River, Chile (Retamal 1993, Guzmán 1999a).

32. *Scintillosergia scintillans* (Burkenroad, 1940)

Distribution in SEP. 400 miles off Caldera Harbour (27°00′07″S, 79°05′05″W) (Guzmán 2004a).

33. *Sergestes atlanticus* H. Milne Edwards, 1830

Distribution in SEP. Soldier Seamount, Nazca Ridge; Big, Yarala, Amber and Pearl Seamounts, Salas y Gomez Ridge (Vereshchaka 1990).

34. *Sergestes cornutus* Kroyer, 1855

Distribution in SEP. Pearl (25°33′S, 89°12′W), Ichthyologist (25°07′S, 99°35′W), and Dome (25°04′S, 97°26′W) Seamounts, Salas y Gomez Ridge (Vereshchaka 1990).

35. *Sergia japonica* (Bate, 1881)

Distribution in SEP. off Valparaiso (33°42′ S, 78°18′ W) (Holthuis 1952; Vereshchaka 2000).

36. *Sergia laminata* (Burkenroad, 1940)

Distribution in SEP. Soldier (21°41′S, 81°46′W) and Eclipse Seamounts (22°06′S, 81°19′W), Nazca Ridge; near Big (25°40′S, 85°27′W), Yarala (25°40′S, 86°34′S), Amber (24°58′S, 88°31′W), Pearl (25°33′S, 89°12′W), Communard (24°40′S, 85°28′W), Dome (25°04′S, 97°26′W), Cliff (25°58′S, 100°41′W), and Ichthyologist (25°07′S, 99°35′W) Seamounts, Salas y Gomez Ridge (Vereshchaka 1990).

Pleocyemata Burkenroad, 1963
Caridea Dana, 1852
Family Pasipheidae Dana, 1852

37. *Eupasiphae gilessi* (Wood-Mason, 1892)

Distribution in SEP. 67 miles NW of Pisagua, Iquique (Guzmán and Wicksten 1998).

38. *Glyphus marsupialis* Filhol, 1884

Distribution in SEP. Off Chicama, Peru (07°56′S, 79°31′W) (Mendez 1981).
Remarks. The material deposited at the MSC (USNM 170564) that is mentioned by Mendez (1981), collected by E. Del Solar in February 1971, and sent to F.A. Chace was reviewed. It was collected at a depth of 810 m.

39. *Parasiphae sulcatifrons* Smith, 1884

Distribution in SEP. 77 miles off Arica (18°25′S, 71°43′W) and 57 miles SW of Iquique (20°20′S, 71°11′W), Chile (Guzmán and Wicksten 1998); 120 miles off Puerto Mal Abrigo, Peru (7°46′S, 81°30′W), 683 m, Expedition

USAP Cruise 3, R/V "Eltanin," 1962; 30 miles W of Ilo (17°43′S, 71°53′W), Peru, 901 m.

Remarks. The specimen from W of Ilo was found among five specimens in one lot in the holdings of the IMARPE collection (02–000937) (coll. A. Kameya and Carbajal). This is the first record of *P. sulcatifrons* for Peru.

40. *Pasiphaea acutifrons* Bate, 1888

Distribution in SEP. Restricted to the southern tip of South America, Magellan Strait.

Remarks. Based on review of different collections in Peru, Chile, Argentina, Uruguay, and South of Brazil, we conclude that many references to this species are erroneous. While referring to a contribution by Del Solar and Flores (1972), Mendez (1981) recorded this species for Matarani, Peru (17°11′S), but this specimen was not reviewed in order to confirm this identification. Retamal (1993) record this species in Iquique, and Guzmán and Wicksten (1998) reproduced the same record. A new examination of this material indicated that it is a variation of *Pasiphaea americana*. In the Atlantic, *P. acutifrons* has also been recorded, but all the records north of Patagonia are erroneous. Spivak (1997) reported the presence of *P. acutifrons* in the SW Atlantic, from Uruguay to the Magellan region. After examining the specimens available in the collections of the Oceanographic Institution of the Federal University of Rio Grande do Sul, labelled *Pasiphaea acutifrons* (FURG 2672), we came to the conclusion that this material belongs to *P. merriami* Schmitt, 1931. The specimens in the Museo Nacional de Historia Natural of Montevideo, Uruguay, labelled *P. acutifrons* belong to *P. barnardi* Yaldwin, 1971, a first record for Uruguay. The specimens deposited in the collection of the Institute of Investigation in Fishery (INIDEP), labelled as *P. acutifrons*, belong to *Pasiphaea balssi* Burukovsky and Romensky, 1987, a new record to Argentina.

41. *Pasiphaea americana* Faxon, 1893

Distribution in SEP. From Mancora Bank (03°23′S, 81°01′W) to off Piura (05°54′S, 81°15′W), Peru (Mendez 1981); Arica (18°25′S, 71°01′W), Chile (Guzmán and Wicksten 1998); Seamounts Amber (24°58′S, 88°31′W) and Pearl (25°33′S, 89°12′W), Salas y Gomez Ridge, 530–3000 m (Vereshchaka 1990; Burukovsky 1990).

42. *Pasiphaea barnardi* Yaldwin, 1971

Distribution in SEP. Baker Channel (47°56′S, 74°29′W) to Wide Channel (49°58′S, 74°26′W), CIMAR 14 cruise (Guzmán 2014), and to Drake Passage, Chile (56°58′S, 69°08′W), 4295 to 4310 m, USAP Expedition, R/V "Eltanin," cruise 5, 1962; Bellingshausen Sea (63°43′30″S, 83°02′00″W), 3660–4099 m, USAP Expedition Station 785, cruise 10, 1963, USARP/EL/10/785/USC, R/V "Eltanin."

Remark. The material from the Expedition USAP R/V "Eltanin," cruise 5, 1962, 4295–4310 m, represents a new record.

43. *Pasiphaea chacei* Yaldwyn, 1962

Distribution in SEP. 80 miles W of Arica (18°25′S, 71°43′W) to 80 miles NW of Loa River mouth (Guzmán and Wicksten 1998); 280 miles W of Caldera Harbour (27°03′37″S, 76°05′28″W) to 500 miles E of Easter Island (27°04′02″S, 100°03′56″W), about 700 m, CIMAR 5 cruise (Guzmán 2004); off San Felix and San Ambrosio Islands (26°20′S, 80°00′W), CIMAR 6 cruise (Guzmán 2004); 50 miles off Valparaiso (33°11′S to 33°14′S and 72°42′W to 72°38′W), R/V "Eltanin," cruise 25, sta. 303, 400–425 m; off Las Cruces (34°35′S, 72°15′W) and 30 miles south of Constitución (35°31′S, 73°04′W), Expedition "Talud Continental I" (1981), Natural History National Museum of Santiago.

Remarks. Several new records are reported for this species, with the material from S of Constitución (35°31′S, 73°04′W) representing its new southernmost distribution limit.

44. *Pasiphaea cristata* Bate, 1888

Distribution in SEP. Seamounts Amber (24°58′S, 88°31′W) and Cliff (25°58′S, 100°41′W), Sala y Gomez Ridge (Vereshchaka 1990).

45. *Pasiphaea dofleini* Schmitt, 1932

Distribution in SEP. Punta Arenas, Magellan Strait, Chile (Holuthuis 1852). W of Penas Gulf (47°14′S, 76°28′W), 732 m, USAP Expedition, cruise 4, station 165, August 1962, R/V "Eltanin"; Taitao Peninsula, Chile (46°25′S, 76°21′W), 1830 m, USAP Expedition, cruise 4, July 1962, R/V "Eltanin."

46. *Pasiphaea flagellata* Rathbun, 1906

Distribution in SEP. Close to Long Seamount (25°47′S, 85°17′W), Salas y Gomez Ridge (Burukovsky 1990).

47. *Pasiphaea kaiwiensis* Rathbun, 1906

Distribution in SEP. Close to Communard Seamount (24°40′S, 85°28′W), Salas y Gomez Ridge (Vereshchaka 1990).

48. *Pasiphaea magna* Faxon, 1893

Distribution in SEP. Lobos de Tierra Island (06°25′S 81°00′W), Peru (Mendez 1981); 30 miles off Arequipa (15°56′S, 74°39′W), IMARPE (02–000935); 30 miles off Ilo Harbour (17°43′S, 71°53′W), IMARPE (02–000937); 75 miles off Iquique, Chile (Guzmán and Wicksten 1998); 13 miles SW of Tocopilla (22°13′S, 70°23′W), Chile (Wehrtmann and Carvacho 1997).

Remarks. The material from the IMARPE collection represent new records.

49. *Pasiphaea rathbunae* (Stebbing, 1914)

Distribution in SEP. 93 miles NW of Diego Ramirez Island, Chile (56°07S, 71°25′W), 983 m, Expedition USAP Cruise 5, R/V "Eltanin," 07 November 1962; 350 miles off SW Tierra del Fuego, Chile (59°16′S, 78°15′W), 2776 m, Expedition USAP, R/V "Eltanin," cruise 10, 23 November 1963.
Remarks. These are the first records for Chile.

50. *Pasiphaea scotiae* (Stebbing, 1914)

Distribution in SEP. Drake Passage (58°29′S, 65°30′W), 2105 m, R/V "Eltanin," cruise 4, 12 August 1962.

Family Oplophoridae Dana, 1852

51. *Acanthephyra brevirostris* Smith, 1885

Distribution in SEP. Ecuador (Faxon 1895). One specimen collected off Chile during the ATACAMEX expedition to Atacama Trench (Guzmán and Escribano unp. data).

52. *Acanthephyra carinata* Bate, 1888

Distribution in SEP. Sarmiento Channel, Patagonia, Chile (Holthuis 1852); off Iquique (21°04′S, 71°31′W), 374 m, 11 September 1988 (Guzmán 2004b).

53. *Acanthephyra cucullata* Faxon, 1893

Distribution in SEP. Seamount Soldier, Nazca Ridge (21°41′S, 81°46′W) (Vereshchaka 1990).

54. *Acanthephyra curtirostris* Wood-Mason, 1891

Distribution in SEP. Off Mollendo (17°05′S, 72°16′W) (Mendez 1981); off Chicama point (7°46′S, 81°30′W), Peru, 683 m, USAP Expedition, cruise 3, station. 34, 7 June 1962, R/V "Eltanin"; Chile-Peru Trench (24°S) (Retamal 1981); 77 miles off Arica (18°25′S, 71°43′W), 18 miles off Pisagua (19°09′S, 70°36′W), and 60 miles SW of Iquique (21°04′S, 70°51′W), 300 to 450 m (Guzmán 2004b).
Remarks. The material from off Chicama Point, Peru, is a new Record for Peru.

55. *Acanthephyra eximia* Smith, 1884

Distribution in SEP. Seamounts Dorofeeva (25°54′S, 84°22′W), Pearl (25°33′S, 89°12′W), and Amber (24°58′S, 88°31′W), Salas y Gomez Ridge (Burukovsky 1990).

56. *Acanthephyra faxoni* Calman, 1939

Distribution in SEP. Tumbes (03°48′S, 81°18′W) to SW of Mollendo (17°05′S, 72°16′W), 800 to 1000 m; SW Peru (Mendez 1981, Moscoso 2012); 120 miles off Chicama (7°46′S, 81°30′W), 683 m, USAP Expedition, cruise 3, station 34,. 7 June 1962, R/V "Eltanin"; 56 miles SW Atico (16°30′S, 74°34′W),

1162 m, USAP Expedition USAP, cruise 3, station 52, 15 June 1962, R/V "Eltanin."

Remarks. These records are new for Peru.

57. *Acanthephyra media* Bate, 1888

Distribution in SEP. Chile-Peru Trench (Retamal 1981).

Remarks. One specimen in the holdings of the Museum of Zoology of Concepción University (MZUDEC31720) is badly damaged and could not be identified.

58. *Acanthephyra pelagica* (Risso, 1816)

Distribution in SEP. Chile-Peru Trench (Retamal 1981), 280 miles off Caldera and San Felix Island (Guzmán 2004b) and Chiloe (42°35′S, 74°48′W), PUCK-156 Expedition, R/V SONNE (Guzmán and Quiroga 2005); 43 miles SW of Valparaiso (33°18′S, 72°27′W to 33°10′S, 72°13′W), 1830 m; USARP Expedition, Eltanin R/V, cruise 9, station 743, 26 September 1963; Drake Passage (58°29′S, 65°30′W), 2105 m, USARP Expedition, USARP/EL/4/149A/USC, station 149A, R/V Eltanin, cruise 4, 12 August 1962.

Remark. The material collected in the Drake Passage represents a new record. *Acanthephyra pelagica* is widely distributed in the SE Pacific, from off Valparaiso to Drake Passage (Wasmer 1986).

59. *Acanthephyra stylorostratis* (Bate, 1888)

Distribution in SEP. Off Concepcion (95°W), Chile (Wasmer 1986).

60. *Acanthephyra trispinosa* Kemp, 1939

Distribution in SEP. Off Arica (10°25′S, 70°40′W), off Iquique (20°20′S, 71°11′W), and off Loa River (21°04′S, 71°03′W), Chile (Guzmán 2004b); Seamounts Soldier (21°41′S, 81°46′W) and Eclipse (22°06′S, 81°19′W), Nazca Ridge; Seamounts Communard (24°40′S 85°28′W), Long (25°47′S; 85°17′W), Yarala (25°40′S, 86°34′W), and Cliff (25°58′S, 100°41′W), Salas y Gomez Ridges (Vereshchaka 1990).

61. *Ephyrina hoskynii* Wood-Mason, 1891

Distribution in SEP. Chile-Peru trench (Retamal 1981).

Remarks. The material of the Retamal (1981) report is no longer available in the Museum of Zoology, Concepción University.

62. *Ephyrina ombango* Crosnier and Forest, 1973

Distribution in SEP. 123 miles off Chicama, Peru (7°46′S, 81°30′W), 683 m, USAP cruise 3, Station 34, R/V Eltanin, 7 June 1962, off Iquique (21°04′S, 71°31′W) (Guzmán 2004b); Seamount Soldier (21°41′S, 81°46′W), Nazca Ridge (Vereshchaka 1990).

Remarks. The record from off Chicama is the first for Peru.

63. *Hymenodora glacialis* (Buchholz, 1874)

Distribution in SEP. Sub-Antarctic (Chace 1986; Wasmer 1986; Hendrickx and Estrada-Navarrete 1989); 415 miles E of Easter Island (26°59′54″S, 101°36″3″W), CIMAR cruise 21; 50 miles W Caldera Harbour (27°00′00″S, 71°46′12″W), off San Felix Island (26°18′00″S, 80°15′00″W), 100 miles N of Juan Fernandez Archipelago (31°48′00″S, 80°00′00″W), and 240 miles off Valparaiso (33°24′00″S, 76°30′00″W), CIMAR cruise 22 (Guzmán and Escribano unp. data).

64. *Hymenodora gracilis* Smith, 1889

Distribution in SEP. Widely distributed in the Southeast Pacific off Chile, from Valparaiso to Drake Passage (Chace 1986; Wasmer 1986).

65. *Meningodora mollis* Smith, 1882

Distribution in SEP. Seamount Soldier (21°41′S, 81°46′W), Nazca Ridge (Vereshchaka 1990).

66. *Notostomus elegans* A. Milne-Edwards, 1881

Distribution in SEP. NW Lobos de Tierra Island (06°13′S, 81°17′W), Peru (Mendez 1981); 123 miles W off Chicama, Peru (7°46′S, 81°30′W), 683 m, USAP Expedition USAP, cruise 3, station 34, R/V "Eltanin," 7 June 1962; off Valparaiso, Chile (33°11′S, 72°40′W) (Wasmer 1986); Seamount Soldier (21°41′S, 81°46′W), Nazca Ridge (Vereshchaka 1990).

Remark. These are new records for Peru.

67. *Oplophorus gracilirostris* A. Milne-Edwards, 1881

Distribution in SEP. Seamount Pearl (25°33′S, 89°12′W), Salas y Gomez ridge (Vereshchaka 1990).

68. *Oplophorus novaezeelandiae* de Man, 1931

Distribution in SEP. Off Chile, from 33° to 40°S (Wasmer 1986); from Juan Fernandez Archipelago (33°20′S, 78°53′W) to San Felix and San Ambrosio Islands (26°10′S, 80°00′W), CIMAR cruise 6 (Guzmán 2004b).

69. *Oplophorus spinosus* (Brulle, 1839)

Distribution in SEP. 140 miles of Caldera (27°00′52″S, 73°37′26″W) to Easter Island and Salas y Gomez Island (Guzmán 2004b); Seamounts Soldier and Eclipse, Nazca Ridge, and Seamounts Communard, Yarala, Pearl, Amber, Dome, and Cliff, Salas y Gomez Ridge (Vereshchaka 1990; Burukovsky 1990).

70. *Systellaspis braueri* Crosnier, 1987

Distribution in SEP. 45 miles off Iquique (20°20′S, 70°58′W), 450 m (Guzmán 2004); 100 miles off Tierra del Fuego (55°22′S, 74°43′W), Chile (Wasmer 1986).

71. *Systellaspis cristata* (Faxon, 1893)

Distribution in SEP. NW of Isla Lobos de Tierra Island (6°13′S, 81°17′W), 120 miles off Chicama (7°46′S, 81°30′W), 683 m, USAP Expedition, cruise 3, station 34, 7 June 1962, R/V "Eltanin"; Chile-Peru Trench (Retamal 1981); 60 miles W off Iquique and 40 miles W off Punta Lobos (21°04′S, 70°51′W), Iquique (Guzmán 2004).

72. *Systellaspis debilis* (A. Milne-Edwards, 1881)

Distribution in SEP. Cliff Seamount, Salas y Gomez Ridge (25°58′S, 100°41′W) (Vereshchaka 1990).

73. *Systellaspis eltanini* Wasmer, 1986

Distribution in SEP. 34 miles off Valparaiso (33°18′S, 72°21′W) and 200 miles SW of Tierra del Fuego, Chile (57°52′S, 74°43′W) (Wasmer 1986).

Family Pandalidae Haworth, 1825

74. *Plesionika martia* (A. Milne-Edwards, 1883)

Distribution in SEP. Seamounts Amber (24°58′S, 88°31′W), Dome (25°04′S, 97°26′W), and Ichthyologist (25°07′S, 99°35′W) of Salas y Gomez Ridge (Burukovsky 1990).

75. *Plesionika sanctaecatalinae* Wicksten, 1983

Distribution in SEP. 12° to 18°S off Peru (Wicksten 1983); off Patache Point (20°46′S, 70°34′W), Iquique, and off Chipana Bay (21°19′S, 70°26′W), Chile (Retamal 1995); from Arica (18°25′S, 70°40′W) to off Loa River mouth (21°41′S, 70°31′W).

Remarks. The material from off Arica and the Loa River mouth represents new records for Chile.

76. *Plesionika trispinus* Squires & Barragan, 1976

Distribution in SEP. Mancora Bank (3°35′S); off Paita (5°04′S), Aguja Point (5°54′S), N of Lobos de Tierra Island (6°23′S), and off Salaberry (8°11′S), Peru (Mendez 1981).

77. *Stylopandalus richardi* (Coutière, 1905)

Distribution in SEP. Off Caldera Harbour to Easter Island, Chile (Guzmán and Rivera 2002).

Family Physetocarididae Chace, 1940

78. *Physetocaris microphthalma* Chace, 1940

Distribution in SEP. Mid SW Pacific (37°06′S, 92°20′W; 38°31′S95°20′W) (Wasmer 1984); off Arica (18°25′S, 71°43′W), Chile (Guzmán 1999).

12.4 Discussion

A total of 78 species were registered in this work, 36 of which belong to the Dendrobranchiata and 42 to the Pleocyemata. We reported on many new specific data, new records, and actualization of taxonomic status. Of these 78 species, 43 are associated with the Seamounts of the Salas y Gomez Ridge. Many records refer to a single sample in which the species was collected. The high geographic dispersion of data is a problem when efforts are made to try to analyze the biogeographical connections of this group of species. Some species have a very wide geographical distribution range, as in the case of *Acanthephyra pelagica* that was found repeatedly during different surveys. The opposite occurred with species considered rare, for example, *Physetocaris microphthalma* and *Glyphus marsupialis,* both collected in some very distant localities and in few numbers.

Some species reported herein for the Southeastern Pacific have been collected in localities in the Arabian Sea and the Bay of Bengal, thus showing a very wide biogeographical pattern which is difficult to explain from a taxonomic viewpoint. *Ephyrina hoskini* was reported by Retamal (1981) in Chilean waters, but there is no complementary information on its morphology, and no specimens are available in collections for a detailed revision of their taxonomic status. The record of *Acanthephyra media* is yet another case, but unlike the previous case, the specimen is available in the collection of the Museo de Zoologia of the Concepción University, but it is severely damaged, and a reliable identification is therefore impossible.

In some cases the records are unique due to the scarcity of sampling or because of very low abundance. Ten species have been registered in one opportunity only. *Aristeomorpha foliacea*, for example, has been recorded only near the Dome Seamount (Burukovsky 1990). The same situation occur with *Pasiphaea flagellata*, *P. kaiwiensis* (Burukovsky 1990), *Acanthephyra cucullata* (Vereshchaka 1990), *A. stylorostratis* (Wasmer 1986), and *Physetocaris microphthalma* (Guzmán 1999), among others. *Physetocaris microphthalma* is another species with only one record and one specimen. These records have been repeatedly published in checklist, biogeographic studies, and review, giving the false idea that they are common or abundant.

Considering our current knowledge, the distribution of species in the southeast Pacific seems to indicate a clear influence of the Seamounts in the Nazca and Salas y Gomez Ridges on the diversity of pelagic shrimps. The main information related to this environment originated from the Russian expeditions published in 1990 (Mironov and Rudjakov 1990). Here we included the results of the Chilean efforts associated with a series of Research Cruises, the CIMAR cruises, during which samples were taken near these Seamounts. A total of 43 species were registered in these habitats, 17 of which have exclusively been recorded near of Seamounts of these ridges (Guzmán 2008). The major diversity is associated with Soldier and Amber Seamounts, with 19 and 18 species, respectively. This number is relatively similar to those observed in other Seamounts habitats. In the south and the north mid-Atlantic ridge, 34 and 47 species of pelagic shrimps have been reported, respectively (Cardoso et al. 2014). Our results seem to coincide with the observations of Letessier et al. (2017) who refer to the Seamounts environment as "oasis" in the middle of the Southeast Pacific.

In a latitudinal sense, the influence of the oceanographic conditions prevailing in the southeast Pacific has a strong impact on the biogeographical patterns observed in this section of the Pacific Ocean (Brattstrom and Johanssen 1983; Thiel et al. 2007). The distribution of pelagic shrimps observed in this contribution shows a relative concordance with the proposal of different authors (see Camus 2001). More information is needed, however, in order to confirm our observations, including the validation of some taxonomic identifications and additional records to verify the distributions.

Acknowledgments We wish to acknowledge the Comité Oceanográfico Nacional of Chile (CONA) by the supports of the CIMAR program. This chapter was supported partially by VRIIP0093-17 project of VRIIP of the Universidad Arturo Prat, Iquique, Chile. To the Atacamex Expedition of Millennium Institute of Oceanography of the Universidad of Concepción, IMO. To the PUCK-156 expedition. To the Museum of Zoology of Concepción University, the National Natural History Museum of Santiago, the millennium core of Ecology and Sustainable Management of Oceanic Islands, ESMOI. To the Museum of Support Center of Smithsonian Institution, the Sanduishe program of the Oordenação de Aperfeiçoamento de Pessoal de Nível Superior, Capes, the Museu de Zoologia da University of Sao Paulo, USP, and the Benthos Laboratory of the Pontifical Catholic University of Valparaiso. Our special thanks to Michel E. Hendrickx for the invitation to be part of this book.

References

Arana PM, Ahumada M, Guerrero A (2003) Distribution y abundancia de la gamba *Haliporoides diomedeae* (Crustacea: Decapoda: Penaeidae) frente a la costa central de Chile. Investig Mar 31(2):57–71. https://doi.org/10.4067/S0717-71782003000200006

Bate S (1888) Report of the Crustacea Macrura dredged by "H.M.S Challenger" during the years 1873–1876. Report on the scientific results of the voyage challenger 1873–76. Zoology 24:1–942

Brattstrom H, Johanssen A (1983) Ecological and regional zoogeography of the marine benthic fauna of Chile. Report 49 of the Lund University Expedition 1948–1949. Sarsia 68:289–339

Burukovsky RN (1990) Shrimps from the Sala y Gomez and Nazca Ridges. Tru Inst Okeanol Akad Nauk SSR 124:187–217

Chace FA (1986) The caridean shrimps (Crustacea: Decapoda) of the Albatross Philippine Expedition, 1907–1910, part 4: families Oplophoridae and Nematocarcinidae. Smithson Contrib Zool 432:1–82

Cardoso I, Falkenhaug T, Loureiro-Fernandes LT (2014) Comparison between the shrimp species richness (Caridea and Dendrobranchiata, Decapoda, Crustacea) of the south and north mid Atlantic ridge. Braz J Oceanogr 62(1):47–56

Camus, PA (2001) Marine biogeography of continental Chile. Rev. Chile. Hist. Nat.74:587–617

Del Solar E, Flores LA (1972) Exploraciones de Crustáceos (Zona Sur), Crucero SNP-1 7201. Instituto del Mar del Perú, Ser Inf Esp, IM-207:1–8.

Gorny M (1999) On the biogeography and ecology of the Southern Ocean decapod fauna. Sci Mar 63(supl. 1):367–382

Guzmán G (1999a) Nuevos registros de camarones sergéstidos (Crustacea; Decapoda; Dendrobranchiata) en el norte de Chile. Investig Mar 27:87–92. https://doi.org/10.4067/S0717-71781999002700009

Guzmán G (1999b) Primer registro de *Physetocarys microphthalmus* Chace, 1940 para el Pacifico Sur oriental (18°25′S–71°43′W) (Crustacea: Decapoda: Physetocaridae). Notic Mens MNHN Santiago 337:3–5

Guzmán G (2003) Mesopelagic shrimps of the family Sergestidae (Decapoda; Dendrobranchiata): new records for the Southeastern Pacific. Crustaceana 75(9):1033–1045. https://doi.org/10.1163/156854002763270446

Guzmán G (2004a) Decápodos mesopelágicos capturados durante los proyectos CIMAR 5 y 6, Islas Oceánicas Chilenas. Cienc Tecnol Mar 27(1):69–78

Guzmán G (2004b) New records of mesopelagic shrimps of the family Oplophoridae (Decapoda; Caridea) in southeastern Pacific waters. With a review of Chilean species. Gayana 68(1):70–75

Guzmán G, Quiroga E (2005) New records of shrimps (Decapoda: Caridea and Dendrobranchiata) in deep waters off Chile. Gayana 69(2):285–290

Guzmán G, Rivera J (2002) *Stylopandalus richardi* (Coutière, 1905) primer registro para aguas del Pacifico sur este (Decapoda, Pandalidae). Investig Mar 30(1):67–70

Guzmán G, Wicksten MK (1998) Nuevos registros de camarones de la familia Pasiphaeidae (Crustacea, Decapoda) en el norte de Chile (18° a 22° Lat Sur, 70° a 72 Long W). Gayana Zool 62(2):203–210

Guzmán G, Wicksten MK (2000) The subfamily Benthesicyminae Bouvier, 1908 (Decapoda, Dendrobranchiata) in northern Chile (18° to 22°S). Crustaceana 73(8):925–931

Guzmán G. (2008). Camarones pelágicos (Crustacea: Decapoda) en aguas del Pacifico sureste. Contribuciones al Estudio de los Crustáceos del Pacífico Este 5(1):27–45

Guzmán G., R. Escribano (2019) Extension of the distributional range of Funchalia woodwardi to the Southeast Pacific (Decapoda, Dendrobranchiata, Penaeidae). Rev Biol Mar Oc 53(3):343–346

Hanamura Y (1983) Pelagic shrimps (Penaeidea and Caridea) from Baja California and its adjacent region with description of a new species. Bull Biogeogr Soc Jpn 38(8):51–85

Hendrickx ME, Estrada-Navarrete F (1989) A checklist of the species of pelagic shrimps (Penaeoidea and Caridea) from the Eastern Pacific, with notes on their geographic and depth distribution. CalCOFI Rep 30:104–121

Hendrickx ME, Wicksten MK (1989) Los Pandalidae (Crustacea: Caridea) del Pacifico Mexicano, con una clave para su identificacion. Caldasia 16:71–86

Holthuis LB (1952) Reports of the Lund University Chile Expeditions, 1948–49. 5. The Crustacea Decapoda Macrura of Chile. Lunds Univ Årsskrift NF Avd 2 47(10):1–110

Letessier B, De Grave S, Boersch-Supan PH, Kemp KM, Brierley AS, Rogers AD (2017) Seamount influences on mid-water shrimps (Decapoda) and gnathophausiids (Lophogastridea) of the south-west Indian ridge. Deep-Sea Research II Top Res Oceanog. https://doi.org/10.1016/j.dsr2.2015.05.009

Mendez M (1981) Claves de identificación y distribución de los langostinos y camarones (Crustacea: Decapoda) del mar y los rios de la costa del Perú. Bol Inst Mar Perú 5:1–170

Moscoso V (2012) Catalogo de crustaceos decapodos y estomatopodos del Peru. Bol Inst Mar Perú 27(1-2):1–207

Mironov AN, Rudjakov JUA, (1990). Plankton and Benthos from the Nazca and Sala-y-GomezSubmarine Ridges. Moscow: Nauka, 269–278 [in Russian]

Nielsen C (2013) Life cycle evolution: was the eumetazoan ancestor a holopelagic, planktotrophic gastraea? Evol Biol 13:171

Noziglia L, Arana P (1976) Observaciones biológico-pesqueras de la gamba, *Hymenopeneaeus diomedeae* Faxon 1893 en el Pacífico Suroriental (Crustacea: Decapoda: Penaeidae). Rev Com Perm Pacífico Sur 5:63–83

Retamal MA (1981) Catálogo ilustrado de los decápodos de Chile. Gayana Zool 44:1–110

Retamal MA (1993) Crustáceos decápodos abisales de la zona Arica-Iquique. Estud Oceanol 12:1–8

Retamal MA (1995) Plesionika santaecatalinae Wicksten, 1983 en aguas Chilenas. Gayana Oceanol 3(1):1–2

Retamal MA (2000) Decápodos de Chile. CD Rom. ETI (Amsterdan) - Universidad de Concepción

Spivak ED (1997) Los crustáceos decápodos del Atlántico sudoccidental (25°-55°S): distribución y ciclos de vida. Investigaciones Marinas, Valparaiso 25:69–91.

Thiel M, Macaya EC, Acuña E, Arntz WE, Bastias H, Brokordt K, Camus PA, Castilla JC, Castro LR, Cortés M, Dumont CP, Escribano R, Fernandez M, Gajardo JA, Gaymer CF, Gomez I, Gonzalez AE, Gonzalez HE, Haye PA, Illanes JE, Iriarte JL, Lancellotti DA, Luna-Jorquera G, Luxoro C, Manriquez PH, Marin V, Muñoz P, Navarrete SA, Perez E, Poulin E, Sellanes J, Hito Sepuveda H, Stotz W, Tala F, Thomas A, Vargas CA, Vasquez JA, Vega JMA (2007) The Humboldt current system of Northern and Central Chile oceanographic processes, ecological interactions and socioeconomic feedback. Oceanogr Mar Biol Annu Rev 45:195–344

von Tiefenbacher L (1991) Anmerkungen zu einigen mesopelagischen Garnelen und ihrer Verbreitung in den Gewässern der westlichen Antarktis (Crustacea, Decapoda, Natantia). Spixiana 14(2):153–158

von Tiefenbacher L (1994) Decapode crustaceen aus westantarktischen Gewässern gesammelt von der R.V. "John Biscoe", Reise 11. Spixiana 17(1):3–19

Vereshchaka AL (1990) Pelagic decapods from Seamounts of Nazka and Sala y Gomez ridges. In: Mironov AN, Rudjakov JA (eds) Plankton and benthos from the Nazka and Sala y Gomez submarine ridges. Academy of the Sciences of the USSR, Moscow, pp 129–155. Trans P.P. Shirshov Inst Oceanol

Vereshchaka AL (2000) Revision of the genus *Sergia* (Decapoda: Dendrobranchiata: Sergestidae): taxonomy and distribution. Galathea Rep 18:69–207

Wasmer RA (1984) New record for *Physetocaris microphthalma* Chace (Decapoda, Caridea, Physetocarididae) from the South Pacific. Crustaceana 49(3):315–317

Wasmer RA (1986) Pelagic shrimps of the family Oplophoridae (Crustacea: Decapoda) from the Pacific sector of the Southern Ocean: USNS Eltanin Cruises 10, 11, 14–16, 19–21, 24 and 25. In: Korniker LS (ed) Biology of Antartic Seas XVII. Antarctic Res Ser 44:29–68

Wehrtmann I, Carvacho A (1997) New records and distribution ranges of shrimps (Crustacea: Decapoda: Penaeoidea and Caridea) in Chilean waters. Proc Biol Soc Wash 110(1):49–57

Wicksten MK (1983) *Plesionika sanctaecatalinae*: a new species of deep-sea shrimp from the Eastern Pacific (Caridea: Pandalidae). Bull South Calif Acad Sci 82(3):138–143

Chapter 13
Deep-Sea Lobsters (Polychelidae and Nephropidae) from the Continental Slope of the Southern Gulf of Mexico: Distribution and Morphometric Relationships

P. Briones-Fourzán, E. Lozano-Álvarez, A. R. Vázquez-Bader, and A. Gracia

Abstract Information on biodiversity from the continental slope of the southern half of the Gulf of Mexico is scarce. Deep-sea lobsters were collected from depths of 300–1090 m (upper continental slope) in all sectors of the southern Gulf of Mexico during several research cruises aimed to survey the benthic biodiversity from this slope. Individuals were sexed, and their carapace length (CL, mm), total length (TL, mm), and weight (W, g) were measured. In all, 3343 lobsters were collected from nine species, four of the family Polychelidae (from more to less abundant: *Stereomastis sculpta*, *Polycheles perarmatus*, *P. typhlops*, *Cardus crucifer*) and five of the family Nephropidae (*Nephropsis aculeata*, *N. rosea*, *Acanthacaris caeca*, *N. neglecta*, *Thaumastocheles zaleucus*). Multivariate analyses revealed differences in the lobster assemblage among 200-m depth strata, mostly due to the depth distribution of the two most abundant species: *N. aculeata* between 300 and 500 m and *S. sculpta* below 700 m. Within families, the median size varied significantly with species. In species with sufficient specimens, length-length and length-weight relationships were compared between sexes and tested for departure from isometry. This information is useful for comparing life history traits of individual species between regions and the relative condition of local populations within metapopulations. Deep-sea fisheries are currently nonexistent in the southern Gulf of Mexico and may be particularly difficult to develop over its generally rugged

P. Briones-Fourzán (✉) · E. Lozano-Álvarez
Universidad Nacional Autónoma de México, Instituto de Ciencias del Mar y Limnología, Unidad Académica de Sistemas Arrecifales, Puerto Morelos, Quintana Roo, Mexico
e-mail: briones@cmarl.unam.mx

A. R. Vázquez-Bader · A. Gracia
Universidad Nacional Autónoma de México, Instituto de Ciencias del Mar y Limnología, Unidad Académica de Biodiversidad y Ecología Marina, Ciudad de México, Mexico

M. E. Hendrickx (ed.), *Deep-Sea Pycnogonids and Crustaceans of the Americas*, https://doi.org/10.1007/978-3-030-58410-8_13

continental slope. Therefore, this area may be conserved as a spatial refugium for local populations of these deep-sea lobsters.

Keywords Allometry · Distribution · Megacrustaceans · Morphometric relationships · Nephropidae · Polychelidae · Southern Gulf of Mexico

13.1 Introduction

Lobsters are a diverse group of decapod crustaceans that have representatives in four infraorders of the suborder Pleocyemata, two of which contain exclusively deep-water species (Glypheidea, Polychelida), and the other two contain both shallow and deep-water species (Astacidea, Achelata) (Chan 2010). There is plenty of biological information on shallow-water lobster species, many of which are economically important, but information on deep-water species is more scant. In some areas, the small number of scientific cruises studying the deep-sea megafauna and the small samples usually obtained in those cruises limit the amount of knowledge of the population biology and ecology of the local species (Abelló and Cartes 1992; Cabiddu et al. 2008). One such area is the continental slope of the Mexican portion of the Gulf of Mexico.

Wicksten and Packard (2005) divided the upper continental slope (200–1500 m in depth) of the entire Gulf of Mexico into seven polygons based on bottom topography and examined the diversity of deep-water decapod crustaceans in each polygon. However, the continental slope of northern Yucatan was not included in any polygon because data on decapod crustaceans from this steep escarpment were not available at the time (Wicksten and Packard 2005, p. 1748). More recently, Lozano-Álvarez et al. (2007), Escobar-Briones et al. (2008), Briones-Fourzán et al. (2010), Gracia et al. (2010), and Vázquez-Bader and Gracia (2013, 2016) have reported on the composition and abundance of several groups of benthic megacrustaceans from the upper continental slope of the southern Gulf of Mexico.

Deep-sea lobsters in the Gulf of Mexico include several species in the families Polychelidae (Polychelida) and Nephropidae (Astacidea). Polychelids are known as deep-sea blind lobsters because all extant forms live in deep water and have strongly reduced eyes (Ahyong 2009). Polychelids are distinguished among reptant decapods by the possession of chelae on pereopods 1–4 and sometimes pereopod 5. In nephropids, pereopods 1–3 are chelate; pereopod 4 is never chelate, and pereopod 5 is chelate only in the genera *Thaumastocheles* and *Thausmastochelopsis* (Wahle et al. 2012). Previous studies on polychelids and nephropids of the Gulf of Mexico have been mostly taxonomic, with a few including some ecological notes (e.g., Firth Jr and Pequegnat 1971; Holthuis 1974), but there is little information on size distribution, morphometric relationships, and allometry in any one species. This information, in particular length-weight relationships, is useful for comparing life history traits of individual species between regions and the relative condition of local populations within metapopulations (Färber-Lorda 1994; Anger and Moreira 1998; Hendrickx 2003a; Lozano-Álvarez et al. 2007). Morphometric relationships also allow testing the relationship between two variables and predicting the value of one (e.g., weight) from the other (e.g., length) (Robinson et al. 2010), whereas allometry is useful to examine how one variable scales with another (Warton et al., 2006; Martínez-Calderón et al. 2018). In the present study, we examined the horizontal and vertical distribution, morphometric relationships, and allometry of deep-sea lobsters of the families Nephropidae (Astacidea) and Polychelidae (Polychelida) from the upper slope of the southern half of the Gulf of Mexico.

13.2 Materials and Methods

13.2.1 Study Area

The upper continental slope (300–1100 m in depth) of the Mexican (Southern) portion of the Gulf of Mexico encompasses, from northwest to southeast, the physiographic provinces known as the Mexican Slope and Mexican Ridges (along the state of Tamaulipas and Veracruz), the Tabasco-Campeche Knolls (from southern Veracruz to Tabasco), the Campeche Escarpment, and the East Campeche Slope (around the Yucatan peninsula) (Fig. 13.1) (Bergantino 1971). The latter two provinces form a steep escarpment with a highly complex topography, particularly to the west and north of the peninsula (Uchupi 1975; Wicksten and Packard 2005). North of the Mexico-USA border, terrigenous sediments dominate due to the influence of the Río Grande, Mississippi, and other rivers, but south of the border, the carbonate content of the sediment gradually increases. The Yucatan–Campeche shelf is dominated by carbonates, many of which are derived from reefs (Balsam and Beeson 2003). More recently, the Gulf of Mexico has been divided for practical reasons into four quadrants of approximately equal areas, with the 90°W meridian dividing the Gulf into western and eastern halves, and the 25°N parallel dividing the Gulf into northern and southern halves. Each quadrant was further divided into two sectors, resulting in eight sectors that could be used to report species with detailed distribution within the Gulf (Felder and Camp 2009; Ellis et al. 2011). The Mexican portion of the Gulf of Mexico, from the Mexico-USA border to the Mexican Caribbean Sea, is contained in five of these sectors (WNW, WSW, SSW, SSE, ESE) (Fig. 13.1).

13.2.2 Lobster Collection

The deep-sea lobsters examined in this study were collected during a series of 21 research cruises of the R/V *Justo Sierra* (National Autonomous University of Mexico) aimed to survey the benthic communities from the upper continental slope

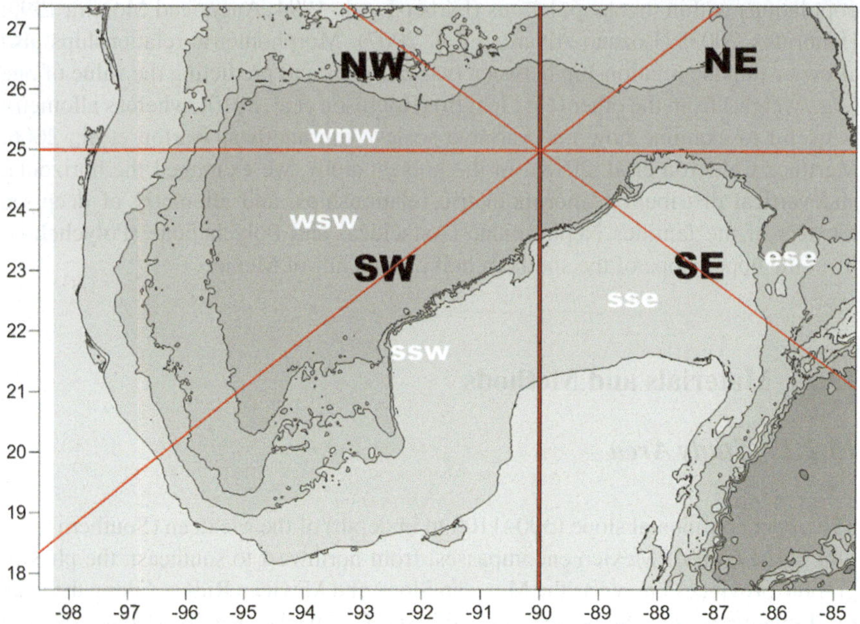

Fig. 13.1 Map of the Gulf of Mexico, showing the eight sectors into which it was divided for practical reasons (see Felder and Camp 2009)

(300–1100 m in depth) of the Mexican (Southern) Gulf of Mexico. The research cruises were BATO (May 1999), BIOREPES (August 2005), BIOREPES 2 (May–June 2007), BIOREPES 3 (November 2008), COBERPES (August 2009), SIGSBEE 9 (August 2005), COBERPES 2011 (April 2011), COBERPES 3 (November 2011), COBERPES 4 (August 2012), COBERPES 5 (May 2012), COBERPES 6 (August 2014), COBERPES 7 (April 2016), COBERPES 8 (October 2016), COBERPES 9 (July–August 2017), SOGOM (June 2015), SOGOM 2 (September 2016), SOGOM 3 (May 2017), and SOGOM 4 (September 2018). A few lobsters were also collected during three previous cruises [PROIBE-IV (October 1985), ARCOMM (August 1986), and TUPICC-III (August 1989)] in which only a few stations deeper than 200 m were sampled. Most specimens were caught with bottom trawl nets (18 m mouth aperture, 4.5 cm stretched mesh, 1.5 cm stretched mesh cod-end). Each tow lasted 30 min at a speed of 2.5–3.0 knots (see Lozano-Álvarez et al. 2007, Gracia et al. 2010). A few specimens were collected in baited traps or with a benthic skimmer (see Barradas-Ortiz et al. 2003).

Nephropids were identified following Holthuis (1974, 1991) and polychelids following Galil (2000) and Ahyong (2009). Individuals were sexed by observation of dimorphic characters. Carapace length (CL, mm) was measured as the shortest distance from the rear margin of the eye orbit to the median posterior edge of the carapace and total body length (TL, mm) from the rear margin of the eye orbit to the distal edge of the telson. All linear measurements were taken to the nearest 0.1 mm with a Vernier caliper. Total weight (W) was measured to the nearest 0.1 g after blotting excess moisture.

13.2.3 Statistical Analyses

13.2.3.1 Horizontal and Bathymetric Distribution

The distribution of deep-sea lobster species within the Gulf of Mexico was examined based on the sectors proposed by Felder and Camp (2009), as well as on the physiographic provinces of Bergantino (1971). The bathymetric distribution was compared among species within families with a Kruskal-Wallis nonparametric ANOVA (Zar 1999), based on the depth at which each specimen was caught.

Multivariate analyses were used to examine the composition of the deep-sea lobster assemblage among sampling stations grouped by Gulf sector, by depth strata (separated by 100 m or 200 m), and by a combination of sector × depth stratum. Data from the three cruises conducted in the 1980s were not included in this analysis. Differences in the taxonomic composition of samples thus categorized were analyzed with nonmetric multidimensional scaling (nMDS) on fourth-root transformed abundance data, using the Bray-Curtis similarity measure (Clarke 1993). For each analysis, the statistical significance of the observed differences in lobster assemblages was further tested with a one-way analysis of similarity (ANOSIM), which provides an R-value indicative of the degree of difference between samples. R values close to 0 are indicative of little difference, whereas values close to 1 are indicative of a large difference in sample composition (Clarke and Warwick 2001). Finally, we did a similarity percentage analysis (SIMPER, Clarke 1993) to identify those species responsible for the observed differences in assemblage composition between group categories. The multivariate analyses were done with the software PRIMER v.6 (Clarke and Gorley 2006).

13.2.3.2 Morphometric Relationships and Allometry

CL was compared among species and sexes within each family with a Kruskal-Wallis nonparametric ANOVA, followed by a pairwise median comparison test (Zar 1999). Size distribution plots were constructed, and morphometric relationships and tests for allometry were performed only for those species with sufficiently large sample sizes. Ordinary least squares regression (OLR) was used to examine morphometric relationships (Warton et al. 2006). In lobsters, CL can be measured with less error than TL because the carapace is rigid, whereas the abdomen is flexible; therefore, TL and W were regressed against CL. However, W was also regressed against TL to facilitate comparisons with other works. Individuals that were damaged upon collection or lacked any of the measurements were not used to derive morphometric equations. For each relationship, the slopes of regressions were compared between sexes with Student's t-tests; if the slopes did not differ significantly, the elevations were compared (Zar 1999). Statistical results were considered significant if $p < 0.05$.

We used the log-transformed data of all dimensions to test for allometry, which involves testing if the slope equals a specific value (i.e., isometry: $b = 1$ for length-length relationships and $b = 3$ for length-weight relationships) (Hartnoll 1982). The appropriate method to estimate slopes for this purpose is the reduced major axis regression (RMA, also known as standardized major axis regression) (Warton et al. 2006). Slopes from these regressions were tested for departures from isometry based on their confidence intervals, i.e., if the confidence interval of the slope covered the hypothesized parameter value ($b = 1$ or $b = 3$, depending on the type of relationship), it was indicative of isometry. If the entire confidence interval was below the hypothesized parameter value, it was indicative of negative allometry, and if it was entirely above the hypothesized parameter value, it was indicative of positive allometry. All regressions were done with PAST v.3.26b (Hammer et al. 2001).

13.2.3.3 Relationships Between Lobster Size and Depth

In some studies (e.g., Firth Jr and Pequegnat 1971; Abelló and Cartes 1992), it was suggested that reproductively mature polychelid females perform upslope migrations to release their eggs, but in others (e.g., Wenner 1979), no evidence of this type of migration was found. Furthermore, Sardà and Cartes (1993) found that in some deep-sea crustaceans (including *P. typhlops*) of the western Mediterranean, size of individuals tended to decrease with increasing depths, whereas in others, it tended to increase, and in others, it showed no trend. Therefore, we used OLRs to test for a potential relationship between depth and size (CL) of lobsters of each species by sex.

13.3 Results

In total, 3343 deep-sea lobsters were caught in 318 sampling stations during the 21 cruises. Of the total catch, 2034 lobsters (60.8%) were Polychelidae, and 1309 (39.2%) were Nephropidae. Within Polychelidae, there were four species in three genera. The most abundant polychelid was *Stereomastis sculpta* (Smith, 1880) with 1265 individuals (62.2% of the total catch of polychelids), followed by *Polycheles perarmatus* Holthuis, 1952 with 614 individuals (30.2%), *Polycheles typhlops* Heller, 1862 with 150 individuals (7.4%), and *Cardus crucifer* (Thomson, 1873) with only five individuals (0.25%) (Fig. 13.2). Within the Nephropidae, there were five species in three genera: *Nephropsis aculeata* Smith, 1881, which was the most abundant with 877 individuals (67% of the total catch of nephropids), followed by *Nephropsis rosea* Bate, 1888 with 337 individuals (10.1%), *Acanthacaris caeca* (A. Milne-Edwards, 1881) with 87 individuals (2.6%), *Nephropsis neglecta*

Fig. 13.2 Species of Polychelidae collected in the southern Gulf of Mexico. (**a**) *Stereomastis sculpta*, (**b**) *Cardus crucifer*, (**c**) *Polycheles typhlops*, and (**d**) *P. perarmatus*. (**a**), (**b**), and (**c**) are fresh specimens; scales are in centimeters. (**d**) is fixed specimen

Holthuis, 1974 with six individuals (0.18%) (Fig. 13.3), and *Thaumastocheles zaleucus* (Thomson, 1873) with only two individuals (0.15%).

13.3.1 Horizontal and Bathymetric Distribution

Within the Polychelidae, *S. sculpta* and *P. typhlops* were broadly distributed, occurring in all five sectors of the Mexican portion of the Gulf of Mexico (Fig. 13.4). *P. perarmatus* occurred in the WNW, WSW, and SSW sectors (corresponding to the Mexican Slope, Mexican Ridges, and Tabasco-Campeche Knolls provinces), but not in the SSE or ESE sectors (Campeche Escarpment and East Campeche Slope provinces). In contrast, *C. crucifer* only occurred in the SSE and ESE sectors (Fig. 13.4). Within the Nephropidae, *N. aculeata*, *N. rosea*, and *A. caeca* occurred in all five sectors, whereas *N. neglecta* was collected in the SSE and ESE sectors. Of the two specimens of the rare species *T. zaleucus*, one was collected in the SSE sector and another one in the SSW sectors, both on the Campeche Escarpment (Fig. 13.5).

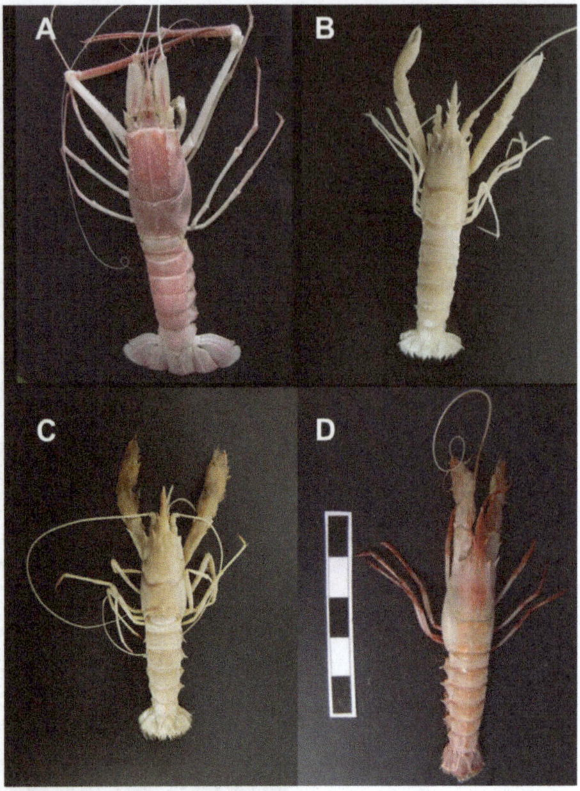

Fig. 13.3 Four of the five species of Nephropidae collected in the southern Gulf of Mexico. (**a**) *Acanthacaris caeca*, (**b**) *Nephropsis neglecta*, (**c**) *N. aculeata*, and (**d**) *N. rosea*. (**a**) and (**d**) are fresh specimens; scale in (**d**) is in centimeters. (**b**) and (**c**) are fixed specimens

A summary of the bathymetric distribution of each lobster species is provided (Table 13.1). In Polychelids, the depth distribution differed significantly among species (Kruskall-Wallis test, $H = 1388$, $df = 3$, $N = 2034$, $p < 0.0001$) (Fig. 13.6a), with the median depths at which these species occurred forming three groups. One group consisted of *P. perarmatus*, which occurred at the shallowest median depth; the second group consisted of *P. typhlops*, with an intermediate median depth, and the third group included *C. crucifer* and *S. sculpta*, with a deeper, but similar, median depth. The bathymetric distribution of Nephropids also differed significantly with species ($H = 731.2$, $df = 4$, $N = 1309$, $p < 0.0001$) (Fig. 13.6b), which formed three groups of medians as well. *N. aculeata* formed one group with the shallowest median depth, followed by *N rosea* and *A. caeca*, which formed a second group with similar intermediate depths. The third group encompassed *N. neglecta* and *T. zaleucus*, which appeared to have a similar bathymetric distribution, although the small number of individuals of both species renders this result uncertain.

Of all the multivariate analyses conducted on horizontal and bathymetric distribution of lobster assemblages grouped by Gulf sector, depth strata, or a combination of sector and strata, only the categorization by 200-m depth strata yielded significant results. The lobster assemblage differed significantly between the 300–499 m (shallow), 500–699 m (intermediate), and >700 m (deep) depth strata (Fig. 13.7),

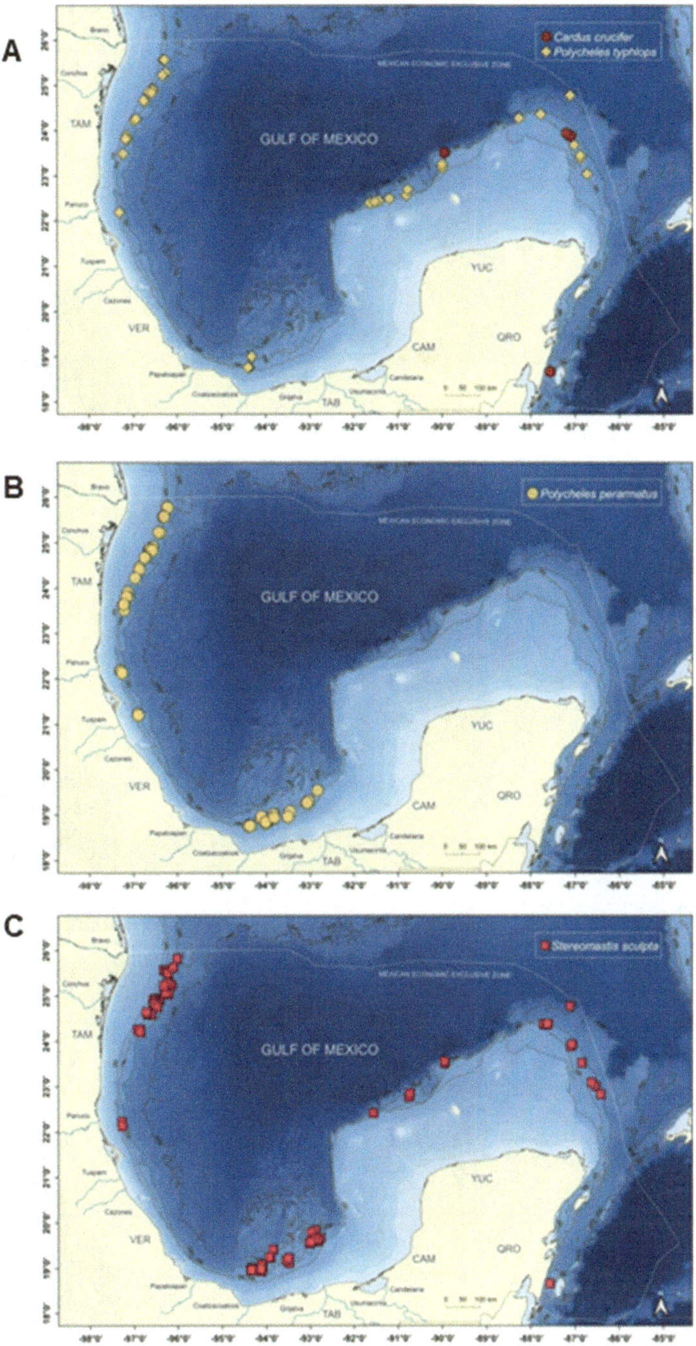

Fig. 13.4 Distribution of species of Polychelidae collected in the southern Gulf of Mexico. (**a**) *Polycheles typhlops* and *Cardus crucifer*, (**b**) *P. perarmatus*, (**c**) *Stereomastis sculpta*

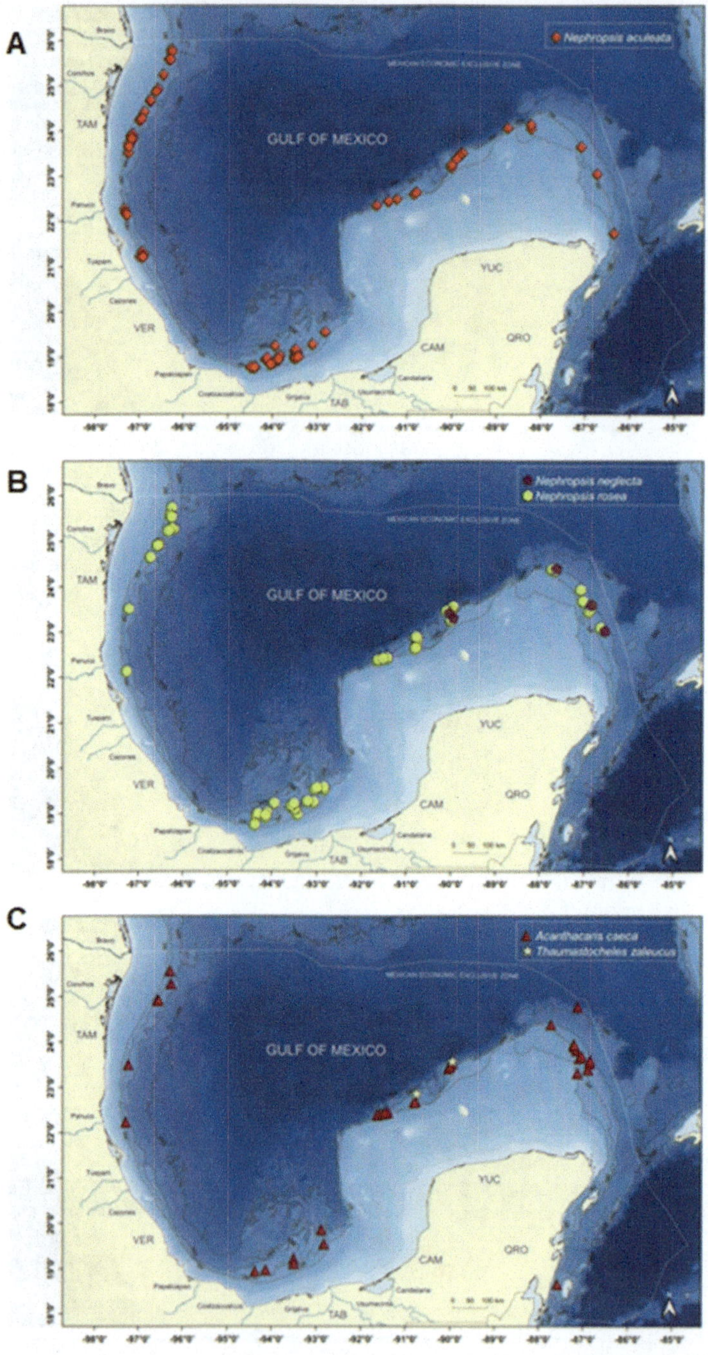

Fig. 13.5 Distribution of species of Nephropidae collected in the southern Gulf of Mexico. (**a**) *Nephropsis aculeata*, (**b**) *N. rosea* and *N. neglecta*, (**c**) *Acanthacaris caeca* and *Thaumastocheles zaleucus*

Table 13.1 Summary of depth (m) statistics for four species of Polychelidae (*Polycheles perarmatus*, *P. typhlops*, *Stereomastis sculpta*, and *Cardus crucifer*) and five species of Nephropidae (*Nephropsis aculeata*, *N. rosea*, *Acanthacaris caeca*, *N. neglecta*, and *Thaumastocheles zaleucus*) collected on the upper continental slope throughout the southern Gulf of Mexico. As for *T. zaleucus* only two individuals were caught, only the minimum and maximum depths are given

	Polychelidae				Nephropidae				
	P. perarm.	*P. typhlops*	*S. sculpta*	*C. crucifer*	*N. aculeata*	*N. rosea*	*A. caeca*	*N. neglecta*	*T. zaleucus*
N	614	149	1247	5	878	338	86	6	2
Min	308	392	401	585	305	314	498	539	828
Max	617	971	1148	953	761	971	1090	1024	971
Mean	437	599	768	739	445	644	621	831	
SD	49	77	112	140	88	101	85	167	
Median	409	616	765	755	442	615	609	826	
25 percentile	401	585	696	615	352	578	568	743	
75 percentile	489	647	813	854	522	703	647	976	

with virtually no overlap between the shallow and deep strata, but with much overlap between the intermediate stratum and the other two depth strata. This was confirmed by ANOSIM, which yielded an overall R = 0.517, but in pairwise comparisons, the R value was much higher (i.e., indicating greater dissimilarity) between the shallow and deep strata (R = 0.867), than between the shallow and intermediate strata (R = 0.275) or between the intermediate and deep strata (R = 0.396).

Results of SIMPER showed a higher average similarity among sampling stations at the deep stratum (64%) than among sampling stations at the shallow (53.5%) or intermediate strata (30%). A cumulative similarity of ~90% among sampling stations was accounted for by five species in the intermediate stratum (*N. rosea*, *N. aculeata*, *A. caeca*, *S. sculpta*, and *P. typhlops*) but only by two species in the shallow stratum (*N. aculeata* and *P. perarmatus*) and by a single species in the deep stratum (*S. sculpta*). Average dissimilarity was far greater between the shallow and deep strata (98.7%), driven mainly by *S. sculpta* and *N. aculeata*, than between the other pairwise comparisons of strata (shallow vs. intermediate: 76%; intermediate vs deep: 78%).

13.3.2 Size Distribution and Morphometric Relationships

A summary of CL statistics for each species by sex is provided (Table 13.2). Three polychelid species had sufficient data for morphometric analyses: *P. typhlops* (size range: 14.2–56.3 mm CL), *P. perarmatus* (14.8–66.2 mm CL), and *S. sculpta* (16.5–70.4 mm CL). In all cases, females reached larger sizes than males (Fig. 13.8). Sufficient data were also available for three nephropid species: *N. aculeata*

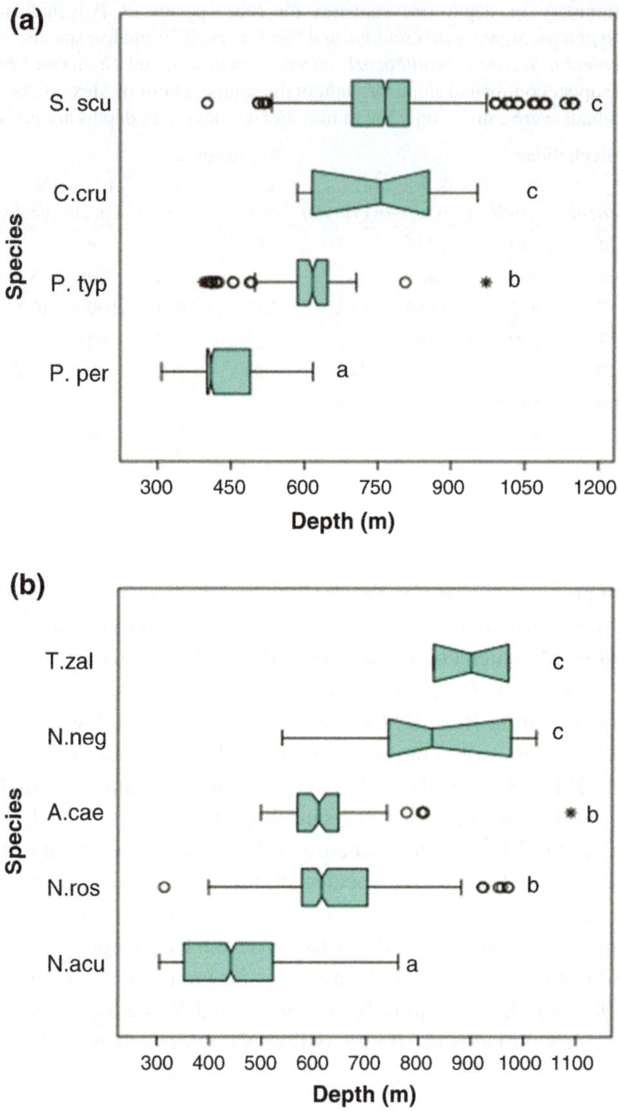

Fig. 13.6 Box plot of the bathymetric distribution of (**a**) polychelid [*Polycheles typhlops* (P. typ), *P. perarmatus* (P. per) and *Stereomastis sculpta* (S. scu)] and (**b**) nephropid species [*Nephropsis aculeata* (N. acu), *N. rosea* (N. ros) and *Acanthacaris caeca* (A. cae)]. The black line in each box is the median, the "notches" are the confidence interval for median, the boxes define the hinge (25% and 75% quartiles), and the line is 1.5 times the hinge. Outliers (points outside the interval) are represented as dots, and extreme outliers as asterisks. In each panel, different letters denote significantly different medians

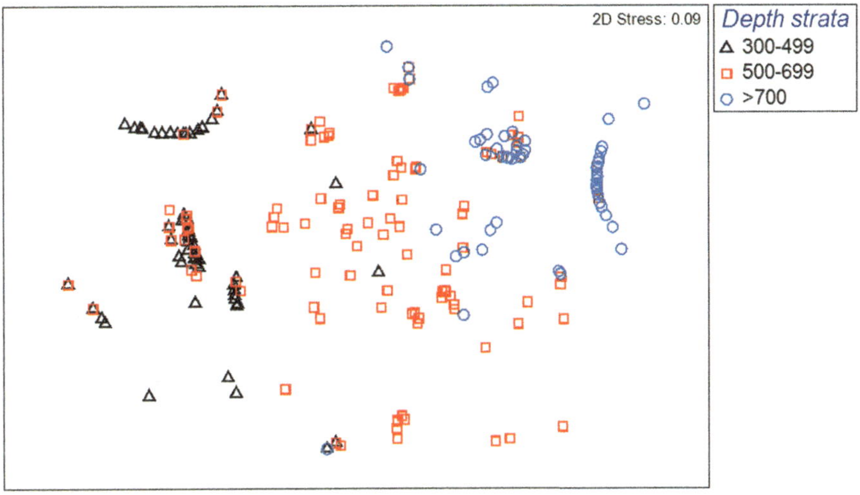

Fig. 13.7 Nonmetric multidimensional (nMDS) ordination of lobster assemblage structure in samples from three depth strata: 300–499 m (gray open triangles), 500–699 m (red open squares), and ≥700 m (blue open circles) throughout the southern Gulf of Mexico, based on species abundances. Each symbol denotes a sampling station

(8.9–42.2 mm CL), *N. rosea* (11.2–50.2 mm CL), and *A. caeca* (14.8–121.3 mm CL). In these species, males typically reached larger sizes than females (Fig. 13.9).

A comparison of the median size of males and females among the three species of polychelids revealed significant differences ($H = 369$, $df = 5$, $p < 0.0001$). Males of *P. typhlops* had the smallest median size, followed by a group conformed by females of *P. typhlops*, males of *P. perarmatus*, and males of *S. sculpta*. The largest median sizes corresponded to females of *P. perarmatus* and females of *S. sculpta* (Fig. 13.10a). In nephropids, the median size also varied significantly with species ($H = 196$, $df = 5$, $p < 0.0001$). The group with the smallest median sizes included males and females of *N. aculeata* and females of *N. rosea*. Males of *N. rosea* formed an intermediate size group, whereas males and females of *A. caeca* had the largest median sizes (Fig. 13.10b).

In polychelid species, there was variation in the similarity of slopes of morphometric relationships between males and females. In *P. typhlops*, the CL vs TL slope differed significantly with sex, but this was not the case for either the slopes or the intercepts of both length-weight relationships (Table 13.3). In *P. perarmatus*, by contrast, the slope of the Ln W vs Ln TL relationship varied with sex; the CL vs TL slope and intercept did not vary with sex, and the Ln W vs Ln CL slope did not vary with sex, but the intercept did. Finally, the slopes of all three relationships differed significantly between sexes in *S. sculpta* (Table 13.3).

Nephropids also showed variation in the similarity of slopes of morphometric relationships between males and females (Table 13.4). In *N. aculeata*, only the CL vs TL slopes differed significantly between sexes, but neither the slopes nor the intercepts of the two length-weight relationships did. In *N. rosea* and *A. caeca*, there

Table 13.2 Summary of size statistics (carapace length, mm) for females and males of three species of Polychelidae (*Polycheles perarmatus*, *P. typhlops*, and *Stereomastis sculpta*) and three species of Nephropidae (*Nephropsis aculeata*, *N. rosea*, and *Acanthacaris caeca*) collected from the upper continental slope throughout the southern Gulf of Mexico

	Polychelidae			Nephropidae		
	P. perarm.	*P. typhlops*	*S. sculpta*	*N. aculeata*	*N. rosea*	*A. caeca*
Females						
N	292	58	695	383	139	50
Min	14.8	17.5	16.5	8.9	11.2	15.4
Max	66.2	56.3	75.8	42.4	49.9	97.0
Mean	46.9	36.2	46.3	25.4	26.8	57.7
SD	10.2	10.8	11.9	6.1	8.6	23.0
Median	47.7	36.1	48.1	25.0	25.6	58.7
25 percentile	41.1	27.1	36.4	21.6	20.3	36.3
75 percentile	54.8	45.6	55.5	29.6	33.6	78.4
Males						
N	320	86	517	477	197	33
Min	17.7	14.2	17.7	10.3	12.6	14.8
Max	53.9	46.6	60.4	36.9	50.2	121.3
Mean	40.5	31.0	37.7	26.0	30.8	60.3
SD	6.3	9.5	7.0	5.1	8.5	30.2
Median	40.7	31.7	38.6	25.9	31.0	57.4
25 percentile	37.0	22.0	33.6	22.4	24.7	38.7
75 percentile	47.1	39.6	42.4	30.0	36.7	84.3

was no effect of sex on either the slopes or intercepts of the CL vs TL and Ln CL vs Ln W relationships, but the Ln TL vs Ln W slopes differed significantly between males and females (Table 13.4).

13.3.3 Tests of Allometry

Tests of allometry for both length-weight relationships revealed a positive allometry in both sexes of all three polychelid species (Table 13.5), but there was more intra- and interspecific variation in the Ln CL vs. Ln TL relationship. In *P. typhlops*, males showed positive allometry in this relationship, which did not depart from isometry in females. Both sexes of *P. perarmatus* showed isometry in this relationship, whereas in *S. sculpta*, females showed negative allometry, and males showed positive allometry (Table 13.5).

Allometry in the three relationships varied broadly among nephropid species (Table 13.6). Both sexes of *N. aculeata* showed positive allometry in the Ln CL vs Ln TL and the Ln CL vs Ln W relationships but isometry in the slopes of the Ln TL vs Ln W relationships. In *N. rosea*, both sexes showed isometry in the Ln CL vs Ln TL relationship and positive allometry in the Ln CL vs Ln W relationship, whereas

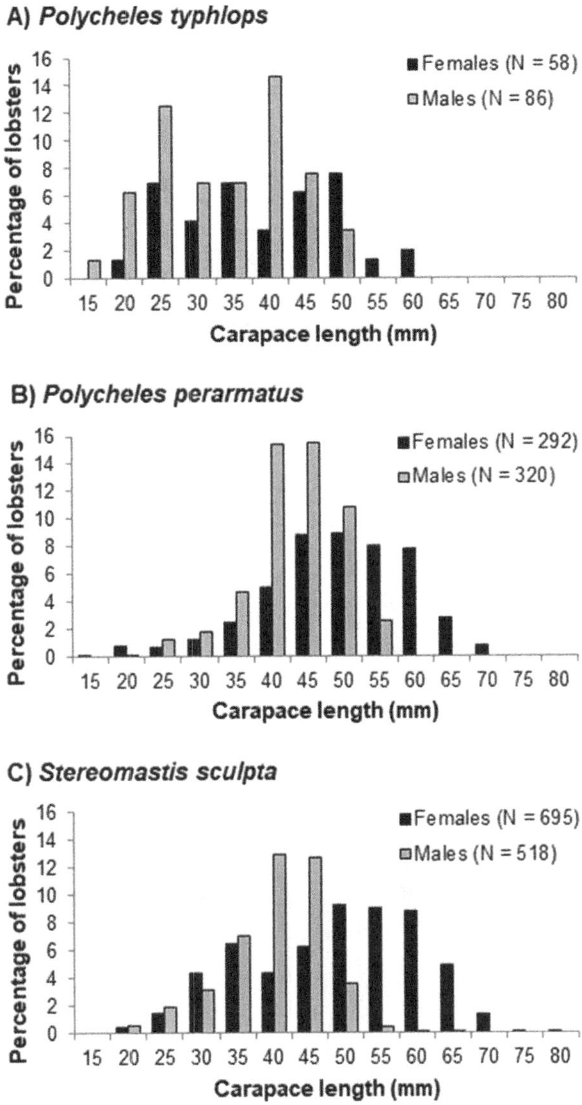

Fig. 13.8 Size distribution of females (black columns) and males (gray columns) of (**a**) *Polycheles typhlops*, (**b**) *P. perarmatus*, and (**c**) *Stereomastis sculpta*

females showed isometry and males positive allometry in the Ln TL vs Ln W relationship. Finally, males and females of *A. caeca* showed negative allometry in the Ln CL vs Ln TL relationship, isometry in the Ln CL vs Ln W relationship, and positive allometry in the Ln TL vs Ln W relationship (Table 13.6).

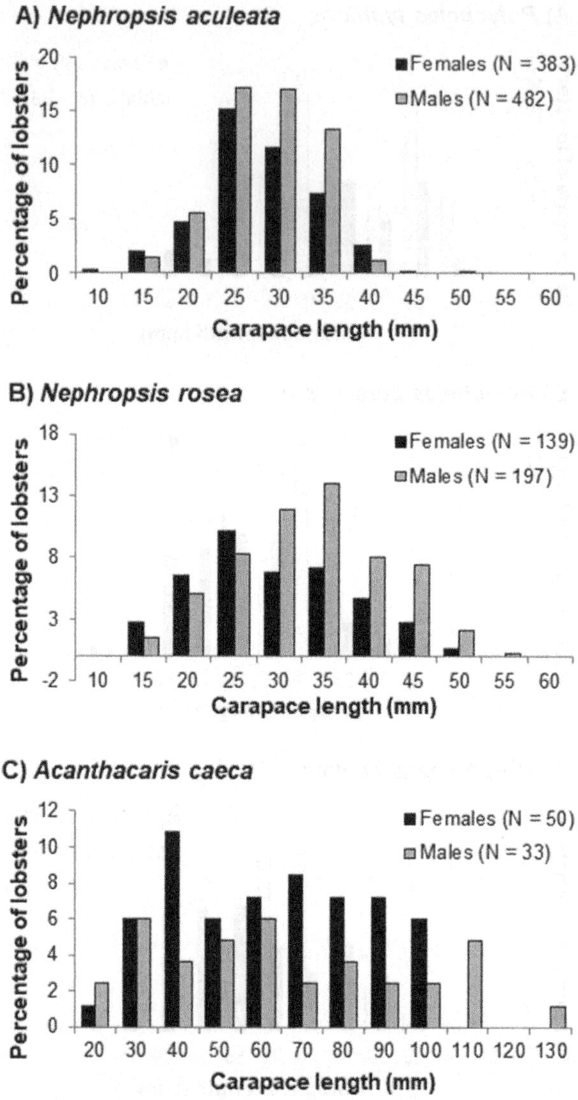

Fig. 13.9 Size distribution of females (black columns) and males (gray columns) of (**a**) *Nephropsis aculeata*, (**b**) *N. rosea*, and (**c**) *Acanthacaris caeca*. Note the different X-axis in (**c**)

13.3.4 Relationships Between Lobster Size and Depth

Within the polychelids, size of both males and females significantly tended to decrease with increasing depth in *P. typhlops* and *P. perarmatus* (Table 13.7). This was also the case for females of *S. sculpta*, but not for males, in which there was no relationship between size of individuals and depth. In the nephropids, size of

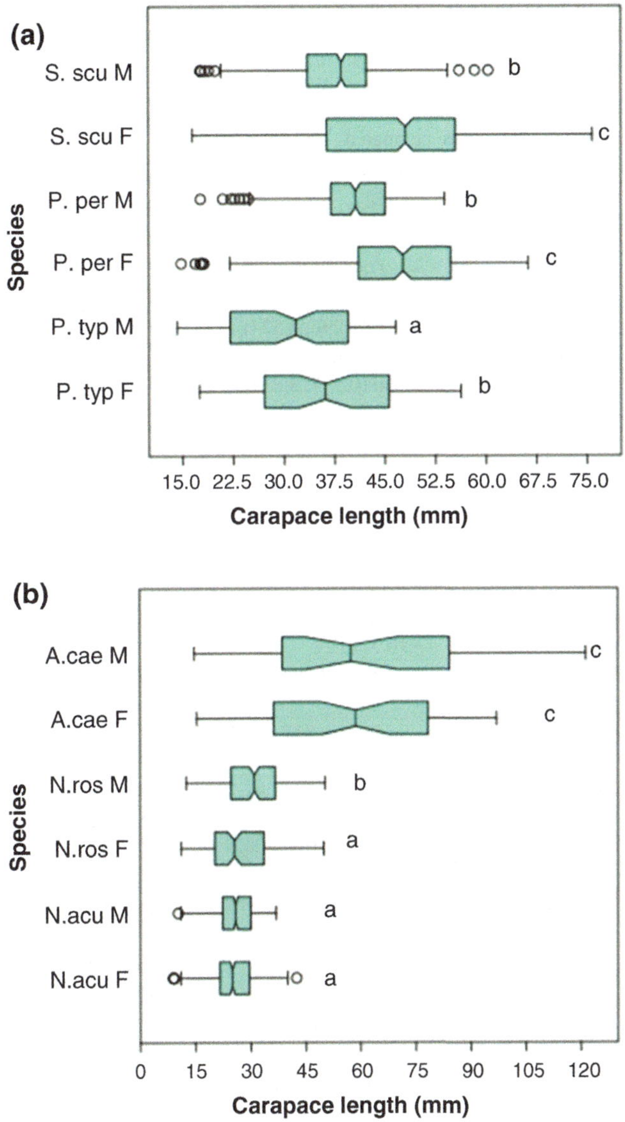

Fig. 13.10 Box plot of size distribution of males (M) and females (F) of (**a**) *Polycheles typhlops* (P. typ), *P. perarmatus* (P. per) and *Stereomastis sculpta* (S. scu), and (**b**) *Nephropsis aculeata* (N. acu), *N. rosea* (N. ros) and *Acanthacaris caeca* (A. cae). The black line in each box is the median, the "notches" are the confidence interval for median, the boxes define the hinge (25% and 75% quartiles), and the line is 1.5 times the hinge. Outliers (points outside the interval) are represented as dots. In each panel, different letters denote significantly different medians

Table 13.3 Parameters (with 95% confidence intervals, CI) of ordinary least squares regressions between body dimensions, with values of dependent (Y) and independent (X) variables for females (F) and males (M) of three polychelid species: *Polycheles typhlops*, *P. perarmatus*, and *Stereomastis sculpta*, from the southern Gulf of Mexico. Within each species, regressions were computed separately for each sex, and the slopes statistically compared between sexes; when slopes did not differ significantly, intercepts were compared

Y	X	Species	Sex	Intercept *a* (95% CI)	Slope *b* (95% CI)	N	r^2
TL	CL	*P. typhlops*	F	−0.228 (−1.488, 0.966)	2.316 (2.277, 2.352)*	58	0.997
			M	−1.978 (−3.204, −0.729)	2.398 (2.356, 2.439)*	86	0.995
		P. perarmatus	F	3.016 (1.296, 4.524)	2.409 (2.376, 2.444)	292	0.988
			M	2.260 (0.597, 3.817)	2.420 (2.381, 2.460)	320	0.982
		S. sculpta	F	3.963 (2.751, 5.033)	2.157 (2.132, 2.186)*	695	0.986
			M	−0.428 (−1.417, 0.555)	2.297 (2.270, 2.324)*	517	0.990
Ln W	Ln CL	*P. typhlops*	F	−10.283 (−11.381, −9.306)	3.488 (3.217, 3.795)	58	0.930
			M	−11.052 (−11.854, −10.322)	3.731 (3.524, 3.959)	86	0.948
		P. perarmatus	F	−8.539 (−8.924, −8.214)	3.079 (2.994, 3.181)	292	0.914
			M	−9.004 (−9.827, −8.204)	3.219 (3.005, 3.440)	320	0.860
		S. sculpta	F	−8.653 (−9.046, −8.274)	2.981 (2.879, 3.085)*	695	0.845
			M	−10.001 (−10.603, −9.382)	3.379 (3.212, 3.544)*	517	0.827
Ln W	Ln TL	*P. typhlops*	F	−13.022 (−14.237, −11.925)	3.448 (3.197, 3.720)	58	0.935
			M	−13.661 (−14.552, −12.825)	3.603 (3.411, 3.806)	86	0.953
		P. perarmatus	F	−11.392 (−11.803, −10.933)	3.093 (2.996, 3.181)*	292	0.926
			M	−12.160 (−13.066, −11.302)	3.270 (3.085, 3.466)*	320	0.884
		S. sculpta	F	−11.459 (−11.922, −11.002)	3.066 (2.967, 3.168)*	695	0.854
			M	−12.498 (−13.156, −11.799)	3.313 (3.158, 3.459)*	517	0.834

Note: CL: carapace length, TL: total length, W: body weight, *significant ($\alpha = 0.05$)

Table 13.4 Parameters (with 95% confidence intervals, CI) of ordinary least squares regressions between body dimensions, with values of dependent (Y) and independent (X) variables for females (F) and males (M) of three nephropid species: *Nephropsis aculeata*, *N. rosea*, and *Acanthacaris caeca*, from the southern Gulf of Mexico. Within each species, regressions were computed separately for each sex, and the slopes statistically compared between sexes; when slopes did not differ significantly, intercepts were compared

Y	X	Species	Sex	Intercept *a* (95% CI)	Slope *b* (95% CI)	N	r^2
TL	CL	*N. aculeata*	F	−0.806 (−1.842, 0.191)	2.878 (2.840, 2.916)*	383	0.980
			M	−3.118 (−4.656, −1.683)	2.994 (2.931, 3.055)*	477	0.958
		N. rosea	F	2.142 (−1.723, 6.477)	2.557 (2.372, 2.716)	139	0.879
			M	7.215 (3.656, 11.323)	2.441 (2.290, 2.574)	197	0.839
		A. caeca	F	12.133 (7.914, 16.231)	2.542 (2.460, 2.623)	50	0.992
			M	9.430 (5.486, 13.501)	2.589 (2.503, 2.669)	33	0.996
Ln W	Ln CL	*N. aculeata*	F	−7.818 (−8.102, −7.530)	3.047 (2.960, 3.132)	383	0.949
			M	−8.000 (−8.498, −7.486)	3.122 (2.964, 3.274)	477	0.920
		N. rosea	F	−8.019 (−8.415, −7.590)	3.061 (2.926, 3.184)	139	0.927
			M	−8.066 (−8.397, −7.696)	3.104 (2.996, 3.203)	197	0.928
		A. caeca	F	−7.745 (−8.428, −6.967)	3.008 (2.816, 3.174)	50	0.978
			M	−7.371 (−7.765, −7.039)	2.909 (2.822, 3.009)	33	0.994
Ln W	Ln TL	*N. aculeata*	F	−10.406 (−10.768, −10.022)	2.907 (2.818, 2.994)	383	0.943
			M	−10.255 (−10.984, -9.509)	2.882 (2.713, 3.052)	477	0.897
		N. rosea	F	−10.649 (−11.049, −10.225)	2.983 (2.887, 3.076)*	139	0.963
			M	−11.467 (−12.009, −11.009)	3.182 (3.080, 3.303)*	197	0.940
		A. caeca	F	−12.121 (−12.928, −11.404)	3.267 (3.129, 3.421)*	50	0.989
			M	−11.347 (−11.891, −10.781)	3.108 (2.994, 3.215)*	33	0.993

Note: CL: carapace length, TL: total length, W: body weight, *significant ($\alpha = 0.05$)

lobsters tended to decrease with increasing depth in *N. rosea*, but the inverse pattern occurred in *N. aculeata*, in which size of both males and females tended to increase with increasing depth. Finally, size showed no relationship with depth in females and males of *A. caeca*. In most cases where the relationship was significant, there was substantial dispersal of points, as indicated by the relatively low values of the correlation coefficients (Table 13.7).

Table 13.5 Parameters (with 95% confidence intervals, CI) of reduced major axis regressions between body dimensions and tests for allometry with Ln-transformed values of dependent (Y) and independent (X) variables for females (F) and males (M) of three polychelid species: *Polycheles typhlops*, *P. perarmatus*, and *Stereomastis sculpta*, from the southern Gulf of Mexico

Y	X	Species	Sex	Intercept *a* (95% CI)	Slope *b* (95% CI)	N	r^2	Allometry
Ln TL	Ln CL	*P. typhlops*	F	0.786 (0.722, 0.840)	1.014 (0.999, 1.032)	58	0.997	Isometry
			M	0.716 (0.636, 0.799)	1.038 (1.014, 1.061)	86	0.995	Positive
		P. perarmatus	F	0.897 (0.817, 0.989)	1.002 (0.978, 1.023)	292	0.990	Isometry
			M	0.914 (0.851, 0.977)	0.998 (0.981, 1.015)	320	0.984	Isometry
		S. sculpta	F	0.896 (0.866, 0.926)	0.977 (0.969, 0.985)	695	0.991	Negative
			M	0.738 (0.689, 0.792)	1.024 (1.010, 1.038)	517	0.991	Positive
Ln W	Ln CL	*P. typhlops*	F	−10.737 (−11.765, −9.782)	3.616 (3.361, 3.899)	58	0.930	Positive
			M	−11.393 (−12.173, −10.666)	3.831 (3.627, 4.052)	86	0.948	Positive
		P. perarmatus	F	−9.083 (−9.387, −8.741)	3.222 (3.132, 3.301)	292	0.914	Positive
			M	−9.934 (−10.78, −9.120)	3.471 (3.254, 3.696)	320	0.860	Positive
		S. sculpta	F	−9.648 (−10.027, −9.280)	3.243 (3.144, 3.341)	695	0.845	Positive
			M	−11.220 (−11.835, −10.596)	3.717 (3.547, 3.884)	517	0.827	Positive
Ln W	Ln TL	*P. typhlops*	F	−13.541 (−14.597, −12.458)	3.566 (3.323, 3.807)	58	0.935	Positive
			M	−14.036 (−14.878, −13.195)	3.691 (3.501, 3.881)	86	0.953	Positive
		P. perarmatus	F	−11.966 (−12.396, −11.395)	3.214 (3.093, 3.304)	292	0.926	Positive
			M	−13.114 (−14.050, −12.236)	3.478 (3.288, 3.680)	320	0.884	Positive
		S. sculpta	F	−12.620 (−13.093, −12.149)	3.318 (3.217, 3.424)	695	0.854	Positive
			M	−13.898 (−14.614, −13.156)	3.629 (3.465, 3.788)	517	0.834	Positive

Note: CL: carapace length, TL: total length, W: body weight

Table 13.6 Parameters (with 95% confidence intervals, CI) of reduced major axis regressions between body dimensions and tests for allometry with Ln-transformed values of dependent (Y) and independent (X) variables for females (F) and males (M) of three nephropid species: *Nephropsis aculeata*, *N. rosea*, and *Acanthacaris caeca*, from the southern Gulf of Mexico

Y	X	Species	Sex	Intercept *a* (95% CI)	Slope *b* (95% CI)	N	r^2	Allometry
Ln TL	Ln CL	*N. aculeata*	F	0.900 (0.849, 0.950)	1.045 (1.029, 1.061)	383	0.982	Positive
			M	0.825 (0.762, 0.890)	1.070 (1.050, 1.090)	477	0.969	Positive
		N. rosea	F	0.819 (0.659, 0.989)	1.045 (0.994, 1.095)	139	0.898	Isometry
			M	1.046 (0.910, 1.201)	0.982 (0.936, 1.021)	197	0.880	Isometry
		A. caeca	F	1.320 (1.227, 1.442)	0.926 (0.896, 0.949)	50	0.995	Negative
			M	1.281 (1.181, 1.382)	0.936 (0.911, 0.960)	33	0.997	Negative
Ln W	Ln CL	*N. aculeata*	F	−8.079 (−8.366, −7.786)	3.129 (3.039, 3.216)	383	0.949	Positive
			M	−8.432 (−8.973, −7.877)	3.256 (3.086, 3.422)	477	0.920	Positive
		N. rosea	F	−8.400 (−8.830, −7.979)	3.178 (3.049, 3.308)	139	0.927	Positive
			M	−8.467 (−8.807, −8.081)	3.222 (3.108, 3.325)	197	0.928	Positive
		A. caeca	F	−7.879 (−8.655, −6.981)	3.041 (2.821, 3.228)	50	0.978	Isometry
			M	−7.408 (−7.826, −7.069)	2.919 (2.828, 3.020)	33	0.994	Isometry
Ln W	Ln TL	*N. aculeata*	F	−10.774 (−11.183, −10.376)	2.994 (2.901, 3.089)	383	0.943	Isometry
			M	−10.943 (−11.661, −10.216)	3.043 (2.873, 3.208)	477	0.897	Isometry
		N. rosea	F	−10.888 (−11.277, −10.476)	3.040 (2.945, 3.131)	139	0.963	Isometry
			M	−11.901 (−12.346, −11.470)	3.282 (3.183, 3.382)	197	0.940	Positive
		A. caeca	F	−12.216 (−13.030, −11.447)	3.285 (3.138, 3.442)	50	0.989	Positive
			M	−11.404 (−11.941, −10.868)	3.119 (3.008, 3.226)	33	0.993	Positive

Note: CL: carapace length, TL: total length, W: body weight

Table 13.7 Parameters (with 95% confidence intervals, CI) of ordinary least squares regressions to test for a relationship between depth (m) (variable X) and carapace length (mm) (variable Y) for females (F) and males (M) of three polychelid species (*Polycheles typhlops*, *P. perarmatus*, and *Stereomastis sculpta*) and three nephropid species (*N. aculeata*, *N. rosea*, *Acanthacaris caeca*), from the southern Gulf of Mexico

Species	Sex	Intercept *a* (95% CI)	Slope *b* (95% CI)	N	r^2	p
P. typhlops	F	76.736 (56.480, 96.992)	−0.069 (−0.103, −0.035)	61	0.218	<0.001
	M	55.358 (40.476, 70.241)	−0.040 (−0.065, −0.016)	87	0.113	0.001
P. perarmatus	F	86.262 (76.847, 95.677)	−0.090 (−0.111, −0.069)	293	0.191	<0.001
	M	47.840 (41.595, 54.085)	−0.017 (−0.031, −0.003)	321	0.017	0.020
S. sculpta	F	56.359 (50.349, 62.370)	−0.014 (−0.021, −0.006)	710	0.016	<0.001
	M	37.586 (33.283, 41.890)	0 (−0.006, 0.005)	528	0	0.988
N. aculeata	F	15.133 (12.238, 18.028)	0.023 (0.017, 0.029)	384	0.117	<0.001
	M	13.470 (11.274, 15.666)	0.028 (0.024, 0.033)	486	0.214	<0.001
N. rosea	F	41.415 (31.973, 50.858)	−0.022 (−0.037, −0.008)	140	0.062	0.002
	M	43.066 (35.905, 50.228)	−0.019 (−0.030, −0.008)	198	0.057	<0.001
A. caeca	F	76.450 (27.128, 125.772)	−0.031 (−0.111, 0.050)	50	0.012	0.445
	M	81.084 (−3.505, 165.673)	−0.030 (−0.162, 0.112)	34	0.007	0.646

13.4 Discussion

The present work provides new information on deep-sea lobsters in the southern Gulf of Mexico, where knowledge on the local deep- sea ecosystems and biodiversity is limited, as well as on the individual species that inhabit these ecosystems (Felder and Camp 2009). Over much of this area, the marked steepness and highly complex topography of the continental slope, featuring many canyons, channels, and escarpments, make it difficult to sample the benthic fauna. This is particularly true for the Campeche Escarpment, off the Yucatan Peninsula (Briones-Fourzán et al. 2010; Gracia et al. 2010). Therefore, it is not surprising that information on deep-water megacrustaceans from this particular province (located in the SSW sector) is scarce (Wicksten and Packard 2005). In addition, as most deep-sea lobsters live in burrows or bury themselves in sediment (Holthuis 1974; Galil 2000; Ahyong 2009), areas with adequate habitats for these species (e.g., soft bottoms) are patchily distributed over the continental slope. This patchy distribution of soft bottoms was observed during the echosounder surveys conducted to find suitable areas for sampling throughout the several cruises.

The nine species of deep-sea lobsters collected in our cruises have a wide distribution. Along the Western Atlantic, they have all been reported from Florida and the Bahamas to Brazil (Firth Jr and Pequegnat 1971; Felder et al. 2009; Tavares and Young 2002; Dall'occo and Tavares 2004; Dall'occo et al. 2007; Silva et al. 2013; Bezerra and Ribeiro 2015; Alves et al. 2016). Within the Gulf of Mexico, *A. caeca*, *N. aculeata*, *N. rosea*, *P. typhlops*, and *S. sculpta* have been collected in all sectors. *C. crucifer* has also been collected in all sectors of the Gulf except the SSW. There are previous reports of *P. perarmatus* only from the northern half of the Gulf, of

T. zaleucus from the NNE and ESE sectors, and of *N. neglecta* exclusively from the ESE sector (Felder et al. 2009). Therefore, the present study extends the known distribution of *P. perarmatus* to the WSW and SSW Gulf sectors, of *T. zaleucus* to the SSE and SSW sectors, and of *N. neglecta* to the SSE sector.

Across the southern Gulf of Mexico, the lobster assemblage did not vary substantially among 100-m depth strata, Gulf sectors, or a combination of depth and sector but varied substantially among 200-m depth strata. This result mainly reflects the different bathymetric distribution of the two most abundant species: *N. aculeata* in the 300–499 m stratum and *S. sculpta* in the >700 m stratum. The OLS regressions that we obtained for morphometric relationships of the most abundant species of polychelids (*P. perarmatus*, *P. typhlops*, and *S. sculpta*) and nephropids (*N. aculeata*, *N. rosea*, and *A. caeca*) in our samples will allow estimating total length and weight from a broad range of CL values in future studies on these species.

13.4.1 Polychelid Lobsters

Despite their wide distribution, comparatively little is known about the biology and ecology of polychelid lobsters, except for *S. sculpta* and *P. typhlops*, which have been extensively studied in the Mediterranean Sea. There, both species reach similar sizes and appear to occupy a similar ecological niche as benthopelagic feeders, but *P. typhlops* is more abundant above 1200 m depth and *S. sculpta* below this depth (Cartes and Abelló 1992; Cartes and Carrassón 2004). Although Galil (2000) synonymized the genus *Stereomastis* with the genus *Polycheles*, Ahyong (2009) further reestablished both genera and provided a thorough analysis of their morphological differences. Both *S. sculpta* and *P. typhlops* abound in the northern Gulf of Mexico where, according to Firth Jr and Pequegnat (1971), the former is probably one of the most numerically important polychelids on the continental slope. This is consistent with our results, with *S. sculpta* being the most abundant polychelid on the upper slope of the southern Gulf of Mexico. Ahyong (2009) suggested that polychelids are ambush predators, striking from a buried position with the chelipeds folded against the lateral margins of the carapace. However, a study in the NW Atlantic using stable isotopes and fatty acid analyses found that the tissues of *S. sculpta* had highly negative values of $\delta^{13}C$ and high levels of EPA and DHA, which are known phytoplankton biomarkers, suggesting a closer link to the pelagic food web (Parzanini et al. 2018).

In areas of the Western North Atlantic where *P. typhlops* does not occur, *S. sculpta* was recorded from 486 to 2257 m and co-occurred with *S. nana* (Wenner 1979), which has not been recorded in the Gulf of Mexico (Felder et al. 2009). The latter species occurred from 1400 to 2599 m (Wenner 1979). In the Caribbean Sea, off French Guyana, Guéguen (2000) captured *P. typhlops* from 342 to 479 m and *S. sculpta* (reported as *P. sculptus*) from 426 to 855 m, whereas in our study area, the median depth distribution differed significantly between *P. typhlops* and *S. sculpta*, with the latter occurring at greater depths. Therefore, where these two species co-occur, *P. typhlops* tends to occur at shallower depths and *S. sculpta* at deeper depths but with some overlap. *P. perarmatus* is closely related to *P. typhlops* but differs from this species in several characteristics that include a rounded second pleuron and the presence of three carinae ventrally on the uropodal expopod (Galil 2000). In the present study, *P. perarmatus* was captured at shallower depths than the other polychelid species and exclusively west of the Yucatan Peninsula. The depth and distribution data of Polychelids strongly suggest that these lobsters are partitioning the habitat to avoid competition in an area that may be highly limited in food resources.

The monospecific genus *Cardus* was established by Galil (2000) to accommodate the species originally described as *Deidamia crucifer*. *Cardus* is the only polychelid in which the fifth pereopodal dactyl is simple (i.e., not chelate) in both sexes. Despite previous records of *C. crucifer* in most sectors of the Gulf of Mexico over a depth range of 549–2195 m (Galil 2000; Felder et al. 2009), this species was very scarce in our samples and was only caught in the SSE sector. Therefore, we could only examine size distribution and morphometric relationships in *P. perarmatus*, *P. typhlops*, and *S. sculpta*. Polychelid females attain larger sizes than males (Follesa et al. 2007; Gastoni et al. 2010), and this was the case for the three examined species in the present study. Female polychelids reproduce year-round, producing several thousands of small eggs (0.6–0.85 mm in diameter, Firth Jr and Pequegnat 1971, Wenner 1979) per spawn (Abelló and Cartes 1992). These features reflect the extended larval development of polychelids, which have a characteristic larva called eryoneicus, a type of zoea with an inflated carapace (Anger 2001). In *P. typhlops*, there are three zoea and several decapodid stages. In a study conducted in the Mediterranean Sea, the three zoea stages and one decapodid stage of *P. typhlops* were caught below the 200 m depth (Torres et al. 2014).

The size of *P. typhlops* and *P. perarmatus* decreased with increasing depth, as was also found for *P. typhlops* on the upper slope (350–750 m) of the Ionian Sea (Maiorano et al. 1998) and the deep slope (926–1824 m) of the Western Mediterranean Sea (Sardà and Cartes 1993). The latter authors, however, did not find any relationship between size of *S. sculpta* and depth, whereas we found a decrease of size with increasing depth for females, but not for males, in which the slope of the regression was virtually zero. Potential explanations for a decrease in size with increasing depth include a lower abundance of both food resources in the meio- and macrobenthos and potential predators of crustaceans with increasing depth, which could favor the survival of smaller crustaceans at greater depths (Thiel 1979; Sardà and Cartes 1993; Stefanescu et al. 1993).

The length-weight relationships exhibited positive allometry in both sexes of all three polychelid species examined, indicating that in individuals of these species, the increase in weight is disproportionally greater as length increases. Allometry in the CL vs TL relationship was more variable. Both males and females of *P. perarmatus* showed isometry in this relationship, as did females of *P. typhlops*, whereas males of *P. typhlops* and *S. sculpta* exhibited positive allometry, indicating a proportionally greater increase in abdomen (tail) length with increasing CL in males relative to females. In contrast, negative allometry in CL vs TL relationship occurred in females of *S. sculpta*, indicating a smaller increase in abdomen length with increasing CL relative to males.

13.4.2 Nephropid Lobsters

Three of the five species of deep-water nephropids collected in our cruises (*A. caeca*, *N. aculeata*, and *N. rosea*) occur across a broad latitudinal distribution in the Western Atlantic (Holthuis 1974, 1991; Silva et al. 2013). *N. rosea* was considered a synonym of *N. aculeata* until Manning (1969) demonstrated that it was a different species. We found *N. aculeata* between depths of 305 and 761 m and *N. rosea* between 314 and 971 m, with the median depths differing significantly between these two congeners. In French Guyana, *N. aculeata* was found from 321 to 491 m and *N. rosea* from 547 to 854 m (Guéguen 2000), and in Brazil, Dall'occo et al. (2007) reported *N. aculeata* from 268 to 576 m and *N. rosea* from 600 to 800 m. Therefore, similar to *P. typhlops* and *S. sculpta*, wherever *N. aculeata* and *N. rosea* co-occur, the latter tends to occur at deeper depths than the former but with some overlap. Compared with *N. aculeata* and *N. rosea*, *N. neglecta* is a rather small species (Holthuis 1974). *N. neglecta* was very scarce in our samples, probably because it appears to be distributed more toward the Caribbean and Brazil than in the Gulf of Mexico and at greater depths (Holthuis 1974, 1991). For example, it has been recorded around Brazil between 800 and 1300 m (Tavares and Young 2002; Alves et al. 2016). As previously suggested for Polychelidae, these data also suggest habitat partitioning among Neprophidae congeners to reduce competition in an area where food resources may be highly limited.

In several deep-sea nephropid species, males generally grow to larger sizes than females (Ivanov and Krylov 1980; Hendrickx 2003b; Dineshbabu 2008); however, in the present study, the size range of males and females was similar in *N. rosea* and *N. aculeata*. This was also the case for *N. occidentalis* off Western Mexico (Papiol et al. 2016). Nephropid lobsters have an abbreviated larval development, usually with only three zoeas and one decapodid (Goy 2014). Therefore, unlike in polychelids, the eggs of *N. aculeata* and *N. rosea* are relatively large (2–2.5 mm in diameter, Holthuis 1974), and females produce only a few hundred per spawn (Roe 1966; Reid and Corey 1991). Interestingly, the relationship between depth and size was negative in *N. rosea* but positive in *N. aculeata*. The causes for this pattern remain to be determined, but it could further reduce competition for food resources between these two congeners.

A. caeca is quite common on the continental slope of the Gulf of Mexico and Caribbean Sea (Holthuis 1974, 1991) and has also been reported off Brazil (Silva et al. 2013). Using submersible video surveys across the Miami Terrace and nearby areas (NW Atlantic), numerous burrows of *A. caeca* were observed in muddy environments, more often below ~600 m in depth. Some burrows were 10–20 cm deep with near-vertical walls (Correa et al. 2012), and some were described as short tunnels with a ramp-like entrance crater at one end and a narrower escape opening at the other (Messing et al. 2006). In our study area, *A. caeca* had a similar depth distribution as *N. rosea*. *A. caeca* reaches much larger sizes than the other deep-sea nephropids. However, although the size range was broader for males than for females, the median size did not differ with sex, and there was no apparent relationship between depth and size for either females or males.

Unlike *A. caeca*, *T. zaleucus* appears to be quite rare despite being the type species of its genus. These lobsters exhibit strongly dimorphic first chelipeds, with the right side extremely elongate and pectinate (Chang et al. 2014). Previous records of *T. zaleucus* amount to seven individuals, all of them females (Holthuis 1974, 1991; Chan and de Saint Laurent 1999), as were the two specimens collected in our study. To the extent of our knowledge, these two females, caught in the SSE and SSW Gulf sectors, constitute the first report in the Gulf of Mexico outside Polygon 2 of Wicksten and Packard (2005), which is located in the NNE sector. A similar situation occurred for *T. japonicus* from the Pacific, which was believed to be the only other species in the genus, until further expeditions produced more individuals of *T. japonicus* and two new species of the genus (Chang et al. 2014). Of the two females of *T. zaleucus* that we caught, one was complete, but her cephalothorax was separated from the abdomen, whereas only the cephalothorax of the other one was recovered. Similarly, three of the seven previously known specimens of *T. zaleucus* were "decapitated" (Holthuis 1974), and eight of 11 specimens assigned to *T. massonktenos* were represented only by a cephalothorax or a major cheliped (Chang et al. 2014). More recently, Poupin and Corbari (2016) reported on the capture by a dredge of only one chela of *T. zaleucus* off the Caribbean island of Guadeloupe. According to Holthuis (1974), this type of catches, obtained with trawling gear, suggests that individuals of *T. zaleucus* occupy holes on the bottom with their carapace protruding.

Results of tests of allometry for the three nephropid species with sufficiently large sample sizes were more variable than those for polychelids. In the CL vs TL relationship, allometry was positive in both sexes of *N. aculeata*, isometric in both sexes of *N. rosea*, and negative in both sexes of *A. caeca*, suggesting a proportionally greater increase in abdomen length with increasing CL in the former, and less increase in abdomen length with increasing CL in the latter.

13.4.3 Should Deep-Sea Lobsters Be Fished?

Many species of shallow-water lobsters sustain important fisheries, but this is not necessarily the case for deep-sea lobsters (Holthuis 1991). For example, polychelids have no commercial value despite some species attaining relatively large sizes (Holthuis 1991; Lavalli and Spanier 2010). In contrast, several deep-water nephropids, including some species of *Nephropsis* (generally called "lobsterettes"), are either exploited or considered as potential fishing resources in many parts of the world (Holthuis 1991; Hendrickx 2003b, Dineshbabu 2008), including the Gulf of Mexico (Roe 1966, Holthuis 1991). However, deep-sea fisheries are more susceptible to overexploitation because deep-water species generally have life history characteristics that result in less population resilience or productivity than shallow-water species (Norse et al. 2012). Moreover, there is growing concern about the negative impacts on biodiversity of fishing activities in the deep sea (Roberts 2002; Norse et al. 2012, Clark et al. 2016; da Ros et al. 2019), where many local species exist as metapopulations whose regional distribution depends on a balance among multiple-scale dynamics (Levin et al. 2001).

In addition, deep-sea ecosystems are more fragile than shallower ecosystems, especially in the context of ocean warming, expanding hypoxia, and acidification associated with climate change (Koslow et al. 2000; Levin et al. 2001; Levin and Dayton 2009). Trawling activities, in particular, can destroy biological structures that provide habitat for many species (Buhl-Mortensen et al. 2010; da Ros et al. 2019). For example, in Florida, *N. aculeata*, *N. rosea*, *P. typhlops*, and *S. sculpta* have been observed associated with deep coral ecosystems (Lutz and Ginsburg 2007), which are especially vulnerable to trawling activities (Maynou and Cartes 2011; Clark et al. 2016). Deep coral ecosystems have also been reported at 500–600 m of depth on the Campeche Escarpment (Hebbeln et al. 2014) and should be protected in the case of eventual fishery exploitation. Currently, deep-sea fisheries for lobsters or any other species are nonexistent in the southern (Mexican) portion of the Gulf of Mexico and may be particularly costly and difficult to develop over the rugged continental slope of much of this portion. Therefore, the potential development of a deep-sea lobster fishery should carefully consider the cost/benefit balance of exploitation in both economic and ecological terms, in order to reach a compromise between eventual utilization and the establishment of conservation areas as spatial refugia for these lobsters and other deep-sea species.

Acknowledgments We thank the officers and crew of the R/V *Justo Sierra* for their skill during sampling operations. We appreciate the invaluable technical support of Magaly Galván Palmerín, Hermelinda Trejo Rosas, Sandra Antonio Bueno, León F. González Morales, and Brenda Barbosa Nieto and Ingrid Antillón Zaragoza, Fernando Negrete-Soto, and Cecilia Barradas-Ortiz, during onboard and laboratory activities. We also thank the many graduate and undergraduate students that collaborated in the research cruises and processing of samples. Universidad Nacional Autónoma de México supported this study with institutional funds and research vessel time and also through research project UNAM-DGAPA-PAPIIT IN223109. The SOGOM cruises were developed with funds of project D1/CH2012-01 "Implementación de redes de observaciones oceanográficas (Físicas, Químicas, Ecológicas) para la generación de escenarios ante posibles contingencias relacionadas a la exploración y producción de hidrocarburos en aguas profundas del Golfo de México", provided by Fondo Sectorial CONACyT-SENER-Hidrocarburos, Consorcio CIGoM.

References

Abelló P, Cartes JE (1992) Population characteristics of the deep-sea lobsters *Polycheles typhlops* and *Stereomastis sculpta* (Decapoda: Polychelidae) in a bathyal mud community of the Mediterranean Sea. Mar Biol 114:109–117

Ahyong ST (2009) The Polychelidan lobsters: phylogeny and systematics (Polychelida: Polychelidae). In: Martin JL, Crandall KA, Felder DL (eds) Crustacean issues, vol 18. CRC Press, Boca Raton, pp 369–396

Alves FA, de Araújo MSC, Souza-Filho JF (2016) Distribution of two species of *Nephropsis* Wood-Mason, 1872 (Crustacea, Decapoda, Nephropidae) from northeastern Brazil. Zootaxa 4114:90–94

Anger K (2001) The biology of decapod crustacean larvae. Crustacean issues, vol 14. Balkema, Lisse

Anger K, Moreira GS (1998) Morphometric and reproductive traits of tropical caridean shrimps. J Crust Biol 18:823–838

Balsam WL, Beeson JP (2003) Sea-floor sediment distribution in the Gulf of Mexico. Deep-Sea Res I 50:1421–1444

Barradas-Ortiz C, Briones-Fourzán P, Lozano-Álvarez E (2003) Seasonal reproduction and feeding ecology of giant isopods *Bathynomus giganteus* from the continental slope of the Yucatan peninsula. Deep Sea Res I 50:495–513

Bergantino RN (1971) Submarine regional geomorphology of the Gulf of Mexico. Geol Soc Am Bull 82:741–752

Bezerra LEA, Ribeiro FB (2015) Primitive decapods from the deep sea: first record of blind lobsters (Crustacea: Decapoda: Polychelidae) in northeastern Brazil. Nauplius 23:125–131

Briones-Fourzán P, Barradas-Ortiz C, Negrete-Soto F, Lozano-Álvarez E (2010) Reproductive traits of tropical deep-water pandalid shrimps (*Heterocarpus ensifer*) from the SW Gulf of Mexico. Deep Sea Res I 57:978–987

Buhl-Mortensen L, Vanreusel A, Gooday AJ, Levin LA, Priede IG, Buhl-Mortensen P, Gheerardyn H, King NJ, Raes M (2010) Biological structures as a source of habitat heterogeneity and biodiversity on the deep ocean margins. Mar Ecol 31:21–50

Cabiddu S, Follesa MC, Gastoni A, Porcu C, Cau A (2008) Gonad development of the deep-sea lobster *Polycheles typhlops* (Decapoda: Polychelidae) from the central western Mediterranean. J Crust Biol 28:494–501

Cartes JE, Abelló P (1992) Comparative feeding habits of polychelid lobsters in the Western Mediterranean deep-sea communities. Mar Ecol Prog Ser 84:139–150

Cartes JE, Carrassón M (2004) Influence of trophic variables on the depth-range distributions and zonation rates of deep-sea megafauna: the case of the Western Mediterranean assemblages. Deep Sea Res I 51:263–279

Chan TY (2010) Annotated checklist of the world's marine lobsters (Crustacea: Decapoda: Astacidea, Glypheidea, Achelata, Polychelida). Raffles Bull Zool Suppl 23:153–181

Chan TY, de Saint Laurent M (1999) The rare lobster genus *Thaumastocheles* (Decapoda: Thaumastochelidae) from the Indo-Pacific, with description of a new species. J Crust Biol 19:891–901

Chang SC, Chan TY, Ahyong ST (2014) Two new species of the rare lobster genus *Thaumastocheles* Wood-Mason, 1874 (Reptantia: Nephropidae) discovered from recent deep-sea expeditions in the Indo-West Pacific. J Crust Biol 34:107–122

Clark MR, Althaus F, Schlacher TA, Williams A, Bowden DA, Rowden AA (2016) The impacts of deep-sea fisheries on benthic communities: a review. ICES J Mar Sci 73:i51–i69

Clarke KR (1993) Non-parametric multivariate analyses of changes in community structure. Aust J Ecol 18:117–143

Clarke KR, Gorley RN (2006) PRIMER v6: user manual/tutorial. PRIMER-E, Plymouth

Clarke KR, Warwick RM (2001) Change in marine communities: an approach to statistical analysis and interpretation. PRIMER-E, Plymouth

Correa TBS, Eberli GP, Grasmueck M, Reed JK, Correa AMS (2012) Genesis and morphology of cold-water coral ridges in a unidirectional current regime. Mar Geol 326–328:14–27

da Ros Z, Dell'Anno A, Morato T, Sweetman AK, Carreiro-Silva M, Smith CJ, Papadopoulou N, Corinaldesi C, Bianchelli S, Gambi C, Cimino R, Snelgrove P, Van Dover CL, Danovaro R (2019) The deep sea: the new frontier for ecological restoration. Mar Policy 108:103642

Dall'occo PL, Tavares M (2004) New and additional records of deep-water blind lobsters from Brazil (Decapoda: Polychelida). Nauplius 12:143–149

Dall'occo PL, Bento RT, de Melo GAS (2007) Range extensions for lobsters off the Brazilian coast (Crustacea, Decapoda, Palinurida, Astacidea). Biociências (Porto Alegre) 15:47–52

Dineshbabu AP (2008) Morphometric relationship and fishery of Indian Ocean lobsterette, *Nephropsis stewarti* Wood-Mason 1873 along the southwest coast of India. J Mar Biol Assoc India 50:113–116

Ellis SL, Incze LS, Lawton P, Ojaveer H, MacKenzie BR, Pitcher CR, Shirley TC, Eero M, Tunnell JW Jr, Doherty PJ, Zeller BM (2011) Four regional marine biodiversity studies: approaches and contributions to ecosystem-based management. PLoS One 6(4):e18997

Escobar-Briones EG, Gaytán-Caballero A, Legendre P (2008) Epibenthic megacrustaceans from the continental margin, slope and abyssal plain of the Southwestern Gulf of Mexico: factors responsible for variability in species composition and diversity. Deep Sea Res II 55:2667–2678

Färber-Lorda J (1994) Length-weight relationships and coefficient of condition of *Euphausia superba* and *Thysanoessa macrura* (Crustacea: Euphausiacea) in southwest Indian Ocean during summer. Mar Biol 118:645–650

Felder DL, Camp DK (eds) (2009) Gulf of Mexico: origins, waters, and biota, vol 1. Texas A&M Univ Press, College Station

Felder DL, Álvarez F, Goy JW, Lemaitre R (2009) Decapoda (Crustacea) of the Gulf of Mexico, with comments on the Amphionidacea. In: Felder DL, Camp DK (eds) Gulf of Mexico: origins, waters, and biota, vol 1. Texas A&M Univ Press, College Station, pp 1019–1104

Firth Jr RW, Pequegnat WE (1971) Deep-sea lobsters of the families Polychelidae and Nephropidae (Crustacea, Decapoda) in the Gulf of Mexico and Caribbean Sea. Texas A&M Univ, Dept Oceanogr, A&M Project 700-15, Reference 71-1 IIT

Follesa MC, Cabiddu S, Gastoni A, Cau A (2007) On the reproductive biology of the deep-sea lobster, *Polycheles typhlops* (Decapoda, Palinura, Polychelidae), from the Central-Western Mediterranean. Crustaceana 80:839–846

Galil B (2000) Crustacea Decapoda: review of the genera and species of the family Polychelidae Wood-Mason, 1874. In: Crosnier A (ed) Résultats des campagnes MUSORSTOM, vol 21. Mém Mus Natl Hist Nat 184:285–387

Gastoni A, Follesa MC, Mulas A, Porcu C, Cau M (2010) Observations on *Polycheles sculptus* S.I. Smith, 1880 (Decapoda, Palinura, Polychelidae) from Sardinian waters (central western Mediterranean). Crustaceana 83:443–456

Goy JW (2014) Astacidea. In: Martin JW, Olesen J, Hoeg JT (eds) Atlas of crustacean larvae. Johns Hopkins University Press, Baltimore, pp 256–264

Gracia A, Vázquez-Bader AR, Lozano-Álvarez E, Briones-Fourzán P (2010) Deep-water shrimp (Crustacea: Penaeidoea) off the Yucatan Peninsula (southern Gulf of Mexico): a potential fishing resource? J Shellfish Res 29:37–43

Guéguen F (2000) Distribution et abondance des crustacés décapodes tu talus continental (200–900 m) de Guyane Française. Crustaceana 73:685–703

Hammer Ø, Harper DAT, Ryan PD (2001) PAST: paleontological statistics software package for education and data analysis. Palaeontol Electron 4(1):1–9

Hartnoll RG (1982) Growth. In: Abele LG (ed) The biology of Crustacea, vol 2. Academic Press, New York, pp 11–196

Hebbeln D, Wienberg C, Wintersteller P, Freiwald A, Becker M, Beuck L, Dullo C, Eberli GP, Glogowski S, Matos L, Forster N, Reyes-Bonilla H, Taviani M et al (2014) Environmental forcing of the Campeche cold-water coral province, southern Gulf of Mexico. Biogeosciences 11:1799–1815

Hendrickx ME (2003a) Distribution and estimation of body size and weight of four species of deep water shrimps in the Southeastern Gulf of California, Mexico. Crustaceana 76:1025–1036

Hendrickx ME (2003b) Distribution and size of the Pacific deep water lobsterette, *Nephropsis occidentalis* Faxon, 1893 (Decapoda, Astacidea, Nephropidae) in the S.E. Gulf of California, Mexico. Crustaceana 76:207–216

Holthuis LB (1974) The lobsters of the superfamily Nephropoidea of the Atlantic Ocean (Crustacea: Decapoda). Bull Mar Sci 24:723–884

Holthuis LB (1991) Marine lobsters of the world. An annotated and illustrated catalogue of species of interest to the fisheries known to date, FAO Fisheries Synopsis No. 125, vol 13. FAO, Rome

Ivanov BG, Krylov VV (1980) Length-weight relationship in some common prawns and lobsters (Macrura, Natantia and Reptantia) from Western Indian Ocean. Crustaceana 38:279–289

Koslow JA, Boehlert GW, Gordon JDM, Haedrich RL, Lorance P, Parin N (2000) Continental slope and deep-sea fisheries: implications for a fragile ecosystem. ICES J Mar Sci 57:548–557

Lavalli KL, Spanier E (2010) Infraorder Palinura Latreille, 1802. In: Schram FR, von Vaupel Klein JC (eds) Treatise on zoology–anatomy, taxonomy, biology: the Crustacea, vol 9, part A. Brill, Leiden, pp 425–532

Levin LA, Dayton PK (2009) Ecological theory and continental margins: where shallow meets deep. Trends Ecol Evol 24:606–617

Levin LA, Etter RJ, Rex MA, Gooday AJ, Smith CR, Pineda J, Stuart CT, Hessler RR, Pawson D (2001) Environmental influences on regional deep-sea species diversity. Annu Rev Ecol Syst 132:51–93

Lozano-Álvarez E, Briones-Fourzán P, Gracia A, Vázquez-Bader AR (2007) Relative growth and size at first maturity of the deep water shrimp *Heterocarpus ensifer* (Decapoda, Pandalidae) from the southern Gulf of Mexico. Crustaceana 80:555–568

Lutz SJ, Ginsburg RN (2007) State of the US deep coral ecosystems in the United States Caribbean region: Puerto Rico and US Virgin Islands. In: Lumsden SE, Hourigan TF, Bruckner AW (eds) The state of deep coral ecosystems of the United States, NOAA Tech Memorandum CRCP-3. NOAA, Silver Spring, pp 307–365

Maiorano P, Pastore M, D'Onghia G, Latorre F (1998) Note on the population structure and reproduction of *Polycheles typhlops* (Decapoda: Polychelidae) on the upper slope of the Ionian Sea. J Nat Hist 32:1609–1618

Manning RB (1969) A new genus and species of lobster (Decapoda, Nephropidae) from the Caribbean Sea. Crustaceana 17:303–309

Martínez-Calderón R, Lozano-Álvarez E, Briones-Fourzán P (2018) Morphometric relationships and seasonal variation in size, weight, and a condition index of post-settlement stages of the Caribbean spiny lobster. Peer J 6:e5297

Maynou F, Cartes JE (2011) Effects of trawling on fish and invertebrates from deep-sea coral facies of *Isidella elongata* in the western Mediterranean. J Mar Biol Assoc UK 92:1501–1507

Messing CG, Walker BK, Dodge RE, Reed J (2006) Calypso U.S. Pipeline, LLC, Mile Post (MP) 31-MP 0 deep-water marine benthic video survey. Final report submitted to Calypso U.S. Pipeline, LLC

Norse EA, Brooke S, Cheung WWL, Clark MR, Ekeland I, Froese R, Gjerde KM, Haedrich RL, Heppell SS, Morato T, Morgan LE, Pauly D, Sumaila UR, Watson R (2012) Sustainability of deep-sea fisheries. Mar Policy 36:307–320

Papiol V, Hendrickx ME, Serrano D (2016) Distribution and ecology of the Pacific lobsterette *Nephropsis occidentalis* Faxon, 1893 (Crustacea, Decapoda, Astacidea), on the continental slope off western Mexico. In: Riosmena-Rodríguez R (ed) Marine benthos: biology, ecosystem functions and environmental impact. Nova Science, New York, pp 197–223

Parzanini C, Parrish CC, Hamel JF, Mercier A (2018) Trophic relationships of deep-sea benthic invertebrates on a continental margin in the NW Atlantic inferred by stable isotope, elemental, and fatty acid composition. Prog Oceanogr 168:279–295

Poupin J, Corbari L (2016) A preliminary assessment of the deep-sea Decapoda collected during the KARUBENTHOS 2015 Expedition to Guadeloupe Island. Zootaxa 4190:1–107

Reid DM, Corey S (1991) Comparative fecundity of decapod crustaceans, III. The fecundity of fifty-three species of Decapoda from tropical, subtropical, and boreal waters. Crustaceana 61:308–316

Roberts CM (2002) Deep impact: the rising toll of fishing in the deep sea. Trends Ecol Evol 17:242–245

Robinson LA, Greenstreet SPR, Reiss H, Callaway R, Craeymeersh J, de Boois I, Degraer S, Ehrich S, Fraser HM, Goffin A et al (2010) Length–weight relationships of 216 North Sea benthic invertebrates and fish. J Mar Biol Assoc UK 90:95–104
Roe RB (1966) Potentially commercial Nephropsids from the western Atlantic. Trans Am Fish Soc 95:92–98
Sardà F, Cartes JE (1993) Relationship between size and depth in decapod crustacean populations on the deep slope in the Western Mediterranean. Deep Sea Res I 40:2389–2400
Silva KCA, Cruz R, Cintra IHA, Abrunhosa FA (2013) Structure and diversity of the lobster community on the Amazon continental shelf. Crustaceana 86:1084–1102
Stefanescu C, Lloris D, Rucabado J (1993) Deep-sea fish assemblages in the Catalan Sea (Western Mediterranean) below a depth of 1000 m. Deep-Sea Res I 40:695–707
Tavares CR, Young PS (2002) Nephropidae (Crustacea, Decapoda) collected by the Revizee Score-Central program from off Bahia to Rio de Janeiro states, Brazil. Arq Mus Nac 60:77–88
Thiel H (1979) Structural aspects of the deep-sea benthos. Ambio Special Rep 6:25–31
Torres AP, Palero F, dos Santos A, Abelló P, Blanco E, Boné A, Guerao G (2014) Larval stages of the deep-sea lobster *Polycheles typhlops* (Decapoda, Polychelida) identified by DNA analysis: morphology, systematic, distribution and ecology. Helgol Mar Res 68:379–397
Uchupi E (1975) Physiography of the Gulf of Mexico and Caribbean Sea. In: Nairn AEM, Stehli FG (eds) The Gulf of Mexico and the Caribbean. Plenum, New York, pp 1–64
Vázquez-Bader AR, Gracia A (2013) Crangonidae and Glyphocrangonidae (Decapoda; Caridea) of the Southern Gulf of Mexico. Zootaxa 3669(3):367–383
Vázquez-Bader AR, Gracia A (2016) Diversity and distribution of Chyrostiloidea and Galatheoidea (Decapoda, Anomura) in the Southern Gulf of Mexico. Zookeys 612:1–30
Wahle R, Tshudy D, Cobb JS, Factor J, Jaini M (2012) Infraorder Astacidea Latreille, 1802 p.p.: the marine clawed lobsters. In: Schram FR, von Vaupel Klein JC (eds) Treatise on zoology–anatomy, taxonomy, biology: the Crustacea, vol 9. Part B. Brill, Leiden, pp 3–108
Warton DI, Wright IJ, Falster DS, Westoby M (2006) Bivariate line-fitting methods for allometry. Biol Rev 81:259–291
Wenner EL (1979) Some aspects of the biology of deep-sea lobsters of the family Polychelidae (Crustacea, Decapoda) from the Western North Atlantic. Fish Bull 77:435–444
Wicksten M, Packard JM (2005) A qualitative zoogeographic analysis of decapod crustaceans of the continental slopes and abyssal plain of the Gulf of Mexico. Deep Sea Res I 52:1745–1765
Zar JH (1999) Biostatistical analysis, 4th edn. Prentice Hall, Upper Saddle River

Chapter 14
New Molecular Data on Squat Lobster from the Coast of São Paulo State (Brazil) (Anomura: *Munida* and *Agononida*) and Insights on the Systematics of the Family Munididae

I. Miranda, P. A. Peres, M. D. S. Tavares, and F. L. Mantelatto

Abstract The squat lobsters *Munida* Leach, 1820 and *Agononida* Leach, 1820 are part of the most speciose genera in the diverse family Munididae. Despite the considerable diversity (>240 species), the Brazilian waters encompass 18 species so far (17 *Munida* and 1 *Agononida*), only 7 (6 and 1, respectively) of which recorded from the coast of São Paulo. The decapod fauna along the coast of São Paulo has been studied in the recent past, mostly using classical alpha morphology. In the present study, we carried out a molecular analysis to phylogenetically contextualize the species of *Munida* and *Agononida* and address future directions on the systematics of the group. The current investigation results from a long-term multidisciplinary taxonomic project that combined analyses of adult specimens for accurate and detailed identification of the biodiversity of marine decapod crustaceans from São Paulo state. Sampling was carried out in five major regions along the São Paulo coast from 2011 to 2018: Ubatuba, Caraguatatuba, São Sebastião and Ilhabela,

I. Miranda
Departamento de Biologia Aplicada, Laboratório de Morfologia de Invertebrados (IML) e Centro de Aquicultura da Unesp (CAUNESP), Universidade Estadual Paulista "Júlio de Mesquita Filho" (UNESP), FCAV, São Paulo, Brazil

Laboratório de Bioecologia e Sistemática de Crustáceos (LBSC), Departamento de Biologia, Faculdade de Filosofia, Ciências e Letras de Ribeirão Preto (FFCLRP), Universidade de São Paulo (USP), São Paulo, Brazil

P. A. Peres · F. L. Mantelatto (✉)
Laboratório de Bioecologia e Sistemática de Crustáceos (LBSC), Departamento de Biologia, Faculdade de Filosofia, Ciências e Letras de Ribeirão Preto (FFCLRP), Universidade de São Paulo (USP), São Paulo, Brazil
e-mail: pasperes@usp.br; flmantel@usp.br

M. D. S. Tavares
Museu de Zoologia da Universidade de São Paulo (MZUSP), São Paulo, Brazil
e-mail: mdst@usp.br

M. E. Hendrickx (ed.), *Deep-Sea Pycnogonids and Crustaceans of the Americas*, https://doi.org/10.1007/978-3-030-58410-8_14

Santos and São Vicente, and Cananéia and Ilha Comprida. Additional material from the MZUSP collections was used to complete the analysis when fresh material was not obtained during the surveys. Previous to molecular analysis, the material was identified by classical literature. We obtained six species out of seven recorded, with sequences of cytochrome oxidase subunit I – barcode region and 16S generated from six species. We include additional genera and close species to run the analysis to better contextualize the phylogenetic positioning of the target species. Our tree shows a clear recognition of some of São Paulo species and points out systematics inconsistencies in Munididae. Based on the present results, and pending future more complete analyses, Munididae should be revised.

Keywords DNA markers · 16S rRNA · Cytochrome oxidase subunit I · Molecular phylogeny · Western Atlantic

14.1 Introduction

The advances of molecular analyses have helped to refine the understanding on the evolutionary history of decapod crustaceans (Timm and Bracken-Grissom 2015). However, many groups are still surrounded by uncertain taxonomy and species resolution (e.g., within Anomura – Bracken-Grissom et al. 2013). The squat lobster family Munididae Ahyong et al. 2010, for instance, have passed through great strides in their knowledge in the last decade (Machordom and Macpherson 2004; Cabezas et al. 2011; Rodríguez-Flores et al. 2019). There are examples of cryptic species (Poore and Andreakis 2012), description of new genera (Cabezas et al. 2008), and new species (Macpherson et al. 2017) mostly due to worldwide efforts combining morphological and molecular data, but also surveys in new locations not previously explored.

The family Munididae is highly speciose, composed of 23 genera and more than 400 recognized species so far (Ahyong et al. 2010; WoRMS 2020a). Despite this tremendous advance on documenting the diversity of the group in all marine habitats from shallow to deep waters, our knowledge on Brazilian Munididae is still poorly explored. Taxonomic contributions were restricted to the 1990s and early 2000s, mostly using morphological data (see Melo-Filho and Melo 1994, 1998, 2001b; Tavares and Campinho 1998; Melo 1999; Melo-Filho 2006; Serejo et al. 2007; Baba et al. 2008). Recently, three new species of *Munidopsis* Witheaves, 1874 were described by Poore (2014), two of them from Amapá and Espirito Santo states and one by Cardoso et al. (2014) also from Espirito Santo. Despite the high diversity of Munididae, only 18 species occur in Brazil so far (17 *Munida* Leach, 1820 and one *Agononida* Leach, 1820), and only 7 (6 and 1, respectively) are recorded for the coast of São Paulo (compiled from the above references).

This scenario evidenced potential perspectives for studies on the taxonomy of the group, in particular those using molecular markers to solving taxonomic inconsistencies and provide phylogenetic contextualization (see Rodríguez-Flores et al. 2019 for literature review). As part of a long-term multidisciplinary project to expand and improve knowledge about the marine biota of the state of São Paulo (Mantelatto et al. 2018), several efforts are being made in this direction and the improvement of the degree of taxonomic refinement, mainly to generate a genomic library for species recorded in that region and to serve for documentation of biodiversity and future studies (FLM pers. com.). During this project, we were faced with some taxonomic uncertainties on some squat lobster species reported for the coast of São Paulo. Here, we combined molecular and morphological data to phylogenetically contextualize the species of *Munida* and *Agononida* reported to this area and address potential cryptic species and future directions.

14.2 Material and Methods

14.2.1 Sampling and Identification of Specimens

Sampling was performed during multiple collecting trips from 2011 to 2015, during the day and/or at night, in different coastal environments of São Paulo State. The collected animals were stored in 95% ethanol and deposited in the Crustacean Collection of the Biology Department at the Faculty of Philosophy, Sciences and Letters in Ribeirão Preto (CCDB/FFCLRP/USP) (CCDB 5806). Additional material was obtained from the Museum of Zoology of the University of São Paulo (MZUSP) (MZUSP 13655, 15360, 16244, 16717, 20501, 28422, 28424).

Previously to DNA extraction, the identification of all individuals was confirmed based on diagnostic morphological characters from appropriate taxonomic studies (Melo 1999; Melo-Filho and Melo 2001a, b; Melo-Filho 2006).

14.2.2 Molecular Protocols

DNA was extracted from the muscle tissue of pereopods, chelae, or pleon using three distinct methods, salting-out method (Miller et al. 1988), Chelating Ion Exchange Resin (Chelex® 100) (Estoup et al. 1996), or Qiagen DNeasy Blood and Tissue Kit (Qiagen), following the manufacturer's protocol. The extracted DNA final concentration was measured using a spectrophotometer (NanoDrop® 2000/2000c).

Approximately 650 base pairs (bp) of COI and 500 bp of 16S genes were amplified using a polymerase chain reaction (PCR) using a thermal cycler (Veriti 96 Well Thermal Cycler Applied Biosystems®). Gene fragments were amplified using the following thermal profiles: initial denaturing for 2–5 min at 94 °C; annealing for 35–40 cycles, 30–45 s at 94/95 °C, 30 s at 38–50 °C (depending on the taxon and primers used), and 1 min at 72 °C; final extension 2–3 min at 72 °C, with the primers 16S–16 L2 (5'-TGC CTG TTT ATC AAA AAC AT-3′) (Schubart et al. 2002) and 1472 (5′-AGA TAG AAA CCA ACC TGG-3′) (Crandall and Fitzpatrick 1996); and cytochrome oxidase subunit I – COL6b (5'-ACA AAT CAT AAA GAT ATY GG-3′) and COH6 (5′-TAD ACT TCD GGR TGD CCA AAR AAY CA-3′) (Schubart and Huber 2006) and COIAL2o (5'-ACG CAA CGA TGA TTA TTT TCT AC-3′) and COIAH2m (5'-GAC CRA AA AAT CAR AAT AAA TGT TG-3′) (Mantelatto et al. 2016).

PCR products were observed in electrophoresis with 1.5% agarose gel and photographed with digital camera Olympus C-7070® and SX520 Canon® on a UV transilluminators M20 UVP® and 302 Kasvy®. Successful PCR products were purified using the SureClean Plus® kit, following the manufacturer's protocol.

Purified samples were sent for sequencing to the Department of Technology at the Faculty of Agrarian and Veterinary Sciences (Jaboticabal) at São Paulo State University (Unesp). All sequences were confirmed by sequencing both strands (forward and reverse directions). A consensus sequence was obtained using the Bioedit 7.0.5 computer program (Hall 1999). Primer regions and non-readable parts at the beginning of the sequences were omitted. All obtained sequences were deposited in the GenBank database.

14.2.3 Data Analysis

In addition to the data we generated, we used GenBank sequences of this and other species of the genera *Agononida* and *Munida*, as well as other Munididae (*Babamunida* Cabezas et al., 2008, *Paramunida* Baba, 1988, and *Plesionida* Baba & Saint Laurent, 1996), Galatheidae (*Fennerogalathea* Baba, 1988), and Porcellanidae [*Porcellana sayana* (Leach, 1820)] (Table 14.1) in order to phylogenetically contextualize our target group and allow comparison with other intrageneric genetic distances.

Sequences were proofread with Chromas 2.23 (Technelysium Pty Ltd., 2005), automatically aligned with Clustal W (Thompson et al. 1994) implemented in BioEdit 7.0.5 (Hall 1999), and unspecific readings were manually corrected when required. Primer regions, poorly aligned regions, and large indels in sequence data were removed following recommendations by GBlocks (Castresana 2000), making the final alignment suitable for phylogenetic analysis (Castresana 2000). The absence of stop codons in COI was confirmed using the software Artemis (Rutherford et al. 2000) to reduce the possibility of including pseudogenes (Song et al. 2008). We tested each marker's substitution saturation (Xia et al. 2003) in the software DAMBE 5 (Xia 2013). Sequences were finally blasted in GenBank and compared with the available assemble.

Alignments of the two mitochondrial markers were concatenated into a single dataset; missing data were labeled as question marks in the alignment. Phylogenetic reconstructions were carried using ML with RAxML, as proposed by Stamatakis

Table 14.1 Squat lobster species (families Galatheidae and Munididae) used for the molecular phylogenetic reconstruction (16S rRNA and COI mtDNA) with respective GenBank accession numbers

Species	Family	16S	COI
Agononida africerta Poore & Andreakis, 2012	Munididae	_	KM281842
Agononida alisae Macpherson, 2000	Munididae	AY351064	_
Agononida incerta (Henderson, 1888)*	Munididae	AY351066	_
Agononida indocerta Poore & Andreakis, 2012	Munididae	_	KM281838
Agononida laurentae (Macpherson, 1994)	Munididae	AY351068	_
***Agononida longipes* (A. Milne-Edwards, 1880) MZUSP 20501**	**Munididae**	**MT320096 - Present work**	_
Agononida marini (A. Milne-Edwards, 1880)	Munididae	AY351071	_
Agononida ocyrhoe (Macpherson, 1994)	Munididae	AY351073	_
Agononida pilosimanus (Baba, 1969)	Munididae	AY351076	_
Agononida procera Ahyong & Poore, 2004	Munididae	_	AY351077
Agononida sphecia (Macpherson, 1994)	Munididae	AY351079	_
Babamunida callista (Macpherson, 1994)	Munididae	AY351110	AY350939
Munida acantha Macpherson, 1994	Munididae	AY351097	AY800033
Munida alonsoi Macpherson, 1994	Munididae	AY351105	AY350936
Munida armilla Macpherson, 1994	Munididae	AY351107	AY350938
Munida asprosoma Ahyong & Poore, 2004	Munididae	JF727283	_
Munida benguela de Saint Laurent & Macpherson, 1988	Munididae	KY230468	KY230453
Munida caeli Cabezas, Macpherson & Machordom, 2009	Munididae	EU417977	EU418002
Munida clinata Macpherson, 1994	Munididae	AY351113	AY350942
Munida compressa Baba, 1988	Munididae	AY351114	AY350944
Munida congesta Macpherson, 2000	Munididae	AY351115	AY350945
***Munida constricta*A. Milne-Edwards, 1880 MZUSP 28424**	**Munididae**	**MT479219 - Present work**	_
***Munida constricta*A. Milne-Edwards, 1880 MZUSP 15360**	**Munididae**	_	**MT479218 - Present work**
Munida delicata Macpherson, 2004	Munididae	EU417976	EU418001
Munida devestiva Macpherson, 2006	Munididae	EU417989	_
Munida distiza Macpherson, 1994	Munididae	AY351119	AY35095
Munida eclepsis Macpherson, 1994	Munididae	AY351122	AY350953
Munida euripa Macpherson, Rodriguez-Flores & Machordom, 2017	Munididae	KY230469	KY230454
Munida flinti Benedict, 1902	Munididae	MF490158	_
***Munida flinti* Benedict, 1902 MZUSP 13655**	**Munididae**	**MF490158 - Present work**	**MF490054 - Present work**
***Munida forceps* A. Milne-Edwards, 1880 MZUSP 16244**	**Munididae**	**MT320100 - Present work**	**MT479221 - Present work**

(continued)

Table 14.1 (continued)

Species	Family	16S	COI
Munida gracilis Henderson, 1885	Munididae	_	KJ544250
Munida gregaria (Fabricius, 1793)[a]	Munididae	KX929697	KX929634
		AY050075	_
		AY700161	_
Munida gordoae Macpherson, 1994	Munididae	AY351127	AY350958
Munida guttata Macpherson, 1994	Munididae	AY351131	AY350962
Munida hoda Macpherson, Rodriguez-Flores & Machordom, 2017	Munididae	_	KY230455
Munida intermedia A. Milne-Edwards & Bouvier, 1899	Munididae	_	JQ348884
Munida iris A. Milne-Edwards, 1880	Munididae	KF182521	KX022442
Munida lailai Cabezas, Macpherson & Machordom, 2009	Munididae	EU417988	EU418012
Munida lanciaria Cabezas, Lin & Chan, 2011	Munididae	JF727289	_
Munida leagora Macpherson, 1994	Munididae	AY351140	AY350971
Munida lenticularis Macpherson & de Saint Laurent, 1991	Munididae	_	AY350972
Munida leptosyne Macpherson, 1994	Munididae	AY351141	AY350973
Munida mendagnai Cabezas, Macpherson & Machordom, 2009	Munididae	EU417974	EU417999
Munida mesembria Macpherson, Rodriguez-Flores & Machordom, 2017	Munididae	KY230471	KY230456
Munida microphthalma A. Milne-Edwards, 1880	Munididae	MF490159	KX022458
***Munida microphthalma* A. Milne-Edwards, 1880 MZUSP 16717**	**Munididae**	**MF490159 - Present work**	**MF490055**
Munida militaris Henderson, 1885	Munididae	AY351143	AY350975
Munida nesiotes Macpherson, 1999	Munididae	_	KY230460
Munida notata Macpherson, 1994	Munididae	AY351147	AY350979
Munida oblongata Cabezas, Macpherson & Machordom, 2009	Munididae	EU417978	_
Munida offella Macpherson, 1996	Munididae	_	KY230460
Munida ommata Macpherson, 2004	Munididae	AY351151	AY350984
Munida pagesi Macpherson, 1994 *Munida leptosyne*	Munididae	AY351152	AY350985
Munida parca Macpherson, 1996	Munididae	EU417987	EU418010
Munida proto Macpherson, 1994	Munididae	AY351153	AY350986
Munida psamathe Macpherson, 1994	Munididae	AY351158	AY350991
Munida psylla Macpherson, 1994	Munididae	AY351159	AY350992
Munida pusilla Benedict, 1902	Munididae	KF182522	_
Munida remota Baba, 1990	Munididae	KY230472	_
Munida quadrispina Benedict, 1902	Munididae	_	MG320382
Munida rhodonia Macpherson, 1994	Munididae	AY351161	AF283886

(continued)

Table 14.1 (continued)

Species	Family	16S	COI
Munida rogeri Macpherson, 1994	Munididae	_	AY350993
Munida rosula Macpherson, 1994	Munididae	AY351162	AY350994
Munida rubridigitalis Baba, 1994[b]	Munididae	AY351163	AF283887
Munida rufiantennulata Baba, 1969	Munididae	AY351164	AY350995
Munida rugosa (Fabricius, 1775)*	Munididae	_	JQ306225
Munida rutllanti Zariquiey Álvarez, 1952	Munididae	_	JQ305920
Munida shaula Macpherson & de Saint Laurent, 2002	Munididae	KY230474	KY230464
Munida spilota Macpherson, 1994	Munididae	AY351165	AY350997
***Munida spinifrons* Henderson, 1885 MZUSP 28422**	**Munididae**	**MT320101 - Present work**	_
Munida spinosa Henderson, 1885	Munididae	AY700162	AY700179
Munida stomifera Macpherson, Rodriguez-Flores & Machordom, 2017	Munididae	KY230475	_
Munida stia Macpherson, 1994	Munididae	AY351170	_
Munida taenia Macpherson, 1994	Munididae	AY351177	AY351008
Munida thoe Macpherson, 1994	Munididae	AY351182	DQ011205
Munida tiresias Macpherson, 1994	Munididae	AY351183	AY351014
Munida tuberculata Henderson, 1885	Munididae	AY351184	AY351015
Munida tyche Macpherson, 1994	Munididae	AY351186	AY351017
Munida valida Smith, 1883	Munididae	JN800548	KX022479
Munida zebra Macpherson, 1994	Munididae	AY351192	AY800054
Outgroup			
Paramunida pictura Macpherson, 1993	Munididae	AY351210	AY351039
Plesionida aliena (Macpherson, 1996)	Munididae	_	AY351061
Munidopsis senticosa Rodriguez-Flores, Macpherson & Machordom, 2018	Munidopsidae	MG979476	_
Fennerogalathea ensifera Rodriguez-Flores, Machordom & Macpherson, 2017	Galatheidae	KY230489	KY230489
Porcellana sayana (Leach, 1820)	Porcellanidae	MF490163	MF490063

*Type species. Species with register to São Paulo state are shown in bold. The following sequences are on GenBank as [a]*Munida subrugosa*, [b]*Munida rubrodigitalis*

(2006), implemented at the CIPRES Science Gateway (Miller et al. 2010), using the GTR + G + I substitution model. The consistency of topologies was measured by the bootstrap method (1000 bootstrap pseudoreplicates), showing all confidence values above 50%. We also conducted maximum likelihood (ML) phylogenetic analyses for each gene fragment separately to reveal any possible discordance in the relationships among the studied lineages (data not shown). To estimate intra- and interspecific divergence rates, genetic distances were calculated for each gene by pairwise comparisons using uncorrected p-distances with the software Mega 6.0 (Tamura et al. 2013).

14.3 Results

Both analyzed genes showed little saturation [Index of substitution saturation (Iss) < Critical Index of substitution saturation (Iss.c), $P < 0.001$, data not shown], indicating that the data are robust enough for phylogenetic reconstruction. Similar phylogenetic topologies were generated with both genes, 16S rRNA and COI mtDNA, individually (data not shown) or concatenated (Fig. 14.1). In general, most external branches showed no bootstrap support, while internal branch supports varied from 50% to 100%.

Based on the dataset we produced and that were able to access, neither *Munida* nor *Agononida* species formed a monophyletic clade. Most of *Agononida* species were recovered as single lineage with low bootstrap (62%). This clade includes

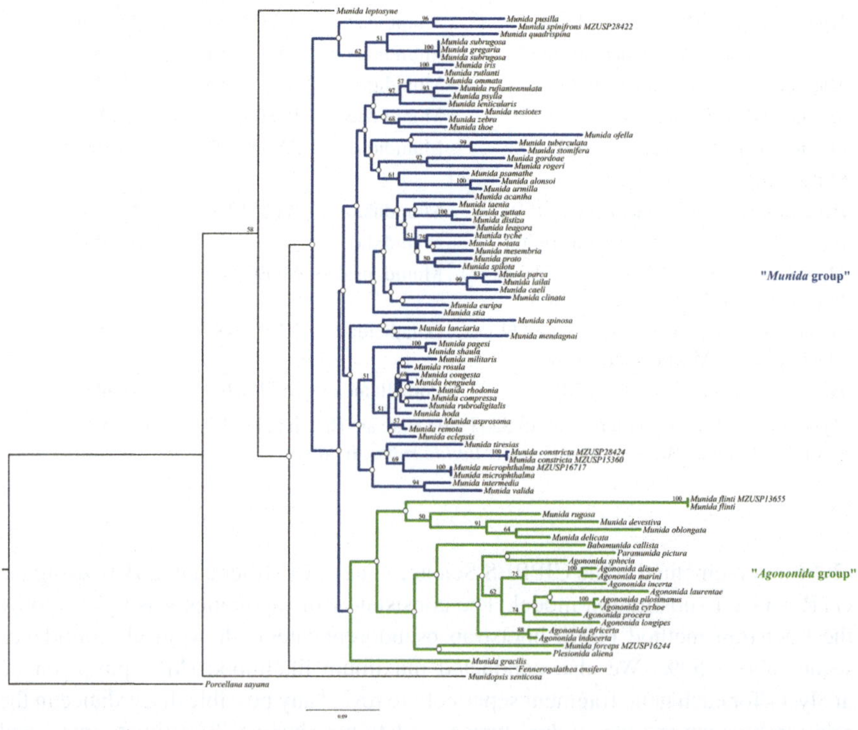

Fig. 14.1 Phylogram based on maximum likelihood analysis of 16S rRNA and COI mtDNA concatenated dataset. Sequences generated by the present work are represented by an asterisk (*). Numbers are significance values after 1000 bootstraps. Values below 50% are marked with a white circle

Paramunida pictura and is close related to *Babamunida callista*, *Plesionida aliena*, and six of the *Munida* species. Hereafter we will name these groups "*Munida* group" and "*Agononida* group".

Overall genetic distance among the total 16S rRNA dataset was 11.5%, while pairwise distances ranged from 10.3% (*Agononida* versus *Paramunida*) to 21.3% (*Babamunida* versus *Munida*) (Table 14.2). Overall genetic distance estimated within *Agononida* specimens analyzed was 8.3%, ranging from 1.4% (*A. alisae* versus *A. marini*) to 9.6% (*A. marini* versus *A. ocyrhoe*) (data not shown). A similar overall genetic distance within *Munida* specimens was observed (8.8%), ranging from 1.4% (*M. rhodonia* versus *M. benguela* and *M. compressa*) to 16.8% (*M. alonsoi* versus *M. flinti* and *M. flinti* versus *M. pusilla*) (data not shown).

Considering the COI mtDNA dataset, the overall genetic distance among the Munididae analyzed was 16.5%, while pairwise distances ranged from 10.3% (*Agononida* versus *Paramunida*) to 20.4% (*Babamunida* versus *Paramunida*) (Table 14.3). The genetic distance estimated between the two species of *Agononida* analyzed was 3% (*A. africerta* versus *A. indocerta*) (data not shown). Genetic distance within *Munida* specimens was higher (15.4%), ranging from 1.9% (*M. benguela* versus *M. congesta* and *M. rosula*) to 23.4% (*M. acantha* versus *M. forceps*) (data not shown).

Regarding the species with distribution registered to São Paulo coast, we generated new sequences for the species *Agononida longipes*, *Munida flinti*, *M. constricta*, *M. forceps*, *M. spinifrons*, and the first South Western Atlantic sequence of *M. microphthalma*. Within the "*Munida* group", *M. microphthalma* form a sister clade to *M. constricta*. *M. spinifrons* specimens cluster together with *M. pusilla* (16S rRNA pairwise genetic divergence between 6.2%). Both *M. flinti* specimens are placed in the "*Agononida* group", so as *M. forceps*. These relations were also present in the single gene topologies (data not shown).

Table 14.2 Genetic divergence matrix (p-distances) of the 16S rRNA gene between Munididae (*Babamunida*, *Paramunida*, *Agononida*, and *Munida*, highlighted) and among each target genera (*Agononida* and *Munida*) and other selected outgroups (Porcellanidae, Galatheidae, and Munidopsidae). Upper and lower values are shown underlined

	1	2	3	4	5	6	7
1. *Porcellana sayana*	-	-	-	-	-	-	-
2. *Fennerogalathea ensifera*	0.181	-	-	-	-	-	-
3. *Munidopsis senticosa*	0.180	0.174	-	-	-	-	-
4. *Babamunida callista*	0.216	0.182	0.193	-	-	-	-
5. *Paramunida pictura*	0.214	0.150	0.171	0.133	-	-	-
6. *Agononida*	0.207	0.172	0.184	0.115	<u>0.103</u>	-	-
7. *Munida*	0.185	0.137	0.170	<u>0.138</u>	0.130	0.135	-

Table 14.3 Genetic divergence matrix (p-distances) of the COI mtDNA gene between Munididae (*Babamunida, Paramunida, Plesionida, Agononida*, and *Munida*, highlighted) and other selected outgroups (Porcellanidae and Galatheidae). Upper and lower values are shown underlined

	1	2	3	4	5	6	7
1. *Porcellana sayana*	-	-	-	-	-	-	-
2. *Fennerogalathea ensifera*	0.220	-	-	-	-	-	-
3. *Babamunida callista*	0.235	0.210	-	-	-	-	-
4. *Paramunida picta*	0.215	0.201	0.204	-	-	-	-
5. *Plesionida aliena*	0.232	0.212	0.187	0.176	-	-	-
6. *Agononida*	0.217	0.193	0.167	0.136	0.148	-	-
7. *Munida*	0.218	0.189	0.198	0.171	0.177	0.171	-

14.4 Discussion

Molecular phylogenies has been made focusing on Pacific Munididae species, and less focus have been given to other areas (Machordom and Macpherson 2004; Cabezas et al. 2011; Palero et al. 2017). This is the first molecular analysis focused specifically in munidids squat lobster from Brazil and São Paulo, generating molecular sequences and integrating it with publicly available data. Our results indicate a clear taxonomy inconsistency in the family. We were able to readily confirm some recent crown groups, but others showed some intriguing positioning of specific species that were pointed but not fully discussed due the scope of our research, but we pointed guidance for future studies. Our tree shows a clear recognition of São Paulo species accessed herein (*Agononida longipes*, *Munida flinti*, *M. microphthalma*, *M. spinifrons*, *M. constricta*, *M. forceps*).

Molecular markers, such as mitochondrial DNA(mtDNA), are commonly used to differentiate species in crustaceans and other invertebrates (Hebert et al. 2003; Lefébure et al. 2006). A common genetic divergence threshold to delimit species in decapod crustaceans is 3% COI, but it can vary depending on the group (Lefébure et al. 2006; Silva et al. 2011). In squat lobsters, these values are around 6% for COI (Machordom and Macpherson 2004; Cabezas et al. 2011).

Even though *Munida* is the most diverse genus of squat lobsters worldwide (Baba et al. 2008), with 353 accepted species (WoRMS 2020b), there is extreme morphological stasis and a high degree of intraspecific variations (Melo-Filho and Melo 1998). Thus, morphological studies are challenging, but also have led to the detection of cryptic species (Macpherson and Machordom 2005). In this sense, it is not surprising that morphological descriptions may assign a specimen to a species, but showing later disagreement with molecular data. Here, individuals preliminarily assigned to *M. spinifrons* using morphology were different species based on genetic data. One of them, after intense morphological analysis, revealed its real identity as *M. constricta*. As an example of phenotypical similarity and intraspecific variation on Brazilian species, previous morphological comparison made by Melo-Filho and Melo (2001a) revealed similarities of *M. spinifrons* to *M. angulata* and *M. petronioi*. Populations of *M. spinifrons* at the southern and northern ends of distribution (Florida, Espirito Santo, Rio de Janeiro, and São Paulo) are very similar (little spine-like rostrum and most individuals with two spines in the anterior branchial region), and, on the other hand, the populations of northeastern Brazil have individuals with a strongly spinous face and, generally, with a thorn in the anterior branchial region. In this case, intraspecific variation coupled with overlapping distribution with other similar species may lead to wrong identification. Subtle morphological differences might be unnoticed until molecular data leads to taxonomic uncertain (Poore and Andreakis 2012). Nevertheless, there are cases like the recognized species *M. gregaria* and *M. subrugosa* that show no differences based on molecular data, but species are phenotypically distinct (Perez-Barros et al. 2008). Integrating different data types may also result in synonym as in the case of *M. rutllanti* and *M. speciosa* (Rodríguez-Flores et al. 2019).

According to morphological revision by Melo-Filho and Melo (2001a), *M. constricta* occurs in the São Paulo coast, and this species is morphologically similar to *Munida miles* A. Milne-Edwards, 1880 from Gulf of Mexico. Besides, the clade formed between *M. spinifrons* and *M. pusilla* (sequence from GenBank) is supported by morphological similarities previously pointed by Melo-Filho and Melo (2001a), being the second one differentiated by a shorter rostrum with distinct spinulation, post-cervical spines, and spinulation of third maxilliped, and fingers are distinctly shorter than the palms and distribution in the northeastern of Brazil (Pará). We confirmed the distribution of *M. microphthalma* in the Gulf of Mexico (GOM) and South Atlantic. Our individual from the coast of São Paulo falls in a clade with another individual from GOM and presents low genetic divergence between them, which supports affirming that this species presents a wide range of distribution.

Our analysis places *Munida flinti* and *M. forceps* at the "*Agononida* group", together with *Babamunida*, *Paramunida*, and *Plesionida*, a relation that deserves further investigation. Based on morphological evidence, *M. flinti* is closely related to *M. benedict* Chace, 1942, *M. stimpsoni* A. Milne-Edwards, 1880 and *M. striata* Chace, 1942 (see Melo-Filho and Melo 2001a), but those could not be confirmed due to the absence of sequences from the above-mentioned species. *Munida forceps* was herein phylogenetically contextualized, and the proximity with *Plesionida* has no morphological support.

Our data points to a major revision within Munididae. Previous studies have indicated a lack of resolution in deep relationships (Machordom and Macpherson 2004; Cabezas et al. 2011; Rodríguez-Flores et al. 2019) and also indicated poly of paraphyletic groups within the group (Bracken-Grissom et al. 2013). Specifically, the genus *Munida* has been under debate since it was split into different genera, e.g. *Agononida* Baba and de Saint Laurent, 1996, *Crosnierita* Macpherson, 1998, *Enriquea* Baba, 2005, *Munida* Leach, 1820, *Paramunida* Baba, 1988, and *Raymunida* Macpherson and Machordom, 2000. Even though some of them are recovered as monophyletic (Machordom and Macpherson 2004; Cabezas et al. 2008), questions remain on the evolutionary relationships of the genera within Munididae (Cabezas et al. 2008, 2010, 2012; Poore and Andreakis 2012). Considering the current intense work on squat lobster, further studies should compile all this data and explore the phylogeny of Munididae using all datasets available.

This study represents the initial results on documenting some squat lobsters' biodiversity under a long-term multidisciplinary project in the state of São Paulo and Brazil. Combining morphological and molecular data, we shed light on possible taxonomic inconsistencies in Munididae that can be used as a starting point to further studies, particularly among the Brazilian members that are poorly known regarding molecular contextualization.

Acknowledgements This paper is part of the multidisciplinary research biodiversity projects supported by Coordenação de Aperfeiçoamento de Pessoal de Nível Superior – CAPES – Finance Code 001 (Ciências do Mar II Proc. 2005/2014 – 23038.004308/2014-14) and São Paulo Research Foundation - FAPESP (Temáticos BIOTA 2010/50188-8 and INTERCRUSTA 2018/13685-5; Coleções Científicas 2009/54931-0; PROTAX 2016/50376-5) and Conselho Nacional de Desenvolvimento Científico e Tecnológico – CNPq (PQ 304968/2014-5 and 302253/2019-0 to FLM). IM and PAP thanks fellowships from CAPES (PD Ciências do Mar II Proc. 1989/2014) and FAPESP (DR 2017/12376-6), respectively. We are thankful to many students and researchers of the BIOTA-FAPESP and CAPES-CIMar II projects, who helped with sampling and making fresh material available for morphological and molecular procedures. We thank anonymous reviewers for their corrections during review process.

References

Ahyong ST, Baba K, Macpherson E, Poore GC (2010) A new classification of the Galatheoidea (Crustacea: Decapoda: Anomura). Zootaxa 2676(1):57–68

Baba K, Macpherson E, Poore GC, Ahyong ST, Bermudez A, Cabezas P, Lin CW, Nizinski M, Rodrigues C, Schnabel KE (2008) Catalogue of squat lobsters of the world (Crustacea: Decapoda: Anomura families Chirostylidae, Galatheidae and Kiwaidae). Zootaxa 1905(1):1–220

Bracken-Grissom HD, Cannon ME, Cabezas P, Feldmann RM, Schweitzer CE, Ahyong ST, Felder DL, Lemaitre R, Crandall KA (2013) A comprehensive and integrative reconstruction of evolutionary history for Anomura (Crustacea: Decapoda). BMC Evol Biol 13(1):128

Cabezas P, Macpherson E, Machordom A (2008) A new genus of squat lobster (Decapoda: Anomura: Galatheidae) from the South West Pacific and Indian Ocean inferred from morphological and molecular evidence. J Crustac Biol 28(1):68–75

Cabezas P, Macpherson E, Machordom A (2010) Taxonomic revision of the genus Paramunida Baba, 1988 (Crustacea: Decapoda: Galatheidae): a morphological and molecular approach. Zootaxa 2712(1):1–60

Cabezas P, Lin CW, Chan TY (2011) Two new species of the deep-sea squat lobster genus *Munida* Leach, 1820 (Crustacea: Decapoda: Munididae) from Taiwan: morphological and molecular evidence. Zootaxa 3036(1):26–38

Cabezas P, Sanmartín I, Paulay G, Macpherson E, Machordom A (2012) Deep under the sea: unraveling the evolutionary history of the deep-sea squat lobster Paramunida (Decapoda, Munididae). Evolution 66(6):1878–1896

Cardoso IA, Serejo CS, Rodrigues C (2014) A new Munidopsis species (Galatheoidea, Munidopsidae) from the Southwestern Atlantic. Zootaxa 3815(3):441–446

Castresana J (2000) Selection of conserved blocks from multiple alignments for their use in phylogenetic analysis. Mol Biol Evol 17(4):540–552

Crandall KA, Fitzpatrick JF Jr (1996) Crayfish molecular systematics: using a combination of procedures to estimate phylogeny. Syst Biol 45(1):1–26

Estoup ACRL, Largieader CR, Perrot E, Chorrout D (1996) Rapid one-tube DNA extraction for reliable PCR detection of fish polymorphic markers and transgenes. Mol Mar Biol Biotechnol 5:295–298

Hall TA (1999) BioEdit: a user-friendly biological sequence alignment editor and analysis program for Windows 95/98/NT. Nucleic Acids Symp Ser, London, Information Retrieval Ltd., c1979–c2000 41(41):95–98

Hebert PD, Cywinska A, Ball SL, Dewaard JR (2003) Biological identifications through DNA barcodes. Proc R Soc B Biol Sci 270(1512):313–321

Lefébure T, Douady CJ, Gouy M, Gibert J (2006) Relationship between morphological taxonomy and molecular divergence within Crustacea: proposal of a molecular threshold to help species delimitation. Mol Phylogenet Evol 40(2):435–447

Machordom A, Macpherson E (2004) Rapid radiation and cryptic speciation in squat lobsters of the genus *Munida* (Crustacea, Decapoda) and related genera in the South West Pacific: molecular and morphological evidence. Mol Phylogenet Evol 33(2):259–279

Macpherson E, Machordom A (2005) Use of morphological and molecular data to identify three new sibling species of the genus *Munida* Leach, 1820 (Crustacea, Decapoda, Galatheidae) from New Caledonia. J Nat Hist 39(11):819–834

Macpherson E, Beuck L, Roder C, Voolstra CR (2017) A new species of squat lobster of the genus *Munida* (Galatheoidea, Munididae) from the Red Sea. Crustaceana 90(7–10):1005–1014

Mantelatto FL, Carvalho FL, Simões SM, Negri M, Souza-Carvalho EA, Terossi M (2016) New primers for amplification of cytochrome c oxidase subunit I barcode region designed for species of Decapoda (Crustacea). Nauplius 24:e2016030

Mantelatto FL, Terossi M, Negri M, Buranelli RC, Robles R, Magalhães T, Tamburus AF, Rossi N, Miyazaki MJ (2018) DNA sequence database as a tool to identify decapod crustaceans on the São Paulo coastline. Mitochondrial DNA A 29(5):805–815

Melo GAS (1999) Manual de identificação dos Crustacea Decapoda do litoral brasileiro: Anomura, Thalassinidea, Palinuridea, Astacidea. Ed. Plêidae, São Paulo

Melo-Filho GAS (2006) Reports on the results of the N. Oc. “Prof. W. Besnard” expeditions to the southern coast of Brazil under the Revizee Program: Chirostylidae and Galatheidae (Crustacea: Decapoda: Anomura). Zootaxa 1238(1):1–22

Melo-Filho GAS, Melo GAS (1994) Three new species of *Munida* (Crustacea: Anomura: Galatheidae) from the Brazilian coast. Pap Avulsos Zool 39:49–60

Melo-Filho GAS, Melo GAS (1998) Redescrição e designação de lectótipos para duas especies do gênero *Munida* Leach, 1820 (Crustacea: Decapoda: Galatheidae) coletadas pelo USFC “Albatross” (1885) no Golfo do México. Pap Avulsos Zool 40(12):189–202

Melo-Filho GAS, Melo GAS (2001a) Espécies do gênero *Munida* Leach (Crustacea, Decapoda, Galatheidae), distribuídas na costa do Brasil. Rev Bras Zool 18(4):1135–1176

Melo-Filho GAS, Melo GAS (2001b) Taxonomia e zoogeografia das espécies do gênero *Munida* Leach, 1820 (Crustacea: Decapoda: Galatheidae) distribuídas ao longo da costa temperada-quente do Atlântico-Sul Ocidental. Trop Oceanogr 29(1):37–57

Miller S, Dykes D, Polesky H (1988) A simple salting out procedure for extracting DNA from human nucleated cells. Nucleic Acids Res 16:1215

Miller MA, Pfeiffer W, Schwartz T (2010) Creating the CIPRES Science Gateway for inference of large phylogenetic trees. In: Proceedings of the gateway computing environments workshop (GCE), New Orleans, LA, 14 November 2010, pp 1–8

Palero F, Robainas-Barcia A, Corbari L, Macpherson E (2017) Phylogeny and evolution of shallow-water squat lobsters (Decapoda, Galatheoidea) from the Indo-Pacific. Zool Scr 46(5):584–595

Perez-Barros P, D'Amato ME, Guzman NV, Lovrich GA (2008) Taxonomic status of two South American sympatric squat lobsters, Munida gregaria and Munida subrugosa (Crustacea: Decapoda: Galatheidae), challenged by DNA sequence information. Biol J Linn Soc 94(2):421–434

Poore GC (2014) Three new American species of Munidopsis (Crustacea: Anomura: Munidopsidae). Nauplius 22(1):53–62

Poore GC, Andreakis N (2012) The Agononida incerta species complex unravelled (Crustacea: Decapoda: Anomura: Munididae). Zootaxa 3492(1):1–29

Rodríguez-Flores PC, Machordom A, Abelló P, Cuesta JA, Macpherson E (2019) Species delimitation and multi-locus species tree solve an old taxonomic problem for European squat lobsters of the genus *Munida* Leach, 1820. Mar Biodivers 49(4):1751–1773

Rutherford K, Parkhill J, Crook J, Horsnell T, Rice P, Rajandream MA, Barrell B (2000) Artemis: sequence visualization and annotation. Bioinformatics 16(10):944–945

Schubart CD, Huber MGJ (2006) Genetic comparisons of German populations of the stone crayfish, Austropotamobius torrentium (Crustacea: Astacidae). Bull Fr Pêche Pisci 380–381:1019–1028

Schubart CD, Cuesta JA, Felder DL (2002) Glyptograpsidae, a new brachyuran family from Central America: larval and adult morphology, and a molecular phylogeny of the Grapsoidea. J Crustac Biol 22(1):28–44

Serejo CS, Young PS, Cardoso IC, Tavares C, Rodrigues C, Almeida TC (2007) Abundância, diversidade e zonação dos crustáceos no talude da costa central do Brasil (11°–22° S) coletados pelo programa revizee/score central: prospecção pesqueira. In: Costa PAS, Olavo G, Martins AS (eds) Biodiversidade da fauna marinha profunda na costa central brasileira. Museu Nacional do Rio de Janeiro, Rio de Janeiro, pp 133–162

Silva JM, Creer S, Dos Santos A, Costa AC, Cunha MR, Costa FO, Carvalho GR (2011) Systematic and evolutionary insights derived from mtDNA COI barcode diversity in the Decapoda (Crustacea: Malacostraca). PLoS One 6(5):e19449

Song H, Buhay JE, Whiting MF, Crandall KA (2008) Many species in one: DNA barcoding overestimates the number of species when nuclear mitochondrial pseudogenes are coamplified. Proc Natl Acad Sci U S A 105(36):13486–13491

Stamatakis A (2006) RAxML-VI-HPC: maximum likelihood-based phylogenetic analyses with thousands of taxa and mixed models. Bioinformatics 22(21):2688–2690

Tamura K, Stecher G, Peterson D, Filipski A, Kumar S (2013) MEGA6: molecular evolutionary genetics analysis version 6.0. Mol Biol Evol 30:2725–2729

Tavares M, Campinho P (1998) Three new records of deep-sea squat lobsters of the genus *Munidopsis* Whiteaves from the southwestern Atlantic Ocean (Decapoda: Galatheidae). Crustac Res 27:88–100

Thompson JD, Higgins DG, Gibson TJ (1994) CLUSTAL W: improving the sensitivity of progressive multiple sequence alignment through sequence weighting, position-specific gap penalties and weight matrix choice. Nucleic Acids Res 22(22):4673–4680

Timm L, Bracken-Grissom HD (2015) The forest for the trees: evaluating molecular phylogenies with an emphasis on higher-level Decapoda. J Crustac Biol 35(5):577–592

WoRMS Editorial Board (2020a) Munididae Ahyong, Baba, Macpherson & Poore 2010. http://marinespecies.org/aphia.php?p=taxdetails&id=562645. Accessed 8 Feb 2020

WoRMS Editorial Board (2020b) *Munida* Leach, 1820. http://www.marinespecies.org/aphia.php?p=taxdetails&id=106835. Accessed 9 Feb 2020

Xia X (2013) DAMBE5: a comprehensive software package for data analysis in molecular biology and evolution. Mol Biol Evol 30(7):1720–1728

Xia X, Xie Z, Salemi M, Chen L, Wang Y (2003) An index of substitution saturation and its application. Mol Phylogenet Evol 26(1):1–7

Chapter 15
Biology and Distribution of *Agononida longipes* (Crustacea, Decapoda, Munididae) in the Colombian Caribbean Sea

M. Fierro-Rengifo, N. H. Campos-Campos, and C. M. López-Orozco

Abstract A total of 806 specimens of *Agononida longipes* (A. Milne-Edwards, 1880), munidid crustaceans of the order Decapoda, were collected in 1998 and 2001 along the Colombian Caribbean coast. These samples were used to determine biological aspects and distribution patterns of the species. A population analysis was conducted based on the length of the carapace, sex ratio, number of ovigerous females, average number of eggs, and occurrence of parasitism. These data were compared with the geographic distribution and bathymetry. Results showed differences in sex ratio and higher abundance levels north of the Magdalena River mouth. This could result from abiotic factors such as differences in salinity and temperature due to the prevalence of seasonal upwelling in the northern Colombian Caribbean Sea, as well as the origin and type of sediments and the amount of organic material. Morphometric evidence revealed sexual dimorphism in size, with females being larger than males and the largest individuals found at the greatest depths. The smallest ovigerous female had 10.9 mm of carapace length, and 65.1% were carrying eggs. The number of eggs per female varied between 100 and 5953, with an average of 1360.3. The eggs were elliptical, with an average major axis of 0.56 and an average minor axis of 0.52 mm. Parasitism was low at 3.1% for isopods and 1.0% for rhizocephalan.

Keywords Sex ratio · Fecundity · Parasitism · Eco-region · Continental slope

M. Fierro-Rengifo
Universidad del Magdalena, Programa de Psicología, Santa Marta, Colombia

N. H. Campos-Campos (✉)
Universidad Nacional de Colombia, Instituto de Estudios en Ciencias del Mar, Santa Marta, Colombia
e-mail: nhcampos@unal.edu.co

C. M. López-Orozco
Universidad de Cartagena, Campus San Pablo, Programa de Biología, Grupo de Investigación Hidrobiología, Cartagena de Indias, Colombia

M. E. Hendrickx (ed.), *Deep-Sea Pycnogonids and Crustaceans of the Americas*, https://doi.org/10.1007/978-3-030-58410-8_15

15.1 Introduction

Decapod crustaceans of the family Munididae are one abundant and rich anomuran mega-epibenthic group found on the seafloor and continental slope of the Colombian Caribbean coast (Navas et al. 2003; Campos et al. 2005). The species of this family are important benthic predators and detritivores. They are potential transporters of organic carbon to nearby communities, performing a stabilizing role in the benthic food chain (Chevaldoneé and Olu 1996). Little is known about the biological characteristics of the munidid populations (Hartnoll et al. 1992; Creasey et al. 2000), and what is known focuses on species of economic interest (Gutiérrez and Zuñiga 1977; Serrano-Padilla and Aurioles-Gamboa 1992; Roa 1993; Wolff and Aroca 1995; Roa and Tapia 1998).

Benedict (1902) was one of the earliest researchers to study the biological aspects and development of munidids belonging to the genera *Munida* Leach, 1820, and *Munidopsis* Whiteaves, 1874, in which differences in egg size were observed. Rayner (1935) studied the distribution, growth, larval development, and occurrence of parasites in species of *Munida*. Wenner and Windsor (1979) reported cases of parasitism in three species of Munididae: *Munida iris* A. Milne Edwards, 1880, *Munida microphthalma* A. Milne Edwards, 1880, and *Agononida longipes* (A. Milne-Edwards, 1880), infested with bopyrid isopods. Van Dover and Williams (1991) found a correlation between the cephalothorax size of females and egg size. Sanz-Brau et al. (1998) studied biological aspects such as individual size, abundance of ovigerous females, and sex ratio of *Munida rutllanti* Zariquiey Álvarez, 1952, collected off the coast of Valencia. Studies carried out in North America by Williams and Brown (1972) on *M. iris* and by Wenner (1982) for two other species of *Munida* revealed information on size, sexual dimorphism, sex ratios, reproductive seasonality, and parasitism.

Studies on the biology of *A. longipes* are scarce. Wenner and Windsor (1979) noted the presence of the bopyrid isopod parasite *Munidion* sp. and Wenner (1982) reported on the geographic and depth distribution, sexual composition, and maturity of *A. longipes* as well as occurrence of parasitism along the eastern North American coast. Studies conducted on the continental Colombian Caribbean coast revealed new data about the biology of the species and presence of at least two different populations, suggesting that the Guajira and Darién eco-regions populations have important morphological and biological differences (Bermúdez-Tobón 2007). This study contributes to the knowledge of the life history of *A. longipes* on the continental Colombian Caribbean by providing information about abundance, distribution, and population characteristics. Also, data on size and sex ratio, incidence of parasitism, of ovigerous females, size, and number of eggs are provided.

15.2 Methods

According to Díaz and Acero (2003), the Colombian Caribbean Coast was divided into seven eco-regions: (1) Guajira (GUA), an area with seasonal upwelling and wide continental shelf covered with extensive seagrass and algal beds in the shallower parts; (2) Palomino (PAL), influenced by several streams and small rivers that drain the northern ranges of the Sierra Nevada de Santa Marta; (3) Tayrona (TAY), hilly and mountainous coast, with a very narrow continental shelf and a minor seasonal upwellings; (4) Magdalena (MAG), highly influenced by the Magdalena River discharge that incudes heavy sediment loads; (5) Golfo de Morrosquillo (MOR), influenced by discharges from the Canal del Dique and the Sinú River; (6) Archipiélagos Coralinos (ARCO), with clear waters, high species, and habitat diversity, including coral reefs, seagrass beds, bioclastic sand flats, and fringing mangroves along some islands; and (7) Darién (DAR), with wide continental shelf, mainly terrigenous muds and high influence of mainland discharges, predominantly turbid waters.

We collected 160 trawl samples in 1998 (1 October – 7 December), 1999 (8–15 April), 2000 (1–3 September), and 2001 (14–31 march), at 80 different localities, ranging from 300 to 500 m depth, in three stations at each eco-region (Fig. 15.1), as part of Macrofauna I and II expeditions of the Museo de Historia Natural Marina de Colombia (MHNMC), aboard the R/V *"Ancón"* at each eco-region in the Colombian Caribbean.

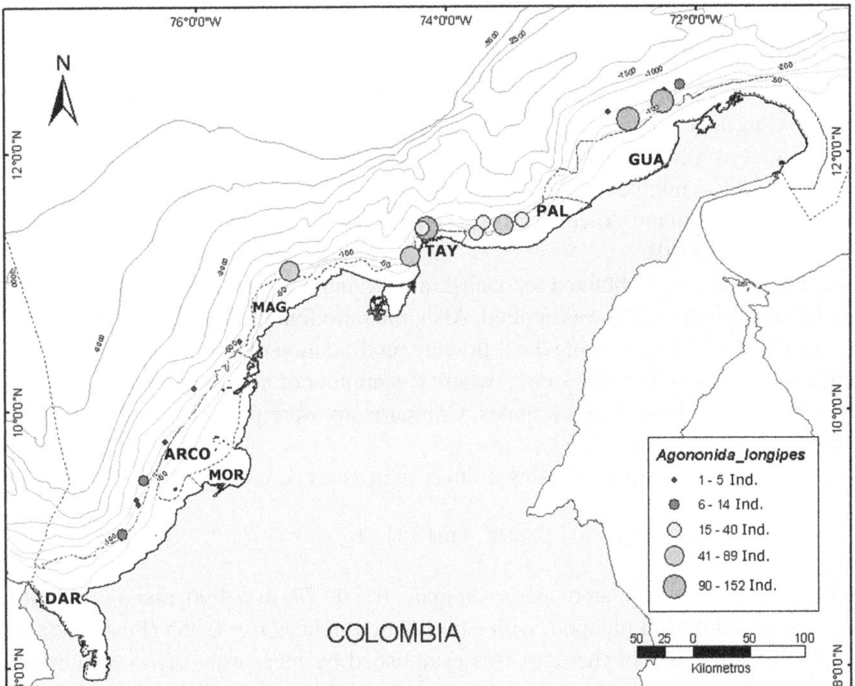

Fig. 15.1 Distribution and abundance of *Agononida longipes* individuals estimated as catch per unit of effort at each station on a 20-minute trawling along the Colombian Caribbean coast. *GUA* Guajira, *PAL* Palomino, *TAY* Tayrona, *MAG* Magdalena, *ARCO* Archipiélagos Coralinos and *DAR* Darien

The collections were made using a semi-balloon bottom trawl with two "V" type metallic doors (91 cm × 63 cm) and a 9 m (estimated 70% during the trawling) mouth aperture. At each station, trawling was performed twice (back and forth), lasting 10 min each, at about 5.6 km/h. We sorted and counted *A. longipes* specimens in each sample and indirectly estimated the abundance as catch per unit of effort at each station as individuals on a 20-minute trawl.

Sampling data (locality, depth, sediments, and abundance data) were summarized in a georeferenced map using ESRI ArcGIS ver. 8.2 software. Sediment type was assigned according to a map produced by the Centro de Investigaciones Oceanográficas e Hidrográficas del Caribe (CIOH 1990).

The specimens were separated by sex (male, non-ovigerous females, and ovigerous females), counted, measured, and deposited at the Museo de Historia Natural Marina de Colombia.

Carapace length without rostrum (CL) was measured to the nearest 0.01 mm from the posterior margin of the orbit to the posterior margin of the carapace. These data were used to establish the median, minimum, maximum, average, and standard deviation of CL for males, females, and ovigerous females. The normal distribution of the data was evaluated using the Kolmogorov-Smirnov test. Comparisons of CL between depths (300–500 m) and eco-regions were conducted among groups of specimens using Analysis of Variance (ANOVA) or Kruskal-Wallis tests with an α of 0.05 by the software Systat 9 (SPSS) and Statistica 6.0 (Statsoft Inc.).

The female proportion was determined for the entire survey (all samples), in 300 and 500 m depth stations, and in the eco-regions, using the equation proposed by Creasey et al. (2000), SO = (M – F)/(M + F), where M = number of males in the sample and F = number of females in the sample. To determine if the proportion of sexes differs from the expected ratio (1:1), we used X^2 tests with a significance level $\alpha = 0.05$ (Zar 2010).

The sex ratio was obtained for each 2 mm CL interval, and the equation proposed by Creasey et al. (2000) was applied. Also, the ratio females/ovigerous females was categorized by CL size, using the following modification of the Creasey et al. (2000) equation, $S_0 = (F - Fov)/(F + Fov)$, where F = number of non-ovigerous females and Fov = number of ovigerous females. Comparisons were performed using ANOVA ($\alpha = 0.05$).

Fecundity was estimated using a linear regression equation:

$$Y = \mathrm{m}(3\,1) + \mathrm{b}$$

where Y = total eggs in an ovigerous female, b = 99.77, m = 4.90, and 3 l = number of eggs in a third left pleopod, with a correlation indices $r^2 = 0.955$ (Fierro 2004).

The average size of the eggs was established by measuring the maximum and minimum diameters of 15 eggs from the third left pleopod of each randomly selected female. The mean and standard deviations of the longest and shortest diameters were established.

The incidence of parasites (%) was determined by direct observation, either as bopyrids or Rhizocephala. For the specimens that were infested with bopyrids, sex of the specimen was recorded. The length of bopyrids was measured from the cephalic region to the posterior edge of uropod's according to the methodology proposed by Wenner and Windsor (1979). CL was calculated for both the host and the parasite. Because the bopyrids produce morphological changes in the carapace and rhizocephala induce sex reversion in infested specimens, we excluded them from the comparative analysis between depth, eco-regions, and sex groups.

15.3 Results

15.3.1 Abundance

A total of 806 specimens of *A. longipes* were collected, 743 individuals at 300 m depth sector (364 M, 123 F, 256 Fov), and 63 individuals at 500 m depth (26 M, 22F, 15Fov), from predominately muddy seafloor (Fig. 15.1).

The highest estimated abundance (as catch per unit of effort at each station as individuals on a 20-minute trawl) occurred in the northern eco-regions, GUA (36%), PAL (23%), and TAY (21%). MAG eco-region, with the influence of the Magdalena River, accounted for 17%. The southern eco-regions had less number of specimens, ARCO 1%, and DAR 2% (Table 15.1).

15.3.2 Carapace Length

We measured the carapace length (CL) of 793 individuals of *Agononida longipes* collected along de Colombian Caribbean Sea in 80 stations. Ovigerous females had the longest CL with a median of 15.6 mm (min 10.9 – max 20.2, avg. 15.6 ± 1.8), followed by the males 14.4 mm (min 5.5 – max 21.0, avg. 13.6 ± 3.1), and non-ovigerous females 12.2 mm (min 6.7 – max 20.6, avg. 12.1 ± 3.1).

Table 15.1 Abundance of sex groups of *Agononida longipes* by depth (300 and 500 m depth) and for eco-regions (*GUA* Guajira, *PAL* Palomino, *TAY* Tayrona, *MAG* Magdalena, *ARCO* Archipiélagos Coralinos, and *DAR* Darien) estimated as catch per unit of effort (individuals for a 20-minute trawl)

	GUA	PAL	TAY	MAG	ARCO	DAR	300 m	500 m	Total
M	142	93	84	59	4	8	364	26	**390**
F	55	40	21	26	1	2	123	22	**145**
FOV	93	56	62	50	0	10	256	15	**271**
%	36	23	21	17	1	2			
300 m	**290**	**154**	**167**	**134**	**4**	**18**			**743**
500 m	**21**	**35**	**0**	**1**	**1**	**2**			**63**
Total	**290**	**189**	**167**	**135**	**5**	**20**			**806**

The smaller sizes were found in the non-ovigerous female group, with the group from 300 m depth sector with a CL average significantly smaller than the specimens from 500 m (F = 12.17 $p = 0.001$). The smallest sizes were dominant at 300 m deep (n = 114), mainly in the CL-class intervals between 8 and 14 mm. The largest sizes were dominant at 500-m-deep sector (n = 22), with the greatest frequency between 12 and 16 mm CL-class intervals (Fig. 15.2).

Males had non-normal distribution (d = 0.10820, $p < 0.01$), and normal distributions were found in non-ovigerous and ovigerous females CL (d = 0.11108, p < 0.1; d = 0.03703, $p < 0.20$) (Fig. 15.3). The LC media in males for all ecoregions was found in the range of 12–16 mm without significant differences. Non-significant differences in LC were found in non-ovigerous females between ecoregions, and the media was found in the range of 14–16 mm. Significant differences were determined in the LC of the ovigerous females between ecoregions, separating the GUA ecoregion from the others (F = 12,768 $p = 0,0$) (Table 15.2). Ovigerous females in GUA with LC distributed between 10 and 20 mm, unlike those that were collected to the south, where the data is concentrated in larger sizes, mainly in the MAG ecoregion. This, together with PAL and TAY, have ranges of similar sizes, and TAY presents the narrowest interquartile range in the area (GUA 13.63–15.73 mm, PAL 14.91–17.51 mm and TAY 14.43–16.89 mm). The medians in this group of ecoregions are in the range of 14.5–17.5 mm. Smaller ovigerous females were found in the GUA ecoregion, where sizes were mainly concentrated between the 12–17 mm intervals, while in the other ecoregions the highest frequencies of ovigerous females

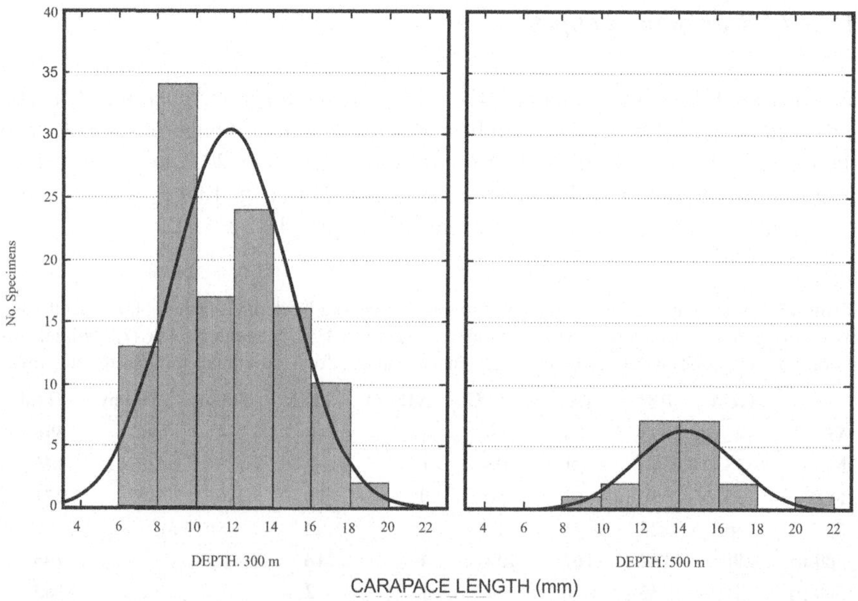

Fig. 15.2 Distribution of abundances of the carapace length (CL) intervals of 2 mm for individuals in stations of 300 and 500 m depth

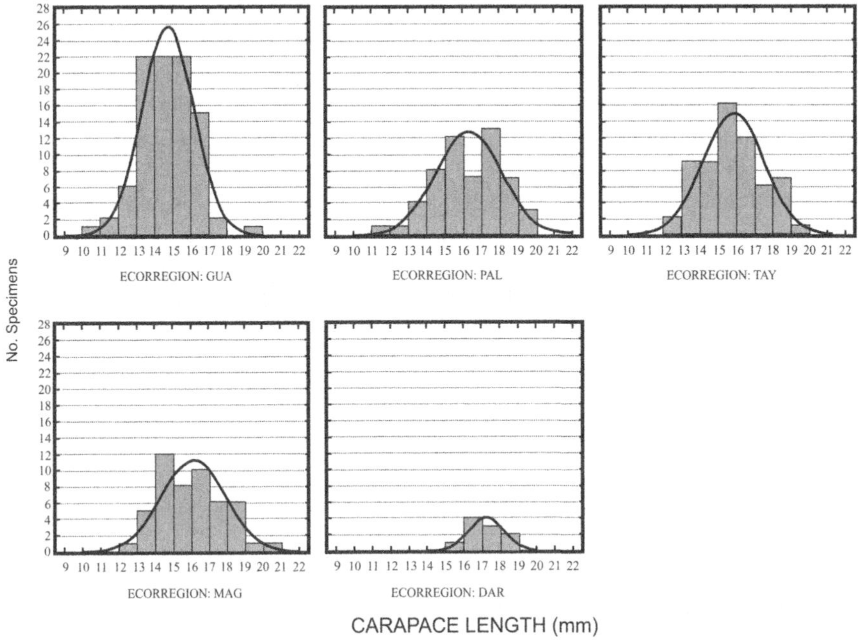

Fig. 15.3 Distribution of abundances of the carapace length of individuals in stations of the eco-regions *GUA* Guajira, *PAL* Palomino, *TAY* Tayrona, *MAG* Magdalena, *ARCO* Archipiélagos Coralinos, and *DAR* Darien

Table 15.2 Comparison of carapace length (CL) for ovigerous females of *Agononida longipes* between eco-regions (three station by eco-region)

Eco-regions	GUA (14.7 mm)	PAL (16.3 mm)	TAY (15.8 mm)	MAG (16.1 mm)	DAR (17.2 mm)
GUA		0.000*	0.003*	0.000*	0.006*
PAL			0.501	0.974	0.719
TAY				0.896	0.302
MAG					0.546

Average CL (mm) in parenthesis. (*) significant differences. Data are p – values

are found from approximately 15–19 mm. Due to the small number of individuals collected in DAR (10 female ovigerous), the size range may possibly widen (Fig. 15.3).

15.3.3 Sex Ratio

The F:M sex ratio of *A. longipes* indicates a deviation (−0.03) of the expected tendency 1:1 in favor of females (51% female and 49% male), but this was not significant between eco-region or depth categories. GUA and PAL, however, had the largest number of females compared to males (sex ratio deviation of −0.3 and -0.1, respectively), increasing slightly in favor of males from north to south (Table 15.3), reaching its highest value in the ARCO eco-region (0.5). Predominance of males was observed in small sizes and up to approximately 11 mm CL (Fig. 15.4). Beyond that size value, females dominated and reached a maximum of 75% of the population in the 18–20 mm size range.

Of the females considered at the 300 m depth (n = 382), 67.8% were ovigerous females (Table 15.4). The smallest proportion was found in the DAR eco-region (−0.8), followed by the TAY eco-region (−0.5) with negative values suggesting a tendency of deviation from the 1:1 ratio in favor of ovigerous females in the largest size range (Fig. 15.5).

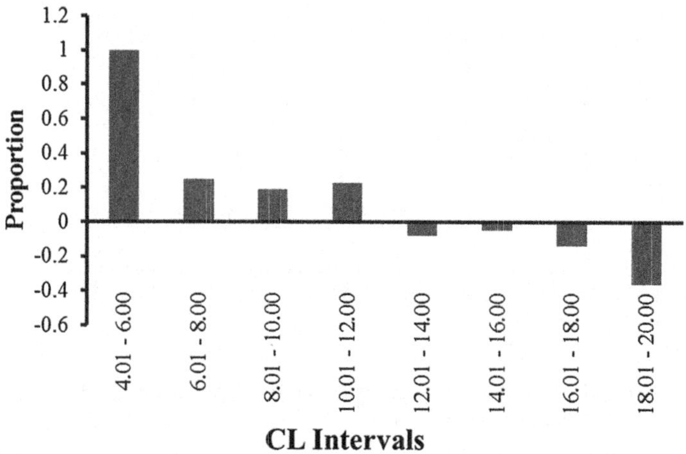

Fig. 15.4 Sex ratio of *Agononida longipes* for each size class (carapace length, CL). The negative values indicate a deviation of the expected tendency 1:1 in favor to females

Table 15.3 Male/female proportion of *Agononida longipes* by eco-region and depth

	Proportion	
Eco-region	300 m	500 m
GUA	0.0	−0.3
PAL	0.0	−0.1
TAY	0.0	0.0
MAG	−0.1	−1.0
ARCO	1.0	0.5
DAR	−0.1	−1.0
Total N	**742**	**67**

Negative values indicate a deviation of the expected tendency 1:1 in favor to females. *GUA* Guajira, *PAL* Palomino, *TAY* Tayrona, *MAG* Magdalena, *ARCO* Archipiélagos Coralinos, and *DAR* Darien

Table 15.4 Proportion of non-ovigerous females/ ovigerous females of *Agononida longipes* by eco-region and depth

	Proportion	
Eco-region	300 m	500 m
GUA	−0.4	0.7
PAL	−0.2	−0.1
TAY	−0.5	0.0
MAG	−0.4	−1.0
ARCO	0.0	1.0
DAR	−0.8	−1.0
Total N	**382**	**37**

Negative values indicate a deviation of the expected tendency 1:1 in favor to ovigerous females. *GUA* Guajira, *PAL* Palomino, *TAY* Tayrona, *MAG* Magdalena, *ARCO* Archipiélagos Coralinos, and *DAR* Darien

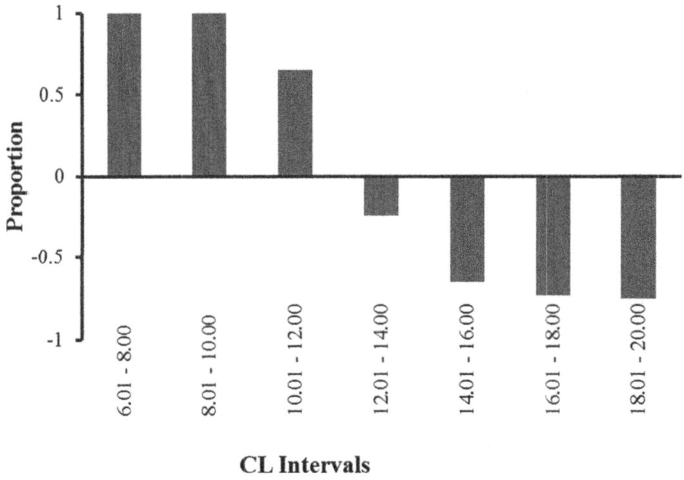

Fig. 15.5 Proportion of non-ovigerous females: ovigerous females vs. carapace length (CL) intervals of 2 mm. The negative values indicate a deviation of the expected tendency 1:1 in favor to ovigerous females

15.3.4 Fecundity

Of 416 females, 271 were ovigerous. At 300-m-depth sector, most of the females (379 ind) were carrying eggs (256). Thirty-seven females were collected at 500 m depth, of which 40.5% were ovigerous (Table 15.1). The smallest ovigerous female (GUA, 300 m depth) had CL 10.9 mm. The number of eggs per female varied between 100 and 5953, with an average of 1360.3 ± 53.6. CL values and numbers of eggs were poorly correlated ($r^2 = 0.30$). The eggs of *A. longipes* were elliptical, with an average major axis of 0.56 ± 0.003 mm and an average minor axis of 0.52 ± 0.002 mm.

15.3.5 Parasite Infestation

Twenty-five specimens of *A. longipes* (CL 10.6–19.3 mm) were infested by bopyrids (3.1% of total) and eight by Rhizocephala (CL 13.8–21.0 mm, 1.0% of total). Fourteen bopyrids were located in the left branchial cavity and three in the right branchial cavity. The length of bopyrids was determined in only nine females (CL between 0.7–1.33 cm) and two males (CL 0.19 and 0.39 cm, respectively). A deformed cephalothorax in several samples suggested the presence of parasites that were not found attached to the specimen. We found specimens parasitized by bopyrids in GUA, PAL, and DAR and by rhizocephala in all eco-regions except DAR. Statistical differences between ecoregions and depth sectors were not calculated because of the number of specimens was not enough.

15.4 Discussion

15.4.1 Abundance

Similar to what has been reported for the tropical western Atlantic, *A. longipes* was one of the most common species of galatheids along the continental Colombian Caribbean (Wenner 1982; Pequegnat and Pequegnat 1970). This species has been commonly collected with *Munida flinti* Benedict, 1902 and *Munida stimpsoni* A. Milne-Edwards, 1880 in the continental slope (Melo-Filho and Melo 2001). No specimens were found at such shallow depths as those recorded by Takeda and Okutani (1983) for Suriname and French Guiana, where it occurs at 40 m. This species usually prefers the upper continental slope, between 270 and 729 m (Pequegnat and Pequegnat 1970; Melo-Filho and Melo 2001).

The relative abundance of *A. longipes* (as catch per unit of effort at each station as individuals on a 20-minute trawl) was higher in stations in the north of the study area (GUA, PAL, and TAY), region under the influence of seasonal upwellings (Lozano-Duque et al. 2010). In PAL eco-region, most of the specimens of *A. longipes* were collected in 300 m in an area where large amounts of decomposing *Thalassia testudinum* K.D.Koenig, 1805, and pieces of wood were recovered, which might favor the presence of *A. longipes* considering that species of Munididae are known to be omnivorous (Garm and Høeg 2000). The high degree of physicochemical disturbance and instability of seabeds caused by continental waters and the vast amounts of silt deposited by the rivers Sinú and Magdalena may cause the low abundance of this species in the eco-regions MAG, MOR, and DAR.

Based on morphometric and genetic data, Bermúdez-Tobón (2007) postulated that one of the main factors affecting the distribution of *A. longipes* is the oceanic current pattern. Indeed, the Caribbean currents and countercurrents disperse the planktonic larvae of *A. longipes* and other species along the coast and offshore in the north.

15.4.2 Carapace Length

The size variation between sexes could be a factor that defines the differences in the abundance of males and females in the *A. longipes* populations in the Colombian Caribbean Sea as was shown by Wenner (1972), Palma and Arana (1996), and Creasey et al. (2000) in other regions. In the present study, males smaller than 11 mm were proportionally more abundant than females, the latter being dominant in larger sizes. As seen in some other crustaceans (Wenner 1972; Hartnoll et al. 1993), it is possible that the male proportions were relatively uniform among the smaller sizes but became smaller with growth.

Hartnoll et al. (1993) postulated that mortality rates could influence the ratio towards larger females since the reduction in male survival in the adult stage caused by the increase in susceptibility to predation would reduce the total male population (Wenner 1972; Sastry 1983; Hartnoll et al. 1993; Turra and Souza 2003). Susceptibility to predation is linked to the development of secondary sexual characters in males, such as the large, more prominent claws, limiting the ability to be easily camouflaged and escape predators. This would result in higher survival rates for females in the adult stage (Hartnoll et al. 1993).

The average sizes for ovigerous females, non-ovigerous females, and males showed significant differences, with ovigerous females having the largest average size followed by males and non-ovigerous females. This tendency of females to dominate the largest size classes, beginning at approximately 11 mm, suggests different growth rates among males and females, similar to what has been reported for other species (Wenner 1972; Petriella and Boschi 1997; Creasey et al. 2000).

The abiotic factors also have an effect on biomass and size in crustaceans. Petriella and Boschi (1997) demonstrated that specimens of the same age have different weights and sizes resulting from different environmental conditions and the availability of food, among other factors. The growth rate varies with the animal age. In early stages of development such as postlarva, juveniles, and subadults, the growth is variable, moulting is frequent, and intermoulting is similar between males and females (Hartnoll 1983; Petriella and Boschi 1997; Gramitto and Froglia 1998).

15.4.3 Sex Ratio

Because crustaceans have a pair of heteromorphic sexual chromosomes, it is expected to observe a 1:1 sex ratio in eggs produced (Ginsburger-Vogel and Charniaux-Cotton 1982; Creasey et al. 2000). However, the results of the present study show that the abundance and sex ratio can vary between regions, depending on their oceanic and ecological characteristics and the adaptations of the species. Likewise, in other studies, we found the same tendency, such as in the middle Atlantic Bight, where Wenner (1982) reports a significant trend for a higher number of females of galatheids. Conversely, when we analyze the sex ratio for *A. longipes* by depth, no significant differences were found from the expected ratio 1:1; this could indicate homogenous environmental conditions, which are common in deep-sea environments (Sastry 1983). The variations between eco-regions were less than changes by depth.

15.4.4 Fecundity

At 300 m in the northern region of the Colombian Caribbean Sea, *A. longipes* showed a major concentration of ovigerous females in GUA and TAY but without significant difference, suggesting the presence of a "nursery" zone. The greater abundance of ovigerous females is associated with high productivity zones, which would assure feeding benefits for the parents and the larvae (Aurioles-Gamboa and Pérez-Flores 1997).

The number of eggs in *A. longipes* was below those reported by Wenner (1982). Studies on *Munida rugosa* suggest that a kind of strategy of reproduction is to produce incomplete masses of eggs to attract males by a chemical signal. Beforehand, when the males are ready to mate, the females produce a complete egg mass at the same time for fecundation (Tapella et al. 2002). Otherwise, the number of eggs can vary due to accidental losses caused by incomplete fertilization (Chacur and Negreiros 1999) or by spawning before collection. It is expected that females with larger CL bear larger egg masses (Tapella et al. 2002) because they have a larger abdominal volume, which implies a higher capacity to carry egg masses (Clarke 1993; Chacur and Negreiros 1999). However, in this study, no significant relationship was found between CL and the number of eggs. Although Sastry (1983) justifies this by reporting that these relations are not very clear in some species, it is possible that the collection process and handling of specimens caused a significant loss of eggs.

Considering the morphology of the egg and the development of the larva (Thorson 1950), *A. longipes* has a planktotrophic reproductive mechanism with small eggs forming large masses, similar to species of the genus *Munida* (Gore 1979; Konishi and Saito 2000). This mechanism is efficient when food availability is optimal (Wilkens et al. 1990; Gómez-Gutierrez and Sánchez-Ortiz 1997), and predation is low (Vance 1973a, b). It is simultaneously determined by external factors such as temperature and salinity (Sulkin 1978; Christiansen and Anger 1990).

The size of the eggs is the first variable that allows the prediction of fecundity patterns, dispersion, and general ecology of the larvae (Vance 1973a, b; Hines 1986) and is not determined by the habitat (Van Dover and Williams 1991). For spherical eggs, the approximate egg diameter of species with planktotrophic development is

0.6 mm, and minimum conditions are required to carry out the biochemical processes needed to form embryos (Van Dover and Williams 1991). In the present study, we found that the average diameter of eggs was less than 0.6 mm, and the number of eggs that could reach 5000 per female indicates that the dispersion strategy could by planktotrophic. Similar data were found for *A. longipes* of the middle Atlantic, where the egg diameters are between 0.49–0.68 mm and 2693–9327 eggs per female (Wenner 1982).

15.4.5 Parasite Infestation

The species of *Munida* are known to be mainly infested by bopyridae (Malacostraca, Isopoda, Bopyridae) and rhizocephala (Cirripedia, Rhizocephala) (Reinhard 1958; Bourdon 1972; Williams and Brown 1972; Wenner and Windsor 1979; Wenner 1982; Gore 1983; Markham 1988; Boyko and Williams 2011). Williams and Brown (1972) reported that 10% of specimens of *Munida iris* were infested by bopyridae on the continental platform off North Carolina. In contrast, the percentage of parasite infestation found in the Colombian Caribbean Sea is low for *A. longipes* (2.1%). Additionally, the infested specimens of *A. longipes* showed a wide CL range, suggesting that the presence of parasites does not interrupt, at least not completely, host development, as indicated by Creasey et al. (2000) for *Munidopsis scobina* Alcock, 1894.

The presence of rhizocephala has been recorded in various species of galatheids such as *Munida irrasa* A. Milne Edwards, 1880, and *G. rostrata* A. Milne Edwards, 1880 (Ritchie and Høeg 1981; Wenner 1982). For *Agononida*, the low level of infestation reported in this study (0.9%) was less than that recorded for *Munidopsis* spp., 1–5% (Wenner 1982; Creasey et al. 2000). It is possible that the rhizocephala interrupted the sexual maturity process in the host specimens because the infested females showed sizes of reproductive maturity, yet had no eggs. This coincides with Creasey et al. (2000), who suggested that rhizocephala-infested smaller-sized specimens (juveniles) induced maturity in the host, thereby allowing the parasite to reproduce first precisely in one of the most vulnerable moments in the host life cycle (ecdysis). The presence of rhizocephala induces the development of female characteristics in the host, directing parental care towards the parasite due to the imitation of the eggs mass (Ritchie and Høeg 1981; Creasey et al. 2000). This causes the inhibition of moulting and the delay or atrophy of the gonads (Petriella and Boschi 1997).

In this study the biological data of *Agononida longipes* from the continental slope of the Colombian Caribbean Sea are congruent with a population in good condition when compared to other areas at the Atlantic Ridge and with another species of galatheids. This is important, due to the relevant role that galatheids play in benthic environments of the deep sea as detritivores, and also prey in the neotropical trophic nets.

Acknowledgments The authors appreciate the contributions of Drs. Adriana Bermúdez and Gabriel R. Navas S. (Programa de Biología, Universidad de Cartagena) and acknowledge the participation of the following research groups: Hidrobiología (Universidad de Cartagena); Museo de Historia Natural Marina de Colombia (INVEMAR); Fauna Marina Colombiana, Biodiversidad y Usos (Universidad Nacional de Colombia); and Ecological Genetics and Conservation (Manchester Metropolitan University). The authors would like to thank the anonymous reviewers for their comments. Contribución No. 503 del Instituto de Estudios en Ciencias del Mar, CECIMAR, de la Universidad Nacional de Colombia, Sede Caribe.

References

Aurioles-Gamboa D, Pérez-Flores R (1997) Seasonal and bathymetric changes in feeding habits of the benthic crab *Pleuroncodes planipes* (Decapoda, Anomura, Galatheidae) off the Pacific coast of Baja California Sur, México. Crustaceana 70(3):272–287. https://doi.org/10.1163/156854097X00573

Benedict J (1902) Description of a new genus and forty-six new species of crustaceans of the family Galatheidae with a list of the known marine species. Proc Biol Soc Wash 26(1311):243–334. https://doi.org/10.5479/si.00963801.26-1311.243

Bermúdez-Tobón A (2007) Populationsstruktur und genetische Isolierung der amerikanischen Springkrebs-Arten *Munida flinti*, *Agononida longipes* und *Munidopsis riveroy* (Crustacea, Decopoda, Anomura, Galatheidae). Dissertation, Justus Liebig University, Gieβen

Bourdon R (1972) Sur quelques Bopyridae (Crustacea: Isopoda) parasites de Galathéides. Bull Mus Nat Hist Nat 3e ser 66:817–838

Boyko CB, Williams JD (2011) Parasites and other symbionts of squat lobsters. In: Poore GCB, Ahyong ST, Taylor J (eds) The biology of squat lobsters. CSIRO Publising, Collingwood, pp 271–296

Campos NH, Navas GR, Bermúdez A, Cruz N (2005) Los crustáceos decápodos de la franja superior del talud continental (300–500 m) del mar Caribe Colombiano. Monografías de Fauna de Colombia Instituto de Ciencias Naturales, Facultad de Ciencias, Universidad Nacional de Colombia, Bogotá

Chacur M, Negreiros M (1999) Aspectos biológicos do Camarão-espinho *Exhippolysmata oplophoroides* (Holthius, 1948) (Crustacea, Caridea, Hyppolytidae). Rev Bras Biol 59(1):173–181. https://doi.org/10.1590/S0034-71081999000100022

Chevaldoneé P, Olu K (1996) Occurrence of anomuran crabs (Crustacea: Decapoda) in hydrothermal vent and cold-seep communities: a review. Proc Biol Soc Wash 109(2):286–298

Christiansen M, Anger K (1990) Complete larval development of *Galathea intermedia* Lilljeborg reared in laboratory culture (Anomura: Galatheidae). Jour Crust Biol 10(1):87–111. https://doi.org/10.2307/1548672

CIOH (Centro de Investigaciones Oceanográficas e Hidrológicas del Caribe) (1990) Mapa de repartición de facies sedimentarias y perfiles batimétricos escala 1:300000. Centro de Investigaciones Oceanográficas e Hidrológicas del Caribe, Cartagena

Clarke A (1993) Reproductive trade-offs of caridean shrimps. Funct Ecol 7:411–419. https://doi.org/10.2307/2390028

Creasey S, Rogers A, Tyler P, Gage J, Jolliver D (2000) Genetic and morphometric comparisons of squat lobster, *Munidopsis scobina* (Decapoda: Anomura: Galatheidae) populations, with notes on the phylogeny of the genus *Munidopsis*. Deep Sea Res Part II Top Stud Oceanogr 47(1):87–118. https://doi.org/10.1016/S0967-0645(99)00098-3

Díaz JM, Acero A (2003) Marine biodiversity in Colombia: achievements, status of knowledge and challenges. Gayana 67(2):261–274. https://doi.org/10.4067/S0717-65382003000200011

Fierro M (2004) Estudio de los principales aspectos biológicos y de distribución de las especies dominantes de *Agononida, Munida* y *Munidopsis* en el Caribe colombiano (Crustacea:

Decapoda: Galatheidae). Trabajo de Grado, Departamento de Biología, Universidad Nacional de Colombia, Bogotá

Garm A, Høeg JT (2000) Functional mouthpart morphology of the squat lobster *Munida sarsi*, with comparison to other Anomurans. Mar Biol 137:123–138. https://doi.org/10.1007/s002270000318

Ginsburger-Vogel T, Charniaux-Cotton H (1982) Sex determination. In: Abele LG (ed) The biology of crustacea, embryology, morphology and genetics, vol 2. Academic Press, New York, pp 257–281

Gómez-Gutierrez J, Sánchez-Ortiz CA (1997) Larval drift and population structure of the pelagic phase of *Pleuroncodes planipes* (Stimpson) (Crustacea: Galatheidae) off the southwest coast of Baja California, Mexico. Bull Mar Sci 61(2):305–325

Gore R (1979) Larval development of *Galathea rostrata* under laboratory conditions, with a discussion of larval development in the Galatheidae (Crustacea: Anomura). Fish Bull 76(4):781–806

Gore R (1983) Notes on rare species of *Munidopsis* (Anomura: Galatheidae) and *Ethusina* (Brachyura: Dorippidae) collected by the USNS Bartlett in the Venezuela Basin, Caribbean Sea. Proc Acad Nat Scie Phil 135:200–217

Gramitto M, Froglia C (1998) Notes on the biology and growth of *Munida intermedia* (Anomura: Galatheidae) in the western Pomo Pit (Adriatic Sea). J Nat Hist 32:1553–1566. https://doi.org/10.1080/00222939800771091

Gutiérrez J, Zuñiga O (1977) *Pleuroncodes monodon* H. Milne Edwards en la bahía de Mejillones del sur, Chile (Crustacea, Decapoda, Anomura). Rev Biol Mar 16(2):161–169

Hartnoll R (1983) Strategies of crustacean growth. In: Lowry J (ed) Papers from the conference on the biology and evolution of crustacea, vol 18. Mem Aust Mus, Sydney, pp 121–131

Hartnoll RG, Rice AL, Attrill MJ (1992) Aspects of the biology of the galatheid genus *Munida* (Crustacea, Decapoda) from the Porcupine Seabight, Northeast Atlantic. Sarsia 76:231–246. https://doi.org/10.1080/00364827.1992.10413479

Hartnoll RG, Bryant AD, Gould P (1993) Size distribution in spider crab populations –spatial and temporal variation. Jour Crust Biol 13(4):647–655. https://doi.org/10.1163/193724093X00219

Hines A (1986) Larval problems and perspectives in life histories of marine invertebrates. Bull Mar Sci 39(2):506–525

Konishi K, Saito T (2000) Larvae of the deep-sea squat lobsters, *Agononida incerta* (Henderson, 1888) and *Munida striola* Macpherson and Baba, 1993 with notes on larval morphology of the family (Crustacea: Anomura: Galatheidae). Zoolog Sci 17:1021–1029. https://doi.org/10.2108/zsj.17.1021

Lozano-Duque Y, Medellín-Mora J, Navas GR (2010) Contexto climatológico y oceanográfico del mar Caribe colombiano. In: Navas GR, Segura-Quintero C, Garrido-Linares M, Benavides-Serrato M, Alonso D (eds) Biodiversidad del margen continental del Caribe colombiano. Serie de publicaciones especiales del Invemar n 20, Santa Marta, pp 53–84

Markham J (1988) Descriptions and revisions of some species of Isopoda Bopyridae of the north western Atlantic Ocean. Zool Verhandel 246:3–63

Melo-Filho GAS, Melo GAS (2001) Espécies do gênero *Munida* Leach (Custacea, Decapoda, Galatheidae) coletadas entre Torres (Rio Grande do Sul, Brasil) e Maldonado (Uruguay) pelo Programa Rio Grande do Sul (PRS I-II, GEDIP). Rev Bras Zool 18(4):1177–1196. https://doi.org/10.1590/S0101-81752001000400014

Navas GR, Bermúdez A, Cruz N, Campos NH (2003) Galatéidos (Decapoda, Anomura, Galatheidae) del Caribe colombiano, incluyendo doce primeros registros. Bol Invest Mar Cost 32:183–218

Palma S, Arana P (1996) Método rápido de determinación del sexo en el erizo comestible *Loxechinus albus* (Molina, 1792) y su aplicación en estudios biológico-pesqueros. Investig Mar 24:123–130. https://doi.org/10.4067/S0717-71781996002400010

Pequegnat WE, Pequegnat LH (1970) Deep-Sea Anomurans of superfamily Galatheoidea with descriptions of two new species. In: Pequegnat W, Chace F (eds) Contribution on the biology of the Gulf of México. University Oceanographic Studies, Texas A&M, pp 125–170

Petriella A, Boschi E (1997) Crecimiento en crustáceos decápodos: resultados de investigaciones realizadas en Argentina. Investig Mar 25:135–157. https://doi.org/10.4067/S0717-71781997002500010

Rayner GW (1935) The Falkland species of the crustacean genus *Munida*. Discov Rep 10:212–245

Reinhard E (1958) Rhizocephala of the family Peltogastridae parasitic on West Indian species of Galatheidae. Proc US Nat Mus 108(3403):295–307. https://doi.org/10.5479/si.00963801.108-3403.295

Ritchie L, Høeg JT (1981) The life history of *Lernaeodiscus porcellanae* (Cirripedia: Rhizocephala) and co-evolution with its porcellanid host. Jour Crust Biol 1(3):334–347. https://doi.org/10.2307/1547966

Roa R (1993) Annual growth and maturity function of squat lobster *Pleuroncodes monodon* in Central Chile. Mar Ecol Prog Ser 97:157–166. https://doi.org/10.3354/meps097157

Roa R, Tapia F (1998) Spatial differences in growth and sexual maturity between branches of a large population of the squat lobster *Pleuroncodes monodon*. Mar Ecol Prog Ser 167:185–196. https://doi.org/10.3354/meps167185

Sanz-Brau A, Mezquita F, Sanz S (1998) Seasonality in the reproductive biology of a shelf-dwelling galatheid (Crustacea: Decapoda): an example from a western Mediterranean population of *Munida rutllanti*. Cah Biol Mar 39:185–195

Sastry AN (1983) Ecological aspects of reproduction. In: Vernberg FJ, Vernberg WB (eds) The biology of crustacea, environmental adaptations, vol 8. Academic Press, New York, pp 179–270

Serrano-Padilla AV, Aurioles-Gamboa D (1992) Dimorfismo sexual en la langostilla, *Pleuroncodes planipes* Stimpson, 1860 (Crustacea: Decapoda: Galatheidae). Proc San Diego Soc Nat Hist 13:1–5

Sulkin S (1978) Nutritional requirements during larval development of the portunid crab, *Callinectes sapidus* Rathbun. J Exp Mar Biol Ecol 34:29–41. https://doi.org/10.1016/0022-0981(78)90055-2

Takeda M, Okutani T (1983) Crustaceans and mollusks trawled off Suriname and French Guiana. Japan Marine Fishery Resource Research Center, Tokio

Tapella F, Lovrich G, Romero C, Thatje S (2002) Reproductive biology of the crab *Munida subrugosa* (Decapoda: Anomura: Galatheidae) in the Beagle Channel, Argentina. J Mar Biol Assoc UK 82:589–595. https://doi.org/10.1017/S0025315402005921

Thorson G (1950) Reproductive and larval ecology of marine bottom invertebrates. Biol Rev 25:1–45. https://doi.org/10.1111/j.1469-185X.1950.tb00585.x

Turra L, Souza E (2003) Population biology and distribution of the tanaid *Kalliapseudes schubarti* Mañé-Garzón, 1949, in an intertidal flat in southeastern Brazil. Braz J Biol 63(3):469–479. https://doi.org/10.1590/S1519-69842003000300013

Van Dover C, Williams A (1991) Egg size in squat lobsters (Galatheoidea): constraint and freedom. In: Wenner A, Kuris AA (eds) Crustacean egg reproduction. Balkema, Rotterdam/Brookfield, pp 143–156

Vance R (1973a) On reproductive strategies in marine benthic invertebrates. Am Nat 107(955):339–352

Vance R (1973b) More on reproductive strategies in marine benthic invertebrates. Am Nat 107(955):353–361

Wenner AW (1972) Sex ratio as a function of size in marine crustacea. Am Nat 106(949):231–350

Wenner E (1982) Notes on the distribution and biology of Galatheidae and Chyrostilidae (Decapoda: Anomura) from the middle Atlantic Bight. Jour Crust Biol 2(3):360–377

Wenner E, Windsor N (1979) Parasitism of galatheid crustaceans from the Norfolk Canyon and Middle Atlantic bight by Bopyrid Isopods. Crustaceana 37(3):293–303. https://doi.org/10.1163/156854079X01176

Wilkens H, Parzefall J, Ribowski A (1990) Population biology and larvae of the anchialine crab *Munidopsis polymorpha* (Galatheidae) from Lanzarote (Canary Islands). Jour Crust Biol 10(4):667–675. https://doi.org/10.2307/1548411

Williams AB, Brown WS (1972) Notes on structure and parasitism of *Munida iris* A. Milne Edwards (Decapoda: Galatheidae) from North Carolina, USA. Crustaceana 22:303–308. https://doi.org/10.1163/156854072X00598

Wolff M, Aroca T (1995) Population dynamics and fishery of the Chilean squat lobster *Cervimunida johni* Porter, (Decapoda: Galatheidae) off the coast of Coquimbo, Northern Chile. Rev Biol Mar 30(1):57–70

Zar JH (2010) Biostatistical analysis, 5th edn. Prentice Hall, Upper Saddle River

Chapter 16
King Crabs of Peruvian Waters During 2003–2004: New Insights

J. Arguelles, P. Larriviere, S. Thatje, and M. Pérez

Abstract Important fisheries of lithodid crabs have been conducted mainly at high latitudes of both hemispheres. In waters of Peru, research and exploratory fishing began only in the 1970s, showing the existence of several species of lithodids, also known as king crab. An offshore exploratory fishery along the coast of Peru was developed in 2003–2004, and this study shows biological and fisheries aspects, which are potentially important for the development of a regional deep-water fishery of king crab. The results show the predominance of *Paralomis longipes* contrary to the dominance of *L. panamensis* reported during a previous survey in 1997–1998. The king crab species found showed a wide bathymetric (500–1500 m) and latitudinal distribution range. However, it is unknown how many stocks or genetic populations exist in waters off Peru. Higher relative abundances of *P. longipes* were estimated between March and October and in deeper waters (>1000 m), which might indicate seasonal bathymetric migrations as part of the life cycle. Males were larger than females, and a size at gonadal maturity of 88.4 mm carapace width was estimated for *P. longipes* females. The presence of parasitic rhizocephalans was observed in *P. longipes*, but not in *L. panamensis* and *L. wiracocha*, probably because of a preference of these parasites for *P. longipes*.

Keywords Lithodidae · Distribution · Fisheries · Size structure · Parasites

---

J. Arguelles (✉) · M. Pérez
Instituto del Mar del Perú, Esq. Gamarra y Gral. Valle s/n Chucuito, Callao, Lima, Peru
e-mail: jarguelles@imarpe.gob.pe; mperez@imarpe.gob.pe

P. Larriviere
Pascual de Andagoya 270 San Miguel, Lima, Peru

S. Thatje
National Oceanography Center Southampton, Ocean and Earth Science,
University of Southampton, Southampton, UK

M. E. Hendrickx (ed.), *Deep-Sea Pycnogonids and Crustaceans of the Americas*, https://doi.org/10.1007/978-3-030-58410-8_16

16.1 Introduction

Lithodid crabs have a wide distribution worldwide and are found from the sublittoral, 50–60 m for *Cryptolithodes expansus* (Kim and Hong 2000) to 3500 m for deep-sea *Neolithodes* sp. (Hall and Thatje 2011). They are typical inhabitants of cold and temperate-cold waters in both hemispheres (del Solar 1981). Their distribution in shallow waters is limited by their thermal tolerance up to 13 °C, and deep and cold water species are excluded from the waters on the continental shelf (Hall and Thatje 2009; Hall and Thatje 2011). Of the at least 121 known species of lithodids (Hall and Thatje 2009a), the greatest diversity and most species (70%) are found in the North Pacific (Zaklan 2002; Stevens and Lovrich 2014). In South America, 34 species have been reported in 4 genera (23 *Paralomis* spp., 7 *Lithodes* spp., 3 *Neolithodes* spp., 1 *Glyptolithodes* sp. (Stevens and Lovrich 2014).

In Peruvian waters, investigations on lithodids began in 1971, when the first specimen of lithodid, a female of *Glyptolithodes cristatipes* (del Solar 1987), was captured on board the RV SNP-1 off Chicama port, at 693 m depth. Subsequently, other cruises were conducted aboard the trawlers "Wiracocha" and "Challwa japic N°1", identifying the species *Lithodes wiracocha* for first time. So far, the presence of nine species of lithodids has been recorded of Peru: *Glyptolithodes cristatipes*, *Lopholithodes diomedeae*, *Paralomis aspera*, *P. longipes*, *P. papillata*, *P. inca*, *Lithodes panamensis*, *L. wiracocha*, and *Neolithodes* sp. (Haig 1974; del Solar 1981; del Solar 1987). The highest concentrations were reported off Tumbes (03°30′S latitude), on a clay mud loaded with organic debris but without a sulfurous odor, and show an abundance decrease from north to south in the archibenthic area of Peru (del Solar 1981). In these areas where lithodids thrive, the temperature varies between 5 °C at 1100 m and 8 °C at 500 m depth, and the oxygen level is greater than 0.70 mL/L at depths greater than 500 m (del Solar 1981).

From September 1997 to March 1998, the experimental fishing for alternative marine resources as "eels," "king crabs," "octopus," "snail," and "sea lamprey" was carried out on board the Korean ship "Moresko 1" from 8°40′S (Chimbote) to 17°58′S (Ilo), between 18.5 and 90.7 km from the coast and from 628 to 1147 m depth. The captured king crabs were *Lithodes panamensis*, *Paralomis diomedeae*, *Paralomis papillata*, *P. longipes*, *Lithodes wiracocha*, and *Glyptolithodes cristatipes*. A total of 507 kg of king crab were captured. The size range of crabs by species is shown in Table 16.1. The monthly relative abundance varied from 0 to 0.81 kg/pot with highest values in October (Juarez et al. 1998).

In December 2005, a deep-water crab survey aboard research vessel IMARPE VI was conducted from 12°S to 15°S using a long-line with trap between 700 and 1400 m. Four species were captured: *Paralomis longipes*, *L. wiracocha*, *L. panamensis*, and *P. diomedeae*, with the dominance of *P. longipes*. The highest abundances indexes were estimated at 14°S at two depth strata (700–1000 and 1000–1400 m depth), while at 12°S the highest abundances occurred at 1000–1400 m

Table 16.1 Size range of crabs reported during the experimental fishing on board the Korean ship "Moresko 1" from September 1997 to March 1998 in Peruvian waters

Specie	Sex	Caparace width range (mm)
Lithodes panamensis	Female	59–176
	Male	105–181
Paralomis diomedeae	Female	74–164
	Male	58–174
Paralomis papillata	Female	97–129
	Male	105–137
Paralomis longipes	Female	84–120
	Male	90–126
Lithodes wiracocha	Female	97–133
	Male	88–148
Glyptolithodes cristatipes	Female	55–89
	Male	50–100

depth. Of the four captured species, *P. longipes* represented the 93,9% followed by *L. panamensis* (5.5%), *P. diomedeae*, and *L. wiracocha*. The size ranged from 111.8 to 175.8 mm CW for *L. panamensis* and 83.4 to 147.6 CW for *P. longipes*, with largest size at deeper waters. The size for *P. diomedeae* and *L. wiracocha* were 99.6 and 129.6 mm CW, respectively. The reported prevalence of parasite infestation was highest in *P. longipes* (7.2%) (Arguelles et al. 2014).

A deep-sea exploratory fishing of king crabs of the longest duration in Peruvian waters was conducted from December 2003 to October 2004. This was carried out aboard long-line vessels equipped with baited pots, and its objective was to determine the distribution, relative abundance, and biological characteristics of the crabs south of 7°S in Peruvian waters. The present paper presents the results of this exploratory fishing, monthly landings, size structure, and distribution of the main species of lithodids in the Peruvian deep sea.

16.2 Materials and Methods

16.2.1 Area and Period of Exploratory Fishing

Exploratory fishing was conducted using long-lines with baited pots off the Peruvian coast from 9°S to 16°S from December 2003 to October 2004. There was no established sampling. The fishing was carried out in areas with good fishing conditions, that is, the presence of the resource and favorable fishing areas where the lines could be launched without running the risk of losing them due to entanglement with the rocky bottom.

16.2.2 Biological and Fisheries Data on Board

On board each long-line vessel, a scientific observer of the IMARPE registered the fishing and biological information. The catch was estimated for each set and/or fishing day. The catch was estimated in number, while the catch in weight was estimated by multiplying the weight (kg) of the fresh processed catch by the corresponding conversion factor. The conversion factor was estimated by dividing the total weight of ten whole individuals by the total weight of the legs of these individuals. This procedure was repeated several times, estimating an average. The geographical position and depth of each set were recorded. The effort was estimated in number of pots by set and soak time per set.

On board, crabs were identified at species level, and their size, as carapace width, was measured using a Vernier caliper to the nearest 1.0 mm. Their mass was weighed to the nearest 5 g. Samples of whole specimens were frozen for a later analysis in the laboratory.

16.2.3 Sample and Laboratory

In the laboratory, the frozen collected specimens were slowly thawed. After taxonomic identification, the cephalothorax width was recorded using a caliper, and total mass, sex, gonad mass, presence of eggs, epibionts, and parasites were registered. The degree of fullness of the egg mass was determined using an arbitrary scale, I, empty with virginal setae; II, partially full; III, full; and IV, spawned.

16.2.4 Statistical Analysis

Statistical analyzes and some figures were made with the RStudio (R Core Team, 2017), while the tables and some figures were performed using Microsoft Office Word and Microsoft Excel, respectively.

The mean size of sexual maturity in females was determined based on the presence of egg masses attached to the pleopods with fullness degree stage II and III. Maturity was estimated by fitting a logistic regression model to the proportion of mature females at a given CW (mm):

$$N_i = \frac{100}{\left(1 + e^{a-bi}\right)}$$

where Ni is the predicted proportion of mature females at the *i*-size-interval and *a* and *b* are estimated parameters.

The capture per unit effort (cpue) was standardized using a GLM model.

A Kruskal-Wallis test was performed in order to estimate differences of the mean size by latitude, month, and depth.

16.3 Results

16.3.1 Distribution and Relative Abundance

During the exploratory fishing, three species were recorded, *Lithodes wiracocha*, *Lithodes panamensis*, and *Paralomis longipes*. *Paralomis longipes* were distributed from 09°S to 16°S, *Lithodes panamensis* between 11°S and 14°S, and *Lithodes wiracocha* between 11°S and 12°S (Fig. 16.1). The CPUE by latitude, month, and depth is shown in Fig. 16.2. The highest median CPUEs were bigger than 2 crabs/pot and occurred between 11°S and 15°S. The lowest mean average CPUEs were recorded from January to February, while from March to October, average CPUEs were bigger than 2.5 crabs/pot. Crab relative abundance increased with depth and peaked between 1000 and 1200 m (Fig. 16.2).

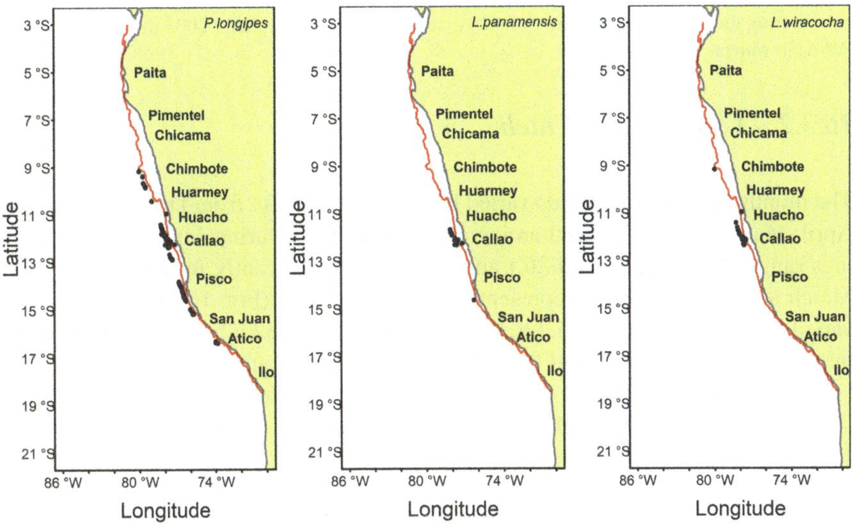

Fig. 16.1 Distribution of three king crab species, *Paralomis longipes Lithodes panamensis* and *Lithodes wiracocha*, during exploratory fishing carried out from December 2003 to October 2004 in Peruvian waters. The red line indicates the isobath of 200 m as the limit of the continental shelf. The black point indicates the presence of king crabs

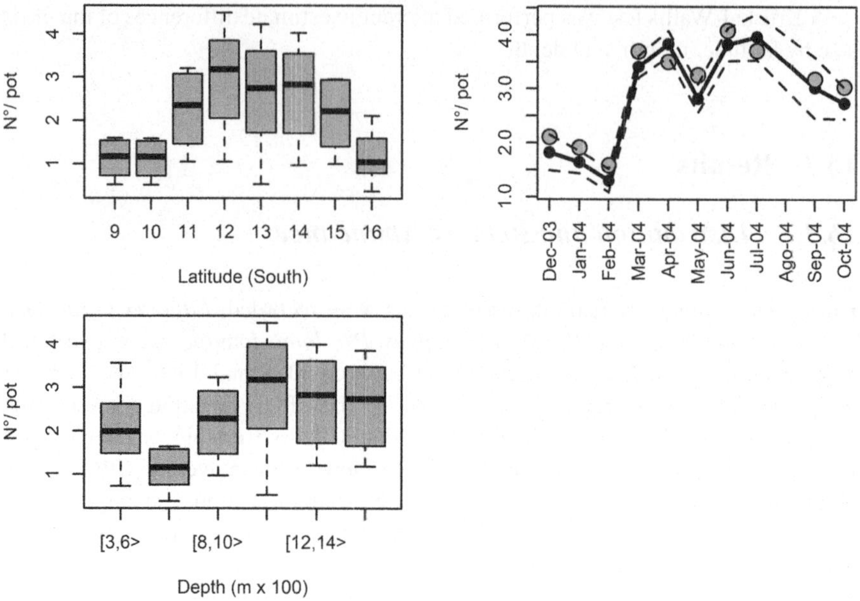

Fig. 16.2 Cpue estimates (Number of individuals/pots by latitude, month, and depth) of *P. longipes* during the king crab exploratory fishery carried out from December 2003 to October 2004 in Peruvian waters

16.3.2 Landing and Catch

The monthly catch of lithodids varied between 10 and 50 t from December 2003 to April–May 2004 and showed an increasing tendency. During June and July, landings ranged between 24 and 26 t and decreased significantly in September. From March to July, the landing represented the 68% of total (Fig. 16.3). The soak time varied between 12.3 and 125.7 h, with a mean value of 32.6 h. The number of pots by set varied between 13 and 91 with an average value of 48 pots. Most sets used between 20 and 80 pots. The fishing depth varied between 345 and 1579 m, with an average value of 1125 m. Most sets were performed at depths between 1000 and 1200 m (Fig. 16.4).

16.3.3 Size Structure

The size of *P. longipes* ranged between 58 and 148 mm CW. Size structure by sex showed differences; males were larger than females as they had an average size and maximum sizes greater than females. Both females and males presented a unimodal size distribution pattern (Fig. 16.5). The sizes by month did not show differences in females and males (Female Chi-square = 62.5, Males Chi-square = 884.6; $p < 0.05$).

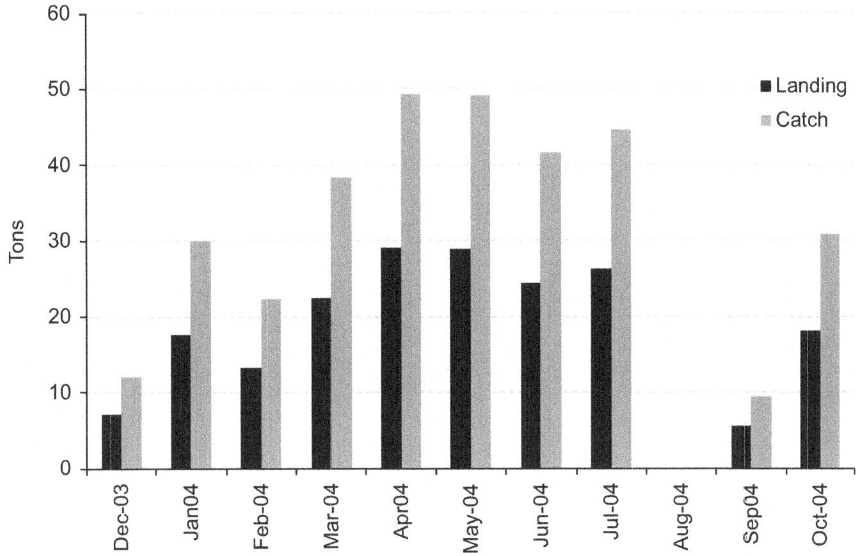

Fig. 16.3 Landing and estimated catch of king crabs in Peruvian waters during exploratory fishing from December 2003 to October 2004

In both females and males, the largest size ranges were observed between 11°S and 14°S. However, no differences of the mean size by latitude were detected in females and males (Female Chi-square = 32.996, male Chi-square = 293,6; $p < 0.05$). The largest range size of crabs of both sexes was observed at depths greater than 900 m; however, no significant differences in size were recorded (Female Chi-square = 20.561, male Chi-square = 56.8; $p < 0.05$) (Fig. 16.6).

The sizes of *L. wiracocha* varied between 33 and 148 mm CW; the largest average sizes and maximum sizes were observed in males. A bimodal size distribution pattern was observed in males and a unimodal size distribution pattern in females. Some females and males smaller than 70 mm were captured by the traps (Fig. 16.7). The monthly sizes in females were similar (Chi-square = 68.5; $p < 0.05$), while the smallest males occurred in May (Fig. 16.8). Females were similar in size along the sampled latitudinal gradient (Chi-square = 24.3, $p < 0.05$). The largest males were found between 11° and 12°S. Sizes of all individuals were similar through the studied depth range (Female Chi-square = 11.9, male Chi-square = 12.9: $p < 0.05$) (Fig. 16.8).

The size of *Lithodes panamensis* varied between 22 and 178 mm CW, with average sizes of 139.2 and 106.8 mm in males and females, respectively. Like in the other sampled lithodids, the males presented larger sizes (Fig. 16.9). In both females and males, the size structure presented different distribution patterns. In females, the modal size was 98 mm CW, while in males the modal size was 143 mm CW. Nevertheless, there was a secondary modal size in both sexes: 133–143 mm CW and 108 mm CW for females and males, respectively. The smallest females

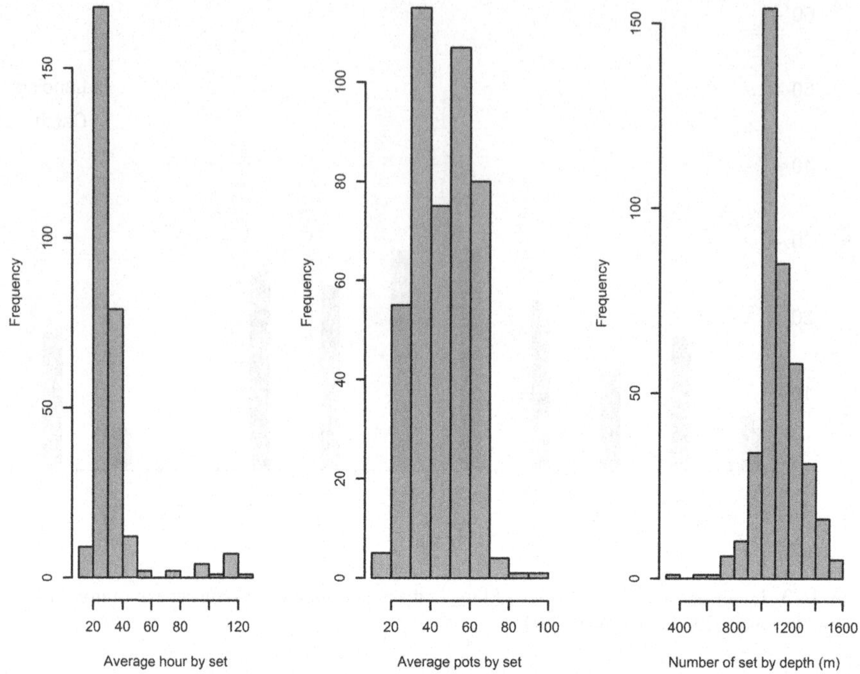

Fig. 16.4 Average number of hours by set, average number of pots by set and number of sets by depth range registered during the king crab exploratory fishing developed from December 2003 to October 2004 in Peruvian waters

occurred between February and May, whereas male sizes were variable and similar throughout our sampling period (Fig. 16.10). In both females and males, sizes were similar by latitude and depth (Fig. 16.10). Only males occurred at 14°S.

16.3.4 Size-Wet Mass Relationship

The size-mass relationship of the *P. longipes*, *L. wiracocha*, and *L. panamensis* species was fitted to potential regression (Fig. 16.11).

The length-weight relationship in *P. longipes* was significantly different by sex; the males presented more weight at larger sizes. While in *L. panamensis* and *L. wiracocha*, no differences were observed (Table 16.2).

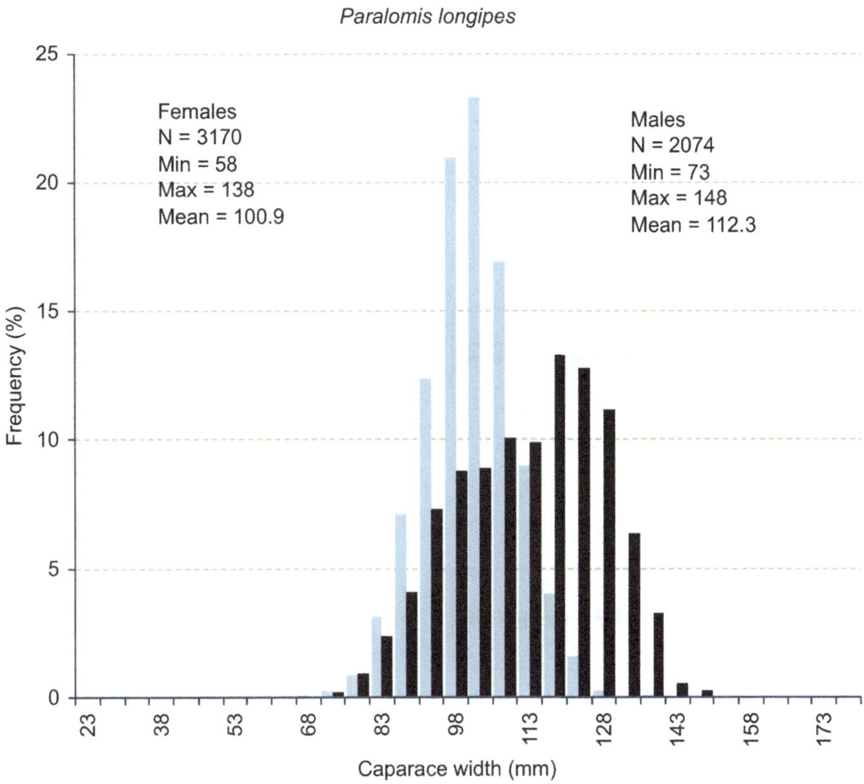

Fig. 16.5 *Paralomis longipes* size frequency distribution of both females and males during the exploratory fishing in December 2003 to October 2004 in Peruvian waters. *Gray bars* female, *black bars* male. Axis x shows class mark

16.3.5 Female Maturity Size

In *Paralomis longipes,* the presence of egg masses was observed from 83 mm CW. The size at 50% of ovigerous females was estimated at 88 mm CW (Fig. 16.12). For the other two species, the few females with eggs masses did not allow to estimate female maturity size. A total of 11 females of *L. panamensis* and 30 females of *L. wiracocha* with egg masses were observed. The size of these ranged from 94.5 to 123.1 mm of CW and from 72.2 to 115 mm of CW, respectively.

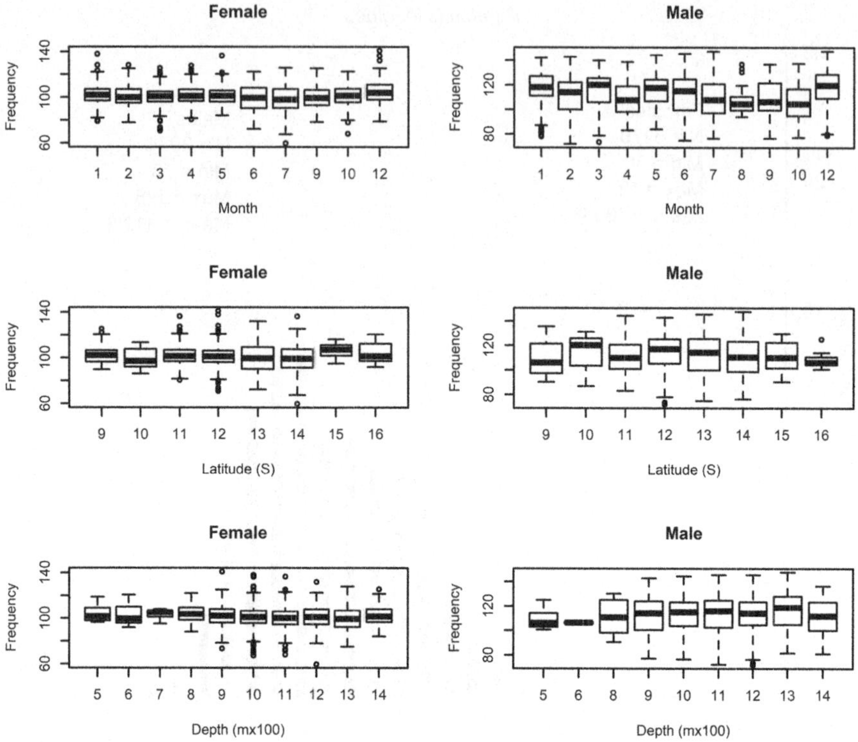

Fig. 16.6 Box-plot of size of *Paralomis longipes* by month, latitude, depth, and sex in Peruvian waters

16.3.6 Parasites

In all cases rhizocephalan externa were found attached to the abdomen of females and males, usually in a number of one per individual, but in some individuals, up to three were observed (Fig. 16.13).

Of 1249 sampled *P. longipes*, 6.7% presented rhizocephalan infestation. In females the prevalence of rhizocephalan was 7.8% (n = 55) and in males 5.4% (n = 29). The highest prevalences were recorded between 14°S and 15°S (Table 16.3, Fig. 16.14). In all cases of parasitized females, egg masses were absent, which evidences a negative effect on reproduction. The sizes of the parasitized specimens varied between 70 and 115 mm CW for females and between 80 and 125 mm CW for males. The highest prevalence in females was observed at smaller sizes, decreasing to a minimum at 105 mm CW. However, a prevalence similar to small sizes was observed in the 115 mm CW size interval. In males, the highest prevalence of rhizocephalans peaked between 90 and 95 mm CW, with no parasitized specimens smaller than 75 mm CW (Fig. 16.14).

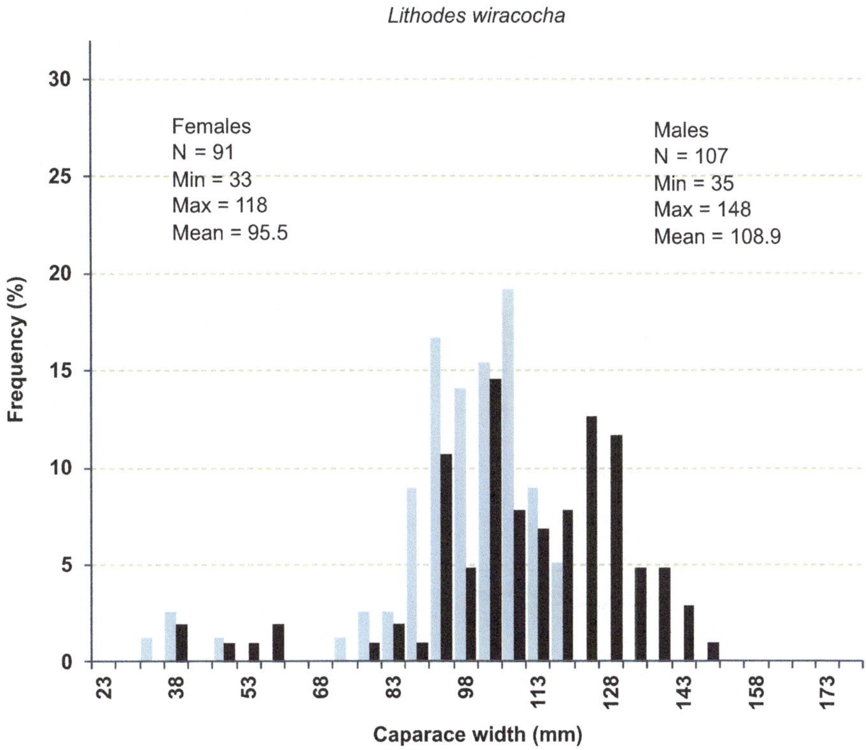

Fig. 16.7 *Lithodes wiracocha* size frequency distribution of both females and males during the exploratory fishery in December 2003 to October 2004 in Peruvian waters. *Gray bars* females, *black bars* male. Axis x shows class mark

16.4 Discussion

The lithodid fishery in the Southeast Pacific has been carried out mainly in Chile since 1928 on *Paralomis granulosa* and *Lithodes santolla* (Cárdenas et al. 1987; Lovrich 2014; Barrera 2016; Daza et al. 2017). In Peru, king crabs landings are not discriminated by species; however, from the sampling here reported and carried out on board artisanal long-line vessels during exploratory fishing, we deduce that *P. longipes* is the most frequent species between 09°S and 16°S in Peruvian waters.

Data about the spatial, bathymetric distribution and population structure of *P. longipes*, *L. panamensis*, *and L. wiracocha* are scarce. For instance, *P. longipes* has been reported in the Northeast Pacific at 05°26′N, 86°55′W (Haig 1974), and in the Southeast Pacific in 07°59′S, 80°22′W and 16°29′S, 73°33′W (del Solar, 1972) in Peruvian waters and in Chilean waters at 20°15′S (Retamal 1994). Juarez et al. (1998) also previously registered *P. longipes* in Peruvian waters from 12°S to 17°S, coincidentally with our findings, where *P. longipes* was the most abundant species with a continuous distribution between the 9°S and 16°S. Also, *P. longipes* was the

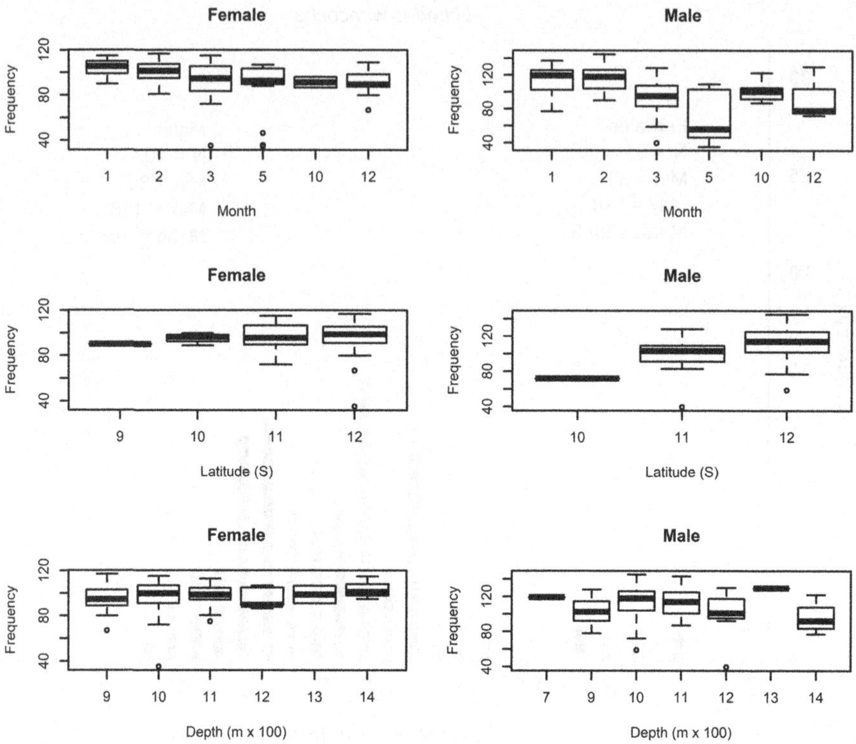

Fig. 16.8 Box-plot of size of *Lithodes wiracocha* by month, latitude, depth, and sex in Peruvian waters

most abundant species in the catches during 2003–2004, while Juarez et al. (1998) report *L. panamensis* as the most representative species in the catches, followed by *P. longipes*. This variation in the specific composition (in 1997–1998 dominance of *L. panamensis* vs 2003–2004 dominance of *P. longipes*) could be due to changes in the food supply from the pelagic zone associated with decadal changes of primary production in the pelagic zone that are determinants in the biogechemical processes of the seabed (Smith et al. 2001).

Lithodes panamensis has been reported in Colombia (07°31′N, 79° 14′W) by Faxon (1895), in northern and southern of Peru (03°48′S,81°22′W, 07°59′S, 80°22′W, 17°34′S, 71°55′W) by del Solar (1972), and in Iquique, Chile (20°10′S) (Retamal 1994). During our exploratory fishing in Peruvian waters, *L. panamensis* was captured between 11°S and 14°S, coinciding with previous data. However, the presence of this species in Chilean waters would indicate this species presents a wider latitudinal distribution in the Southeast Pacific.

During our exploratory fishing, *L. wiracocha* was registered between 11°S and 12°S. However, records further north in Peruvian waters (03°48′S, 81°22′W and 07°59′S, 80°22′W; del Solar 1972) and further south in Chilean waters (Brito 2002), indicate a wider latitudinal distribution. In general, the records of *P. longipes*,

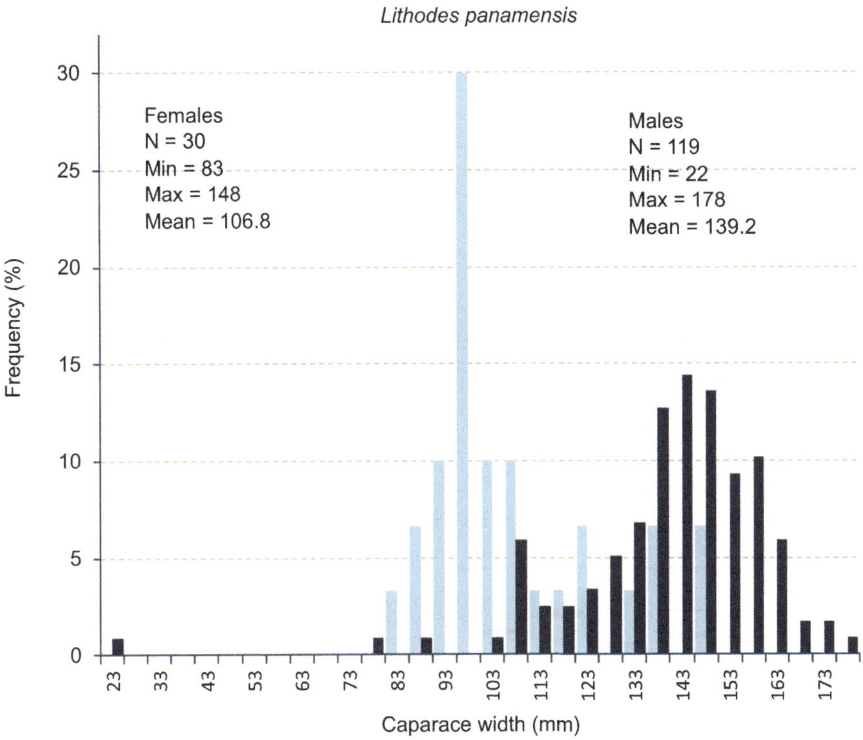

Fig. 16.9 *Lithodes panamensis* size frequency distribution of both females and males during the exploratory fishery in December 2003 to October 2004 in Peruvian waters. *Gray bars* female, *black bars* = male. Axis x shows class mark

L. panamensis, and *L. wiracocha* indicate a wider distribution range. However, it is important to determine the spatial structure of these species (genetic units – population stocks) for fishery management purposes (Barrera 2016).

Paralomis longipes was previously recorded between 700 and 1400 m (Faxon 1895; del Solar 1972; Retamal 1994), while, in the present study, this species was captured between 549 and 1480 m, extending its vertical distribution to shallower waters. *Lithodes panamensis* was previously recorded at depths between 620 and 800 m (del Solar 1972; Retamal 1994) and *L. wiracocha* between 680 and 800 m (del Solar 1972). Here, these two species were captured down to 1480 m depth, indicating that this species can thrive at greater depths. According to Hall and Thatje (2009) and Hall and Thatje (2011), king crabs are distributed from intertidal to 3500 m and that their distribution in waters on the continental shelf is limited by their thermal tolerance up to 13 °C. In the same sense, del Solar (1981) indicates that the minimum depth at which lithodids are known to thrive in the Peruvian sea is 500 m where the temperature is 8 °C. In addition, considering that the mean depth of the 15 °C isotherm not exceed the 180 m (Flores et al. 2013), we conclude that it is unlikely that king crabs are able to inhabit the continental shelf off Peru.

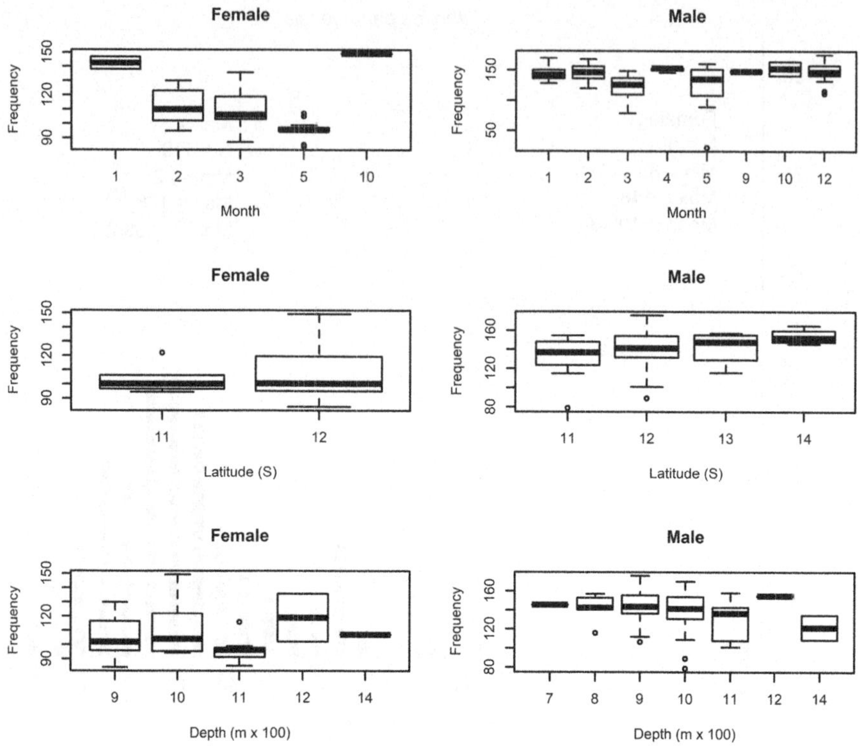

Fig. 16.10 Box-plot of size of *Lithodes panamensis* by month, latitude, depth, and sex in Peruvian waters

In other king crab species, the spatial (depth) and temporal (months) variations are related to reproductive migrations (Lovrich 2014). It is known that some species migrate to shallow waters to reproduce, with juvenile specimens found in these areas (Lovrich 2014). However, it has also been postulated that recruitment in *Paralomis formosa* can occur in deeper areas (Purves et al. 2003). In this work juvenile individuals were not captured probably due to the pots' selectivity. However, there is a possibility that juvenile specimens are in other less explored areas or depths. However, if one considers that the estimated abundance patterns show monthly variations (higher values between March and October) and by depth (higher values at depths greater than 1000 m), it is possible to infer that *P. longipes* shows a gregarious behavior between March and October and at depths greater than 1000 m a behavior that allows higher yields. In addition, monthly and depth variations of the size frequency distributions have not been observed in *P. longipes*, which may indicate that this gregarious behavior is carried out by specimens of all ranges of observed sizes.

Males of *L. panamensis*, *L. wiracocha*, and *P. longipes* are larger than females, which is consistent with observations in other lithodids (Boschi et al. 1984; Macpherson 1988). This difference is important in the copulation in brachyuran and

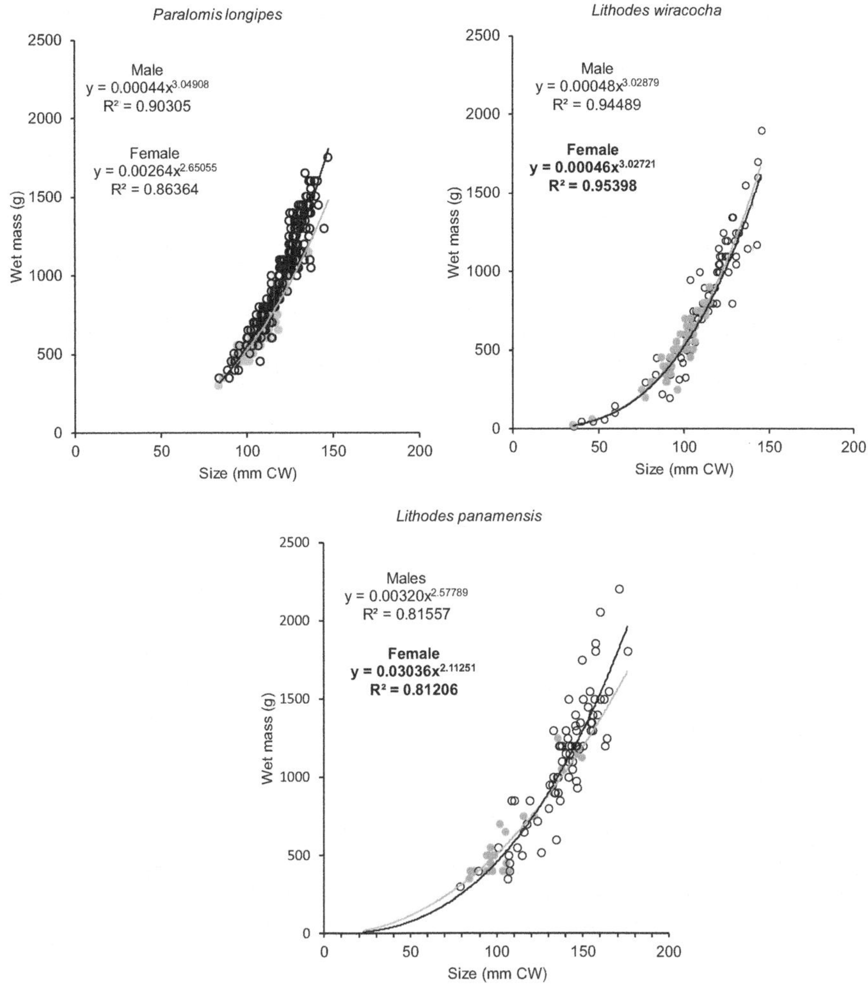

Fig. 16.11 Size-wet mass relationship of *Paralomis longipes* (Left), *Lithodes wiracocha* (center), and *Lithodes panamensis* (right) in Peruvian waters. Males, open circles; females, solid circles

anomuran crabs since males only mate with smaller females (Paul and Paul 1990; Lovrich et al. 2002).

The size at maturity of *P. longipes* females was estimated at 88.4 mm CW. This should be considered valid for Peruvian waters only, because this species has a wide latitudinal distribution in the Eastern Pacific. Size at maturity in other lithodids shows latitudinal variations and smaller sizes at maturity at higher latitude, in response to changes in environmental temperatures or food supply (Olson et al. 2018).

In this work rhizocephalan barnacles have been recorded only in *P. longipes* and where not observed in *L. panamensis* and *L. wiracocha*. These parasites have been

Table 16.2 Log mass vs log size relationship of king crabs in Peruvian waters during an exploratory fishing 2003–2004

Specie	Sex	Log mass (g) vs log size (mm)	n	t-test value
Paralomis longipes	Female	y = 2.6506 × − 2.578	301	2.714
	Male	y = 3.0491 × − 3.355		
Lithodes wiracocha	Female	y = 3.0272 × − 3.334	150	0.013
	Male	y = 3.0288 × − 3.321		
Lithodes panamensis	Female	y = 2.1125 × −1.517	108	1.807
	Male	y = 2.5779 × − 2.494		

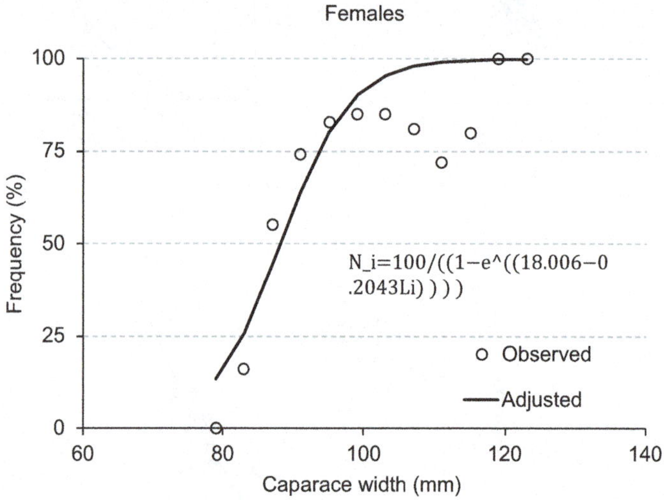

Fig. 16.12 Maturity ogives for females *Paralomis longipes* caught in Peruvian waters during the exploratory fishery between December 2003 and October 2004. The ogive was fitted to observed frequency of eggs masses by size range (4 mm)

Fig. 16.13 Rhizocephala externae infesting females of *Paralomis longipes* (ventral view) in Peruvian waters. Pleopods without eggs and one externa (left) and three externae (right)

Table 16.3 Frequency of presence and absence of rhizocephalan externae parasitizing *Paralomis longipes* by sex, latitude, longitude, and depth range

Latitude/Longitude	Sex	Rhizocephalan presence	Frequency N°	Frequency %	Depth (m)
12°00′–12°59′S	Female	Yes	1	0.6	950
		Not	166	99.4	837–1222
	Male	Yes	2	3.4	1090–1093
		Not	56	96.6	837–1222
13°00′–13°59′S	Female	Yes	2	5.4	1058
		Not	35	94.6	958–1059
	Male	Yes	1	5.6	1059
		Not	17	94.4	958–1059
14°00′–14°59′S	Female	Yes	48	10.4	822–1470
		Not	412	89.6	817–1470
	Male	Yes	25	5.8	817–1470
		Not	405	94.2	817–1470
15°00′–15°59′S	Female	Yes	3	9.1	1083–1136
		Not	30	90.9	990–1153
	Male	Yes	1	3.7	1136
		Not	26	96.3	990–1153
16°00′–16°59′S	Female	Yes	1	8.3	659
		Not	11	91.7	549–659
	Male	Yes	0	0.0	
		Not	8	100.0	549–659
Total	Female	Yes	55	7.8	659–1470
		Not	653	92.2	549–1470
	Male	Yes	29	5.4	659–1470
		Not	512	94.6	549–1470
	Total	Yes	84	6.7	659–1470
		Not	1165	93.3	549–1470

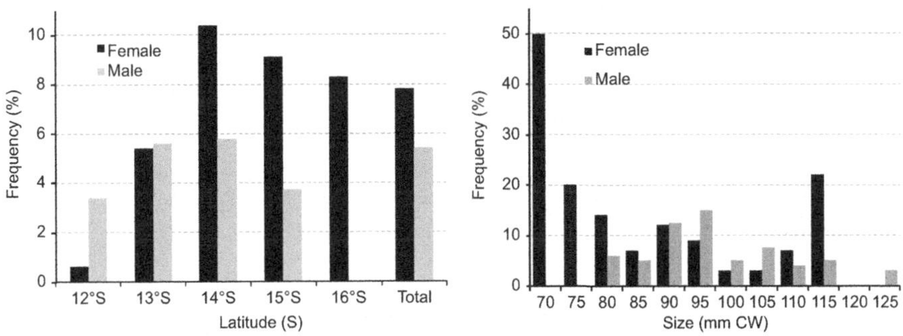

Fig. 16.14 Prevalence of rhizocephalan externae in *P. longipes* by sex and latitude (left) or size (right) in Peruvian waters during an exploratory fishing between December 2003 and October 2004

found in many species of lithodids (Lovrich 2014), and based on morphological characteristics, they have been attributed to *Briarosaccus callosus* (Boschma 1930 in Noever et al. 2016). Following this practice, many specimens of rhizocephalans have been assigned as *B. callosus* (Noever et al. 2016), including rhizocephalan found in Peruvian waters (Pino et al. 2010). However, Noever et al. (2016) used genetic and morphological comparisons to determine that *Briarosaccus* specimens infesting three king crab species are not *B. callosus* and furthermore identified two new cryptic and sympatric species of *Briarosaccus*. The infestation of rhizocephalan in lithodid crabs is variable. For instance, in *P. granulosa* a maximum of 10% has been reported at small sizes (Lovrich et al. 2004), while in *P. spinosissima*, occurring in Southern Ocean waters off South Georgia, an 80% infestation rate has been reported (Watters 1998). Also the lack of occurrence of *B. callosus* in some specimens of king crabs could be due to differences in habitat, greater presence of rhizocephalans in closed waters in contrast to open ocean habitats (Sparks and Morado 1986). In this work, the prevalence of rhizocephalans on *P. longipes* was 7%. It has also been observed that parasitized females did not have an ovigerous mass, confirming the negative effect of the presence of rhizocephalans on the reproductive process (Nagler et al., 2017). Also, the highest prevalence of Rhizocephala has been observed at smaller sizes, according to other observations in other species of lithodids (Lovrich et al. 2004). However, at higher sizes of *P. longipes*, the high prevalence of parasites was also observed. Although *L. panamensis* and *L. wiracocha* have been captured in the same areas where *P. longipes* was captured, they did not present rhizocephalans, which could indicate the preference of these parasites for some species.

We conclude that future work needs to focus on analyzing the life cycle of all species under investigation, such as reproductive behavior and larval development, growth and size at maturity under varying thermal regimes found at different depths, as well as growth, mortality, and individual age. This information is essential for constituting a sustainable fishery of these species otherwise vulnerable to overexploitation.

Acknowledgments Special thanks go to M.E. Hendrickx for allowing us to participate in this project. We thank Pablo Marin for his help in the R Sofware and to Gustavo A. Lovrich for thoroughly commenting on the draft manuscript. The information collected during exploratory fishing was used by P. Larriviere for her thesis to obtain a bachelor's degree in biology at the National University of San Marcos, Lima-Peru.

References

Arguelles J, Ganoza F, Aliaga A, Chacon G, Rubio J, Tafur R, Baldeon A, Castañeda W (2014) Investigación de crustáceos de profundidad en la zona de Pisco – Callao. Inf Inst Mar Peru 41(1–4):36–44

Barrera MG (2016) Análisis de la variación y la estructura genética de la centolla (Lithodes santolla, Molina, 1782) en la Región de Magallanes y Antartica Chilena mediante marcadores moleculares. Tesis para obtener el grado de Maestro en Ciencias en Uso, manejo y preservación de los recursos naturales. Centro de Investigaciones Biológicas del Noroeste S.C. La Paz, Baja California Sur

Boschi E, Bertuche DA, Wyngaard JG (1984) Estudio biológico pesquero de la centolla (*Lithodes antarcticus*) del Canal Beagle, Tierra del Fuego, Argentina. Contribuciones INIDEP 441:7–72

Brito JL (2002) Lithodidae registrados frente a San Antonio, Chile central (Crustacea, Anomura). Invest Mar Valparaiso 30(1):57–62

Cárdenas JC, Gibbons J, Oporto J, Stutzin M (1987) Impacto de la pesquería de centolla y centollon sobre las poblaciones de mamíferos marinos de Magallanes, Chile. Amb y Des 3(1–2):111–119

Daza E, Almonacid E, Hernández R (2017) Boletín de difusión IFOP. Crustáceos bentónicos XII Región, 2016: centolla y centollón. IFOP, Valparaiso

del Solar E (1972) Addenda al catálogo de crustáceos del Perú. Inst Mar Perú 38:1–21

del Solar E (1981) Lithodidae, nueva familia de cangrejos gigantes en el Perú. Bol Lima 14(3):68–81

del Solar E (1987) Recursos marinos de la zona arquibentónica peruana. Bol Lima 50:77–85

Faxon W (1895) Reports on an exploration off the west coasts of Mexico, Central and South America, an off the Galapagos Islands by the U.S. Fish Commission steamer "Albatross" during 1891. The stalk-eye Crustacea. Mem Mus Comp Zool Harvard Univ 18:1–292

Flores R, Espino M, Luque G, Quispe J (2013) Patrones de variabilidad ambiental en el mar peruano. In: Csirke J, Guevara-Carrasco R, Espino M (eds) Ecología, pesquería y conservación del jurel (*Trachurus murphyi*) en el Perú. Rev Peru Biol 20(1):21–28

Haig J (1974) Observations on the lithodid crabs of Peru, with description of two new species. Bull S Calif Acad Sci 73(3):152–164

Hall S, Thatje S (2009) Global bottlenecks in the distribution of marine Crustacea: temperature constraints in the family Lithodidae. J Biogeogr 36:2125–2135

Hall S, Thatje S (2009a) Four new species of the family Lithodidae (Crustacea: Decapoda) from the collections of the National Museum of Natural History, Smithsonian Institution. Zootaxa 2302:31–47

Hall S, Thatje S (2011) Temperature-driven biogeography of the deep-sea family Lithodidae (Crustacea: Decapoda: Anomura) in the Southern Ocean. Polar Biol 34:363–370

Juarez L, Llactarina E, Cochatoma A, Sanchez J (1998) Pesca experimental de los recursos "Anguilas", "Centollas", "Pulpo", "Caracol" y "Lamprea de mar" a bordo del buque coreano Moresko 1 (13 de septiembre de 1997 al 13 de marzo de 1998). Informe Interno Instituto del Mar del Peru, Lima

Kim MH, Hong SY (2000) Larval development of *Cryptolithodes expansus* Miers (Decapoda: Anomura: Lithodidae) reared in the laboratory. Proc Biol Soc Wash 113(1):54–65

Lovrich G (2014) Lithodidae. In: Calcagno JA (ed) Los invertebrados del Mar Argentino. Fundación de Historia Natural Félix de Azara, Buenos Aires, pp 241–261

Lovrich GA, Vinuesa JH, Smith BD (2002) Male growth, maturity and mating of *Lithodes santolla* in the Beagle Channel, Argentina. In: Paul AJ, Dawe EG, Elner R, Jamieson GS, Kruse GH, Otto RS, Sainte-Marie B, Shirley TC, Woodby D (eds) Crabs in cold water regions: biology, management and economics. Sea Grant, University of Alaska, Fairbanks, pp 147–168

Lovrich GA, Roccatagliata D, Peresan L (2004) Hyperparasitism by the cryptoniscid isopod *Liriopsis pygmaea* and parasitism by the rhizocephalan *Briarosaccus callosus* on the lithodid *Paralomis granulosa* from the Beagle Channel, Argentina. Dis Aquat Org 58:71–77

Macpherson E (1988) Revision of the family Lithodidae Samouelle, 1819 (Crustacea, Decapoda, Anomura) in the Atlantic Ocean. Monog Zool Mar 2:9–153

Nagler C, Hornig M, Haug J, Noever C, Hoeg J, Glenner H (2017) The bigger, the better? Volume measurements of parasites and hosts: parasitic barnacles (Cirripedia, Rhizocephala) and their decapod host. PlosOne 12(7):e0179958. https://doi.org/10.1371/journal.pone.0179958

Noever C, Olson A, Glenner H (2016) Two new cryptic and sympatric species of the King crab parasite *Briarosaccus* (Cirripedia: Rhizocephala) in the North Pacific. Zool J Linnean Soc 176:3–14

Olson AP, Siddon CE, Eckert GL (2018) Spatial variability in size at maturity of Golden King crab (*Lithodes aequispinus*) and implications for fisheries management. R Soc Open Sci 5:171802

Paul JM, Paul AJ (1990) Reproductive success of sublegal size male Red King Crab. In: Proceedings International Symposium of King and Tanner crabs. Alaska Sea Grant College Program, University of Alaska, AK-SG 90-04, Anchorage, pp 37–50

Pino J, Larriviere P, Shiga B, Arguelles J (2010) Histology of the pedicle of the "externa" of *Briarosaccus callosus* (Cirripedia, Peltogastridae) parasite of *Paralomis longipes* (Crustacea, Decapoda). Biologist 8:73–78

Purves M, Agnew D, Moreno G, Yau C, Pilling G (2003) Distribution, demography, and discard mortality of crabs caught as bycatch in an experimental pot fishery for toothfish (*Dissostichus eleginoides*) in the South Atlantic. Fish Bull 101:874–888

Retamal M (1994) Los Lithodidae chilenos. Ans Inst Pat Ser Cs Nts Punta Arenas (Chile) 21:111–129

Smith KL, Kaufmann RS, Baldwin RJ, Carlucci AF (2001) Pelagic benthic coupling in the abyssal eastern North Pacific: a 8-year time-series study of food supply and demand. Limnol Oceanogr 46(3):543–556

Sparks A, Morado F (1986) Histopathology and host response in lithodid crabs parasitized by *Briarosaccus callosus*. Dis Aquat Org 2:31–38

Stevens BG, Lovrich GA (2014) King crabs of the world: species and distribution. In: Stevens BG (ed) King crabs of the world: biology and fisheries management. CRC Press, Boca Raton, pp 1–30

Watters G (1998) Prevalences of parasitized and hyperparasitized crabs near South Georgia. Mar Ecol Prog Ser 170:215–229

Zaklan SD (2002) Review of the family Lithodidae (Crustacea: Anomura: Paguroidea): distribution, biology, and fisheries. In: Paul AJ, Dawe G, Elner RW, Jamieson G, Kruse GH, Otto RS, Sainte-Marie B, Shirley TC, Woodby D (eds) Crabs in cold water regions: biology, management, and economics. University of Alaska Sea Grant College Program, Fairbanks, pp 751–845

Chapter 17
Lower Slope and Abyssal Benthic Decapods of the Eastern Pacific

M. K. Wicksten

Abstract A total of 119 species of decapod crustaceans have been reported from depths of 700 m or more in the eastern Pacific. From 1875 to 2000, decapods were caught mostly by trawls and dredges on muddy sea floors. Sampling has not been consistent and has been concentrated in certain geographic areas. Taxonomic confusion and potential misidentifications continue to create problems. Of the species in the area, 14 are considered to occur worldwide. The area from Alaska to northern California has the fewest species (8); the area from southwestern Mexico to Peru or Chile has the most (27). Of the taxonomic groups, the most speciose are the family Munidopsidae (31 species). Adaptations to the environment include loss of functional eyes, capacious carapaces, lack of strict food preferences, and antipredator behavior including ability to swim and association with other invertebrates. Reproduction is poorly known but seems to be asynchronous. Largely due to costs and difficulties in processing the catch, there are no extensive fisheries for deepwater decapods in the area.

Keywords Deep sea · Benthic crustaceans · Eastern Pacific · Decapoda

17.1 Introduction

A brief historical overview indicates that eastern Pacific decapods were collected as early as 1825–1827, when the HMS *Blossom* visited Monterey Bay. Further studies have continued on nearshore subtidal species, using scuba diving, trawls, dredging, and traps, and commercial fisheries for crabs and shrimps on the continental shelf. Records of decapod crustaceans below 700 m (the approximate lower limit of the oxygen minimum zone, OMZ) are scarce and almost always based on collections from trawls or dredges. In the eastern Pacific, the British expedition's ship HMS *Challenger* included a short stop at Valparaiso, Chile, and a visit to the Juan Fernández Islands in 1875; and later samples were taken in the Strait of Magellan.

M. K. Wicksten (✉)
Department of Biology, Texas A&M University, College Station, TX, USA
e-mail: m-wicksten@tamu.edu

M. E. Hendrickx (ed.), *Deep-Sea Pycnogonids and Crustaceans of the Americas*, https://doi.org/10.1007/978-3-030-58410-8_17

The most comprehensive study of deep-sea life was by the US fisheries steamer *Albatross* in 1890–1910. The participants used trawls to collect marine life from the area around the Galápagos Islands to San Francisco Bay, California. Reports of decapods from these collections included those by Faxon (1893, 1895) and Benedict (1910). Decapods of the Peru-Chile Trench and adjacent areas were studied by the cruises of the *Anton Bruhn* (Garth and Haig 1971). Other expeditions that collected decapods in that area included the cruise of the research vessel *Akademik Kurchatov* (Zarenkov 1976, carideans) and work by the Instituto del Mar del Perú (del Solar 1972; Haig 1974; Méndez 1981; Wicksten and Méndez 1982). McCauley (1972) trawled deep decapods to a depth of more than 3000 m as part of a survey of the effects of low-level radioactive wastes from the Hanford Site to the Columbia River and nearby ocean waters. Ambler (1980) described new species of squat lobsters taken off Oregon. Various deep-sea decapods were collected off Baja California during cruises by Scripps Institution of Institution of Oceanography, including specimens of *Glyphocrangon* (Wicksten 1979), *Bathystylodactylus* (Wicksten and Martin 2004), and new species of *Munidopsis* (Jones and Macpherson 2007). The University of Southern California received decapods collected during the California Department of Fish and Game's study of catches in sablefish traps (Wicksten 1982; Baba and Haig 1990).

Ongoing studies include works by M.E. Hendrickx and his colleagues in western Mexico including the Gulf of California, I. Wehrtmann and colleagues in Costa Rica, and G. Guzmán and his colleagues off Chile. The biologists of the Monterey Bay Aquarium Institute (MBARI) have concentrated on the biota of the Monterey Submarine Canyon and nearby seamounts. Using the remotely operated vehicle (ROV)*Ventana*, they have obtained high-definition photographs of deep decapods showing living colors, behavior, and new ranges (Wicksten and Kuhnz 2015). Jeffrey Drazen and his colleagues at the University of Hawaii at Manoa have investigated the fishes and larger invertebrates in areas of manganese nodules along the Clarion-Clipperton Zone and the Peru Basin. New studies from the Exploration Vessel *Nautilus*, operated by the US National Oceanographic and Administration and the Ocean Exploration Trust and with the cooperation of the Charles Darwin Foundation, have combined use of a remotely operated vehicle with collections to provide new information on the deep fauna off the Galápagos Islands.

In 1989, I attempted a zoogeographic analysis of ranges of crustaceans living at 200 m or more in the entire eastern Pacific (Alaska to Cape Horn). Since that paper was published, new species (especially galatheoid anomurans) have been described, new records have been published, and additional information on color, habitat, physiology, and predator-prey relationships of these species have been reported. In situ color photographs now are available for some of the species. This new information has been compiled and included in the present contribution, updating the list of species occurring below 700 m depth.

17.2 Methods

Based on the contribution by Wicksten (1989), an updated list of species was elaborated containing new species and new records made available during the last 31 years. Aspects related to taxonomy, distributional range, habitat, functional anatomy, reproduction, larval stage, population densities, natural history, exploitation, and environmental disturbance are addressed.

17.3 Results

17.3.1 Species in the Eastern Pacific

A total of 120 species, not including species currently suspected of being undescribed, has been recorded (Table 17.1). For the purposes of this report, species are included for which the majority of their records were at 700 m or deeper. The species mentioned in this paper generally are confined to depths below the local oxygen minimum zone (OMZ). Some records are suspect. Rathbun (1904) reported *Pagurus capillatus* from only 4 m. This unlikely record could be the result of a misidentification or perhaps a specimen thrown overboard from a fishing vessel. Lithodid crabs of the genus *Paralomis* generally come from deep water, but the records of *Paralomis tuberipes* do not provide a depth. The area in which the holotype was collected was shallow, perhaps 15 m (F. Palero, pers. comm.) Identifications based on photographs or video stills from remotely operated vehicles often can be identified to genera, not species, because distinguishing features cannot be seen. Some species are known only from a single sample and thus there is no depth "range," while others (e.g., *Pandalus platyceros*) have enormous ranges (0–1846 m, Wicksten 1989), which may reflect north-south distributional patterns or migration by age or sex. Depths in this chapter are given for settled adults and not pelagic larval stages.

Genera that contain species consistently found at 700 m or deeper include *Cerataspis*, *Benthesicymus*, *Bathypalaemonella*, *Bathystylodactylus*, *Nematocarcinus*, *Glyphocrangon*, *Parapagurus*, *Eiconaxius*, *Munidopsis*, and *Paralomis*. The deepest record from the eastern Pacific is for *Munidopsis verrucosus* at 4880 m (Jones and Macpherson 2007).

17.3.2 Taxonomy

Most shrimps have been identified on the basis of morphology. "Trans-ocean" (Atlantic and Pacific; eastern and western Pacific) species may in fact be separate species, but at present there is little or no genetic information for the vast majority of deep benthic decapods. Vereshchaka et al. (2019) revised the species assigned to

Table 17.1 List of species with geographic ranges and depths. If no source is provided, source is Wicksten (1989). Changes in names since 1989 are cited "as (old name) in (author, date)." The synonymies are not complete. See the latest reference for each species for further information

Order Decapoda
Suborder Dendrobranchiata
Infraorder Penaeoidea
Family Aristaeidae
1. *Cerataspis monstrosa* (Gray, 1828). Cosmopolitan, "north-west coast of U.S.A." (Farfante and Kensley 1997 as *Plesiopenaeus armatus*); in eastern Pacific, Monterey Bay, Clarion-Clipperton Zone, 3477–5400 m, Peru Basin, 4120–4200 m (Wicksten and Kuhnz 2015; Amon et al. 2017; Drazen et al. 2019)
Family Benthesicymidae
2. *Benthesicymus laciniatus* (Rathbun, 1906). Azores, Canary Islands, Madagascar, Japan to Hawaiian Islands; in eastern Pacific, off Santa Catalina Island, California, off Baja California, Clarion-Clipperton Fracture Zone; 1471–4028 m (Farfante and Kensley 1997; Wicksten 2004, 2012)
2. *Dalicaris altus* (Bate, 1881). Indian Ocean, Philippines, Japan, to Fiji; in eastern Pacific, San Nicolas Island, California to Galápagos Islands, 916–4089 m (Farfante and Kensley 199, as *Benthesciymus altus*)
3. *Trichocaris tanneri* (Faxon, 1893). Off San Diego, California, to northern Chile, 484–2010 m (Hendrickx and Hernandez Payán 2017 as *Benthesciymus tanneri*)
Family Solenoceridae
Haliporoides diomedeae (Faxon, 1893). Gulf of Panama to southern Chile, 240–3455 m (Wicksten 1989)
Hymenopenaeus doris (Faxon, 1893). Mexico to northern Peru, 549–4802 m (Wicksten 1989)
Hymenopenaeus nereus (Faxon, 1893). Costa Rica to Ecuador; Clarion-Clipperton Zone, 330–4001 m; Peru Basin, 4120–4200 m (Hendrickx and Wicksten 2016; Drazen et al. 2019)
Suborder Pleocyemata
Suborder Pleyceomata
Infraorder Caridea
Family Acanthephyridae
Acanthephyra eximia (Smith, 1884): Cosmopolitan, in East Pacific at Erben Seamount off California, 200–3700 m (Crosnier and Forest 1973; Chace 1986); Costa Rica (Pequegnat and Wicksten 2006)
Family Nematocarcinidae
Nematocarcinus agassizii (Faxon, 1893). Gulf of California, Mexico to Peru including Cocos and Malpelo Island and Galápagos Islands, 230–1800 m (Hendrickx 1995a; Hernandez-Payán and Hendrickx 2016)
Nematocarcinus faxoni (Burukovsky, 2001). Southwestern Mexico to northern Peru, 799–2055 m (Hernandez-Payán and Hendrickx 2016)
Nematocarcinus proximatus (Bate, 1888). Southern Indian Ocean, south of New Guinea, off Japan, off Juan Fernández Islands, southern Chile, 2514–2651 m (Chace 1986)
Nematocarcinus tenuipes (Bate, 1888). East Pacific Rise, 2558–2619 (Komai and Segonzac 2005 as *Nematocarcinus ovalis*)
Segonzackomaius burukovskyi (Komai and Segonzac, 2005). East Pacific Rise, 2330–2612 m.

(continued)

Table 17.1 (continued)

Family Bathypalaemonellidae
Bathypalaemonella delsolari (Wicksten and Méndez, 1983). SW of Lobos de Tierra, Peru, 712–714 m
Family Stylodactylidae
Bathystylodactylus echinus (Wicksten and Martin, 2004). Off Magdalena Bay and Patton Escarpment, 3427–3689 m
Family Pandalidae
Heterocarpus hostilis (Faxon, 1893). NW of Cabo San Lucas, Baja California; SW Mexico; Isla del Coco, Costa Rica; to Supe, Peru, 890–1895 m (Wicksten and Hendrickx 2016)
Heteronika nesisi (Burukovsky, 1986). Western Baja California Mexico (Hendrickx 2019a, b).
Pandalus amplus (Bate, 1888). Washington to Gulf of California, 553–1986 m (Wicksten 2012, as *Pandalopsis amplus*)
Pandalus tridens (Rathbun, 1902). Pribilof Islands to San Nicolas Island, 5–1984 m (Wicksten 2012)
Family Thoridae
Eualus biunguis (Rathbun, 1902). Sea of Japan north to Bering Sea, south to Oregon, 90–2090 m (Butler 1980)
Eualus macropthalmus (Rathbun, 1902). Unalaska to Point Sur, California, 110–1163 m (Butler 1980)
Heptacarpus yaldwyni (Wicksten, 1984). Off Salina Cruz, Mexico, 1052–1145 m
Lebbeus bidentatus (Zarenkov, 1976). Off Peru, 1680 m
Lebbeus carinatus (Zarenkov, 1976). Off Peru, 1850 m
Lebbeus curvirostris (Zarenkov, 1976). Off Peru, 1680–1860 m
Lebbeus laurentae (Wicksten 2010). West of Costa Rica (12°49′N 103°57′W), 2630 m
Lebbeus scrippsi (Wicksten and Méndez, 1982). SE Gulf of California, Peru to Chile, 768–1240 m (Hendrickx 2001)
Lebbeus splendidus (Wicksten and Méndez, 1982). Off Peru, 712–1100 m
Lebbeus vicinus montereyensis (Wicksten and Méndez, 1982). Monterey Bay, California to Gulf of California, 954–2086 m
Lebbeus washingtonianus (Rathbun, 1902). Northern Japan, across north Pacific to off San Clemente Island, California, 820–1808 (Komai and Takeda 2004)
Family Crangonidae
Neocrangon abyssorum (Rathbun, 1902). Bering Sea to Cortez Bank, California 97–2975 m (Wicksten 2012)
Metacrangon procax (Faxon, 1893). San Miguel Island, California to southern Peru, 830–1658 m (Wicksten 1989 as *Crangon lomae*, Wicksten 2012)
Paracrangon areolata (Faxon, 1893). SE Gulf of California; off Santa María Bay, Tres Marías Islands, Mexico to Peru, 1016–1650 m (Hendrickx 1996, 2001)
Parapontophilus occidentalis (Faxon, 1893). Off San Clemente Island (Wicksten 2012 as *Pontophilus gracilis occidentalis*), central Gulf of California to Peru, 1789–4082 m (Komai 2008; Hendrickx 2012b)
Sclerocrangon atrox (Faxon, 1893). Off Sinaloa (25° 15′N), Mexico to Peru, 1209–1238 m (Hendrickx 2012b)
Family Glyphocrangonidae
Glyphocrangon alata (Faxon, 1893). North of state of Michoacán, off Acapulco, Mexico to off Valparaiso, Chile, 600–1300 m (Hendrickx 2012b)

(continued)

Table 17.1 (continued)

Glyphocrangon rimapes (Bate, 1983). Near Juan Fernández Islands, 2500 m
Glyphocrangon sicaria (Faxon, 1893). Costa Rica and Gulf of Panama, 1454–3310 m
Glyphocrangon spinulosa (Faxon, 1893). Cortez Basin, California, SE Gulf of California, to off Mariato Point, Panama, 956–1374 m (Hendrickx 2001, 2012b)
Glyphocrangon taludensis (Hendrickx, 2010). SW Mexico, 780–1879 m
Glyphocrangon vicaria (Faxon, 1893). San Clemente Basin, California, off Cedros Island and Cabo San Lucas, Baja California, Mexico; to Galápagos Islands, 1374–2441 m (Hendrickx 2012b)
Infraorder Astacidea
Family Nephropidae
Nephropsis occidentalis (Faxon, 1893). West coast of Baja California, Mexico to Chile, 550–1238 m (Manning 1970)
Infraorder Polychelida
Family Polychelidae
Pentacheles laevis (Bate, 1878). Cosmopolitan, in eastern Pacific, Mexico, Gulf of Panama, off Colombia, off Galápagos Islands, Nasca Ridge and Arica, Chile, 347–2505 m (Luke 1972 as *Polycheles granulatus* Galil 2000 (Hendrickx and Serrano 2012)
Pentacheles validus (A. Milne-Edwards, 1880). Cosmopolitan, off Juan Fernández Islands in eastern Pacific, 914–3365 m (Galil 2000)
Polycheles tanneri (Faxon, 1893). Gulf of Panama, off Galápagos Islands, Ecuador, Peru, 540–830 m (Galil 2000)
Stereomastis nana (Smith, 1884). Cosmopolitan, 300–4000 m (Galil 2000 as *Polycheles nanus*)
Stereomastis pacifica (Faxon, 1893). Noyo Canyon, California to off Valparaiso, Chile, 600–3380 m (Galil 2000; Wicksten 2012 as *Stereomastis sculptus pacificus*)
Stereomastis suhmi (Bate, 1878). Off Valparaiso, Chile, circum-Antarctic, 293–4000 m (Galil 2000 as *Polycheles suhmi*)
Willemoesia inornata (Faxon, 1893). Off Magdalena Bay, Baja California (Luke 1992); off Costa Rica, Gulf of Panama, Colombia, Ecuador, Chile, Galápagos and Juan Fernández Islands, 2380–4005 m (Wicksten 1989 as *W. challenger*, Galil 2000)
Willemoesia pacifica (Sund, 1920). Off Juan Fernández Islands, Kermadec Trench, off New Zealand, 2745–5000 m (Galil 2000)
Infraorder Axiidea
Family Axiidae
Eiconaxius albatrossae (Kensley, 1996). Pacific Panama, 851–1016 m (Komai and Tsuchida 2012). *Eiconaxius baja* Kensley 1996. Between Cortes Bank and San Clemente island, California; off northwestern Baja California (Pacific), 1098–1252 m (Komai and Tsuchida 2012)
Family Ctenochelidae
Callianopsis goniophthalma (Rathbun, 1902) Clarence Strait, Alaska to off Palos Verdes Peninsula, Los Angeles County, California; off Ahome Point, Sinaloa, Mexico, 483–1920 m (Hart 1982 as *Callianassa goniophthalma*; Hendrickx 1995b; Wicksten 2012)
Family Calocarididae
Calocaris investigatoris (Anderson, 1896). Arabian Sea, Aleutian Islands to San Diego, California, 549–1733 m (Hart 1982)

(continued)

Table 17.1 (continued)

Infraorder Anomura
Family Chirostylidae
Heteroptychus galapagos (Baba and Wicksten, 2019). East Darwin Seamount, Galápagos Islands, 1012 m
Heteroptychus nautilus (Baba and Wicksten, 2019). East Wolf Seamount, 1049.4 m
Uroptychus occidentalis (Faxon, 1893). Gulf of Panama, 839 m; East Wolf Seamount, Galápagos Islands, 873 m (Baba and Wicksten 2019)
Family Sternostylidae
Sternostylus defensus (Benedict, 1902). Off Galápagos Islands, 717–873 m (Baba and Wicksten 2019)
Sternostylus iaspis (Baba and Haig, 1990). Southern Vancouver Island to Jasper Seamount, California, 600–1189 m (Wicksten, 2012 as *Gastroptychus iaspis*)
Family Munididae
Munida curvipes (Benedict, 1902). Off Chonos Archipelago, 1924 m
Munida perlata (Benedict, 1902). Southern Gulf of California, 1920–3292 m
Munida propinqua (Faxon, 1893). Gulf of Panama to Peru, 1290–1713 m
Family Munidopsidae
Galacantha diomedeae (Faxon, 1893). San Clemente Island, California to off Constitución, Chile, 768–3790 m (Wicksten 1989 as *Munidopsis diomedeae*; Guzmán and Sellanes 2015)
Galacantha rostrata (Milne-Edwards, 1880). Off Acapulco, off Galápagos and Juan Fernández Islands, off Antofagasta, 1775–2492 m (Wicksten 1989 as *Munidopsis rostrata*; Guzmán and Sellane, 2015
Munidopsis agassizii (Faxon, 1893). Gulf of Panama, Peru, to Iquique, Chile, 384–1000 m (Guzmán and Sellanes 2015)
Munidopsis albatrossae (Pequegnat and Pequegnat, 1973). Oregon to central America, East Pacific Rise, 2550–2891 (Ambler 1980 as *M. aries*; Jones and Macpherson 2007)
Munidopsis alfredolaguardai (Hendrickx and Ayon-Parente, 2013). Gulf of California to Chiloé, Chile, 480–1225 m (Guzmán and Sellanes 2015)
Munidopsis antonii (Filhol, 1884). Cosmopolitan, in eastern Pacific, from Bering Sea, off Oregon, central California, Costa Rica, Gulf of Panama, off Juan Fernández Islands, 3134–4100 m (Jones and Macpherson 2007 as *M. beringana*)
Munidopsis barrerai (Bahamonde, 1964). Off Peru to Los Vilos, Chile, 280–800 m (Guzmán and Sellanes 2015)
Munidopsis bracteosa (Jones and Macpherson, 2007). Mendocino Fracture Zone, Monterey Bay, California; 2441–2891 m
Munidopsis cascadia (Ambler, 1980). Cascadia Basin, off Oregon; Monterey Bay, California, 2743–2926 m (Jones and Macpherson 2007)
Munidopsis cochlearis (Khodkina, 1973). Southwest of Antofagasta, Chile, 4550 m
Munidopsis follirostris (Khodkina, 1973). North of Juan Fernández Islands, Chile, 1280 m
Munidopsis granosicorium (Williams and Baba, 1989). Off Strait of Juan de Fuca, 2020 m
Munidopsis hamata (Faxon, 1893). Baja California, Mexico to Chile, 390–1337 m (Guzmán and Sellanes 2015)
Munidopsis hendersoniana (Faxon, 1893). SW Coast of Mexico; Gulf of Panama, 1101–1869 m (Hendrickx 2017)
Munidopsis hirsuta (Jones and Macpherson, 2007). Off cental California, 34 deg. 50 min. N, 123 deg. 00 min. W, 4100 m

(continued)

Table 17.1 (continued)

Munidopsis hystrix (Faxon, 1893). Anacapa Island, California to Peru, 552–1243 m (Wicksten 2012)
Munidopsis kensmithi (Jones and Macpherson, 2007). Cental California, 34 deg. 50 min. N, 123 deg. 00 min. W, 4100 m; photograph off Monterey Bay, California (MBARI photo files); Lamont Guyot, East Mariana Basin, 4833 m (Dong et al. 2017)
Munidopsis lignaria (Williams and Baba, 1989): Cascadia Basin off Oregon, East Pacific Rise off south central Mexico,2030–2875 m (Ambler 1980 as *Munidopsis ciliata*)
Munidopsis nitida (A. Milne-Edwards, 1880). Gulf of Mexico, Caribbean, Indian Ocean, Japan; in eastern Pacific, Gulf of Panama and off Cocos Island, 1245–2363 m (Ambler 1980 as *Munidopsis ciliata*, Baba 2005)
Munidopsis opalescens (Benedict, 1902). Straits of Magellan, Chilean Patagonia, subantarctic islands, 922 m (Guzmán and Sellanes 2015)
Munidopsis palmatus (Khodkina, 1973). Gulf of California, Chile, 1225–1240 m (Hendrickx 2001)
Munidopsis panamae (Baba, 2005). Gulf of Panama, 3800 m
Munidopsis producta (Baba, 2005). Bay of Panama, Mariato Point to Cocos Island, 3260–3680 m (Faxon 1895 as *Munidopsis subsquamosa aculeata*; Henderson 1888 as *Munidopsis aculeata*; Guzmán and Sellanes 2015 as *M. subsquamosa*)
Munidopsis quadrata (Faxon, 1893). Queen Charlotte Islands, Canada to off Tres Marías Islands; Antofagasta to Constitución, Chile, 245–1574 m (Guzmán and Sellanes 2015)
Munidopsis scotti (Jones and Macpherson, 2007). Juan de Fuca Ridge, 2715 m
Munidopsis segonzaci (Jones and Macpherson, 2007). Cental California, 34°50′N, 123°W, 4100 m.
Munidopsis tiburon (Jones and Macpherson, 2007). Oregon (Ambler 1980 as *Munidopsis* sp.), Vance Seamount and Monterey Bay Canyon, California, 1829–2029 m
Munidopsis tuftsi (Ambler, 1980). Tufts Abyssal Plain, northern Pacific, 3500–3858 m
Munidopsis verrilli (Benedict, 1902). Oregon, Monterey Bay to "Cerros" (= Cedros) Island; western Pacific in Makassar Strait and Tasmania, 732–4169 m (Baba 2005)
Munidopsis verrucosus (Khodkina, 1973). Aleutian Islands to Antofagasta, Chile, 3932–4880 m (Jones and Macpherson 2007)
Munidopsis vicina (Faxon, 1893). Off Alaska Peninsula, Gulf of Panama, Cocos Island, 936–3885 m (Wicksten 2012; Baba 2005)
Munidopsis yaquinensis (Ambler, 1980): Off Oregon, 2763–2377 m
Family Paguridae
Pagurus capillatus (Benedict, 1892). Chukchi Sea, Bering Sea of off Santa Cruz, California, 4–1189 m (Wicksten 2012).
Family Parapaguridae
Parapagurus foraminosus (Lemaitre, 1999). Off Baja California to Ecuador, Cocos and Galápagos Islands, 915–2807 m
Parapagurus holthuisi (Lemaitre, 1989). Gulf of California to Chile, 1410–3340 m (Garth and Haig 1971 as *Parapagurus abyssorum*)
Parapagurus benedicti (de St. Laurent, 1972). Alaska to off Juan Fernández Islands, 415–2012 m (McLaughlin 1974; Wicksten 2012 as *Parapagurus pilosimanus benedicti*)
Probeebei mirabilis (Boone, 1926): Costa Rica to Peru, 1145–3995 (Wicksten 1989)
Family Lithodidae
Lithodes couesi (Benedict, 1895). Japan, Okhotsk Sea, Bering Sea, Alaska south to Tortugas Bay, Baja California, Mexico, 384–1125 m (Martin et al. 1997)

(continued)

Table 17.1 (continued)

Lithodes panamensis (Faxon, 1893). Costa Rica, Gulf of Panama, off Peru, 760–838 m. (Macpherson and Wehrtmann 2010). (Unconfirmed record off Cabo San Lucas, Mexico, Hendrickx and Harvey 1999)
Lithodes wiracocha (Haig, 1974). Off Costa Rica, SW of Banco de Mancora, Peru, 620 m (Macpherson and Wehrtmann 2010)
Neolithodes diomedeae (Benedict, 1894). Monterey Bay and Sur Ridge, California; off Panama to Peru, 830–1300 m (Barry et al. 2016)
Paralomis aspera (Faxon, 1893). West coast of Baja California and SW of Manzanillo, Mexico, Panama to northern Peru, 560–1397 m (Hendrickx 2019a, b)
Paralomis chilensis (Andrade, 1980). Off Coquimbo to off Los Vilos, Chile, 400–420 m.
Paralomis diomedeae (Faxon, 1893). California, Costa Rica to northern Peru, 458–830 m (Macpherson and Wehrtmann 2010)
Paralomis inca (Haig, 1974). Peru to Chile, 620–744 m
Paralomis longipes (Faxon, 1893). Off Cocos Island to Peru, 760–1409 m
Paralomis multispina (Benedict, 1895). Japan, Alaska to San Diego, California, 500–1665 m (Sakai 1971)
Paralomis otsuae (Wilson, 1990). West coast of Baja California, off Jalisco, western Mexico; Peru to Mejillones del Sur, Chile, 80–2054 m (Hendrickx 2019a, b)
Paralomis papillata (Benedict, 1895). "Off Lower California, or perhaps south of the region" (Benedict 1895), off Costa Rica, Peru 712–744 m (Macpherson and Wehrtmann 2010)
Paralomis phrixa (Macpherson, 1991). Northern Peru and off Tasmania, 1815–1860 m
Paralomis sonne (Guzmán, 2009). Off Antofagasta, Chile, 1775 m
Paralomis verrilli (Benedict, 1895). Sea of Okhotsk to Cortez Bank, California, 1238–2379 m
Infraorder Brachyura
Family Homolodromiidae
Homolodromia robertsi (Garth, 1973). off Peru, 800 m
Family Dorippidae
Ethusina faxonii (Rathbun, 1933). Western Mexico to Peru, 2999–4081 m
Ethusina robusta (Miers, 1886). Bay of Panama to Galápagos Islands and coast of Ecuador, 1618–3334 m
Family Cymonomidae
Cymonomus menziesi (Garth in Garth and Haig, 1971). Peru, 1005–1124 m
Family Epialtidae
Rochinia cornuta (Rathbun, 1898). Off Galápagos Islands, 718–1160 m
Family Trichopeltariidae
Trichopeltarion corallinum (Faxon, 1893). SE Gulf of California; off Acapulco to Peru, 834–1280 m (Hendrickx 2001)
Trichopeltarion hystricosus (Garth in Garth and Haig, 1971). Peru to Chile, 907–935 m

Benthesicymus, changing *B. altus* to *Dalicaris altus* and *B. tanneri* to *Trichocaris tanneri* but a more recent revision changed the name of the latter species to *Benthoecetes tanneri* (WoRMS editorial board (2020). Pequegnat and Wicksten (2006) used morphology to compare specimens of *Acanthephyra eximia* taken in the Gulf of Mexico with those from western Costa Rica, but did not find any observable differences between them, nor did Chace (1986) in his treatment of Pacific

Oplophoridae. *Pandalus ampla*, reported from both the Atlantic and Pacific Oceans, may be a species complex (Komai 1994). Burukovsky (2001) redefined the species of *Nematocarcinus* so that some supposedly cosmopolitan species have since been divided into related species. Krygier and Pearcy (1981) reported *Nematocarcinus exilis* off Oregon, but this specimen needs to be re-examined. *Lebbeus washingtonianus* may be a highly variable species or a species complex (Komai and Takeda 2004). Komai et al. (2004) noted that *L. washingtonianus* had been reported from the eastern Pacific and from a single location in the Okinawa Trough, but the habitat and depth of these records were different. They suggested that the identification was uncertain and that the specimens be re-examined. There are photographs of what probably are *Lebbeus* spp. taken by ROV's off Monterey Bay, Costa Rica, and the Galápagos Islands, but these cannot be identified to species without specimens (MKW unpublished). A shrimp identified as *Bathypalaemonella serratipalma* was photographed off the Galápagos Islands (L. Watling pers. comm.), but it was not compared to the eastern Pacific *B. delsolari*.

Records of lobster-like decapods need further confirmation. A previous report of *Eiconaxius acutifrons* by Wicksten (1982) from California is now identified as *E. baja* Kensley, 1996 (Komai and Tsuchida 2012). If confirmed, a record of *Willemoesia leptodactyla* off Punta San Juan, Peru, would be the first record of this species from the eastern Pacific (Luke 1992). Photographs in the "benthic invertebrate guide" for the ship *Okeanos Explorer* (NOAA 2019) in the central Pacific show a squat lobster identified as *Gastroptychus* cf. *iaspis*. I examined a paratype of *Sternostylus iaspis* and concluded that the pattern of spines on the carapace and chelae of *S. iaspis* are markedly different from those of the (as yet unidentified) squat lobster in the central Pacific. As of this writing, *S. iaspis* is known only from the eastern Pacific.

There is general agreement that many members of the superfamilies Galatheoidea and Chirostyloidea need to be studied further and may in fact be species complexes. The taxonomy of species of *Munidopsis* is particularly confusing. For example, *Munidopsis bairdii* and *M. antonii* currently are considered to be cosmopolitan species, but there has been no genetic comparison of specimens from different oceans (E. Macpherson pers. comm.). *Munidopsis subsquamosa*, considered to be a cosmopolitan species, recently has been determined in the eastern Pacific to be a different species, *M. producta* (Baba 2005). Specimens reported as *M. subsquamosa* from Oregon (Ambler 1980) and off Chile (Guzmán and Sellanes 2015) need to be re-examined (Baba 2005). Presumably the same species has been reported in the eastern Pacific as *Munidopsis subsquamosa aculeata* by Faxon (1895) and *Munidopsis aculeata* Henderson, 1888. There has been extensive recent revision of species of the Chirostylidae based on morphological work (Baba 2005; Baba et al. 2018) and new genetic studies (Jones and Macpherson 2007). Some species formerly assigned to the genus *Gastroptychus* have been re-examined and assigned to the genus *Sternoptychus* (Baba et al. 2018), including *Sternoptychus iaspis* in the eastern Pacific.

The nomenclature of the various crab-like decapods has remained relatively stable since 1989, although there are new records and range extensions. Lemaitre (1989, 1999) revised the deep-sea hermit crabs of the genus *Parapagurus*.

17.3.3 Ranges

Ranges of eastern Pacific decapods known from more than a single location are provided (Table 17.2). If a species only has been reported from latitudinally isolated locations (e.g., from Monterey Bay, California, and off Chile), then those two reports are listed separately instead of giving the range as "Monterey Bay to Chile." For the majority of the species, there are few or no numerical data, so the system largely is based on the presence or absence of specimens. Records from the Clarion-Clipperton Zone, Peru Basin, Eastern Pacific Rise, and Juan de Fuca Ridge are included.

The huge spread of the Indo-Pacific region has many more species than the eastern Pacific, especially in the Chirostylidae (Baba and Schnabel 2018) and other members of the Galatheoidea (as Galatheidae in older literature). The biota of the eastern Pacific is isolated and less diverse than that of the western and central Pacific (Macpherson et al. 2010). The eastern Pacific is separated from the central Pacific by as much as 8000 km: Easter Island lies over 1850 km west of the Juan Fernandez Islands; Hawaii is nearly 5000 km from California. This vast distance is known as the "Eastern Pacific Barrier" and probably limits the dispersal of species between these two areas. The lack of records in abyssal and lower slope areas may be due to a lack of study. Wicksten and Kuhnz (2015) first reported the common and widespread abyssal shrimp *Cerataspis monstrosa* off California, when photographs and video stills from ROV's and improved cameras became available. Species reported to be worldwide or trans-Pacific tend to be abyssal species (living as deep as 2000 m or more), as are species that range from Alaska or California to Peru or Chile. At least four species range across the north Pacific from Japan to California, perhaps able to "island hop" along submerged spurs or slopes across the length of the Aleutian Archipelago.

Previous work suggests differentiation in species assemblages by latitude (Wicksten 1989; Macpherson et al. 2010; Guzmán and Sellanes 2015). The area from southwestern Mexico to Peru is particularly rich in species. These ranges seem to correspond to patterns of modern currents or possibly ancient geologic evens or climate changes, such as the closing of the seaway across Central America. Ranges of many species are based only on a single specimen or a few crustaceans from one isolated locality: for example, Jones and Macpherson (2007) described three new species of *Munidopsis* from a single site in California. Some areas (south of the Columbia River, Oregon to Monterey Bay, California; much of western Central America, or most areas 1500 m deep or more) remain poorly studied or not sampled at all. At present, there is no definitive evidence that any deep benthic decapods are endemic to a particular island chain, vent system, or seamount.

Table 17.2 Range of decapod crustaceans with more than one record in different sectors of the eastern Pacific

World-wide, trans-Pacific or in more than one ocean	Alaska or Canada to Chile	Oregon or California to Western Mexico, Costa Rica or Panama	Japan and Russia into Eastern Pacific
Cerataspis monstrosa (as *Plesiopenaeus armatus*)	*Parapagurus benedicti*	*Lebbeus vicinus montereyensis*	*Eualus biunguis*
Benthesicymus laciniatus	*Munidopsis quadrata*	*Glyphocrangon spinulosa*	*Lebbeus Washingtonianus*
Acanthephyra eximia	*Munidopsis verrucosus*	*Glyphocrangon vicaria*	*Lithodes couesi*
Nematocarcinus proximatus	*Munidopsis vicina*	*Munidopsis albatrossae*	
Pandalus ampla		*Munidopsis challengeri*	
Pentacheles laevis	**Alaska, Washington or Oregon to California**	*Munidopsis lignaria*	**California to Peru or Chile**
Pentacheles validus	*Pandalus tridens*	*Paralomis multispina*	*Bentheocetes tanneri*
Polycheles nanus	*Eualus biunguis*	*Munidopsis hystrix*	*Metacrangon procax*
Calocaris investigataris	*Paralomis verrilli*	*Neocrangon abyssorum*	*Parapontophilus occidentalis*
Munidopsis antonii		*Sternostylus iaspis*	*Polycheles pacificus*
Munidopsus kensmithi		*Munidopsis cascadia*	*Galacantha diomedae*
Munidopsus nitida		*Pagurus capillatus*	*Neolithodes diomedeae*
Munidopsis subsquamosa species complex		*Munidopsis tiburon*	*Paralomis diomedeae*
Munidopsis verrilli			

Southwestern Mexico and Gulf of California to Panama and Ecuador	Southwestern Mexico to Peru or Chile	Costa Rica or Panama to Ecuador, Peru or Chile	Circum-Antarctic or Southern Chile
Munidopsis hendersoniana	*Hymenopenaeus doris*	*Haliporides diomedeae*	*Willemoesia pacifica*
Parapagurus foraminatus	*Glyphocrangon alata*	*Glyphocrangon sicario*	*Polycheles suhmi*
	Sclerocrangon atrox	*Hymenopenaeus nereus*	*Munidopsis opalescens*
	Nematocarcinus agassizi	*Polycheles tanneri*	
	Nematocarcinus faxoni	*Munida propinqua*	
	Lebbeus scrippsi	*Munidopsis agassizi*	
	Paracrangon areolata	*Munidopsis alfredolaguardai*	
	Willemoesia inornata	*Paralomis phrixa*	

(continued)

Table 17.2 (continued)

Southwestern Mexico and Gulf of California to Panama and Ecuador	Southwestern Mexico to Peru or Chile	Costa Rica or Panama to Ecuador, Peru or Chile	Circum-Antarctic or Southern Chile
	Nephropsis occidentalis	*Munidopsis barrerei*	
	Galacantha rostrata	*Munidopsis producta*	
	Munidopsis hamata	*Probeebei mirabilis*	
	Munidopsis palmatus	*Ethusina robusta*	
	Parapagurus holthuisi	*Trachycarcinus hystricosus*	
	Paralomis aspera	*Paralomis inca*	
	Paralomis otsuae	*Paralomis longipes*	
	Paralomis papillata		
	Trachycarcinus corallinus		
	Ethusina faxoni		

The eastern Pacific, Indo-Pacific, and Atlantic contain species that seem to be cosmopolitan or belong to closely related clades, such as the *Munidopsis subsquamosa* complex. Such clades suggest interconnectivity until fairly recently. The closing of the Isthmus of Panama between the Caribbean and the eastern Pacific may have occurred in stages. The latest estimate gives an estimated time of complete closing by an estimated 2.8 million years ago (O'Dea et al. 2017). This closing resulted in many sibling species in the nearshore Caribbean and tropical eastern Pacific. How this closing affected the deep biota is less well understood. In comparison to the western Atlantic and Caribbean, the eastern Pacific exceeds the western Atlantic in species of the Lithodidae and Thoridae but has only one of the Nephropidae and fewer of the Glyphocrangonidae and Polychelidae than the western Atlantic-Caribbean. The large crabs of the family Geryonidae, common inhabitants of the western Atlantic, are not known from the eastern Pacific. Thirty-nine species of the Munidopsidae (including *Galacantha*) have been reported for the entire eastern Pacific, as opposed to 25 in the Gulf of Mexico alone (Wicksten and Packard 2005). Whether these totals reflect differences in sampling effort or habitat diversity remain unknown. Recent descriptions of many new species by Jones and Macpherson (2017), among others, have added greatly to the number of eastern Pacific species and suggest that species diversity is poorly known and likely to be underestimated.

17.3.4 Habitats

Most early studies of deep-sea benthic animals were done by trawling, which usually is conducted on soft substrates. Shrimps and lobsters that usually are caught or seen on muddy or silty sea floors include the abyssal species *Cerataspis monstrosa*,

Benthesicymus laciniatus, and *Willemoesia* spp.; *Haliporoides diomedeae*, which buries itself in soft substrates (Hendrickx 1995a, b); *Nephropsis occidentalis* (Manning 1970; Hendrickx 2003); and *Glyphocrangon* spp. Most records of *Nematocarcinus* spp. come from muddy areas, but recent ROV photographs also show them among rocks or manganese nodules (J. Drazen, pers. comm.). *Nematocarcinus burukovskyi* and *N. ovalis* have been photographed at or near hydrothermal vents but also at a distance from them and are not considered to be obligate vent associates (Komai and Segonzac 2005). Recent photographs by Drazen et al. (2019) show *Cerataspis monstrosa*, *Hymenopenaeus nereus*, *Benthesicymus* sp., *Probeebei mirabilis*, and an unidentified species of *Munidopsis* in a field of manganese nodules.

Eastern Pacific decapods are affected by the oxygen minimum zone (OMZ), a consistent feature along much of the eastern Pacific. The upper depth of the OMZ varies considerably, from less than 50 m off Peru to 200–600 m off California and Oregon. The lower limit also is variable, from 600–700 m off Chile and Peru to about 1100 m off North America (Helly and Levin 2004). Hypoxia is most severe off western Mexico and Central and South America, to less than 0.2 ml oxygen per liter, with North America usually above the 0.2 ml limit. The El Niño climatic condition also can have a great influence, deepening the OMZ off Peru and Chile by more than 100 m but also having severe effects on pelagic organisms at lesser depths (Arntz et al. 2006). In the southern Gulf of California, there is no macrofauna inhabiting areas between approximately 150–500 m due to severe hypoxic or anaerobic conditions (Hendrickx 2003). Hendrickx and Hernandez Payán (2017) noted that both *Nematocarcinus agassizii* and *N. faxoni* are restricted to zones deeper than the core of the OMZ, as are *Nephropsis occidentalis* and *Stereomastis pacificus* (Papiol et al. 2016; Hendrickx and Serrano 2012).

17.3.5 Functional Anatomy

As in shallow-water species, decapods living on muddy sea floors generally have elongate dactyls that serve as stilts. Species of *Nematocarcinus* are rightly called "thread leg" shrimps, for their hairlike legs provide a minimum support for these shrimps at depth. Species of the Benthesicymidae have very thin legs as well as poorly calcified exoskeletons. Squat lobsters and lithodids living on rocks usually have sturdy appendages and may have spinules on their dactyls. The spinules may aid in gripping a host coral or the substrate. Lithodid crabs and burrowing species of *Glyphocrangon* and the Munidopsidae generally have sturdy exoskeletons. Deep-sea crabs generally do not have swimming paddles or flat digging appendages, which can be characteristic of decapods of the continental shelf (e.g., the swimming crabs, family Portunidae, and the frog crabs, family Raninidae).

Garth and Haig (1971) noted that the crab *Lophorochinia parabranchia*, living within the OMZ of the Peru-Chile Trench, had greatly swollen gill chambers and enlarged excurrent openings. *Lithodes couesi* also has inflated branchial chambers and large exhalent openings and scaphognathites (Somerton 1981). Species of *Munidopsis* represent the largest group of decapod crustaceans below the OMZ (Hendrickx 2012a). *Bentheocetes tanneri*, *Glyphocrangon spinulosa*, and *Nephropsis occidentalis* also have been reported from conditions of low oxygen (Hendrickx 2003). Their particular adaptations to these conditions remain unstudied.

Many deep decapods have no or reduced vision. Members of the Polychelidae and Munidopsidae do not have pigmented eyes. *Cymonomus menziesii* has long eyestalks but no cornea. Crustaceans living in dim light may have superpositional eyes, in which screening pigments in the ommatidia are concentrated and allow light to pass through to more than one ommatidium before striking the rhabdome, which contains rhodopsin. These eyes are adapted to very low light but have less visual acuity and capability for image formation than is found in light-adapted eyes. In the Atlantic lobster, *Nephrops norvegicus*, the superposition compound eyes are large and able to adjust their sensitivity to spectral and temporal changes through movement of pigments. As in some other deep crustaceans, exposure to ambient surface light intensities can damage the retina layer, but such damage does not cause a noticeable impact on survival (Gaten et al. 2013).

In contrast, the abyssal anomurans *Probeebei mirabilis* and *Parapagurus holthuisi* (as *P. abyssorum*) have pigmented eyes that seem to be functional (Garth and Haig 1971). Hiller-Adams and Case (1988) stated benthic crustaceans living at greater depths tend to have larger eyes than those in lesser pelagic depths, and Warrant and Locket (2004) suggested that the better nutrition of the sea floor would favor agile crustaceans with larger eyes. Frank et al. (2012) found that the Atlantic chirostylid squat lobster *Gastroptychus spinifer*, which has large eyes, had greatest sensitivity in the blue region of the spectrum and could detect greenish bioluminescence. Although this squat lobster itself is not bioluminescent, it might be able to detect the difference between bioluminescence produced by zooplankton which it eats and that of the pennatulaceans and zoanthids on which it lives.

When illuminated or brought to the surface, deep-sea decapods may be bright red. Faxon (1895) illustrated six red benthic species by watercolors immediately after collection. Photographs show the same coloration in *Cerataspis monstrosa* (https://www.ncddc.noaa.gov/okeanosanimalguide/Dendrobranchiata005.html)) and *Acanthephyra eximia* (Baba et al. 1985). Many species of *Munidopsis* are white. Deep-sea fishes may be able to detect the colors from pigmentation at depth, so red colors, invisible at that depth, could serve in visual camouflage. Carotenoid (red) pigments also may be involved in physiological functions that remain unstudied.

Deep decapods, like shallow-water ones, have sensory setae of a wide variety of shapes and functions. Many have tactile setae on the pereopods and chemosensory setae (aesthetascs) on the first antennae. Decapods in general have gustatory setae on the inner mouthparts. The function of small peg setae, spinules, and other sensors remains unknown but might include detection of vibrations or currents. Many have long, whiplike second antennae that may serve in determining individual distance from conspecifics or moving through tight spaces.

17.3.6 Reproduction, Larval Stages, and Population Densities

Reproduction in deep decapods seems to be year-round. Photographs of lithodid crabs (*Paralomis* sp. and *Neolithodes agassizii*) show that adult females may be much smaller than the males holding them in the "mating embrace" (MBARI photo files). Some species produce large numbers of small eggs. The eastern Pacific *Nematocarcinus agassizii* and *N. faxoni* can carry up to 10,000 eggs (Hendrickx and Hernandez Payán 2017). Wenner (1979a) found that western Atlantic *Nematocarcinus* spp. produced as many as 15,573 eggs per female. *Pandalus platyceros* females carry a mean egg count of 2028 eggs per female (Butler 1970). *Stereomastis pacificus* had eggs approximately 1 mm in largest dimension. This size is similar to those of S. *sculpta*, a related Atlantic species, which carried as many as 19,080 eggs (Wicksten 1981). A female *Lithodes couesi* can carry as many as 5000 eggs but usually close to 4000 (Somerton 1981). *Parapagurus foraminosus* carries as many as 2716 eggs that increase in diameter during development (Ayon Parente and Hendrickx 2009). Penaeoid shrimp females do not carry eggs ventrally, and so reproduction must be estimated by examination of the gonads.

Other decapods carry fewer, larger eggs. *Lebbeus scrippsi* females carried 32–66 eggs (Hendrickx 2001). *Glyphocrangon alata* females carry 28–51 eggs; *G. spinulosa*, 11–16; and *G. vicaria*, 10–29 (Wicksten 1979). The few species of *Munidopsis* that have been studied have few large eggs and abbreviated development, but the larvae may be able to float for extended periods in nutritionally poor water (Jones and Macpherson 2007). Female *Galacantha diomedeae* carry from 11 to 126 large eggs (Hendrickx and Papiol 2019). A specimen of *Uroptychus nitidus* from the Gulf of Mexico carried 35 eggs (MKW, unpublished data).

Larval stages of deep decapods are poorly known and may be unidentifiable. Size distribution of larvae of *S. pacificus* taken in an Isaacs-Kidd midwater trawl (mesh size 3 mm) suggested that there were at least five larval stages, previously called the "Eryoneicus" larvae. The largest larval stages had a total length of 60 mm or more, as much as 0.8 times the total length of reproducing adults (Wicksten 1981). *Cerataspis monstrosa* produces a bizarre larval form that only recently was matched with the adult (Bracken-Grissom et al. 2012). *Pandalus ampla* has five zoeal stages and one postlarva (Park et al. 2004). In *Galacantha diomedeae*, the large eggs contain advanced embryos at stage 5, consistent with extended, lecithotrophic embryonic development and a reduced pelagic larval stage (Hendrickx and Papiol 2019).

Decapods have few indicators of lifespan other than the obvious—big ones probably are older than smaller ones. Brachyurans show obvious sexual dimorphism in the shape of the abdomen and often the length of the chelipeds relative to the body and the gape between the fingers of the chela. *Pandalus dispar* and *Pandalus platyceros* from the northern Pacific generally live no more than 4 years, which may be typical of pandalids (Butler 1980).

Populations of deep benthic decapods can be patchy. Arana et al. (2003) found that groups of *Haliporoides diomedae* were separated by a submarine canyon.

Hendrickx and Hernandez Payán (2017) found that populations of *Nematocarcinus faxoni* varied from 1 to 800 animals per hectare in the Gulf of California and Pacific side of Baja California. *Nematocarcinus agassizii* was encountered less frequently but also varied in density, from 1 to 50 animals per hectare. The densities of both species declined with depth, with *N. agassizii* having its lowest values at the 1701–2100 m interval and *N. faxoni* at 1301–500 m. *Galacantha diomedeae* occurred at maximum densities of 71 squat lobsters per hectare (Hendrickx and Papiol 2019).

Drazen et al. (2019) reported on scavenging decapods seen form baited camera traps and in camera transects in the Peru Basin. The shrimp *Hymenopenaeus nereus* was present in 63% of the images, showing up to 15 shrimp at once (average of 9 shrimp per view). This abundance was similar to that seen for the species in the eastern Clarion-Clipperton Zone (Leitner et al. 2017). The hermit crab *Probeebei mirabilis* appeared in 29% of the images and in all deployments, with 1 to 9 crabs per view. The large number of these crabs seemed to be "unique" among abyssal scavenger studies. This large number of hermit crabs may not be an isolated occurrence. More than 300 *Sympagurus pictus* were taken in a single trawl in the Gulf of Mexico (MKW unpublished data). Such great numbers were not collected at any other station during 1964–2004 and seem likely to represent a short-term aggregation, perhaps at a sunken food source.

17.3.7 Natural History

There is very little natural history information available for deep slope and abyssal decapods. I have included published information on related species living in the Gulf of Mexico and western Atlantic, unpublished information from communications with colleagues, my own field notes, and information from preserved specimens in the Biodiversity Research and Teaching Collections of Texas A&M University.

There is a general assumption that deep decapods feed on small benthic invertebrates, debris, mucus, bacterial mats, or other nutritional sources that they glean from the substrate. Two squat lobster species (*Munidopsis verrilli* and *Munidopsis bracteosa*) and the lithodid crab *Neolithodes diomedeae* have been collected or photographed at carcasses of dead whales (Jones and Macpherson 2007, MBARI photo files). Papiol and Hendrickx (2015) reported that *Bentheocetes tanneri* often fed on benthic or benthopelagic prey, especially the squat lobster *Pleuroncodes planipes* and oplophorid shrimps. Wenner (1979a, b) reported that *Nematocarcinus ensifer* and *N. rotundus* from the western Atlantic were omnivorous, feeding on fish parts, foraminiferans, polychaetes, and detritus. *Glyphocrangon* spp. from the western Atlantic also were omnivores, feeding on gastropods, fish scales, polychaetes, and unidentified detritus (Thompson 1963; Gore 1985). *Glyphocrangon sculpta* off Ireland was photographed more or less passively moving with the current across the sea floor, where it fed on bivalves, foraminiferans, and small infaunal crustaceans

(Lampitt and Burnham 1983). In situ photographs of *Bathystylodactylus* cf. *inflatus* from the western Pacific and an unidentified *Bathystylodactylus* from the Clarion-Clipperton Zone show the shrimps with their setose second appendages extended laterally, probably filter-feeding on small particles carried by a current (Amon et al. 2017; Wicksten et al. 2017). Krygier and Pearcy (1981) captured a single *Neocrangon abyssorum* (as *Crangon abyssorum*) in a midwater trawl at a distance of more than 1500 m from the sea floor and suggested that it might forage off the sea floor. *Stereomastis pacificus* has been taken in baited sablefish traps (Wicksten 1981), and *Cerataspis monstrosa*, *Hymenopenaeus nereus*, and an unidentified species of *Benthesicymus* were attracted to baited traps in the Peru Basin (Drazen et al. 2019). Lithodid crabs in general are detritivores or feed on mollusks, barnacles, and smaller crustaceans. Species of *Munida* may scavenge, feed on smaller crustaceans, or use setose maxillipeds to sweep detritus from the sea floor. A squat lobster (*Sternostylus* sp.) was photographed gleaning small particles from the water (L. Kuhnz MBARI pers. comm.). A large lithodid crab, probably *Neolithodes agassizii*, was photographed feeding on an ophiuroid in the western Atlantic (D. Wagner, NOAA, *Okeanos Explorer* expedition 18-06).

Deep decapods may have consistent associations with other invertebrates. *Bathypalaemonella serratipalma* associates with soft corals (Wicksten and Heathman 2015). Unidentified *Lebbeus* species live on hexactinellid sponges (Fig. 17.1). Species of the small lobster *Eiconaxius* are commensals of hexactinellid sponges (Kensley 1996; Komai and Tsuchida 2012). Members of the family Chirostylidae seem consistently to associate with antipatharians or alcyonaceans. *Uroptychus occidentalis*, like the Atlantic *U. nitidus*, lives on *Chrysogorgia* spp. (Baba and Wicksten 2017, 2019); *Heteroptychus nautilus*, on bamboo corals (Isididae) (Baba and Wicksten 2019); *Sternostylus iaspis*, on gorgonians and antipatharians (Wicksten 2012); and *S. defensus*, on antipatharians (Baba and Wicksten 2019). The nature of the associations between these shrimps, lobsters, or anomurans

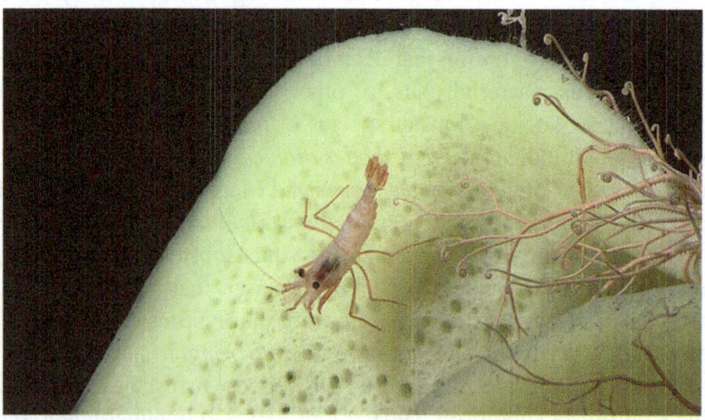

Fig. 17.1 *Lebbeus* sp. on hexactinellid sponge *Staurocalyptus* sp., Davidson Seamount, 35.721 degrees N, 122.724 degrees W. 1315 m. Photo courtesy of MBARI 2006

and their sponge or cnidarian hosts is uncertain. Crustaceans may use their host as a perch off the sea floor, from which they can capture zooplankton, a safe refuge from predators, or a source of mucus and tissue on which they feed. They might also drive off potential predators on their host or remove debris adhering to it (Wagner et al. 2012). The hermit crab *Parapagurus benedicti*, like the related Atlantic *P. pilosimanus*, can inhabit a shell overgrown by zoanthids (M. Wicksten, field notes). *Parapagurus foraminosus* occupies shells of *Bathybembix bairdii*, often covered by an unidentified sea anemone (Actiniaria) (Hendrickx and Ayón Parente 2009). These cnidarians may discourage potential predators of the crab and at the same time gain locomotion. Juvenile *Neolithodes diomedeae* are found on or beneath the elasipodid holothurian *Scotoplanes* sp., where they may take shelter from predators (Barry et al. 2016). The caprellid amphipod *Caprella ungulina* clings to the legs of *Paralomis multispina* (Wicksten 1982). *Paralomis longipes* and other lithodid crabs may have stalked barnacles attached to the exoskeleton (Haig 1974).

Deep decapods can be are parasitized. *Nematocarcinus* spp. and *Glyphocrangon* spp. may be parasitized by isopods (Wenner 1979a; Markham 2016). *Munidopsis antonii*, *M. depressa*, and *M. beringana* are parasitized by bopyrid isopods (Román-Contreras 2008). *Lithodes couesi* can be parasitized by the rhizocephalan cirriped *Briarosaccus callosus* (Somerton 1981). *Lithodes couesi* can be parasitized by a dinoflagellate, *Hematodinium* sp. (Jensen et al. 2010).

Predators on deep decapods probably include each other, especially larval stages. Squids and octopuses live at depths and are likely to prey on decapods. Small sharks and benthic fishes of the families Macrouridae, Ophidiidae, and others prey on crustaceans and whatever else they can catch (Fitch and Lavenberg 1968). Deep decapods can escape from predators by a rapid tail flip and swimming backward, as is found in many shrimps and squat lobsters (superfamily Galatheoidea), digging into the sea floor or hiding. Species of *Glyphocrangon* have locking ball and socket joints on the last three abdominal somites and telson as well as sharp teeth, ridges, and nodules on the exoskeleton (Rice 1981). These armored shrimps can arch the body so that the carapace and abdomen touch (Anderson and Bullis 1970). Large lithodid crabs may escape predation by means of a "size refugium"—they are too big for most predators to attack. *Galacantha rostrata* and *G. diomedeae* have formidable dorsal spines, as do some juvenile lithodids.

17.3.8 Exploitation and Environmental Disturbance

Most fishing for decapods occurs on the continental shelf and upper slope, at depths at much less than 700 m. The cost of operating gear below 500 m has been a major factor in limiting fisheries, and so many deepwater species either are incidental catch or part of exploratory fisheries (Hendrickx 1995a, b; Hendrickx and Hernandez Payán 2017). Wehrtmann et al. (2012) compiled a comprehensive report on deepwater fisheries of the Central and South America. Most of these fisheries were for shrimps of the family Pandalidae (*Heterocarpus affinis* from

Mexico to Peru, *H. reedei* in Chile). Of these, *H. affinis* seems to have the greatest potential for a fishery but off Costa Rica; *H. affinis* is fished "only in negligible quantities." The lower slope species considered for fisheries in western Mexico include the carideans *Pandalus amplus* and *Nematocarcinus* spp. and the penaeoid *Bentheocetes tanneri*. The penaeoid *Haliporus diomedeae* gradually increased in landings in Chile from the 1970s and then decreased as of 2012. The lithodid crab *Paralomis longipes* has been taken in traps off Peru, but as of 2012, the landings had declined considerably. *Bentheocetes tanneri*, with a thin exoskeleton, was deemed to be too fragile to be easily caught and transported. *Pandalus platyceros* has been caught in traps and trawls from Alaska to southern California and has been considered for a potential fishery off western Mexico (Butler 1970; Sunada 1984; Flores et al. 2004). Prized as seafood, the "spot prawn" currently sells for at least $37.95/pound (454 g) and thus is a luxury item. Lost fishing gear, including traps used to catch prawns, is a concern because it may become entangled with corals or sponges on the sea floor (Etnoyer et al. 2013).

The effects of manganese mining can include removal of hard substrate, creation of large sediment plume, and immediate mortality of resident species. Drazen et al. (2019) visited experimentally plowed areas in the Peru Basin. Using camera transects and baited traps, they found that fish density was lower in the first years following disturbance but increased over time. Twenty-six years after disturbance, there are no differences in overall fish densities between reference and experimental areas, but the dominant fish species still exhibited lower densities in the plowed habitat. The scavenging community was dominated by fishes, shrimps, and the hermit crab *Probeebi mirabilis*. Almost nothing is known about the lifespans and recruitment of these decapods, and one can only speculate that extensive disturbance in their habitat would have a negative effect on them.

New evidence points to the damaging effects of plastics on marine animals, even deep-sea crustaceans. Jamieson et al. (2019) found that 72% of the deep-sea amphipods (Lyssianassidae) collected in deep-sea trenches (including the Peru-Chile Trench) at 7000–10,890 m contained microplastics. Amphipods being common scavengers and likely to enter the food chain for larger crustaceans, fishes, and cephalopods, it seems likely that these pollutants also pose a threat to deep-sea decapods.

Pandalid shrimps, lithodid crabs, and *Munidopsis* spp. have been photographed clinging to shipwrecks and sunken cargo containers (MBARI file photographs, M. Wicksten unpublished.) The photographs from ROV's generally do not show the interior of these structures, where rust, obstructions, and layers of anoxic sediments tend to accumulate. These metal structures at least for a time can offer a raised hard area that acts as an "artificial reef," but their lifespan is limited by the time it takes for them to rust, fall apart, and sink down into a silty sea floor.

Acknowledgments I thank Les Watling, University of Hawaii at Manoa, and Shane Ahyong, Australian Museum, for information on records of specimens from off the Galápagos Islands; Chris Kelley, University of Hawaii, for a record of a chirostyloid off Hawaii; Ferran Palero, Centro de Estudios Avanzados de Blanes, Spain, for providing information on *Paralomis tuberipes*;

Guillermo Guzmán, Universidad Arturo Pratt, Chile, for providing records of *Munidopsis*; Enrique Macpherson, Instituto de Ciencias del Mar, Barcelona, for information on supposedly cosmopolitan species of *Munidopsis*; and especially Keiji Baba, Kumamoto University, Japan for his considerable help with chirostyloids. Linda Kuhnz, Monterey Bay Aquarium Research Institute (MBARI), provided video footage and still photographs of decapods in the Gulf of California and off Monterey County, California. Daniel Wagner of the National Oceanographic and Atmospheric Administration Office of Ocean Exploration and Research was an enormous help in obtaining data from photographs taken by the ship *Okeanos Explorer*. Studies by the vessel *Nautilus* off the Galápagos Islands were conducted under the auspices of the Ocean Exploration Trust, the Charles Darwin Foundation, and the Galápagos National Park Directorate.

References

Ambler J (1980) Species of *Munidopsis* (Crustacea, Galatheidae) occurring off Oregon and in adjacent waters. U S Fish Bull 78:13–34

Amon D, Ziegler AF, Drazen JV, Grischenko AV, Leitner AB, Lindsay DJ, Voight JR, Wicksten MK, Young CM, Smith CR (2017) Megafauna of the UKSRL exploration contract area and eastern Clarion-Clipperton Zone in the Pacific Ocean: Annelida, Arthropoda, Bryozoa, Chordata, Ctenophora, Mollusca. Biodiv Data J 5:e14598

Anderson W, Bullis JH (1970) Searching the shrimp beds by sub. Sea Frontiers 16:112–119

Andrade H (1980) Nueva especie de *Paralomis* en aguas de Chile: *Paralomis chilensis* n. sp. Bol Mus Nac Hist Nat, Santiago, Chile 37:269–273

Arana P, Ahumada M, Guerrero A (2003) Distribución y abundancia de la gamba *Haliporoides diomedeae* (Crustacea: Decapoda: Penaeidae) frente a la costa central de Chile. Invest Mar Valparaíso 31:57–71

Arntz W, Gallardo V, Gutiérrez D, Isla E, Levin L, Mendo J, Neira C, Rowe G, Tarazona J, Wolff M (2006) El Niño and similar perturbation effects on the benthos of the Humboldt, California, and Benguela current upwelling ecosystems. Adv Geosci 6:243–265

Baba K (2005) Deep-sea chirostylid and galatheid crustaceans (Decapoda: Anomura) from the Indo-West Pacific, with a list of species. Galathea Rep 20:1–319

Baba K, Haig J (1990) A new species of chirostylid crustacean (Decapoda: Anomura) from off the west coast of North America. Proc Biol Soc Wash 103:854–860

Baba K, Wicksten M (2017) *Uroptychus nitidus* (A. Milne-Edwards, 188) and related species (Crustacea: Decapoda: Anomura: Chirostylidae) from the western Atlantic. Zootaxa 4221:251–290

Baba K, Wicksten M (2019) Chirostyloidean squat lobsters (Crustacea: Decapoda: Anomura) from the Galapagos Islands. Zootaxa 4564(2):391–421

Baba K, Hayashi K, Masahiro T (1985) Decapod crustaceans from continental shelf and slope around Japan. Jap Fish Res Cons Ass, Tokyo:1–336

Baba K, Ahyong S, Schnabel K (2018) Rediagnosis of the squat lobster genus *Gastroptychus* Caullery, 1896, with a new genus *Sternostylus* and a new family Sternostylidae (Crustacea: Decapoda: Anomura: Chirostyloidea). Zootaxa 4524:077–086

Barry J, Taylor J, Kuhnz L, De Vogelaere A (2016) Symbiosis between the holothurian *Scotoplanes* sp. A and the lithodid crab *Neolithodes diomedeae* on a featureless abyssal plain. Mar Ecol 2017:38e12396

Benedict J (1895) Descriptions of new genera and species of crabs of the family Lithodidae with notes on the young of *Lithodes camtschaticus* and *Lithodes brevipes*. Proc US Nat Mus 17:79–488

Benedict J (1902) Descriptions of a new genus and forty-six new species of crustaceans of the family Galatheidae, with a list of the known marine species. Procs U S Nat Mus 26:243–334

Bracken-Grissom H, Felder D, Volimer N, Martin J, Crandall K (2012) Phylogenetics links monster larva to deep-sea shrimp. Ecol Evol 2012. https://doi.org/10.1002/ece3.347
Burukovsky R (2001) Taxonomy of *Nematocarcinus* (Decapoda: Nematocarcinidae). Description of *Nematocarcinus* from waters of the American continent. Zool Zh 80:1429–1443 (In Russian)
Butler T (1970) Synopsis of biological data on the prawn *Pandalus platyceros* Brandt, 1851. FAO Fish Rep 57:521–526
Butler T (1980) Shrimps of the Pacific coast of Canada. Can Bull Fish Aquat Sci 202:1–280
Chace FA Jr (1986) The caridean shrimps (Crustacea: Decapoda) of the *Albatross* Philippine Expedition, 1907-1910, Part 4: Families Oplophoridae and Nematocarcinidae. Smiths Contr Zool 432:1–82
Crosnier A, Forest J (1973) Faune Tropicale XIX. Les crevettes profondes de l'Atlantique oriental tropical. Office de la Recherche Scientifique et Technique Outre Mer, Paris, 409 pp
Del Solar E (1972) Addenda al catalogo de Crustaceos del Peru. Inf Inst Mar Peru 38:1–21
Dong D, Li X, Wang C (2017) Three squat lobsters (Crustacea: Decapoda: Anomura) from tropical West Pacific seamounts, with description of a new species of *Uroptychus* Henderson, 1888. Zootaxa 4311:389–398
Drazen J, Leitner A, Morningstar S, Marcon Y, Greinert J, Purser A (2019) Observations of deep-sea fishes and mobile scavengers from the abyssal DISCOL experimental mining area. Biogeosci Discuss 2019(5):1–31. https://doi.org/10.5194/bg-2019-5
Etnoyer PJ, Stierhoff KL, Nehasil S (2013) Spatial overlap of commercial bottom-fishing activity with deep-water coral and sponge observations in the Southern California Bight: a report to NOAA's Deep-Sea Coral Research and Technology Program. Silver Spring, Maryland
Farfante I, Kensley B (1997) Penaeoid and sergestoid shrimps and prawns of the world. Mém Mus Nat Hist Nat Paris 175:1–233
Faxon W (1893) Reports on the dredging operations off the west coast of Central America to the Galapagos, to the west coast of Mexico, and in the Gulf of California in charge of Alexander Agassiz, by the U.S. Fish Commission steamer "Albatross" during 1891, Lieut. Commander Z.L. Tanner, U.S.N. commanding. VI. Preliminary descriptions of new species of Crustacea. Bull Mus Comp Zool Harv Univ 24:149–220
Faxon W (1895) Reports on an exploration off the west coast of Mexico, central and South America, and off the Galapagos Islands, in charge of Alexander Agassiz, by the U.S. Fish Commission steamer "Albatross" during 1891, Lieut. Commander Z.L. Tanner, U.S.N. commanding. 15. The stalk-eyed Crustacea. Mem Harvard Univ Mus Comp Zool 18. 292 pp
Flores J, Hendrickx ME, Chapa-Saldaña H (2004) Capturas del camarón mancha blanca, *Pandalus platyceros* Brandt, 1851 (Caridea: Pandalidae), en la costa oeste de Baja California, México, con una breve revisión de su biología y su pesca. In: Hendrickx ME (ed) Contributions to the study of East Pacific crustaceans 3. Instituto de Ciencias del Mar y Limnología, Universidad Nacional Autónoma de México, Mexico, pp 199–211
Frank T, Johnsen S, Cronin T (2012) Light and vision in the deep-sea benthos: II. Vision in deep-sea crustaceans. J Exp Biol 215(19):3344–3353
Fitch J, Lavenberg R (1968) Deep-water teleostean fishes of California. Univ Calif Berkeley. 155 pp
Galil B (2000) Crustacea Decapoda: review of the genera and species of the family Polychelidae Wood-Mason, 1874. In: Crosnier A (ed) Résultats des Campagnes MUSORSTOM, vol 21. Mém Mus nat Hist nat, Paris 184:285–387
Garth J, Haig J (1971) Decapod Crustacea (Anomura and Brachyura) of the Peru-Chile Trench. Anton Bruun Rep 6:3–19
Gaten E, Moss S, Johnson M (2013) The reniform superposition compound eyes of *Nephrops norvegicus*: optics, susceptibility to light induced damage, electrophysiology and a ray tracing model. Adv Mar Biol 64:107–148
Gore R (1985) Some rare species of abyssobenthic shrimp (Families Crangonidae, Glyphocrangonidae and Nematocarcinidae) from the Venezuela Basin, Caribbean Sea (Decapoda, Caridea). Crustaceana 48:269–285

Guzmán G (2009) Paralomis sonne nueva especie de Litodido (Decapoda: Paguroidea: Lithodidae) en agua profundas frente a Antofagasta, Chile. Lat Am J Aquat Res 37:253–257

Guzmán G, Sellanes J (2015) A review of the Munidopsidae Ortmann, 1898 (Decapoda, Galatheoidea) in Chilean waters, including new records for the southeastern Pacific. Zootaxa 4021:282–306

Haig J (1974) Observations on the lithodid crabs of Peru, with description of two new species. Bull S Calif Acad Sci 73:152–164

Hart J (1982) Crabs and their relatives of British Columbia. British Columbia Provincial Museum, Victoria

Helly J, Levin L (2004) Global distribution of naturally occurring marine hypoxia on continental margins. Deep Sea Res, Part I: Ocean Res Pap 51:1159–1168

Hendrickx ME (1995a) Camarones. In: Fischer W, Krupp F, Schneider W, Sommer C, Carpenter KE, Niem VH (eds) Guía para la Identificación de especies para los fines del la Pesca, Pacíco Centro-Oriental, vol 1. FAO, Rome, pp 417–537

Hendrickx ME (1995b) Langostas (Langostas espinosa, bogavantes, cigarras y zapateras, langostas de lodo, etc.). In: Fischer W, Krupp F, Schneider W, Sommer C, Carpenter KE, Niem VH (eds) Guía para la Identificación de especies para los fines del la Pesca, Pacifico Centro-Oriental, vol 1. FAO, Rome, pp 383–415

Hendrickx ME (2001) Occurrence of a continental slope decapod crustacean community along the edge of the minimum oxygen zone in the south eastern Gulf of California, Mexico. Bel J Zool 131(Supplement 2):95–110

Hendrickx ME (2003) Distribution and size of the Pacific deep water lobsterette, *Nephropsis occidentalis* Faxon 1893 (Decapoda, Astacidea, Nephropidae) in the S.E. Gulf of California, Mexico. Crustaceana 76:207–216

Hendrickx ME (2010) A new species of *Glyphocrangon* (Decapoda: Caridea: Glyphocrangonidae) from off the coast of western Mexico. In: De Grave S, Fransen C (eds) Contributions to shrimp taxonomy. Zootaxa 2372:358–366

Hendrickx ME (2012a) Crustáceos decápodos (Arthropoda: Crustacea: Decapoda) de aguas profundas del Pacífico mexicano: Lista de especies y material recolectado durante el proyecto TALUD. In: Zamorano P, Hendrickx ME, Caso M (eds) Biodiversidad y comunidades del talud continental del Pácifico mexicano. Secretaría del Medio Ambiente y Recursos Naturales, Instituto Nacional de Ecología, Mexico, pp 283–317

Hendrickx ME (2012b) Los Glyphocrangonidae y Crangonidae (Crustacea: Decapoda: Caridea) recolectados durante los cruceros TALUD en el Pácifico mexicano. In: Zamorano P, Hendrickx ME, Caso M (eds) Biodiversidad y comunidades del talud continental del Pacifico mexicano. Secretaría del Medio Ambiente y Recursos Naturales, Instituto Nacional de Ecología, Mexico, pp 319–356

Hendrickx ME (2017) First western Mexico record of the deep-water squat lobster *Munidopsis hendersoniana* Faxon, 1893 (Crustacea: Decapoda: Galatheoidea: Munidopsidae). Bol Inst Oceanogr Venezuela 56:13–15

Hendrickx ME (2019a) Redescription of the rare shrimp *Heterocarpus nesisi* (Burukovsky, 1986) (Crustacea: Caridea: Pandalidae) rediscovered off western Mexico, with the proposal of a new genus. Zootaxa 4565:49–60

Hendrickx ME (2019b) Deep-water lithodids of the genus *Paralomis* White, 1856 (Decapoda, Anomura, Lithodidae) off western Mexico. Crustaceana 92:1257–1264

Hendrickx ME, Ayón-Parente M (2009) Notes on distribution and ecology of *Parapagurus foraminosus* Lemaitre, 1999 (Decapoda, Anomura, Parapaguridae) in the eastern Pacific. Crustaceana 82:1339–1344

Hendrickx ME, Harvey A (1999) Checklist of anomuran crabs (Crustacea: Decapoda) from the Eastern Tropical Pacific. Belg J Zool 129:327–352

Hendrickx ME, Hernandez-Payán J (2017) Distribution and abundance of *Nematocarcinus* spp. (Crustacea: Decapoda: Caridea: Nematocarcinidae) off western Mexico, eastern Pacific. Reg Stud Mar Sci 2017. https://doi.org/10.1016/j.rsma.2017.10.012

Hendrickx ME, Papiol V (2019) Distribution, relative growth, ecology and fecundity of the deep-water squat lobster *Galacantha diomedeae* (Crustacea: Decapoda: Galatheoidea: Munidopsidae) in the Mexican Pacific. Invertebr Biol 2019:1–12

Hendrickx ME, Serrano D (2012) Influencia de la zona del mínimo de oxígeno sobre la distribución de *Steromastis pacifica* (Faxon, 1893) y primer registro para Mexico de *Pentacheles laevis* Spence Bate, 878 (Crustacea; Decapoda: Polychelidae). Cienc Mar 16:29–36

Hendrickx ME, Wicksten M (2016) New records of decapod crustaceans in the eastern Pacific. Crustaceana 89:603–610

Hiller-Adams P, Case J (1988) Eye size of pelagic crustaceans as function of habitat, depth and possession of photophores. Vis Res 28:667–680

Jamieson A, Brooks L, Reid W, Piertney S, Narayanaswamy B, Linley T (2019) Microplastics and synthetic particles ingested by deep-sea amphipods in six of the deepest marine ecoystems on Earth. Royal Society Open Science. https://doi.org/10.1098/rsos.180667

Jensen P, Califf K, Lowe V, Hauser L, Morado J (2010) Molecular detection of *Hematodinium* sp. in Northeast Pacific *Chionoecetes* spp. and evidence of two species in the northern hemisphere. Dis Aquat Org 89:55–166

Jones W, Macpherson E (2007) Molecular phylogeny of the East Pacific squat lobsters of the genus *Munidopsis* (Decapoda: Galatheidae) with the descriptions of seven new species. J Crustac Biol 27:77–501

Kensley B (1996) New thalassinidean shrimp from the Pacific Ocean (Crustacea: Decapoda: Axiidae and Calocarididae). Bull Mar Sci 59:69–489

Khodkina I (1973) New species of the genus *Munidopsis* (Decapoda, Anomura) from the East Pacific. Zool Zh Moscow 52:1156–1167 (In Russian)

Komai T (1994) Deep-sea shrimps of the genus *Pandalopsis* (Decapoda: Caridea: Pandalidae) from the Pacific coast of eastern Hokkaido, Japan, with the description of two new species. J Crustac Biol 14:538–559

Komai T (2008) A world-wide review of species of the deep-water crangonid genus *Parapontophilus* Christoffersen, 1988 (Crustacea, Decapoda, Caridea) with descriptions of ten new species. Zoosystema 30:261–332

Komai T, Segonzac M (2005) Two new species of *Nematocarcinus* A. Milne-Edwards, 1881 (Crustacea, Decapoda, Caridea, Nematocarcinidae) from hydrothermal vents on the North and South East Pacific Rise. Zoosystema 27:343–364

Komai T, Takeda T (2004) A new species of the hippolytid shrimp genus *Lebbeus* White, 1847 from Sagami-Nada Sea, Central Japan, and further records of two little-known species (Crustacea: Decapoda: Caridea). Bull Natl Mus Nat Sci (A) 30:77–86

Komai T, Tsuchida S (2012) Rediscovery and redescription of a sponge-associated axiid shrimp, *Eiconaxius acutifrons* Bate, 1888 (Crustacea: Decapoda: Axiidae). Zootaxa 3393:27–40

Komai T, Hayashi K, Kohtsuka H (2004) Two new species of the shrimp genus *Lebbeus* White from the Sea of Japan, with redescription of *Lebbeus koboi* Hayash (Decapoda: Caridea; Hippolytidae). Crustac Res 33:103–125

Krygier E, Pearcy W (1981) Vertical distribution and biology of pelagic decapod crustaceans off Oregon. J Crustac Biol 1:70–95

Lampitt R, Burnham M (1983) A free fall time lapse camera and current system "Bathysnap" with notes on the foraging behaviour of a bathyal decapod shrimp. Deep-Sea Res 30:1009–1017

Leitner A, Neuheimer A, Donlon E, Samith C, Drazen J (2017) Environmental and bathymetric influences on abyssal bait-attending communities of the Clarion-Clipperton Zone. Deep-Sea Res I 125:65–80

Lemaitre R (1989) Revision of the genus *Parapagurus* (Anomura, Paguroidea, Parapaguridae), including redescriptions of the western Atlantic species. Zool Verhandel Leiden 253:1–106

Lemaitre R (1999) Crustacea Decapoda: a review of the species of the genus *Parapagurus* Smith, 1879 (Parapaguridae) from the Pacific and Indian Oceans. In: Crosnier A (ed) Résultats des Campagnes MUSORSTOM 20. Mém Mus nat Hist nat Paris 180:303–378

Luke S (1992) The marine Shrimps of the Scripps Institution of Oceanography Benthic Invertebrate Collection. SIO, La Jolla

Macpherson E, Wehrtmann I (2010) Occurrence of lithodid crabs (Decapoda, Lithodidae) on the Pacific coast of Costa Rica, Central America. Crustaceana 83:143–151

Macpherson E, Richer de Forges B, Schnabel K, Samadi S, Boisselier M, Garcia-Rubies A (2010) Biogeography of the deep-sea galatheid squat lobsters of the Pacific Ocean. Deep-Sea Res I 57:228–238

Manning R (1970) Notes on the west American nephropidean lobster, *Nephropsis occidentalis* Faxon. Proc Biol Soc Wash 82:865–870

Markham J (2016) Redescription of *Bathygyge grandis* Hansen, 1897 (Crustacea, Isopoda, Bopyridae) from southern California with erection of a new subfamily, Bathygyginae, for it. Bull S Calif Acad Sci 115:72–80

Martin J, Sánchez C, Pereyra R (1997) Notes on the distribution of two lithodid crabs (Crustacea: Decapoda; Anomura) from off the coast of Baja California Sur, Mexico. Bull S Calif Acad Sci 96:78–86

McCauley J (1972) A preliminary checklist of selected groups of invertebrates from otter-trawl and dredge collections off Oregon. In: Pruter A, Alvers, University of Washington Press, Seattle, pp 409–443

McLaughlin P (1974) The hermit crabs (Crustacea Decapoda, Paguroidea) of northwestern North America. Zool Verhandl Leiden 130:1–396

Méndez M (1981) Claves de identificación y distribución de los langostinos y camarones (Crustacea: Decapoda) del mar y rios de la coast del Peru. Inst Mar Peru Bol 5:1–170

NOAA (2019) NOAA Office of Ocean Exploration and Research benthic deepwater animal identification Guide. Available from http://oceanexplorer.noaa.gov/okeanos/animalguide/animal guide.html. Consulted 14 August 2019

O'Dea A, Lessios H, Coates A, Eytan R, Restrepo-Moreno S, Clone A, Collins L, De Queiroz A, Farris D, Norris R, Stallard R, Woodburne M, Aguilera O, Aubry M, Berggren W, Budd A, Cozzuol M, Coppard S, Duque-Caro H, Finnegan S, Gasparini G, Grossman E, Johnson K, Keigwin L, Knowlton N, Leigh E, Leonard-Pingel J, Marko P, Pyenson N, Rachello-Dolman P, Soibelzon E, Soibelzon K, Todd J, Vermeij G, Jackson J (2017) Formation of the Isthmus of Panama. Sci Adv 2016(2):e16900883

Papiol V, Hendrickx ME (2015) Ecology of *Benthesicymus tanneri* Faxon, 1893 (Dendrobranchiata: Benthesicymidae) from the Mexican Pacific slope. J Crustac Biol 36:50–60

Park W, Perry R, Hong S (2004) Larval development of the sidestriped shrimp *Pandalopsis dispar* Rathbun (Çrustacea, Decapoda, Pandalidae) reared in the laboratory. U.S. Fish Bull 102:118–126

Pequegnat L, Wicksten M (2006) Oplophorid shrimps (Decapoda: Caridea: Oplophoridae) in the Gulf of Mexico and Caribbean Sea from the collections of the research vessels *Alaminos*, *Oregon* and *Oregon II*. Crustac Res 35:92–107

Pequegnat W, Pequegnat L (1973) Munidopsis albatrossae, a new species of deep-sea Galatheidae (Decapoda, Anomura) from the eastern Pacific Ocean. Crustaceana 24:163–168

Rathbun M (1904) Decapod crustaceans of the northwest coast of North America. Harriman Alaska Ser 10:1–190

Rice A (1981) The abdominal locking mechanism in the deep-sea shrimp genus *Glyphocrangon* (Decapoda, Glyphocrangonidae). Crustaceana 40:316–391

Román-Contreras R (2008) A new genus of bopyrid isopod infesting *Munidopsis beringana* (Decapoda: Galatheidae) from deep waters. J Crustac Biol 28:378–384

Sakai T (1971) Illustrations of 15 species of crabs of the family Lithodidae, two of which are new to science. Res Crustac 4(5):1–49

Somerton D (1981) Contribution to the life history of the deep-sea king crab, *Lithodes couesi*, in the Gulf of Alaska. U S Fish Bull 79:259–269

Sunada J (1984) Spot prawn (*Pandalus platyceros*) and ridgeback prawn (*Sicyonia ingentis*) fisheries in the Santa Barbara Channel. CalCOFI Rep 25:100–104
Thompson J (1963) The bathyalbenthic caridean shrimps of the southwestern North Atlantic. Ph.D. Dissertation, Duke University, Durham, North Carolina, pp 1–504
Vereshchaka A, Corbari L, Kulagin D, Lunina A, Olesen J (2019) A phylogeny-based revision of the shrimp genera *Altelatipes*, *Bentheonectes* and *Benthesicymus* (Crustacea: Decapoda: Benthesicymidae). Zool J Linnean Soc 20:1–12. https://doi.org/10.1093/zoolinnean/zlz125
Wagner D, Luck D, Toonen R (2012) Chapter Two. The Biology and ecology of black corals (Cnidaria: Anthozoa: Hexacorallia: Antipatharia). In: Lesser M (ed) Advances in marine biology, vol 63. Elsevier, Amsterdam, pp 67–132
Warrant E, Locket N (2004) Vision in the deep sea. Biol Rev 79:671–671
Wehrtmann I, Arana P, Barriga E, Gracia D, Pezzuto P (2012) Deep-water shrimp fisheries in Latin America: a review. Latin Am J Aquat Res 40:497–535
Wenner E (1979a) Distribution and reproduction of nematocarcinid shrimp (Decapoda: Caridea) from the northwestern North Atlantic. Bull Mar Sci 29:380–393
Wenner E (1979b) Some aspects of the biology of deep-sea lobsters of the family Polychelidae (Crustacea, Decapoda) from the western North Atlantic. US Fish Bull 77:435–444
Wicksten M (1979) New records of the species of *Glyphocrangon* in the northeastern Pacific Ocean (Caridea: Glyphocrangonidae). Proc Biol Soc Wash 92:217–224
Wicksten M (1981) New records of *Stereomastis sculpta pacifica* (Faxon) (Decapoda: Polychelidae) in the eastern Pacific Ocean. Proc Biol Soc Wash 93:914–919
Wicksten M (1982) Crustaceans from baited traps and gill nets off southern California. Calif Fish Game 68:244–248
Wicksten M (1989) Ranges of offshore decapod crustaceans in the eastern Pacific Ocean. Trans San Diego Soc Nat Hist 21:291–316
Wicksten M (2004) The status of *Benthesicymus laciniatus* Rathbun (Decapoda, Penaeoidea, Benthesicymidae) in the northestern Pacific. Bull S Calif Acad Sci 10:93–94
Wicksten M (2010) *Lebbeus laurentae*: a replacement name for *Lebbeus carinatus* de Saint Laurent, 1984 (Decapoda: Caridea: Hippolytidae) and a redescription of the species. Proc Biol Soc Wash 123:196–203
Wicksten M, Heathman T (2015) New host records for *Bathypalaemonella serratipalma* (Caridea: Bathypalaemonellidae) in the northern Gulf of Mexico. Mar Biodiv Rec 8:1–4
Wicksten M, Kuhnz L (2015) A swimming deep-sea penaeoid shrimp photographed off California. Calif Fish Game 101:146–148
Wicksten M, Méndez M (1982) New records and new species of the genus *Lebbeus* (Caridea: Hippolytidae) in the eastern Pacific Ocean. Bull S Calif Acad Sci 81:106–120
Wicksten M, Packard J (2005) A qualitative zoogeographic analysis of decapod crustaceans of the continental slopes and abyssal plain of the Gulf of Mexico. Deep-Sea Res I 52:1745–1765
Wicksten M, De Grave S, France S, Kelley C (2017) Presumed filter-feeding in a deep-sea benthic shrimp (Decapoda, Caridea, Stylodactylidae), with records of the deepest occurrence of carideans. ZooKeys 646:17–23
WoRMS editorial board (2020) Benthoecetes Smith, 1884. Accessed at: http://www.marinespecies.org/aphia.php?p=taxdetails&id=584912 on 2020-10-01.
Wicksten M (1984) Distributions of some common decapod crustaceans and a pycnogonid from the continental shelf of northern California. Calif Fish Game 70:132–139
Wicksten M (2012) Decapod Crustacea of the Californian and Oregonian Zoogeographic Provinces. Zootaxa 3371:1–307
Wicksten M, Martin J (2004) A new species of caridean shrimp of the family Stylodactyidae from the eastern Pacific Ocean. Proc Biol Soc Wash 117:377–384
Zarenkov N (1976) On the fauna of decapods of the waters adjacent to South America. Biol Morya 5:8–18 (in Russian)

Chapter 18
Conservation Strategies for Potential New Deep-Sea Crustacean Fisheries in the Colombian Caribbean under an Ecosystem Approach

J. Paramo, M. Grijalba-Bendeck, D. Pérez, A. Espinoza-Tenorio, and M. Wolff

Abstract The further development of the fisheries in Colombia should consider the identification of potential new resources based on knowledge of characteristics of the deep-sea habitats and the organisms which highlight the need for broad exploratory surveys and an ecosystem approach to fisheries management (EAF). The objective of this work was thus to advice management and conservation strategies for the possible new deep-sea crustacean fisheries in the Colombian Caribbean based on an ecosystem approach to fisheries. The management of both the shallow water shrimp fishery and the potential new deep-sea crustacean fishery should be based on baseline information of the population dynamics and size structure of target species and the optimum level of fishing effort and catch volumes and should

J. Paramo (✉)
Universidad del Magdalena, Tropical Fisheries Science and Technology Research Group (CITEPT), Santa Marta, Colombia
e-mail: jparamo@unimagdalena.edu.co

M. Grijalba-Bendeck
Faculty of Natural Sciences and Engineering, Doctorate in Marine Sciences Program, Research Group: Dynamics and Management of Marine-Coastal Ecosystems (DIMARCO), Universidad de Bogotá Jorge Tadeo Lozano, Santa Marta, Colombia

D. Pérez
Universidad del Magdalena, Tropical Fisheries Science and Technology Researh Group (CITEPT), Santa Marta, Colombia

Faculty of Natural Sciences and Engineering, Doctorate in Marine Sciences Program, Research Group: Dynamics and Management of Marine-Coastal Ecosystems (DIMARCO), Universidad de Bogotá Jorge Tadeo Lozano, Santa Marta, Colombia

A. Espinoza-Tenorio
El Colegio de la Frontera Sur, Unidad Campeche. Av. Rancho Polígono 2-A, Col. Ciudad Industrial, Campeche, México

M. Wolff
Universität Bremen, Leibniz Zentrum für Marine Tropenforschung (ZMT), Bremen, Germany

M. E. Hendrickx (ed.), *Deep-Sea Pycnogonids and Crustaceans of the Americas*, https://doi.org/10.1007/978-3-030-58410-8_18

also consider the development of strategies for constant resource and ecosystem monitoring. Open ocean marine protected areas for the protection of nursery and spawning areas should also be considered.

Keywords Colombian Caribbean · Deep-sea crustacean · Ecosystem approach to fisheries · Management

18.1 Introduction

Fisheries have always been important to humanity (Christensen 2011) and are the source of a significant amount of food produced for human consumption (Hart and Pearson 2011). However, intensive fisheries are depleting marine resources in many parts of the ocean, and fisheries management has often been ineffective (Daan et al. 2011). Declining fish stocks, combined with the indirect effects of fishing on the marine ecosystems, demonstrate that fisheries management in a great part of the world has failed to achieve sustainability (Worm et al. 2006, 2011; Hutchings et al. 2010; Longhurst 2010). This failure is mainly the result of increasing fishing effort in response to the intense social-political pressure to obtain larger catches in the short term, despite limited knowledge of fisheries systems (Pauly et al. 2003) and disregarding the complexity of wider ecological interactions and possible impacts on fragile marine ecosystems (Longhurst 2010).

Over the past decades, intensive exploitation of the fishing resources, mainly on the continental shelves, has led to the progressively declining catches of many fish and crustacean stocks (Pauly et al. 2003). As a response, new fishing areas in the high seas and in deeper waters are being searched for, taking advantage of recent advances in capture technologies (Pauly et al. 2003). Nevertheless, deep-sea ecosystems (defined here as >200 m; Cavanagh and Kyne 2006) and their fisheries are not considered highly productive and are known to be especially vulnerable to overexploitation due to the life history characteristics of deep-sea species, including extreme longevity, slow growth rate, late maturity, and low fecundity (Morato et al. 2006; Follesa et al. 2011). The potential effects of the fishery on deep-sea resources include the extensive restructuring of entire ecosystems, changes in the geographical ranges of many species, large-scale elimination of taxa, and a decline in biodiversity at all scales (Robison 2009). Stocks of deep waters thus tend to collapse much more rapidly, and their recuperation is slower, compared to resources from shallow environments (Roberts 2002). The United Nations (UN) General Assembly had declared 2021–2030 as Decade on “Ecosystem Restoration” (ONU 2020), and within 17 Sustainable Development Goals (SDGs) for 2030, the SDG 14 (life below water) is related to the conservation and sustainable use of the oceans, seas, and marine resources for sustainable development (www.sustainabledevelopment.un.org/sdg14). Therefore, management on new deep-sea fisheries should avoid actions of irreversible degradation, advocate for mitigation of the impacts caused on deep-sea ecosystems, and restore the degraded deep-sea habitats (Da Ros et al. 2019).

Among the results of anthropogenic impact, which affect these ecosystems, are (i) the removal of predators by fishing and the removal of habitat-forming species (such as gorgonians and stony corals), (ii) the modification of the food webs as a response of the bycatch and catching only commercial important species, (iii) the accumulation of heavy metals and toxins, and (iv) global climate changes that alter the quantity and quality of food that reaches the deep waters (WWF/IUCN 2004). Therefore, the sustainable use of new deep-sea fishery resources should include knowledge of the life history of the target species, of their ecology and bio-economic potential, as well as of the associated biodiversity in deep-sea ecosystems (FAO 2003; Munro 2011).

In the Colombian Caribbean, most of the coastal fisheries are currently being exploited at or above maximum sustainable level (Paramo et al. 2009; Paramo and Saint-Paul 2010). Since tropical fisheries are complex multispecies systems, which intensively exploit a wide range of species and sizes, a generalized overexploitation of fisheries resources may easily occur (Munro 2011). If the fisheries of Colombia are to be further developed, potential new resources in areas yet not accessed should be identified based on knowledge of characteristics of the deep-sea habitats and the organisms they inhabit. A potential sustainable use of those resources needs an ecosystem approach to fisheries management (EAF).

EAF seeks to balance diverse societal objectives, by taking account of biotic, abiotic, and human components of ecosystems and their interactions and applying a holistic approach to fisheries management (Garcia et al. 2003; Bianchi 2008). Since for the development of an adaptive management agenda the monitoring and evaluation of sustainability indicators is a key step (Espinoza-Tenorio et al. 2014), we used several of those indicators (Shin et al. 2010) to advice management and conservation strategies for the possible new deep-sea crustacean fisheries in the Colombian Caribbean.

18.2 Materials and Methods

Data were obtained by trawling in depths between 200 and 550 m (100 m strata intervals) in the Colombian Caribbean. Sampling was carried out in August and December 2009 as well as in March and May 2010 using a commercial shrimp trawler through a bottom trawl with a cod-end mesh size of 44.5 mm from knot to knot. The location of samples depended on the existence of trawlable bottoms, determined by an echosounder Furuno FCV 1150 with a transducer at a frequency of 28 kHz. A total of 87 stations were sampled, with at least two hauls per 100 m depth stratum (Fig. 18.1). The haul duration was 30 min, and the distance traveled by the net was estimated using a GPS Garmin MAP 76CSx. The deep-sea crustacean and fish catch from each haul was analyzed for biomass (kg/km^2). The swept

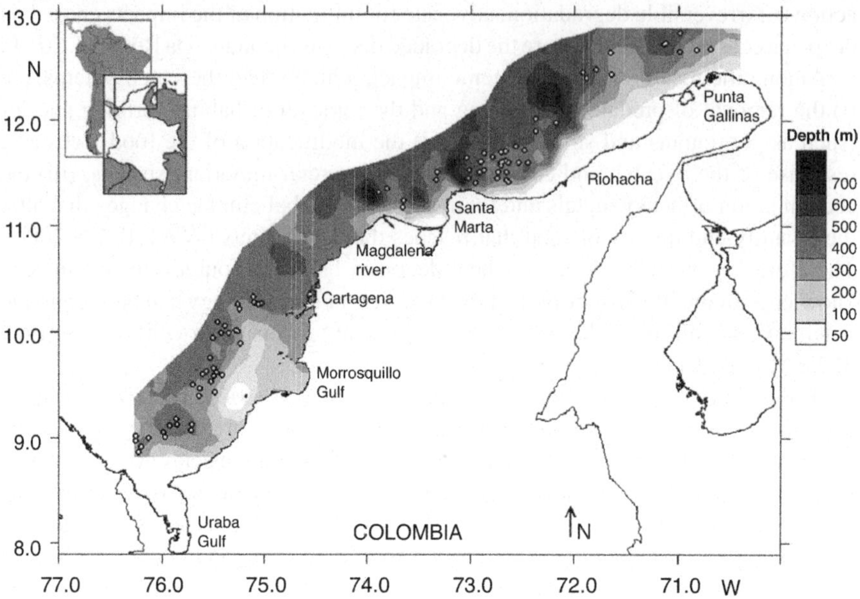

Fig. 18.1 Study area in the Colombian Caribbean; circles indicate the sampled stations

area was estimated from the spread of the net (11.58 m) and the speed of the vessel (average 2.5 knots) (Gunderson 1993; King 2007).

18.3 Results

18.3.1 Potential New Crustacean Fishery in the Colombian Caribbean

Penaeid shrimps of shallow water are an important fisheries resource in the Colombian Caribbean, generating large amounts of direct and indirect employment and foreign currency through exports due to their high value on international markets. However, there is little biological information on this shallow water shrimp fishery, and stock assessment to determine the abundance and spatial distribution of the species is required to establish artisanal and industrial fishing zones and solve conflicts between stakeholders in accordance with the code of conduct for responsible fisheries (FAO 1995). In the Colombian Caribbean, the shallow water shrimp fishery targets the species *Penaeus notialis* (Pérez-Farfante, 1967), *Penaeus brasiliensis* (Latreille, 1817), *Penaeus subtilis* (Pérez-Farfante, 1967), and *Litopenaeus schmitti* (Burkenroad, 1936), but *P. notialis* constitute around 70% of the total shrimp catch. This fishery has gone through different stages of development from growth, fully exploited, overexploited, to collapse and is probably currently at a recovery stage, with the result of a significant depletion in shrimp stocks (Paramo and Saint-Paul 2010). Therefore, in the Colombian Caribbean Sea, fisheries management measures, additional to those used traditionally, are necessary to protect fisheries resources and to improve the sustainability of the fisheries. These measures should be based on an EAF, including the establishment of marine protected areas, which have recently emerged as a tool for marine conservation and fisheries management (Paramo et al. 2009). Most shrimps are benthic organisms, inhabiting a variety of bottom habitats such as sandy, muddy, rocky, or a mixture of these, that are exploited in shallow waters at depths above 100 m (Carpenter 2002). However, in FAO Fishing Area 31 (Caribbean region), there is a deep-sea shrimp fishery targeting the commercial species *Aristaeomorpha foliacea* (Risso, 1827) and *Pleoticus robustus* (Smith, 1885) at depths greater than 200 m (Carpenter 2002), and the deep-sea Caribbean lobster *Metanephrops binghami* (Boone, 1927) is reported to have potential for economic exploitation in Venezuelan waters (Gómez et al. 2000, 2005) (Table 18.1). Nevertheless, at present there is no deep-sea crustacean fishery in the Colombian Caribbean. Due to the substantial marketability of these deep-sea crustaceans in international markets, these deep-sea crustaceans represent a potential new economic resource in the Colombian Caribbean Sea. Recent research revealed the potential of the deep-sea giant red shrimp (*Aristaeomorpha foliacea*), the royal red shrimp (*Pleoticus robustus*) (Paramo and Saint-Paul 2012a), the pink speckled deep-sea shrimp (*Penaeopsis serrata* Burkenroad, 1936) (Paramo and Saint-Paul 2012b), and the deep-sea lobster (*Metanephrops binghami*) (Paramo and Saint-Paul 2012c) as new fishing resources. However, the authors of the mentioned studies recommend further scientific assessment to be conducted to determine the population life cycle characteristics of those deep-sea crustaceans and to estimate the associated biodiversity before initiating a new commercial fishery.

In a survey deep-sea fish assemblages were studied across a depth range of 200–550 m from the Colombian Caribbean Sea, and concerns were raised for the need of an ecosystem approach to fisheries management (Paramo et al. 2012). The authors recommended that any management of potential new deep-sea crustacean fishery need to allow an appropriate level of biodiversity and the habitat quality to be maintained. Specifically, they argue that population dynamics and size structure of the target and non-target species, the optimum allocation of catches and effort, and the protection of nursery and spawning areas should be elements of the EAF, as well as strategies for monitoring the health of those deep-sea ecosystems.

Highest biomass values of deep-sea crustacean and fish were found in the northern zone of the Colombian Caribbean, mainly between Santa Marta and Riohacha. In addition to this, high biomass values in the southern zone were found in front of Cartagena and Morrosquillo Gulf (Fig. 18.2).

Table 18.1 Indicators for monitoring the ecological status of marine ecosystems

Indicator	Potential deep-sea implications	Source
1. Total biomass of surveyed species	Alternatives to complete biodiversity monitoring include the assessment of functional ecological groupings at each trophic level, such as guilds of predatory fishes or large, omnivorous siphonophores	Robison (2009)
2. ABC method	Indicator of disturbance for antropic and environmental impacts in different communities, macrozoobenthos, macrobenthos, and demersal fish. Based on the comparison of biomass respect to the abundance of individuals in a community at the same time and space	Warwick (1986), Clarke and Warwick (1994), Stenton-Dozey et al. (1999), Yemane et al. (2005)
3. Size spectra	Evaluate the state of communities and their degree of disturbance, caused mainly by the fisheries, by characterizing the behavior of abundance or biomass with respect to the variation in body size	Shin et al. (2005), Edwards et al. (2017)
4. Trophic level of landings	Commercial fishing pressures may affect deep pelagic biodiversity by "fishing down the food web." As the numbers of top predators have declined, fishing effort has shifted to species at lower trophic levels	Pauly et al. (1998)
5. Proportion of predatory fish	Deep-sea top predators are pinnipeds, whales, tunas, and swordfish that feed in midwater as deep as 1000 m or more. Life history characteristics of these deep-sea species include extreme longevity, slow growth rate, late maturity, and low fecundity	Morato et al. (2006), Follesa et al. (2011), Robison (2009)
6. Proportion of under- and moderately exploited stocks	Number of under- and moderately exploited species/ number of target species	Shin et al. (2010)
7. Mean life span	Turnover or retention rates for each compartment: How long is the mean, characteristic life span of the group?	Håkanson (2004)
8. Reciprocal of the coefficient of variation of total biomass	1/coefficient of variation (CV) of total biomass	Robison (2009)

18.3.2 *Potential Fishery Target Species*

18.3.2.1 The Giant Red Shrimp (*Aristaeomorpha foliacea*)

Aristaeomorpha foliacea is found in deep waters from 250 to 1300 m over mud bottom and is actively fished outside Colombian waters because of its high commercial value (Tavares 2002). It has a wide geographical distribution from the Mediterranean Sea, the eastern and western Atlantic, the Indian Ocean, and the western Pacific from Japan to Australia, New Zealand, and the Fiji Islands (Tavares 2002). An exploration of deep waters off the Yucatan Peninsula in Mexico showed that *A.*

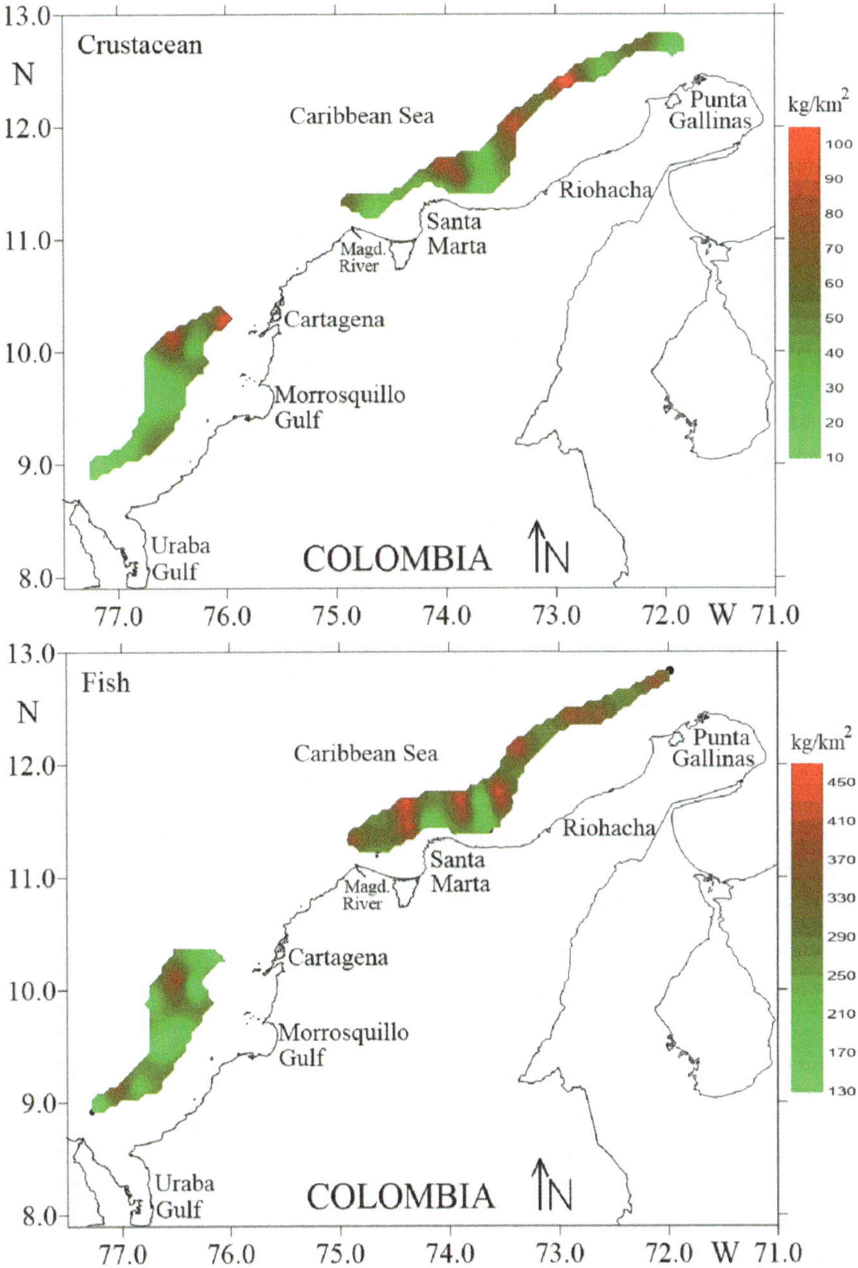

Fig. 18.2 Spatial distribution of biomass (kg/km^2) of deep-sea crustacean and fish in the Colombian Caribbean Sea

foliacea represents a potential fisheries resource (Gracia et al. 2010). *A. foliacea* constitutes a valuable deep-sea shrimp fishery of the south-eastern and southern sectors of the Brazilian coast (Pezzuto et al. 2006; Dallagnolo et al. 2009) and is a commercially important shrimp species in the deep waters of the Mediterranean Sea (D'Onghia et al. 1998; Figueiredo et al. 2001; Papaconstantinou and Kapiris 2003; Politou et al. 2004; Mouffok et al. 2008).

18.3.2.2 The Royal Red Deep Shrimp (*Pleoticus robustus*)

Pleoticus robustus is captured in high abundances off the coast of northeastern Florida, near the Dry Tortugas Islands, in the Mississippi delta and off the coast of Venezuela (Tavares 2002). However, recent exploration of deep waters off the Yucatan Peninsula in Mexico showed that *P. robustus* represents a potential fisheries resource (Gracia et al. 2010). This species inhabits continental slopes from 180 to 730 m depth and is most abundant at depths between 250 and 475 m over mud, sand, muddy sand, or white calcareous mud (Tavares 2002).

18.3.2.3 The Pink Speckled Shrimp (*Penaeopsis serrata*)

Penaeopsis serrata is of potential commercial importance in the western and eastern Atlantic (Holthuis 1980). The species of genus *Penaeopsis* are benthic, occurring in the upper part of the continental and insular slopes of tropical and subtropical regions (Pérez-Farfante 1980). *P. serrata* is distributed in the eastern Atlantic off north-west Africa (Morocco, Río de Oro), the western Atlantic in North Carolina (USA) to Surinam, the Bahamas Islands, the Gulf of Mexico, the Caribbean Sea (Holthuis 1980; Pérez-Farfante 1980), and the Mediterranean Sea off Alborón Island and in the Sardinian Channel (Mura et al. 2002).

18.3.2.4 The Caribbean Lobster (*Metanephrops binghami*)

Metanephrops binghami is distributed from the Bahamas and southern Florida to French Guiana, including the Gulf of Mexico and the Caribbean Sea (Tavares 2002; Holthuis 1991). Crustaceans of the genus *Metanephrops* are an important economic resource (Holthuis 1991; Chan 1998; Chan et al. 2009) in some tropical and subtropical regions. Three deep-water crustaceans of the species *Metanephrops* are exploited commercially on the continental slopes of north-west Australia: *Metanephrops boschmai* (Holthuis, 1964), *M. andamanicus* (Wood-Mason, 1892), and *M. australiensis* (Bruce, 1966) (Ward and Davis 1987; Wassenberg and Hill 1989). New Zealand has developed a deep-water lobster fishery, targeting scampi (*Metanephrops challenger* Balss, 1914) (Smith 1999). An economically important commercial fishery exists in Taiwan and East China Sea for *M. thomsoni* (Bate, 1888), *M. japonicus* (Tapparone-Canefri, 1873), and *M. formosanus* (Chan and Yu,

1987) (Choi et al. 2008). *Metanephrops binghami* has potential for an economic exploitation in waters of Venezuela (Gómez et al. 2000, 2005).

18.3.3 Monitoring the Ecological Status of Resources and Ecosystem

Fisheries management practices are failing to protect individual stocks and ecosystems; therefore, fisheries management should move from single species management towards an ecosystem approach to fisheries (EAF) (Hilborn 2010). Nevertheless, approaches based on species are also necessary to understand and to analyze how they contribute to the communities and ecosystems (Guijarro et al. 2011). A fishery consists of an ecosystem with the embedded resources, a fishing fleet, and a management system. An EAF should be implemented with stakeholder cooperation, provide good governance and effective enforcement (Hilborn 2010), and be guided by key indicators of ecosystem condition (Levin et al. 2009). The code of conduct for responsible fisheries (FAO 2009) calls for the use of best scientific evidence for management and the need of comprehensive knowledge of resource dynamics and habitat conditions to advice for the conservation and management of fisheries. However, in developing countries, there is a lack of institutional capacity building and funding to carry out the research related to fisheries and their ecosystems, required for an EAF (Mathew 2011). Additionally, deep-sea fisheries are usually data-poor, with only landing records and rarely scientific survey data being available. Many deep-water species are also difficult to age reliably (Lorance et al. 2011). Nevertheless, to overcome the data limitation and in the absence of modelling support, simple indicators have proven to be useful to assess the state of the fisheries resources and its ecosystem and to monitor changes as they occur (Ye et al. 2011). Such indicators should be based on easily obtainable and reliable data, should adequately reflect the condition of the resource, and should allow to define associated reference values and responsive management measures.

To develop baseline knowledge for a new deep-sea fishery in the Colombian Caribbean, the following fundamental questions must be addressed: Which resources can we use sustainably? Where can we fish them? How is the associated biodiversity to be affected and how can we achieve a sustainable fishery under an ecosystem approach to fisheries management? What are the ecologically spatial and temporal scales that should be measured? Given the marked population reductions of crustaceans and deep-sea fish in many parts of the oceans, voices around the world are raised seeking to protect these environments (Dallagnolo et al. 2009; Pérez et al. 2013), considering even the prohibition of fishing below a depth of 1000 m (Fisheries Commission for the Mediterranean GFCM). Therefore, it is necessary to understand and evaluate how human activities could modify the complex community structures that these stocks are embedded in the Colombian Caribbean and to take decisions within a context of development based on strategies of

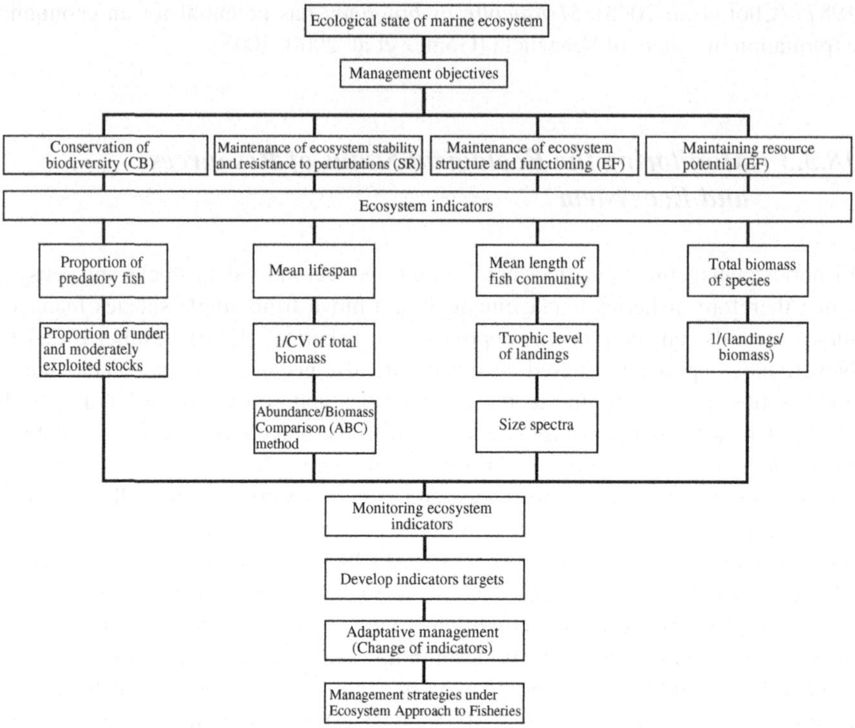

Fig. 18.3 Proposal of management strategies under an ecosystem approach to fisheries (EAF) for deep-sea fisheries in the Colombian Caribbean

management with an ecosystem approach (Bensch et al. 2008), which permits the healthy permanence of these environments and their resources. According to Shin et al. (2010), four management objectives can be distinguished as a means of evaluating the ecological state of marine ecosystems (Fig. 18.3): conservation of biodiversity (CB), maintenance of ecosystem stability and resistance to perturbations (SR), maintenance of ecosystem structure and functioning (EF), and maintaining resource potential (RP). Eight indicators were selected to attain these objectives (Table 18.1).

To assess the resource potential (RP) of an ecosystem, the total biomass of surveyed species is a useful indicator of changes over time, which measures the production capacity and potential contribution of the ecosystem as an exploitable marine resource. The inverse measure of the level of exploitation or total fishing pressure on an ecosystem 1/(landing/biomass) reflects the proportion of the community production that is taken by the fishery and serves as a proxy for exploitation rate. With the mean length of fish in the community, we can assess the direct effects of fishing on an ecosystem (EF). Accordingly, the mean trophic level (TL) of species exploited by the fishery represents the trophic position of the whole catch, which is expected to decrease with fishing intensity (EF). Fishing can thus change the structure of marine food webs by reducing the mean TL and thereby affecting ecosystem functioning by shortening the length of food chains and releasing predation on low-trophic-level organisms. For instance, species with short generation times and high fecundity (e.g., gelatinous animals, squids) are particularly suited to the opportunistic replacement of vertebrates with late maturity and fewer young, such as top predators (Lynam et al. 2006).

The proportion of predatory fish is a measure of the diversity of fish in the community (CB) and reflects the potential effects of fishing on the functioning of marine food webs. The removal of larger and more fecund fish could compromise population productivity. Also, the removal of larger species changes the size structure of the community and ecosystem functioning.

How the fishery of deep-sea species will affect deep-sea pelagic biodiversity and ecosystem functioning is difficult to predict, however, because too little is as yet known about the trophic webs and because there is insufficient baseline data to reveal changes (Robison 2009). The proportion of under- and moderately exploited stocks (CB) is a measure of the success of fisheries management and is used to compare the state of ecosystems. The initial content of a baseline data set would be a list of commercial species and their relative abundances, both in catch statistics and ecosystem fishery surveys. The mean life span of a community will reflect the relative abundances of species with different turnover rates and is a proxy for the mean turnover rate of species and communities and is intended to reflect the stability of a system. This is also considered a measure of ecosystem resistance to perturbations (SR). A measure of the stability of the ecosystem (SR) is the 1/coefficient of variation (CV) of total biomass, with a low 1/CV indicating low biomass stability and thus low ecosystem stability in response to perturbations. As total biomass decreases, the area occupied by the various stocks may decrease, the stocks may be more patchily distributed, or they may occupy the same area at a lower density.

The abundance/biomass comparison (ABC) method is widely used as an indicator of disturbance in different communities, macrozoobenthos, macrobenthos, and demersal fish (Warwick 1986; Stenton-Dozey et al. 1999; Yemane et al. 2005). This technique is based on the comparison of biomass respect to the abundance of individuals in a community at the same time and place, without the need for spatial or a temporal control as a reference (Warwick 1986). An undisturbed community is dominated by slow-growing, large-sized, late-maturing species, and therefore the biomass curve tends to be above the abundance curve. Whereas in a disturbance scenario that can be caused by anthropic activities such as fishing, oil and gas extraction, or environmental impacts, opportunistic species of small sizes, with rapid growth and early maturation, are dominated, where the biomass curve is below the abundance curve (Clarke and Warwick 1994). The ABC method was used to identify the degree of alteration of a shallow water demersal fish community in the Colombian Caribbean, indicating a moderately altered assemblage (Paramo et al. 2009). Something very interesting about the ABC method is that a spatial analysis can also be made of the degree of alteration related to fishing activity and to identify marine protected areas for fisheries management (Paramo et al. 2009; Pérez et al. 2019). Pérez et al. (2019) used the ABC method to study mega-invertebrate assemblage in a pristine ecosystem in the Colombian Caribbean, showing an undisturbed community. The authors also made a spatial analysis of the degree of alteration, showing possible nursey areas for fishing protection.

Another indicator commonly used to determine the state of ecosystems or disturbances mainly due to fishing activity is the size spectrum (Rice and Gislason 1996; Bianchi et al. 2000; Blanchard et al. 2005; Shin et al. 2005; Law et al. 2012; Jacobsen et al. 2014; Thorpe et al. 2015; Edwards et al. 2017). Body size is linked to the physiological characteristics of individuals and their life history (Peters 1983; West et al. 1997; Gillooly et al. 2001; Brown et al. 2004). This indicator is based on the comparison of the slopes or b parameter of the spectra of abundance or biomass distributions of the communities (Edwards et al. 2017). The steepness of the slope of the abundance size spectrum reflects the removal of large species from the system and possibly increased predation pressure on smaller sizes. On the other hand, when the slope decreases, it indicates a general reduction in the abundance of the community (Blanchard et al. 2005; Daan et al. 2005). The data to develop this indicator can be obtained from various sources including samples by trawling, underwater visual census survey for fish, benthic invertebrates in sediments, and remote sensors to determination of chlorophyll, which makes it a useful tool for management and conservation of the ecosystem (Petchey and Belgrano 2010; Edwards et al. 2017).

18.3.4 Bycatch Reduction

Discards are described as the proportion of the total organic material of animal origin in the catch that is thrown away or dumped at sea, for whatever reason, and are one of the most important topics in fisheries management (FAO 2010). Bycatch is the part of a catch that is taken incidentally in addition to the target species towards which fishing effort is directed, but some or all of it may be returned to the sea as discards, usually dead or dying (FAO 2010). Notwithstanding, global fishery discards have significantly declined (Zeller and Pauly 2005; Davies et al. 2009), due to improved selectivity of fishing technology and greater utilization of the bycatch for aquaculture and human consumption (Bellido et al. 2011). According to this, Heymans et al. (2011) modelled the deep-sea ecosystem of the Rockall area (200 miles off the west of Scotland) using Ecopath with Ecosim and identified the lack of discard data from deep-water fisheries in the area as an important limitation and potentially a substantial source of error in the model. This emphasizes the importance of having a good knowledge and quantification of discards to assess ecosystem status, as required for the implementation of EAF.

Zhou et al. (2010) mention that less selective fishing gears may help to maintain diversity and functioning in certain marine ecosystems, through a "balanced" exploitation, in which a reduced fishing effort is combined with less selective fishing strategies allowing for better use of the multispecies catch to achieve sustainable yields while maintaining healthy ecosystems. Diversifying the harvest and learning to utilize a wider variety of products were also recommended by Hall and Mainprize

(2005). Fisheries production could increase through better use of non-target species while reducing unsustainably high catches of high-trophic-level target species, thereby helping to meet the challenge of increasing global food demand (Zhou et al. 2010) and maintaining the ecosystem size spectrum.

18.3.5 Spatial Management Through Marine Protected Areas

The implementation of spatial management, with zoning for different kinds of fishing activity and use of seasonal or temporary closures, is one important measure that should be included in the management of the proposed new deep-sea fishery in the Colombian Caribbean. Those spatial management measures must be underpinned by a good knowledge of the biology, spatial distribution, and abundance of both resource species and other species impacted by fisheries, including protected species (Bellido et al. 2011). Marine protected areas (MPA) have recently emerged as a tool for marine conservation and fisheries management following an ecosystem-based approach (Worm et al. 2006; Paramo et al. 2009; Fraser et al. 2009; Jackson and Jacquet 2011). The general concept of an open-ocean international reserve that includes the full water column and the deep seafloor was proposed by Mills and Carlton (1998). They envisioned restrictions on shipping, fishing, mining, dumping, weapons, and floating cities.

The assemblages of fish and invertebrates in deep-sea are complex and are comprised by a high number of species and individuals (Paramo et al. 2012) of comparable numbers. In the deep sea of the Colombian Caribbean, species abundances have been described that are even superior to those that have been described for other deep regions of the North Atlantic and the Gulf of Mexico (Haedrich et al. 2001; Powell and Haedrich 2003; Priede et al. 2011; Godbold et al. 2013). The results confirm that the Caribbean Sea of Colombia, as part of the southeast of the Great Caribbean Basin, contributes as a center of high biodiversity and a nucleus of origin and adaptive evolution (Briggs 2007). Several studies have demonstrated that the greatest species abundance and biodiversity are concentrated in the north of the Colombian Caribbean, along the length of the department of the Guajira, with a marked preponderance off the coast of Riohacha (Paramo et al. 2009; Paramo et al. 2012). The elevated productivity in this zone is attributed to events of seasonal upwelling that enrich the deepest ecosystems with organic material of phytodetritus from the epipelagic zone (Paramo et al. 2009; Rice et al. 1986; Paramo et al. 2011, Correa-Ramírez et al. 2020), which also influences neighboring zones. In this way, the supply of food, which proceeds from the productive shallow waters, acts as a mediating seasonal agent of the reproductive processes and of the trophic interactions that occur in the different strata of the water column down to the sea floor (Fernández-Arcaya et al. 2013). This zone of upwelling, like others in the world, permits the enrichment of the waters and with it the development, establishment, and renovation of the innumerable populations of organisms, counteracting the impact that human activities have on them.

In the same manner as in the case of deep-sea corals, the fish resources in these strata are subject to a high risk due to the growth of the offshore hydrocarbon industry and the modernization of the fishing technology to reach deeper ecosystems; their protection would be imperative which involves objectives of conservation:

1. Conserve fishing resources both commercial and non-commercial (fish and crustaceans), which are found between 200 and 600 m of depth, which comprise highly complex and dynamic assemblages that represent one of the nuclei of the diversity of the southwestern Caribbean and a center of origin and adaptive evolution; protection is necessary.
2. Contribute towards the conservation of deep-sea communities and their habitats which supply resources for fishing, by considering these the focal point of exportation of materials and energy towards neighboring ecosystems.
3. Contribute to the processes of vertical and horizontal connection in the northern zone of the Colombian Caribbean.
4. Favor the conservation of the deep-sea resources considering their individual conditions of low resistance and high vulnerability.
5. Do not allow fishing of any kind beyond 600 m depth to protect the reproductive adult population and biodiversity.

Upon considering some of the proposed criteria in the literature (http://www.aida.ngo), policies of protection of the marine waters, which could be applied in this case, are the **Marine Protected Areas** (**MPA**) and the **Fisheries Protected Areas** (**FPA**). In the latter two types of areas are included: (**i**) areas closed to fishing and (**ii**) fisheries reserves that propose a code of conduct for the management of the zone, the techniques used, and the resources targeted. By the same, various modalities are considered in the areas closed to fishing: (**a**) a regulated fishing zone that prohibits determined methods of fishing, at least for certain periods of time; (**b**) a protected fishing zone, also called a conservation zone, which seeks the protection of one or various pelagic resources and/or specific demersal zones; (**c**) a zone in which fishing is prohibited, which restricts some type of fishing and other extractive measures; (**d**) a zone in the process of restoration, which seeks to restore marine habitats on which certain resources of fisheries depend; and (**e**) marine refuges, which are marine areas permanently closed to the fishing of one or various species in order to promote the concentration of breeders or nursery areas (http://www.aida.ngo).

Understanding the connections between the constituents and the ecological communities is fundamental for defining strategies of conservation (Halpern and Agardy 2014). Nevertheless, while much of the structure and function of the deep-sea environments is unknown, these environments provide numerous ecosystemic services that must be conserved for the future. In this case, an applicable strategy of conservation seems to be the FPA, with "The spatial and/or temporal prohibition of fishing in a given zone, whether totally or partially, and in this case, permitting determined fishing activities that do not affect negatively the resource that needs to be protected" (AIDA 2015). This would be an adequate strategy, given its applicability to the continental shelf, its orientation towards the protection of species of commercial importance, and its final objective of the protection of fishing grounds, rather than the protection of biodiversity, although it favors the latter directly (https://aida-americas.org/).

Some countries are considering implementing a depth limit to bottom trawling to manage deep-sea fisheries. Clarke et al. (2015) showed evidence that biodiversity and the ratio of Elasmobranchii (sharks and rays) to commercial biomass increased between 600 and 800 m depth, while commercial value decreases. Therefore, limiting bottom trawling to a maximum depth of 600 m could be an effective management strategy. In fact, the bathymetric distribution of the ratios of bycatch to commercial crustaceans (CC), teleostean to CC, and non-CC to CC decreased in relation to increasing depth among the main biological categories (crustaceans, teleostean fish, chondrichthyes, and mollusks) of a potentially new deep-sea resource in the Colombian Caribbean (Grijalba-Bendeck et al. 2019). For that reason, the beginning of this potentially new deep-sea fishery must consider sustainable use only up to 600 m depth.

18.4 Conclusions

The shallow water shrimp fishery in the Colombian Caribbean is a typical case in which high exploitation, combined with non-existing fisheries management, has resulted in the significant depletion of stocks (Paramo and Saint-Paul 2010). For that reason, possible new fishing areas were investigated on the deep-sea habitats in the Colombian Caribbean, to determine the potential for a viable deep-sea crustacean fishery (Paramo and Saint-Paul 2012a, b, c). Potential new fishing resources of substantial biomass in the area are the deep-sea giant red shrimp (*Aristaeomorpha foliacea*), the royal red shrimp (*Pleoticus robustus*), the pink speckled deep-sea shrimp (*Penaeopsis serrata*), and the deep-sea lobster (*Metanephrops binghami*). The profitability of these deep-sea crustaceans' species, both locally and in numerous international markets, underlines the potential of these species as a new economic resource in the Colombian Caribbean. Nevertheless, more scientific assessment is necessary to determine the life cycle and population characteristics of deep-sea crustaceans and the associated biodiversity, before initiating a new commercial fishery. The management of the proposed new deep-sea crustacean fishery should be based on an ecosystem approach that considers population dynamics and structure and function of the ecosystem, the optimum allocation of catches and effort, protection of nursery and spawning areas, as well as the development of monitoring strategies by means of ecosystem indicators (Shin et al. 2010). Moreover, ecosystem protection through the implementation of MPAs is also indicated. Finally, survey-based indicators that are independent of the fishery should be used to study the ecosystem responses to fishing pressure. In this way a real evaluation of changes in the marine ecosystem subjected to fishing activities can be achieved following the needs for EAF. This holistic approach will allow an appropriate level of biodiversity and the habitat quality to be maintained while implementing a sustainable fishery.

Acknowledgments We thank the crew of vessel "Tee Claude" and members of the research group Tropical Fisheries Science and Technology–CITEPT of Universidad del Magdalena (Colombia) that participated in the laboratory and research survey. The work was sponsored by Colciencias grant number 1117-452-21288, Universidad del Magdalena, and the Autoridad de Acuicultura y Pesca (AUNAP) through cooperation agreement number 153-2020 with the research project "Reproductive biology of deep-sea crustaceans of commercial importance in the Colombian Caribbean." Marcela Grijalba-Bendeck and Daniel Pérez also thank to Doctorate in Marine Sciences Program, Facultad de Ciencias Naturales e Ingeniería, Universidad de Bogotá Jorge Tadeo Lozano.

References

AIDA (2015) Herramientas para la pesquería sostenible y el manejo costero Capítulo 7. Figuras de Protección del Medio Marino. Available from: http://www.aida.ngo. Accessed 21 Sept 2017

Bellido JM, Santos MB, Pennino MG et al (2011) Fishery discards and bycatch: solutions for an ecosystem approach to fisheries management? Hydrobiologia 670:317–333. https://doi.org/10.1007/s10750-011-0721-5

Bensch A, Gianni M, Gréboval D et al (2008) Worldwide review of bottom fisheries in the high seas, FAO fisheries and aquaculture technical paper, Rome, p 522

Bianchi G (2008) The concept of the ecosystem approach to fisheries. In: Bianchi G, Skjoldal HR (eds) The ecosystem approach to fisheries. CAB and FAO, Wallingford, pp 20–38

Bianchi G, Gislason H, Graham K et al (2000) Impact of fishing on size composition and diversity of demersal fish communities. ICES J Mar Sci 57:558–571. https://doi.org/10.1006/jmsc.2000.0727

Blanchard JL, Dulvy NK, Jennings S et al (2005) Do climate and fishing influence size-based indicators of Celtic Sea fish community structure? ICES J Mar Sci 62:405–411. https://doi.org/10.1016/j.icesjms.2005.01.006

Briggs JC (2007) Marine longitudinal biodiversity: causes and conservation. Divers Distrib 13(5):544–555. https://doi.org/10.1111/j.1472-4642.2007.00362.xl

Brown JH, Gillooly JF, Allen AP et al (2004) Toward a metabolic theory of ecology. Ecology 85(7):1771–1789. https://doi.org/10.1890/03-9000

Carpenter KE (2002) The living marine resources of the Western Central Atlantic. Volume 1: Introduction, molluscs, crustaceans, hagfishes, sharks, batoid fishes, and chimaeras, FAO species identification guide for fishery purposes and American Society of Ichthyologists and Herpetologists Special Publication no. 5. FAO, Rome

Cavanagh R, Kyne PM (2006) The conservation status of deep-sea chondrichthyans fishes. In: Shotton R (ed) Deep Sea 2003: conference on the governance and management of deep-sea fisheries. Part 2: conference poster papers and workshop papers, vol 3/2. FAO Fish. Proceed, Rome, pp 366–378

Chan TY (1998) Shrimps and prawns. In: Carpenter KE, Niem VH (eds) Species identification guide for fisheries purposes. The living marine resources of the Western Central Pacific. FAO, Rome, pp 851–971

Chan TY, Yu HP (1987) *Metanephrops formosanus* sp. Nov., a new species of lobster (Decapoda, *Nephropidae*) from Taiwan. Crustaceana 52(2):172–186

Chan TY, Ho KC, Li CP, Chu KH (2009) Origin and diversification of clawed lobster genus *Metanephrops* (Crustacea: Decapoda: *Nephropidae*). Mol Phylogenet Evol 50:411–422. https://doi.org/10.1016/j.ympev.2008.11.020

Choi JH, Kim JN, Kim MH et al (2008) Population biology and feeding habits of the nephropid lobster *Metanephrops thomsoni* (Bate 1888) in the East China Sea. J Environ Biol 29(4):453–456

Christensen V (2011) Introduction: toward ecosystem-based management. In: Christensen V, Maclean J (eds) Ecosystem approaches to fisheries, Cambridge/New York, pp 1–8
Clarke KR, Warwick RM (1994) Change in marine communities: an approach to statistical analysis and interpretation. Plymouth Marine Laboratory, Plymouth
Clarke J, Milligan RJ, Bailey DM, Neat FC (2015) A scientific basis for regulating deep-sea fishing by depth. Curr Biol 25(18):2425–2429. https://doi.org/10.1016/j.cub.2015.07.070
Correa-Ramírez M, Rodríguez-Santana A, Ricaurte-Villota C et al (2020) The Southern Caribbean upwelling system off Colombia: water masses and mixing processes. Deep-Sea Res Part I Oceanogr Res Pap 155:103–145. https://doi.org/10.1016/j.dsr.2019.103145
D'Onghia G, Tursi A, Maiorano P et al (1998) Distribution, biology, and population dynamics of *Aristaeomorpha foliacea* (Risso, 1827) (Decapoda, Natantia, Aristeidae) in the north-western Ionian Sea (Mediterranean Sea). Crustaceana 71(5):518–544
Da Ros Z, Dell'Anno A, Morato T et al (2019) The deep-sea: the new frontier for ecological restoration. Mar Policy 108:103642. https://doi.org/10.1016/j.marpol.2019.103642
Daan N, Gislason H, Pope JG et al (2005) Changes in the North Sea fish community: evidence of indirect effects of fishing? ICES J Mar Sci 62:177–188. https://doi.org/10.1016/j.icesjms.2004.08.020
Daan N, Gislason H, Pope JG, Rice JC (2011) Apocalypse in world fisheries? The reports of their death are greatly exaggerated. ICES J Mar Sci 68(7):1375–1378. https://doi.org/10.1093/icesjms/fsr069
Dallagnolo R, Perez JAA, Pezzuto PR et al (2009) The deep-sea shrimp fishery off Brazil (Decapoda: Aristeidae) development and present status. Lat Am J Aquat Res 37(3):327–346. https://doi.org/10.3856/vol37-issue3-fulltext-5
Davies RWD, Cripps SJ, Nickson A et al (2009) Defining and estimating global marine fisheries bycatch. Mar Pol 33:661–672. https://doi.org/10.1016/j.marpol.2009.01.003
Decade on Restauration (2020). http://www.decadeonrestauration.org. Accessed 10 May 2020
Edwards AM, Robinson PW, Plank MJ et al (2017) Testing and recommending methods for fitting size spectra to data. Methods Ecol Evol 8(1):57–67. https://doi.org/10.1111/2041-210X.12641
Espinoza-Tenorio A, Moreno-Báez M, Pech D et al (2014) El ordenamiento ecológico marino en México: un reto y una invitación al quehacer científico. Lat Am J Aquat Res 42(3):386–400. https://doi.org/10.3856/vol42-issue3-fulltext-1
FAO (1995) Precautionary approach to fisheries. Part 1: Guidelines on the precautionary approach to capture fisheries and species introductions, FAO Fisheries technical paper. No. 350, Part 1, Rome
FAO (2003) Fisheries management 2. The ecosystem approach to fisheries, FAO technical guidelines for responsible fisheries No 4, Suppl. 2, Rome
FAO (2009) Progress in the implementation of the code of conduct for responsible fisheries, related international plans of action and strategy, COFI/2009/2. FAO, Rome
FAO (2010) Report of the technical consultation to develop international guidelines on bycatch management and reduction of discards, Rome, 6–10 December 2010, FAO Fisheries and aquaculture report No. 957. FAO, Rome
Fernández-Arcaya U, Ramírez-Llodra X, Rotlant E et al (2013) Reproductive biology of two macrourid fish, *Nezuima aequalis* and *Coelorinchus mediterraneus*, inhabiting the NW Mediterranean continental margin (400–2000 m). Deep Sea Res II 92:63–72. https://doi.org/10.1016/j.dsr2.2013.03.003
Figueiredo MJ, Figueiredo I, Machado PB (2001) Deep-water penaeid shrimps (Crustacea: Decapoda) from off the Portuguese continental slope: an alternative future resource? Fish Res 51:321–326. https://doi.org/10.1016/S0165-7836(01)00255-7
Follesa MC, Porcu C, Cabiddu S et al (2011) Deep-water fish assemblages in the central-western Mediterranean (south Sardinian deep-waters). J Appl Ichth 27:129–135. https://doi.org/10.1111/j.1439-0426.2010.01567.x

Fraser HM, Greenstreet SPR, Piet GJ (2009) Selecting MPAs to conserve groundfish biodiversity: the consequences of failing to account for catchability in survey trawls. ICES J Mar Sci 66:82–89. https://doi.org/10.1093/icesjms/fsn181

Garcia SM, Zerbi A, Aliaume C et al (2003) The ecosystem approach to fisheries. Issues, terminology, principles, institutional foundations, implementation and outlook, FAO fisheries technical paper. No. 443, Rome

Gillooly JF, Brown JH, West GB et al (2001) Effects of size and temperature on metabolic rate. Science 293:2248–2251. https://doi.org/10.1126/science.1061967

Godbold J, Bailey D, Collins M et al (2013) Putative fishery-induced changes in biomass and population size structures of demersal deep-sea fishes in ICES Sub-area VII, Northeast Atlantic Ocean. Biogeosciences 10:529–539. https://doi.org/10.5194/bg-10-529-2013

Gómez G, Guzmán R, Marcano L (2000) Contribución al conocimiento de la Cigala del Caribe, *Metanephrops binghami*, en aguas de Venezuela: aspectos biométricos y fecundidad. Zool Trop 18(1):107–118

Gómez G, Marcano L, Guzmán R, Ramírez T (2005) Aspectos biométricos y reproductivos de la Cigala, *Metanephrops binghami*, en aguas de Venezuela. Proc Gulf Carib Fish Inst 47:537–547

Gracia A, Vázquez-Bader AR, Lozano-Álvarez E, Briones-Fourzán P (2010) Deep-water shrimp (Crustacea: Penaeoidea) off the Yucatan peninsula (Southern Gulf of Mexico): a potential fishing resource. J Shellfish Res 29(1):37–43. https://doi.org/10.2983/035.029.0124

Grijalba-Bendeck M, Paramo J, Wolff M (2019) Catch composition of deep-sea resources of commercial importance in the Colombian Caribbean. Rev Biol Mar Oceanog 54(2):194–203. https://doi.org/10.22370/rbmo.2019.54.2.1891

Guijarro B, Tserpes G, Moranta J et al (2011) Assessment of the deep water trawl fishery off the Balearic Islands (western Mediterranean): from single to multi-species approach. Hydrobiologia 670:67–85. https://doi.org/10.1007/s10750-011-0670-z

Gunderson DR (1993) Surveys of fisheries resources. Wiley, New York

Haedrich RL, Merrett NR, O'Dea NR (2001) Can ecological knowledge catch up with deep-water fishing? A North Atlantic perspective. Fish Res 51(2–3):113–122. https://doi.org/10.1016/S0165-7836(01)00239-9

Håkanson L (2004) Break-through in predictive modelling opens new possibilities for aquatic ecology and management – a review. Hydrobiologia 518:135–157. https://doi.org/10.1023/B:HYDR.0000025064.26497.53

Hall SJ, Mainprize BM (2005) Managing by-catch and discards: how much progress are we making and how can we do better? Fish Fish 6(2):134–155. https://doi.org/10.1111/j.1467-2979.2005.00183.x

Halpern B, Agardy T (2014) Ecosystem-based approaches to marine conservation and management. In: Bertness M, Bruno J, Siliman B et al (eds) Marine community ecology and conservation, 1st edn. Sinauer, Sunderland, pp 477–493

Hart PJB, Pearson E (2011) An application of the theory of island biogeography to fish speciation on seamounts. Mar Ecol Prog Ser 430:281–288. https://doi.org/10.3354/meps08948

Heymans JJ, Howell KL, Ayers M et al (2011) Do we have enough information to apply the ecosystem approach to management of deep-sea fisheries? An example from the West of Scotland. ICES J Mar Sci 68(2):265–280. https://doi.org/10.1093/icesjms/fsq065

Hilborn R (2010) Future directions in ecosystem based fisheries management: a personal perspective. Fish Res 108(2–3):235–239. https://doi.org/10.1016/j.fishres.2010.12.030

Holthuis LB (1980) Shrimps and prawns of the world. An annotated catalogue of species of interest to fisheries. (FAO species catalogue, 1). FAO Fish Synopsis 125(1):1–271

Holthuis LB (1991) Marine lobsters of the world, FAO species catalogue, vol 13. FAO Fisheries Synopsis, Rome

Hutchings JA, Minto C, Ricard D et al (2010) Trends in the abundance of marine fishes. Can J Fish Aquat Sci 67(8):1205–1210. https://doi.org/10.1139/F10-081

Jackson J, Jacquet J (2011) The shifting baselines syndrome: perception, deception, and the future of our oceans. In: Christensen V, Maclean J (eds) Ecosystem approaches to fisheries, Cambridge/New York, pp 128–141
Jacobsen NS, Gislason H, Andersen KH (2014) The consequences of balanced harvesting of fish communities. Proc R Soc Ser B 281:20132701
King M (2007) Fisheries biology, assessment and management. Blackwell Publishing, Oxford
Law R, Plank MJ, Kolding J (2012) On balanced exploitation of marine ecosystems: results from dynamic size spectra. ICES J Mar Sci 69(4):602–614. https://doi.org/10.1093/icesjms/fss031
Levin PS, Fogarty MJ, Murawski SA et al (2009) Integrated ecosystem assessments: developing the scientific basis for ecosystem-based management of the Ocean. PLoS Biol 7(1):e1000014. https://doi.org/10.1371/journal.pbio.1000014
Longhurst A (2010) Mismanagement of marine fisheries. University Press, Cambridge
Lorance P, Agnarsson S, Damalas D et al (2011) Using qualitative and quantitative stakeholder knowledge: examples from European deep-water fisheries. ICES J Mar Sci 68(8):1815–1824. https://doi.org/10.1093/icesjms/fsr076
Lynam CP, Gibbons MJ, Axelsen BE et al (2006) Jellyfish overtake fish in a heavily fished ecosystem. Curr Biol 16(13):R492–R493. https://doi.org/10.1016/j.cub.2006.06.018
Mathew S (2011) Fishery-dependent information and the ecosystem approach: what role can fishers and their knowledge play in developing countries? ICES J Mar Sci 68(8):1805–1808. https://doi.org/10.1093/icesjms/fsr113
Mills CE, Carlton JT (1998) Rationale for a system of international reserves for the open ocean. Conserv Biol 12(1):244–247
Morato T, Watson R, Pitcher TJ et al (2006) Fishing down the deep. Fish Fish 7(1):24–34. https://doi.org/10.1111/j.1467-2979.2006.00205.x
Mouffok S, Massuti E, Boutiba Z et al (2008) Ecology and fishery of the deep-water shrimp, *Aristeus antennatus* (Risso, 1816) off Algeria (South-western Mediterranean). Crustaceana 81(10):1177–1199. https://doi.org/10.1163/156854008X374513
Munro JL (2011) Assessment of exploited stocks of tropical fishes: an overview. In: Christensen V, Maclean J (eds) Ecosystem approaches to fisheries. New York, Cambridge, pp 145–1170
Mura M, Murenu M, Cau A (2002) The occurrence of *Penaeopsis serrata* Bate 1881 (Decapoda, Penaeidae) in the Middle-West Mediterranean Sea. Crustaceana 75(10):1263–1269. https://doi.org/10.1163/156854002321518180
ONU (2020) Sustainable development http://www.sustainabledevelopment.un.org/sdg14. Accessed 10 May 2020
Papaconstantinou C, Kapiris K (2003) The biology of the giant red shrimp (*Aristaeomorpha foliacea*) at an unexploited fishing ground in the Greek Ionian Sea. Fish Res 62(1):37–51. https://doi.org/10.1016/S0165-7836(02)00254-0
Paramo J, Saint-Paul U (2010) Morphological differentiation of southern pink shrimp *Farfantepenaeus notialis* in Colombian Caribbean Sea. Aquat Liv Res 23(1):95–101. https://doi.org/10.1051/alr/2010004
Paramo J, Saint-Paul U (2012a) Deep-sea shrimps *Aristaeomorpha foliacea* and *Pleoticus robustus* (Crustacea: Penaeoidea) in the Colombian Caribbean Sea as a new potential fishing resource. J Mar Biol Assoc UK 92(4):811–818. https://doi.org/10.1017/S0025315411001202
Paramo J, Saint-Paul U (2012b) Spatial structure of the pink speckled deep-sea shrimp *Penaeopsis serrata* (Bate, 1881) (Decapoda, Penaeidae) during November-December 2009 in the Colombian Caribbean Sea. Crustaceana 85(1):103–116. https://doi.org/10.1163/156854012X623665
Paramo J, Saint-Paul U (2012c) Spatial structure of deep-sea lobster (*Metanephrops binghami*) in the Colombian Caribbean Sea. Helgol Mar Res 66:25–31. https://doi.org/10.1007/s10152-011-0243-6
Paramo J, Guillot L, Benavides S et al (2009) Aspectos poblacionales y ecológicos de peces demersales de la zona norte del Caribe colombiano en relación con el hábitat: una herramienta para identificar Áreas Marinas Protegidas (AMPs) para el manejo pesquero. Caldasia 31(1):123–144

Paramo J, Correa M, Núñez S (2011) Evidencias de desacople físico-biológico en el sistema de surgencia en La Guajira, Caribe colombiano. Rev Biol Mar Oceanog 46(3):421–430. https://doi.org/10.4067/S0718-19572011000300011

Paramo J, Wolff M, Saint-Paul U (2012) Deep-sea fish assemblages in the Colombian Caribbean Sea. Fish Res 125–126:87–98. https://doi.org/10.1016/j.fishres.2012.02.011

Pauly D, Christensen V, Dalsgaard J et al (1998) Fishing down marine food webs. Science 279:860–863. https://doi.org/10.1126/science.279.5352.860

Pauly D, Alder J, Bennett E et al (2003) The future for fisheries. Science 21:1359–1361. https://doi.org/10.1126/science.1088667

Pérez JA, Pereira NB, Pereira DA et al (2013) Composition and diversity patterns of megafauna discards in the deep-water shrimp trawl fishery off Brazil. J Fish Biol 83(4):804–825. https://doi.org/10.1111/jfb.12141

Pérez D, Paramo J, Wolff M (2019) Distribution, abundance and fishing potential of mega-invertebrates in the sub-euphotic zone (150–535 m) in the Colombian Caribbean. Reg Stud Mar Sci 32:100868. https://doi.org/10.1016/j.rsma.2019.100868

Pérez-Farfante I (1980) Revision of the penaeid shrimp genus Penaeopsis (Crustacea: Decapoda). Fish Bull 77(4):721–763

Petchey OL, Belgrano A (2010) Body-size distributions and size-spectra: universal indicators of ecological status? Biol Lett 6:434–437. https://doi.org/10.1098/rsbl.2010.0240

Peters RH (1983) The ecological implications of body size. Cambridge University Press, Cambridge

Pezzuto R, Pérez JAA, Wahrlich R (2006) Deep-sea shrimps (Decapoda: Aristeidae): new targets of the deep-water trawling fishery in Brazil. Braz J Oceanogr 54(2–3):123–134. https://doi.org/10.1590/S1679-87592006000200003

Politou C-Y, Kapiris K, Maiorano P et al (2004) Deep-sea Mediterranean biology: the case of *Aristaeomorpha foliacea* (Risso, 1827) (Crustacea: Decapoda: Aristidae). Sci Mar 68(S3):129–139

Powell SM, Haedrich RL (2003) The deep-sea demersal fish fauna of the northern gulf of Mexico. J Northwest Atl Fish Sci 31:19–33

Priede IG, Godbold JA, Niedzielski T et al (2011) A review of the spatial extent of fishery effects and species vulnerability of the deep-sea demersal fish assemblage of the Porcupine Seabight, Northeast Atlantic Ocean (ICES subarea VII). ICES J Mar Sci 68(2):281–289. https://doi.org/10.1093/icesjms/fsq045

Rice J, Gislason H (1996) Patterns of change in the size spectra of numbers and diversity of the North Sea fish assemblage, as reflected in surveys and models. ICES J Mar Sci 53(6):1214–1225. https://doi.org/10.1006/jmsc.1996.0146

Rice AL, Billett DS, Fry J et al (1986) Seasonal deposition of phytodetritus to the deep-sea floor. Proc R Soc Edinburgh 88:265–279. https://doi.org/10.1017/S0269727000004590

Roberts CM (2002) Deep impact: the rising toll of fishing in the deep sea. Trends Ecol Evol 17(5):242–245. https://doi.org/10.1016/S0169-5347(02)02492-8

Robison BH (2009) Conservation of deep pelagic biodiversity. Conserv Biol 23:847–858. https://doi.org/10.1111/j.1523-1739.2009.01219.x

Shin YJ, Rochet MJ, Jennings S et al (2005) Using size-based indicators to evaluate the ecosystem effects of fishing. ICES J Mar Sci 62(3):384–396. https://doi.org/10.1016/j.icesjms.2005.01.004

Shin YJ, Shannon LJ, Bundy A et al (2010) Using indicators for evaluating, comparing, and communicating the ecological status of exploited marine ecosystems. 2. Setting the scene. ICES J Mar Sci 67(4):692–716. https://doi.org/10.1093/icesjms/fsp294

Smith PJ (1999) Allozyme variation in scampi (*Metanephrops challengeri*) fisheries around New Zealand. New Zeal J Mar Fresh Res 33(3):491–497. https://doi.org/10.1080/00288330.1999.9516894

Stenton-Dozey JMF, Jackson IF, Busby AJ (1999) Impact of mussel culture of macro-enthic community structure in Saldahana Bay, South Africa. Mar Pollut Bull 39(1–12):357–366. https://doi.org/10.1016/S0025-326X(98)00180-5

Tavares M (2002) Shrimps. In: Carpenter KE (ed) The living marine resources of the western central Atlantic, species identification guide for fisheries purposes. FAO, Rome, pp 251–291
Thorpe RB, Le Quesne WJF, Luxford F et al (2015) Evaluation and management implications of uncertainty in a multispecies size structured model of population and community responses to fishing. Methods Ecol Evol 6(1):49–58. https://doi.org/10.1111/2041-210X.12292
Ward TJ, Davis TLO (1987) Diel periodicity of *Metanephrops australiensis* and other deep-water Crustaceans of northwest Australia. Fish Res 5(1):91–97. https://doi.org/10.1016/0165-7836(87)90018-X
Warwick RM (1986) A new method for detecting pollution effects on marine macrobenthic communities. Mar Biol 92:557–562
Wassenberg TJ, Hill BJ (1989) Diets of four decapod crustaceans (*Linuparus trigonus*, *Metanephrops andamanicus*, *M. australiensis* and *M. boschmai*) from the continental shelf around Australia. Mar Biol 103:161–167
West GB, Brown JH, Enquist BJ (1997) A general model for the origin of allometric scaling laws in biology. Science 276(5309):122–126. https://doi.org/10.1126/science.276.5309.122
Worm B, Barbier EB, Beaumont N et al (2006) Impacts of biodiversity loss on ocean ecosystem services. Science 314(3):787–790. https://doi.org/10.1126/science.1132294
Worm B, Hilborn R, Baum JK et al (2011) Rebuilding global fisheries. Science 325:578–584. https://doi.org/10.1126/science.1173146
WWF/IUCN (2004) The Mediterranean deep-sea ecosystems: an overview of their diversity, structure, functioning and anthropogenic impacts, with a proposal for conservation. WWF IUCN, Malaga
Ye Y, Cochrane K, Qiu Y (2011) Using ecological indicators in the context of an ecosystem approach to fisheries for data-limited fisheries. Fish Res 112(3):108–116. https://doi.org/10.1016/j.fishres.2011.06.004
Yemane D, Field JG, Leslie RW (2005) Exploring the effects of fishing on fish assemblages using Abundance Biomass Comparison (ABC) curves. ICES J Mar Sci 62(3):374–379. https://doi.org/10.1016/j.icesjms.2005.01.009
Zeller D, Pauly D (2005) Good news, bad news: global fisheries discards are declining, but so are total catches. Fish Fish 6:156–115. https://doi.org/10.1111/j.1467-2979.2005.00177.x
Zhou S, Smith ADM, Punt AE et al (2010) Ecosystem-based fisheries management requires a change to selective fishing philosophy. Proc Natl Acad Sci U S A 107(21):9485–9489. https://doi.org/10.1073/pnas.0912771107

Chapter 19
Diversity, Abundance, and Biomass of Deep-Sea Decapod Crustaceans of the Uruguayan Continental Slope in the Southwestern Atlantic Ocean

G. Rotllant, A. Verdi, R. Santos-Bethencourt, N. Bahamón, and J. B. Company

Abstract The megafauna of deep continental margin of the exclusive economic zone (EEZ) of Uruguay have been little studied. The present study includes deep-sea trawling operations and represents the first analysis in detail of the deep-sea community of decapod crustaceans. Cluster analysis of bottom trawl data indicated that benthic megafauna are grouped in four bathymetric ranges along the continental margin: A, 250 > 1100 m; B, 1100 > 2000 m; C, 2000 > 3000 m; and D, 3000–3800 m, while pelagic species are not grouped by depth strata. The decapod individuals belong to 79 different species from which 64% correspond to shrimps (suborder Dendrobranchiata and infraorder Caridea), and the third most important group corresponds to the infraorder Anomura (18%). From those, previously 67% were not reported off Uruguay and 47% in the southwestern Atlantic Ocean. The bathymetric range of the identified decapods was enlarged by 32%. Moreover, the frequency distribution of species occurrences was rare since 56% of the species were only sampled once, thus indicating that the decapod community of this area is still far unknown. The benthic species were also more diverse than the pelagic decapods, and their abundance and biomass were higher. The biomass of the decapod community was dominated by the geryonid crab, *Chaceon notialis*, mainly located in the shallowest depth strata (representing 97% of the total biomass). To detect possible changes in the structure, biomass, and diversity of benthic assemblages, we recommend the implementation of long-term monitoring programs in the continental slope off Uruguay before fishing or mining exploitation is developed.

G. Rotllant (✉) · R. Santos-Bethencourt · N. Bahamón · J. B. Company
Institut de Ciències del Mar (ICM-CSIC), Barcelona, Spain
e-mail: guio@icm.csic.es; rsantos@icm.csic.es; n.bahamon@csic.es; batista@icm.csic.es

A. Verdi
Facultad de Ciencias, Universidad de la República, Montevideo, Uruguay
e-mail: averdi@fcien.edu.uy

M. E. Hendrickx (ed.), *Deep-Sea Pycnogonids and Crustaceans of the Americas*, https://doi.org/10.1007/978-3-030-58410-8_19

Keywords Crustaceans · Deep sea · Shannon index · Equity index of Pielou Species richness · Conservation

19.1 Introduction

A large part of biodiversity on the world is still unknown, and it is estimated that, at the current pace, it will take several centuries to complete description of living species (Fontaine et al. 2012). The deep-sea environment represents over 95% of the characteristic volume on earth, remaining as one of the least known ecosystems. The topographical heterogeneity of the ocean increases the probability of species diversity (Buhl-Mortensen et al. 2010). The knowledge of the occurrence, abundance, and biomass of megafauna species from the deep sea is scarce and remains difficult to study diversity patterns, in terms of either species richness or functional diversity.

Such big expeditions as HMS Challenger, R/V Vema, R/V Atlantis II, and R/V Calypso explored the Uruguayan slope where decapods were earlier identified. Only a few taxonomic groups were studied in detail, Alpheoida (Christoffersen 1979) and Lithodidae (Macpherson 1988). In 2010, the cruise "Uruguay 0110" on board of the RV Miguel Oliver studied the EEZ waters of Uruguay from 200 to 4500 m depth including decapod crustaceans. In the cruise report, it was indicated that the diversity of decapod was important with a large biomass of non-identified sergestid shrimps. Some species of Pinnotheridae and Majidae families and *Propagurus gaudichaudi*, *Peltarion* sp., *Munida* sp., and *Chaceon notialis* were the most representative (Carranza 2010). On the other hand, a list of marine and estuarine decapods of Uruguay appeared in 1995 that has been later updated (Zolessi and Philippi 1995; Scarabino 2006; Spivak et al. 2019). In the most recent update, from the 211 species of decapod crustaceans listed only 51 species corresponded to deep-sea (<200 m) species off Uruguayan waters.

The objective of this study was to check if the exceptional diversity of decapods recorded in the Uruguayan continental slope before mining or fishing exploitation reaches this pristine deep environment. We perform a comprehensive examination at different taxonomic levels (suborders or infraorders and families), life strategies (pelagic and benthic), depth strata levels, and areas (north, central, and south). We also investigate decapod diversity patterns with different ecological characteristics (proportional abundance of species, evenness, and species richness).

19.2 Material and Methods

Biological material was collected during April–May 2016 in an oceanographic cruise on board the F/V Marianne, exploring the deep continental margin of the exclusive economic zone (EEZ) of Uruguay (Table 19.1) in the southwestern Atlantic Ocean (SWAO). We sampled a number of transects located in front the north (N), central (Ce), and south (S) coast of Uruguay in direction south-east from 250 m to 3800 m depth (Fig. 19.1). Up to 26 trawling lanes positioned as parallel to isobaths as possible to maintain a constant depth during the hauls were carried out

Table 19.1 Characteristics of the fishing maneuvers and calculation of the swept area of the campaign carried out in the EEZ of Uruguay

TN	Area	TST	Latitude S (initial)	Longitude W (initial)	TFT	TD	Latitude S (final)	Longitude W (final)	Depth (m)	Date	**SA (km²)**
P1	S	0:03	36°37.56′	53°50.35′	0:40	0:37	36°36.52′	53°49.47′	250	4/19/2016	0.006
P2	Ce	1:03	35°36.92′	52°36.28′	1:43	0:40	35°35.97′	52°35.00′	700	4/21/2016	0.006
P3	Ce	4:52	35°48.28′	52°24.70′	5:35	0:43	35°47.52′	52°23.04′	1680	4/21/2016	0.007
P4	Ce	9:10	35°47.00′	52°23.03′	9:50	0:40	35°46.61′	52°22.01′	1680	4/21/2016	0.005
P5	Ce	15:07	36°01.99′	52°14.20′	16:06	0:59	36°00.80′	52°12.42′	2600	4/21/2016	0.009
P6	Ce	22:10	36°12.51′	52°00.48′	23:10	1:00	36°11.18′	51°59.79′	3100	4/21/2016	0.009
P7	Ce	12:18	36°36.00′	51°36.71′	14:58	2:40	36°34.51′	51°33.06′	3600	4/22/2016	0.025
P8	S	2:39	36°49.37′	52°37.13′	4:11	1:32	35°47.48′	52°35.39′	3000	4/23/2016	0.011
P9	S	14:12	37°36.34′	52°37.62′	15:10	0:58	37°35.35′	52°41.74′	3800	4/23/2016	0.009
P10	S	21:05	37°36.04′	52°36.84′	22:35	1:30	37°34.95′	52°33.76′	3800	4/23/2016	0.0139
P11	S	7:51	36°56.73′	53°09.76′	8:55	1:04	36°54.73′	53°08.76′	2600	4/24/2016	0.009
P12	S	12:00	36°53.57′	53°07.62′	13:35	1:35	36°56.18′	53°08.85	2540	4/24/2016	0.014
P13	S	17:35	36°47.24′	53°22.42′	18:50	1:15	36°48.39′	53°24.80′	1660	4/24/2016	0.011
P14	S	0:00	36°35.88′	53°35.81′	0:42	0:42	36°36.61′	53°37.05′	1050	4/24/2016	0.006
P15	S	2:40	36°40.54′	53°48.01′	3:07	0:27	36°41.25′	53°48.56′	500	4/25/2016	0.004
P16	Ce	15:00	35°34.07′	52°35.48′	15:30	0:30	35°34.74′	52°36.18′	500	4/25/2016	0.004
P17	Ce	16:46	35°34.01′	52°40.00′	17:20	0:34	35°34.76′	52°40.72′	250	4/25/2016	0.005
P18	N	4:33	34°42.40′	52°00.99′	10:03	5:30	34°43.12′	52°01.63′	250	5/1/2016	0.004
P19	N	11:15	34°41.95′	51°58.18′	11:45	0:30	34°42.64′	51°58.78′	500	5/1/2016	0.008
P20	N	14:20	34°48.20′	51°48.97′	15:20	1:00	34°49.73′	51°50.93′	1160	5/1/2016	0.009
P21	N	20:03	34°48.96′	51°50.49′	21:05	1:02	34°47.47′	51°49.43′	1160	5/1/2016	0.009
P22	N	22:14	35°01.86′	51°37.94′	23:45	1:31	34°59.67′	51°36.20′	1900	5/1/2016	0.015
P23	N	4:14	35°11.68′	51°23.41′	5:45	1:31	35°14.11′	51°25.12′	2500	5/2/2016	0.013
P24	N	10:45	35°20.34′	51°11.72′	12:20	1:35	35°22.29′	51°14.18′	2800	5/2/2016	0.0140
P25	N	20:17	35°41.07′	50°43.69′	22:19	2:02	35°44.41′	50°46.43′	3600	5/2/2016	0.019
P26	S	19:12	37°11.80′	52°59.19′	21:12	2:00	37°14.94′	53°02.12′	3300	5/3/2016	0.018

Abbreviations: *TN* trawl number, area (*Ce* central, *N* North, *S* South), *TST* trawling starting time, *TFT* trawling finishing time, *TD* trawling duration, *SA* swept area

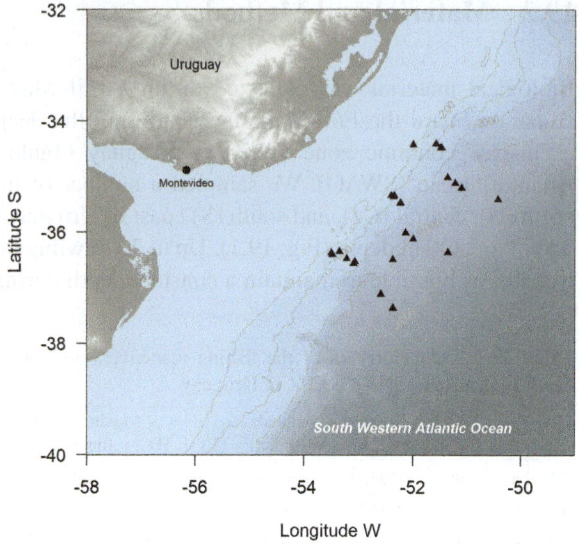

Fig. 19.1 Map of the deep continental margin of the EEZ of Uruguay showing the average locations of bottom trawl fishing

for 30 min to 2 h, being the deeper trawls the longest to be able to catch enough number of individuals. Megafauna sampling was carried out with an Agassiz trawl, with a 2.5 m horizontal opening, 1.2 m vertical opening, and a net mesh size of 12 mm, dragged at 2.0 knots.

The megafauna (fishes, cnidarians, echinoderms, crustaceans, mollusks) samples from each trawl were sorted on board and identified to the lowest possible taxonomic level using the available guides of the closer studied area (southern Brazil, Uruguay, and/or northern Argentina) of the most important taxonomic groups (Cervigón et al. 1992; Melo 1996, 1999; Scarabino 2003a, b, c; Kitahara et al. 2009; Benavides-Serrato et al. 2011; Smith and Kroh 2011; Scarabino 2004; Nion et al. 2016; Scarabino et al. 2019; Spivak et al. 2019). Color images of fresh samples of all species were obtained at each trawling lane to keep taxonomic characteristics that could be lost during preservation. Then, animals were counted and weighted by species in each haul. Finally, specimens were preserved in 70% ethanol. When more than one individual was captured, another sample was fixed in 10% formaldehyde for further detailed taxonomic classification. Abundance and biomass of every species were standardized for a swept area (in km^2), which was previously calculated using the average speed of the boat (in knots, s), the horizontal opening of the net (in meters, bt), and the sampling time (in hours, h): swept area = bt x s x h x 1852/10^6.

Data on abundance and biomass of the species were log-transformed, and similarity among all samples was calculated using the Raup-Crick method, since community data are treated as presence/absence data (Oksanen et al. 2019 and references therein). This method was used for both benthic and pelagic megafauna. Life strategies in each species were determined using SeaLifeBase (www.sealifebase.ca) or literature cited above. Depth ranges obtained in the cluster were used to study the

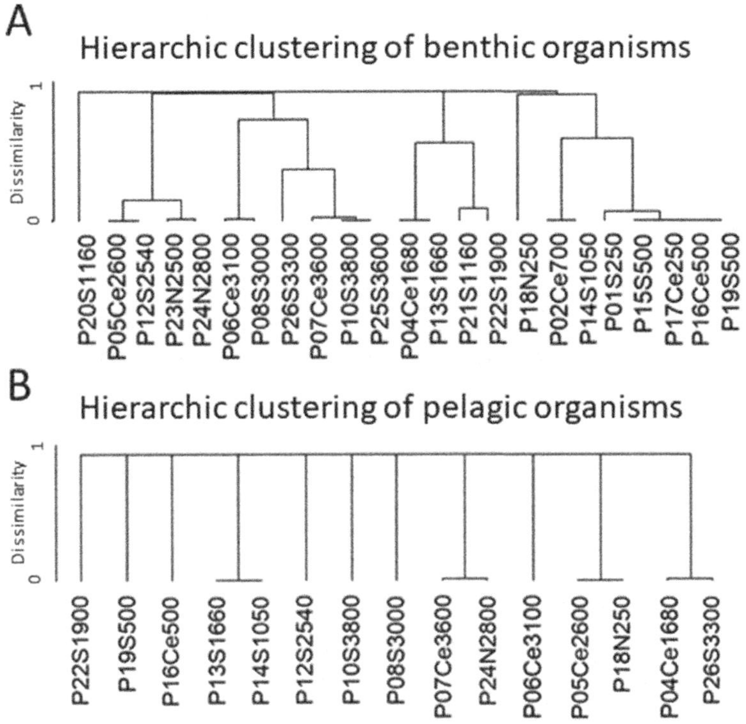

Fig. 19.2 Dissimilarity among all megafauna samples collected in the deep continental slope of Uruguay estimated using the Raup-Crick method. Depth ranges of similarity were only obtained for benthic megafauna (**a**), while no grouping was estimated for pelagic megafauna (**b**). Full trawling code includes trawling number, area code, and depth (m). Abbreviations: Ce = central area, P = trawling lane, N = north area, S = south area

diversity of decapods for benthic species only (Fig. 19.2a), since no strata were found for pelagic species (Fig. 19.2b).

In the Institute of Marine Sciences at Barcelona (ICM-CSIC), the classification of all decapod species was reviewed using the available taxonomic literature (see Table 19.2). In brief, to reach family level, we used the following literature by subcategory: suborder Dendrobranchiata (Pérez Farfante and Kensley 1997) and infraorders Caridea (Holthuis 1993), Astacidea (Melo 1999), Polychelida (Ahyong 2009), Anomura (Ahyong et al. 2010), and Brachyura (Melo 1996). Whenever available, keys to family from the area were used to reach species level, and species first description was checked. Species names were crosschecked with the World Register of Marine Species, available online (WoRMS Editorial Board 2020). Several samples were damaged when sampling or preservation was not identify to species level. All identified decapod specimens are deposited in the collections of the ICM-CSIC.

Ecological diversity indices Shannon (*H′*), species richness (*S*), and Pielou (*J'*) were calculated using the vegan package (Oksanen et al. 2019). In brief, Shannon

Table 19.2 Abundance and biomass of deep-sea crustaceans fished in the EEZ of Uruguay

TN	Area	Depth	H_ DR	Species	N	W (g)	Ab (N/Km2)	Bi_g_km2
P01	S	250	B_A	*Libidoclaea granaria*	16	16	2667	2667
P01	S	250	B_A	*Peltarion spinulosum*	4	28	667	4667
P02	Ce	700	B_A	*Chaceon notialis*	65	28400	10833	4733333
P03	Ce	1680	P	*Gennadas elegans*	1	1	151	151
P03	Ce	1680	B_B	Nematocarcinidae sp. A	3	7	452	1055
P03	Ce	1680	P	Pasiphaeidae sp. A	1	2	151	301
P03	Ce	1680	P	*Deosergestes henseni*	1	1	151	151
P03	Ce	1680	P	Sergestidae sp. A	1	4	151	603
P04	Ce	1680	B_B	Nematocarcinidae sp. A	99	375	19800	75000
P04	Ce	1680	B_B	*Stereomastis sculpa*	7	25	1400	5000
P05	Ce	2600	B_C	*Benthonectes filipes*	1	4	111	444
P05	Ce	2600	B_C	*Neolithodes agassizii*	2	5	222	556
P05	Ce	2600	B_C	*Nematocarcinus tenuipes*	4	9	444	1000
P05	Ce	2600	B_C	*Parapontophilus gracilis*	6	11	667	1222
P05	Ce	2600	P	*Parasergestes armatus*	1	1	111	222
P05	Ce	2600	B_C	*Pentacheles validus*	9	79	1000	8778
P05	Ce	2600	B_C	Porcellanidae sp. A	1	5	111	556
P05	Ce	2600	P	*Prehensilosergia prehensilis*	1	1	111	222
P06	Ce	3100	P	*Acanthephyra pelagica* var. *sica*	1	15	111	1667
P06	Ce	3100	B_D	*Aristaeopsis edwardsiana*	1	10	111	1111
P06	Ce	3100	B_D	*Bathicaris brasiliensis*	2	15	222	1667
P06	Ce	3100	B_D	*Ethusina abyssicola*	1	1	111	111
P06	Ce	3100	B_D	Nematocarcinidae sp. A	1	5	111	556
P07	Ce	3600	B_D	*Aristaeopsis edwardsiana*	4	120	160	4800
P07	Ce	3600	B_D	*Bathicaris brasiliensis*	3	70	120	2800
P07	Ce	3600	B_D	*Cerataspis monstrosus*	4	190	160	7600
P07	Ce	3600	P	*Gennadas kempi*	1	1.0	40	40
P07	Ce	3600	P	*Gennadas* sp. A	2	2.0	80	80
P07	Ce	3600	B_D	*Hepomadus tener*	2	60	80	2400
P07	Ce	3600	B_D	*Munidopsis abyssicola*	1	55	40	2200
P07	Ce	3600	P	Sergestidae sp. B	3	8	120	320
P08	S	3000	B_D	*Acanthephyra quadrispinosa*	1	10	91	909
P08	S	3000	B_D	*Bathicaris brasiliensis*	3	40	227	3636
P08	S	3000	B_D	*Ethusina abyssicola*	3	11	273	1000
P08	S	3000	B_D	*Glyphocrangon atlantica*	1	2	91	182
P08	S	3000	B_D	*Munida subcaeca*	2	6	182	545
P08	S	3000	B_D	*Munidopsis* sp. A	1	1	91	91
P08	S	3000	B_D	*Munidopsis aries*	1	250	91	22727
P08	S	3000	B_D	*Parapagurus abyssorum*	1	7	91	636
P08	S	3000	B_D	*Willemoesia forceps*	1	40	91	3636

(continued)

Table 19.2 (continued)

TN	Area	Depth	H_ DR	Species	N	W (g)	Ab (N/Km2)	Bi_g_km2
P10	S	3800	P	*Phorcosergia* sp. A	1	8	143	571
P10	S	3800	P	Sergestidae sp. B	1	8	143	571
P12	S	2540	B_C	*Acanthephyra quadrispinosa*	1	6	71	429
P12	S	2540	P	*Eusergestes antarcticus*	1	1	71	71
P12	S	2540	B_C	*Nematocarcinus* sp. A	1	1.5	71	107
P13	S	1660	B_B	*Chaceon notialis*	13	3400	1182	309091
P13	S	1660	B_B	Nematocarcinidae sp. A	195	750	17727	68182
P13	S	1660	P	*Phorcosergia phorca*	2	4	182	364
P13	S	1660	B_B	Polychelidae sp. A	1	2	91	182
P13	S	1660	P	*Sicyonella* sp. A	2	3	17727	68182
P13	S	1660	B_B	*Stereomastis suhmi*	1	3	17727	68182
P13	S	1660	B_B	*Systellaspis guillei*	4	5	17727	68182
P13	S	1660	B_B	*Thymops birsteini*	2	400	182	36364
P14	S	1050	B_A	*Chaceon notialis*	27	1400	4500	233333
P14	S	1050	P	*Eusergestes antarcticus*	13	16	2250	2666
P14	S	1050	B_A	*Pandalus amplus*	1	6	167	1000
P14	S	1050	P	*Pasiphaea alcocki*	13	14	2167	2334
P14	S	1050	P	*Petalidium foliaceum*	14	16	2250	2666
P14	S	1050	P	*Phorcosergia phorca*	58	200	9667	33333
P15	S	500	B_A	*Chaceon notialis*	3	1500	750	375000
P15	S	500	B_A	Munidopsidae sp. B	1	1	250	250
P16	Ce	500	B_A	*Chaceon notialis*	108	34500	27000	8625000
P17	Ce	250	B_A	*Chaceon notialis*	57	16000	11400	3200000
P17	Ce	250	B_A	*Tetraxanthus rathbunae*	1	2	200	400
P18	N	250	B_A	*Alpheus pouang*	1	1	250	250
P18	N	250	B_A	*Artemesia longinaris*	7	200	1750	50000
P18	N	250	B_A	*Catapagurus cunhai*	8	4	2000	1000
P18	N	250	B_A	*Latreillia williamsi*	2	3	500	750
P18	N	250	B_A	*Lysmata stenolepis*	1	2	250	500
P18	N	250	B_A	*Mesopenaeus tropicalis*	1	1	250	250
P18	N	250	B_A	*Munida benedicti*	2	3	500	750
P18	N	250	B_A	*Myropsis quinquespinosa*	2	13	500	2750
P18	N	250	B_A	*Notopandalus magnoculus*	1	1	250	250
P18	N	250	B_A	*Palicus gracilipes*	2	2	500	500
P18	N	250	B_A	*Parapontocaris caribbaea*	1	2	250	500
P18	N	250	B_A	*Plesionika brevipes*	3	1.5	750	375
P18	N	250	P	Pylochelidae sp. A	1	1	250	250
P18	N	250	P	*Spinolambrus pourtalesii*	4	20	1000	5000
P18	N	250	B_A	*Tetraxanthus rathbunae*	18	22	4500	5500
P19	S	500	P	*Alainopasiphaea australis*	6	15	750	1875
P19	S	500	B_A	*Chaceon notialis*	15	3600	1875	450000

(continued)

Table 19.2 (continued)

TN	Area	Depth	H_ DR	Species	N	W (g)	Ab (N/Km2)	Bi_g_km2
P19	S	500	B_A	*Merhippolyte americana*	1	5	125	625
P19	S	500	B_A	Penaeoidea sp. A	1	1	125	125
P20	S	1160	B_B	*Systellaspis debilis*	1	4	111	444
P21	S	1160	B_B	*Nematocarcinus exilis*	5	7	556	778
P21	S	1160	B_B	*Stereomastis sculpa*	2	30	222	3333
P22	S	1900	P	*Gennadas kempi*	1	1	67	67
P22	S	1900	P	*Parasergestes armatus*	1	1	67	67
P22	S	1900	P	*Parasergestes armatus*	1	1	67	67
P22	S	1900	P	*Pasiphaea oshoroae*	2	3	133	200
P22	S	1900	B_B	*Pentacheles validus*	2	35	133	2333
P23	N	2500	B_C	*Aristaeopsis edwardsiana*	8	36	615	2769
P23	N	2500	B_C	Crangonidae sp. A	24	44	1846	3385
P23	N	2500	B_C	*Munidopsis bermudezi*	2	44	154	3385
P23	N	2500	B_C	*Parapagurus abyssorum*	2	40	154	3077
P23	N	2500	B_C	*Parapontophilus gracilis*	70	80	5385	6154
P23	N	2500	B_C	Polychelidae sp. A	28	340	2154	26154
P24	N	2800	B_C	*Bathicaris iridescens*	11	40	786	2857
P24	N	2800	B_C	*Ethusina abyssicola*	9	3	643	214
P24	N	2800	P	*Eusergestes antarcticus*	2	1	143	71
P24	N	2800	P	*Gennadas kempi*	4	6	286	429
P24	N	2800	B_C	*Glyphocrangon sculpa*	1	10	71	714
P24	N	2800	B_C	*Nematocarcinus* sp. B	30	47	2143	3375
P24	N	2800	B_C	*Nematocarcinus tenuipes*	10	16	714	1125
P24	N	2800	B_C	*Parapontophilus gracilis*	22	6	1571	429
P24	N	2800	B_C	Polychelidae sp. A	5	54	357	3857
P24	N	2800	B_C	*Systellaspis debilis*	1	2	71	143
P25	S	3600	B_D	*Bathicaris brasiliensis*	1	30	53	1579
P25	S	3600	B_D	*Cerataspis monstrosus*	4	220	211	11579
P25	S	3600	P	*Gennadas gilchristi*	1	0.5	53	26
P25	S	3600	B_D	Munidopsidae sp. A	1	20	53	1053
P26	S	3300	B_D	*Cerataspis monstrosus*	1	25	53	1316
P26	S	3300	P	*Gennadas kempi*	7	10	368	526
P26	S	3300	B_D	*Hemipenaeus spinidorsalis*	1	8	53	421
P26	S	3300	B_D	*Munida gregaria*	2	3	105	158
P26	S	3300	B_D	*Nematocarcinus sp.* B	30	47	2143	3375
P26	S	3300	B_D	*Nematocarcinus tenuipes*	10	16	714	1125
P26	S	3300	B_D	*Parapasiphae* sp. A	1	2	53	105
P26	S	3300	P	*Sergestes atlanticus*	2	2	105	105

Abbreviations: *TN* trawl number, area (*Ce* central, *N* North, *S* South), *H* habitat (*B* benthic, *P* pelagic), *DR* depth range (A, 250 > 1100 m; B, 1100 > 2000 m; C, 2000 > 3000 m; and D, 3000–3800 m), *W* weight, *Ab* abundance, *Bi* biomass index is defined as: $H' = -\sum_i p_i \ln(p_i)$ $H' = -\sum_i pi \ln pi$ where *p*i is the proportional abundance of species *i*. It expresses the uniformity of important values through all species present in the sample choices au hazard. Species richness is the total number of individuals in a concrete community. Pielou's evenness (equitability) assesses relation between Shannon diversity and the maximal diversity observed in a concrete community as: $J' = H'/(\ln S)$. These indices were estimated by habitat (pelagic and benthic, depth strata), sampling area, and taxon.

19.3 Results and Discussion

A total of 1166 individuals of decapod crustacean specimens, representing a biomass of 93 kg, were collected in the deep-sea continental slope of the EEZ off Uruguay (Table 19.2). Decapods were collected at all trawl lines. Benthic species showed the highest overall occurrence (85% of sampled trawl lines), while pelagic species occurrence was 54%. Benthic taxa were more diverse than pelagic taxa (57 vs 22 species). Abundance and biomass of benthic taxa were higher (82% and 99%, respectively). We recorded 79 decapod species (Figs. 19.3, and 19.4), 64% of which were shrimps (Dendrobranchiata and Caridea), and the third most important group was Anomura (18%). Only five species of Polychelida and one species of Astacidea were caught (Table 19.3). Similar proportions of decapod groups were previously described in Uruguay (Spivak et al. 2019 and references therein).

Penaeoidea presented 14 species distributed across 4 families. Benthesicymidae were represented by seven species belonging and three genera. Among four species

Fig. 19.3 Deep-sea Decapoda from the continental slope of Uruguay. (**a**) *Cerataspis monstrosus*, Dendrobranchiata. (**b**) *Eusergestes antarcticus*, Dendrobranchiata. (**c**) *Pasiphaea alcocki*, Caridea. (**d**) *Systellaspis debilis*, Caridea

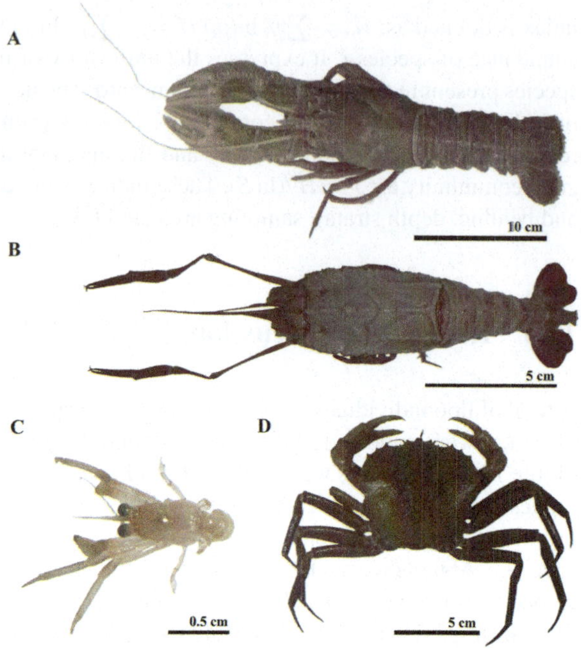

Fig. 19.4 Deep-sea Decapoda from the continental slope of Uruguay. (**a**) *Thymops birsteini*, Astacidea. (**b**) *Stereomastis sculpa*, Polychelida. (**c**) Glaucothoe of Pylochelidae, Anomura. (**d**) *Chaceon notialis*, Brachyura

of *Gennadas*, only *G. gilchristi* was previously reported in Uruguay (Spivak et al. 2019), but they were captured above a depth of 3600 m that is deeper than the previously reported by Calman (1925) in the South African coast. *Gennadas kempi* and *G. elegans* were not previously reported in Uruguayan waters neither in the SWAO, but their range of distribution was already known (Smith 1882; Stebbing 1914; Zariquiey Alvarez 1968). Both recorded species of the genus *Bathicaris*, *B. brasiliensis* and *B. iridescens*, were sampled in the same depth range and in the SWAO but not in Uruguayan waters (Bate 1881; Pérez Farfante and Kensley 1997; Vereshchaka et al. 2019). The benthesicymid *Benthonectes filipes* was reported for the first time in the SWAO at slightly deeper depths than previously reported (Pérez Farfante and Kensley 1997; Smith 1884).

Within the Aristeidae only *Aristaeopsis edwardsiana* was previously reported in Uruguay (Spivak et al. 2019), but it was recorded deeper at 3600 m while previously was found down at 1850 m in the southeastern and northern Atlantic Ocean (Zariquiey Alvarez 1968; Dore and Frimodt 1987). The other three species of the Aristeidae, *Cerataspis monstrosus* (Fig. 19.3a), *Hemipenaeus spinidorsalis*, and *Hepomadus tener*, were not previously reported in Uruguay neither in the SWAO, and in the case of *H. tener*, we also increased the depth range to 3600 m (Pérez Farfante and Kensley 1997). The Penaeidae were represented by two species previously reported in the SWAO, *Artemesia longinaris* and *Mesopenaeus tropicalis*; the former is already cited in Uruguay at the same depth range (Spivak et al. 2019; Alves-Júnior et al. 2017).

Table 19.3 Taxonomic classification of deep-sea crustaceans fished in the EEZ of Uruguay, keys used, and previous distribution (√)

Subcategory	Family	Species	Key 1	**Original name**	**U**	**SWAO**	**DR**
Dendrobranchiata		Penaeoidea sp. A	Pérez Farfante and Kensley (1997)				
Dendrobranchiata	Aristeidae	*Aristaeopsis edwardsiana* (Johnson, 1868)	Pérez Farfante and Kensley (1997)	*Penaeus edwarsianus*	√	√	
Dendrobranchiata	Aristeidae	*Cerataspis monstrosus* Gray, 1828	Pérez Farfante and Kensley (1997)				√
Dendrobranchiata	Aristeidae	*Hemipenaeus spinidorsalis* Spence Bate, 1881	Pérez Farfante and Kensley (1997)				√
Dendrobranchiata	Aristeidae	*Hepomadus tener* Smith, 1884	Pérez Farfante and Kensley (1997)				
Dendrobranchiata	Benthesicymidae	*Bathicaris brasiliensis* (Spence Bate, 1881)	Pérez Farfante and Kensley (1997), Vereshchaka et al. (2019)	*Benthesicymus brasiliensis*		√	√
Dendrobranchiata	Benthesicymidae	*Bathicaris iridescens* Spence Bate, 1881	Pérez Farfante and Kensley (1997)			√	√
Dendrobranchiata	Benthesicymidae	*Benthonectes filipes* Smith, 1885	Pérez Farfante and Kensley (1997)				
Dendrobranchiata	Benthesicymidae	*Gennadas* sp. A	Pérez Farfante and Kensley (1997)				
Dendrobranchiata	Benthesicymidae	*Gennadas elegans* (Smith, 1882)	Pérez Farfante and Kensley (1997)	*Amalopenaeus elegans*			√
Dendrobranchiata	Benthesicymidae	*Gennadas gilchristi* Calman, 1925	Pérez Farfante and Kensley (1997)		√	√	
Dendrobranchiata	Benthesicymidae	*Gennadas kempi* Stebbing, 1914	Pérez Farfante and Kensley (1997)				√
Dendrobranchiata	Penaeidae	*Artemesia longinaris* Spence Bate, 1888	Pérez Farfante and Kensley (1997)		√	√	√
Dendrobranchiata	Penaeidae	*Mesopenaeus tropicalis* (Bouvier, 1905)	Pérez Farfante and Kensley (1997), Alves Junior et al. (2017)	*Parartemesia tropicalis*		√	
Dendrobranchiata	Sergestidae	Sergestidae sp. A	Pérez Farfante and Kensley (1997)				
Dendrobranchiata	Sergestidae	Sergestidae sp. B	Pérez Farfante and Kensley (1997)				

(continued)

Table 19.3 (continued)

Subcategory	Family	Species	Key 1	**Original name**	**U**	**SWAO**	**DR**
Dendrobranchiata	Sergestidae	*Eusergestes antarcticus* (Vereshchaka, 2009)	Pérez Farfante and Kensley (1997), Vereshchaka (2000), Vereshchaka et al. (2014)	*Sergestes antarcticus*	√	√	
Dendrobranchiata	Sergestidae	*Sergestes atlanticus* H. Milne-Edwards, 1830	Pérez Farfante and Kensley (1997), Vereshchaka (2009), Vereshchaka et al. (2014)		√	√	
Dendrobranchiata	Sergestidae	*Deosergestes henseni* (Ortmann, 1893)	Pérez Farfante and Kensley (1997), Vereshchaka (2009), Vereshchaka et al. (2014)	*Sergia henseni*	√	√	
Dendrobranchiata	Sergestidae	*Parasergestes armatus* (Krøyer, 1855)	Pérez Farfante and Kensley (1997), Vereshchaka (2009), Vereshchaka et al. (2014)	*Sergestes armatus*	√	√	
Dendrobranchiata	Sergestidae	*Phorcosergia* sp. A	Pérez Farfante and Kensley (1997), Vereshchaka (2000), Vereshchaka et al. (2014)				
Dendrobranchiata	Sergestidae	*Phorcosergia phorca* (Faxon, 1893)	Pérez Farfante and Kensley (1997), Vereshchaka (2000), Vereshchaka et al. (2014)	*Sergestes phorcus*			
Dendrobranchiata	Sergestidae	*Prehensilosergia prehensilis* (Spence Bate, 1881)	Pérez Farfante and Kensley (1997), Vereshchaka (2000), Vereshchaka et al. (2014)	*Sergestes prehensilis*	√	√	
Dendrobranchiata	Sergestidae	*Petalidium foliaceum* Spence Bate, 1881	Pérez Farfante and Kensley (1997), Vereshchaka (2000), Vereshchaka and Lunina (2015)		√	√	√
Dendrobranchiata	Sergestidae	*Sicyonella* sp. A	Pérez Farfante and Kensley (1997), Vereshchaka et al. (2014), Vereshchaka et al. (2016)				
Caridea	Pasiphaeidae	Pasiphaeidae sp. A	Holthuis (1993)				

Subcategory	Family	Species	Key 1	**Original name**	U	SWAO	DR
Caridea	Pasiphaeidae	*Alainopasiphaea australis* (Hanamura, 1989)	Holthuis (1993), Burukovsky (1996)	*Pasiphaea australis*			√
Caridea	Pasiphaeidae	*Pasiphaea alcocki* (Wood-Mason in Wood-Mason & Alcock, 1891)	Holthuis (1993), Burukovsky (1996), Komai and Chan (2012), Rodrigues and Cardoso (2019)	*Parapasiphae alcocki*		√	√
Caridea	Pasiphaeidae	*Pasiphaea oshoroae* Komai & Amaoka, 1993	Holthuis (1993), Burukovsky (1996)				
Caridea	Pasiphaeidae	Parapasiphae sp. A	Holthuis (1993)				
Caridea	Hippolytidae	*Merhippolyte americana* Holthuis, 1961	Holthuis (1993)		√	√	√
Caridea	Acanthephyridae	*Acanthephyra pelagica* (Risso, 1816) var. *Sica*	Holthuis (1993), Cardoso (2013), Vereshchaka (pers. comm.)	*Alpheus pelagicus*	√	√	√
Caridea	Acanthephyridae	*Acanthephyra quadrispinosa* Kemp, 1939	Holthuis (1993), Cardoso (2013)		√	√	
Caridea	Oplophoridae	*Systellaspis debilis* (A. Milne-Edwards, 1881)	Holthuis (1993), Crosnier (1987)	*Acanthephyra debilis*	√	√	√
Caridea	Oplophoridae	*Systellaspis guillei* Crosnier, 1988	Holthuis (1993), Chace Jr (1986), Sha and Wang (2015)				√
Caridea	Nematocarcinidae	Nematocarcinidae sp. A	Holthuis (1993)				
Caridea	Nematocarcinidae	*Nematocarcinus* sp. A	Holthuis (1993)				
Caridea	Nematocarcinidae	*Nematocarcinus* sp. B	Holthuis (1993)				
Caridea	Nematocarcinidae	*Nematocarcinus exilis* (Spence Bate, 1888)	Holthuis (1993), Crosnier and Forest (1973), Chace Jr (1986), Cardoso and Burukovsky (2014)	*Stochasmus exilis*			√
Caridea	Nematocarcinidae	*Nematocarcinus tenuipes* Spence Bate, 1888	Holthuis (1993), Chace Jr (1986)				

(continued)

Table 19.3 (continued)

Subcategory	Family	Species	Key 1	**Original name**	**U**	**SWAO**	**DR**
Caridea	Alpheidae	*Alpheus pouang* Christoffersen, 1979	Holthuis (1993), Soledade and Almeida (2013)		√	√	√
Caridea	Lysmatidae	*Lysmata stenolepis* Crosnier & Forest, 1973	Holthuis (1993)				√
Caridea	Pandalidae	*Notopandalus magnoculus* (Spence Bate, 1888)	Holthuis (1993)	*Pandalus magnoculus*			√
Caridea	Pandalidae	*Pandalus amplus* (Spence Bate, 1888)	Holthuis (1993), Boschi (1973)	*Pandalopsis ampla*	√	√	√
Caridea	Pandalidae	*Plesionika brevipes* (Crosnier & Forest, 1968)	Holthuis (1993), Cruz and Fransen (2004)	*Parapandalus brevipes*			
Caridea	Crangonidae	Crangonidae sp. A	Holthuis (1993)				
Caridea	Crangonidae	*Parapontocaris caribbaea* (Boone, 1927)	Holthuis (1993), Dardeau and Heard Jr (1983), Alves-Junior et al. (2018)	*Aegeon caribbaeus*		√	√
Caridea	Crangonidae	*Parapontophilus gracilis* (Smith, 1882)	Holthuis (1993), Crosnier and Forest (1973)	*Pontophilus gracilis*			√
Caridea	Glyphocrangonidae	*Glyphocrangon atlantica* Chace Jr, 1939	Holthuis (1993), Vazquez-Bader and Gracia (2013), Holthuis (1971)				√
Caridea	Glyphocrangonidae	*Glyphocrangon sculpta* (Smith, 1882)	Holthuis (1993), Crosnier and Forest (1973)	*Rhachocaris sculpta*			√
Astacidea	Nephropidae	*Thymops birsteini* (Zarenkov & Semenov, 1972)	Melo (1999), Holthuis (1974)		√	√	√
Polychelida	Polychelidae	Polychelidae sp. A	Ahyong (2009)				
Polychelida	Polychelidae	*Willemoesia forceps* A. Milne-Edwards, 1880	Ahyong (2009), Galil (2000)				√
Polychelida	Polychelidae	*Pentacheles validus* A. Milne-Edwards, 1880	Ahyong (2009)		√	√	√
Polychelida	Polychelidae	*Stereomastis suhmi* (Bate, 1878)	Ahyong (2009), Farias et al. (2015)			√	√

Subcategory	Family	Species	Key 1	**Original name**	**U**	**SWAO**	**DR**
Polychelida	Polychelidae	*Stereomastis sculpta* (Smith, 1880)	Ahyong (2009)	*Polycheles sculptus*		√	√
Anomura	Munidopsidae	Munidopsidae sp. A	Ahyong et al. (2010)				
Anomura	Munidopsidae	Munidopsidae sp. B	Ahyong et al. (2010)				
Anomura	Munidopsidae	*Munidopsis* sp. A	Ahyong et al. (2010)				
Anomura	Munidopsidae	*Munidopsis abyssicola* Baba, 2005	Ahyong et al. (2010), Pequegnat and Pequegnat (1971)				√
Anomura	Munidopsidae	*Munidopsis aries* (A. Milne Edwards, 1880)	Ahyong et al. (2010), Pequegnat and Pequegnat (1971)	*Orophorhynchus aries*			√
Anomura	Munidopsidae	*Munidopsis bermudezi* Chace Jr, 1939	Ahyong et al. (2010), Pequegnat (1970), Pequegnat and Pequegnat (1971)				√
Anomura	Munididae	*Munida benedicti* Chace Jr, 1942	Ahyong et al. (2010), Pequegnat (1970)				√
Anomura	Munididae	*Munida gregaria* (Fabricius, 1793)	Ahyong et al. (2010), Pequegnat (1970), Tapella and Lovrich (2006), Zeldis (1985)	*Galathea gregaria*		√	
Anomura	Munididae	*Munida subcaeca* Bouvier, 1922	Ahyong et al. (2010), Pequegnat (1970)				
Anomura	Porcellanidae	Porcellanidae sp. A	Haig (1966), Werding et al. (2003)				
Anomura	Lithodidae	*Neolithodes agassizii* (Smith, 1882)	Melo (1999), MacPherson (1988)				
Anomura	Paguridae	*Catapagurus cunhai* Nucci & Schmidt de Melo, 2012	McLaughlin (2003), Nucci and Melo (2012)			√	√
Anomura	Parapaguridae	*Parapagurus abyssorum* Filhol, 1885	McLaughlin (2003), Lemaitre (1989), Forest and McLaughlin (2000)				√
Anomura	Pylochelidae	Pylochelidae sp. A	McLaughlin (2003), McLaughlin and Lemaitre (2009), Forest (1987)				
Brachyura	Trichopeltariidae	*Peltarion spinulosum* (White, 1843)	Melo (1996)	*Atelecyclus spinulosum*	√	√	√

(continued)

Table 19.3 (continued)

Subcategory	Family	Species	Key 1	**Original name**	**U**	**SWAO**	**DR**
Brachyura	Ethusinae	*Ethusina abyssicola* Smith, 1884	Melo (1996), Rathbun (1937)			√	√
Brachyura	Leucosiidae	*Myropsis quinquespinosa* Stimpson, 1871	Melo (1996), de Melo (2010)		√	√	√
Brachyura	Epialtidae	*Libidoclaea granaria* H. Milne Edwards & Lucas, 1842	Melo (1996)			√	√
Brachyura	Majidae	*Latreillia williamsi* Melo, 1990	Melo (1996), Castro et al. (2003)		√	√	√
Brachyura	Palicidae	*Palicus gracilipes* (A. Milne-Edwards, 1880)	Melo (1996), Castro (2000a, b)	*Cymopolia gracilipes*			√
Brachyura	Parthenopidae	*Spinolambrus pourtalesii* (Stimpson, 1871)	Melo (1996), Tan and Ng (2007)	*Lambrus pourtalesii*		√	√
Brachyura	Geryonidae	*Chaceon notialis* Manning & Holthuis, 1989	Melo (1996), Tavares and Pinheiro (2011)		√	√	
Brachyura	Xanthidae	*Tetraxanthus rathbunae* Chace Jr, 1939	Melo (1996)		√	√	√

DR depth range, *PC* personal communication, *SWAO* southwestern Atlantic Ocean, *U* Uruguay

The Sergestoidea was represented by 11 species in 7 samples. *Phorcosergia phorca*, *Prehensilosergia prehensilis*, and *Deosergestes henseni* were not previously reported neither in the Uruguay area nor in the SWAO (Vereshchaka 2000, 2009). *Eusergestes antarcticus* (Fig. 19.3b), *Sergestes atlanticus*, *Parasergestes armatus*, and *Petalidium foliaceum* have already been reported in the Uruguay area (Spivak et al. 2019). All these sergestids were captured deeper than previously reported (Vereshchaka 2000, 2009; Vereshchaka and Lunina 2015).

Caridean shrimps were the next important group in number of species caught in the continental slope of Uruguay. Species from ten different families were identified. We were able to identify five species of Pasiphaeidae, but only three were identified to species level. *Pasiphaea alcocki* (Fig. 19.3c) was recently reported in deep Brazilian waters (Rodrigues and Cardoso 2019), and the other two species, *Alainopasiphaea australis* and *Pasiphaea oshoroae*, were never reported in the SWAO and in shallower depths (Komai and Amaoka 1993; Burukovsky 1996; Poore et al. 2008). Nevertheless, the depth at which these species inhabit may be less, since species were probably caught during the retrieval of the net in the water column.

Only one species of Hippolytidae, *Merhippolyte americana*, was identified and had already been reported in Uruguay and in the same depth range in Chile (Christoffersen 1979; Guzmán and Quiroga 2005). Two species of Acanthephyridae, *Acanthephyra pelagica* and *A. quadrispinosa*, were earlier reported in Uruguay, and the later presented a deeper depth range (Cardoso 2013; Spivak et al. 2019). Two species of Oplophoridae were caught. *Systellaspis debilis* (Fig. 19.3d) was previously reported in Uruguay at 1800 m and deeper in Mar del Plata (Farias et al. 2015; Spivak et al. 2019). The second species, *Systellaspis guillei*, was not cited previously in the SWAO although it was reported in the northwestern Atlantic, the eastern Atlantic, and some areas of the Indian and Pacific oceans close to 3000 m depth (Smith 1882; Chace Jr 1986). Five species of Nematocarcinidae were found, although we were only able to get to species level in two cases and both had not been reported before in the SWAO. *Nematocarcinus exilis* occurs in the eastern Atlantic from SW Ireland to Morocco and off the Canary Islands, including the Mediterranean Sea, as deep as 4765 m (Burukovsky 2012); *N. tenuipes* was cited in some areas of the Pacific and Indian Oceans, close to 2000 m depth (Spence Bate 1888; Crosnier and Forest 1973).

Alpheus pouang was the single species of Alpheidae caught at 250 m that has been previously cited in Uruguay at a similar depth (Christoffersen 1979). Lysmatidae was represented by *Lysmata stenolepis* previously reported in similar depth in Senegal but not in Uruguay (Crosnier and Forest 1973). From the Pandalidae shrimps, only *Pandalus amplus* was previously reported in Uruguay deeper than 1000 m (Bate 1888), while *Notopandalus magnoculus* and *Plesionika brevipes* that were caught at 250 m were not previously cited in the SWAO, although *P. brevipes* was reported in the South East Atlantic coast (Bate 1888; Crosnier and Forest 1968). Three species of Crangonidae were caught, and we were able to reach species level in two of them. The shallow *Parapontocaris caribbaea* and the deeper *Parapontophilus gracilis* were recently reported at the same depth in Brazil but not in Uruguay (Cardoso 2012; Alves-Júnior et al. 2018). The two species of

Glyphocrangonidae, *Glyphocrangon atlantica* and *Glyphocrangon sculpta*, caught at deep waters in Uruguay, were reported previously in the Atlantic but not in the SWAO also in deep waters (Chace Jr 1939; Holthuis 1971).

In contrast to the large number of caridean shrimps, the infraorder Astacidea was only represented by the Nephropidae and a single species, *Thymops birsteini* (Fig. 19.4a). This commercial lobster species has been previously reported in Uruguay at deep as 1660 m, as for in the present study (Laptikhovsky and Reyes 2009; Spivak et al. 2019). In the blind and abyssal lobster Polychelidae, five species were caught and four identified at species level. *Pentacheles validus* presents a worldwide distribution, including Uruguayan continental slope, and was reported deeper in Brazilian waters than in the present study (Serejo et al. 2007; Chang et al. 2013). The two species of the genus *Stereomastis*, *S. suhmi* and *S. sculpa* (Fig. 19.4b), were previously reported in the SWAO, close to Uruguay and in a similar depth (Galil 2000; Serejo et al. 2007; Farias et al. 2015). *Willemoesia forceps*, although reported in the Atlantic waters as deep as 4064 m, was not earlier cited in the SWAO (Gore 1984; Galil 2000).

Anomuran crabs were more represented that the lobster species in the Uruguayan continental slope including 7 families and 14 different species. The Munidopsidae was represented by six species, although we were only able to reach species level in three of them. *Munidopsis aries* and *Munidopsis bermudezi* were previously reported at deeper depths and in the Atlantic but not the SWAO (Milne-Edwards 1880; Chace Jr 1939; Pequegnat and Pequegnat 1971). *Munidopsis abyssicola* was previously reported close to New Zealand at a deeper depth too (4520 m vs 3600 m) (Baba 2005). From the Munididae three species were collected in the present study. The shallower one, *Munida benedicti*, was previously reported only in the Caribbean waters down to 430 m depth (Chace Jr 1942). *Munida gregaria* and *Munida subcaeca* were caught down to 3000 m. *M. gregaria* was previously cited in deep-sea waters off Buenos Aires province as depth as 1100 m (Boschi et al. 1992; Diez et al. 2016). *Munida subcaeca* has been reported in Tenerife and Madeira down to 1700 m, while in this study, the specimen was collected at 3000 m (Bouvier 1922). *Neolithodes agassizii* was the only species from the Lithodidae collected in the continental slope of Uruguay at 2600 m, while it was only reported previously in the northeastern Atlantic coast down to 1900 m (Smith 1882; Macpherson 1988). *Catapagurus cunhai* was the single species found of Paguridae in shallow waters off Uruguayan coast previously reported in Brazil (Nucci and Melo 2012). The Parapaguridae were represented by *Parapagurus abyssorum* collected at 3000 m off Montevideo. This species was cited previously in the North Atlantic at similar depth range (Lemaitre 1989). The Pylochelidae was represented by an unidentified glaucothoe (Fig. 19.4c). The specimen characteristics were shield two-thirds to three-fifths total carapace length, equal and paired pereiopods in somites 2 to 5, abdominal tergites well separated and calcified, distinct from the telson, subquadrangular telson broader than longer, and uropods biramous, with exopods longer than endopods. These intermediated characters are typical of glaucothoe stages in anomurans as reported in other species (Reese and Kinzie III 1968; Roberts 1973; Stuck and Truesdale 1986; McLaughlin et al. 1992; McLaughlin and Gore 1992).

The infraorder Brachyura was represented by nine species belonging to nine different families. *Ethusina abyssicola* belonging to the Ethusidae was the deepest true crab captured and previously cited in a similar depth in Mar del Plata Canyon (Argentina) and South Brazil but not in Uruguay (Ocampo et al. 2014; Spivak et al. 2019). Several individuals of *Chaceon notialis* (Fig. 19.4d) belonging to the Geryonidae were collected between 250 and 1660 m in the present study. This important commercial species has been previously reported in Uruguay down to 1200 m (Boschi et al. 1992; Spivak et al. 2019). The other species were all collected at 250 m depth, and their bathymetric distribution has not been enlarged in this study. *Peltarion spinulosum* (Trichopeltariidae), *Myropsis quinquespinosa* (Leucosiidae), *Latreillia williamsi* (Majidae), and *Tetraxanthus rathbunae* (Xanthidae) have been previously cited in Uruguay, while *Libidoclaea granaria* (Epialtidae) and *Spinolambrus pourtalesii* (Parthenopidae) were cited in SWAO and *Palicus gracilipes* (Palicidae) in the Caribbean Sea (Melo 1996; Felder et al. 2009; de Melo 2010; Spivak et al. 2019).

To summarize, in the present study, 79 different species of decapods were collected from the continental slope of Uruguay between 250 and 3800 m, and from those only 56 were previously reported (Spivak et al. 2019). Our results indicated that the waters off the Uruguayan deep-sea coast are well represented by decapod crustaceans when comparing with other studied deep-sea areas in the Atlantic. In the Madeira Archipelago, 186 species were reported, but 126 species were found in shallow waters (< 50 m) and 81 between 200 and 500 m (Rosa et al. 2012). In the continental slope off Guinea-Bissau, 122 species were reported, but 47 species were found at <50 m and 41 between 200 and 500 m (Muñoz et al. 2012). Moreover, 67% of the decapod species collected in this study were not previously reported off Uruguay coast and 47% in the SWAO. In relation to the bathymetric range, it was enlarged for 32% of the identified decapods. In addition, the frequency distribution of species occurrence was considered as "rare" since 56% of the species were only sampled once and often only one or few specimens were caught. This rarity was higher when compared to other areas of the world where few studies has been conducted as the deep-sea communities of French Polynesia (Delavenne et al. 2019), clearly indicating that the decapod community of this area has been poorly studied and the presence of other non-cited species should be expected (Spivak et al. 2019).

These decapod specimens correspond to an abundance of 214,324 individuals km^{-2} and a biomass of 18,599,146 g km^{-2} (Table 19.2). Caridean shrimp (36%), brachyuran crabs (32%), and Dendrobranchiata shrimps (19%) were the groups presenting the highest abundance (Fig. 19.5). However, at family level, the most abundant was Geryonidae (27%), Nematocarcinidae (21%), and Sergestidae (16%). Only 1 species, *C. notialis* (Fig. 19.4d), represented the Geryonidae, while Nematocarcinidae presented 5 species and Sergestidae 11 species. The high abundance and individual weight of *C. notialis* induced that brachyuran crabs represented 97% of the biomass and Geryonidae 96%. Without taking in account, the Brachyura, Caridea, and Dendrobranchiata shrimps were the most representative groups in biomass, and at family level, Geryonidae was followed by Nematocarcinidae, Polychelidae, and Sergestidae. The five species of Polychelidae

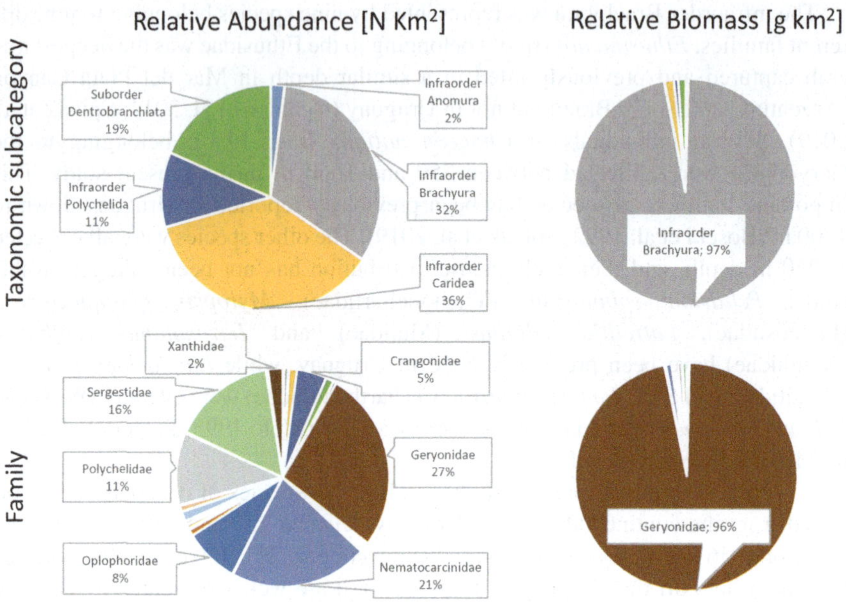

Fig. 19.5 Abundance and biomass of decapods collected in the deep continental slope of Uruguay by taxonomic subcategory and family. For complete list of taxonomic groups, see Table 19.3

were representative in biomass by their individual weight since they were only 23 individuals versus 388 individual of Nematocarcinidae and the 105 of Sergestidae. Regarding depth strata (Fig. 19.6), the shallowest (250 > 1100 m) was dominated by Brachyurans either in abundance and biomass due to the high number of *C. notialis*. In fact, in this strata were 85% of the individuals of this species. The abundance of the other deep benthic strata was dominated by caridean shrimps, and this abundance was decreasing with depth. The abundance in the second depth benthic strata (1100 > 2000 m) presented 73% of caridean shrimps and 25% of Polychelida lobsters. In terms of biomass, these depth benthic strata did not present a clear representation of any group. Regarding pelagic decapods, most of them corresponded to the Dendrobranchiata, representing the 91% in abundance and 93% in biomass. Two specimens of *Thymops birsteini* presenting a weight of 400 g were the only representatives of the Nephropidae. Brachyuran crabs also accounted for nearly the 90% of the biomass in the shallowest stratum of the coastal shelf of Guinea-Bissau (Muñoz et al. 2012), although our stratum was much deeper. In this former study and other from the North Atlantic waters (Fariña et al. 1997; Cartes et al. 2007), the increase of decapod biomass with increasing depth has been described, though in the present study the highest biomass was observed in the shallower strata and no clear pattern was shown in the deeper strata. Regarding the geographical distribution (Fig. 19.7), the highest abundance was observed in the transect located in the south maybe due to the highest nutrient concentration offered by the Río de la Plata.

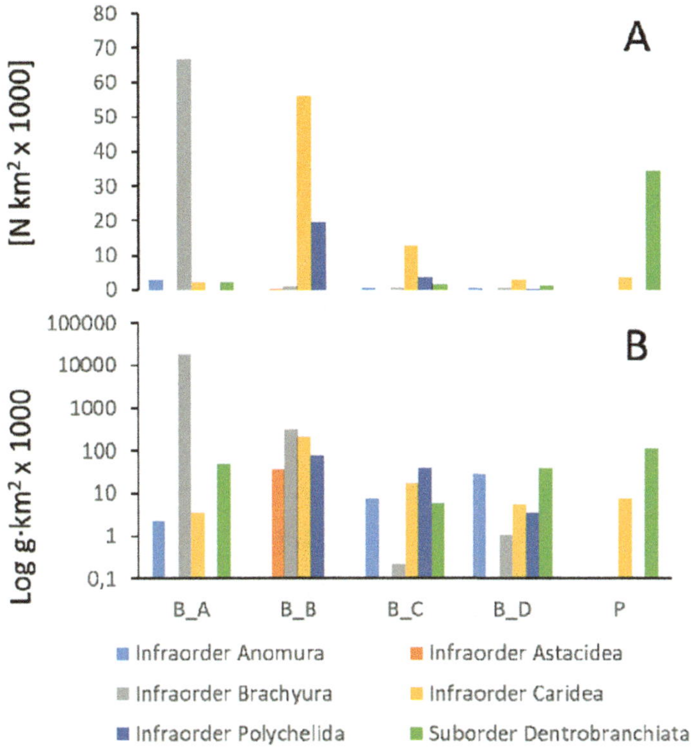

Fig. 19.6 Abundance and biomass of decapods collected in the deep continental slope of Uruguay by habitat (pelagic vs benthic, depth for benthic decapods) in each taxonomic subcategory. Abbreviations: B-A = benthic at depth range A = 250 > 1100 m, B-B = benthic at depth range B = 1100 > 2000 m, B-C = benthic at depth range C = 2000 > 3000 m, B-D = benthic at depth range D = 3000–3800 m, P = pelagic

However, abundance was higher in the central transect, but those high values were also due to the presence of *C. notialis*.

In the present study, species richness (Table 19.4) in shrimps was the highest ($S = 26$ in Dendrobranchiata and $S = 25$ in Caridea), followed by Anomura (S = 14). However, their diversity indices, Shannon-Wiener diversity index (H′) and Pielou (J), were higher in Anomura ($H' = 1.95$ vs 1.83 and 1.66; $J = 0.74$ vs 0.56 and 0.52), maybe because less single occurrence species was shown for anomurans. Regarding diversity by families, species richness was higher in Sergestidae ($S = 12$) followed by Benthesicymidae ($S = 7$) and Munidopsidae ($S = 6$). Also for category level, the diversity indices were higher for Munidopsidae ($H' = 1.61$ vs 1.26 and 1.56; $J = 0.90$ vs 0.51 and 0.80). Studying the diversity by habitat, the highest richness was observed in pelagic shrimps ($S = 23$) closely followed by our shallowest strata up to 1000 m depth ($S = 22$). The diversity indices for pelagic species ($H' = 1.68$ and $J' = 0.54$) was intermediate among benthic depth strata and in benthic strata diversity increase by depth ($H' = 1.14$ to 2.17 and $J = 0.37$ to 0.75). Regarding the

Fig. 19.7 Abundance and biomass of decapods collected in the deep continental slope of Uruguay by area (Ce = central, N = north, S = south) in each taxonomic subcategory

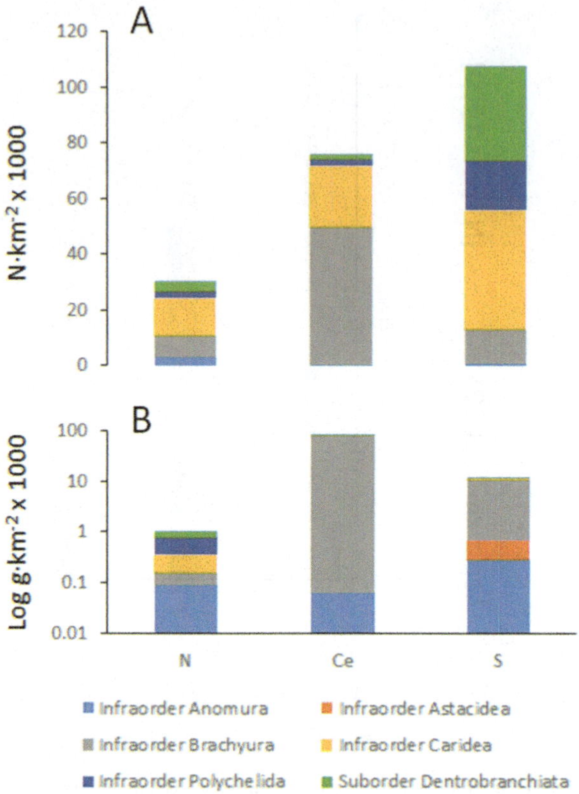

geographical transects, the species richness was nearly twofold in the south ($S = 47$) compared to central ($S = 26$) and north ($S = 29$) transects, as we could expect by the release of nutrients from the Río de la Plata. However, the diversity indices were higher in the north close to the Brazilian border ($H' = 2.72$ vs 1.07 and 2.41; $J = 0.81$ vs 0.33 and 0.63). When comparing with data from Guinea-Bissau (Muñoz et al. 2012), importance of families in terms of species richness was different, being the most diverse Pandalidae ($S = 14$), followed by Oplophoridae ($S = 9$), Portunidae ($S = 8$), Pasiphaeiade ($S = 7$), Inachidae ($S = 6$), and Scyllaridae ($S = 5$). Average values of species richness and diversity were higher at their deepest stratum (500–1000 m depth) being the one with the highest number of decapod species ($S = 59$) and the greatest diversity ($H' = 2.1$). This stratum corresponded to our shallower strata where species richness and diversity indices were half of the ones obtained in Guinea-Bissau. In our study, Dendrobranchiata represented 32% of all decapod diversity observed versus the 12% recorded in the Madeira Archipelago (Rosa et al. 2012). In benthic species, they reported a significant exponential decline with increasing depth and comprised seven infraorders: Caridea, Stenopodidea, Polychelida, Achelata, Astacidea, Anomura, and Brachyura; that for our data from Uruguayan waters was only evident for Caridea and Anomura in the three deeper

Table 19.4 Diversity indices (see text) for deep-sea crustaceans in the EEZ of Uruguay by subcategory, family, habitat and depth range, and area

Subcategory	H′	J	S
Dendrobranchiata	1.83	0.56	26.00
Caridea	1.66	0.52	25.00
Astacidea	0.00	NA	1.00
Polychelida	0.81	0.50	5.00
Anomura	1.95	0.74	14.00
Brachyura	0.74	0.34	9.00
Family			
Aristeidae	0.94	0.68	4.00
Benthesicymidae	1.56	0.80	7.00
Penaeidae	0.38	0.54	2.00
Sergestidae	1.26	0.51	12.00
Pasiphaeidae	0.95	0.59	5.00
Hippolytidae	0.00	NA	1.00
Acanthephyridae	0.68	0.97	2.00
Oplophoridae	0.06	0.08	2.00
Nematocarcinidae	0.56	0.35	5.00
Alpheidae	0.00	NA	1.00
Lysmatidae	0.00	NA	1.00
Pandalidae	0.89	0.81	3.00
Crangonidae	0.60	0.55	3.00
Glyphocrangonidae	0.69	0.99	2.00
Nephropidae	0.00	NA	1.00
Polychelidae	0.81	0.50	5.00
Munidopsidae	1.61	0.90	6.00
Munididae	0.90	0.82	3.00
Porcellanidae	0.00	NA	1.00
Lithodidae	0.00	NA	1.00
Paguridae	0.00	NA	1.00
Parapaguridae	0.00	NA	1.00
Pylochelidae	0.00	NA	1.00
Trichopeltariidae	0.00	NA	1.00
Ethusinae	0.00	NA	1.00
Leucosiidae	0.00	NA	1.00
Epialtidae	0.00	NA	1.00
Majidae	0.00	NA	1.00
Palicidae	0.00	NA	1.00
Parthenopidae	0.00	NA	1.00
Geryonidae	0.00	NA	1.00
Xanthidae	0.00	NA	1.00

(continued)

Table 19.4 (continued)

Subcategory	H′	J	S
H_DR			
B_A	1.14	0.37	22.00
B_B	1.25	0.54	10.00
B_C	2.03	0.72	17.00
B_D	2.17	0.75	18.00
P	1.68	0.54	23.00
Area			
N	2.72	0.81	29
Ce	1.07	0.33	26
S	2.41	0.63	47

Habitat and depth (H_DR): *H* habitat (*B* benthic, *P* pelagic), *DR* depth range (A, 250 > 1100 m; B, 1100 > 2000 m; C, 2000 > 3000 m; and D, 3000–3800 m). Area: *Ce* central, *N* north, *S* south

strata. They also showed that richness was maximum between 1000 and 2000 m and then declined sharply towards lower bathyal (20 species, from a total of 175, between 2500 and 3000 m depth) and more steadily towards abyssal depths (10 species between 4500 and 5000 m depth). In this study, there was a decrease in the second stratum (10 species between 1100 and < 2000 m) while then increased in deepest strata (17 species between 2000 and < 3000 m and 18 species between 3000 and 3800 m). This tendency of increasing decapod diversity with depth, reaching maxima values between 1000 and 2000 m, was also described for other areas in the northeastern Atlantic (Fariña et al. 1997; Cartes et al. 2007) but was not observed in the present study in the SWAO.

19.4 Conclusions

In the continental slope of Uruguay, we collected 79 different species of decapods from 250 to 3800 m depth, and 56% of the species were only sampled once and often with only 1 or few specimens. Among the 51 decapod species described previously below 200 m depth (Spivak et al. 2019), only 16 species were collected by us and other 5 species previously cited by these authors in the SWAO, and their distribution was expanded to Uruguay plus other 3 that were reported in shallowest waters. These results might indicate that the data off Uruguayan waters are still scarce and therefore it remains difficult to study diversity patterns, in terms of either occurrence, abundance, rarity, species richness, or functional diversity. For maintenance of the biodiversity in the deep waters off Uruguayan coast and before fishing or mining exploitation is developed, we recommend the implementation of long-term biological and environmental monitoring programs that could enlarge the knowledge in this poor studied area and detect possible changes in the structure, biomass, and diversity of benthic assemblages.

Acknowledgments We would like to thank the crew of the F/V Marianne and to Alejandro V. Ariza, Marc Baeta, Claudio Barría, Ernesto Chiesa, Anabel Colmenero, Susana Díez, Alejandro Escánez, Josep Maria Gili, Valentina Leoni, Pablo José López González, Ariadna Mecho, Carmen Morant, Javier Sellanes, Anna Soler, Carlo Turi, and Rebeca Zapata for their participation in the CUCAGUAY cruise and/or the classification of other megafauna than crustaceans. We are also thankful to Anastasia A. Lunina and Alexander L. Vereshchaka from the Institute of Oceanology (Russian Academy of Sciences) and Enrique MacPherson from Center of Advanced Studies of Blanes (CSIC) for their good disposition and their advices in the revision of several decapod species identification. Moreover, we are grateful to Laure Corbari for her reception in the Muséum national d'Histoire naturelle, Ulla Fernández-Arcaya from the Instituto Español de Oceanografía for their support during the preparation of the sea mission and the data pre-treatment, and Ignacio Castaño from the library of the Marine Sciences Institute (CSIC) for his support with the bibliographic search. Funding for cruising was supported by "Administración Nacional de Combustibles, Alcohol y Portland" (ANCAP) center in Uruguay.

References

Ahyong S (2009) The polychelidan lobsters: phylogeny and systematics (Polychelida: Polychelidae). Decapod Crustacean Phylogenetics 18:369–396

Ahyong ST, Baba K, Macpherson E, Poore GC (2010) A new classification of the Galatheoidea (Crustacea: Decapoda: Anomura). Zootaxa 2676(1):57–68

Alves-Júnior FDA, Câmara De Araújo MDSL, Souza-Filho JF (2017) New records of deep-sea shrimps of family Solenoceridae Wood-Mason & Alcock, 1891 (Crustacea: Decapoda: Dendrobranchiata) from Southwestern Atlantic. Zootaxa 4254(4):473–484

Alves-Júnior FDA, Viana GFS, De AraújoCâmara MDSL, Souza-Filho JF (2018) New records of the family Crangonidae (Decapoda: Caridea) from Southwestern Atlantic. Zootaxa 4369(1):046–062

Baba K (2005) Deep-sea chirostylid and galatheid crustaceans (Decapoda: Anomura) from the Indo-Pacific, with a list of species. Galathea Rep 20:1–317

Bate CS (1881) On the Penaeidae. Ann Mag Nat Hist 5(8):169–196

Bate CS (1888) Report on the Crustacea Macrura collected by the challenger during the years 1873–76. Report on the scientific results of the voyage of HMS challenger during the years 1873–76. Zoology 24(52):1–942

Benavides-Serrato M, Borrero-Pérez G, Díaz-Sánchez CM (2011) Equinodermos del Caribe colombiano I: Crinoidea, Asteroidea y Ophiuroidea. Serie de Publicaciones Especiales de Invemar 22:1–384

Boschi E (1973) Aportes al conocimiento de tres especies de camarones de aguas profundas del Atlántico Sur (Crustacea Caridea). [Contribution to the knowledge of three deep sea South Atlantic species of shrimps (Crustacea, Caridea)]. Physis 32(85):233–241

Boschi EE, Fischbach CE, Iorio MI (1992) Catálogo ilustrado de los crustáceos estomatópodos y decápodos marinos de Argentina. Comisión Técnica Mixta del Frente Marítimo, Montevideo

Bouvier EL (1922) Observations complémentaires sur les Crustacés décapodes (Abstraction faite des Carides) provenant des Campagnes de S.A.S. le Prince de Monaco. Résultats des Campagnes Scientifiques Accomplies sur son Yacht par Albert Ier, Prince Souverain de Monaco 62:1–106

Buhl-Mortensen L, Vanreusel A, Gooday AJ, Levin LA, Priede IG, Buhl-Mortensen P, Gheerardyn H, King NJ, Raes M (2010) Biological structures as a source of habitat heterogeneity and biodiversity on the deep ocean margins. Mar Ecol 31(1):21–50

Burukovsky R (1996) Shrimps of the genus *Pasiphaea*: systematics and some remarks on new findings (Decapoda, Caridea). Zoologiske Zhurnal 75(6):841–847. [In Russian with English abstract]

Burukovsky R (2012) Deep sea shrimps of the family Nematocarcinidae (history of study, systematic, distribution, and biology). Prospekt Nauki, St Petersburg

Calman WT (1925) On macrurous decapod Crustacea collected in South African waters by the S.S. "Pickle". Reports of the Fisheries and Marine Biological Surveys of South Africa 4:1–26

Cardoso I (2012) The Southwestern Atlantic Crangonidae (Caridea, Decapoda) including first records of two species. Cah Biol Mar 54:85–92

Cardoso I (2013) Sobre algunos Oplophoridae (Caridea, Decapoda) raros de la Cordillera Meso-Atlántica Sur. Lat Am J Aquat Res 41(2):209–216

Cardoso IA, Burukovsky RN (2014) *Nematocarcinus* Milne Edwards, 1881 (Crustacea, Decapoda) from southwestern Atlantic, including the southern mid-Atlantic ridge area. Zootaxa 3887(3):437–458

Carranza A (2010) Informe campaña Uruguay 0110. B/O Miguel Oliver (SGM). Tragsatec, Madrid

Cartes JE, Huguet C, Parra S, Sanchez F (2007) Trophic relationships in deep-water decapods of Le Danois bank (Cantabrian Sea, NE Atlantic): trends related with depth and seasonal changes in food quality and availability. Deep-Sea Res Part I Oceanogr Res Pap 54(7):1091–1110

Castro P (2000a) Crustacea Decapoda: a revision of the indo-West Pacific species of palicid crabs (Brachyura Palicidae). Résultats des campagnes MUSORSTOM 21:437–610

Castro P (2000b) Crustacea Decapoda: a revision of the Indo-west Pacific species of palicid crabs (Brachyura Palicidae). In: Crosnier A (ed) Résultats des campagnes MUSORSTOM, Mém Mus natl Hist nat 184, vol 21, pp 437–610

Castro P, Williams AB, Cooper LL (2003) Revision of the family Latreilliidae Stimpson, 1858 (Crustacea, Decapoda, Brachyura). Zoosystema, Paris 25(4):601–634

Cervigón F, Cipriani R, Garibaldi L, Hendrickx ME, Lemus AJ, Márquez R, Poutiers JM, Robaina G, Rodríguez B (1992) Guía de campo de las especies comerciales marinas y de aguas salobres de la costa septentrional de Sur América. Fichas F.A.O. de Identificación de especies para los fines de pesca. F.A.O, Roma

Chace FA Jr (1939) Reports on the scientific results of the first Atlantis expedition to the West Indies, etc.. Preliminary descriptions of one new genus and seventeen new species of decapod and stomatopod crustacea. Mem Soc Cubana Hist Nat 13(1):31–54

Chace FA Jr (1942) Reports on the scientific results of the Atlantis expeditions to the West Indies, under the joint auspices of the University of Havana and Harvard University. The Anomuran Crustacea I Galatheidea Torreia 11:1–106

Chace FA Jr (1986) The caridean shrimps (Crustacea: Decapoda) of the "Albatross" Philippine expedition 1907–1910, part 4: Oplophoridae and Nematocarcinidae. Smithsonian Contributions to Zoology 432:1–82

Chang S-C, Ahyong ST, Chan T-Y (2013) New records of deep-sea blind lobsters (Crustacea: Decapoda: Polychelidae) from Taiwan. J Mar Sci Technol 21:8–14

Christoffersen ML (1979) Decapod Crustacea: Alpheoidea. Campagnes de la Calypso au large des côtes atlantiques de l'Amérique du Sud (1961–1962) I Résultats Scientifiques de la Campagne de la Calypso 11:297–377

Crosnier A (1987) *Systellaspis intermedia* (Crustacea Decapoda Oplophoridae), espèce nouvelle de l'Indo-Pacifique. Bul Mus natl Hist nat (4) section A, Zoologie, Biologie et Écologie animales 9:947–959

Crosnier A, Forest J (1968) Note préliminaire sur les carides recueillis par l' "Ombango" au large du plateau continental, du Gabon à l'Angola (Crustacea Decapoda Natantia). Bull Mus Natl Hist Nat 39(6):1123–1147

Crosnier A, Forest J (1973) Les crevettes profondes de l'Atlantique oriental tropical. Cah ORSTOM 19:1–409

Cruz N, Fransen CHJM (2004) Addition of three species of the genus *Plesionika* (Crustacea: Caridea: Pandalidae) to the known Atlantic marine fauna of Colombia. Zool Meded 78:131–146

Dardeau MR, Heard R Jr (1983) Crangonid shrimps (Crustacea: Caridea), with a description of a new species of *Pontocaris*. Mem Hourglass Cruises VI(II):1–39

de Melo GAS (2010) Os Brachyura (Crustacea: decapoda) coletados pelo projeto GEDIP entre Torres, Rio Grande do Sul (Brasil) e Maldonado (Uruguai). Atlântica (Rio Grande) 32(1):39–58

Delavenne J, Keszler L, Castelin M, Lozouet P, Maestrati P, Samadi S (2019) Deep-sea benthic communities in the largest oceanic desert are structured by the presence of polymetallic crust. Sci Rep 9(1):6977

Diez F, Tapella F, Romero MC, Madirolas A, Lovrich GA (2016) La langostilla *Munida gregaria* en el Mar Argentino: biología e interés económico. In: Boschi E (ed) El Mar Argentino y sus recursos pesqueros, 6, Los crustáceos de interés pesquero y otras especies relevantes en los ecosistemas marinos, pp 213–228

Dore I, Frimodt C (1987) *Penaeus latisulcatus*. In: An illustrated guide to shrimp of the world. Van Nostrand Reinhold, New York, p 229

Farias NE, Ocampo EH, Luppi TA (2015) On the presence of the deep-sea blind lobster *Stereomastis suhmi* (Decapoda: Polychelidae) in Southwestern Atlantic waters and its circum-Antarctic distribution. New Zeal J Zool 42(2):119–125

Fariña A, Freire J, González-Gurriarán E (1997) Megabenthic decapod crustacean assemblages on the Galician continental shelf and upper slope (north-west Spain). Mar Biol 127(3):419–434

Felder DL, Álvarez F, Goy JW, Lemaitre R (2009) Decapoda (Crustacea) of the Gulf of Mexico, with comments on the Amphionidacea. In: Felder DLaDKC (ed) Gulf of Mexico origin, waters, and biota, vol 1. Texas A&M University Press, Texas, pp 1019–1104

Fontaine B, Perrard A, Bouchet P (2012) 21 years of shelf life between discovery and description of new species. Curr Biol 22(22):R943–R944

Forest J (1987) Les Pylochelidae ou "Pagures symétriques"(Crustacea Coenobitoidea). Résultats des Campagnes MUSORSTOM, Volume 3. Mém Mus Natl Hist Nat Zool 137:1–254

Forest J, McLaughlin PA (2000) The marine fauna of New Zealand: Paguridea (Decapoda: Anomura) exclusive of the Lithodidae. NIWA Biod Mem 114:1–250

Galil BS (2000) Crustacea Decapoda: review of the genera and species of the family Polychelidae Wood-Mason, 1874. In: Crosnier A (ed) Résultats des Campagnes MUSORSTOM, Mém Mus natl Hist nat 21 (184), vol 21, pp 285–387

Gore RH (1984) Abyssal lobsters, genus *Willemoesia* (Palinura, Polychelidae), from the Venezuela Basin. Caribbean Sea Proc Acad Nat Sci Phila:1–11

Guzmán G, Quiroga E (2005) New records of shrimps (Decapoda: Caridea and Dendrobranchiata) in deep waters off Chile. Gayana 69(2):285–290

Haig J (1966) Campagne de la Calypso au large des côtes atlantiques de l'Amérique du Sud (1961–1962). 1. Porcellanid crabs (Crustacea Anomura). Ann Inst Océanogr Monaco 44:351–358

Holthuis LB (1971) Biological results of the University of Miami deep-sea expeditions. 75. The Atlantic shrimps of the deep-sea genus *Glyphocrangon* A. Milne Edwards, 1881. Bull Mar Sci 21(1):267–373

Holthuis LB (1974) Biological results of university of miami deep-sea expeditions.106. Lobsters of superfamily Nephropidea of Atlantic ocean (Crustacea, Decapoda). Bull Mar Sci 24(4):723–884

Holthuis LB (1993) The recent genera of the Caridean and Stenopodidean shrimps (Crustacea, Decapoda): with an appendix on the order Amphionidacea. National Natuurhistorish Museum, Leiden

Kitahara MV, Capítoli RR, Horn Filho NO (2009) Distribuição das espécies de corais azooxantelados na plataforma e talude continental superior do sul do Brasil. Iheringia (Zoology Series) 99:223–236

Komai T, Amaoka K (1993) A new species of the genus *Pasiphaea* (Crustacea, Decapoda, Pasiphaeidae) from the North Pacific. Zool Sci 10:367–367

Komai T, Chan T-Y (2012) A new species of the bathypelagic shrimp genus *Pasiphaea* Savigny, 1816 (Crustacea: Decapoda: Caridea) from off Hawaii, Central Pacific. Bull Mar Sci 88(1):105–112

Laptikhovsky V, Reyes PR (2009) Distribution and reproductive biology of a subantarctic deep-sea lobster, the Patagonian lobsterette Thymops birsteini (Zarenkov and Semenov, 1972) (Decapoda, Astacidea, Nephropidae). J Nat Hist 43(1–2):35–46. https://doi.org/10.1080/00222930802567099

Lemaitre R (1989) Revision of the genus *Parapagurus* (Anomura, Paguroidea, Parapaguridae), including redescriptions of the western Atlantic species. Zool Verh (Leiden) 253(21–vii):1–106

Macpherson E (1988) Revision of the family Lithodidae Samouelle, 1819 (Crustacea, Decapoda, Anomura) in the Atlantic Ocean, Monografías de Zoología Marina, vol 2. Consejo Superior de Investigaciones Científicas (España), Barcelona

McLaughlin PA (2003) Illustrated keys to families and genera of the superfamily Paguroidea (Crustacea: Decapoda: Anomura), with diagnoses of genera of Paguridae. Mem Mus Victoria 60(1):111–144

McLaughlin PA, Gore RH (1992) Studies on the Provenzanoi and other pagurid groups.7. The zoeal and megalopal stages of *Pagurus armatus* (Dana) (Decapoda, Anomura, Paguridae), reared in the laboratory. J Crust Biol 12(3):448–463. https://doi.org/10.2307/1549039

McLaughlin PA, Lemaitre R (2009) A new classification for the Pylochelidae (Decapoda: Anomura: Paguroidea) and descriptions of new taxa. Raffles Bull Zool

McLaughlin P, Crain J, Gore R (1992) Studies on the provenzanoi and other pagurid groups: VI. Larval and early juvenile stages of *Pagurus ochotensis* Brandt (Decapoda; Anomura; Paguridae) from a northeastern Pacific population, reared under laboratory conditions. J Nat Hist 26(3):507–531

Melo GAS (1996) Manual de identificação dos Brachyura (caranguejos e siris) do litoral brasileiro. Plêiade, São Paulo

Melo GAS (1999) Manual de identificação dos Crustacea Decapoda do litoral brasileiro: Anomura; Thalassinidea; Palinuridea e Astacidea. Plêiade, São Paulo

Milne-Edwards A (1880) Reports on the results of dredging, under the supervision of Alexander Agassiz, in the Gulf of Mexico and in the Caribbean Sea, 1877,'78,'79, by the US Coast Survey steamer "Blake", Lieut.-Commander CD Sigsbee, USN, and Commander JR Bartlett, USN, Commanding. VIII. Etudes preliminaires sur les Crustaces. Bull Mus Compar Zool, Harvard 8(1):1–68

Muñoz I, García-Isarch E, Sobrino I, Burgos C, Funny R, González-Porto M (2012) Distribution, abundance and assemblages of decapod crustaceans in waters off Guinea-Bissau (north-west Africa). J Mar Biol Assoc UK 92:475–494. https://doi.org/10.1017/S0025315411001895

Nion H, Ríos C, Meneses P (2016) Peces del Uruguay. Lista sistemática y nombres comunes. Dinara, Montevideo

Nucci PR, Melo G (2012) A new species of *Catapagurus* A. Milne-Edwards (Crustacea: Decapoda: Anomura: Paguridae) from Brazil, with a key to the western Atlantic species. Zootaxa 3499(1):81–85

Ocampo EH, Farías NE, Luppi TA (2014) New record of the deep-sea crab *Ethusina abyssicola* from the Mar del Plata Canyon, Argentina. New Zeal J Zool 41(3):218–221

Oksanen J, Blanchet FG, Friendly M, Kindt R, Legendre P, McGlinn D, Minchin PR, O'Hara R, Simpson GL, Solymos P, Stevens MHH, Szoecs E, Wagner H (2019) Vegan: community ecology package. R package version 2.5–6. http://www.cranr-project.org. Acesso em:296

Pequegnat LH (1970) Deep-sea anomurans of superfamily Galatheoidea with descriptions of three new species. Texas A&M University Oceanographic Studies, Contrib Biol Gulf of Mexico 1:125–170

Pequegnat W, Pequegnat L (1971) New species and new records of Munidopsis (Decapoda: Galatheidae) from the Gulf of Mexico and Caribbean Sea. Texas A&M University Oceanographic Studies. Contrib Biol Gulf of Mexico:1–24

Pérez Farfante I, Kensley BF (1997) Penaeoid and sergestoid shrimps and prawns of the world. Keys and diagnosis for the families and genera. Mém Mus Natl Hist Nat 175:1–233

Poore G, McCallum A, Taylor J (2008) Decapod Crustacea of the continental margin of southwestern and Central Western Australia: preliminary identifications of 524 species from FRV southern surveyor voyage SS10-2005. Mus Victoria Sci Rep 11:1–106

Rathbun MJ (1937) The oxystomatous and allied crabs of America. Bull US Natl Mus 166:1–272

Reese ES, Kinzie RA III (1968) The larval development of the coconut or robber crab *Birgus latro* (L.) in the laboratory (Anomura, Paguridea). Crustaceana Supplement:117–144

Roberts PE (1973) Larvae of *Munida subrugosa* (White), 1847, from Perseverance Harbour, Campbell Island. J Roy Soc New Zeal 3(3):393–408

Rodrigues T, Cardoso I (2019) First record of four *Pasiphaea* Savigny, 1816 (Crustacea, Decapoda, Pasiphaeidae) species from the southwestern Atlantic. Zootaxa 4560(2):321–330. https://doi.org/10.11646/zootaxa.4560.2.5

Rosa R, Boavida-Portugal J, Trübenbach K, Baptista M, Araújo R, Calado R (2012) Descending into the abyss: Bathymetric patterns of diversity in decapod crustaceans shift with taxonomic level and life strategies. Deep Sea Res. Part I Oceanogr Res Pap 64:9–21

Scarabino F (2003a) Lista sistemática de los Aplacophora, Polyplacophora y Scaphopoda de Uruguay. Comunicaciones de la Sociedad Malacológica del Uruguay 8(79):191–196

Scarabino F (2003b) Lista sistemática de los Bivalvia marinos y estuarinos vivientes de Uruguay. Comunicaciones de la Sociedad Malacológica del Uruguay 8(81):227–258

Scarabino F (2003c) Lista sistemática de los Cephalopoda vivientes de Uruguay. Comun Soc Malacol Urug 8(79):197–202

Scarabino F (2004) Lista sistemática de los Gastropoda marinos y estuarinos vivientes de Uruguay. Comun Soc Malacol Urug 8(84):85–86

Scarabino F (2006) Faunística y taxonomía de invertebrados bentónicos marinos y estuarinos de la costa uruguaya. In: Menafra R, Rodríguez-Gallego L, Scarabino F, Conde D (eds) Bases para la conservación y el manejo de la costa uruguaya. Vida Silveste Uruguay, Montevideo, pp 113–142

Scarabino F, Lucena RA, Munilla T, Soler-Membrives A, Ortega L, Schwindt E, Lopez G, Orensanz JM, Christoffersen ML (2019) Pycnogonida (Arthropoda) from Uruguayan waters (Southwest Atlantic): annotated checklist and biogeographic considerations. Zootaxa 4550(2):185–200

Serejo CS, Young PS, Cardoso IC, Tavares C, Rodrigues C, Almeida TC (2007) Abundance, diversity and zonation of crustaceans on the central Brazilian slope (11 degrees–22 degrees S) collected by the REVIZEE Program – Central region: fishery survey, vol 24. Biodiversidade Da Fauna Marinha Profunda Na Costa Central Brasileira

Sha Z, Wang Y (2015) A new deep-sea species of the genus *Systellaspis* (Decapoda, Caridea, Oplophoridae) in the western Pacific. Crustaceana 88(10–11):1181–1192

Smith SI (1882) Reports on the results of dredging under the supervisión of Alexander Agassiz, on the east coast of the United States during the summer of 1880, by the U.S. Coast Survey Steamer "Blake", Commander J.R. Bartlett, U.S.N., commanding. Bull Mus Compar Zool, Harvard 10(1):1–108

Smith SI (1884) Report on the decapod Crustacea of the albatross dredgings off the east coast of the United States in 1883, vol 10. US Government Printing Office, Washington, DC

Smith A, Kroh A (2011) The echinoid directory. WorldWide Web electronic publication. http://www.nhmacuk/research-curation/projects/echinoid-directory

Soledade GO, Almeida AO (2013) Snapping shrimps of the genus *Alpheus* Fabricius, 1798 from Brazil (Caridea: Alpheidae): updated checklist and key for identification. Nauplius 21(1):89–122

Spivak E, Farías N, Ocampo E, Lovrich G, Luppi T (2019) Annotated catalogue and bibliography of marine and estuarine shrimps, lobsters, crabs and their allies (Crustacea: Decapoda) of Argentina and Uruguay (Southwestern Atlantic Ocean). Frente Marít 26:1–164

Stebbing TRR (1914) Stalk-eyed Crustacea Malacostraca of the Scottish National Antarctic Expedition. Trans R Soc Edinburgh 50:253–307

Stuck KC, Truesdale FM (1986) Larval and early postlarval development of *Lepidopa benedicti* Schmitt, 1935 (Anomura: Albuneidae) reared in the laboratory. J Crust Biol 6(1):89–110

Tan S, Ng PK (2007) Descriptions of new genera from the subfamily Parthenopinae (Crustacea: Decapoda: Brachyura: Parthenopidae). Raffles Bull Zool 16:95–119

Tapella F, Lovrich GA (2006) Morphological differences between '*subrugosa*'and '*gregaria*'morphs of adult *Munida* (Decapoda: Anomura: Galatheidae) from the Beagle Channel, southern South America. J Mar Biol Assoc UK 86(5):1149–1155

Tavares M, Pinheiro AP (2011) A new species of *Chaceon* Manning & Holthuis, 1989, from the southwestern Atlantic, with a key to the western Atlantic species (Crustacea, Decapoda, Geryonidae). Zootaxa 3086(1):57–68

Vázquez-Bader A, Gracia A (2013) Crangonidae and Glyphocrangonidae (Decapoda; Caridea) of the Southern Gulf of Mexico. Zootaxa 3669(3):367–383

Vereshchaka AL (2000) Revision of the genus *Sergia* (Decapoda: Dendrobranchiata: Sergestidae): taxonomy and distribution. Galathea Rep 18:69–207

Vereshchaka AL (2009) Revision of the genus *Sergestes* (Decapoda: Dendrobranchiata: Sergestidae): taxonomy and distribution. Galathea Report 22:7–104

Vereshchaka AL, Lunina AA (2015) Phylogeny and taxonomy of the enigmatic genus *Petalidium* (Decapoda, Sergestidae), with biological remarks. Zool J Linnean Soc 174(3):459–472

Vereshchaka AL, Olesen J, Lunina AA (2014) Global diversity and phylogeny of pelagic shrimps of the former genera *Sergestes* and *Sergia* (Crustacea, Dendrobranchiata, Sergestidae), with definition of eight new genera. PLoS One 9(11):e112057. https://doi.org/10.1371/journal.pone.0112057

Vereshchaka AL, Lunina AA, Olesen J (2016) Phylogeny and classification of the shrimp genera *Acetes*, *Peisos*, and *Sicyonella* (Sergestidae: Crustacea: Decapoda). Zool J Linnean Soc 177(2):353–377

Vereshchaka AL, Corbari L, Kulagin DN, Lunina AA, Olesen J (2019) A phylogeny-based revision of the shrimp genera *Altelatipes*, *Benthonectes* and *Benthesicymus* (Crustacea: Decapoda: Benthesicymidae). Zool J Linnean Soc 20:1–21

Werding B, Hiller A, Lemaitre R (2003) Geographic and depth distributional patterns of western Atlantic Porcellanidae (Crustacea: Decapoda: Anomura), with an updated list of species. Mem Mus Victoria 60(1):79–85

WoRMS Editorial Board (2020) www.marinespecies.org. Accessed Jan–Apr 2020

Zariquiey Alvarez R (1968) Crustáceos Decápodos Ibéricos. Investigación Pesquera 32:1–510

Zeldis JR (1985) Ecology of *Munida gregaria* (Decapoda, Anomura): Distribution and abundance, population dynamics and fisheries. Marine ecology progress series Oldendorf 22(1):77–99

Zolessi L, Philippi M (1995) Lista sistemática de decápoda del Uruguay (Arthropoda: Crustacea). Com Zool Mus Hist Nat Montevideo 12(183):1–23

Chapter 20
Deep-Sea Megacrustacean Biodiversity (Crustacea, Decapoda) in the South Gulf of Mexico

A. R. Vázquez-Bader and A. Gracia

Abstract One hundred sixty-eight crustacean species were identified in a study of the deep-sea benthic megafauna conducted along the upper continental slope (300–1200 m depth) of the Mexican Gulf of Mexico. A total of 18 cruises were conducted on board the R/V *Justo Sierra* of the Universidad Nacional Autónoma de México from off Tamaulipas to Yucatán. Samples were obtained with a commercial shrimp trawl net (18 m mouth aperture, 4.5 cm stretched mesh, 1.5 cm stretched mesh cod-end). The Decapoda were the dominant taxa and comprised 46 families, 94 genera, and 162 species. Three species of Lophogastrida and two genera with one species each of Stomatopoda were also encountered. The records of the only genus and species of Isopoda, *Bathynomus giganteus*, extended its geographical range. Comparing families, Munidopsidae was the most speciose (20), followed by the Pandalidae (11), Acanthephyridae (9), Munididae (9), Oplophoridae (7), and Paguridae (7). The rest of the families were only presented a low number of species ($\leq$ 5). The south-southwest sector of the Gulf of Mexico exhibited the highest richness and abundance. For several species extensions of their bathymetrical and geographical ranges were recorded. The faunal composition in the southern Gulf of Mexico showed differences compared to northern region.

Keywords Megacrustacean · Diversity · Distribution · Bathymetric ranges

20.1 Introduction

The Gulf of Mexico is the ninth largest body of water in the world, and it is recognized as one of 64 Large Marine Ecosystems by the US National Oceanic and Atmospheric Administration (NOAA). It is located in a subtropical region influenced by the Atlantic Ocean through the Yucatan Channel and the Straits of Florida

A. R. Vázquez-Bader (✉) · A. Gracia
Instituto de Ciencias del Mar y Limnología, Laboratorio de Ecología Pesquera de Crustáceos, Universidad Nacional Autónoma de México, Ciudad de México, Mexico

M. E. Hendrickx (ed.), *Deep-Sea Pycnogonids and Crustaceans of the Americas*, https://doi.org/10.1007/978-3-030-58410-8_20

(Monreal-Gómez et al. 2004) and is one of the most productive and important bodies of water due to its economic activities and fisheries (Cato 2009; Ward and Tunnell Jr 2017). Deep waters constitute the largest part of the Gulf of Mexico as 68% correspond to continental slope (200–3000 m) and abyssal plain (> 3000 m). Also, the Gulf of Mexico abundant deep supplies of oil and gas in shallow-waters, supported for a long time the marine oil industry of Mexico and USA and nowadays deep-water and ultra-deep-water sources (≥ 1500 m water depth) are potentially important (Murawski et al. 2020).

The continental shelf of the southern part of the Gulf of Mexico (hereafter abbreviated SGMx) is narrow and terrigenous in the northwest zone. In the southwest of the GMx, the continental platform is wide with a terrigenous zone and a large carbonate area in the southeast, adjacent to the Yucatan Peninsula. Despite the complexity and variety of habitats of the SGMx, the knowledge of its marine diversity is scarce, especially in deep waters (> 200 m). This is mainly due to the difficulty of exploring deep waters owing to the financial support and the availability of taxonomic experts of different groups.

The available information about deep-water crustacean fauna in the SGMx (from 200 to 1200 m) is mainly based on records by US expeditions (e.g., R/V *Alaminos*, *Oregon*, and *Pillsbury*). Most of the information is dispersed and available in these reports and US collections, although unfortunately much information has not been published. Recently, some studies have been reported on the fauna in Mexican deep waters that substantially increased the knowledge of crustacean diversity in the Mexican part of the GMx (Vázquez-Bader and Gracia 2004; Lozano-Álvarez et al. 2007; Escobar-Briones et al. 2008; Gracia et al. 2010; Briones et al 2010; Vázquez-Bader and Gracia 2013, 2016; Gracia and Vázquez-Bader 2014; Vázquez-Bader et al. 2014; Lemaitre et al. 2014).

The present study, developed with funds of the National Autonomous University of Mexico, is unique because it is the first time that the deep megacrustaceans of the Mexican continental slope (from 200 to 1200 m) were sampled in an extensive and comprehensive way along the whole Mexican Gulf. Furthermore, these results could form the basis for specific surveys aimed at further enhancing the knowledge of megacrustacean diversity, distribution, and abundance, as well as deeper water environmental conservation and potential impact of human activities in the Gulf of Mexico.

20.2 Material and Methods

20.2.1 Sampling Methods

This study analyzed biological material collected in 20 cruises carried out along the upper continental slope off Tamaulipas to Quintana Roo, in water depths of 200–1200 m (Fig. 20.1). The oceanographic cruises, BATO (n = 1); BIOREPES

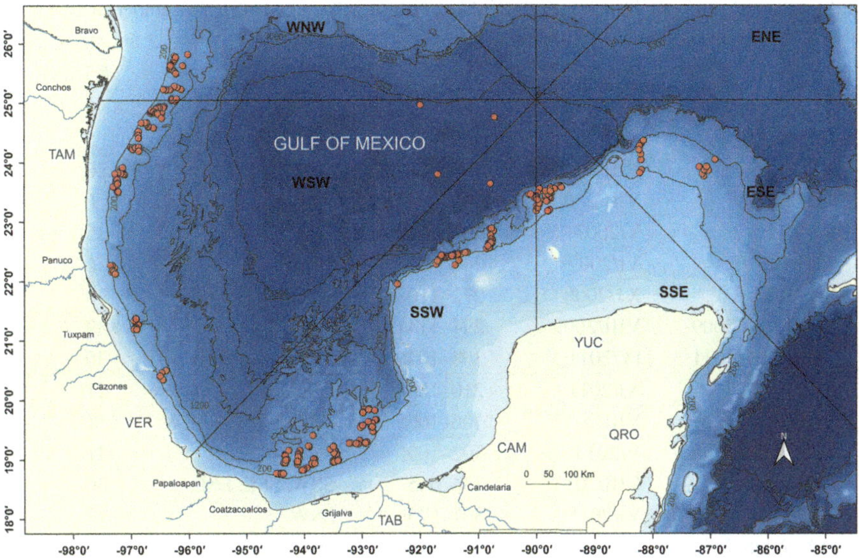

Fig. 20.1 Map of the South Gulf of Mexico showing sampling sites (circles) of surveys in the southwestern Gulf of Mexico. TAM = Tamaulipas, Ver = Veracruz, TAB = Tabasco, CAM = Campeche, YUC = Yucatán, and QRO = Quintana Roo. WNW = west-northwest, WSW = west-southwest, SSW = south-southwest, SSE = south-southeast, and ESE = east-southeast (taken from Felder and Camp 2009)

(n = 3); COBERPES (n = 9); SIGSBEE (n = 2); and SOGOM (n = 4), were conducted on board the R/V *Justo Sierra* of the Universidad Nacional Autónoma de México. DGoMb was carried out on board the R/V G*yre*. Table 20.1 shows all relevant cruise information.

Samples were obtained with a commercial shrimp trawl net (18 m mouth aperture, 4.5 cm stretched mesh, 1.5 cm stretched mesh cod-end). Before each deployment, bottom topography was explored with a Multibeam Echosounder EM 300 and sub-bottom profiler TOPAS PS 18 to find soft bottoms suitable for trawling. Trawl sampling was done for 30 min at an average speed of 2.5–3 knots; details of the procedure are described in Gracia and Vázquez-Bader (Chap. 10, this book). The trawls were carried out at 100 m interval along the upper continental shelf. Due to difficulties in finding suitable trawling sites, data of some depth strata were lacking. Although sampling was specifically directed to epibenthic fauna, pelagic and benthopelagic crustaceans were also caught during trawl net recovery. The retrieval time duration increased with depth, which could be up to 150% longer (45 min) when sampling in deeper locations (>900 m). In this way it could be considered that the crustaceans living in the water column were "indirectly" sampled; however, it is clear that the gear used is not entirely appropriate for this purpose.

Table 20.1 Oceanographic cruises (OC) data: date, depth range (m), and number of trawls by sector of southwestern Gulf of Mexico

OC	Date	Depth	Sector	# Trawls
BATO	V/1999	244–612	SSW	40
DGoMB	VIII/2002	3015–3730	WSW, SSW, ESE	5
BIOREPES	VIII/2005	201–774	SSW, SSE, ESE	20
SIGSBEE 9	VIII/2006	266–398	WSW	6
BIOREPES 2	V/2007	200–1071	SSW, SSE	35
SIGSBEE 10	VI/2007	269–350	WSW	6
BIOREPES 3	X1/2008	272–1108	WNW, WSW	33
COBERPES 2009	VIII/2009	294–1141	SSW	39
COBERPES 2011	IV/2011	308–1040	SSW, SSE, ESE	30
COBERPES 3	XI/2011	210–1068	SSW	21
COBERPES 4	VIII/2012	306–1065	WNW, WSW, SSW	31
COBERPES 5	V/2013	302–1095	SSW	31
COBERPES 6	VIII/2014	336–645	SSW, SSE, ESE	30
SOGOM 1	VI/2015	301–575	SSW	13
COBERPES 7	III–IV/2016	315–1143	SSW	11
SOGOM 2	IX/2016	310–815	SSW	13
COBERPES 8	X/2016	298–1063	SSW	34
SOGOM 3	V/2017	317–832	SSW	16
COBERPES 9	VIII/2017	296–909	WNW, WSW	48
SOGOM 4	IX/2018	294–827	SSW	15

WNW west-northwest, *WSW* west-southwest, *SSW* south-southwest, *SSE* south-southeast, and *ESE* east-southeast

20.2.2 Analysis of Biological Data

All specimens studied are kept in the Crustacean Reference Collection of the Laboratorio de Ecología Pesquera de Crustáceos (LEPC-ICML-UNAM), including the specimens donated by cruises DGoMb, SIGSBEE 9, and SIGSBEE 10 (Table 20.1). All of them were preserved on board in 70% ethanol and, where possible, were identified to species level and then counted, sexed, weighted, and labeled.

The highest taxonomy was accorded to De Grave et al. (2009) and Poore (2016) for Decapoda, Reaka et al. (2009) for Stomatopoda, and Price et al. (2009) for Lophogastrida. As a rule, the latest taxonomic or systematic treatment for each species was followed. In several collected specimens, we noticed that the taxonomic description did not fully match with known species; hence, these are possibly new species. Only species fully identified were included in the analysis.

We also recorded depth range (m) and distribution for each species in the SGMx. Particularly for distribution analysis, we used the Gulf of Mexico sector division proposed by Felder and Camp (2009): west-northwest (WNW), west-southwest (WSW), south-southeast (SSE), and east-southeast (ESE). Although the division of GMx is somewhat arbitrary and potentially does not have any biogeographic significance, it is herein followed for comparative purposes with previous work.

Species distribution and bathymetric ranges were compared with data found in the literature: for decapods (Felder et al. 2009a, b), isopods (Schotte et al. 2009), lophogastrids (Price et al. 2009), and stomatopods (Reaka et al. 2009).

Each species identification was listed with corresponding authority and number of presences in each sector. Species with distribution in the Caribbean were highlighted with **, and species in bold are known only from the Gulf of Mexico (GMx) (Table 20.2).

20.3 Results and Discussion

We recorded 168 crustacean species: Decapoda (162), Isopoda (1), Lophogastrida (3), and Stomatopoda (2) (Table 20.2). The only genus and species of Isopoda, *Bathynomus giganteus* A. Milne-Edwards, 1879, was frequent in all trawls; and its known geographical range in SGMx was extended (Table 20.2). Two taxa of Stomatopoda were recorded in the upper slope of the SGMx. *Eurysquilla chacei* Manning, 1969, constituting was the first record in the southern GMx. Also, we collected *Squilla edentata edentata* (Lunz, 1937) which is common across all sectors of the GMx.

The three species of lophogastridans (*Neognathophausia ingens* (Dohrn, 1870), *Gnathophausia gigas* (Willemoes-Suhm, 1873), and *Gnathophausia zoea* Willemoes-Suhm, 1875) were previously reported as cosmopolitan in abyssal depths.

The decapods were the dominant taxa, comprising of 46 families, 94 genera, and 162 species; almost all of them belonged to taxa typical of the upper continental slope. Among families, Munidopsidae was the most speciose (20), followed by the Pandalidae (11), Acanthephyridae (9), Munididae (9), Oplophoridae (7), and Paguridae (7). The rest of the families presented a low number of species ($\leq$ 5) (Table 20.3).

Burdett (2016) reported for the North Gulf of Mexico a high abundance and biomass for species of the families Acanthephyridae and Oplophoridae, but it is not comparable with the present study since the sampling gear employed was for pelagic trawls (MOCNESS net). The importance of these families is that the majority of their pelagic species contribute greatly to food webs of deep-sea communities.

The number of crustacean genera was similar in each sector (> 60), except for SSW (n = 75). For species, the highest number was registered in sector SSW (n = 122), followed by ESE (n = 106); the rest of the sectors presented a smaller number of species (< 100).

The differences of sector species richness are probably influenced by sea bottom topography and consequently sampling effort herein deployed. It must be reminded that our sampling gear operation was limited to soft substrate bottoms. The upper continental shelf in front of Tamaulipas and north of Veracruz (WNW) is narrow, rugged, and steep which is difficult for sampling with trawl nets and thus has fewer sampling locations, whereas the areas of Campeche Bay and Campeche Bank

Table 20.2 List of species collected in each sector of southwestern Gulf of Mexico

	Higher taxon	Species	Presence				
	Families		**WNW**	**WSW**	**SSW**	**SSE**	**ESE**
Decapoda	**Dendrobranchiata**						
	Penaeoidea						
	Aristeidae	*Aristaeomorpha foliacea* (Risso, 1827)	1	1	1	1	1
		Aristaeopsis edwardsiana (Johnson, 1868)	1	1	1	1	1
		Aristeus antillensis A. Milne-Edwards and Bouvier, 1909	1	1	1	1	1
		Hepomadus tener Smith, 1884	1	0	1	1	0
	Benthesicymidae	*Bentheogennema intermedia* (Spence Bate, 1888)	0	1	0	0	0
		Benthoecetes bartletti (Smith, 1882)	1	1	1	1	1
	Penaeidae	*Funchalia villosa* (Bouvier, 1905)	0	0	1	0	1
		Parapenaeus americanus Rathbun, 1901	1	1	1	1	1
		Parapenaeus politus (Smith, 1881)	0	1	1	1	1
		Penaeopsis serrata Bate, 1881	1	1	1	1	1
	Solenoceridae	*Hymenopenaeus debilis* Smith, 1882	1	1	1	1	1
		Pleoticus robustus (Smith, 1885)	1	1	1	1	1
		Solenocera atlantidis Burkenroad, 1939	0	0	1	0	0
		Solenocera necopina Burkenroad, 1939	0	1	1	1	1
		Solenocera vioscai Burkenroad, 1934	1	1	1	1	0
	Pleocyemata						
	Anomura						
	Chirostyloidea						

	Chirostylidae	*Gastroptychus spinifer* (A. Milne-Edwards, 1880)	0	0	1	1	1
		Uroptychus capillatus Benedict, 1902	0	0	0	0	1
		Uroptychus nitidus (A. Milne-Edwards, 1880)	1	1	1	1	1
		Uroptychus spiniger Benedict, 1902**	1	0	0	0	1
		Uroptychus spinosus (A. Milne-Edwards and Bouvier, 1894)	0	0	0	1	0
		Uroptychus uncifer (A. Milne- Edwards, 1880)	0	0	1	0	1
	Sternostylidae	*Sternostylus salvadori* (Rice & Miller, 1991) **	0	0	1	1	0
Decapoda	**Galatheoidea**						
	Munididae	*Agononida longipes* (A. Milne-Edwards, 1880)	1	1	1	1	1
		Munida constricta A. Milne-Edwards, 1880	0	1	0	1	1
		Munida evermanni Bendedict, 1901	0	1	0	1	1
		Munida forceps A. Milne-Edwards, 1880	1	1	1	1	1
		Munida iris A. Milne-Edwards, 1880	1	1	1	1	1
		Munida irrasa A. Milne-Edwards, 1880	0	0	0	1	1
		Munida microphthalma A. Milne-Edwards, 1880	0	0	1	0	0
		Munida miles A. Milne-Edwards, 1880	0	0	1	1	1
		Munida valida Smith, 1883	1	1	1	1	1
	Munidopsidae	*Galacantha spinosa* A. Milne-Edwards, 1880	1	1	1	1	1
		Munidopsis abbreviata (A. Milne-Edwards, 1880)			1		
		Munidopsis alaminos L.H. Pequegnat and W. E Pequegnat, 1970	1	1	1	1	1
		Munidopsis armata (A. Milne-Edwards, 1880)	1	0	1	0	1
		Munidopsis bermudezi Chace Jr, 1939	0	1	0	0	0
		Munidopsis bradleyi Pequegnat and Pequegnat, 1971 **	0	0	1	0	0
		Munidopsis erinacea (A. Milne-Edwards, 1880)	0	1	1	1	1
		Munidopsis expansa Benedict, 1902	0	0	0	0	1

(countinued)

Table 20.2 (continued)

	Higher taxon	Species	Presence				
		Munidopsis latifrons (A. Milne-Edwards, 1880)	0	0	0	0	1
		Munidopsis longimanus (A. Milne-Edwards, 1880)	0	1	1	0	0
		Munidopsis polita (Smith, 1883)	0	1	1	0	0
		Munidopsis ramahtaylorae Pequegant and Pequegnat, 1971	0	1	1	0	0
Decapoda	Munidopsidae	*Munidopsis riveroi* Chace Jr, 1939 **	1	1	0	0	0
		Munidopsis robusta (A. Milne-Edwards, 1880)	1	1	1	0	1
		Munidopsis serratifrons (A. Milne-Edwards, 1880)	0	0	0	0	1
		Munidopsis serricornis (Lovén, 1852)	0	0	0	0	1
		Munidopsis schulerae Lemaitre, Vázquez-Bader and Gracia, 2014	0	0	1	0	1
		Munidopsis sigsbei (A. Milne-Edwards, 1880)	1	1	1	1	1
		Munidopsis simplex (A. Milne-Edwards, 1880)	0	0	0	0	1
		Munidopsis spinoculata (A. Milne-Edwards, 1880)	0	0	1	0	0
	Porcellanidae	*Porcellana sigsbeiana* A. Milne-Edwards, 1880	0	1	0	1	0
	Lithodoidea						
	Lithodidae	*Neolithodes agassizii* (Smith, 1882)	1	0	0	0	0
	Paguroidea						
	Diogenidae	*Areopaguristes oxyophthalmus* (Holthuis, 1959)	0	0	1	0	0
		Dardanus insignis (de Saussure, 1857)	0	0	1	0	1
	Paguridae	*Anisopagurus bartletti* (A. Milne-Edwards, 1880)	1	0	1	1	1
		Pagurus bullisi Wass, 1963	0	1	1	0	0
		Pagurus curacaoensis (Benedict, 1892)	0	1	1	1	0
		Pylopagurus discoidalis (A. Milne-Edwards, 1880)	0	0	1	1	1
		Tomopaguropsis ahkinpechensis Lemaitre, Vazquez-Bader and Gracia, 2014	0	0	1	0	0
		Tomopaguropsis problematica (A. Milne-Edwards and Bouvier, 1893)	0	0	1	0	1
		Tomopagurus cokeri (Hay, 1917)	1	0	0	0	0

	Parapaguridae	*Oncopagurus gracilis* (Henderson, 1888)	0	0	1	1	1
		Paragiopagurus pilimanus (A. Milne-Edwards, 1880)	0	0	1	1	0
		Parapagurus alaminos Lemaitre, 1986	1	0	1	0	0
		Parapagurus pilosimanus Smith, 1879	1	1	1	0	1
		Sympagurus pictus Smith, 1883	1	1	1	1	1
	Pylochelidae	*Mixtopagurus paradoxus* A. Milne-Edwards, 1880	0	1	1	1	0
Decapoda	**Astacidea**						
	Nephropoidea						
	Nephropidae	*Acanthacaris caeca* (A. Milne-Edwards, 1881)	1	1	1	1	1
		Nephropsis aculeata Smith, 1881	1	1	1	1	1
		Nephropsis neglecta Holthuis, 1974	0	0	0	0	1
		Nephropsis rosea Spence Bate, 1888	1	1	1	1	1
	Brachyura						
	Calappoidea						
	Calappidae	*Acanthocarpus alexandri* Stimpson, 1871	1	1	1	1	1
	Dorippoidea						
	Ethusidae	*Ethusa microphthalma* Smith, 1881	1	1	1	1	1
	Goneplacoidea						
	Euryplacidae	*Frevillea hirsuta* (Borradaile, 1916)	0	0	1	1	1
		Trizocarcinus tacitus Chace Jr, 1940	0	0	1	1	1
	Goneplacidae	*Bathyplax typhla* A. Milne-Edwards, 1880	1	1	1	1	1
	Leucosioidea						
	Leucosiidae	*Iliacantha liodactylus* Rathbun, 1898	0	1	1	0	0
		Myropsis quinquespinosa Stimpson, 1871	0	1	1	1	1
	Majoidea						

(countinued)

Table 20.2 (continued)

	Higher taxon	Species	**Presence**				
	Epialtidae	*Minyorhyncha crassa* (A. Milne-Edwards, 1879)	1	1	1	1	1
		Rochinia hystrix (Stimpson, 1871)	0	0	0	1	0
		Rochinia tanneri (Smith, 1883)	1	0	1	1	1
		Scyramathia umbonata (Stimpson, 1871)	0	1	1	1	1
		Nibilia antilocapra (Stimpson, 1871)					1
		Stenocionops spinimanus (Rathbun, 1892)	0	1	1	1	0
Decapoda	Inachidae	*Anomalothir furcillatus* (Stimpson, 1871)	0	0	1	0	0
	Inachoididae	*Anasimus fugax* A. Milne-Edwards, 1880	0	0	1	0	0
		Anasimus latus Rathbun, 1894	0	0	0	0	1
		Collodes robustus Smith, 1883	0	1	0	0	0
		Collodes trispinosus Stimpson, 1871	0	0	0	1	0
		Pyromaia arachna Rathbun, 1924	1	1	0	1	0
		Pyromaia cuspidata Stimpson, 1871	0	0	1	1	0
	Palicoidea						
	Palicidae	*Palicus floridanus* (Rathbun, 1918)	0	0	0	0	1
		Palicus gracilis (Smith, 1883)	0	1	1	0	1
		Palicus obesus (A. Milne-Edwards, 1880)	1	1	1	1	1
	Parthenopoidea						
	Parthenopidae	*Agolambrus agonus* (Stimpson, 1871)	0	0	1	0	0
		Leiolambrus nitidus Rathbun, 1901	0	1	0	0	0
		Solenolambrus typicus Stimpson, 1871	0	0	1	1	1
	Portunoidea						
	Geryonidae	*Chaceon quinquedens* (Smith, 1879)	1	1	1	1	1
		Raymanninus schmitti (Rathbun, 1931)	1	1	1	0	0
	Polybiidae	*Bathynectes longispina* Stimpson, 1871	0	1	1	1	1

	Portunidae	*Achelous spinicarpus* Stimpson, 1871	0	1	0	1	0
		Achelous spinimanus (Latreille, 1819)	0	1	0	0	0
		Callinectes similis Williams, 1966	1	0	0	0	0
Decapoda	**Trichopeltarioidea**						
	Trichopeltariidae	*Trichopeltarion nobile* A. Milne-Edwards, 1880	1	1	1	0	0
	Xanthoidea						
	Pseudorhombilidae	*Robertsella meridionalis* Tavares and Gouvêa, 2013**	0	0	0	0	1
		Tetraxanthus rathbunae Chace Jr, 1939	0	1	1	1	1
	Xanthidae	*Batodaeus urinator* (A. Milne-Edwards, 1880)	0	0	1	0	0
	Cyclodorippoidea						
	Cyclodorippidae	*Cyclodorippe antennaria* A. Milne-Edwards, 1880	0	0	1	0	1
		Cyclodorippe bouvieri Rathbun, 1934	0	0	1	0	1
		Clythrocerus carinatus Coelho, 1973 **	0	0	0	0	1
		Clythrocerus granulatus (Rathbun, 1898)	0	0	0	0	1
	Homoloidea						
	Homolidae	*Homola minima* Guinot and Richer de Forges, 1995	0	0	0	0	1
	Homolidae	*Homola vigil* A. Milne-Edwards, 1880	0	0	0	0	1
		Homologenus rostratus (A. Milne-Edwards, 1880)	0	0	0	0	1
	Latreilliidae	*Latreillia elegans* P. Roux, 1830	0	0	0	0	1
	Homolodromioidea						
	Homolodromiidae	*Homolodromia paradoxa* A. Milne-Edwards, 1880	1	1	1	1	1
	Raninoidea						
	Raninidae	*Lysirude nitidus* (A. Milne-Edwards, 1880)	1	1	1	0	0
Decapoda	**Caridea**						
	Cangronoidea						
	Crangonidae	*Parapontocaris caribbaea* (Boone, 1927)	1	1	1	0	1
		Parapontocaris vicina (Dardeau and Heard, 1983)	1	1	1	0	0

(countinued)

Table 20.2 (continued)

	Higher taxon	Species	Presence				
	Glyphocrangonidae	*Glyphocrangon aculeata* A. Milne-Edwards, 1881	1	1	1	1	1
		Glyphocrangon alispina Chace Jr, 1939	1	1	1	1	1
		Glyphocrangon haematonotus Holthuis, 1971	0	0	1	1	1
		Glyphocrangon longleyi Schmitt, 1931	0	1	1	1	1
		Glyphocrangon spinicauda A. Milne-Edwards, 1881	0	0	1	1	1
	Nematocarcinoidea						
	Eugonatonotidae	*Eugonatonotus crassus* (A. Milne-Edwards, 1881)	0	0	1	0	0
	Nematocarcinidae	*Nematocarcinus cursor* A. Milne-Edwards, 1881	1	1	1	1	1
		Nematocarcinus rotundus Crosnier and Forest, 1973	1	1	1	1	1
	Oplophoroidea						
	Acanthephyridae	*Acanthephyra acutifrons* Spence Bate, 1888	1	0	0	0	0
		Acanthephyra armata A. Milne-Edwards, 1881	1	0	1	1	1
		Acanthephyra eximia Smith, 1884	1	0	1	1	1
		Acanthephyra pelagica (Risso, 1816)	0	0	1	0	0
		Acanthephyra purpurea A. Milne-Edwards, 1881	0	0	1	0	1
		Ephyrina benedicti Smith, 1885	0	1	0	0	0
		Heterogenys microphthalma (Smith, 1885)	0	1	0	0	0
		Notostomus elegans A. Milne-Edwards, 1881			1		
		Notostomus gibbosus A. Milne-Edwards, 1881	0	0	0	0	1
Decapoda	Oplophoridae	*Janicella spinicauda* (A. Milne-Edwards, 1883)	1	1	1	1	0
		Oplophorus gracilirostris A. Milne-Edwards, 1881	1	1	1	1	1
		Oplophorus spinosus (Brullé, 1839)	1	0	0	0	0
		Systellaspis cristata (Faxon, 1893)	0	0	0	0	1
		Systellaspis debilis (A. Milne-Edwards, 1881)	1	1	1	1	1
		Systellaspis pellucida (Filhol, 1884)	1	1	1	0	1
	Pandaloidea						

	Pandalidae	*Heterocarpus ensifer* A. Milne-Edwards, 1881	1	1	1	1	1
		Heterocarpus oryx A. Milne-Edwards, 1881	1	1	1	1	1
		Plesionika acanthonotus (Smith, 1882)	1	0	1	0	1
		Plesionika edwardsii (Brandt, 1851)	0	1	1	0	0
		Plesionika ensis (A. Milne-Edwards, 1881)	0	1	1	1	1
		Plesionika holthuisi Crosnier and Forest, 1968	0	1	1	1	1
		Plesionika longicauda (Rathbun,1901)	0	1	1	0	0
		Plesionika longipes (A. Milne-Edwards, 1881)	0	1	1	1	1
		Plesionika martia (A. Milne-Edwards, 1883)	1	0	1	1	0
		Plesionika tenuipes (Smith, 1881)	0	1	1	1	1
		Plesionika richardi (Coutière, 1905)	0	1	0	0	0
	Psalidopodoidea						
	Psalidopidae	*Psalidopus barbouri* Chace Jr, 1939	1	1	1	0	1
	Pasiphaeoidea						
	Pasiphaeidae	*Glyphus marsupialis* Filhol, 1884	0	1	1	0	0
		Pasiphaea merriami Schmitt, 1931	1	1	0	1	1
Decapoda	**Polychelida**						
	Polychelidae	*Cardus crucifer* (Thomson, 1873)	0	1	0	1	1
		Polycheles perarmatus Holthuis, 1952	1	1	1	0	0
		Polycheles typhlops Heller, 1862	1	1	1	1	1
		Stereomastis sculpta (Smith, 1880)	1	1	1	0	1
	Hoplocarida						
	Stomatopoda						
	Eurysquilloidea						
	Eurysquillidae	*Eurysquilla chacei* Manning, 1969	0	0	0	1	0
	Squilloidea						
	Squillidae	*Squilla edentata* (Lunz, 1937)	0	1	1	1	0

(countinued)

Table 20.2 (continued)

	Higher taxon	Species	**Presence**				
	Isopoda						
	Cymothooidea						
	Cirolanidae	*Bathynomus giganteus* A. Milne-Edwards, 1879	1	1	1	1	1
	Lophogastrida						
	Gnathophausiidae	*Gnathophausia zoea* Willemoes-Suhm, 1873	0	1	1	1	1
		Neognathophausia gigas (Willemoes-Suhm, 1873)	1	1	1	1	1
		Neognathophausia ingens (Dohrn, 1870)	1	1	1	0	0

WNW west-northwest, *WSW* west-southwest, *SSW* south-southwest, *SSE* south-southeast, and *ESE* east-southeast. (Presence = 1, absence = 0). Bold = species only for the Gulf of Mexico; ** = Caribbean species

Table 20.3 Number of genera (GEN) and species (SPP) by family in the sectors of the southwestern Gulf of Mexico

	SGMX											
Higher taxon	TOTAL		WNW		WSW		SSW		SSE		ESE	
Family	GEN	SPP	GEN	SPP	GEN	SPP	GEN	SPP	GEN	SPP	GEN	SPP
Dendrobranchiata												
Penaeoidea												
Aristeidae	4	4	4	4	3	3	4	4	4	4	3	3
Benthesicymidae	2	2	1	1	2	2	1	1	1	1	1	1
Penaeidae	3	4	2	2	2	3	3	4	2	3	3	4
Solenoceridae	3	5	3	3	3	4	3	5	3	4	3	3
Pleocyemata												
Anomura												
Chirostyloidea												
Chirostylidae	2	6	2	2	1	1	2	3	3	4	1	4
Sternostylidae	1	1	0	0	0	0	1	1	1	1	0	0
Galatheoidea												
Munididae	2	9	2	4	2	6	2	6	2	8	2	8
Munidopsidae	2	20	2	6	2	10	2	13	2	4	2	12
Porcellanidae	1	1	0	0	1	1	0	0	1	1	0	0
Lithodoidea												
Lithodidae	1	1	1	1	0	0	0	0	0	0	0	0
Paguroidea												
Diogenidae	2	2	0	0	0	0	2	2	0	0	1	1
Paguridae	5	7	2	2	1	2	4	6	3	3	3	3
Parapaguridae	4	5	2	3	2	2	4	5	3	3	3	3
Pylochelidae	1	1	0	0	1	1	1	1	1	1	0	0
Astacidea												
Nephropoidea												
Nephropidae	2	4	2	3	2	3	2	3	2	3	2	4
Brachyura												
Calappoidea												
Calappidae	1	1	1	1	1	1	1	1	1	1	1	1
Dorippoidea												
Ethusidae	1	1	1	1	1	1	1	1	1	1	1	1
Goneplacoidea												
Euryplacidae	2	2	0	0	0	0	2	2	2	2	2	2
Goneplacidae	1	1	1	1	1	1	1	1	1	1	1	1
Leucosioidea												
Leucosiidae	2	2	0	0	2	2	3	3	1	1	2	2
Majoidea												
Epialtidae	5	6	2	2	3	3	3	4	4	5	4	4
Inachidae	1	1	0	0	0	0	0	0	0	0	1	1
Inachoididae	3	5	1	1	2	2	2	3	2	3	0	0

(continued)

Table 20.3 (continued)

	SGMX											
Higher taxon	TOTAL		WNW		WSW		SSW		SSE		ESE	
Family	GEN	SPP	GEN	SPP	GEN	SPP	GEN	SPP	GEN	SPP	GEN	SPP
Palicoidea												
Palicidae	1	3	1	1	1	2	1	2	1	1	1	3
Parthenopoidea												
Parthenopidae	3	3	0	0	1	1	2	2	1	1	1	1
Portunoidea												
Geryonidae	2	2	2	2	2	2	2	2	1	1	1	1
Polybiidae	1	1	1	1	1	1	1	1	1	1	1	1
Portunidae	2	3	1	1	1	2	0	0	1	1	0	0
Trichopeltarioidea												
Trichopeltariidae	1	1	1	1	1	1	1	1	0	0	0	0
Xanthoidea												
Pseudorhombilidae	3	4	0	0	1	1	2	2	3	3	2	2
Xanthidae	1	1	0	0	0	0	1	1	0	0	0	0
Cyclodorippoidea Cyclodorippidae	2	4	0	0	0	0	1	2	0	0	2	4
Homoloidea												
Homolidae	2	3	0	0	0	0	0	0	0	0	2	3
Latreilliidae	1	1	0	0	0	0	0	0	0	0	1	1
Homolodromoidea												
Homolodromiidae	1	1	0	0	1	1	1	1	0	0	1	1
Raninoidea												
Raninidae	1	1	1	1	1	1	1	1	0	0	0	0
Caridea												
Cangronoidea												
Crangonidae	1	2	1	2	1	2	1	2	0	0	1	1
Glyphocrangonidae	1	5	1	2	1	3	1	5	1	5	1	5
Nematocarcinoidea												
Eugonatonotidae	1	1	0	0	0	0	1	1	0	0	0	0
Nematocarcinidae	1	2	1	2	1	2	1	2	1	2	1	2
Psalidopodoidea												
Psalidopidae	1	1	1	1	1	1	1	1	0	0	1	1
Oplophoroidea							0	0	0	0	0	0
Acanthephyridae	4	9	1	3	2	2	1	5		2	2	3
Oplophoridae	3	6	3	5	3	4	3	4	3	3	2	4
Pandaloidea												
Pandalidae	2	11	2	4	2	9	2	10	2	7	2	7
Pasiphaeoidea												
Pasiphaeidae	2	2	1	1	2	2	1	0	1	1	1	1
Polychelida												
Polychelidae	3	4	2	3	3	4	2	3	2	2	3	3

(continued)

Table 20.3 (continued)

	SGMX											
Higher taxon	TOTAL		WNW		WSW		SSW		SSE		ESE	
Family	GEN	SPP	GEN	SPP	GEN	SPP	GEN	SPP	GEN	SPP	GEN	SPP
Hoplocarida												
Stomatopoda												
Eurysquilloidea												
Eurysquillidae	1	1	0	0	0	0	0	0	1	1	0	0
Squilloidea												
Squillidae	1	1	0	0	1	1	1	1	1	1	0	0
Isopoda												
Cymothooidea												
Cirolanidae	1	1	1	1	1	1	1	1	1	1	1	1
Lophogastrida												
Gnathophausiidae	2	3	1	0	2	3	2	3	2	2	2	2

WNW west-northwest, *WSW* west-southwest, *SSW* south-southwest, *SSE* south-southeast, and *ESE* east-southeast

(SSW) are characterized by a broad shelf and smooth slope, presenting more opportunities for trawling. In addition, the upper continental shelves of Campeche and Yucatán (SSE and ESE) are characterized by the presence of ridges and canyons and several deep coral reef formations, making sampling difficult.

Data obtained about crustacean species occurrence in the SGMx showed that at least 50% of species were not recorded before in each sector. In the WSW and ESE, there were more species present; however, it does not reach even 50% of the total species registered in the present study. An extension of the bathymetric range, either in the lower or in the upper limit, was documented in 26 species (21%) of the total collected in SSW sector. In the other sectors, this percentage was as follows: ESE 14 (13%), SSE 12 (14%), WNW 11 (16%), and WSW 9 (10%) (Table 20.4). It is expected that further studies could give more information about the species depth distribution ranges as well as its seasonal variations.

The high percentage of species with geographic and bathymetric extension ranges, in addition, to species not recorded in the South Gulf of Mexico, reflects the scarce knowledge about crustaceans in this region prior to this survey, particularly in depths beyond to 200 m. The lack of information was pointed out by Wicksten and Packard (2005) who discussed the difference in total number of sampled sites between the North Gulf of Mexico compared to the southern area (396 and 42, respectively). A similar case was observed in the study of demersal fish species in the upper slope in SGMx, which contributed with several new distribution and bathymetrical range extensions (Ramírez et al. 2019).

The present surveys revealed that crustaceans were a dominant group in genera and species number, frequency, and abundance among other megainvertebrates like echinoderms and mollusks. Total crustaceans collected in our cruises carried out in the SGMx amounted to 39,000 specimens. The families Aristeidae, Nematocarcinidae, Benthesicymidae, Penaeidae, Parapaguridae, and Solenoceridae

Table 20.4 Numbers of species in each family, record, and range extensions (B, bathymetric; D, distribution) by sector in the southwestern Gulf of Mexico

		SGMX									
	Higher taxon	WNW		WSW		SSW		SSE		ESE	
	Family	B	D	B	D	B	D	B	D	B	D
Decapoda	**Dendrobranchiata**										
	Penaeoidea										
	Aristeidae	3	2	1	2	4	3	4	3	3	1
	Benthesicymidae	1	0	0	0	1	0	1	0	1	0
	Penaeidae	0	2	0	1	0	2	0	1	0	2
	Solenoceridae	1	1	0	1	1	1	0	1	1	1
	Anomura										
	Chirostyloidea										
	Chirostylidae	1	2	0	2	1	5	1	4	2	4
	Sternostylidae	0	0	0	0	1	1	1	1	1	1
	Galatheoidea										
	Munididae	0	1	0	1	0	4	1	3	1	1
	Munidopsidae	0	2	1	6	3	8	0	1	1	6
	Porcellanidae	0	0	0	0	0	1		1	0	0
	Lithodoidea										
	Lithodidae	0	1	0	1	1	1	0	1	0	1
	Paguroidea										
	Diogenidae	0	1	1	0	3	1	0	1	0	1
	Paguridae	0	0	0	2	0	3	0	1	0	1
	Parapaguridae	0	0	0	0	0	0	0	0	0	0
	Pylochelidae	0	0	0	0	0	0	0	0	0	0
	Astacidea										
	Nephropoidea										
	Nephropidae	0	3	0	3	3	3	0	3	0	4
	Brachyura										
	Calappoidea										
	Calappidae	0	0	0	0	0	1	0	1	0	0
	Dorippoidea										
	Ethusidae	0	0	0	0	0	1	0	0	0	0
	Goneplacoidea										
	Euryplacidae	0	0	0	0	0	0	0	0	0	1
	Goneplacidae	1	0	0	0	0	1	0	2	0	0
	Leucosioidea										
	Leucosiidae	0	0	1	0	1	0	0	0	0	0
	Majoidea										
	Epialtidae	0	2	0	2	0	4	0	3	0	2
	Inachidae	0	0	0	0	0	1	0	0	0	0
	Inachoididae	0	0	0	0	2	0	0	0	0	0
	Palicoidea										

(continued)

Table 20.4 (continued)

		SGMX									
	Higher taxon	WNW		WSW		SSW		SSE		ESE	
	Family	B	D	B	D	B	D	B	D	B	D
	Palicidae	1	1	1	2	1	2	1	1	1	1
	Parthenopoidea										
	Parthenopidae	0	0	0	1	0	2	0	1	0	1
	Portunoidea										
	Geryonidae	0	1	1	1	0	1	0	1	0	0
	Polybiidae	0	0	0	1	0	1	0	1	0	1
	Portunidae	1	1	1	1	1	1	0	0	0	0
	Trichopeltarioidea										
	Trichopeltariidae	1	1	1	1	1	1	0	0	0	0
	Xanthoidea										
	Pseudorhombilidae	0	1	0	1	0	1	0	2	0	0
	Xanthidae	0	1	0	1	0	1	0	1	0	1
	Cyclodorippoidea										
	Cyclodorippidae	0	0	0	0	0	0	0	1	0	0
	Homoloidea										
	Homolidae	0	1	0	1	0	2	0	2	0	0
	Latreilliidae	0	0	0	0	0	0	0	0	0	0
	Homolodromioidea										
	Homolodromiidae	0	1	0	1	0	2	0	2	0	0
	Raninoidea										
	Raninidae	0	1	0	0	0	1	0	1	0	1
	Caridea										
	Cangronoidea										
	Crangonidae	0	1	0	1	0	1	0	1	0	0
	Glyphocrangonidae	1	0	1	0	2	0	1	0	1	0
	Nematocarcinoidea										
	Eugonatonotidae	0	0	0	0	0	1	0	0	0	0
	Nematocarcinidae	0	1	0	1	0	1	0	1	0	1
	Psalidopodoidea										
	Psalidopidae	0	0	0	1	0	1	0	1	0	1
	Oplophoroidea										
	Acanthephyridae	0	0	0	1	0	0	0	0	0	
	Oplophoridae	0	2	0	1	0	2	0	1	1	2
	Pandaloidea										
	Pandalidae	0	2	1	4	1	8	1	5	1	3
	Pasiphaeoidea										
	Pasiphaeidae	0	0	0	0	0	0	0	1	0	1
	Polychelida										
	Polychelidae	0	2	0	2	0	2	0	1	0	1

(continued)

Table 20.4 (continued)

		SGMX									
	Higher taxon	WNW		WSW		SSW		SSE		ESE	
	Family	B	D	B	D	B	D	B	D	B	D
	Hoplocarida Stomatopoda										
	Eurysquilloidea										
	Eurysquillidae	0	1	0	0	0	1	0	1	0	1
	Squilloidea										
	Squillidae	0	0	0	0	0	0	0	0	0	0
	Isopoda										
	Cymothooidea										
	Cirolanidae	0	0	0	1	0	1	0	0	0	0
	Lophogastrida										
	Gnathophausiidae	0	3	0	2	0	2	0	2	0	2

WNW west-northwest, *WSW* west-southwest, *SSW* south-southwest, *SSE* south-southeast, and *ESE* east-southeast

(in descending order) presented the highest abundance in all sectors. The following species were the most abundant and frequent: *Aristaeomorpha foliacea* (Risso, 1827); *Nematocarcinus rotundus* Crosnier and Forest, 1973; *Benthoecetes bartletti* (Smith, 1882); *Penaeopsis serrata* Bate, 1881; *Sympagurus pictus* Smith, 1883; and *Pleoticus robustus* (Smith, 1885). A comparative analysis of abundance by sector showed that SSW presented the highest abundance, followed by ESE, WSW, SSE, and WNW. Besides the sampling effort, the difference in abundance was probably influenced by the freshwater discharge of the main rivers (Grijalva-Usumacinta system) and cyclonic gyres in Campeche Bank that may enhance primary production through the input of nutrients to deep-water ecosystem.

The distributional analysis showed that the majority of crustacean species (50) occurred in only 1 sector (SSW) and only 38 species occur throughout all 5 sectors; the rest of the species were collected in 2–4 sectors.

The fauna of deep-water crustaceans of the SGMx contains a few endemic species. Two of these species were recently described: *Munidopsis shulerae* Vázquez-Bader et al., 2014 (SSW, ESE); *Tomopaguropsis ahkinpechensis* Lemaitre et al., 2014 (SSW); and the pagurid *Pagurus bullisi* Wass, 1973 (SSW).

Six species were previously only known from the Caribbean Sea and are new records for the Gulf of Mexico: two species of Munididae, *Munidopsis bradleyi* Pequegnat and Pequegnat, 1971 (SSW), and *Munidopsis riveroi* Chace Jr, 1939 (WNW and WSW); two Chirostylidae, *Sternostylus salvadori* (Rice and Miller, 1991) (SSW and SSE) and *Uroptychus spiniger* Benedict, 1902 (WNW and ESE); one species of Cyclodorippidae, *Clythrocerus carinatus* Coelho, 1973 (ESE); and one species of Pseudorhombilidae, *Robertsella meridionalis* Tavares and Gouvêa, 2013 (ESE).

Fifteen species show a distribution range that encompass the Gulf of Mexico and waters beyond the Atlantic Ocean (Pacific and Indo-Pacific), and nine can be considered as cosmopolitan. The majority of species (86 %) range from south of Cape Hatteras, North Carolina, to the Gulf of Mexico and the Caribbean and further western and eastern Atlantic.

The faunal composition in the SGMx showed differences compared to the northern region, particularly in species not recorded before. For example, Wicksten and Packard (2005) mentioned that the parapagurid *Paragiopagurus pilimanus* (A. Milne-Edwards, 1880) has a distribution in the western Atlantic off Florida and in the Caribbean but not entering the Gulf of Mexico. However, we registered this hermit crab in low numbers (<5 specimens) in the SSW and SSE sectors. These authors also pointed out that *Munida iris* A. Milne-Edwards, 1880, was reported only once in the southern Gulf of Mexico, but this species was very frequent and abundant throughout all sectors in our study.

Shrimp species as *Parapontocaris caribbaea* (Boone, 1927) and *Plesionika longipes* (A. Milne-Edwards, 1881) were reported with a low abundance in the northern Gulf of Mexico, considered more common in the Caribbean, and frequently collected in four of the five sectors (50 and more than 400 specimens, respectively). Besides, we recorded a continuous distribution (throughout all sectors of SGMx) for the commercial shrimp *Aristeus antillensis* A. Milne-Edwards and Bouvier, 1909.

In spite of the intensive and systematic sampling carried on the upper slope of the SGMx in this study, it is obvious that the crustacean species list will be updated with increased sampling effort and by using different sampling equipment. Another aspect is related to taxonomic difficulties found in the identification of several specimens collected that do not fit to descriptions available in the literature (e.g., families Chirostylidae, Nematocarcinidae, Pasiphaeidae, Pseudorhombilidae, Sergestidae, Stenopodidae, and Xanthidae).

Our analysis suggests that the deep-water megacrustacean fauna of the northern and southern Gulf of Mexico may have different composition and abundance patterns. However, it is evident that more extensive and exhaustive studies are needed to characterize the crustacean community distribution pattern across the whole Gulf of Mexico.

Acknowledgment The authors are grateful to all the staff of LEPC of the Instituto de Ciencias del Mar y Limnología UNAM, in particular I. Antillón, B. Barbosa, S. Bueno, M. Palmerín, and H. Trejo and especially to L.F. González who made the map; to crew of B/O *Justo Sierra*, UNAM; and to R. Lemaitre for the identification of pagurid specimens of the southern Gulf of Mexico. The specimens of oceanographic cruise DGoMB were identified by M. Wicksten. Sammy de Grave is greatly appreciated for his thorough revision and comments to the manuscript. SOGOM cruises were funded by the Mexican National Council for Science and Technology – Mexican Ministry of Energy – Hydrocarbon Fund, project 201441 as part of the Gulf of Mexico Research Consortium (CIGoM) due to PEMEX's specific request to the Hydrocarbon Fund to address the environmental effects of oil spills in the Gulf of Mexico.

References

Bate CS (1881) On the Penaeidae. The Annals and Magazine of Natural History, series 5, 8:169–196, pp 11–12

Bate CS (1888) Report on the Crustacea Macrura collected by the challenger during the years 1873–76. Report on the scientific results of the voyage of H.M.S. challenger during the years 1873–76. Zoology, 24 (part 52):i–xc,1–942, pp1–150. http://www.19thcenturyscience.org/HMSC/HMSC-Reports/Zool-52/htm/doc.html

Benedict JE (1892) Preliminary descriptions of thirty-seven new species of hermit crabs of the genus *Eupagurus* in the U.S. National Museum. Proc US Natl Mus 15:1–26

Benedict JE (1901) The anomuran collections made by the fish hawk expedition to Porto Rico. Bull U S Fish Commun 20:129–148

Benedict JE (1902) Description of a new genus and forty-six new species of crustaceans of the family Galatheidae with a list of the known marine species. Proc US Natl Mus 26:243–334

Boone L (1927) Scientific results of the first oceanographic expedition of the "Pawnee", 1925. Crustacea from the tropical east American Seas. Bull Bingham Oceanogr Coll 1:1–147

Bouvier EL (1905) Sur les Pénéides et les Sténopides receuillis par les expéditions françaises et monégasques dans l'Atlantique oriental. C R Hebd Séances Acad Sci 140:980–983

Brandt JF (1851) Krebse. In: von Middendorff, AT (ed) Reise in den äussersten Norden und Osten Sibiriens während der Jahre 1843 und 1844 mit allerhöchster Genehmigung auf Veranstaltung der Kaiserlichen Akademie der Wissenschaften zu St. Petersburg ausgeführt und in Verbinding mit vielen Gelehrten herausgegeben 2 (Theil 1): 77–148, Plates 5–6. St. Petersburg. https://www.biodiversitylibrary.org/page/37047789

Brullé A (1839) Crustacés. In: Barker-Webb P, Berthelot S (eds) Histoire naturelle des Iles Canaries:15–18, Planche unique. Béthune, Paris

Burdett EA (2016) Geographic and depth distributions of decapod shrimps (Caridea: Oplophoridae) from the northeastern Gulf of Mexico with notes on ontogeny and reproductive seasonality. Thesis of Master Science. Nova Northeastern University, Halmos College of Natural Sciences and Oceanography, pp 1–42

Cato JC (2009) Gulf of Mexico origin, waters and biota: ocean and coastal economy, vol 2. Texas A & M University Press, College Station, p 136

Chace FA Jr (1939) Reports on the scientific results of the first Atlantis expedition to the West Indies, under the joint auspices of the University of Havana and Harvard University. Preliminary descriptions of one new genus and seventeen new species of decapod and Stomatopod Crustacea. Mem Soc Cub Hist Nat 13:31–54

Chace FA Jr (1940) Reports on the scientific results of the Atlantis expeditions to the West Indies, Under the Joint Auspices of the University of Havana and Harvard University. The brachyuran crabs. Torreia 4:3–67

Coutière H (1905) Sur quelques Crustacés provenant des campagnes de la Princesse-Alice (filet à grande ouverture). CR Hebd Séances Acad Sci 140:1113–1115

Crosnier A, Forest J (1968) Note préliminaires sur les carides receuillis par l "Ombango" au large du plateau continental, du Gabon à l'Angola (Crustacea Decapoda Natantia). Bull Mus Nat Hist Nat 39(2):1123–1147

Crosnier A, Forest J (1973) Les crevettes profondes de l'Atlantique oriental tropical. Faune Trop 19:1–409

De Grave S, Pentcheff ND, Ahyong ST, Chan T-Y, Crandall KA, Dworschak PC, Felder DL, Feldmann RM, Fransen CHJM, Goulding LYD, Lemaitre R, Low MEY, Martin JW, Ng PKL, Schweitzer CE, Tan SH, Wetzer R (2009) A classification of living and fossil genera of decapod crustaceans. Raffles Bull Zool Suppl 21:1–109

de Saussure H (1857) Diagnoses de quelques Crustacés nouveaux de l'Amérique tropicale. Rev Mag Zool Ser 2(9):501–505

Dohrn A (1870) Untersuchungen über Bau und Entwicklung der Arthropoden. 8. Die Ueberreste des Zoëa-Stadiums in der ontogenetischen Entwicklung der verschiedenen Crustaceen-Familien. Jenaische Zeitschr Naturw 5:471–491

Escobar-Briones E, Gaytán A, Legendre P (2008) Epibenthic megacrustaceans from the continental margin, slope and abyssal plain of the southwestern Gulf of Mexico: factors responsible for variability in species composition and diversity. Deep Sea Res Part II Top Stud 55:2667–2678. https://doi.org/10.1016/j.dsr2.2008.07.016

Faxon W (1893) Reports on the dredging operations off the west coast of Central America to the Galapagos, to the west coast of Mexico, and in the Gulf of California, in charge of Alexander Agassiz, carried on by the U.S. Fish Commission steamer "Albatross", during 1891, Lieut. Commander Z.L. Tanner, U.S.N., commanding. VI. Preliminary descriptions of new species of Crustacea. Bull Mus Comp Zool Harvard 24:149–220

Felder DL, Camp D, Tunnell JW Jr (2009a) An introduction to Gulf of Mexico biodiversity assessment. In: Felder DL, Camp DK (eds) Gulf of Mexico: origin, waters, and biota, Biodiversity, vol 1. Texas A & M University Press, College Station, pp 1–13

Felder DL, Álvarez F, Goy JW, Lemaitre R (2009b) Decapoda (Crustacea) of the Gulf of Mexico, with comments on the Amphionidacea. In: Felder DL, Camp DK (eds) Gulf of Mexico: origin, waters, and biota, Biodiversity, vol 1. Texas A & M University Press, College Station, pp 1019–1104

Filhol H (1884) Explorations sous-marines. Voyage du" Talisman". La Nature 12:119–122, 134–138, 147–151, 161–164, 182–186, 198–202, 230–234, 278–282, 326–330, 391–394

Gracia A, Vázquez-Bader AR, Lozano-Álvarez E, Briones-Fourzán P (2010) Deep-water shrimps (Crustacea: Penaeoidea) off the Yucatan peninsula (southern Gulf of Mexico): a potential fishing resource? J Shellfish Res 29(1):37–43. https://doi.org/10.2983/035.029.0124

Guinot D, Richer de Forges B (1995) Crustacea Decapoda Brachyura: Révision de la famille des Homolidae de Haan, 1839. In: Crosnier A (ed) Résultats des Campagnes MUSORSTOM 13, Mém Mus nat Hist nat, Series A, Zool 163, pp 283–517

Hay WP (1917) Preliminary descriptions of five new species of crustaceans from the coast of North Carolina. Proc Biol Soc Wash 30:71–74

Henderson JR (1888) Report on the Anomura collected by H.M.S. Challenger during the years 1873–76. Report on the Scientific Results of the Voyage of H.M.S. Challenger during the years 1873–76. Zool 27 (69):i–xi,1–221,pl. 1–21 http://www.19thcenturyscience.org/HMSC/HMSC-Reports/Zool-69/README.htm

Holthuis LB (1959) The Crustacea Decapoda of Suriname (Dutch Guiana). Zool Verh 44:1–296

Holthuis LB (1971) Biological results of the University of Miami deep-sea expeditions. 75. The Atlantic shrimps of the deep-sea genus *Glyphocrangon* A. Milne-Edwards, 1881. Bull Mar Sci 21:267–373

Johnson JY (1868) Descriptions of a new genus and a new species of macrurous decapod crustacean belonging to the Penaeidae, discovered at Madeira. Proc Zool Soc London 1867:895–901

Lemaitre R (1986) Western Atlantic species of the *Parapagurus pilosimanus* complex (Anomura: Paguroidea: Parapaguridae): description of a new species and morphological variations. J Crust Biol 6(3):525–542

Lemaitre R, Vázquez-Bader AR, Gracia A (2014) An unusual new species of paguroid (Crustacea: Anomura: Paguridae) from deep waters of the Gulf of Mexico. Zookeys 449:57–67. https://doi.org/10.3897/zookeys.449.8541. ISSN 1313–2970

Lovén S (1852) De svenska arterna af slägtet Galathea. Ofversigt Kong Vetensk Akad Förhandl 9:20–23

Lozano-Álvarez E, Briones-Fourzán P, Gracia A, Vázquez-Bader AR (2007) Population characteristics of *Heterocarpus ensifer* from the continental slope of the Yucatan Peninsula, Mexico, with notes on *Heterocarpus oryx* (Decapoda: Pandalidae). Crustaceana 80(5):555–568

Lunz GR (1937) Stomatopoda of the Bingham oceanographic collection. Bull Bingham Oceanogr Coll 5(5):1–19

Manning RB (1969) Stomatopod Crustacea from the Western Atlantic. Stud Trop. Oceanogr 8:pp 380

Milne-Edwards A (1879) Sur un isopode gigantesque, des grandes profondeurs de la mer. C R Hebd Séances Acad Sci, París 88:21–23
Milne-Edwards A (1880) Reports on the results of dredging under the supervision of Alexander Agassiz, in the Gulf of Mexico and in the Caribbean Sea,1877, '78, '79, by the U.S. Coast Survey Steamer "Blake", Lieut.-Commander C.D. Sigsbee, U.S.N., and Commander J.R. Bartlett, U.S.N. commanding VIII. Études préliminaires sur les Crustacés. Bull Mus Comp Zool Harvard Coll 8(2):1–68, Plates 1–2
Milne-Edwards A (1881) Description de quelques crustacés macroures provenant des grandes profondeurs de la Mer des Antilles. Ann Sci Nat 11(6):1–15
Milne-Edwards A (1883) Recueil de figures de Crustacés nouveaux ou peux connus: 1–3, Plates 1–44. Paris
Milne-Edwards A, Bouvier EL (1893) Reports on the results of dredging, under the supervision of Alexander Agassiz, in the Gulf of Mexico (1877–78), in the Caribbean Sea (1878–79), and along the Atlantic coast of the United States (1880), by the U.S. Coast Survey Steamer "Blake", Lieut.-Commander C. D Sigsbee, U.S.N., and Commander J. R. Bartlett, U.S.N., commanding. XXXIII. Description des Crustacés de la famille des paguriens recueillis pendant l'expédition. Mem Mus Comp Zool Harvard 14 (3):1–172, pp I–XII https://www.biodiversitylibrary.org/page/4305631
Milne-Edwards A, Bouvier EL (1894) Considerations générales sur la famille des Galatheides. Ann Sci Nat Zool 16(series 7):191–327
Milne-Edwards A, Bouvier EL (1909) Reports on the results of dredging, under the supervision of Alexander Agassiz, in the Gulf of Mexico (1877-78), in the Caribbean Sea (1878-79), and along the Atlantic coast of the United States (1880), by the U.S.N. Coast Survey Steamer "Blake", Lieut.-Com. C.D. Sigsbee, U.S.N., and Commander J.R. Bartlett, US.N. Commanding. 44. Les Pénéides et Sténopides. Mem Mus Comp Zool Harvard 27:177–274, Plates 1–9
Monreal-Gómez MA, Salas de León DA, Velasco-Mendoza H (2004) La hidrodinámica del Golfo de México. In: Caso M, Pisanty I, Ezcurra E (eds) Diagnóstico Ambiental del Golfo de México. SEMARNAT, Ciudad de México, pp 47–68
Murawski SA, Hollander D, Gilbert S, Gracia A (2020) Deep-water oil and gas production in the Gulf of Mexico, and related global trends. In: Murawski SA, Ainsworth C, Gilbert S, Hollander D, Paris CB, Schluter M, Wetzel D (eds) Scenarios and responses to future deep oil spills: fighting the next war. Springer, Cham, pp 16–32
Pequegnat LH, Pequegnat WE (1970) Deep-sea anomurans of superfamily Galatheoidea with description of three new species. In: Pequegnat WE, Chace FA (eds) Contributions on the biology of the Gulf of Mexico. Texas A and M University, pp 125–170
Poore GCB (2016) The names of the higher taxa of Crustacea Decapoda. J Crust Biol 36(2):248–255
Price WW, Heard RW, Aas P, Meland K (2009) Lophogastrida (Crustacea) of the Gulf of Mexico. In: Felder DL, Camp DK (eds) Gulf of Mexico, origins, waters, and biota, Biodiversity, vol 1. Texas A and M Press, College Station, pp 923–927
Ramírez M, Vázquez-Bader AR, Gracia A (2019) Icthyofaunal list of the continental slope of the southern Gulf of Mexico. Zookeys:1–16. https://doi.org/10.3897/zookeys.@@.31944. **ISSN** 1313–2970
Rathbun MJ (1892) Catalogue of crabs of the family Periceridae in the U. S. National Museum. Proc US Natl Mus 15(901):231–277, pls 28–40
Rathbun MJ (1898) The Brachyura of the biological expedition to the Florida keys and the Bahamas in 1893. Bull Lab Nat Hist State Univ Iowa 3:250–294. https://www.biodiversitylibrary.org/page/6313618#page/125/mode/1up
Rathbun MJ (1901) Investigations of the aquatic resources and fisheries of Porto Rico by the United States fish commission steamer fish hawk in 1899. The Brachyura and Macrura of Porto Rico. Bull US Fish Commun 20:1–127. [for 1900, preprint dated 1901, published in journal in 1902]
Rathbun MJ (1918) The Grapsoid crabs of America. US Natl Mus Bull 97:1–461, Plates 1–161

Reaka ML, Camp DK, Álvarez F, Gracia A, Ortiz M, Vázquez-Bader AR (2009) Stomatopoda (Crustacea) of the Gulf of Mexico. In: Felder DL, Camp DK (eds) Gulf of Mexico-origins, waters, and biota, Biodiversity, vol 1. Texas A&M University Press, College Station, pp 901–921

Rice AL, Miller JE (1991) Chirostylid and galatheid crustaceans associates of coelenterates and echinoderms collected from the Johnson-Sea-Link submersible, including a new species of *Gastroptychus*. Proc Biol Soc Wash 104(2):299–308

Risso A (1816) Histoire Naturelle des Crustacés des Environs de Nice. Librairie Grecque-Latine-Allemande Paris, p 175, Plates 1–3

Risso A (1826–1827) Histoire naturelle des principales productions de l'Europe Méridionale et particulièrement de celles des environs de Nice et des Alpes Maritimes. Paris, Levrault 3(XVI):1–480,14 Plates. http://www.biodiversitylibrary.org/bibliography/58984

Roux P (1828–1830) Crustacés de la Méditerranée et de son littoral. Paris, p 176, Plates 1–45 http://www.biodiversitylibrary.org/page/9283448

Schmitt WL (1931) Some carcinological results of the deeper water trawlings of the Anton Dohrn, including description of two new species of Crustacea. Carnegie Inst Wash Yearbook 30:389–394

Schotte M, Markham JC, Wilson GDF (2009) Isopoda (Crustacea) of the Gulf of Mexico. In: Gulf of Mexico: its origins, waters, and biota. Vol. 1: Biodiversity. In: Felder DL, Camp DK (eds) Gulf of Mexico: origin, waters, and biota. Vol. 1, biodiversity. Texas A&M University Press, College Station, pp 973–986

Smith SI (1881) Preliminary notice on the Crustacea dredged in 64 to 325 fathoms, off the south coast of New England, by the United States fish commission in 1880. Proc US Natl Mus 3:413–452

Smith SI (1882) Reports on the results of dredging under the supervision of Alexander Agassiz, on the east coast of the United States during the summer of 1880, by the U.S. Coast Survey Steamer "Blake", Commander J.R. Bartlett, U.S.N., commanding. Bull Mus Comp Zool Harvard 10(1):1–108, Plates 1–16

Smith SI (1883) Preliminary report on the Brachyura and Anomura dredged in deep water off the south coast of New England by the United States Fish Commission in 1880, 1881, and 1882. Proc US Natl Mus 6(1):1–57, Plates 1–6

Smith SI (1884) Report on the decapod Crustacea of the Albatross dredging off the East coast of the United States in 1883. Rep US Fish Comm 10:345–426, Plates 1–10

Smith SI (1885) On some new or little-known decapod Crustacea, from recent Fish Commission dredging off the east coast of the United States. Proc US Natl Mus 7:493–511 [imprint 1884]

Stimpson W (1871) Preliminary report on the Crustacea dredged in the Gulf Stream in the Straits of Florida, by L.F. de Pourtales, Assist. U.S. Coast Survey. Bull Mus Comp Zool Harvard 2:109–160

Tavares M, Gouvêa A (2013) A new species of *Robertsella* Guinot, 1969 (Crustacea: Brachyura: Panopeidae) from the southwestern Atlantic. Zootaxa 3734(1):72–80. https://doi.org/10.11646/zootaxa.3734.1.8

Vázquez-Bader AR, Gracia A (2004) A new genus and new species of crab of the family Xanthidae MacCleay, 1838 (Crustacea: Decapoda: Brachyura) from the Southwestern Gulf of Mexico. Proc Biol Soc Wash 117(4):56–64

Vázquez-Bader AR, Gracia A (2013) Crangonidae and Glyphocrangonidae (Decapoda: Caridea) of the Southern Gulf of Mexico. Zootaxa 3669(3):367–383. https://doi.org/10.11646/zootaxa.3669.3.1

Vázquez-Bader AR, Gracia A (2016) Diversity and distribution of Chirostyloidea and Galatheoidea (Decapoda, Anomura) in the Southern Gulf of Mexico. ZooKeys 612:1–30. https://doi.org/10.3897/zookeys.612.9492. ISSN 1313–2970

Vázquez-Bader AR, Gracia A, Lemaitre R (2014) A new species of *Munidopsis* Whiteaves, 1874 (Crustacea: Anomura: Galatheoidea: Munidopsidae) from the Gulf of Mexico and Caribbean Sea. Zootaxa 3821(3):354–362 doi.org/10.11646/zootaxa.3821.3.4. ISSN 1175–5334

Von Willemoes-Suhm (1873) In: Thomson W (ed) Notes from the "Challenger", VII. Nature, London, vol. 8:400–403

Ward CH, Tunnell JW Jr (2017) Habitats and Biota of the Gulf of Mexico: an overview. In: Ward CH (ed) Habitats and Biota of the Gulf of Mexico: before the Deepwater Horizon Oil Spill. Springer Nature, p 1–54

Wass ML (1963) New species of hermit crabs (Decapoda, Paguridae) from the western Atlantic. Crustaceana 6(2):133–157

Wicksten M, Packard JM (2005) A qualitative zoogeographic analysis of decapod crustaceans of the continental slopes and abyssal plain of the Gulf of Mexico. Deep-Sea Res Part I Oceanogr Res Pap 52:1745–1765

Williams AB (1966) The Western Atlantic swimming crabs *Calinectes ornatus, C. danae* and a new, related species (Decapoda, Portunidae). Tulane Stud Zool 13(3):83–93

Chapter 21
Catalogue of Typical Deep-Sea Decapod Fauna from Brazilian Waters

I. Cardoso Azevedo, F. Alves-Junior Almeida, and T. Rodrigues Garcia Almeida

Abstract The deep sea is one of the last frontiers to biodiversity knowledge mainly due to the difficulties to access this environment. Additionally, the deep sea is constantly suffering anthropogenic impact (fishery activities, mininger exploration, and pollution). Recent efforts have contributed to knowledge on deep-sea Decapoda in Brazilian waters. However, these inventories are still far from complete. The main goal of this work is to compile the existing information, including these recent inventories of typically deep-sea species from Brazilian waters, >500 m depth (also including the Rio Grande Rise area). We herein list a total of 181 species of the two decapod suborders: Dendrobranchiata (32 species) and Pleocyemata (149 species). From the Pleocyemata eight out of the ten decapod infraorders were represented: 68 species of carideans; 31 of anomurans, 36 of brachyurans, 5 species of Astacidea and Polychelidae, 2 species of Achelata, and 1 of Stenopodidea and Axiidea each. As expected, the shrimp fauna was more diverse than the crab and lobster fauna. This is in contrast with the global diversity which is highest in the Brachyura.

Keywords Decapoda · Deep sea · Checklist · Brazil · Southwestern Atlantic

21.1 Introduction

About 70% of the Earth's surface is covered by sea, of which more than 90% concerns areas with depths below 1000 m (Bouchet 2006; Sumida 2009). It is quite difficult to access these deep regions and many of them remain unexplored. The

I. Cardoso Azevedo (✉) · T. Rodrigues Garcia Almeida
Setor de Carcinologia, Departamento de Invertebrados, Museu Nacional/UFRJ, Quinta da Boa Vista, São Cristóvão, Rio de Janeiro, RJ, Brazil
e-mail: irenecardoso@mn.ufrj.br

F. Alves-Junior Almeida
Laboratório de Carcinologia, Museu de Oceanografia Prof. Dr. Petrônio Alves Coelho da Universidade Federal de Pernambuco (UFPE). Av. Arquitetura, s/n, Cidade Universitária, Recife, Brazil

M. E. Hendrickx (ed.), *Deep-Sea Pycnogonids and Crustaceans of the Americas*, https://doi.org/10.1007/978-3-030-58410-8_21

knowledge of deep-sea biodiversity goes along with the development of sample technologies (ships, drags, nets) and "in situ" observations ("ROV remotely operated vehicle" and "HOV human-occupied vehicle") to access this distant environment.

The first deep-sea hauls in Brazilian waters were carried out by *Challenger Expedition* (1872–1876). Despite recent sampling efforts, the knowledge of Brazilian deep-sea biodiversity is still modest. This is mainly caused by the high cost of deep-sea oceanographic campaigns and their logistic difficulties. In 1987, the oceanographic cruise TAAF MD55/Brasil sampled along the Vitória-Trindade chain (Espírito Santo). The first major effort to increase our knowledge of the Brazilian deep-sea fauna however was the REVIZEE project (Estudo dos Recursos Vivos da Zona Econômica Exclusiva) that sampled along the entire Brazilian coast. The increasing exploration by Brazilian oil companies in deep-sea areas enhanced the development of Brazilian deep-sea research technologies and increased the knowledge of marine biodiversity. Some relevant projects are "Environmental Characterization of Deep waters at Campos Basin" (CENPES/PETROBRAS), "Evaluation of environmental heterogeneity at Campos Basin" (CENPES/PETROBRAS), "Avaliação da Biota Bentônica e Planctônica da Bacia Potiguar e Ceará (BPOT)" (Petrobras), and more recently "Acoustic along the Brazilian Coast" (ABRACOS/IRD France).

Apart from the difficulties to access the deep sea, this environment is suffering from anthropogenic impact mainly due to the fishery activities (dregging and trawling with bottom nets), mining explorations, and pollution (Santos et al. 2013). Knowledge on the impact of natural processes and long-term conservation actions is indispensable to tackle present-day biodiversity issues. Recent efforts are improving the knowledge on deep-sea Decapoda in Brazilian waters. This inventory, however, is still far from complete (Komai 2004; Cardoso and Young 2005; Tavares and Cardoso 2006; Cardoso 2009a, 2013; Cardoso and Burukovsky 2014; Cardoso et al. 2017; Alves-Júnior et al. 2017a, 2018a, 2019a, b; Rodrigues and Cardoso 2019).

21.2 Material and Methods

This paper is a data compilation of papers dealing with the taxonomy of typically deep-sea species from Brazilian waters (also including the Rio Grande Rise area). The Brazilian coastline extends around 7.500 km on the Atlantic coast from Cape Orange at north (5°N) to Chuí (34°S) (Serejo and Siqueira 2018). According to Sumida and Pires-Vanin (1997) and Sumida (2009), the typical deep-sea fauna from Brazilian waters starts below 500 m depth. The upper slope region between 200 and 500 m depth is a transition zone between the shelf fauna and the deep-sea fauna (Sumida 2009). We here only include species living below 500 m depth.

Material listed herein were sampled mainly by the following deep-sea research projects: the TAAF MD55/Brazil (Espírito Santo); the REVIZEE Central project, which collected samples on the continental shelf and slope with the R.V. Thalassa off the Central Brazilian coast (11°S and 21°S) between 1999 and 2001, at depths from 200 to 2000 m, with two net types (ARROW and GOV); the "Campos Basin Deep Sea Environmental Project" (OCEANPROF), coordinated by CENPES/PETROBRAS, sampled in the Campos Basin (between 21°48′S and 22°48′S, Rio de Janeiro), using N/RB Astrogaroupa, two campaigns, Oceanprof I and II, were performed using opening/closing nets, the Oceanprof I campaign was carried out in February 2003 and collected at depths between 1074 and 1649 m, the Oceanprof II was carried out in August 2003 and collected between 1059 and 1640 m; the "Evaluation of Environmental Heterogeneity in the Campos Basin (HABITATS) project," coordinated by CENPES/PETROBRAS, sampled on the continental shelf and slope of the Campos Basin (Rio de Janeiro) in April 2008, with the R.V. Gyre in depths between 400 and 1200 m, using dragnets for fishery; the BPOT "Avaliação da biota bentônica e plânctonica da Bacia Potiguar e Ceará" sampled at Potiguar Basin, located in the northeast of Brazil (03/05°S; 38/35°W), in the states of Ceará (CE) and Rio Grande do Norte (RN) along the continental slope using bottom trawls and semi-balloon otter trawl net, between 150 and 2068 m depth, on board of the R/V Luke Thomas in December 2009 and on board of the R/V Seward Johnson in May 2011; the "Acoustic along the Brazilian Coast" (ABRACOS 2) sampled around the seamounts of the Ceará Chain, Atol das Rocas, and Fernando de Noronha Archipelago, using a micronekton net, between 40 and 1660 m depth, carried out by the R/V Antea in April 2017.

Suborders and infraorders are listed in systematic order following the classification presented by De Grave et al. (2009). Families, genera, and species inside these groups are listed alphabetically. We present here a restricted synonymy for each species, including the original description, references to junior synonyms, and some of the most relevant works with taxonomy of the species, mainly dealing with Brazilian fauna. References included at synonymy of each species were used to elaborate the respective geographic and bathymetric topics. For some species that realize wide vertical migration, all depths of occurrence were considered, including superficial waters.

21.3 Results

21.3.1 Species Richness

A total of 181 species distributed in the two decapod suborders (32 Dendrobranchiata and 149 Pleocyemata) were recorded herein. Eight of the ten Pleocyemata infraorders were represented: 68 species of Caridea, 31 of Anomura, 36 of Brachyura, 5 species of Astacidea and Polychelidea, 2 of Achelata, and 1 species of Stenopodidea and Axiidea.

Herein we present the first record of the pasiphaeid shrimp *Eupasiphae gilesii* (Wood-Mason, 1892) to the western Atlantic. This species was recorded previously at the eastern Atlantic (West of Cape Verde, Madeira, Bermuda) and at the South

Mid-Atlantic Ridge. We present also range extensions of eight caridean species to Brazilian states and oceanic islands (Table 21.1).

21.3.2 Systematic Section

Order Decapoda Sars, 1903.
Suborder Dendrobranchiata Spence-Bate, 1888.
Family Aristeidae Wood-Mason and Alcock, 1891.

1. *Aristaeomorpha foliacea* (Risso, 1827) (Fig. 21.1a).

Penaeus foliacea Risso, 1827: 69, pl. 2, Fig. 6.
Aristeus rostridentatus Spence-Bate, 1888:189.
Aristaeomorpha giglioliana Wood-Mason, 1892: pl. 2, Fig. 2.
Aristaeomorpha mediterranea Adensamer, 1898: 627, unnumbered text figure.
Penaeus meridionalis Hope, 1851: 19.
Aristeus japonicus Yokoya, 1933: 3, Fig. 1.
Aristaeomorpha foliacea– Barnard 1950: 625; Crosnier and Forest 1973: 287, Fig. 96 a; Crosnier 1978: 52–59, Figs. 23–24; Kensley et al. 1987: 279–280; Pérez-Farfante 1988: 6, Fig. 6; Pérez-Farfante and Kensley 1997: 33–36, Figs. 5–6; Dall 2001: 412, Fig. 2; Serejo et al. 2007: 138; Tavares and Serejo 2007: 5, Figs. 2–4; Lira et al. 2017a, b: 32; Alves-Júnior et al. 2019a: 2, Figs. 1, 2, 8a, 9a.

Type locality. Meridional Europe, Nice, "Alpes maritimes." **Diagnosis**. Tavares and Serejo (2007). **Distribution**. Pelagic. Western Atlantic: Massachussets to Florida, Gulf of Mexico, Caribbean Sea to Venezuela, Brazil (Rio Grande do Norte, Bahia, Espírito Santo, Rio de Janeiro, Paraná, Santa Catarina, Rio Grande do Sul). Eastern Atlantic: Bay of Biscay to Western Sahara, Azores, Madeira, Canary Islands,

Table 21.1 First records registered during this study for eight species of decapods

Species	First records
Nematocarcinus gracilipes Filhol, 1884[a]	Rio Grande do Norte, Saint Peter and Saint Paul Archipelago
Heterocarpus ensifer A. Milne-Edwards, 1881[a]	Ceará, Rio Grande do Norte
Plesionika acanthonotus (Smith, 1882[a])	Ceará, Rio Grande do Norte
Plesionika edwardsii (Brandt, 1851[a])	Rio Grande do Norte
Plesionika ensis (A. Milne Edwards, 1881)	Pará, Ceará, Rio Grande do Norte
Plesionika martia (A. Milne-Edwards, 1883)	Ceará, Rio Grande do Norte
Eupasiphae gilesii (Wood-Mason, 1892)	Western Atlantic
Eupasiphae ostrovski Rodrigues and Cardoso, 2018	Rocas Atoll

[a]Species with first records for Brazilian states and oceanic islands

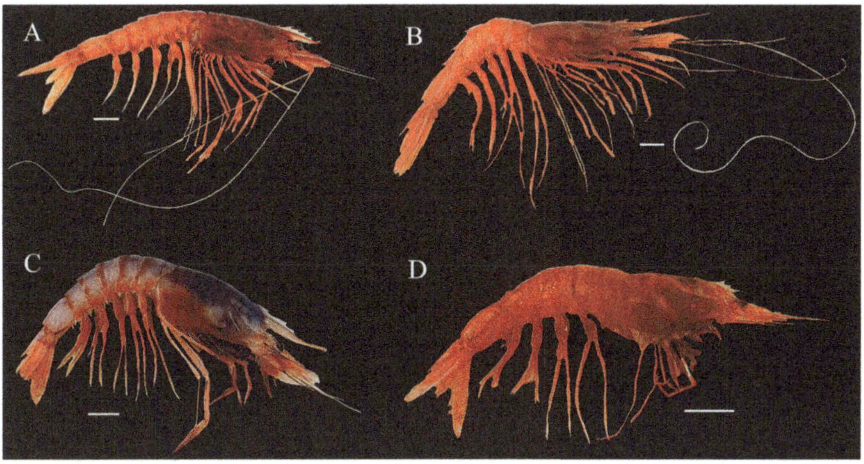

Fig. 21.1 (**a**) *Aristaeomorpha foliacea* (Risso, 1827); (**b**) *Aristaeopsis edwardsiana* (Johnson 1867); (**c**) *Aristeus antillensis* A. Milne-Edwards and Bouvier, 1909; (**d**) *Benthesicymus bartletti* Smith, 1882. All sampled by REVIZEE/Score Central, deposited at Museu Nacional/UFRJ (unknown catalogue number). Scale bars = 1 cm

Mediterranean, Namibia, SW Africa. Indian Ocean: off east coast of South Africa, Mozambique, Madagascar, Réunion, Maldives Islands, Sri Lanka. Pacific Ocean: Philippines, Indonesia, Taiwan, Japan, Eastern Australia, New Caledonia, New Zealand, Wallis and Futuna Islands, Fiji. **Depth range**. From 61 to 1300 m.

2. *Aristaeopsis edwardsiana* (Johnson, 1867) (Fig. 21.1b).

    Penaeus edwarsianus Johnson, 1867: 897.
    Aristeus coralinus Spence-Bate, 1888: xxxii, Fig. X.
    Plesiopenaeus edwardsianus– Alcock 1901a: 35–37; Barnard 1950: 624; Crosnier and Forest 1973: 291, Figs. 98, 99a-b; Crosnier 1978: 86–92, Table 11, Figs. 31a-c, 32a-c, 33a-c; Kensley et al. 1987: 281; Pérez-Farfante 1988: 7, Fig. 8.
    Aristaeopsis edwardsiana– Pérez-Farfante and Kensley 1997: 36–39, Figs. 7–8; Dall 2001: 413–414, Fig. 3; Serejo et al. 2007: 138; Tavares and Serejo 2007: 9, Figs. 5–7; Lira et al. 2017a, b: 32; Alves-Júnior et al. 2019a: 4, Figs. 3, 4, 8b, 9b.

Type locality. Off Madeira, Northeast Atlantic Ocean. **Diagnosis**. Crosnier (1973, 1978), Tavares and Serejo (2007). **Distribution**. Pelagic. Western Atlantic: Gulf of Mexico, Caribbean Sea to French Guiana, Bermuda, Brazil (Pará, Amapá, Ceará and Rio Grande do Norte, Bahia, Espírito Santo, Santa Catarina), Uruguay. Eastern Atlantic: Bay of Biscay, Azores, Madeira, Canary Islands, Portugal, Morocco, Western Sahara to South Africa. Indian Ocean: Madagascar, Arabian Sea, Bay of Bengal, Andaman Sea. Pacific Ocean: Indonesia, Japan, South China Sea, Eastern Australia, Wallis and Futuna Islands. **Depth range**. From 200 to 1850 m.

3. *Aristeus antennatus* (Risso, 1816).

Peneus Antennatus Risso, 1816: 95, pl. u, Fig. 6.
Sicyonia duvernoii Risso, 1844: 95 (nomen nudum).
Penaeus antemarius Costes, 1890: 558 (nomen nudum).
Aristeus antennatus– Crosnier and Forest, 1973: 288; Crosnier 1978: 71; Pérez-Farfante and Kensley 1997: 39; Serejo et al. 2007: 138; Tavares and Serejo 2007: 13, Figs. 8–10.

Type locality. Off Nice, Mediterranean Sea. **Diagnosis**. Tavares and Serejo (2007). **Distribution**. Pelagic. Western Atlantic: Brazil (Bahia, Espírito Santo). Eastern Atlantic: from Portugal to Cape Verde Islands, Azores, Mediterranean, Natal, South Africa. Indian Ocean: Mozambique, Zanzibar, Madagascar, Réunion, Maldives Islands. **Depth range**. From 150 to 1799 m.

4. *Aristeus antillensis* A. Milne-Edwards and Bouvier, 1909 (Fig. 21.1c).

Aristeus antillensis A. Milne-Edwards and Bouvier, 1909: 201–203, pl. I, Fig. 8–13; Crosnier and Forest 1973: 290, 291; 1978: 61; Pérez-Farfante 1988: 6, Fig. 7; Pérez-Farfante and Kensley 1997: 41; Serejo et al. 2007: 138; Tavares and Serejo 2007: 19, Figs. 11–13; Lira et al. 2017a, b: 32; Alves-Júnior et al. 2019a: 4, Figs. 4, 5, 6, 8c, 9c.

Type locality. Nevis Island, Antilles. **Diagnosis**. Pérez-Farfante (1988), Tavares and Serejo (2007). **Distribution**. Pelagic. Western Atlantic: Delaware to Florida, Gulf of Mexico, Caribbean Sea to French Guiana, Brazil (Amapá, Maranhão, Ceará, Rio Grande do Norte, Bahia, Espírito Santo). **Depth range**. From 200 to 2057 m.

5. *Hemipenaeus carpenteri* Wood-Mason, 1891.

Hemipenaeus carpenteri Wood-Mason, 1891: 189; Crosnier 1978: 76–80, Fig. 27c–d, 28a–b, 29a; Pérez-Farfante and Kensley 1997: 46; Serejo et al. 2007: 138; Tavares and Serejo 2007: 22, Figs. 14–16; Lira et al. 2017a, b: 33; Alves-Júnior et al. 2019a: 5, Figs. 7, 8d, 9d.
Hemipenaeus triton Faxon, 1893: 215.
Aristaeus (Hemipenaeus) carpenteri– Alcock, 1901a: 32–33.

Type locality. Bay of Bengal. **Diagnosis**. Crosnier (1978), Tavares and Serejo (2007). **Distribution**. Pelagic. Western Atlantic: Bahamas, Gulf of Mexico, Caribbean Sea, Brazil (Bahia, Rio Grande do Norte, Espírito Santo). Indian Ocean: Madagascar, Réunion, Arabian Sea, Bay of Bengal, Western Australia. Pacific Ocean: Japan, Wallis and Futuna Islands, Northeastern Australia, Gulf of Panama, Galapagos Islands. **Depth range**. From 900 m to 3900 m.

6. *Hepomadus tener* Smith, 1884.

Hepomadus tener Smith, 1884: 409, pl. 9, Figs. 7–8; Burkenroad 1936: 86–89; Pérez-Farfante 1973: 442, Figs. 1–8; Pérez-Farfante and Kensley 1997: 46, Figs. 15–16; Serejo et al. 2007: 138; Tavares and Serejo 2007: 28, Figs. 17–19; Lira et al. 2017a, b: 33.

Type locality. Northwestern Atlantic, off eastern USA. **Diagnosis**. Pérez-Farfante (1973), Tavares and Serejo (2007). **Distribution**. Western Atlantic: Gulf of Mexico, Caribbean Sea, Venezuela, Brazil (Bahia, Espírito Santo, Rio de Janeiro, Paraná, Santa Catarina). Eastern Atlantic Ocean: off Azores, Madeira, Canary Islands, Cape Verde. Indian Ocean: Zanzibar, Madagascar, Reunion Island, Maldives Islands, Bay of Bengal, Northwestern Australia. Pacific Ocean: Japan, Philippines, Tuamotu Islands, Wallis and Futuna Islands, Hawaii. **Depth range**. From 765 to 5400 m.

7. *Cerataspis monstrosus* Gray, 1828.

   Cerataspis monstrosus Gray, 1828: 8, pl. 6, Figs. 5a, b; Lira et al. 2017a, b: 338, Fig. 2.
   Aristaeus armatus Spence-Bate, 1881: 188; 1888: 312–317, pl. 45–46, Figs. 1–2.
   Aristaeus (Aristaeopsis) armatus– Alcock 1901a: 41.
   Plesiopenaeus armatus– Crosnier and Forest 1973: 294–296, Fig. 99c-d; Crosnier 1978: 92–94, Figs. 31d–e, 32d-f, 33b; Pérez-Farfante and Kensley 1997: 50–52, Figs. 19–20; Serejo et al. 2007: 138; Tavares and Serejo 2007: 33, Figs. 20–22.
   Cryptopus Defrancii Latreille, 1829: 100; H Milne Edwards 1837: 439.
   Cerataspis monstruosus– H Milne Edwards 1837: 438.

Type locality. Gray (1828, p. 8) "Found in the stomach of a dolphin off the coast of Brazil." **Diagnosis**. Crosnier (1978), Tavares and Serejo (2007). **Distribution**. Pelagic. Western Atlantic: Gulf of Mexico, Venezuela, Brazil (Rio Grande do Norte, Pernambuco, Fernando de Noronha Archipelago, Bahia, Espírito Santo). Eastern Atlantic: Azores, Madeira, Canary Islands, Cape Verde. Indian Ocean: Zanzibar, Madagascar, Maldives Islands, Bay of Bengal. Pacific Ocean: Japan, Philippines, Tuamotu Islands, Wallis and Futuna Islands, northeast of Australia. Hawaii. **Depth range**. From 752 to 5413 m.

8. *Plesiopenaeus coruscans* (Wood-Mason in Wood-Mason and Alcock, 1891).

   Aristeus coruscans Wood-Mason in Wood-Mason and Alcock, 1891: 280, Fig. 6.
   Aristaeus (Plesiopenaeus) coruscans– Alcock 1901a: 37–38.
   Plesiopenaeus coruscans– Burkenroad 1936: 95–100; Crosnier 1978: 86–96, Figs. 31f–g, 32c–f; Pérez-Farfante and Kensley 1997: 52; Serejo et al. 2007: 138; Tavares and Serejo 2007: 38, Figs. 23–25.

Type locality. Bay of Bengal. **Diagnosis**. Crosnier (1978), Tavares and Serejo (2007). **Distribution**. Western Atlantic: Gulf of Mexico, Bahamas, Brazil (Bahia, Espírito Santo). Indian Ocean: Madagascar, Arabian Sea, Bay of Bengal, Andaman Islands. Pacific Ocean: Northeastern Australia. **Depth range**. From 900 to 2367 m.

Family Benthesicymidae Wood-Mason and Alcock, 1891.

9. *Altelatipes carinatus* (Smith, 1884).

Benthesicymus carinatus Smith, 1884: 396; pl 10, Figs. 6–7; Tavares 2009: 204, Figs. 2–6.
Benthesicymus expansus Kensley, 1977: 22; Figs. 4–5.
Altelatipes carinatus– Crosnier and Vereshchaka 2008: 399.

Type locality. *Albatross* stn 2094, 39°44′30″N, 71°4′W, off Long Island, Northwestern Atlantic. **Diagnosis**. Kensley (1977), Tavares (2009). **Distribution**. Pelagic. Western Atlantic: Caribbean Sea, Brazil (Bahia, Espírito Santo, Rio de Janeiro). Indian Ocean: eastern South Africa, Réunion, Arabian Sea. Pacific Ocean: northwest area. **Depth range**. From 599 to 2076 m.

10. *Bentheogennema intermedia* (Spence-Bate, 1888).

Gennadas intermedius Spence-Bate, 1888: 343; pl. 58, Fig. 3.
Gennadas alicei Bouvier, 1906a: 748.
Bentheogennema intermedia– Burkenroad 1936: 56; Barnard 1950: 634; Roberts and Pequegnat 1970: 39; Coelho and Ramos 1972: 138; Kensley 1972: 10; Crosnier and Forest 1973: 278; Omori 1974: 236; Crosnier 1978: 30; Burukovsky 1982: 63; Freitas 1984: 22; Ramos-Porto et al. 1989: 227; D'Incao 1998: 311.

Type locality. Africa, Sierra Leone, 1°47′N, 24°26′W; between Bermuda and Azores, 35°59′S, 1°34′E. **Diagnosis**. Spence-Bate (1888). **Distribution**. Pelagic. Western Atlantic: USA (Florida), Bermuda, Gulf of Mexico, Brazil (Pernambuco). Eastern Atlantic: Spanish, Portugal (Azores, Canary and Madeira Islands), Morocco, Cape Verde, Sierra Leone, Gabon, Zaire, western South Africa. Indian Ocean: Mozambique, Madagascar, Tanzania, Maldives Islands, Arabian Sea. Pacific Ocean: Japan, Hawaii, Mexico (Baja California). **Depth range**: From 900 to 4360 m.

11. *Benthesicymus bartletti* Smith, 1882 (Fig. 21.1d).

Benthesicymus bartletti Smith, 1882: 82; pl. 14, Figs. 1–7; Burkenroad 1936: 47; Crosnier and Forest 1973: 275, Figs. 92a, b; Serejo et al. 2007: 138; Lira et al. 2017a, b: 33; Alves-Júnior et al. 2019a: 6, Figs. 10, 14a, 15a.
Benthesicymus pleocanthus Spence-Bate 1888: 334, Fig. 48, pl. 57, Fig. 2.

Type locality. *Blake* stn 343, 39°45′40″N, 70°55′W, off Massachusetts, Northwestern Atlantic. **Diagnosis**. Burkenroad (1936). **Distribution**. Pelagic. Western Atlantic: Canada, Virginia, Bahamas, Belize, Caribbean Sea, Honduras, Panama, Lesser Antilles, Barbados, Grenada, Tobago, Colombia, Venezuela, Guyana, Surinam, French Guiana, Brazil (Ceará, Rio Grande do Norte, Bahia, Rio de Janeiro, Santa Catarina). Eastern Atlantic: Azores, Alboran Sea, Cadiz Gulf, Morocco, Canaries, Cape Verde, Mauritania, Gabon, Congo. Indian Ocean: Bay of Bengal. Pacific Ocean: Philippines, North Pacific (37°49′S, 166°47′W). **Depth range**. From 180 to 5777 m.

12. *Gennadas bouvieri* Kemp, 1909.

Gennadas bouvieri Kemp, 1909: 726, pl. 74, Figs. 1–4, pl. 75, Figs. 6–7; D'Incao 1998: 311; Lira et al. 2017a, b: 33; Alves-Júnior et al. 2019a: 8, Figs. 11, 14b, 15b.
Gennadas alcocki Kemp, 1910: 174 (in part, males only).

Type locality. Philippines; West of Manila, 17°54′N, 117°14′E; North of New Guinea, 0°42′S, 147°E. **Diagnosis**. Kemp (1909). **Distribution**. Pelagic. Western Atlantic: Bermuda, Gulf of Mexico, Bahamas, Caribbean Sea, Venezuela, Brazil (Ceará Chain, Rio Grande do Norte, Fernando de Noronha Archipelago, Pernambuco). Eastern Atlantic: west coast of South Africa. Indian Ocean: Tanzania (Zanzibar), Madagascar, Maldives, Gulf of Arabia, Bay of Bengal. Pacific Ocean: Indonesia, Philippines, Eastern Australia, Japan. **Depth range**. From 250 to 4970 m.

13. *Gennadas capensis* Calman, 1925.

Gennadas capensis Calman, 1925: 5, pl. I, Figs. 1, 2; Burkenroad 1936: 67, Figs. 51, 53; Barnard 1950: 630; Kensley 1971: 277, Fig. 3a–e; Dall 2001: 430, Fig. 15; Hendrickx 2015: 423; Lira et al. 2017a, b: 33; Alves-Júnior et al. 2018a: 377, Fig. 1a–c.

Type locality. Off Cape, South Africa. **Diagnosis**. Kensley (1971), Dall (2001), Alves-Júnior et al. (2018a). **Distribution**. Pelagic. Western Atlantic: Gulf of Mexico, Caribbean Sea, Bahamas, Venezuela, Brazil (Seamounts Ceará Chain, Atol das Rocas, and Fernando de Noronha Archipelago). Eastern Atlantic: western South Africa. Indian Ocean: Western Australia. Pacific Ocean: New Caledonia, Wallis and Futuna Islands. **Depth range**. From 505 to 2000 m.

14. *Gennadas scutatus* Bouvier, 1906a.

Gennadas scutatus Bouvier, 1906a: 748; 1908: 42–44, plate VIII; A. Milne-Edwards and Bouvier 1909: 193, 194, Figs. 10–12; Burkenroad 1938: 59, 60; Kensley 1971: 288, 289, Fig. 10; 1972, Figs. 4d, 6 g; Crosnier and Forest 1973: 281–283, Fig. 94a, 95a, b; Kensley et al. 1987: 279; Dall 2001: 434, 435, Fig. 20; Guzmán 2008: 29; Hendrickx 2015: 423; Alves-Júnior et al. 2018a: 381, Fig. 4a–c.

Type locality. "l'Hirondelle," stn 156, between Azores and Newfoundland. **Diagnosis**. Kensley (1971), Dall (2001), Alves-Júnior et al. (2018a). **Distribution**. Pelagic. Western Atlantic: Caribbean Sea, Brazil (Seamounts Ceará Chain and Fernando de Noronha Archipelago). Mid-Atlantic Ridge. Eastern Atlantic: off Cape Peninsula, western South Africa. Indian Ocean: Agulhas Basin. Pacific Ocean: Eastern Australia, Mexico, Peru, Chile. **Depth range**. From 92 to 3400 m.

15. *Gennadas talismani* Bouvier, 1906a.

Gennadas talismani Bouvier, 1906a: 749; Lenz and Strunk 1914: 311–313, plate XVIII, Fig. 1–14; Crosnier and Forest 1969: 549; Kensley 1971: 289, 290, Fig. 11; 1972, Figs. 4j, 6d; Alves-Júnior et al. 2018a: 382, Fig. 5a, b.

Type locality. 16°38′N, 18°24′W, Cape Verde. **Diagnosis**. Alves-Junior et al. (2018a). **Distribution**. Pelagic. Western Atlantic: Gulf of Mexico, Brazil (Seamounts of Ceará Chain and Fernando de Noronha Archipelago). Eastern Atlantic: Walvis Ridge, Cape Verde, Gabon, Angola, western South Africa. **Depth range**. From 100 to 4000 m.

Family Penaeidae Rafinesque, 1815.

16. *Funchalia danae* Burkenroad, 1940.

Funchalia danae Burkenroad, 1940, 36; Alves-Júnior et al. 2019a: 9, Figs. 12, 14c, 15c.

Type locality. Canary Islands, 29°13′N 14°12′W. **Diagnosis**. Burkenroad (1940). **Distribution**. Western Atlantic: Brazil (Ceará, Fernando de Noronha Archipelago, Rio Grande do Norte). Eastern Atlantic: Azores, Madeira, Canary, Morocco, Cape Verde, Guinea, Congo. Indian Ocean: Arabian Sea, South India. **Depth range**. From 60 to 560 m.

17. *Funchalia villosa* (Bouvier, 1905a).

Hemipenaeopsis villosus Bouvier, 1905a: 981 (in part).
Funchalia woodwardi.– Bouvier, 1907: 952 (in part); 1908: 93 (in part), pl. 1, Fig. 7, pl. 15, Fig. 3, 19, pl. 16, Fig. 9–21.
Funchalia vanhöffeni Lenz and Strunck, 1914: 306, pl. 17, Fig. 20–21.
Funchalia villosa– Burkenroad, 1936: 129; Crosnier 1985: 869, Figs. 13a–e, 14b, c; Kensley et al. 1987: 281; Wasmer 1989: 483, Fig. 2c; Serejo et al. 2007: 138; Lira et al. 2017a, b: 33; Alves-Júnior et al. 2019a: 10, Figs. 13, 14d, 15d.

Type locality. Between Canary and Azores. **Diagnosis**. Crosnier (1985). **Distribution**. Pelagic. Western Atlantic: Gulf of Mexico, Mexico, Caribbean Sea, Venezuela, Brazil (Ceará, Rio Grande do Norte, Fernando de Noronha Archipelago, Bahia, Espírito Santo, Rio de Janeiro, Paraná, Santa Catarina, Rio Grande do Sul), Uruguay. Eastern Atlantic: Azores, Madeira, Canary, Turkay, Cape Verde. Mediterranean Sea, Tristan da Cunha, Valdivia Bank. Pacific Ocean: Eastern Australia. **Depth range**. From 36 to 2600 m.

18. *Parapenaeus americanus* Rathbun, 1901.

Parapenaeus americanus Rathbun, 1901: 102, pl. 2; D'Incao 1998: 312; Serejo et al. 2007: 138; Alves-Júnior et al. 2019a: 13, Figs. 17, 20b, 21b.

Type locality. Western Atlantic, Puerto Rico Mayaguez Harbor. **Diagnosis**. Rathbun (1901). **Distribution**. Pelagic. Western Atlantic: USA (Florida), Gulf of Mexico, Bahamas, Cuba, Puerto Rico, Saint Lucia Island, Martinique, Brazil (Rio Grande do Norte, Rio de Janeiro, São Paulo, Paraná, Santa Catarina, Rio Grande do Sul), Uruguay. **Depth range**. From 54 to 613 m.

19. *Penaeopsis serrata* Spence-Bate, 1881 (Fig. 21.2a).

Penaeopsis serratus Spence-Bate, 1881: 183.

Parapenaeus megalops Smith, 1885b: 172; Faxon 1896:163; Alcock 1905: 520.

Artemesia talismani Bouvier, 1905a: 982.

Penaeopsis serratus var. *antillensis* A. Milne-Edwards and Bouvier, 1909: 226; pl. 3, Fig. 10, pl. 4, Fig. 5.

Penaeopsis serrata– Pérez-Farfante 1977a: 297; 1977b: 180; Pérez-Farfante 1980: 748, Figs. 28–38; Serejo et al. 2007: 138; Lira et al. 2017a, b: 34; Alves-Júnior et al. 2019a: 10, Figs. 16, 20a, 21a.

Type locality. Gulf of Mexico, off Barbados, 12°58′33″N, 59°36′45″W. **Diagnosis**. Spence-Bate (1881). **Distribution**. Pelagic. Western Atlantic: USA, New Jersey to Florida, Gulf of Mexico, Bahamas, Cuba, Caribbean Sea, Jamaica, Dominican Republic, Puerto Rico, Lesser Antilles, Dominica, Saint Lucia, Saint Vincent, Barbados, Grenada, Tobago, Trinidad, Belize, Nicaragua, Panama, Colombia, Venezuela, Guiana, Suriname, French Guiana, Brazil (Ceará, Rio Grande do Norte, Paraná, Santa Catarina, Rio Grande do Sul), Uruguay. Eastern Atlantic: Portugal, Mediterranean Sea, Morocco, Sahara, Mauritania. **Depth range**. From 183 to 750 m.

Family Sergestidae Dana, 1852.

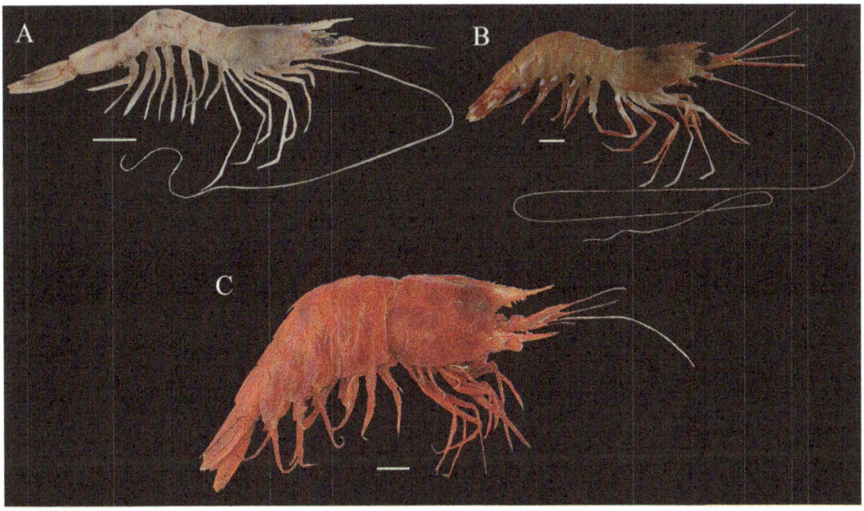

Fig. 21.2 (**a**) *Penaeopsis serrata* Spence-Bate, 1881; (**b**) *Solenocera acuminata* Pérez-Farfante and Bullis, 1973; (**c**) *Acanthephyra eximia* Smith, 1884. All sampled by REVIZEE/Score Central, deposited at Museu Nacional/UFRJ (unknown catalogue number). Scale bars = 1 cm

20. *Deosergestes corniculum* (Køyer, 1855).

Sergestes corniculum Krøyer, 1855: 22, Fig. 10; Cardoso and Serejo 2003; Serejo et al. 2007: 139; Vereshchaka 2009: 38, Figs. 16, 16, pl. 1a.
Sergestes laciniatus Krøyer, 1859: 274, 282, pl. 5, Fig. 15.
Sergestes longirostris Spence-Bate, 1888: 415, pl. 75, Fig. 3.
Sergestes (*Sergestes*) *curvatus* Crosnier and Forest, 1973: 315, Figs. 105i–k, 107c-d, f–g.
Sergestes (Sergestes) corniculum– Yaldwyn 1957: 7; Kensley 1968: 307, Figs. 9a, 10b, 11b; 1971: 236, Fig. 10.
Deosergestes corniculum– Judkins and Kensley 2008: 75; Vereshchaka et al. 2014: 4.

Type locality. 4.5°N, 21.5°W, off Liberia. **Diagnosis**. Kensley (1971), Cardoso and Serejo (2003), Vereshchaka (2009). **Distribution**. Pelagic. Western Atlantic: Brazil (Espírito Santo). Eastern Atlantic: Azores, Canary, Sargassum Sea. Mediterranean. **Depth range**. From 50 to 2000 m.

21. *Deosergestes paraseminudus* (Crosnier and Forest, 1973).

Sergestes (Sergestes) paraseminudus Crosnier and Forest, 1973: 313, Figs. 105e–h, 107a–b, e.
Sergestes paraseminudus– Cardoso and Tavares, 2006: 10, Fig. 1a–e; Serejo et al. 2007: 139; Vereshchaka 2009: 46, Figs. 18, 20.
Deosergestes paraseminudus– Judkins and Kensley 2008: 75; Vereshchaka et al. 2014: 4.

Type locality. 1°55′S 8°30′E, off Gabon. **Diagnosis**. Cardoso and Tavares (2006), Vereshchaka (2009). **Distribution**. Pelagic. Eastern Atlantic: Gabon, Congo, Angola. Western Atlantic: Brazil (Rio Grande do Norte, Bahia). **Depth range**. From 50 to 1170 m.

22. *Parasergestes armatus* Krøyer, 1855.

Sergestes armatus Krøyer, 1855: 10; Barnard 1950: 643, Fig. 120 m–p; Cardoso and Serejo 2003: 4, Fig. 1; Serejo et al. 2007: 139; Vereshchaka 2009: 93, Figs. 45, 46, pl. 92b.
Sergestes incertus Hansen, 1896: 962.
Sergestes diapontus– Illig 1927: 333, Figs. 111–118.
Sergestes extensus Hanamura, 1983: 64, Figs. 7–8.
Sergestes (Sergestes) armatus– Yaldwyn 1957: 8; Kensley 1968: 304; 1971: 232, Fig. 8.
Parasergestes armatus– Judkins and Kensley 2008: 77; Vereshchaka et al. 2014: 4.

Type locality. 7°37′N, 22.5°W, off Guinee. **Diagnosis**. Cardoso and Serejo (2003), Vereshchaka (2009). **Distribution**. Pelagic. Western Atlantic: Bermuda, Brazil (Rio de Janeiro). Eastern Atlantic: Azores, Canary. Mediterranean. Indian

Ocean: southwestern region. Pacific Ocean: northwestern region. **Depth range**. From 30 to 830 m.

23. *Phorcosergia burukovskii* (Vereshchaka, 2000).

Sergia burukovskii Vereshchaka, 2000: 121, Figs. 31–33; Cardoso and Tavares 2006: 13, Fig. 3a–e; Serejo et al. 2007: 139.
Sergestes (Sergia) grandis– Kensley 1971(part): 249, Fig. 17; Crosnier and Forest 1973 (part): 331, Figs. 113–116.
Phorcosergia burukovskii– Vereshchaka et al. 2014: 5; Alves-Júnior et al. 2019a: 14, Figs. 18, 20c, 21c.

Type locality. 23°26′S, 03°56′E, South Atlantic. **Diagnosis**. Vereshchaka (2000), Cardoso and Tavares (2006). **Distribution**. Pelagic. Western Atlantic: off Canada (Terra Nova, Grand Banks), Bermuda, Sargasso Sea, Surinam, French Guiana, Brazil (Ceará Chain, Rio Grande do Norte, Rocas Atoll, Pernambuco, Fernando de Noronha Archipelago, Espírito Santo). Eastern Atlantic: Azores, Canary, Cape Verde, off Namibia, western South Africa. **Depth range**. From 300 to 2478 m.

24. *Prehensilosergia prehensilis* (Spence-Bate, 1881).

Sergestes prehensilis Spence-Bate, 1881: 193; 1888: 385, pl. 71; Gordon 1935: 314, Figs. la, 3b, 6c, d,8a, b, c, 9a, b, c, d.
Sergestes gloriosus Stebbing 1905: 84, pls. 22, 23; 1910: 381; Barnard 1950: 642, Fig. 120 h, i, j.
Sergestes fujiyamaensis Nakazawa, 1932: 32.
Sergestes (Sergia) prehensilis.– Yaldwyn 1957: 9; Kensley 1968: 308.
Sergia prehensilis– Vereshchaka 2000: 160, Figs. 59–61; Cardoso and Serejo 2003: 9, Fig. 4; Serejo et al. 2007: 139.
Prehensilosergia prehensilis– Vereshchaka et al. 2014: 4.

Type locality. 34°58′N, 139°29′E, off Japan. **Diagnosis**. Vereshchaka (2000), Cardoso and Serejo (2003). **Distribution**. Pelagic. Western Atlantic: Brazil (Bahia, Espírito Santo, Rio de Janeiro). Eastern Atlantic: southwestern Africa. Indian Ocean: southwestern region. Pacific Ocean: northwestern region. Antarctic. **Depth range**. From 30 to 1700 m.

25. *Robustosergia regalis* (Gordon, 1939).

Sergestes regalis Gordon, 1939: 498, Figs. 1–4.
Sergestes (Sergia) regalis– Yaldwyn 1957: 9; Kensley 1968: 308, Figs. 9b, 10a, 11a.
Sergestes (Sergia) creber Burkenroad, 1940: 44; Yaldwyn 1957: 9; Kensley 1971: 247, Fig. 16.
Sergia regalis– Vereshchaka 2000: 149, Figs. 52a–d, 53a–c, 54, pl. 2a; Serejo et al. 2007: 139.
Robustosergia regalis– Vereshchaka et al. 2014: 5; Alves-Júnior et al. 2019a: 14, Figs. 19, 20d, 21d.

Type locality. 32°45′S, 8°47′W, South Atlantic. **Diagnosis**. Vereshchaka (2000), Cardoso and Serejo (2003). **Distribution**: Pelagic. Western Atlantic: USA (Florida), Gulf of Mexico, Cuba, Mexico, Caribbean Sea, Lesser Antilles, Guiana, Suriname, French Guiana, Brazil (Ceará Chain, Rio Grande do Norte, Rocas Atoll, Paraíba, Fernando de Noronha Archipelago, Rio de Janeiro, Paraná, Santa Catarina, Rio Grande do Sul). Eastern Atlantic: western South Africa. Indian Ocean: Mozambique, Madagascar, Tanzania, Somalia, Arabian Sea, South India, Bay of Bengal, Singapore, Thailand. Pacific Ocean: Indonesia, Philippines, Eastern China, Papua New Guinea, Eastern Australia, New Zealand, Galapagos. **Depth range**. From 100 to 2500 m.

26. *Sergia tenuiremis* (Krøyer, 1855).

Sergestes tenuiremis Krøyer, 1855: 30, 34; 1856: 39, 62, 67–70, pl. 4, Fig. 11a–b. Spence-Bate 1888: 420; Illig 1914: 349; 1927: 283, Figs. 6–10; Gurney and Lebour 1940: 21.
Sergestes kröyeri Spence-Bate, 1881: 193; 1888: 388, pl. 70, Figs. 3–4. Hansen 1903: 58; 1920: 479; Illig 1914: 354 (part); 1927: 289 (part); Burkenroad 1940: 50; Dennell 1955: 403; Richardson and Yaldwyn 1958: 26.
Sergestes junceus Spence-Bate, 1888: 416, pl. 76, Fig. 1.
Sergestes longicollis Spence-Bate, 1888: 421, Fig. 1.
Sergestes tropicus Sund, 1920a: 18, Figs. 27–28, 30–32.
Sergestes (*Sergia*) *tenuiremis*– Yaldwyn 1957: 9; Donaldson 1975: 45.
Sergestes (*Sergia*) *kröyeri*– Yaldwyn 1957: 9; Crosnier and Forest 1973: 308; Lagardère 1978: 7.
Sergia kröyeri– Krygier and Wasmer 1988: 72.
Sergia tenuiremis– Krygier and Pearcy 1981: 101, Fig. 1; Vereshchaka 1994: 76, Figs. 1–3, 26; 2000: 84, Figs. 3–5; Cardoso and Tavares 2006: 13, Fig. 4a–e; Serejo et al. 2007: 139; Vereshchaka et al. 2014: 5.

Type locality. 2°N, 21°W, Atlantic Ocean. **Diagnosis**. Vereshchaka (2000), Cardoso and Tavares (2006). **Distribution**. Pelagic. Western Atlantic: Brazil (Espírito Santo). Eastern Atlantic: Canary Islands, Azores, Madeira, Gulf of Guinea. Pacific Ocean: Kermadec, New Zealand. Western Pacific (Hawaii, Oregon). **Depth range**. From 330 to 2000 m.

Family Solenoceridae Wood-Mason, 1981.

27. *Hadropenaeus modestus* (Smith, 1885).

Hymenopenaeus modestus Smith, 1885: 183; Burkenroad 1936: 104; Crosnier and Forest 1973: 259.
Haliporus modestus– Bouvier 1905a: 980; A. Milne-Edwards and Bouvier 1909: 209.
Hadropenaeus modestus– Pérez-Farfante 1977a: 323, Fig. 50; Abele and Kim 1986: 8; Ramos-Porto et al. 1987: 224; D'Incao 1998: 316; Coelho et al. 1990: 23; Coelho et al. 2006: 46; Alves-Júnior et al. 2017a: 474, Figs. 1 a–b, 2.

Type locality. Delaware, off Bethany Beach, 38°31′N, 73°21′W. **Diagnosis**. Pérez-Farfante (1977a), Alves-Júnior et al. (2017a). **Distribution**. Pelagic. Western Atlantic: USA (from Delaware to Florida), Bahamas, Gulf of Mexico, Lesser Antilles, Caribbean Sea, Panama, Trinidad and Tobago, Venezuela, Brazil (Ceará, Rio Grande do Norte, Alagoas). **Depth range**. From 146 to 550 m.

28. *Hymenopenaeus chacei* Crosnier and Forest, 1969.

Hymenopenaeus chacei Crosnier and Forest, 1969: 545, Figs. 1–2; Crosnier and Forest 1973: 261, Fig. 82; Cardoso et al. 2014a: 51; Alves-Júnior et al. 2017a: 476, Figs. 3 a–b, 4.

Type locality. Gabon, 8°35′S, 12°51′E. **Diagnosis**. Crosnier and Forest (1969), Alves-Júnior et al. (2017a). **Distribution**. Pelagic. Western Atlantic: Brazil (Ceará, Rio Grande do Norte). Eastern Atlantic: Mid-Atlantic Ridge, Portugal (Madeira Island), West Sahara, Morocco, Mauritanian, Senegal, Gabon, Guinea Bissau, Guinea, Namibia. **Depth range**. From 410 to 2068 m.

29. *Hymenopenaeus debilis* Smith, 1882.

Hymenopenaeus debilis Smith, 1882: 91; 1886a, b: 687; Burkenroad 1936: 111; Holthuis 1962: 108; Crosnier and Forest 1969: 545; Pérez-Farfante 1977a: 268; Serejo et al. 2007: 139.
Haliporus debilis– Faxon 1896: 163; Boone 1927: 78.

Type locality. Savannah Beach, Georgia, 31°57′00″N, 78°18′35″W; Southeast of Cape Fear, North Carolina, 33°19′00″N, 76°12′30″W, East of Cape Fear, North Carolina, 33°42′15″N, 76°00′50″W. **Diagnosis**. Crosnier and Forest (1969). **Distribution**. Pelagic. Western Atlantic: Hudson Canyon, New Jersey, Gulf of Mexico, Caribbean Sea, Guyana, Brazil (Rio de Janeiro, Espírito Santo, Rio Grande do Sul), Rio Grande Rise. Eastern Atlantic: Azores, northwest Africa, Morocco, Cape Verde, Canary Islands, Southwestern England. Mediterranean Sea. **Depth range**. From 300 to 2163 m.

30. *Hymenopenaeus laevis* (Spence-Bate, 1881).

Haliporus laevis Spence-Bate, 1881: 185; Spence-Bate 1888: 289, pl. 42, Fig. 2; Bouvier 1906b: 3.
Hymenopenaeus microps Smith, 1884: 413, pl. 10; Smith 1886a: 189; 1887: 688, pl. 16, Fig. 8; Wood-Mason and Alcock 1891: 188.
Haliporus androgynus Bouvier, 1906c: 253.
Hymenopenaeus laevis– Burkenroad 1936: 106; 1938: 61; Crosnier and Forest 1973: 253, Figs. 82a, 83b; Alves-Júnior et al. 2017a: 478, Figs. 5 a–b, 6.

Type locality. Mid-Atlantic. **Diagnosis**. Pérez-Farfante (1977a), Alves-Júnior et al. (2017a). **Distribution**. Pelagic. Western Atlantic: USA (Massachusetts, Georges Bank), Bahamas, Bermuda, and Brazil (Rio Grande do Norte). Eastern Atlantic: Mid-Atlantic Ridge, Mauritania, Senegal, Equatorial Guinea, Sierra Leone, Cameroon, Liberia. Indian Ocean: Arabian Sea (Laccadive Sea), Bay of

Bengal, Andaman Islands. Pacific Ocean: Philippines. **Depth range**. From 1110 to 4792 m.

31. *Mesopenaeus tropicalis* (Bouvier, 1906a).

Parartemesia tropicalis Bouvier, 1906a: 748.
Haliporus tropicalis– Bouvier 1905b: 4; 1907: 80; A. Milne-Edwards and Bouvier 1909: 217, Figs. 45–54, pl. 3, Figs. 1–19.
Hymenopenaeus tropicalis– Burkenroad 1936:103; Springer and Bullis 1956: 8.
Solenocera weymouthi Lindner and Anderson, 1941:181, Fig. 1a–e.
Mesopenaeus tropicalis– Pérez-Farfante 1977a: 333, Fig. 56; Pérez-Farfante and Kensley 1997; Ramos-Porto et al. 2000: 76; Alves-Júnior et al. 2017a: 481, Figs. 7 a–b, 8.

Type locality. Antilles. **Diagnosis**. Pérez-Farfante (1977a), Alves-Júnior et al. (2017a). **Distribution**. Pelagic. Western Atlantic: USA (North and South Carolina, Georgia, Florida), Gulf of Mexico, Bahamas, Caribbean Sea, Antilhas, Barbados, Dominican Republic, Nicaragua, Panama, Venezuela, Brazil (Amapá, Pará, Maranhão, Ceará, Rio Grande do Norte, Espírito Santo, Rio de Janeiro, São Paulo, Rio Grande do Sul). **Depth range**. From 30 to 915 m.

32. *Solenocera acuminata* Pérez-Farfante and Bullis, 1973 (Fig. 21.2b).

Solenocera acuminata Pérez-Farfante and Bullis, 1973: 8, Figs. 1c, 2, 4–6, 18, 19; Serejo et al. 2007: 139.

Type locality. North of Riviere Organabo, French Guiana, 07° 15' N, 53°35' W. **Diagnosis**. Pérez-Farfante and Bullis (1973). **Distribution**. Pelagic. Western Atlantic: Bahamas, Caribbean Sea, from British Honduras to northeastern Venezuela. Off South America, from the Gulf of Paria to French Guiana, Brazil (Bahia). **Depth range**. From 32 to 622 m.

Suborder Pleocyemata Burkenroad, 1963.
Infraorder Stenopodidea Spence-Bate, 1888.

Family Stenopodidae Claus, 1872.

33. *Odontozona lopheliae* Goy and Cardoso, 2014.

Odontozona lopheliae Goy and Cardoso, 2014: 558, Figs. 4–8.
Odontozona edwardsi– Becker et al. 2009: 792; Lessard-Pilon et al. 2010: 1885; Goy 2010: 251.

Type locality. Campos Basin, Rio de Janeiro, Brazil, 22°24′12"S, 46°06′18"W. **Diagnosis**. Goy and Cardoso (2014). **Distribution**. Benthic. Western Atlantic: off Sapelo Island, Georgia, USA; Green Canyon, Gulf of Mexico; Brazil (Rio de Janeiro). **Depth range**. From 459 to 665 m.

Infraorder Caridea Dana, 1852.

Family Acanthephyridae Spence-Bate, 1888.

34. *Acanthephyra acanthitelsonis* Spence-Bate, 1888.

Acanthephyra acanthitelsonis Spence-Bate, 1888: 745, pl. 125, Fig. 3; Chace 1936: 27; 1947: 16; 1986: 9; Barnard 1950: 668; Holthuis 1951: 27; Crosnier and Forest 1968: 1129; 1973: 31; Cardoso 2013: 210; Cardoso et al. 2014a: 52; Alves-Júnior et al. 2016a: 194; Alves-Júnior et al. 2019b: 435.

Type locality. Sierra Leone. **Diagnosis**. Spence-Bate (1888), Alves-Júnior et al. (2016a). **Distribution**. Bathypelagic. Western Atlantic: USA (off Florida, New Jersey, South Carolina, Virginia), Bermuda, Gulf of Mexico, Bahamas, Caribbean Sea, off French Guiana, Brazil (Rocas Atoll). South Equatorial Mid-Atlantic Ridge. Eastern Atlantic: Iceland, Faroe island, Guinea, Sierra Leone, Gabon, Congo, Angola, Namibia. **Depth range**. From 230 to 4000 m.

35. *Acanthephyra acutifrons* Spence-Bate, 1888.

Acanthephyra acutifrons Spence-Bate, 1888: 749 (part), pl. 126: Fig. 3; Chace 1940: 146, Fig. 23; 1986: 10, Figs. 2b, 4b, 5b; Cardoso and Young 2005: 8, Figs. 3–7; Pequegnat and Wicksten 2006: 95; Serejo et al. 2007: 139; Judkins 2014: 304; Alves-Júnior et al. 2019b: 403, Figs. 1 a–b, 2, 39a.

Type locality. Off Kepulauan Aru, Indonesia, 5°41′S, 134°04′30″E. **Diagnosis**. Chace (1986), Cardoso and Young (2005), Alves-Júnior et al. (2019b). **Distribution**. Bathypelagic. Western Atlantic: Gulf of Mexico, Cuba, Bahamas, Tortuga, Guiana, Suriname, French Guiana, Brazil (Rio Grande do Norte, Rocas Atoll, Pernambuco, Fernando de Noronha Archipelago, Alagoas, Bahia, Espírito Santo). Indian Ocean: western region, Madagascar, Sumatra. Pacific Ocean: Philippines, Eastern Australia, Japan. **Depth range**. From 50 to 4200 m.

36. *Acanthephyra armata* A. Milne-Edwards, 1881.

Acanthephyra armata A. Milne-Edwards, 1881: 12; Spence-Bate 1888: 744; Wood-Mason and Alcock 1892: 359; Kensley 1977: 18; Chace 1986: 10; Crosnier 1987: 697; Pequegnat and Wicksten 2006: 95; Alves-Júnior et al. 2016a: 196, Figs. 3 a–c, 4. Alves-Júnior et al. 2019b: 435.

Type locality. Lesser Antilles, off Saint Lucia. **Diagnosis**. Spence-Bate (1888), Alves-Júnior et al. (2016a). **Distribution**. Bathypelagic. Western Atlantic: USA (Louisiana, Florida, Albany), Gulf of Mexico, West Indies (Natal), Guadeloupe, Lesser Antilles, Brazil (Rio Grande do Norte). Indian Ocean: eastern South Africa, Madagascar, Arabian Sea, Bay of Bengal, Andamam Sea, Thailand, Vietnam. Pacific Ocean: Philippines, Indonesia, Japan, New Caledonia, Fiji Island, Polynesia. **Depth range**. From 37 to 2880 m.

37. *Acanthephyra curtirostris* Wood-Mason and Alcock, 1891.

Acanthephyra acutifrons Spence-Bate, 1888: 749 (part); Kemp 1906: 22.
Acanthephyra curtirostris Wood Mason and Alcock, 1891: 195; 1892: 364, pl. 3, Fig. 5; Faxon 1895: 164, pl. 43, Figs. 2–5; Kemp 1906: 22; Chace 1936: 26; 1940: 143, Fig. 21; Calman 1939: 194; Crosnier and Forest 1968:

1129; 1973: 39, Fig. 8a.; Kensley 1981: 21; Kensley et al. 1987: 283; Pequegnat and Wicksten 2006: 96; Alves-Júnior et al. 2019b: 404, Figs. 3 a–b, 4, 39b.

Type locality. Arabian Sea. **Diagnosis**. Chace (1940), Alves-Júnior et al. (2019b). **Distribution**. Bathypelagic. Western Atlantic: USA (Oregon), Bermuda, Bahamas, Gulf of Mexico, Caribbean Sea, Antilhas, Panama Basin, British Guiana, and Brazil (Pará, Rocas Atoll, Fernando de Noronha, Pernambuco). Eastern Atlantic: Madeira. Indian Ocean: east coast of Africa, Aldabra Atoll, Arabian Sea, Maldives Islands, Gulf of Bengal, Andaman Sea. Pacific Ocean: USA (Northward California, coast of Baja California), Peru. **Depth range**. From 65 to 5900 m.

38. *Acanthephyra eximia* Smith, 1884 (Fig. 21.2c).

Acanthephyra eximia Smith, 1884: 376; Crosnier and Forest 1973: 34, Fig. 7c–d; Chace 1986: 18, Figs. 2j, 4j, 5j, 6 h, 9A; Ramos-Porto et al. 1998: 326; Tavares 1999: 675; Ramos-Porto et al. 2000: 76; Cardoso and Young 2005: 14, Figs. 8–13; Cardoso and Serejo 2007: 44; Serejo et al. 2007: 139; Pequegnat and Wicksten 2006: 96; Poupin 2018: 102; Alves-Júnior et al. 2016a, b, c: 406, Figs. 5 a–b, 6, 39c.
Acanthephyra angusta Spence-Bate, 1888: 737, pl.12, Fig. 6.
Acanthephyra edwardsii Spence-Bate, 1888: 747, pl. 124, Fig. 1; Moreira 1901: 10.
Acanthephyra brachytelsonis Spence-Bate, 1888: 753, pl. 126, Fig. 7.
Acanthephyra pulchra A. Milne-Edwards, 1890: 163.
Acanthephyra frontieri Crosnier, 1987: 699, Fig. 1.

Type locality. 35°09′50″N, 74°57′40″W, off North Carolina. **Diagnosis**. Chace (1986), Cardoso and Young (2005), Alves-Júnior et al. (2019b). **Distribution**. Bathypelagic. Western Atlantic: from Cape Hatteras to Bahamas, Gulf of Mexico, Lesser Antilles (Guadeloupe), Brazil (Amapá, Pará, Ceará, Rio Grande do Norte, Sergipe, Alagoas, Bahia, Espírito Santo, Rio de Janeiro). Eastern Atlantic: Bay of Biscay, Gulf of Cadiz, near Gibraltar, Azores, Madeira, Canary, Morocco, Angola. Mediterranean Sea. Indian Ocean: southeastern Africa. Pacific Ocean: Indonesia, Philippines, Japan, Hawaii, New Zealand. **Depth range**. From 200 to 4700 m.

39. *Acanthephyra kingsleyi* Spence-Bate, 1888.

Acanthephyra kingsleyi Spence-Bate, 1888: 751, pl. 126, Fig. 4; Holthuis 1951: 28; Crosnier and Forest 1973: 37, Fig. 7 e–f; Alves-Júnior et al. 2019b: 408, Figs. 7 a–b, 8, 39d.
Acanthephyra purpurea– Ortmann 1893: 43 (part); Lenz and Strunck 1914: 326; Balss 1925: 252; Chace 1936: 27.
Acanthephyra Kingsleyi– Kemp 1906: 22; De Man 1920: 45; Balss 1925: 251.
Acanthephyra sexspinosa Kemp, 1939: 570; Barnard 1950: 669; Holthuis 1951: 26; Crosnier and Forest 1968: 1129.

Type locality. Southwest of Sierra Leone. **Diagnosis**. Spence-Bate (1888), Alves-Júnior et al. (2019b). **Distribution**. Bathypelagic. Western Atlantic: Brazil (Ceará, Rocas Atoll, Fernando de Noronha Archipelago). Eastern Atlantic: Senegal, Sierra Leone, Gabon, Congo, Angola. Mid-Atlantic Ridge. **Depth range**. From 65 to 4575 m.

40. *Acanthephyra quadrispinosa* Kemp, 1939.

Acanthephyra quadrispinosa Kemp, 1939: 576; Barnard 1950: 668, Fig. 124 g; Chace 1986: 26, Figs. 3 h, 4 t, 7 g, 10c, 14; Kensley 1987: 284; Cardoso and Young 2005: 21, Figs. 14–18; Serejo et al. 2007: 139; Judkins 2014: 304. Alves-Júnior et al. 2019b: 410, Figs. 9 a–b, 10, 39e.
Acanthephyra batei– Stebbing 1905: 107, pl. 24b.

Type locality. Indo-Pacific from the East African coast to 163°W and from 25°N to 42°S; South Atlantic from 32°S to 40°S. **Diagnosis**. Chace (1986), Cardoso and Young (2005), Alves-Júnior et al. (2019b). **Distribution**. Bathypelagic. Western Atlantic: Brazil (Pernambuco, Espírito Santo, Rio de Janeiro), Uruguay, south from 35°S to 40°S. Indian Ocean: eastern South Africa, Madagascar, Arabian Sea, Sumatra. Pacific Ocean: from eastern Africa to 163°W and from 25°N to 44°S, Indonesia, Australia, Papua New Guinea, Japan, Canada. **Depth range**. From 250 to 3716 m.

41. *Acanthephyra stylorostratis* (Spence-Bate, 1888).

Bentheocaris stylorostratis Spence-Bate, 1888: 726, pl. 123, Fig. 4.
Acanthephyra stylorostratis– Calman 1925: 14; Chace 1936: 30; 1940: 144, Fig. 22; 1986: 10; Cardoso and Young 2005: 27, Figs. 19–23; Cardoso and Serejo 2007: 44; Serejo and Cardoso 2010: 194, Fig. 1 a–c; Serejo et al. 2007: 139; Alves-Júnior et al. 2019b: 412, Figs. 11 a–c, 12, 39 f.

Type locality. 21°38′N, 44°39′W, Mid North Atlantic. **Diagnosis**. Chace (1940), Cardoso and Young (2005), Cardoso and Serejo (2007), Alves-Júnior et al. (2019b). **Distribution**. Bathypelagic. Western Atlantic: USA (off New Jersey, Florida), Bermuda, Gulf of Mexico, Tortuga, Brazil (Ceará, Rio Grande do Norte, Alagoas, Espírito Santo, Rio de Janeiro). Eastern Atlantic: Canary Islands, Madeira, Cape Verde, Sahara Occidental. Indian Ocean: South Africa. Pacific Ocean: Tuamotu. **Depth range**. From 700 to 3548 m.

42. *Ephyrina benedicti* Smith, 1885 (Fig. 21.3a).

Ephyrina benedicti Smith, 1885: 506; De Man 1920: 46; Crosnier and Forest 1973: 65, Figs. 18, 19a; Chace 1986: 33; Cardoso and Young 2005: 33: Figs. 24–28; Pequegnat and Wicksten 2006: 98; Serejo et al. 2007: 139; Alves-Júnior et al. 2019b: 414, Figs. 13 a–c, 14.
Tropiocaris planipes Spence-Bate, 1888: 835, pl. 136, Fig. 1.

Type locality. Northwestern Atlantic, off New York, 40°26′40″N 67°5′15″W. **Diagnosis**. Cardoso and Young (2005), Alves-Júnior et al. (2019b).

Fig. 21.3 (**a**) *Ephyrina benedicti* Smith, 1885; (**b**) *Notostomus elegans* A. Milne-Edwards, 1881; (**c**) *Glyphocrangon longirostris* (Smith, 1882); (**d**) *Glyphocrangon neglecta* Faxon, 1896. All sampled by REVIZEE/Score Central, deposited at Museu Nacional/UFRJ (unknown catalogue number). Scale bars = 1 cm

Distribution. Bathypelagic. Western Atlantic: South Greenland, Northwestern Atlantic (40°26′40″N 67°5′15″W), USA, Gulf of Mexico, Brazil (Ceará, Bahia, Espírito Santo). Eastern Atlantic: southwest Ireland, west Portugal, Canary, Gabon, Saint Tomé. Pacific Ocean: northeastern Philippine Sea, Japan, west of Bonin Islands, Hawaii. **Depth range**. From 300 to 5000 m.

43. *Ephyrina ombango* Crosnier and Forest, 1973.

> *Ephyrina ombango* Crosnier and Forest, 1973: 68, Figs. 20a, 21a, 22a; Chace 1986: 36, Figs. 18, 19; Vereshchaka 1990: 139; Alves-Júnior et al. 2019b: 415, Figs. 15 a–c, 16, 40a.

Type locality. Off São Tomé, Gulf of Guinea. **Diagnosis**. Chace (1986), Alves-Júnior et al. (2019b). **Distribution**. Bathypelagic. Western Atlantic: Gulf of Mexico, Brazil (Ceará, Rocas Atoll, Fernando de Noronha, Pernambuco). Eastern Atlantic: Cape Verde, Guinea, Sao Tome. Mid-Atlantic Ridge. Indian Ocean: Cocos-Keeling. Pacific Ocean: Sulu Sea, Philippines, Banda Sea, Indonesia, Panama, Easter Island, Nazca Ridge, Sala y Gòmez. **Depth range**. From 50 to 4000 m.

44. *Meningodora compsa* (Chace, 1940).

> *Notostomus compsus* Chace, 1940:156, Figs. 31, 32a–i.
> *Meningodora compsa*– Crosnier and Forest 1973: 48, Fig. 10e; Kikuchi 1985: 196; 1991: 25, Fig. 2; Chace 1986: 49; Alves-Júnior et al. 2019b: 416, Figs. 17 a–b, 18, 40b.

Type locality. Bermuda. **Diagnosis**. Alves-Júnior et al. (2019b). **Distribution**. Pelagic. Western Atlantic: Bermuda, Brazil (Pernambuco). Eastern Atlantic: Portugal (Azores Island), Senegal. Northern Mid-Atlantic Ridge. **Depth range**. From 680 to 1829 m.

45. *Meningodora longisulca* Kikuchi, 1985.

Meningodora longisulca Kikuchi, 1985: 191, Figs. 1–3; Kikuchi 1991: 27, Fig. 2; Alves-Júnior et al. 2019b: 418, Figs. 19 a–c, 20.

Type locality. Western North Pacific, near Japan. **Diagnosis**. Kikuchi (1985), Alves-Júnior et al. (2019b). **Distribution**. Pelagic. Western Atlantic: Brazil (Fernando de Noronha Archipelago). Pacific Ocean: Philippine Sea, off Japan. **Depth range**. From 0 to 2394 m.

46. *Meningodora mollis* Smith, 1882.

Meningodora mollis Smith, 1882: 74, pl. 11, Figs. 8–9, pl. 12, Figs. 5–9; Crosnier and Forest 1973: 44, Fig. 10c; Kensley et al. 1987: 285; Ramos-Porto et al. 1998: 327; Vereshchaka 1990: 139; Kikuchi 1991: 32, Fig. 6; Pequegnat and Wicksten 2006: 100; Alves-Júnior et al. 2019b: 420, Figs. 21 a–b, 22.
Hymenodora mollis– Spence-Bate 1888: 841, pl. 136, Fig. 5.
Notostomus fragilis Faxon, 1893: 207; 1895: 170, pl. 44, Fig. 2b.
Notostomus mollis– Balss 1925: 266, Fig. 37; Chace 1940: 164, Fig. 38.

Type locality. North Carolina (Cape Lookout). **Diagnosis**. Chace (1940), Alves-Júnior et al. (2019b). **Distribution**. Pelagic. Western Atlantic: Canada (Terra Nova), USA, Bermuda, Gulf of Mexico, Bahamas, Costa Rica, and Brazil (Fernando de Noronha, off Pernambuco). Eastern Atlantic: Spain (Bay of Biscay), Portugal (Canary Island), Gabon, Angola. Indo-Pacific Ocean: Somalia, Philippine Islands, China Sea, Panama, Galapagos Islands, Easter Island, Nazca ridge, seamounts Sala y Gòmez. **Depth range**. From water surface to 5000 m.

47. *Meningodora vesca* (Smith, 1887).

Notostomus viscus Smith, 1886a: 189 (nomen nudum).
Notostomus vescus Smith, 1887: 676; De Man 1920: 46.
Acanthephyra brevirostris Spence-Bate, 1888: 751, plate 126, Figs. 5–6.
Acanthephyra batei Faxon, 1895: 167 (nomen novum for *A. brevirostris* Spence-Bate 1888).
Acanthephyra Batei– Kemp 1906: 22; De Man 1920: 41.
Acanthephyra parvirostris Coutière, 1911: 157.
Notostomus Batei– Balss 1925: 267.
Meningodora vesca– Sivertsen and Holthuis 1956: 13; Crosnier and Forest 1968: 1130; 1973: 46, Fig. 10d; Zariquiey Alvarez 1968: 87; Foxton 1970: 955, Fig. 6; Chace 1986: 50; Kensley et al. 1987: 286; Cardoso 2006: 2, Figs. 1–4; Cardoso and Serejo 2007: 45; Alves-Júnior et al. 2019b: 421, Figs. 23 a–b, 24.

Type locality. 37°12′20″N, 69°39′00″W, off New Jersey. **Diagnosis**. Cardoso (2006), Cardoso and Serejo (2007), Alves-Júnior et al. (2019b). **Distribution**. Pelagic. Western Atlantic: Bermuda, Bahamas, Brazil (Rocas Atoll, Fernando de Noronha, Rio de Janeiro). Eastern Atlantic: Portugal, Azores, Canary, Gabon,

Angola. Indian Ocean: Bay of Bengal. Pacific Ocean: Philippines, Indonesia. **Depth range**. From 510 to 2500 m.

48. *Notostomus elegans* A. Milne-Edwards, 1881 (Fig. 21.3b).

Notostomus elegans A. Milne-Edwards, 1881: 7; Crosnier and Forest 1973: 49; Chace 1986: 56, Figs. 28d–f, 29c–e, 30; Kensley 1987: 287; Cardoso and Young 2005: 46, Figs. 34–38; Pequegnat and Wicksten 2006: 100; Cardoso and Serejo 2007: 46; Serejo et al. 2007: 139; Alves-Júnior et al. 2019b: 423, Figs. 25 a–c, 26, 40c.
Notostomus patentissimus Spence-Bate, 1888: 826, pl. 123, Figs. 1, 1a–c, 2.
Notostomus longirostris Spence-Bate, 1888: 833, pl. 135, Fig. 4.
Notostomus westergreni Faxon, 1893: 208.
Notostomus atlanticus Lenz and Strunck, 1914: 330; De Man 1920: 46.

Type locality. 24°36′N, 84°05′W, Antilles. **Diagnosis**. Chace (1986), Cardoso and Young (2005), Cardoso and Serejo (2007); Alves-Júnior et al. (2019b). **Distribution**. Pelagic. Western Atlantic: Bermuda, Bahamas, Gulf of Mexico, Brazil (Rio Grande do Norte, Rocas Atoll, Fernando de Noronha Archipelago, Espírito Santo, Rio de Janeiro). Eastern Atlantic: Portugal, Gulf of Cadiz, Açores, Western South Africa. Pacific Ocean: Philippines, Indonesia, Australian east coast, Ecuador, Easter Island, Nazca Ridge, seamounts Sala y Gòmez. **Depth range**. From 65 to 5380 m.

49. *Notostomus gibbosus* A. Milne-Edwards, 1881.

Notostomus gibbosus A. Milne-Edwards, 1881: 7; 1883, pl. 32; De Man 1920: 46; Chace 1936: 28; Crosnier and Forest 1973: 49, Fig. 13; Chace 1986: 57; Kensley et al. 1987: 288; Ramos-Porto et al. 1998: 327; Pequegnat and Wicksten 2006: 101; Poupin 2018: 102; Alves-Júnior et al. 2019b: 425, Figs. 27 a–c, 28, 40d.
Notostomus perlatus Spence-Bate, 1888: 831, pl. 135, Fig. 2; Kemp 1913: 66, pl. 7, Fig. 10; Chace 1936: 28; 1940: 170, Fig. 42; Holthuis 1951: 29; Crosnier and Forest 1968: 1130.
Notostonzus brevirostris Spence-Bate, 1888: 832, pl. 135, Fig. 3.
Notostoiniis perlatus– Stebbing 1893: 246, pl. 13.
Notostomus brevirostris– Moreira 1901: 10; De Man 1920: 46.

Type locality. Off Grenada. **Diagnosis**. Chace (198) Alves-Júnior et al. (2019b). **Distribution**. Bathypelagic. Western Atlantic: Bermuda, Gulf of Mexico, off Grenada, Lesser Antilles, and Brazil (Ceará, Fernando de Noronha Archipelago). Eastern Atlantic: Sierra Leone, Ghana, Senegal, Congo. Indian Ocean: Chagos Archipelago, Western Australia, Sumatra. Pacific Ocean: Indonesia, Marquesas Islands. **Depth range**. From 569 to 4000 m.

Family Bathypalaemonellidae de Saint Laurent, 1985.

50. *Bathypalaemonella texana* Pequegnat, 1970a.

Bathypalaemonella texana Pequegnat, 1970a: 81, Figs. 4.7, 4.8; Crosnier and Forest 1973: 156; Wicksten and Méndez 1983: 230; Chace 1997: 31; Cleva 2001: 780; Cardoso 2010a: 1, Fig. 1.

Type locality. Northwestern Gulf of Mexico, 27°01.6′N, 94°42′W. **Diagnosis**. Cardoso (2010a). **Distribution**. Benthic. Western Atlantic: northwestern Gulf of Mexico, Brazil (Rio de Janeiro). **Depth range**. From 620 to 1463 m.

Family Crangonidae Haworth 1825.

51. *Parapontocaris caribbaea* (Boone, 1927).

Aegeon caribbaeus Boone, 1927: 125, Fig. 28.
Parapontocaris caribbaea– Chace 1984: 30; Chan 1996: 319; Cruz et al. 2002: 189; Campos et al. 2005: 86, Figs. 49, 50; Cardoso 2013: 88, Fig. 2; Vázquez-Bader and Garcia 2013: 369; Alves-Júnior et al. 2018b: 47, Figs. 1 a–b, 2.

Type locality. Caribbean Sea. **Diagnosis**. Chan (1996); Cardoso (2013). **Distribution**. Benthic. Western Atlantic: Bahamas, Straits of Florida, Gulf of Mexico, Caribbean Sea, Honduras, Colombia, Surinam, French Guiana, Brazil (Rio Grande do Norte, Bahia, Espírito Santo, Rio de Janeiro). **Depth range**. From 251 to 885 m.

52. *Parapontophilus gracilis* (Smith, 1882).

Pontophilus gracilis Smith, 1882: 36, pl. 7, Figs. 2, 2a–c, 3,3a; Crosnier and Forest 1968: 1145; 1973: 242, Fig. 79e, f; Pequegnat 1970a: 113.
Parapontophilus gracilis– Campos et al. 2005: 89, Figs. 53, 54; Komai 2008: 271, Figs. 2, 20a; Cardoso 2013: 88, Fig. 3; Alves-Júnior et al. 2018b: 49, Figs. 3 a–c, 4.

Type locality. South Carolina, USA, 32°18.20′N, 78°43′W. **Diagnosis**. Komai (2008); Cardoso (2013). **Distribution**. Benthic. Western Atlantic: USA east coast (between 39° 57′ N and 32° 18′ N, New Jersey), Gulf of Mexico, Antilles, Colombia, Brazil (Rio Grande do Norte, Rio de Janeiro). Eastern Atlantic: Morocco, Senegal, Congo, Cabinda, Angola. Indian Ocean: east coast of Africa, Zanzibar, Gulf of Aden, Andaman Sea, Maldives. **Depth range**. From 294 to 3440 m.

53. *Parapontophilus longirostris* Komai, 2008.

Parapontophilus longirostris Komai, 2008: 305, Figs. 18, 19, 21A; Cardoso 2009b: 30, Figs. 1–4; 2013: 91.

Type locality. Tubuai Island, Austral Islands, 23°21.3′S, 149°33.9′W. **Diagnosis**. Komai (2008); Cardoso (2009a, b, 2013). **Distribution**. Benthic. Western Atlantic Ocean, off Brazilian coast. Indian and Pacific Oceans. **Depth range**. From 1070 to 1889 m.

54. *Prionocrangon brasiliensis* Anker et al., 2014.

Prionocrangon brasiliensis Anker et al., 2014: 273, Figs. 7, 8.

Type locality. 19°36′S, 38°53′W, off Espírito Santo, Brazil. **Diagnosis**. Anker et al. (2014). **Distribution**. Benthic. Western Atlantic: Brazil (Espírito Santo). **Depth range**. From 707 to 733 m.

55. *Sabinea hystrix* (A. Milne-Edwards, 1881).

Paracrangon hystrix A. Milne-Edwards, 1881: 6.
Sabinea princeps Smith, 1882: 38, pl. 8, Fig.1, 1a, 1b; 1886a: 189; 1887: 654, pl. 10, Figs. 1, 1a, 1b, 2.
Sabinea hystrix– Hansen 1908: 51; De Man 1920: 256, 302, 303; Holthuis 1955: 132, Fig. 95b; Sivertsen and Holthuis 1956: 40; Crosnier and Forest 1973: 232, Fig. 73c–d; Chace 1984: 58; Squires 1990: 12; Alves-Júnior et al. 2018b: 56, Figs. 10a; 11a–d, 12.

Type locality. Near Guadalupe. **Diagnosis**. Squires (1990); Alves-Júnior et al. (2018b). **Distribution**. Benthic. Western Atlantic: Greenland (Davis Strait, 64°54′ N), Nova Scotia (off La Have Bank to St. Pierre Bank), West Indies, Gulf of Mexico, Caribbean Sea, Guadalupe, Brazil (Ceará, Rio Grande do Norte). Eastern Atlantic: Bank of Galicia, Western Sahara. **Depth range**. From 550 to 3957 m.

Family Glyphocrangonidae Smith, 1884.

56. *Glyphocrangon aculeata* A. Milne-Edwards, 1881.

Glyphocrangon aculeatum A.Milne-Edwards, 1881:5; 1883, pl. 39.
Rhacocaris agassizii Smith, 1882:43, pl. 5, Fig. 2, pl. 6, Fig. 2.
Glyphocrangon aculeata– Spence-Bate 1888: 521, pl.94, Fig. 1; Pequegnat 1970a:104; Holthuis 1971: 323, Fig. 10; Forest and Holthuis 1997: 56; Ramos-Porto et al. 1998: 342; Komai 2004: 32, Fig. 1a, b; Coelho et al. 2006: 55; Serejo et al. 2007: 139; Felder et al. 2009: 1061; Vázquez-Bader and Garcia 2013: 371; Alves-Júnior et al. 2017b: 5, Fig. 2 a–b.

Type locality. Off Saint Vincent, West Indies. **Diagnosis**. Holthuis (1971), Komai (2004). **Distribution**. Benthic. Western Atlantic: from USA (off Cape Hatteras, North Carolina) to off Brazil (Ceará, Rio Grande do Norte, Pernambuco, Bahia, Espírito Santo, Rio de Janeiro, São Paulo), including the entire Gulf of Mexico and Caribbean Sea. **Depth range**. From 443 to 1760 m.

57. *Glyphocrangon alispina* Chace, 1939.

Glyphocrangon alispina Chace, 1939: 39; Pequegnat 1970a: 105; Holthuis 1971: 347, Fig. 15; Komai 2004: 33, Fig. 1c, d; Coelho et al. 2006: 55; Serejo et al. 2007: 139; Felder et al. 2009: 1061; Vázquez-Bader and Gracia 2013: 373; Alves-Júnior et al. 2017b: 6, Fig. 3 a–b.

Type locality. North of Matanzas Province, Cuba, 23°24′N, 81°00.5′W. **Diagnosis**. Holthuis (1971), Komai (2004). **Distribution**. Benthic.

Western Atlantic: USA (Florida), Caribbean Sea, Gulf of Mexico, Cuba from the Florida Straits to British Guiana, Brazil (Amapá, Ceará, Rio Grande do Norte, Bahia). **Depth range**. From 548 to 2094 m.

58. *Glyphocrangon aurantiaca* Holthuis, 1971.

Glyphocrangon aurantiaca Holthuis, 1971: 303, Fig.8; Takeda and Okutani 1983: 68; Komai 2004: 35, Fig. 2A, B; Serejo et al. 2007: 139.

Type locality. Caribbean Sea, off Tobago, 11°37.3′N, 60°59.4′W. **Diagnosis**. Holthuis (1971); Komai (2004). **Distribution**. Benthic. Western Atlantic: from Tobago to French Guiana, Brazil (Rio de Janeiro). **Depth range**. From 410 to 733 m.

59. *Glyphocrangon longirostris* (Smith, 1882) (Fig. 21.3c).

Rhachocaris longirostris Smith, 1882: 51, pl. 5, Fig. 1, pl. 6, Fig. 1.
Glyphocrangon longirostris – Pequegnat 1970a: 106; Holthuis 1971: 330, Figs. 11–13; Crosnier and Forest 1973: 230, Fig. 73a, b; Komai 2004: Fig. 2C, D; Coelho et al. 2006: 55; Cardoso and Serejo 2007: 40, Fig. 1; Serejo et al. 2007: 139; Felder et al. 2009: 1061; Alves-Júnior et al. 2017b: 7, Fig. 4 a–b.

Type locality. Off Cape Hatteras, North Carolina, USA, 35°41.03′N, 74°31.00′W. **Diagnosis**. Holthuis (1971); Komai (2004); Cardoso and Serejo (2007). **Distribution**. Benthic. Western Atlantic: from Massachusetts, USA, to off Suriname, Brazil (Bahia, Espírito Santo, Rio de Janeiro). Eastern Atlantic: from southwest of Ireland to Cape Point, South Africa. **Depth range**. From 1280 to 2500 m.

60. *Glyphocrangon neglecta* Faxon, 1896 (Fig. 21.3d).

Glyphocrangon neglecta Faxon, 1896: 159, pl. 1, Figs. 5, 6; Holthuis 1971: 319, Fig. 9; Takeda and Okutani 1983: 69; Komai 2004: 37, Fig. 3A, B; Serejo et al. 2007: 140.

Type locality. Off Grenada, 12°03.3′N, 61°47.1′W. **Diagnosis**. Holthuis (1971); Komai (2004). **Distribution**. Benthic. Western Atlantic: southern Caribbean Sea, from Panama to Surinam, Brazil (Bahia, Rio de Janeiro). **Depth range**. From 365 to 1050 m.

61. *Glyphocrangon nobilis* A. Milne-Edwards, 1881.

Glyphocrangon nobile A. Milne-Edwards, 1881: 5; 1883, pl. 40, Fig. 2, 2a; Pequegnat 1970a: 107.
Glyphocrangon nobilis– Holthuis 1971: 341, Fig. 14; Forest and Holthuis 1997: 56, pl. 40, Fig. 2, 2a; Komai 2004: 39, Fig. 3C, D; Serejo et al. 2007: 140.

Type locality. Off Dominica, 15°26.36′N, 61°36.45′W. **Diagnosis**. Holthuis (1971), Komai (2004). **Distribution**. Benthic. Western Atlantic: from South Carolina (USA) and the Bahamas Islands to Suriname, Brazil (Bahia). **Depth range**. From 410 to 2150 m.

62. *Glyphocrangon sculpta* (Smith, 1882).

Rhachocaris sculpta Smith, 1882: 49, pl. 5, Fig. 3, pl. 6, Fig. 3–3d.
Glyphocrangon sculptus– Smith 1886b: 608, 655, pl. 8, Fig. 3, pl. 9, Figs. 1, 2.
Glyphocrangon sculpta– Holthuis 1971: 279, Figs. 2, 3: Komai 2004: 39, Fig. 4A, B; Coelho et al. 2006: 55; Serejo et al. 2007: 140; Alves-Júnior et al. 2017b: 8, Fig. 5 a–b; Serejo et al. 2007: 140.

Type locality. Off Delaware, USA, 38°16.45′N, 73°10.30′W. **Diagnosis**. Holthuis (1971), Komai (2004). **Distribution**. Benthic. Western Atlantic: from USA (Massachusetts and Delaware) to Antilles and Bahamas, Brazil (Rio Grande do Norte, Bahia, Rio de Janeiro). Eastern Atlantic: from Iceland to Nigeria. **Depth range**. From 1645 to 3219 m.

63. *Glyphocrangon spinicauda* A. Milne-Edwards, 1881.

Glyphocrangon spinicauda A. Milne-Edwards, 1881: 3; 1883, pl. 40, Fig. 1, 1a; Pequegnat 1970a: 110; Holthuis 1971: 295, Figs. 6–7; Ramos-Porto et al. 1998: 342; Komai 2004: 40, Fig. 4c, d; Coelho et al. 2006: 55; Serejo et al. 2007: 140; Felder et al. 2009: 1061; Vázquez-Bader and Gracia 2013: 378; Alves-Júnior et al. 2017b: 10, Fig. 6 a–b.

Type locality. Off St. Christopher, West Indies, 17°19.27′S, 62°50.30′W. **Diagnosis**. Holthuis (1971), Komai (2004). **Distribution**. Benthic. Western Atlantic: USA (east coast of Florida) from Florida to Barbados, northwest of Cuba, Caribbean Sea, Yucatan, south of Jamaica, Honduras, Nicaragua, Guadalupe, Dominica, Brazil (Amapá, Pará, Maranhão, Ceará, Rio Grande do Norte, Bahia). **Depth range**. From 256 to 692 m.

Family Nematocarcinidae Smith, 1884.

64. *Nematocarcinus gracilipes* Filhol, 1884.

Nematocarcinus gracilipes Filhol, 1884: 232, Fig. 1; Crosnier and Forest 1973: 123; García Raso 1996: 734; Burukovsky 2000a: 4; 2004a: 558; 2009: 81; 2012: 116; Cardoso and Burukovsky 2014: 440, Figs. 2–4.
Nematocarcinus agassizii Faxon, 1893: 204; 1895: 158, pl. 42; Burukovsky 2001: 1432, Fig. 3; 2004: 558.

Type locality. Sampled at 850 m depth, no other indication. **Diagnosis**. Cardoso and Burukovsky (2014). **Distribution**. Benthic. Western Atlantic: Brazil (Rio Grande do Norte, Saint Peter and Saint Paul Archipelago, Bahia). Mid-Atlantic Ridge. Eastern Atlantic: Azores, along the western Spain and African continent, Cape Verde. Mediterranean Sea. Pacific Ocean: Southern California to the Galapagos Islands, Peru. **Depth range**. From 190 to 1860 m.

65. *Nematocarcinus parvus* Burukovsky, 2000b.

Nematocarcinus parvus Burukovsky, 2000b: 1163; 2012: 146; Cardoso et al. 2017: 4.
Nematocarcinus gracilis– Crosnier 1976: 229; Chace 1986: 71 (part).

Type locality. 12°26′S, 42°08′5″E, Madagascar. **Diagnosis**. Burukovsky (2000a, b, c), Burukovsky (2012). **Distribution**. Benthic. Western Atlantic: Rio Grande Rise. Indian Ocean: Madagascar, Tanzania, Mozambique. Pacific Ocean: Solomon Islands, Vanuatu, Samoa Islands. **Depth range**. From 600 to 1050 m. **Color pattern**. Live specimens white, cephalotorax with a pink hepatopancreas; pereopods with white ischium, merus, and carpus; pale pink propods and dactyls; pale pink antennal and antennular flagellum; dorsally pink abdominal somites 1–2 and 5–6; dorsally white abdominal somites 3–4; white telson.

66. *Nematocarcinus rotundus* Crosnier and Forest, 1973.

Nematocarcinus rotundus Crosnier and Forest, 1973: 103, Fig. 29 c, 30f–i, 31e–f; Wenner 1979: 380; Takeda and Okutani 1983: 53; Squires 1990: 105; Burukovsky 2001: 1440; 2003: 144; 2004a: 558; 2012: 173; Cardoso and Burukovsky 2014: 445, Figs. 5–7.
Nematocarcinus cursor– Smith 1886a: 189 (part); 1887: 665 (part), pl. 17, Fig. 1; Agassiz 1888: 46 (part); Thompson 1966: 138, Figs. 4–5; Pequegnat 1970a: 73.

Type locality. Gulf of Mexico, 28°43′N 87°14′30″W. **Diagnosis**: Cardoso and Burukovsky (2014). **Distribution**. Benthic. Western Atlantic: from Long Island to Chesapeake Bay, northwestern Bahamas, South of Florida, Northern Gulf of Mexico, French Guiana, Brazil (Bahia, Espírito Santo, Rio de Janeiro). **Depth range**. From 421 to 1629 m.

67. *Nematocarcinus tenuipes* Spence-Bate, 1888.

Nematocarcinus tenuipes Spence-Bate, 1888: 812 (part, "Challenger" st. 232); Burukovsky 1991: 42; 2000c: 164; 2001: 1302; 2002a: 11; 2002b: 1515; 2003: 159; 2004: 558; 2012: 187; 2013: 182; Cardoso and Burukovsky 2014: 450, Figs. 8–10.
Nematocarcinus productus– Spence-Bate 1888: 811 (part, "Challenger", st. 237).
Nematocarcinus longirostris– Spence-Bate 1888: 806 (part).
Nematocarcinus parvidentatus Spence-Bate, 1888: 814.
Nematocarcinus ensiferus– Rathbun 1906: 926.
Nematocarcinus serratus Spence-Bate, 1888: 819; Zarenkov 1968: 158.
Nematocarcinus serratirostris Burukovsky, 1991: 41.
Nematocarcinus ovalis Komai and Segonzac, 2005: 355.
Nematocarcinus ensifer.– Cardoso and Serejo, 2007: 41.

Type locality. Japan, 35°11′N, 139°28′E. **Diagnosis**: Cardoso and Burukovsky (2014). **Distribution**. Benthic. Western Atlantic: Brazil (Bahia, Espírito Santo, Rio de Janeiro). Eastern Atlantic: Gabon, Angola, Namibia coasts. South Mid-Atlantic Ridge. Indian Ocean: Agulhas Bank, Mozambique Strait. Pacific Ocean: Taiwan, Japan, New Caledonia, Hawaii, Vanuatu, Fiji, Tonga Islands, East Pacific Rise, Chile. **Depth range**. From 630 to 3075 m.

Family Pandalidae Haworth,1825.

68. *Heterocarpus dorsalis* Spence-Bate, 1888.

Heterocarpus dorsalis Spence-Bate, 1888: 630; Chace 1985: 22; Crosnier 1988b: 62; Tavares 1999: 673, Figs. 2, 3.
Heterocarpus alphonsi Spence-Bate, 1888: 632, pl. 112 Fig. 1.
Heterocarpus affinis– Borradaile 1915: 208.

Type locality. 4°34′00″S, 129°57′30″E, off Banda Island, Caribbean Sea. **Diagnosis**. Chace (1985), Crosnier (1988a). **Distribution**. Bathypelagic. Western Atlantic: Brazil (Espírito Santo). Indian Ocean: eastern Africa to Indonesia. Pacific Ocean: Japan, Philippines, New Caleonia, Samoa. **Depth range**: From 185 to 1400 m.

69. *Heterocarpus ensifer* A. Milne-Edwards, 1881.

Heterocarpus ensifer A. Milne-Edwards, 1881: 8; 1883: pl. 27; Pequegnat 1970a: 84; Crosnier and Forest 1973: 189, Fig. 61a; Crosnier 1988b: 67, p. 2, Figs. a–f; Ramos-Porto et al. 1998: 340; Serejo et al. 2007: 139; Rego and Cardoso 2010: 120, Fig. 1.
Pandalus carinatus Smith, 1882: 63, pl. 10.
Atlantocaris gigas Ortmann, 1893: 80, pl. 5, Fig. 2.
Procletes atlanticus Lenz and Strunck, 1914: 334, pl. 22, Figs. 9–14.

Type locality. Western Atlantic, near Barbados. **Diagnosis**. Crosnier (1988a), Rego and Cardoso (2010). **Distribution**. Bathypelagic. Western Atlantic: from USA (North Carolina) to Brazil (Ceará, Rio Grande do Norte, Rio de Janeiro). Eastern Atlantic: from Iberian Peninsula to Congo, Azores, Madeira, Canary Islands, Cape Verde. Mediterranean Sea. Indian Ocean: southwestern region. Pacific Ocean: Hawaii, Kiribati, Marquesas. **Depth range**. From 88 to 885 m.

70. *Heterocarpus inopinatus* Tavares, 1999.

Heterocarpus inopinatus Tavares, 1999: 673, Fig. 1; Cardoso and Serejo 2007: 47, Fig. 3; Serejo et al. 2007: 139; Rego and Cardoso 2010: 121, Fig. 4; Alves-Júnior et al. 2017c: 39, Fig. 1 a–c.

Type locality. Western Atlantic, Brazil, Espírito Santo, 19°38′S, 038°43′W. **Diagnosis**. Cardoso and Serejo (2007); Rego and Cardoso (2010). **Distribution**. Bathypelagic. Brazil (Ceará, Rio grande do Norte, Bahia, Espírito Santo, Rio de Janeiro). **Depth range**. From 150 to 1718 m.

71. *Heterocarpus laevigatus* Spence-Bate, 1888.

Heterocarpus laevigatus Spence-Bate, 1888: 636, pl. 112, Fig. 3; Crosnier and Forest 1973: 195, Fig. 61c; Chace 1985: 33, Fig. 13i; Crosnier 1988b: 74; Viana et al. 2007: 35, Fig. 2; Rego and Cardoso 2010: 124, Fig. 6.

Type locality. 4°34′0″S, 129°57′30″E, off Banda Island. **Diagnosis**. Chace (1985); Rego and Cardoso (2010). **Distribution**. Bathypelagic. Western Atlantic: Brazil (Pernambuco, Bahia, Espírito Santo). Eastern Atlantic: off Rio d'Oro, off Villa Cisneros, Cape Verde. Indian Ocean: Arabian Sea, Malay. Pacific Ocean: Hawaii, French Polynesia, Eastern Australia, Japan, New Caledonia. **Depth range**. From 366 to 966 m.

72. *Heterocarpus oryx* H. Milne-Edwards, 1881.

Heterocarpus oryx H. Milne-Edwards, 1881: 10; Pequegnat 1970a: 85; Chace 1985: 21; Crosnier 1988b: 91; Ramos-Porto et al. 1998: 340; Tavares 1999: 675; Cabral et al. 2000: 246; Ramos-Porto et al. 2003: 91.

Type locality. 24°36′N 84°05′W, Gulf of Mexico. **Diagnosis**. Pequegnat (1970a, b). **Distribution**. Bathypelagic. Western Atlantic: Gulf of Mexico, Lesser Antilles, Guadeloupe, Martinique, Brazil (Amapá, Pará, Maranhão, Ceará, Rio Grande do Norte, Pernambuco, Sergipe, Espírito Santo). **Depth range**. From 118 to 1774 m.

73. *Plesionika acanthonotus* (Smith, 1882).

Pandalus acanthonotus Smith, 1882: 61.
Pandalus parfaitii A. Milne-Edwards, 1883: 21; De Man 1920: 107.
Pandalus geniculatus A. Milne-Edwards, 1883: 25; Coutière 1905: 675.
Nothocaris geniculatus– Bate 1888: 661; Moreira 1901: 8.
Plesionika geniculata– De Man 1920: 106.
Plesionika acanthonotus– De Man 1920: 105; Holthuis 1951: 62; Crosnier and Forest 1968: 661; Pequegnat 1970a: 91; Omori 1971: 241; Kensley 1972; 50; Crosnier and Forest 1973: 203; Ramos-Porto et al. 1998: 341; Cruz and Fransen 2004: 132; Cardoso 2010b: 222, Figs. 1–5; Cardoso et al. 2019: 463, Figs. 2, 3.

Type locality. 32°43′25″N, 77°20′30″W, South Carolina, USA. **Diagnosis**. Crosnier and Forest (1973), Cardoso (2010a, b). **Distribution**. Bathypelagic. Western Atlantic: USA (South Caroline, Florida), Gulf of Mexico, Brazil (Pará, Ceará, Rio Grande do Norte, Pernambuco, Bahia, Espírito Santo). Mediterranean Sea. Eastern Atlantic: Portugal, Spain, Congo, Angola. **Depth range**. From 190 to 1350 m.

74. *Plesionika edwardsii* (Brandt, 1851) (Fig. 21.4a).

Pandalus (Pontophilus) edwardsii Brandt, 1851: 122.
Pandalus narval– H. Milne Edwards 1837: 385.
Pandalus guerinii Risso, 1844: 95 (nomen nudum).
Pandalus (Parapandalus) longirostris Borradaile, 1899: 413, pl.37, Fig.10.

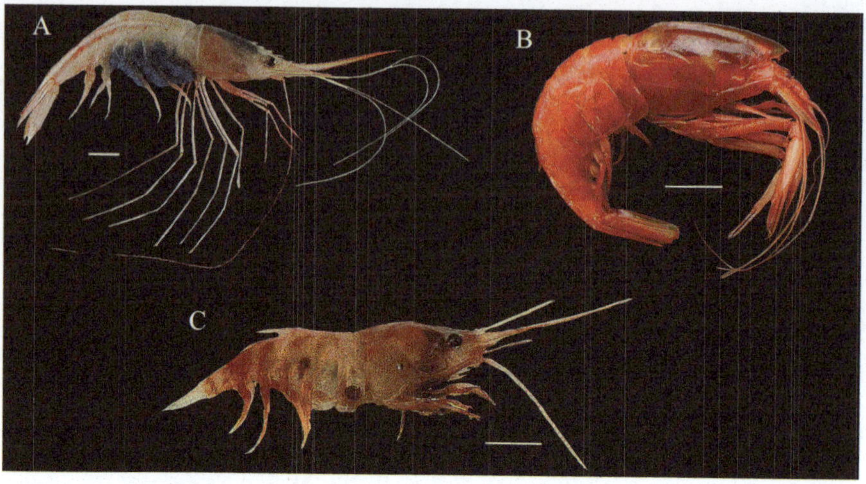

Fig. 21.4 A. *Plesionika edwardsii* (Brandt, 1851); B. *Eupasiphae ostrovski* Rodrigues and Cardoso, 2018; C. *Oplophorus gracilirostris* A. Milne-Edwards, 1881. A and C sampled by REVIZEE/Score Central, deposited at Museu Nacional/UFRJ (unknown catalogue number); B sampled by ABRACOS, deposited at Universidade Federal de Pernambuco (unknown catalogue number). Scale bars = 1 cm

Plesionika edwardsii– Holthuis 1947: 316; Pequenat 1970a, b: 93; Omori 1971: 241; Crosnier and Forest 1973: 202, Fig. 63b, 64b; Chace 1985: 62, Fig. 26; Crosnier 1986: 362; Kensley et al. 1987: 314; Chan and Yu 1991: 550, Figs. 2, 3b; Chan and Crosnier 1997: 193, Fig. 23; Cabral et al. 2000: 246; Cruz and Fransen 2004: 141; Serejo et al. 2007: 139; Viana et al. 2007: 36; Cardoso 2009a: 55, Figs. 1, 2.

Type locality. Mediterranean (designated by Chan and Yu 1991). **Diagnosis**. Chan and Crosnier (1997), Cardoso (2009a, b). **Distribution**. Bathypelagic. Western Atlantic: Virginia, Gulf of Mexico, Bahamas, Brazil (Ceará, Rio Grande do Norte, Pernambuco, Sergipe, Rio de Janeiro). Eastern Atlantic: Spain, Canary Islands, Madeira Islands, Morocco, Senegal, Angola. Mediterranean. Indian Ocean: La Reunion. Pacific Ocean: Indonesia, Philippines, Taiwan, Eastern Australia, New Caledonia, Vanuatu, Fiji, French Polynesia, Tubuai, Society Islands. **Depth range**. From 50 to 680 m.

75. *Plesionika ensis* (A. Milne-Edwards, 1881).

Acanthephyra ensis A. Milne-Edwards, 1881: 14; Young 1900: 476.

Pandalus ensis– A. Milne-Edwards 1883: pl. 18; Faxon 1896: 161; Alcock 1901b: 96; Coutiére 1905: 675; Rathbun 1906: 914.

Plesionika uniproducta Spence-Bate, 1888: 641, pl. 113, Fig. 1 (part); Moreira 1901: 8 (part).

Plesionika semilaevis– Spence-Bate 1888: 664 (part).

Plesionika ensis– De Man 1920: 106; Holthuis 1951: 55, Fig. 10; Crosnier and Forest 1968: 1138; 1973: 209, Figs. 63c, 64f; Pequegnat 1970a: 94; Omori 1971: 241; Ramos-Porto et al. 1998: 341; Cabral et al. 2000: 246; Cruz and Fransen 2004: 141; Viana et al. 2007: 36; Cardoso 2009a: 57, Figs. 3, 4.

Type locality. Near Barbados. **Diagnosis**. Holthuis (1951); Cardoso (2009a, b). **Distribution**. Bathypelagic. Western Atlantic: Florida, Gulf of Mexico, Antilles, Brazil (Pará, Maranhão, Ceará, Rio Grande do Norte, Paraíba, Alagoas, Bahia, Rio de Janeiro). Eastern Atlantic: Senegal, Gabon, Congo, Angola. Indian Ocean: Andaman Sea, Arabian Sea. Pacific Ocean: Hawaii, Fiji. **Depth range**. From 55 to 1251 m.

76. *Plesionika gigliolii* (Senna, 1903).

Pandalus gigliolii Senna, 1903: 315, pl. XVI, Figs. 5–16.
Plesionika gigliolii– De Man 1920: 106, 111; Martin and Hargreaves 1991: 53; Fransen 1991: 173; Biscoito 1993: 324; Cruz and Fransen 2004: 135, Fig. 1a–c; Cardoso 2011: 132, Figs. 1–3.
Pandalus subtilirostris Riggio, 1905: 283.

Type locality. 39°15′37″3‴N, 9°26′37″7‴E, near Carbonara Cape, Sardegna. **Diagnosis**. Cruz and Fransen (2004), Cardoso (2011). **Distribution**. Bathypelagic. Western Atlantic: Caribbean Sea, Colombian coast (Aguja), Brazil (Bahia). Eastern Atlantic: Mediterranean Sea, Morocco; Madeira. **Depth range**. From 334 to 600 m.

77. *Plesionika holthuisi* Crosnier and Forest, 1968.

Plesionika holthuisi Crosnier and Forest, 1968: 1141, Fig. 7b, c; 1973: 206, Fig. 64c, 65b–d; Pequegnat 1970a: 94, Fig. 4–12; Omori 1971: 241; Cardoso 2010b: 222, Figs. 6–10; Cardoso et al. 2019: 463.
Plesionika acanthonotus– Holthuis 1951: 62 (part); 1952: 36, Fig. 9 (part).

Type locality. Off Pointe Noire, Congo. **Diagnosis**. Crosnier and Forest (1973), Cardoso (2010a, b). **Distribution**. Bathypelagic. Western Atlantic: Gulf of Mexico, Brazil (Bahia and Espírito Santo). Eastern Atlantic: Senegal, Congo, Angola. **Depth range**. From 480 to 900 m.

78. *Plesionika longipes* (A. Milne-Edwards, 1881).

Pandalus longipes A. Milne-Edwards, 1881: 15; 1883: pl. 20.
Plesionika longipes– Boone 1927: 114, Figs. 24–26; Chace 1956: 12; Bullis and Thompson 1965: 8; Campos et al. 2005: 81, Fig. 45; Cardoso 2011: 136, Figs. 4–6.

Type locality. Near Barbados. **Diagnosis**. Boone (1927), Cardoso (2011). **Distribution**. Bathypelagic. Western Atlantic: USA, Barbados, Belize, Colombian Caribbean, Brazil (Bahia). **Depth range**. From 196 to 669 m.

79. *Plesionika macropoda* Chace, 1939.

Plesionika macropoda Chace, 1939: 37; Palmier 1993: 19, pls. 19–20; Monterossa 1988: 637, Figs. 4–5; Chan and Crosnier 1997: 205, Figs. 11–13, 30–31; Cardoso 2011: 141, Figs. 7–9.

Type locality. Bahia de Cochinos, Santa Clara Province, Cuba, 22°07′N, 81°08′W. **Diagnosis**. Chan and Crosnier (1997); Cardoso (2011). **Distribution**. Bathypelagic. Western Atlantic: Caribbean Sea, Antilles, Brazil (Bahia). Pacific Ocean: Loyalty Islands, French Polynesia. **Depth range**. From 260 to 623 m.

80. *Plesionika martia* (A. Milne-Edwards, 1883)

Pandalus martius A. Milne-Edwards, 1883, pl. 21.
Plesionika uniproducta Spence-Bate, 1888: 641 (partim); pl. 113 Fig. 1.
Plesionika (Pandalus) sicherii Riggio, 1900: 20.
Plesionika martia– Crosnier and Forest 1973: 112, Figs. 63d, 64, 66; Ramos-Porto et al. 1998: 341; Ramos-Porto et al. 2003: 95; Nunes et al. 2017: 342, Fig. 8; Ahamed et al. 2017: 585.

Type locality. Atlantic Ocean, between 34° and 44°N. **Diagnosis**. Crosnier and Forest (1973), Ahamed et al. (2017). **Distribution**. Bathypelagic. Western Atlantic: USA (off South Carolina to Florida), off Bermuda, Gulf of Mexico, Brazil (Pará, Amapá, Maranhão, Ceará, Rio Grande do Norte, Saint Peter and Saint Paul Archipelago). Eastern Atlantic: off SW Ireland, Portugal (Azores Island), Bay of Biscay, Mediterranean Sea, Gulf of Guinea, Cape of Good Hope. Indian Ocean: Andaman Sea, Bay of Bengal, Arabian Sea. **Depth range**. From 180 to 459 m.

81. *Plesionika miles* (A. Milne-Edwards, 1883).

Pandalus miles A. Milne-Edwards, 1883: pl. 18.
Parapandalus miles– De Man 1920: 107, 138; Crosnier and Forest 1968: 1137; 1973: 228, Fig. 71; Ramos-Porto et al. 1998: 342; Cruz and Fransen 2004: 141; Cardoso 2009a: 61, Figs. 5, 6.

Type locality. Dominica. **Diagnosis**. Cardoso (2009a, b). **Distribution**. Bathypelagic. Western Atlantic: Martinique, Dominica, Brazil (Alagoas, Bahia, Espírito Santo, Rio de Janeiro). **Depth range**. From 54 to 700 m.

82. *Plesionika revizeei* Cardoso, 2011.

Plesionika revizeei Cardoso, 2011: 52, Figs. 1–3.

Type locality. 21°13′S, 40°14′W, off Rio de Janeiro, Brazil. **Diagnosis**. Cardoso (2011). **Distribution**. Bathypelagic. Southwestern Atlantic: Brazil (Rio Grande do Sul, Rio de Janeiro, Espírito Santo). **Depth range**. From 533 to 666 m.

83. *Stylopandalus richardi* (Coutière, 1905).

Pandalus (Stylopandalus) richardi Coutière, 1905: 1115.
Stylopandalus richardi– Richard 1905: 11.
Parapandalus richardi– De Man 1920: 108.

Parapandalus richardi– Chace 1940: 192, Fig. 58–61; Crosnier and Forest 1968: 1138; 1973: 224, Fig. 69b; Pequegnat 1970a: 86.
Plesionika nana Murray and Hjort, 1912: 585.
Pandalus (Plesionica) gracilis Borradaile, 1915: 208.
Parapandalus zur strasseni Balss, 1914: 597; De Man 1920: 108, pl.12, Fig. 32.
Parapandalus Zurstrasseni– Balss, 1925: 273; Calman 1939: 201.
Stylopandalus richardi– Burukovsky 1982: 45; Chace 1985: 136, Fig. 62; Kensley et al. 1987: 319; Cardoso 2009a: Figs. 7, 8.

Type locality. 32°18′N, 23°58′W, west of Madeira; 27°43′N, 18°28′W, Canary Islands. **Diagnosis**. Chace (1940), Cardoso (2009a, b). **Distribution**. Bathypelagic. Western Atlantic: New Foundland, Gulf of Mexico, Bermuda, Cadix, Brazil (Rio de Janeiro). Eastern Atlantic: Azores, Canary, Madeira, Gibraltar, Gulf of Guinea, Adriatic Sea, Gabon, Congo, Angola. Mediterranean. Indian Ocean: Natal, Seychelles, Red Sea, Sumatra, Malay Archipelago, Gulf of Bengal, Banda Sea. Pacific Ocean: Philippines, Eastern Australia, Indonesia, Hawaii, Alaska. **Depth range**. From 7 to 3600 m.

Family Pasiphaeidae Dana, 1852.

84. *Eupasiphae gilesii* (Wood-Mason, 1892).

Parapasiphaë Gilesii Wood-Mason, 1892: pl. 3, Fig. 8; Calman 1939: 187.
Parapasiphae (Eupasiphaë) Gilesii– Wood Mason and Alcock 1893: 166.
Parapasiphaea (Eupasiphaea) gilesii– Alcock and Anderson 1894: 158.
Parapasiphaea Gilesii– Alcock 1901b: 66.
Eupasiphae rhinocerata Burukovsky, 1977: 473, figs. a–c.
Eupasiphae gilesi– Fisher and Goldie 1961: 78; Crosnier and Forest 1973: 150, Fig. 44.
Eupasiphae gilesii– Holthuis 1955: 36; Tirmizi 1969: 213, Figs. 1–4; Kensley 1977: 32, Figs. i, o, b; Hanamura 1983: 78, Fig. 16 a–b; Kikuchi and Nemoto 1986: 55; Kensley et al. 1987: 293; Crosnier 1988b: 786, Figs. 1–5a.

Type locality. Off Cinque Island, Andaman Sea. **Diagnosis**. Crosnier and Forest (1973), Kensley (1977). **Distribution**. Pelagic. Western Atlantic: Brazil (Fernando de Noronha Archipelago, Bahia). Eastern Atlantic: West of Cape Verde, Madeira, Bermuda. South Mid-Atlantic Ridge. Indian Ocean: Gulf of Oman, Andaman Sea, Madagascar, central and northern Arabian Sea. Pacific Ocean: Baja California, New Zealand. **Depth range**. From 350 to 2500 m.

85. *Eupasiphae ostrovski* Rodrigues and Cardoso, 2018 (Fig. 21.4b).

Eupasiphae ostrovski Rodrigues and Cardoso, 2018: 190, Figs. 1–3.

Type locality. Off Brazil, Rio de Janeiro, 21°25.738′S, 39°43.946′W. **Diagnosis**. Rodrigues and Cardoso (2018). **Distribution**. Pelagic. Southwestern Atlantic Ocean, Brazil (Rocas Atoll, Bahia, Espírito Santo, Rio de Janeiro). **Depth range**. From 1374 to 1718 m.

86. *Parapasiphae cristata* Smith, 1884.

Parapasiphae cristata Smith, 1884: 388, Fig. 3; Krygier and Pearcy 1981: 81; Tchesunov 1984: 1170, Figs. 3–4; Hendrickx and Estrada-Navarrete 1989: 112; Wicksten 2002: 133; Wasmer 2005: 167, Fig. 5; Tavares and Cardoso 2006: 33, Figs. 5–6; Serejo et al. 2007: 139.
Parapasiphaë macrodactyla Chace, 1939: 33.

Type locality. 39°22′N, 68°34′30″W, off New Jersey. **Diagnosis**. Wasmer (2005), Tavares and Cardoso (2006). **Distribution**. Pelagic. Western Atlantic: Brazil (Bahia). South Mid-Atlantic Ridge. Pacific Ocean: Oregon, Mexico, North Atlantic. **Depth range**. From 400 to 2870 m.

87. *Parapasiphae sulcatifrons* Smith, 1884.

Parapasiphaë sulcatifrons Smith, 1884: 384, Figs. 1–7; Crosnier and Forest 1973: 142, Fig. 41; Krygier and Pearcy 1981: 81; Tchesunov 1984: 1166–1170; Hendrickx and Estrada-Navarrete 1989: 112; 1996: 99, Fig. 62; Iwasaki 1990: 200; Wicksten 2002: 133; Tavares and Cardoso 2006: 36, Figs. 7–8; Serejo et al. 2007: 139.
Orphania tenuimana Spence-Bate, 1888: 872; pl. 141, Fig. 4.
Dantecia caudani Caullery, 1896: 372, pl. 14, Figs. 1–11.
Pasiphaea metriomma Dohrn, 1908: 9.

Type locality. Off USA, 37°02′42″N, 74°17′36″W; 39°27′10″N, 69°56′20″W; 38°53′00″N, 69°23′30″W; 41°53′00″N, 65°35′00″W; 41°43′00″N, 65°21′50″W; 39°44′30″N, 71°04′00″W; 37°56′20″N, 70°57′30″W; 37°12′20″N, 69°39′00″W; 37°50′00″N, 73°03′50″W; 35°12′10″N, 74°57′15″W. **Diagnosis**. Tavares and Cardoso (2006). **Distribution**. Pelagic. Western Atlantic: Brazil (Ceará, Rio Grande do Norte, Rocas Atoll, Fernando de Noronha Archipelago, Espírito Santo, Rio de Janeiro). Indian Ocean: southeastern Africa. Pacific Ocean: south and east coast of Australia, west coast of North and Central America. **Depth range**. From 500 to 5340 m.

88. *Pasiphaea alcocki* (Wood-Mason, 1891).

Parapasiphaë Alcocki Wood-Mason, 1891: 196.
Pasiphaea (*Phye*) *Alcocki*– Wood-Mason 1892: 3, Fig. 5.
Phye Alcocki– Wood-Mason and Alcock 1893: 164.
Pasiphaea (*Phye*) *alcocki*– Alcock 1901b: 61.
Pasiphaea Alcocki– De Man 1920: 2.
Pasiphaea alcocki– Komai and Amaoka 1993: 371; Hayashi 2006: 196, Figs. 1–2; Komai et al.2012: 300, Fig. 4; Rodrigues and Cardoso 2019: 322, Fig. 1.

Type locality. Bay of Bengal, 16°11′15″N, 82°30′30″E. **Diagnosis**. Hayashi (2006), Rodrigues and Cardoso (2019). **Distribution**. Pelagic. Western Atlantic: Brazil (Espírito Santo). Indian Ocean: Gulf of Manar, Arabian Sea, Southeast Asia. Pacific Ocean: Taiwan, Indonesia, Philippines. **Depth range**. From 185 to 1733 m.

89. *Pasiphaea antea* Rodrigues, Alves-Júnior and Cardoso, 2018.

Pasiphaea antea Rodrigues, Alves-Júnior and Cardoso, 2018: 494, Figs. 1–3.

Type locality. Fernando de Noronha Archipelago (Brazil). **Diagnosis**. Rodrigues et al. (2018). **Distribution**. Pelagic. Fernando de Noronha Archipelago and Rocas Atoll (Brazil). **Depth range**. From 65 to 505 m.

90. *Pasiphaea major* Hayashi, 2006.

Pasiphaea major Hayashi, 2006: 219, Figs. 9–11; Rodrigues and Cardoso 2019: 323, Fig. 2.

Type locality. 25°09′S, 168°53′E, Norfolk Ridge. **Diagnosis**. Rodrigues and Cardoso (2019). **Distribution**. Pelagic. Western Atlantic: Brazil (Bahia, Espírito Santo, Rio de Janeiro). Indian Ocean: Madagascar. Pacific Ocean: Lord Howe Ridge, New Caledonia, Wallis and Futuna, Marquesas, Tonga. **Depth range**. From 226 to 1890 m.

91. *Pasiphaea merriami* Schmitt, 1931.

Pasiphaea merriami Schmitt, 1931: 391; Chace 1956: 11; Bullis and Thompson 1965: 7; Pequegnat 1970a, b: 64; Takeda 1983: 59; Tchesunov 1984: 997; Burukovsky and Romensky 1987: 58; Burukovsky 1996: 843; Hayashi 2004: 344, Fig. 12; Tavares and Cardoso 2006: 28, Figs. 1–2; Serejo et al. 2007: 139.
Pasiphaea nishiei– Iwasaki 1990: 190, Figs. 1–2; Hanamura 1994: 171.

Type locality. South of the Dry Tortugas, North Atlantic Ocean. **Diagnosis**. Hayashi (2004), Tavares and Cardoso (2006). **Distribution**. Pelagic. Western Atlantic: Gulf of Mexico, Caribbean Sea, Suriname, Brazil (Ceará, Rio Grande do Norte, Bahia, Espírito Santo). **Depth range**: From 280 to 3206 m.

92. *Pasiphaea princeps* Smith, 1884.

Pasiphaëa princeps Smith, 1884: 383, Fig. 2; De Man 1920: 2; Iwasaki 1990: 196, Figs. 5, 6; Hanamura 1994: 171; Burukovsky 1996: 843; Hayashi 2004: 353; Tavares and Cardoso 2006: 30, Figs. 3–4; Serejo et al. 2007: 139.

Type locality. 39°29′N 70°58′40″W, North Atlantic Ocean. **Diagnosis**. Iwazaki (1990), Cardoso and Tavares (2006). **Distribution**. Pelagic. Western Atlantic: USA to Cape Verde Islands, Brazil (Bahia, Espírito Santo, Rio de Janeiro). **Depth range**. From 1450 to 4570 m.

93. *Pasiphaea taiwanica* Komai et al., 2012.

Pasiphaea taiwanica Komai et al., 2012: 318; Figs. 18–20; Rodrigues and Cardoso 2019: 323, Fig. 3.

Type locality. 21°47.35′N, 120°29.7′E, Southwestern Taiwan. **Diagnosis**. Rodrigues and Cardoso (2019). **Distribution**. Pelagic. Western Atlantic: Brazil (Bahia). Pacific Ocean: Southwestern Taiwan. **Depth range**. From 226 to 2137 m.

94. *Pasiphaea tarda* Krøyer, 1845.

Pasiphaea tarda Krøyer, 1845: 453; Matthews and Pinnoi 1973: 139; Hayashi 2006: 234; Rodrigues and Cardoso 2019: 324, Fig. 4.
Pasiphaea princeps– Rathbun 1904: 23.
Pasiphaea principalis Sund, 1913: 6; De Man 1920: 2.

Type locality. Not indicated. **Diagnosis**. Hayashi (2006), Rodrigues and Cardoso (2019). **Distribution**. Pelagic. Western Atlantic: South Carolina, Brazil (Espírito Santo). Eastern Atlantic: South Greenland, Canary Islands, Angola. **Depth range**. From 250 to 2400 m.

Family Psalidopodidae Wood-Mason and Alcock, 1892.

95. *Psalidopus barbouri* Chace, 1939.

Psalidopus barbouri Chace, 1939: 36; Bullis and Thompson 1965: 8; Chace and Holthuis 1978: 2; Ramos-Porto et al. 2000, 77; Ramos-Porto et al. 2003: 81.

Type locality. 23°21′N, 79°58′W, south of Cay Sal Bank, Nicholas Channel, Western Atlantic. **Diagnosis**. Pequegnat (1970a, b), Chace and Holthuis (1978). **Distribution**. Benthic. Western Atlantic: USA (Florida), Gulf of Mexico, Caribbean Sea, Belize, Venezuela, Suriname, Brazil (Pará). **Depth range**. From 300 to 626 m.

Family Oplophoridae Dana, 1852.

96. *Janicella spinicauda* (A. Milne-Edwards, 1883).

Oplophorus spinicauda A. Milne-Edwards, 1883; Chace 1940: 184, Fig. 54; Kensley 1972: 38, Fig. 17d, e.
Oplophorus foliaceus Rathbun, 1906: 922, pl. 20, Fig. 8; De Man 1920: 48.
Acanthephyra anomala Boone, 1927: 104, Fig. 21.
Janicella spinicauda– Chace 1986: 44, Figs. 23, 24; Kensley 1987: 285, Figs. 29–33; Cardoso and Young 2005: 39, Figs. 29–33; Cardoso and Serejo 2007: 45; Serejo et al. 2007: 139; Alves-Júnior et al. 2019b: 426, Figs. 29 a–c, 30, 40e.

Type locality. 34°13′30″N, 7°43′W, off Casablanca, Morocco. **Diagnosis**. Chace (1986), Cardoso and Young (2005), Cardoso and Serejo (2007), Alves-Júnior et al. (2019b). **Distribution**. Pelagic. Western Atlantic: east coast of Florida, Bermuda, Bahamas, Caribbean Sea, Honduras, Grenada, Lesser Antilles, Brazil (Rocas Atoll, Pernambuco, Bahia, Espírito Santo, Rio de Janeiro). Indian Ocean: north of Madagascar, southwestern Indian. Pacific Ocean: Philippines, French Polynesia, Hawaii. **Depth range**. From 105 to 3716 m.

97. *Oplophorus gracilirostris* A. Milne-Edwards, 1881 (Fig. 21.4c).

Oplophorus gracilirostris A. Milne-Edwards, 1881: 6; De Man 1920: 48; Chace 1947: 44, Figs. 4–7; 1986: 59, Fig. 32a–e; Ramos-Porto et al. 2000: 77; Cardoso and Young 2005: 52, Figs. 39–43; Serejo et al. 2007: 139; Alves-Júnior et al. 2019b: 428, Figs. 31 a–b, 32, 40 f.

Oplophorus longirostris Spence-Bate, 1888: 765, pl. 127, Fig. 2.
Hoplophorus smithii Wood-Mason and Alcock, 1891: 194.
Oplophorus okitsuensis Yokoya, 1922: 302.

Type locality. Western Atlantic, Dominica. **Diagnosis**. Chace (1986), Cardoso and Young (2005), Alves-Júnior et al. (2019b). **Distribution**. Pelagic. Western Atlantic: Bahamas, Gulf of Mexico, Caribbean Sea, Brazil (Bahia). Indian Ocean: southeastern Africa. Pacific Ocean: Indonesia, Philippines, southern Japan, Fiji Islands, Hawaii. **Depth range**. From 100 to 2400 m.

98. *Oplophorus spinosus* (Brullé, 1839).

Palaemon spinosus Brullé, 1839: 18.
Hoplophorus grimaldii Coutiére, 1905: 1, Fig. 1; Calman 1939: 189; Chace 1940: 187, Fig. 55; Kensley 1972: 38, Fig. 17i, j.
Oplophorus spinosus– Holthuis 1949: 229; Crosnier and Forest 1973: 25; Chace 1986: 59; Kensley 1987: 289; Cardoso and Young 2005: 58, Figs. 44–48; Cardoso and Serejo 2007: 46; Serejo et al. 2007: 139.

Type locality. Canary Islands. **Diagnosis**. Chace (1947), Cardoso and Young (2005), Cardoso and Serejo (2007). **Distribution**. Pelagic. Western Atlantic: USA, Bermuda, Bahamas, Brazil (Bahia, Espírito Santo). Eastern Atlantic: Azores, Madeira, Canary, Senegal, Tristan da Cunha. Indian Ocean: southwestern Indian Ocean, west Australia. Pacific Ocean: Indonesia, south Japan, Hawaii. **Depth range**. From water surface to 2700 m.

99. *Systellaspis curvispina* Crosnier, 1987.

Systellaspis curvispina Crosnier, 1987: 711, Figs. 6–8; Lunina et al. 2018: 3, Fig. 2; Alves-Júnior et al. 2019b: 429, Figs. 33 a–c, 34.
Systellaspis cristata Chace, 1986: 64 (part), Fig. 35c.

Type locality. Magadascar. **Diagnosis**. Crosnier (1987), Alves-Júnior et al. (2019b). **Distribution**. Pelagic. Western Atlantic: Gulf of Mexico, Bahamas, Dominica, Lesser Antilles, Caribbean Sea, Brazil (Ceará, Rio Grande do Norte, Rocas Atoll). Indian Ocean: southeastern Africa. Pacific Ocean: Indonesia, Philippines, southern Japan, Fiji Islands, Hawaii, French Polynesia, Seamounts Sala y Gòmez, Nazca Ridge. **Depth range**. From 140 to 1150 m.

100. *Systellaspis debilis* (A. Milne-Edwards, 1881).

Acanthephyra debilis A. Milne-Edwards, 1881: 13.
Miersia gracilis Smith, 1882: 70.
Systellaspis bouvieri Coutiére, 1905: 8, Fig. 3.
Hoplopasiphaea philippinensis Yokoya and Shibata, 1965: 4, Figs. 4, 5.
Systellaspis debilis– Crosnier and Forest 1973: 87, Figs. 26b, 27b; Chace 1940: 181, Fig. 51; 1986: 67, Figs. 34 m–o, 35 g, h; Cardoso and Young 2005: 64, Figs. 49–53; Pequegnat and Wicksten 2006: 102; Cardoso and Serejo 2007: 47; Felder et al. 2009: 1053; Poupin 2010: 73; Alves-Júnior et al. 2019b: 431, Figs. 35 a–b, 36.

Type locality. Bahamas. **Diagnosis**. Chace (1986), Cardoso and Young (2005), Cardoso and Serejo (2007), Alves-Júnior et al. (2019b). **Distribution**. Pelagic. Western Atlantic: south of Greenland, USA (Virginia, North Carolina), Bermuda, Gulf of Mexico, Bahamas, Caribbean Sea, Brazil (Pernambuco, Rio de Janeiro). Eastern Atlantic: south of Iceland, Faroe, Belgium, Bay of Biscay, Azores, Cape Verde, west African coast, Nigeria, Congo, Angola. Indian Ocean: Mayotte area, Madagascar, South of the Keeling Islands, Malay Archipelago. Pacific Ocean: Philippines, Indonesia, Hawaii, French Polynesia. **Depth range**. From 25 to 4594 m.

101. *Systellaspis pellucida* (Filhol, 1885).

Acanthephyra pellucida Filhol, 1885: 144, 162.
Acanthephyra affinis Faxon, 1896: 162, pl. 2, Figs. 1–3.
Systellaspis affinis– De Man 1920: 43; Chace 1936: 29; Calman 1939: 190; Springer and Bullis 1956: 11; Forest 1964: 621; Bullis and Thompson 1965: 7; Monod 1966: 110; Crosnier and Forest 1968: 1133.
Systellaspis pellucida– Crosnier and Forest 1973: 92, Figs. 26c, 27c; Chace 1986: 67, Figs. 34 m–o, 35 g, h; Cardoso and Young 2005: 70, Figs. 54–58; Pequegnat and Wicksten 2006: 102; Serejo et al. 2007: 139; Poupin 2010: 73; Alves-Júnior et al. 2019b: 431, Figs. 35 a–b, 36.

Type locality. 26°20′N, 14°53′W, Near Canary. **Diagnosis**. Cardoso and Young (2005). **Distribution**. Pelagic. Western Atlantic: Gulf of Mexico, Bahamas, Antilles, Brazil (Bahia). Eastern Atlantic: West African coast, from Guinea to Gabon. Indian Ocean: Madagascar, western Indian Ocean, Zanzibar. Pacific Ocean: Philippines, Indonesia, South China Sea. **Depth range**. From 85 to 3700 m.

Infraorder Astacidea Latreille, 1802.

Family Nephropidae Dana, 1852.

102. *Acanthacaris caeca* (A. Milne-Edwards, 1881) (Fig. 21.5a).

Phoberus caecus A. Milne-Edwards, 1881:1.
Neophoberus caecus– Glaessner 1969: 490.
Acanthacaris caeca– Holthuis 1974: 741, Figs. 4–8; 1991: 5; Melo 1999: 478, Fig. 320; Takeda 1983: 85; Tavares and Young 2002: 85, Fig. 9; Serejo et al. 2007: 140.

Type locality. 12°03′15″N, 61°48′30″W, off Grenada, West Indies. **Diagnosis**. Holthuis (1974), Tavares and Young (2002). **Distribution**. Benthic. Western Atlantic: Gulf of Mexico, Caribbean Sea, Straits of Florida, West Indies, Grenade, Brazil (Amapá, Pará, Bahia). **Depth range**. From 293 to 878 m.

103. *Nephropsis aculeata* Smith, 1881.

Nephropsis aculeata Smith, 1881: 431; Holthuis 1974: 776, Figs. 15–16; 1991: 13; Melo 1999: 484, Fig. 324; Takeda 1983:82; Tavares and Young 2002: 79, Figs. 1–2; Serejo et al. 2007: 140.

Type locality. 40°02′N, 70°57′W, off Massachusetts, USA. **Diagnosis**. Holthuis (1974), Tavares and Young (2002). **Distribution**. Benthic. Western Atlantic: off Massachusetts (USA) to French Guiana, Suriname, Bermuda, Gulf of Mexico, Caribbean Sea, Brazil (Espírito Santo, Rio de Janeiro, São Paulo). **Depth range**. From 137 to 824 m.

104. *Nephropsis agassizii* A. Milne-Edwards, 1880 (Fig. 21.5b).

Nephropsis agassizii A. Milne-Edwards, 1880: 1; Filhol 1885: 144, pl. 1 Fig. 5; De Man 1916: 97, 110, 111.
Nephropsis agassizii– Holthuis 1974: 796, Figs.19–20; 1991:15; Coelho and Ramos-Porto 1985: 65; Melo 1999:486, Fig. 326; Tavares and Young 2002: 81; Serejo et al. 2007: 140; Alves-Júnior et al. 2016b: 90, Figs. 1, 2.

Type locality. 24°01′N, 88°58′W, North of Yucatan Bank. **Diagnosis**. Holthuis (1974), Tavares and Young (2002). **Distribution**. Benthic. Western Atlantic: Bahamas, northern and eastern Gulf of Mexico, southern and eastern Caribbean Sea, Tobago, Brazil (Rio Grande do Norte, Bahia, Espírito Santo, Rio de Janeiro, São Paulo). **Depth range**. From 878 to 2900 m.

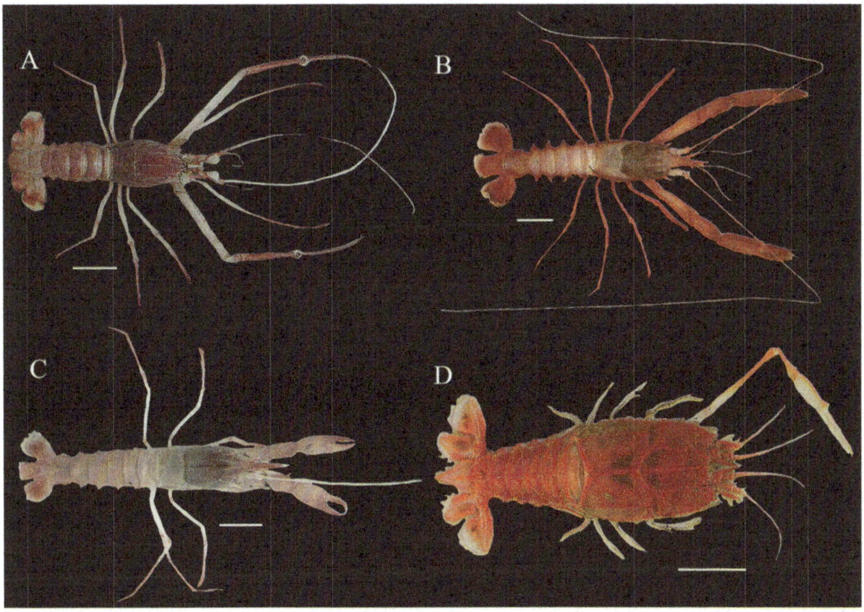

Fig. 21.5 (**a**) *Acanthacaris caeca* A. Milne-Edwards, 1881; (**b**) *Nephropsis agassizii* A. Milne-Edwards, 1880; (**c**) *Nephropsis rosea* Spence-Bate, 1888; (**d**) *Polycheles typhlops* Heller, 1862. All sampled by REVIZEE/Score Central, deposited at Museu Nacional/UFRJ (unknown catalogue number). Scale bars A = 10 cm; B, C and D = 2 cm

105. *Nephropsis neglecta* Holthuis, 1974.

Nephropsis neglecta Holthuis, 1974: 792, Fig. 18; Takeda 1983:83; Tavares and Young 2002: 82, Figs. 5–6; Serejo et al. 2007: 140. Alves-Júnior et al. 2016b: 91, Figs. 2–3.

Type locality. Dry Tortugas, Florida. **Diagnosis**. Holthuis (1974), Tavares and Young (2002). **Distribution**. Benthic. Western Atlantic: Straits of Florida, Dry Tortugas, Lesser Antilles, Guadeloupe, Tobago, Caribbean Sea, Jamaica, Colombia, Venezuela, Guianas, Brazil (Rio Grande do Norte, Espírito Santo). **Depth range**. From 655 to 1300 m.

106. *Nephropsis rosea* Spence-Bate, 1888 (Fig. 21.5c).

Nephropsis rosea Spence-Bate, 1888: 178, Fig. 39, pl.23, Figs. 1–2, pl. 24, Fig. 1; Holthuis 1974: 787, Figs. 16–17; Takeda 1983: 84; Tavares and Young 2002: 84, Figs. 7–8; Serejo et al. 2007: 140.

Diagnosis. Holthuis (1974), Tavares and Young (2002). **Type locality**. 32°11′7″N, 65°3′20″W, off Bermuda. **Distribution**. Benthic. Western Atlantic: between Bermuda and French Guyana, Bahamas, Gulf of Mexico, Caribbean Sea, Brazil (Amapá, Bahia, Espírito Santo). **Depth range**. From 420 to 1260 m.

Infraorder Axiidea De Saint Laurent, 1979.

Family Callianassidae Dana, 1852.

107. *Cheramus profunda* (Biffar, 1973).

Callianassa profunda Biffar, 1973: 225, Figs. 1, 2; Sakai 1999: 28; Sakai 2005: 46.
Callianassa occidentalis Bate, 1888: 29, pl. 2 Fig. 2 k; Young 1900: 425; Borradaile 1903: 548; Balss 1925: 212; De Man 1928a: 115; Schmitt 1935: 3; Biffar 1971: 649.
Cheramus occidentalis Bate, 1888: 32, pl. 2 Fig. 1; Young 1900: 246.
Callianassa (*Cheramus*) *batei* Borradaile, 1903: 546; De Man 1928a: 26, 98; 1928b: 10, pl. 1 Fig. 3.
Callianassa batei– Schmitt 1935: 5; Biffar 1971: 649, 654; Manning 1987: 398.
Cheramus batei– Manning and Felder 1991: 780; Coelho 1997: 150; Melo 1999: 358, Figs. 237–238.
Cheramus profundus– Tudge et al. 2000: 145; Sakai 2011: 371.

Type locality. 18°29.3 N, 63°24.6 W, West Indies, off Sombrero Island. **Diagnosis**. Melo (1999), Sakai (2011). **Distribution**. Benthic. Western Atlantic: USA (Florida), Lesser Antilles, Brazil (Rio Grande do Sul). **Depth range**. From 686 to 820 m.

Infraorder Achelata Scholtz and Richter, 1995.

Family Palinuridae Latreille, 1802.

108. *Palinustus truncatus* A. Milne-Edwards, 1880.

Palinustus truncatus A. Milne-Edwards, 1880: 66; Holthuis 1991: 125; Fausto-Filho 1977: 75; Takeda 1983: 81; Melo 1999: 432; Silva et al. 2003: 29; Tavares 2003: 316.

Type locality. Grenadines. **Diagnosis.** Holthuis (1991), Melo (1999). **Distribution**. Benthic. Western Atlantic: Lesser Antilles, Venezuela, Suriname, Brazil (Amapá, Pará). **Depth range**. From 100 to 1000 m.

109. *Projasus parkeri* (Stebbing, 1902).

Jasus parkeri Stebbing, 1902: 39; Barnard 1950: 540.
Puerulus parkeri– Holthuis 1946: 110, 148.
Projasus parkeri– George and Grindley 1964: 89; Webber and Booth 1988: 82; Holthuis 1991: 159; Griffin and Stoddart 1995: 236; Cardoso et al. 2017: 4.

Type locality. Buffalo River, north 15 miles, Natal, South Africa, about 33°S 28°E. **Diagnosis**. Holthuis (1991), Griffin and Stoddart (1995). **Distribution**. Benthic. Western Atlantic: Rio Grande Rise. Eastern Atlantic: southwest Africa (Valdivia Bank, off Namibia). Indian Ocean: southeast Africa, St. Paul. Pacific Ocean: Southeastern Australia, New Zealand. **Depth range**. From 370 to 880 m.

Infraorder Polychelida De Haan, 1841.

Family Polychelidae Wood-Mason, 1874.

110. *Pentacheles laevis* Spence-Bate, 1878a.

Pentacheles laevis Spence-Bate, 1878a: 278, Fig. 7; 1878b: 484; 1878c: 563; 1888: 144, pl. 15, Figs. 4c, 5; Faxon 1895: 118; Galil 2000: 301, Fig. 7; Dall'Occo and Tavares 2004: 143, Fig. 1a; Serejo et al. 2007: 140.
Pentacheles gracilis Spence-Bate, 1878a: 279; 1878b: 484; 1878c: 563; 1888: 146, pl. 16, Figs. 1–2; Faxon 1895: 118.
Polycheles granulatus Faxon, 1893: 197; 1895: 123, pl. 32, Fig. 1, pl. 33, Figs. 2, 2a; Rathbun 1906: 899, Fig. 54; Stebbing 1910: 378; Calman 1925: 18; Barnard 1950: 569; Squires 1965: 89, Fig. 38; Kensley 1981: 29; Hendrickx 1995: 156; Dawson 1997: 10.
Pentacheles beaumontii Alcock, 1894: 236; 1901b: 175; Wood-Mason and Alcock 1894, pl. 8, Fig. 3.
Polycheles beaumontii– Faxon 1895: 125.
Polycheles dubius Bouvier, 1905b: 4; 1905c: 480.
Polycheles eryoniformis Bouvier, 1905b: 2; 1905d: 644; 1907: 62.
Polycheles gracilis– Firth and Pequegnat 1971: 46.
Polycheles laevis– Firth and Pequegnat 1971: 49.

Type locality. 4°33′N, 127°6′E, off the Moluccas. **Diagnosis**. Galil (2000), Dall'Occo and Tavares (2004). **Distribution**. Benthic. Western Atlantic: Canada

(Nova Scotia), USA, Bahamas, Brazil (Rio Grande do Norte, Bahia, Rio de Janeiro, Espírito Santo, Rio de Janeiro, Santa Catarina), Rio Grande Rise. Eastern Atlantic: southwestern Ireland, Ibero-Morrocan Gulf, Madeira, Canary Islands, Azores, Cape Verde. Indian Ocean: Sri Lanka, Madagascar. Pacific Ocean: Indonesia, Molluca Sea, Philippines, Gulf of Panama, Colombia, Galapagos, Nazca Ridge, Marquesas, Hawaii, Fiji, Wallis and Futuna, Vanuatu, New Caledonia, New Zealand, Tasmania, Australia. **Depth range**. From 347 to 2505 m.

111. *Pentacheles validus* A. Milne-Edwards, 1880.

Pentacheles validus A. Milne-Edwards, 1880: 65; Bouvier 1925: 434; Galil 2000: 308–311, Fig. 10; Ahyong and Brown 2002: 49; Dall'Occo and Tavares 2004: 144, Fig. 1b; Serejo et al. 2007: 140; Ahyong 2009: 383; Chan 2010: 162; Bezerra and Ribeiro 2015: 126, Fig. 1.
Polycheles debilis Smith, 1884: 360.
Polycheles validus– Bouvier, 1905c: 480; 1925: 434, Fig. 10, pl. 5, Figs. 1–2; Firth and Pequegnat 1971: 61.
Polycheles demani Stebbing, 1917: 28, pl. XCII; Firth and Pequegnat 1971: 45.
Polycheles chilensis Sund, 1920b: 226; Firth and Pequegnat 1971: 42.

Type locality. Bequia, Windward Islands, Antilles. **Diagnosis**. Galil (2000), Bezerra and Ribeiro (2015). **Distribution**. Benthic. Western Atlantic: USA, Bahamas Islands, Gulf of Mexico, Caribbean Sea, Brazil (Rio Grande do Norte, Bahia, Espírito Santo, Rio de Janeiro), South Georgia. Eastern Atlantic: Bay of Biscay, Azores, Canary Islands, West Africa, South Africa. Indian Ocean: East Indian Ridge. Pacific Ocean: Vanuatu, Wallis and Futuna, New Caledonia, Australia, New Zealand, Tasmania, Chile. **Depth range**. From 914 to 3365 m.

112. *Polycheles typhlops* Heller, 1862 (Fig. 21.5d).

Polycheles typhlops Heller, 1862: 392, pl. 1, Figs. 1–6; Bouvier 1925: 237; Galil 2000: 354, Fig. 30; Silva et al. 2003: 27–28; Ahyong and Chan 2004: 179–181, Figs. 1d–f, 4 h, 5a, b; Dall'Occo and Tavares 2004: 146–148, Figs. 1d, 2a; Coelho et al. 2007: 7; Galil 2013: 497–498, Fig. 1b; Serejo et al. 2007: 140; Bezerra and Ribeiro 2015: 127, Fig. 2.
Pentacheles agassizii A. Milne-Edwards, 1880: 65.
Pentacheles hextii Alcock, 1894: 237–239.
Polycheles intermedius Balss, 1914: 599.

Type locality. Sicily, Italy. **Diagnosis**. Galil (2000), Bezerra and Ribeiro (2015). **Distribution**. Benthic. Western Atlantic: USA, Bermuda, Gulf of Mexico, Caribbean Sea, Puerto Rico, Suriname, French Guyana, Brazil (Pará, Rio Grande do Norte, Bahia, Espírito Santo, Rio de Janeiro, São Paulo, Paraná, Santa Catarina, Rio Grande do Sul). Eastern Atlantic: North Sea, North Africa, Cape Verde, West Africa, South Africa. Mediterranean Sea (Israel, Spain, France). Indian Ocean: Kenya, Comoro, Mozambique, Madagascar, Gulf of Aden. Pacific Ocean: Japan, East

China Sea, Taiwan, Philippines, Indonesia, Eastern Australia, New Caledonia, Fiji. **Depth range**. From 77 to 2055 m.

113. *Stereomastis nana* (Smith, 1884).

Pentacheles nanus Smith, 1884: 359.
Pentacheles andamanensis Alcock, 1894: 239.
Polycheles nanus– Galil 2000: 329, Fig. 19; Ahyong and Brown 2002: 71; Ahyong and Galil 2006: 765.
Polycheles grimaldii Bouvier, 1905c: 481; 1905b: 4.
Stereomastis nana– Kensley 1968: 293; Yaldwyn and Webber 2011: 206; Ahyong 2012: 3; Alvarenga and Cardoso 2014: 1, Fig. 1.

Type locality. 38°44′N, 72°38′W, off New Jersey. **Diagnosis**. Galil (2000), Alvarenga and Cardoso (2014) **Distribution**. Benthic. Western Atlantic: Canada, USA, Brazil (Rio de Janeiro). Eastern Atlantic: Iceland, Greenland, Ireland, Gulf of Biscay, Portugal, Azores, West Africa, South Africa. Indian Ocean: Gulf of Aden, Arabian Sea, India. Pacific Ocean: Japan, China Sea, Philippines, Indonesia, Eastern Australia, Tasman Sea, Solomon Islands, Vanuatu, New Zealand. **Depth range**. From 300 to 4000 m.

114. *Stereomastis sculpta* (Smith, 1880) (Fig. 21.6a).

Polycheles sculptus Smith, 1880: 346; Bouvier 1925: 438; Galil 2000: 340, Fig. 24; Ramos-Porto et al. 2000: 250; Ahyong and Brown 2002: 75; Ahyong and Chan 2004: 179, Fig. 3e, g; Dall'Occo and Tavares 2004: 146, Fig. 1c; Ahyong and Galil 2006: 765; Coelho et al. 2007: 7; Serejo et al. 2007: 140.
Stereomastis sculpta– Ahyong 2009: 385; Chan 2010: 382; Bezerra and Ribeiro 2015: 128, Fig. 3.

Type locality. Nova Scotia, Canada. **Diagnosis**. Galil (2000), Bezerra and Ribeiro (2015). **Distribution**. Benthic. Western Atlantic: Labrador Sea, Canada, Iceland, USA, Gulf of Mexico, Caribbean Sea, Guyana, Brazil (Pará, Rio Grande do Norte, Bahia, Espírito Santo, Rio de Janeiro, São Paulo, Paraná, Santa Catarina). Eastern Atlantic: Ibero-Moroccan Gulf, Mauritania, Canary Islands, Cape Verde, West Africa, South Africa. Mediterranean Sea. Indian Ocean: Arabian Sea, East Africa, Gulf of Aden, Comoro Islands, and Madagascar. Pacific Ocean: Japan, China Sea, Philippines, Indonesia, Malay Archipelago, Eastern Australia, Tasmania, Vanuatu, New Zealand Canada, USA, Chile. **Depth range**. From 200 to 4000 m.

Infraorder Anomura McLeay, 1838.

Family Chirostylidae Ortmann, 1892.

115. *Uroptychus janiceae* Baba and Wicksten, 2017.

Uroptychus janiceae Baba and Wicksten, 2017: 265, Figs. 9–11, 23C.
Diptychus nitidus– A. Milne-Edwards 1880: 62 (part).
Uroptychus nitidus var. B– Chace 1942: 15, Fig. 5; Pequegnat and Pequegnat 1970: 161; Melo-Filho 2006: 4.

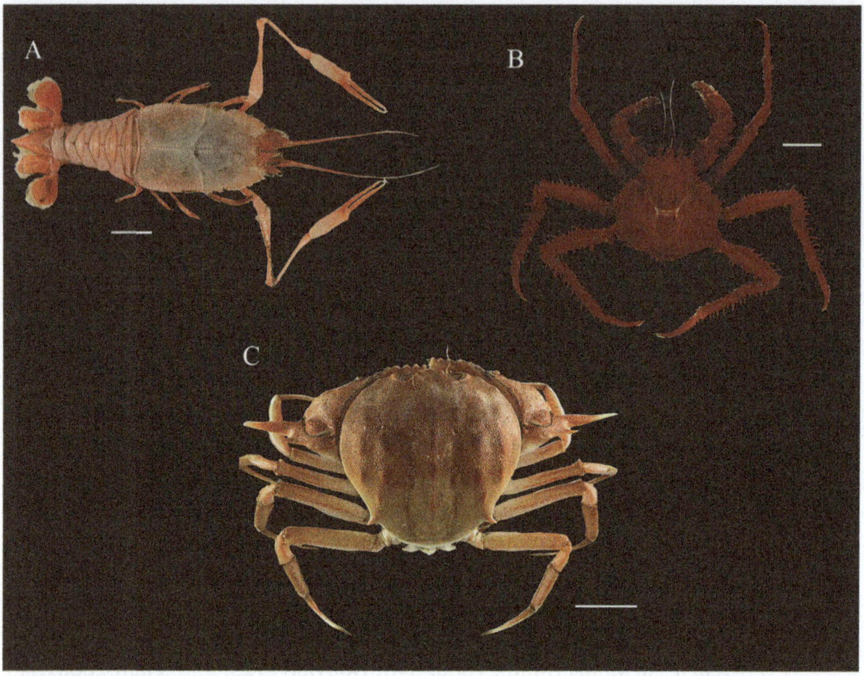

Fig. 21.6 (**a**) *Stereomastis sculpta* (Smith, 1880); (**b**) *Paralomis formosa* Henderson, 1888; (**c**) *Acanthocarpus alexandri* Stimpson, 1871. All sampled by REVIZEE/Score Central, deposited at Museu Nacional/UFRJ (unknown catalogue number). Scale bars A, C = 1 cm; B = 2 cm

Type locality. Nicholas Channel, south of Cay Sal Bank, north coast of Cuba, 23°20′N, 80°00′W. **Diagnosis**. Chace (1942), Baba and Wicksten (2017). **Distribution**. Benthic. Western Atlantic: USA (Florida, East of St. Augustine), north coast of Cuba, northern and southeastern Gulf of Mexico, southern Caribbean Sea, Brazil (São Paulo). **Depth range**. From 458 to 808 m.

116. *Uroptychus nitidus* (A. Milne-Edwards, 1880).

Diptychus nitidus A. Milne-Edwards, 1880: 62 (part); A. Milne-Edwards and Bouvier 1897: 134 (part), pl. 11, Figs. 21–22, pl. 12, Figs. 10–16.
Uroptychus nitidus– Henderson 1888: 174, pl. 21, Figs. 6, 2a; Melo-Filho 1998: 393; Melo 1999: 168, Fig. 102; Melo-Filho 2006: 3; Serejo et al. 2007: 140; Baba et al. 2008; 38; Vazquez-Bader and Gracia 2016: 5; Baba and Wicksten 2017: 253, Figs. 1, 2, 23a.
Uroptychus nitidus (typical form)– Chace 1942: 11, Fig. 3; Pequegnat and Pequegnat 1970: 161, Fig. 5–15.
Uroptychus nitidus nitudus– Melo-Filho 1999: 384.

Type locality. Off Martinique, 14°31′55"N, 61°07′28"W **Diagnosis**. Chace (1942), Baba and Wicksten (2017). **Distribution**. Benthic. Western Atlantic: USA, Straits of Florida, northern and southwestern Gulf of Mexico, north coast of Cuba, Lesser Antilles, northwestern and southern Caribbean Sea, Brazil (Pernambuco,

Espírito Santo, Rio de Janeiro, São Paulo, Rio Grande do Sul). **Depth range**. From 161 to 1362 m.

117. *Uroptychus uncifer* (A. Milne-Edwards, 1880).

Diptychus uncifer A. Milne-Edwards, 1880: 63.
Diptychus nitidus A. Milne-Edwards, 1880: 62 (part); A. Milne-Edwards and Bouvier 1897: 140, pl. 11: Figs. 1–2, pl. 12: Figs. 17–29.
Uroptychus uncifer– Benedict 1901: 148; Chace 1942: 18, Fig. 7 (part); Lemaitre 1984: 427; Melo-Filho 1998: 393; Melo 1999: 170, Fig. 104; Baba et al. 2008; 45; Vazquez-Bader and Gracia 2016: 5; Baba and Wicksten 2017: 281, Figs. 20–22, 3d.

Type locality. 13°5′0″N, 59°39′18″W, off coast of Barbados. **Diagnosis**. Baba and Wicksten (2017). **Distribution**. Benthic. Western Atlantic: Bahamas, Puerto Rico, Guadeloupe, Saint Vincent, Barbados, Brazil (Pernambuco). **Depth range**. From 161 to 453 m.

Family Diogenidae Ortmann, 1892.

118. *Paguristes spinipes* A. Milne-Edwards, 1880.

Paguristes spinipes A. Milne-Edwards 1880: 44; Melo 1999: 86, Fig. 40; Coelho-Filho 2006: 14.

Type locality: 11°25′N, 62°04′15″W, Grenada, Caribbean Sea. **Diagnosis**. Melo (1999). **Distribution**. Benthic. Western Atlantic: USA (North Carolina to Florida), Lesser Antilles, Brazil (Ceará, Paraíba, Pernambuco, Alagoas). **Depth range**. From 70 to 640 m.

Family Lithodidae Samouelle, 1819.

119. *Lithodes confundens* Macpherson, 1988.

Lithodes confundens Macpherson, 1988: 55, Fig. 24, pl. 11, 12; Lianos et al. 2017: 983, Fig. 1.

Type locality. 54°02.07′ S, 58°40.4′ W, off Maldives Island. **Diagnosis**. Macpherson (1988), Lianos et al. (2017). **Distribution**. Benthic. Western Atlantic: Brazil (Rio Grande do Sul), Argentina (Buenos Aires, Patagonia, Strait of Magellan, Burdwood Bank, south of the Falkland Islands). Pacific Ocean: Chile (Punta Arenas). **Depth range**. From 430 to 600 m.

120. *Lithodes manningi* Macpherson, 1988.

Lithodes manningi Macpherson, 1988: 62, Figs. 27, 28, pl. 14; Chevaldonne and Olu 1996: 287; Serejo et al. 2007: 140; Serejo and Cardoso 2010: 218, Fig. 48; Alves-Júnior et al. 2018c: 2, Figs. 1–2.

Type locality. Dominica, French Guiana. **Diagnosis**. Macpherson (1988). **Distribution**. Benthic. Western Atlantic: Gulf of Mexico, Dominica, Caribbean Sea, French Guiana, Brazil (Rio Grande do Norte, Espírito Santo, Rio de Janeiro). **Depth range**. From 640 to 1105 m.

121. *Neolithodes agassizii* (Smith, 1882).

Lithodes agassizii Smith, 1882: 8, pl. 1, Fig. 1 (in part, only adults); Agassiz 1888: 39, Fig. 232.
Neolithodes agassizii– Bouvier 1895: 178; 1896: 8, 22.
Neolithodes agassizii– Takeda and Okutani 1983: 106; Macpherson 1988: 33, Figs. 13, 14, 15a; pl. 2c; Serejo et al. 2007: 140; Serejo and Cardoso 2010: 220; Alves-Júnior et al. 2018c: 3, Figs. 3–4.

Type locality. Virginia, 34°39.4′ N, 75°14.4′ W. **Diagnosis**. Macpherson (1988). **Distribution**. Benthic. Western Atlantic: USA (North and South Carolina, Alabama), Gulf of Mexico, Mexico (Tabasco), Bahamas, Caribbean Sea, Martinique, Colombia, Suriname, French Guiana, Brazil (Rio Grande do Norte, Bahia, Espírito Santo, Rio de Janeiro). **Depth range**. From 650 to 2076 m.

122. *Paralomis formosa* Henderson, 1888 (Fig. 21.6b).

Paralomis formosus Henderson, 1888 p. 46, pl. V. Fig. 2.
Paralomis formosa– Bouvier 1896: 26; Macpherson 1988: 88, Figs. 36b, 40, pl. 20; Tavares and Albuquerque 1990; Serejo et al. 2007; 140.
Paralomis spectabilis Birstein and Vinogradov, 1972: 352.

Type locality. 37°17′ S, 53°52′ W. **Diagnosis**. Macpherson (1988), Tavares and Albuquerque (1990). **Distribution**. Western Atlantic: Brazil (Bahia, Espírito Santo), Uruguay, Argentina, South Georgia Islands, Shag Rocks, South Orkney Islands. **Depth range**. From 400 to 2075 m.

Family Munididae Ahyong et al., 2010.

123. *Agononida longipes* (A. Milne-Edwards, 1880).

Munida longipes A. Milne-Edwards, 1880: 50; Chace 1942: 47; Pequegnat and Pequegnat 1970: 132, Fig. 5/3; Williams 1984: 235, Fig. 170; Abele and Kim 1986: 35, figs. c, p. 405; Melo-Filho and Melo 1992a: 514; Poupin 1994: 36; Melo-Filho 1998: 395; Melo 1999: 192, Fig. 121, 122 a–e; Melo-Filho 1999: 388, Fig. 13; Melo-Filho and Melo 2001a: 1190, Fig. 9; 2001b:1155, Fig. 20, 21; 2001c: 47; Serejo et al. 2007: 140.
Munida paynei Boone, 1927: 53, Fig. 11.
Agononida longipes– Baba and de Saint Laurent 1996: 442; Melo-Filho 2006: 5; Baba et al. 2008: 49; Melo-Filho 2008: 34; Vazquez-Bader and Gracia 2016: 7.

Type locality. Off Barbados. **Diagnosis**. Melo-Filho and Melo (2001a), Melo-Filho (2006). **Distribution**. Benthic. Western Atlantic: USA (Virginia, North Carolina), Gulf of Mexico, Bahamas, Cuba, Cay Sal Banks, Puerto Rico, Lesser Antilles, Mexico, Venezuela, Guiana, and Brazil (Pernambuco, Bahia, São Paulo, Santa Catarina, Rio Grande do Sul). **Depth range**. From 129 to 1089 m.

124. *Munida constricta* A. Milne-Edwards, 1880.

Munida constricta A. Milne-Edwards, 1880: 52; Chace 1942: 34, Fig. 14; Pequegnat and Pequegnat 1970: 127; Melo-Filho 1998: 394; Melo-Filho and Melo 1992b: 766, Fig.17; Melo 1999: 180, Figs. 109, 110 a-d; Melo-Filho 1999: 389, Fig. 8; Melo-Filho and Melo 2001b: 1144, Figs. 8, 9; 2001c: 42; Melo-Filho 2006: 5; Serejo et al. 2007: 140; Baba et al. 2008: 91; Melo-Filho 2008: 34; Vazquez-Bader and Gracia 2016: 11.
Munida miles Henderson, 1888: 126 (part).

Type locality. Saint Lucie, 13° 54′N, 61°06′W. **Diagnosis**. Chace (1942), Melo-Filho (2001b). **Distribution**. Benthic. Western Atlantic: Cuba, Lesser Antilles, Brazil (Alagoas, Bahia, Espírito Santo, Rio de Janeiro, São Paulo, Santa Catarina, Rio Grande do Sul). **Depth range**. From 232 to 835 m.

125. *Munida flinti* Benedict, 1902.

Munida flinti Benedict, 1902: 258, Fig. 9; Chace 1942: 57; Pequegnat and Pequegnat 1970: 130; Takeda 1983: 87; Melo-Filho and Melo 1992b: 765, Figs. 15–16; Melo-Filho and Melo 1997: 193, Fig. 2a–f; Melo-Filho 1998: 394; Melo 1999: 182, Figs. 111, 112a–d; Melo-Filho and Melo 2001b: 1146, Figs. 10–11; 2001c: 43; Navas et al. 2003: 195; Campos et al. 2005: 149; Serejo et al. 2007: 140; Baba et al. 2008: 95.

Type locality. 28°44′00″N, 85°16′00″W, between Delta of Mississippi and Cedar Keys, Florida, USA. **Diagnosis**. Melo (1999). **Distribution**. Benthic. Western Atlantic: Gulf of Mexico, Lesser Antilles, Grenade, Guiana, Brazil (Alagoas, Bahia, Espírito Santo, Rio de Janeiro, São Paulo, Paraná, Santa Catarina and Rio Grande do Sul), Uruguay. **Depth range**. From 11 to 630 m.

126. *Munida forceps* A. Milne-Edwards, 1880.

Munida forceps A. Milne-Edwards, 1880: 49; A. Milne-Edwards and Bouvier 1894: 256; A. Milne-Edwards and Bouvier 1897: 28, pl. 2, Fig. 8; Pequegnat and Pequegnat 1970: 131, Fig. 5/2; Williams 1988: 69, 71, 74, Fig. 3; Melo-Filho and Melo 1992b: 768, Figs. 18–24; Melo-Filho 1998: 394; Melo-Filho and Melo 2001b: 1148, Figs. 12, 13; Navas et al. 2003: 197, Figs. 9–10; Melo-Filho 2006: 8; Serejo et al. 2007: 140; Baba et al. 2008: 96; Vazquez-Bader and Gracia 2016: 13.

Type locality. Gulf of Mexico. **Diagnosis**. A. Milne-Edwards and Bouvier (1894), Melo-Filho (2006). **Distribution**. Benthic. Western Atlantic: USA (Virginia and Florida), Gulf of Mexico, Cuba, Venezuela, Guiana, Brazil (Alagoas, Bahia, Espírito Santo, Rio de Janeiro, Santa Catarina, Rio Grande do Sul), Uruguay. **Depth range**. From 73 to 950 m.

127. *Munida iris* A. Milne-Edwards, 1880.

Munida iris A. Milne-Edwards, 1880: 49; Chace 1942: 33; Pequegnat and Pequegnat 1970: 131; Coelho et al. 1986: 137, 140, 149; Poupin 1994: 35;

Melo-Filho 1998: 394; Melo 1999: 188, Figs. 117, 118 a-e; Melo-Filho 1998: 394; Melo-Filho 1999: 395, Fig. 11; Melo-Filho and Melo 2001a: 1184, Fig. 6; 2001b: 1150, Figs. 16, 17; 2001c: 45; Melo-Filho 2006: 9; Serejo et al. 2007: 140; Baba et al. 2008: 101; Melo-Filho 2008: 36; Vazquez-Bader and Gracia 2016: 13.

Type locality. Off Barbados, 13°00′N, 59°36′W. **Diagnosis**. Melo-Filho and Melo (2001b), Melo-Filho (2006). **Distribution**. Benthic. Western Atlantic: Virginia, Carolinas, Gulf of Mexico, Cuba, Cay Sal Banks, Lesser Antilles, Mexico, Guiana, Brazil (Alagoas, Bahia, São Paulo, Santa Catarina, Rio Grande do Sul), Uruguay. **Depth range**. From 45 to 1303 m.

128. *Munida irrasa* A. Milne-Edwards, 1880.

Munida caribaea Stimpson, 1860: 244; A. Milne-Edwards 1880: 49; Smith 1881: 428; 1883: 40, pl. 3, Fig. 11; Benedict 1902: 306.
Munida irrasa A. Milne-Edwards, 1880: 49; Benedict 1902: 310; Chace 1942: 46; Williams 1965: 105; 1984: 234; Pequegnat and Pequegat 1970: 132; Coelho and Ramos 1972: 171; Melo-Filho and Melo 1992a: 513; Melo-Filho and Melo 2001a: 1153, Figs. 18–19; Melo-Filho 2006: 11; Serejo et al. 2007: 140; Baba et al. 2008: 107; Vazquez-Bader and Gracia 2016: 14.

Type locality. Grenada. **Diagnosis**. Melo-Filho and Melo (2001a). **Distribution**. Benthic. Western Atlantic: USA (Carolinas), Gulf of Mexico, Bahamas, Antilles, Caribbean Sea, Brazil (Amapá, Pará, Maranhão, Espírito Santo, Rio de Janeiro, São Paulo, Rio Grande do Sul), and Uruguay. **Depth range**. From 38 to 829 m.

129. *Munida microphthalma* Milne-Edwards, 1880.

Munida microphthalma Milne-Edwards, 1880: 51; Chace 1942: 40, Fig. 16; Pequegnat and Pequegnat 1970: 135, Fig. 4–5; Melo-Filho and Melo 1992a: 515; Melo-Filho 1998: 395; Melo-Filho 1998: 394; Melo-Filho and Melo 2001a: 1157, Figs. 22–23; Serejo et al. 2007: 140; Baba et al. 2008: 107; Vazquez-Bader and Gracia 2016: 14.

Type locality. 13°10′N, 61°18′W, off Saint Vincent. **Diagnosis**. Chace (1942), Melo-Filho and Melo (2001a). **Distribution**. Benthic. Western Atlantic: Virginia, Gulf of Mexico, Cuba, Lesser Antilles, Brazil (Espírito Santo, São Paulo). Eastern Atlantic: Iceland, Porcupine abyssal plain, Bay of Biscay, Morocco, Cape Verde, Ascension, South African coast. **Depth range**. From 667 to 2165 m.

130. *Munida valida* Smith, 1883.

Munida valida Smith, 1883: 42, pl. 1; Chace 1942: 32; Pequegnat and Pequegnat 1970: 137; Williams 1984: 237, Figs. 172–173; Melo-Filho and Melo 1992b: 770, Figs. 25–31; Melo-Filho 1998: 395; Melo-Filho and Melo 2001a: 1165, Figs. 32–33; 2001c: 49; Serejo et al. 2007: 140; Baba et al. 2008: 127.
Munida miles– Henderson 1888: 26.

Type locality. 39°56′N, 70°35′W/40°01′N, 68°54′W. **Diagnosis**. Melo-Filho and Melo (1992a, b, 2001a). **Distribution**. Benthic. Western Atlantic: Virginia, Carolinas, Gulf of Mexico, Bahamas, Caribbean Sea, Colombia, Curacao, Venezuela, Guianas, Brazil (Alagoas, Bahia, Espírito Santo, Rio de Janeiro, Rio Grande do Sul). **Depth range**. From 90 to 2297 m.

131. *Munida victoria* Melo-Filho, 1996.

Munida victoria Melo-Filho, 1996: 272, Figs. 1–7; Serejo et al. 2007: 140; Baba et al. 2008: 127.

Type locality. Espírito Santo, Brazil. **Diagnosis**. Melo-Filho (1996). **Distribution**. Benthic. Western Atlantic: Brazil (Espírito Santo). **Depth range**. From 910 to 927 m.

Family Munidopsidae Ortmann, 1898.

132. *Galacantha rostrata* (A. Milne-Edwards, 1880).

Galacantha rostrata A. Milne-Edwards, 1880: 52; Smith 1884: 355; Macpherson 2007: 18, Figs. 10i, j; Ahyong 2007: 4, Figs. 2c, d; Baba et al. 2008: 62.
Munidopsis rostrata– Smith 1885: 493; Smith 1886a, b: 45, pl. 6, Figs. 1, 1; Chace 1942: 75; Baba and Poore 2002: 239, Fig. 5; Ingle and Christiansen 2004: 144, Figs. 117, 120; Poore 2004: 237, Fig. 65f; Macpherson and Segonzac 2005: 41; Serejo et al. 2007: 140.
Galacantha talismanii Filhol, 1885: pl. 3.
Galacantha investigatoris Alcock and Anderson, 1894: 173.

Type locality. Lesser Antilles, Bequia, Windward. **Diagnosis**. Poore (2004), Ahyong (2007), Macpherson (2007). **Distribution**. Benthic. Western Atlantic: New Jersey, Antilles, Colombia, Tobago, Gulf of Mexico, Cuba, Brazil (Bahia). Eastern Atlantic: Morocco, South Africa. Indian Ocean: Arabian Sea, Gulf of Aden, Zanzibar, Bay of Bengal, Moluccas, Indonesia. Pacific Ocean: Japan, Australia, Galapagos, Valparaiso, Chile. **Depth range**. From 1600 to 3800 m.

133. *Munidopsis erinacea* (A. Milne-Edwards, 1880).

Galathodes erinaceus A. Milne-Edwards, 1880: 53.
Munidopsis erinacea– Henderson 1888: 149, pl. 16, Figs. 4, 4a; Chace 1942: 90; Melo-Filho 1998: 396; Tavares and Campinho 1998: 88, Figs. 1–2; Serejo et al. 2007: 140; Baba et al. 2008: 141; Vazquez-Bader and Gracia 2016: 17.
Munidopsis erinaceus– Navas et al. 2003: 205; Campos et al. 2005: 160, Figs. 122, 123.

Type locality. Near Saint Lucia, West Indies. **Diagnosis**. Tavares and Campinho (1998), Campos et al. (2005). **Distribution**. Benthic. Western Atlantic: Gulf of Mexico, Cuba, Lesser Antilles, Colombia, Suriname; Guianas; Honduras, Brazil (Pernambuco, Bahia, Espírito Santo, Rio de Janeiro, São Paulo). **Depth range**. From 279 to 1107 m.

134. *Munidopsis nitida* (A. Milne-Edwards, 1880).

Orophorhynchus nitidus A. Milne-Edwards, 1880: 59.
Orophorhynchus spinosus A. Milne-Edwards, 1880: 58.
Munidopsis nitida– A. Milne-Edwards and Bouvier 1894: 275, 278; Faxon 1895: 84; Chace 1942: 73; Pequegnat and Pequegnat 1970: 153, Figs. 5–12; 1971: 6; Pequegnat et al. 1971: 5, 8; Williams and Baba 1989: 902, Fig. 2 h; Tavares and Campinho 1998: 91, Figs. 3, 4; Baba et al. 2008: 151.
Munidopsis nitidus– Doflein and Balss 1913: 177.

Type locality. Guadeloupe, Dominica, Lesser Antilles. **Diagnosis**. Pequegnat (1970a, b), Tavares and Campinho (1998). **Distribution**. Benthic. Western Atlantic: Bahamas, Guadeloupe, Dominica, Gulf of Mexico, Brazil (Bahia). Eastern Atlantic: Gulf of Guinea. Indian Ocean: Mozambique Channel, Madagascar, Bay of Bengal, Gulf of Aden, Philippines, Indonesia, New Caledonia, Vanuatu and Solomon Islands. Pacific Ocean: Taiwan, Admiralty Islands, Gulf of Panama, Japan, Gulf of California. **Depth range**. From 1350 to 2149 m.

135. *Munidopsis polita* (Smith, 1883).

Anoplonotus politus Smith, 1883: 50, pl. 2, Fig. 1, pl. 3: Fig. 1–5a.
Munidopsis polita– Benedict 1902: 324; Chace 1942: 75; Pequegnat and Pequegnat 1970: 155, Fig. 5.1; Poupin 1994: 39; Melo-Filho 1998: 396; 2006: 5; 2008: 36; Baba et al. 2008: 155; Vazquez-Bader and Gracia 2016: 19.

Type locality: off Martha's Vineyard. **Diagnosis**. Melo-Filho (2006), Melo-Filho (2008). **Distribution**: Benthic. Western Atlantic: Massachusetts, Virginia, Florida, Gulf of Mexico, Nicaragua, Lesser Antilles, Guadeloupe, Colombia, Brazil (São Paulo, Santa Catarina). **Depth range**. From 129 to 860 m.

136. *Munidopsis riveroi* Chace, 1939.

Munidopsis riveroi Chace, 1939: 48; 1942: 93, Figs. 31–32; Pequegnat and Pequegnat 1970: 140; Campos et al. 2005: 166, Figs. 133–134; Serejo et al. 2007: 140; Baba et al. 2008: 157; Vazquez-Bader and Gracia 2016: 19.

Type locality. Cuba, Nicholas Channel, Santa Clara. **Diagnosis**. Chace (1942), Pequegnat and Pequegnat (1970). **Distribution**. Benthic. Western Atlantic: Caribbean Sea, Honduras, Colombia, Venezuela, Cuba, Dominica, Brazil (Bahia). **Depth range**. From 260 to 659 m.

137. *Munidopsis sigsbei* (A. Milne-Edwards, 1880).

Galathodes sigsbei A. Milne Edwards, 1880: 56.
Munidopsis sigsbei– Henderson 1888: 150, pl. 18, Fig. 2, 2a; Chace 1942: 82; Tavares and Campinho 1998: 95, Figs. 5–8; Ahyong and Poore 2004: 53, Fig. 11 b–f; Serejo et al. 2007: 140; Baba et al. 2008: 160; Vazquez-Bader and Gracia 2016: 20.
Munidopsis Sigsbei– A. Milne Edwards and Bouvier 1897: 83.
Munidopsis sigsbey– Navas et al. 2003: 201.

Type locality. Guadeloupe, North of Yucatan Bank, off Frederickstadt, Santa Cruz, Martinique. **Diagnosis**. Tavares and Campinho (1998), Ahyong and Poore (2004). **Distribution**. Benthic. Western Atlantic: Gulf of Mexico, Yucatan Bank, Cuba, Jamaica, Lesser Antilles, Guadeloupe, Martinique, Brazil (Espírito Santo, Bahia). **Depth range**. From 677 to 1784 m.

138. *Munidopsis transtridens* Pequegnat and Pequegnat, 1971.

Munidopsis transtridens Pequegnat and Pequegnat, 1971: 15; McLaughlin et al. 2005: 239; Baba et al. 2008: 166; Fierro Rengifo et al. 2008: 9; Tavares et al. 2008: 95, Figs. 1a–c, 2, 3; Rodrigues and Serejo 2010: 216.

Type locality. Southeastern Gulf of Mexico. **Diagnosis**. Tavares et al. (2008). **Distribution**. Benthic. Western Atlantic: Florida, Guyana, southeastern Gulf of Mexico, Brazil (22°25′4 4.211"S-45°57′32.305"W). **Depth range**. From 1048 to 1446 m.

Family Paguridae Latreille, 1802.

139. *Anisopagurus bartletti* (A. Milne-Edwards, 1880).

Eupagurus bartletti A. Milne-Edwards, 1880: 41.
Pylopagurus Bartletti– A. Milne-Edwards and Bouvier 1893: 91, pl. 7, Figs. 1–9.
Pylopagurus bartletti– Alcock 1905: 189; Gordan 1956: 340.
Anisopagurus bartletti– McLaughlin 1981a: 6; Abele and Kim 1986: 31, 361, 369, fig. b; Lemaitre and McLaughlin 1996: 92, Figs. 1–4; Melo 1999: 102, Fig. 52; McLaughlin et al. 2010: 27; Nucci and Melo 2011: 29, Figs. 1b, 2b, 3b.

Type locality. Saint Vincent, Lesser Antilles. **Diagnosis**. Lemaitre and McLaughlin (1996), Nucci and Melo (2011). **Distribution**. Benthic. Western Atlantic: USA, Florida, Gulf of Mexico, Antilles, Colombia, Brazil (Pará). **Depth range**. From 50 to 550 m.

140. *Catapaguroides microps* A. Milne-Edwards and Bouvier, 1892.

Catapaguroides microps A. Milne-Edwards and Bouvier, 1892: 221; 1899: 63; 1900: 207, pl. 204, Figs. 17–20; Holthuis 1962: 243; De Saint Laurent 1968: 935, Figs. 17, 21, 22, 24; McLaughlin 2002: 497; Lins and Cardoso 2010: 51, Figs. 1–2; McLaughlin et al. 2010: 28.

Type locality. 33°9′N, 11°58′W, Morocco. **Diagnosis**. De Saint Laurent (1968), Lins and Cardoso (2010). **Distribution**. Benthic. Western Atlantic: Brazil (Rio de Janeiro). Eastern Atlantic: Azores, Cape Finisterre, Spain, Morocco. Pacific Ocean: Indonesia. **Depth range**. From 1067 to 1626 m.

141. *Iridopagurus iris* (A. Milne-Edwards, 1880).

Spiropagurus iris A. Milne-Edwards, 1880: 44; Perrier 1893; A. Milne-Edwards and Bouvier 1893: 112, pl. 8, Figs. 14–25; Alcock 1905: 188; Gordan 1956: 341; Rabaud 1941: 190, Fig. 1; Hazlett 1966: 88.

Iridopagurus iris– De Saint Laurent 1966: 159, Figs. 1, 3–9, 17–20, 24, 29, 34; García-Gómez 1983: 16; Coelho and Ramos-Porto 1986: 44; Rieger 1998: 417; Melo 1999: 110, Fig. 58; McLaughlin et al. 2010: 30; Nucci and Melo 2011: 33, Figs. 1e, 2e, 3e.

Type locality. Barbados. **Diagnosis**. Melo (1999), Nucci and Melo (2011). **Distribution**. Benthic. Western Atlantic: USA, Florida, Antilles, Brazil (Amapá, Alagoas, Rio de Janeiro), Uruguay. **Depth range**. From 60 to 700 m.

142. *Phimochirus occlusus* (Henderson, 1888).

Eupagurus occlusus Henderson, 1888: 70, pl. 7, Fig. 6.
Pagurus occlusus– Gordan 1956: 332.
Pylopagurus occlusus– Forest and De Saint Laurent 1968: 145, Figs. 113, 115–119.
Phimochirus occlusus– McLaughlin 1981a: 5; 1981b: 360, Figs. 4 g, 9c, 10c; Coelho and Ramos-Porto 1986: 42; Rieger 1998: 416; Melo 1999: 142, Fig. 84; McLaughlin et al. 2010: 34; Nucci and Melo 2011: 36, Figs. 1 k, 2 k, 3 k.

Type locality. 9°5′ S, 34°50′W, off Pernambuco. **Diagnosis**. McLaughlin (1981b), Nucci and Melo (2011). **Distribution**. Benthic. Western Atlantic: Antilles, Brazil (Pernambuco, Alagoas, Sergipe, Bahia, Espírito santo, Rio de Janeiro, São Paulo). **Depth range**. From 100 to 640 m.

143. *Pylopagurus discoidalis* (A. Milne-Edwards, 1880).

Eupagurus discoidalis A. Milne-Edwards, 1880: 41.
Pylopagurus discoidalis– A. Milne-Edwards and Bouvier 1893: 76, pl.6, Figs. 7–14; Williams 1965: 134, Fig. 109; 1984: 226, Fig. 162; McLaughlin 1981a, b: 2; Coelho and Ramos-Porto 1986: 43; Rieger 1998: 416; Melo 1999: 144, Fig. 86; McLaughlin and Lemaitre 2001: 451, Figs. 4–6; McLaughlin et al. 2010: 35; Nucci and Melo 2011, Figs. 1 m, 2 m, 3 m.

Type locality. Montserrat, Lesser Antilles. **Diagnosis**. Nucci and Melo (2011). **Distribution**. Benthic. Western Atlantic: USA (North Carolina to Florida), Gulf of Mexico, Antilles, and Brazil (from Amapá to Santa Catarina). **Depth range**. From 55 to 930 m.

Family Parapaguridae Smith, 1882.

144. *Sympagurus dimorphus* (Studer, 1883).

Eupagurus dimorphus Studer, 1883: 20, Figs. 11–12.
Sympagurus dimorphus– Lemaitre 1989: 71, Figs. 36–38, 40E–H; 1990: 229; Lemaitre and Mclaughlin 1992: 747, Figs. 1–5; Melo 1999: 154, Figs. 92–93.

Type locality. Ascencion Island. **Diagnosis**. Lemaitre (1989), Melo (1999). **Distribution**. Benthic. Western Atlantic: Brazil (Rio Grande do Sul), Argentina. Eastern Atlantic: Assencion, western South Africa. Pacific Ocean: Australia, Tasmania, New Zaeland. **Depth range**. From 70 to 1995 m.

145. *Strobopagurus gracilipes* (A. Milne Edwards, 1891).

Sympagurus gracilipes A. Milne Edwards, 1891: 132; Forest 1954: 167. *Parapagurus gracilipes*– Forest 1954: 103; Kensley 1973: 287. *Strobopagurus gracilipes*– Lemaitre 1989: 36; 1996: 167; 2004: 364; Cardoso et al. 2017: 4.

Type locality. 38°26′25″N, 30°59′10″W, Azores. **Diagnosis**. Lemaitre (1989). **Distribution**. Benthic. Western Atlantic: Rio Grande Rise. Eastern Atlantic: from Portugal to Morocco, Azores, Canary, Cape Verde Islands. Pacific Ocean: New Caledonia, Vanuatu, Solomon Islands, Taiwan, Hawaii, French Polynesia. **Depth range**. From 75 to 1200 m.

Infraorder Brachyura Latreille, 1802.

Family Calappidae De Haan, 1833.

146. *Acanthocarpus alexandri* Stimpson, 1871 (Fig. 21.6c).

Acanthocarpus alexandri Stimpson, 1871: 153; Rathbun 1937: 224, pl. 69, Figs. 1–2; Melo 1996: 121; Melo et al. 1998: 445; Ramos-Porto et al. 2002: 100; Rodrigues and Young 2003: 2, Figs. 1–2; Serejo et al. 2007: 140; Alves-Júnior et al. 2018d: 61, Fig. 1.

Type locality. Off the Quicksands, Florida Keys, USA. **Diagnosis**. Rathbun (1937), Melo (1996), Alves-Júnior et al. (2018c). **Distribution**. Benthic. Western Atlantic: Canada, USA (Massachusetts, North Carolina to Florida), Gulf of Mexico, Cuba, Porto Rico, Lesser Antilles, Dry Tortugas, Brazil (Piauí, Rio Grande do Norte, Bahia, Espírito Santo, Rio de Janeiro, Rio Grande do Sul). **Depth range**. From 20 to 550 m.

147. *Acanthocarpus bispinosus* A. Milne Edwards, 1880.

Acanthocarpus bispinosus A. Milne-Edwards, 1880; Powers 1977; Rathbun 1937: pl. 68, Figs. 1–3; Ramos-Porto et al. 2002: 101; Rodrigues and Young 2003: 5, Figs. 3–4; Serejo et al. 2007: 140; Alves-Júnior et al. 2018d: 63, Fig. 2.

Type locality. Reefs of the Grenadines, Caribbean Sea. **Diagnosis**. Rathbun (1937), Alves-Júnior et al. (2018a, b, c, d). **Distribution**. Benthic. Western Atlantic: USA (Florida), Gulf of Mexico, Lesser Antilles, Grenadines Island, Dry Tortugas, Brazil (Rio Grande do Norte, Pernambuco, Sergipe, Bahia). **Depth range**. From 200 to 552 m.

Family Chasmocarcinidae Serène, 1964.

148. *Chasmocarcinus cylindricus* Rathbun, 1901.

Chasmocarcinus cylindricus Rathbun, 1901: 10; Melo 1996: 420.

Type locality. Mayaguez Harbor, Puerto Rico. **Diagnosis**. Melo (1996). **Distribution**. Benthic. Western Altantic: Gulf of Mexico, Lesser Antilles, Brazil (Rio de Janeiro, São Paulo). **Depth range**. From 15 to 1900 m.

Family Cyclodorippidae Ortmann, 1892.

149. *Cyclodorippe antennaria* A. Milne-Edwards, 1880.

Cyclodorippe antennaria A. Milne-Edwards, 1880: 25; Rathbun 1937: 104; Melo 1996: 91; Melo et al. 1998: 441.

Type locality. Gulf of Mexico. **Diagnosis**. Melo (1996). **Distribution**. Benthic. Western Atlantic: Gulf of Mexico, Lesser Antilles, Brazil (Rio de Janeiro). **Depth range**. From 40 to 650 m.

150. *Clythrocerus granulatus* (Rathbun, 1898).

Cyclodorippe granulata Rathbun, 1898: 293, pl. 9, Fig. 1.
Clythrocerus granulatus– Melo 1996: 87; Melo et al. 1998: 441.

Type locality. off Trinidad, Caribbean Sea. **Diagnosis**: Rathbun (1898), Melo (1996). **Distribution**. Benthic. Western Atlantic: USA (Florida), Lesser Antilles, Venezuela, Brazil (Amapá, Espírito Santo, Rio de Janeiro, Rio Grande do Sul). **Depth range**. From 120 to 600 m.

Family Cymonomidae Bouvier, 1898.

151. *Cymonomoides guinotae* (Tavares, 1991).

Cymonomus guinotae Tavares, 1991: 640, Figs. 7c, 8b, 9c, 10a, b, d.
Cymonomoides guinotae– Melo 1996: 98; Melo et al. 1998: 442.

Type locality. Brazil, 23°46′S, 42°09′W. **Diagnosis**. Tavares (1991), Melo (1996). **Distribution**. Benthic. Western Atlantic: Brazil (Espírito Santo, Rio de Janeiro). **Depth range**. From 500 to 900 m.

152. *Cymonomus guillei* Tavares, 1991.

Cymonomus guillei Tavares, 1991: 639, Figs. 7b, 8d, 9b, 11d; Melo 1996: 100; Melo et al. 1998: 442.

Type locality. 23°46′S, 42°09′W, Brazil. **Diagnosis**. Tavares (1991), Melo (1996). **Distribution**. Western Atlantic: Brazil (Espírito Santo, Rio de Janeiro). **Depth range**. From 590 to 730 m.

153. *Cymonomus magnirostris* Tavares, 1991.

Cymonomus magnirostris Tavares, 1991: 635, Figs. 7a, 8e, 9a, 10d–f; Melo 1996: 101; Melo et al. 1998: 443.

Type locality. 23°46′S, 42°09′W, Brazil. **Diagnosis**. Tavares (1991), Melo (1996). **Distribution**. Benthic. Western Atlantic: Brazil (Espírito Santo, Rio de Janeiro). **Depth range**. From 590 to 730 m.

154. *Cymonomus quadratus* A. Milne-Edwards, 1880.

Cymonomus quadratus A. Milne-Edwards, 1880: Rathbun 1937: 98; 26; Melo 1996: 102; Melo et al. 1998: 443.

Type locality. Gulf of Mexico. **Diagnosis**. Melo (1996). **Distribution**. Benthic. Western Atlantic: USA (Florida), Gulf of Mexico, Lesser Antilles, Brazil (Amapá, Rio de Janeiro, Paraná, Santa Catarina, Rio Grande do Sul). **Depth range**. From 190 to 930 m.

Family Dromiidae De Haan, 1833.

155. *Dromia bollorei* Forest, 1974.

Dromia bollorei Forest, 1974: 91, Fig. 1d, 2, 3d, 5, 6b, 7c, d; pl. 2, Fig. 1,2; pl. 3, Fig. 4; pl. 5, Fig. 1; Manning and Holthuis 1981: 11; Guinot and Tavares 2003: 94: Fig. 27b; Cleva et al. 2007: 240, Fig. 8c; Ng et al. 2008: 33; Nunes et al. 2017: 336, Fig. 4.

Type locality. Mauritania. **Diagnosis**. Nunes et al. (2017). **Distribution**. Benthic. Western Atlantic: Brazil (Saint Peter and Saint Paul Archipelago). Eastern Atlatic: Mauritania, Ivory Coast. **Depth range**. From 100 to 546 m.

Family Epialtidae MacLeay, 1838.

156. *Herbstia depressa* Stimpson, 1860.

Herbstia depressa Stimpson, 1860: 185; Melo 1996: 255.

Type locality. Saint Thomas, Caribbean Sea. **Diagnosis**. Stimpson (1860), Melo (1996). **Distribution**. Benthic. Western Atlantic: Lesser Antilles, Venezuela, Brazil (Alagoas). **Depth range**. From 60 to 700 m.

157. *Holoplites armatus* (A. Milne-Edwards, 1880).

Nibilia armata A. Milne-Edwards, 1880.
Holoplites armata– Melo 1996: 256.
Holoplites armatus– Ng et al. 2008: 103.

Type locality. Gulf of Mexico. **Diagnosis**. Melo (1996). **Distribution**. Benthic. Western Atlantic: Gulf of Mexico, Lesser Antilles, Brazil (Pará). **Depth range**. From 160 to 800 m.

158. *Minyorhyncha crassa* (A. Milne-Edwards, 1879).

Amathia crassa A. Milne-Edwards, 1879: 203, pl. 28, Fig. 2.
Rochinia crassa– Rathbun 1925: 210, pls. 68, 69, 226; Pequegnat 1970b: 183; Williams 1984: 322, Figs. 256, 260a; Ng et al. 2008: 105.
Minyorhyncha crassa– Tavares and Santana 2018: 213, Figs. 4, 12.

Type locality. Florida, USA. **Diagnosis**. Tavares and Santana (2018). **Distribution**. Benthic. Western Atlantic: Nova Scotia, Canada, USA (from Martha's Vineyard to Florida), Gulf of Mexico, Caribbean Sea, Brazil (Amapá, Pará, Maranhão, Piauí, Ceará, Rio Grande do Norte, Paraíba, Pernambuco, Alagoas, Sergipe, Bahia, Espírito Santo, Rio de Janeiro, São Paulo, Paraná, Santa Catarina). **Depth range**. From 66 to 1216 m.

159. *Scyramathia umbonata* (Stimpson, 1871).

Scyra umbonata Stimpson, 1871: 115; A. Milne-Edwards 1879: pl. 31, Figs. 5, 5a, 5b; Smith 1886b, 625.
Rochinia umbonata– Rathbun 1925: 222, pl. 72, pl. 73, Fig. 1; Chace 1940, 63; Williams et al. 1968: 61, Fig. 16; Pequegnat 1970b: 183; Williams 1984: 323, Fig. 258; Abele and Kim 1986: 42; Griffin and Tranter 1986: 175; Poupin 1994: 43, pl. 4 g; McLaughlin et al. 2005, 253; Wicksten and Packard 2005: 1762; Serejo et al. 2007: 141; Coelho et al. 2008: 17; Tavares et al. 2015: 1, Figs. 1–6.
Scyramathia umbonata– A. Milne-Edwards and Bouvier 1923: 381.
Amathia modesta Stimpson, 1871: 124; Miers 1886: 26.
Anamathia modesta– Smith 1885: 493; Faxon 1895: 10.
Anamathia umbonata– Rathbun 1894: 61, pl. 1, Figs. 1–3; Faxon 1895: 10.
Rochinia confusa Tavares, 1991: 162; Melo 1996: 266; Melo 1998: 471; Coelho et al. 2008: 17.

Type locality. Off Sand Key, Florida. **Diagnosis**. Williams (1984), Tavares (1991). **Distribution**. Benthic. USA (from North Carolina to Gulf of Mexico), Nicaragua, West Indies, and Brazil (Amapá, Pará, Pernambuco, Bahia, Espírito Santo, Rio de Janeiro). **Depth range**. From 161 to 900 m.

Family Ethusidae Guinot, 1977.

160. *Ethusa microphthalma* Smith, 1881.

Ethusa microphthalma Smith, 1881: 418; Melo 1996: 107; Melo et al. 1998: 443.

Type locality. Long Island, from 39°46′ to 40°06′N, 70°22′ to 71°10′W. **Diagnosis**. Melo (1996). **Distribution**. Benthic. Western Atlantic: USA (Massachussets to Florida), Gulf of Mexico, Lesser Antilles, Brazil (São Paulo). **Depth range**. From 110 to 750 m.

161. *Ethusina abyssicola* Smith, 1884.

Ethusina abyssicola Smith, 1884: 349, pl. 2, Figs. 1, la; Melo 1996: 109; Melo et al. 1998: 443.

Type locality. 38°52′40″N, 09°24′40″W; 38°53′00″N, 00°23′30″W; 37°41′20″N, 73°03′20″W. **Diagnosis**. Melo (1996). **Distribution**. Benthic. Western Atlantic: USA (Massachussets to North Carolina), Gulf of Mexico, Brazil (Rio de Janeiro). Eastern Atlantic: Spain; Mediterranean Sea. **Depth range**. From 850 to 4050 m.

Family Geryonidae Colosi, 1923.

162. *Chaceon gordonae* (Ingle, 1985).

Geryon gordonae Ingle, 1985: 90, Figs. 1, 2, 5a.
Chaceon gordonae– Manning and Holthuis 1989: 8; Afonso-Dias et al. 2008: 1, Fig. 2a; Nunes et al. 2017: 333, Fig. 2.

Type locality. West Africa. **Diagnosis**. Ingle (1985), Afonso-Dias et al. (2008), Nunes et al. (2017). **Distribution**. Benthic. Western Atlantic: Brazil (Saint Peter and Saint Paul Archipelago). Eastern Atlantic: Iceland, Sierra Leone, São Tomé, and Príncipe. **Depth range**. From 100 to 2000 m.

163. *Chaceon notialis* Manning and Holthuis, 1989.

Chaceon notialis Manning and Holthuis, 1989: 59, Figs. 14, 15; Tavares and Pinheiro 2011: 66.
Geryon quinquedens– Juanico 1973:145; Scelzo and Valentini 1974: 561, Figs. 1–2 (part); Boschi 1976: 66; Barea and Defeo 1985:189, Fig. 1.

Type locality. 38°55′S, 55°35′W, Argentina. **Diagnosis**. Manning and Holthuis (1989). **Distribution**. Benthic. Western Atlantic: Brazil (Espírito santo, Rio de Janeiro, São Paulo), Uruguay, Argentina. **Depth range**. From 120 to 800 m.

164. *Chaceon ramosae* Manning et al., 1989.

Chaceon ramosae Manning et al., 1989: 646, Figs. 2–3; Melo 1998: 480; Serejo et al. 2007: 140; Tavares and Pinheiro 2011: 66.
Geryon quinquedens– Rathbun 1937: 270 (part); Scelzo and Valentini 1974: 561 (part).

Type locality. 21°31′S, 40°07′W, Espírito Santo, Brazil. **Diagnosis**. Manning et al. (1989). **Distribution**. Benthic. Western Atlantic: Brazil (Espírito Santo, Rio de Janeiro, São Paulo). **Depth range**. From 601 to 1228 m.

165. *Chaceon sanctaehelenae* Manning and Holthuis, 1989.

Chaceon sanctaehelenae Manning and Holthuis, 1989: 71; Afonso-Dias et al. 2008: 1; Fransen 2014: 302; Tavares and Pinheiro 2011: 66; Cardoso et al. 2017: 4.

Type locality. 15°58′S, 05°43′W, Sandy Bay, St. Helena. **Diagnosis**. Manning and Holthuis (1989). **Distribution**. Benthic. Western Atlantic: Rio Grande Rise. Eastern Atlantic Ocean: St. Tome, St. Helena. **Depth range**. From 500 to 1200 m.

166. *Chaceon linsi* Tavares and Pinheiro, 2011.

Chaceon linsi Tavares and Pinheiro, 2011: 58, Figs. 1a, 2a, 3a–c, 4a, d, f.
Chaceon fenneri– Oliveira et al. 1999: 50; Cunha et al. 1999: 531; Sankarankutty et al. 2001: 649; Carvalho et al. 2009: 572.

Type locality. 01°45.231′S, 38°15.444′W, Ceará, Brazil. **Diagnosis**. Tavares and Pinheiro (2011). **Distribution**. Benthic. Western Atlantic: Brazil (Ceará, Rio Grande do Norte). **Depth range**. From 529 to 709 m.

Family Goneplacidae Macleay, 1838.

Fig. 21.7 (**a**) *Bathyplax typhla* A. Milne-Edwards, 1880; (**b**) *Homolodromia monstrosa* Martin et al., 2001; (**c**) *Myropsis quinquespinosa* Stimpson, 1871. All sampled by REVIZEE/Score Central, deposited at Museu Nacional/UFRJ (unknown catalogue number). Scale bars = 1 cm

167. *Bathyplax typhla* A. Milne-Edwards, 1880 (Fig. 21.7a).

Bathyplax typhlus A. Milne-Edwards, 1880: 16; Coelho and Coelho-Filho 1993: 560.

Bathyplax typhlus var. *oculiferus* Miers, 1886: 230; Tavares 1996: 414.

Bathyplax typhla– Rathbun 1918: 19; Chace 1940: 43; Pequegnat 1970a, b: 192; Powers 1977: 112; Melo 1996: 399; Melo 1998: 491; Serejo et al. 2007: 140.

Type locality. Frederickstadt, Saint Croix. **Diagnosis**. Chace (1940), Melo (1996). **Distribution**. Benthic. Western Atlantic: North Caroline, Florida, Gulf of Mexico, Cuba, Guadelupe, St. Croix, Saint Lucia, Brazil (Pernambuco, Alagoas, Bahia, Espírito Santo, Rio de Janeiro, São Paulo). **Depth range**. From 220 to 1100 m.

Family Homolidae De Haan, 1839.

168. *Homola minima* Guinot and Richer de Forges, 1995.

Homola minima Guinot and Richer de Forges, 1995: 326, Figs. 8c, d, f, 9b; Nizinski 2003: 123; Felder et al. 2009: 1072; Tavares and Lemaitre 2014: 514, Figs. 6–9; Nunes et al. 2017: 339, Fig. 6.

Thelxiope barbata– Rathbun 1937: 63, Fig. 16, Table 17, pl. 15, Figs. 1–2; Chace 1940: 8.
Homola barbata– Williams 1984: 261, Fig. 193; Melo 1996: 75; 1999: 440, Fig. 2.

Type locality. Off Delaware. **Diagnosis**. Guinot and Richer de Forges (1995), Tavares and Lemaitre (2014), Nunes et al. (2017). **Distribution**. Benthic. Western Atlantic: USA (Massachusetts to Florida), Gulf of Mexico, Caribbean Sea, Suriname, Brazil (Rio Grande do Norte, Saint Peter and Saint Paul Archipelago), Uruguay. **Depth range**. From 55 to 700 m.

169. *Homologenus rostratus* (A. Milne-Edwards, 1880).

Homolopsis rostratus A. Milne-Edwards, 1880: 34.
Homologenus rostratus– Rathbun 1937: 70; Chace 1940: 9; Manning and Holthuis 1981: 25; Guinot and Richer de Forges 1995: 471; Cleva et al. 2007: 250; Serejo et al. 2007: 141; Felder et al. 2009: 1072; Almeida et al. 2010: 360; Tavares and Lemaitre 2014: 520, Fig. 6c.

Type locality. Antilles, between St. Thomas and St. Croix. **Diagnosis**. A. Milne-Edwards (1880). **Distribution**. Benthic. Western Atlantic: USA (Delaware to Florida), Straits of Florida, Gulf of Mexico, Cuba, Venezuela, and Brazil (Rio Grande do Norte, Bahia). Eastern Atlantic: Azores, Madeira, Morocco. **Depth range**. From 600 to 2195 m.

Família Homolodromiidae Alcock, 1899.

170. *Homolodromia monstrosa* Martin et al., 2001 (Fig. 21.7b).

Homolodromia monstrosa Martin et al., 2001: 319, Figs. 4–7; Serejo et al. 2007: 141; Tavares and Lemaitre 2014: 512, Fig. 1c, d.

Type locality. Paramaribo, Suriname, and Georgetown, French Guiana. **Diagnosis**. Martin et al. (2001). **Distribution**. Benthic. Western Atlantic: USA, Florida, Gulf of Mexico, Caribbean Sea, Honduras, Nicaragua, Panama, Colombia, Suriname, French Guiana, Brazil (Rio Grande do Norte). **Depth range**. From 375 to 918 m.

171. *Homolodromia paradoxa* A. Milne-Edwards, 1880.

Homolodromia paradoxa A. Milne-Edwards, 1880: 33; Takeda 1983: 112; Martin et al. 2001: 314; Tavares and Lemaitre 2014: 511, Fig. 5.

Type locality. Nevis, Lesser Antilles. **Diagnosis**. A. Milne-Edwards (1880). **Distribution**. Benthic. Western Atlantic: Gulf of Mexico, Caribbean Sea, Suriname, Brazil (Bahia). **Depth range**: From 549 to 933 m.

Family Inachidae MacLeay, 1838.

172. *Anomalothir furcillatus* (Stimpson, 1871).

Anomalopus furcillatus Stimpson, 1871: 125.
Anomalothir furcillatus– Melo 1996: 181; Melo 1998: 457.

Type locality. The Samboes, Straits of Florida, USA. **Diagnosis** Stimpson (1871), Melo (1996). **Distribution**. Benthic. Western Atlantic: USA (North Carolina to Florida), Gulf of Mexico, Lesser Antilles, Brazil (Bahia, Espírito Santo, Rio Grande do Sul, São Paulo, Paraná, Santa Catarina, Rio de Janeiro). **Depth range**. From 50 to 690 m.

Family Inachoididae Dana, 1851.

173. *Euprognatha acuta* A. Milne-Edwards, 1880.

Euprognatha acuta A. Milne-Edwards, 1880; Melo 1996: 204; Melo 1998: 460.

Type locality. Gulf of Mexico. **Diagnosis**. Melo (1996). **Distribution**. Benthic. Western Atlantic: USA (Massachusetts to Florida), Gulf of Mexico, Lesser Antilles, Guianas, Brazil (Amapá to Rio Grande do Sul). **Depth range**. From 15 to 710 m.

Family Leucosiidae Samouelle, 1819.

174. *Myropsis quinquespinosa* Stimpson, 1871 (Fig. 21.7c).

Myropsis quinquespinosa Stimpson, 1871: 157; Melo 1996: 149; Melo et al. 1998: 450; Torres et al. 2002: 109, Fig.1; Serejo et al. 2007: 140; Coelho et al. 2008: 14.

Type locality. Off Tennessee Reef. **Diagnosis**. Melo (1996). **Distribution**. Benthic. Western Atlantic: Massachusetts, North Carolina, Florida, Gulf of Mexico, Antilles, Colombia, Venezuela, Suriname, Brazil (Bahia, Rio de Janeiro, São Paulo, Santa Catarina, Paraná, Rio Grande do Sul), Uruguay, Argentina. **Depth range**. From 90 to 1047 m.

Family Mathidellidae Karasawa and Kato, 2003.

175. *Neopilumnoplax americana* (Rathbun, 1898).

Pilumnoplax americanus Rathbun, 1898: 283.
Neopilumnoplax americana– Rathbun 1918: 21; Melo 1998: 492.

Type locality. Georgia. **Diagnosis**. Rathbun (1898). **Distribution**. Benthic. Western Atlantic: USA (North Carolina to Florida), Gulf of Mexico, Cuba, Lesser Antilles, Brazil (Bahia, Espírito Santo, São Paulo). Indian Ocean: Arabian Sea. **Depth range**. From 130 to 800 m.

Family Parthenopidae MacLeay, 1838.

176. *Solenolambrus typicus* Stimpson, 1871.

Solenolambrus typicus Stimpson, 1871: 133; Melo 1996: 293; Melo 1998: 474.

Type locality. Off The Samboes and off Alligator Reef, Straits of Florida, USA. **Diagnosis**. Stimpson (1871), Melo (1996). **Distribution**. Benthic. Western Atlantic: USA (North Carolina to Florida), Gulf of Mexico, Lesser Antilles, Brazil (Rio de Janeiro). **Depth range**. From 90 to 620 m.

Family Palicidae Bouvier, 1898.

177. *Palicus acutifrons* (A. Milne-Edwards, 1880)

Cymopolia acutifrons A. Milne-Edwards, 1880.
Palicus acutifrons Melo, 1996: 501; Melo 1998: 507; Serejo et al. 2007: 141.

Type locality. Bahia, Brazil. **Diagnosis**. Melo (1996). **Distribution**. Benthic. Western Atlantic: Brazil (Bahia, Espírito Santo). **Depth range**. From 30 to 900 m.

Family Plagusiidae Dana, 1851.

178. *Euchirograpsus americanus* A. Milne-Edwards, 1880.

Euchirograpsus americanus A. Milne-Edwards, 1880; A. Milne-Edwards and Bouvier 1894; Rathbun 1918; Türkay 1975; Melo 1996: 476; Melo 1998: 503; Alves-Júnior et al. 2016c: 3, Fig. 1a, c.

Type locality. Gulf of Mexico. **Diagnosis**. A. Milne-Edwards and Bouvier (1894), Rathbun (1918), Türkay (1975), Melo (1996). **Distribution**. Benthic. Western Atlantic: Canada (Bay of Fundy), USA (North and South Carolina, Massachusetts, New Jersey, Louisiana, Florida), Gulf of Mexico, Dry Tortugas, Cuba, Caribbean Sea, Antilles, West Indies, Barbados, Colombia, Venezuela and Brazil (Rio Grande do Norte, Rio Grande do Sul). **Depth range**. From 10 to 510 m.

Family Portunidae Rafinesque, 1815.

179. *Achelous spinicarpus* Stimpson, 1871.

Achelous spinicarpus Stimpson, 1871: 148.
Portunus spinicarpus– Williams 1965: 167 (part), Fig. 150; Holthuis 1969: 415, Fig. 1; Coelho and Ramos 1972: 187; Felder 1973: 60, pl. 8, Fig. 13; Williams 1984: 392, Fig. 308; Melo 1998: 480; Serejo et al. 2007: 141; Rodrigues et al. 2017: 5, Fig. 3a.

Type locality. 24°23′N, 82°57′W, Florida Straits, USA. **Diagnosis**. Williams (1984), Rodrigues et al. (2017). **Distribution**. Benthic. Western Atlantic: USA (North Carolina, South Carolina, Florida), Gulf of Mexico, Antilles, Colombia, Venezuela, Guiana, Brazil (Amapá, Pará, Maranhão, Piauí, Ceará, Rio Grande do Norte, Paraíba, Pernambuco, Alagoas, Sergipe, Bahia, Espírito santo, Rio de Janeiro, São Paulo, Paraná, Santa Catarina, Rio Grande do Sul). **Depth range**. From shallow waters to 910 m.

180. *Bathynectes longispina* Stimpson, 1871.

Bathynectes longispina Stimpson, 1871: 146; Abele and Kim 1986: 52; Tavares 2003: 1, Figs. 1, 2; Torres et al. 2006: 134, Fig. 1; Nunes et al. 2017: 335, Fig. 3.

Bathynectes superba– Milne-Edwards and Bouvier 1900: 65; 1923: 311; Rathbun 1930: 28, pls. 9, 10; Powers 1977: 72.

Type locality. USA, Florida. **Diagnosis**. Tavares (2003), Torres et al. (2006). **Distribution**. Benthic. Western Atlantic: USA (Massachusetts to Florida), Gulf of Mexico, Cuba, Brazil (Rio Grande do Norte, Saint Peter and Saint Paul Archipelago, Pernambuco, Alagoas, Bahia, Espírito Santo). **Depth range**. From 260 to 546 m.

Family Xanthidae MacLeay, 1838.

181. *Allactaea lithostrota* Williams, 1974.

Allactaea lithostrota Williams, 1974: 19, Figs. 1–3; Melo 1996: 340; Melo 1998: 481.

Type locality. 43′N, 76°40′W SE, Cape Lookout, North Carolina, USA. **Diagnosis**. Williams (1974), Melo (1996). **Distribution**. Benthic. Western Atlantic: USA (North Carolina to Florida), Gulf of Mexico, Lesser Antilles, Brazil (Rio de Janeiro, São Paulo, Paraná, Santa Catarina, Rio Grande do Sul). **Depth range**. From 50 to 640 m.

21.4 Discussion

Among the 181 species reported herein, 6 have economic relevance for fisheries and have been directly exploited in the Brazilian deep sea: *Aristaeomorpha foliacea*, *Aristaeopsis edwardsiana*, *Aristeus antillensis*, *Chaceon notialis*, *C. ramosae*, and *C. linsi*. According to Dallagnolo et al. (2009a), the three Aristaeidae shrimps are fishery targets since 2002 in Brazilian deep waters, from 18°S and 34°S and 700 and 800 m depth. *Aristaeopsis edwardsiana* is called "carabinero" shrimp, and it has been the most caught Aristeidae species (456,710 kg) in this area between 2002 and 2007; *Aristaeomorpha foliacea* (also called the "carabinero" shrimp) is the second most caught Aristeidae species (121,497 kg), while *Aristeus antillensis* ("alistado" shrimp) is the third most caught Aristeidae species (27,919 kg). Dallagnolo et al. (2009b) published a management plan for shrimp fisheries in south and southeastern Brazilian slope waters and affirm that it is clear that the stock of these species is in biologic danger due to overexploitation.

According to Tavares and Pinheiro (2011), *Chaceon notialis* and *C. ramosae* have been fished in the southwestern Atlantic since the 1980s. According to Perez et al. (2009), landings of *C. notialis* and *C. ramosae* from Brazilian waters were about 6 and 4 tons between 2000 and 2006. These species are now overexploited and their fishing is not allowed anymore. Also according to Tavares and Pinheiro (2011), "between 2003 and 2007 *Chaceon linsi* was fished in northeastern Brazil either as *C. fenneri* or *Chaceon* sp., where captures fluctuated between about 23 (2003) and 3 tons (2007)" (Lessa 2006; Carvalho et al. 2009).

Interestingly, some species reported herein are rare, like the caridean shrimp *Bathypalaemonella texana* recorded only twice in literature, in the original description (Gulf of Mexico; Pequegnat 1970a) and by Cardoso (2010a, b) in Brazilian waters. Other species are very difficult to sample as in the case of *Odontozona lopheliae*, a stenopodid from deep-water coral reefs that was sampled using submersibles (ROV at Campus Basin, Brazil, and Johnson Sea Link at primarily two locations, Green Canyon and Viosca Knoll, Gulf of Mexico) (Goy and Cardoso 2014).

Eupasiphae gilesii is recorded herein for the first time in the western Atlantic. The examined material fits very well with the descriptions of Tirmizi (1969) and Crosnier (1988b). Other eight caridean species had their geographic ranges extended in Brazilian waters mainly to Ceará and Rio Grande do Norte states and Rocas Atoll and Fernando de Noronha Archipelago, thanks to samples obtained during the BPOT and ABRAÇOS projects (Table 21.1).

A total of 181 species were listed in this catalogue and belong to the following major groups of Decapoda: 32 Dendrobranchiata, 68 Caridea, 31 Anomura, 36 Brachyura, 5 Astacidea and 5 Polychelida, 2 species of Achelata, and 1 species each of Stenopodidea and Axiidea. As expected, the shrimp fauna was more diverse than the crab and lobster fauna, although the global diversity of Brachyura exceeds that of the Dendrobranchiata and Caridea. The same pattern has been reported by other authors working with Decapoda in relevant deep-sea oceanographic projects. Company et al. (2004), for instance, worked with central Mediterranean material sampled by DESEAS project survey and found 32 Decapoda species, including 21 species of shrimps (Dendrobranchiata and Caridea) and 11 species each of crabs, lobsters, and anomurans. Serejo et al. (2007) working with material from Brazilian deep-sea region sampled during the REVIZEE Central project found 65 species of shrimp, 15 of lobsters, 17 of anomurans, and 19 of brachyurans. Aspects related with vertical migration ability, reproduction, and use of resources should be related with this diversity pattern, but more studies on deep-sea Decapoda biology are necessary to understand it.

Acknowledgments Authors would like to aknowledge the reviewers, Charles H. J. M. Fransen and Enrique Macpherson, for all comments and corrections to this work, many thanks for your time dedicated to work on it. We aknowledge also the editor, Michel Hendrickx, for all attention and work dedicated to this catalogue.

References

Abele LG, Kim W (1986) An illustrated guide to the marine decapod crustaceans of Florida. Florida, Department of Environmental Regulation. Technical Ser 8(1):1–760

Adensamer T (1898) Berichte der Commission für Erfoschung des östlichen Mittelmeeres. XXII. Zoologische Ergebnisse. XI. Decapoden gesammelt auf S. M. Schiff Pola in den Jahren 1890-1894. Denkschriften der Kaiserlichen Akademie der Wissenschaften. Mathematisch-Naturwissenschaftliche Classe 65:597–628

Afonso-Dias MS, Pires A, Clark PF (2008) Occurrence of *Chaceon gordonae* and *C. sanctae-helenae* (Crustacea: Brachyura: Geryonidae) off the Island of São Tomé. Mar Biol Rec 1:1–2
Agassiz A (1888) Three cruises of the United States coast and survey steamer "Blake" in the Gulf of Mexico, in the Caribbean Sea, and along the Atlantic Coast of the United States from 1877 to 1880. Bull Mus Comp Zool 2(15):1–220
Ahamed F, Cardoso IA, Ahmed ZF, Hossan MDY, Ohtomi JUN (2017) An overview of the genus *Plesionika* Bate, 1888 (Decapoda: Caridea: Pandalidae) in Asian waters. Zootaxa 4221:575
Ahyong ST (2007) Decapod Crustacea collected by the NORFANZ expedition: Galatheidae and Polychelidae. Zootaxa 1593:1–54
Ahyong ST (2009) The Polychelidan lobsters: phylogeny and systematics (Polychelida: Polychelidae). In: Martin JW, Crandall KA, Felder L (eds) Decapod crustacean Phylogenetics. Crustacean issues, vol 18. CRC Press/Taylor, Francis Group, Boca Raton/London/New York, pp 369–396
Ahyong ST (2012) Polychelid lobsters (Decapoda: Polychelida: Polychelidae) collected by the CIDARIS expeditions off Central Queensland, with a summary of Australian and New Zealand distributions. Mem Queensl Mus 56(1):1–8
Ahyong ST, Brown DE (2002) New species and new records of Polychelidae from Australia (Decapoda: Crustacea). Raffles B Zool 50:53–79
Ahyong ST, Chan TY (2004) Polychelid lobsters of Taiwan (Decapoda: Polychelidae). Raffles B Zool 52(1):171–182
Ahyong ST, Galil BS (2006) Polychelidae from the southern and western Pacific (Decapoda: Polychelida). Zoosystema 28:757–767
Ahyong ST, Poore GCB (2004) Deep-water Galatheidae (Crustacea: Decapoda: Anomura) from southern and eastern Australia. Zootaxa 472:3–7
Alcock A (1894) Natural history notes from H.M. Indian marine survey steamer "Investigator", Commander R.F. Hoskyn, R.N., commanding. Ser. II. No. 1. On the results of deep-sea dredging during the season 1890-91. Ann Mag Nat Hist 13(6):225–245
Alcock A (1901a) Zoological gleanings from the Royal Indian Marine Survey Ship "investigator". Sci Mem Med Off Army India Calcutta 12:1–76
Alcock A (1901b) A descriptive catalogue of the Indian deep-sea Crustacea Decapoda Macrura and Anomala, in the Indian Museum. Being a revised account of the deep-sea species collected by the Royal Indian Marine Survey Ship Investigator. Trustees of the Indian Museum, Calcutta
Alcock A (1905) Catalogue of the Indian decapod Crustacea in the collection of the Indian Museum. Part II. Anomura. Fascicle I, Pagurides. Trustees of the Indian Museum, Calcutta
Alcock A, Anderson AR (1894) Natural history notes from H.M. Indian Marine Survey steamer 'Investigator,' Commander CF Oldham, RN, commanding. 14. An account of the recent Collection of deep-sea Crustacea from Bay of Bengal and Laccadive Sea. J Asiat Soc Bengal 63(2):141–185
Almeida AO, Souza GBG, Boehs G, Bezerra LEA (2010) Shallow-water anomuran and brachyuran crabs (Crustacea: Decapoda) from southern Bahia, Brazil. Lat Am J Aquat Res 38:329–376
Alvarenga FM, Cardoso IA (2014) First record of *Stereomastis nana* (Polychelidae: Crustacea: Decapoda) from the southwestern Atlantic. Mar Biodivers Rec 7(e80):1–4
Alves-Júnior FA, Araújo MSLC, Bertrand A, Souza-Filho JF (2016a) First report of two deep-sea shrimps of the genus *Acanthephyra* A Milne-Edwards, 1881 (Crustacea: Decapoda: Acanthephyridae) from southwestern Atlantic. Zootaxa 4184:193–200
Alves-Júnior FA, Araújo MSLC, Souza-Filho JF (2016b) Distribution of two species of *Nephropsis* Wood-Mason, 1872 (Crustacea: Decapoda: Nephropidae) from northeastern Brazil. Zootaxa 4114:90–94
Alves-Júnior FA, de Santama JL, Araújo MSLC, Souza-Filho JF (2016c) First record of *Euchirograpsus americanus* A. Milne-Edwards, 1880 (Crustacea: Decapoda: Plagusiidae) from northeastern Brazil. Nauplius 24:1–4
Alves-Júnior FA, Araújo MSLC, Souza-Filho JF (2017a) New records of deep-sea shrimps of family Solenoceridae Wood-Mason & Alcock, 1891 (Crustacea: Decapoda: Dendrobranchiata) from Southwestern Atlantic. Zootaxa 4254:473–484

Alves-Júnior FA, Araújo MSLC, Souza-Filho JF (2017b) New records and bathymetric distribution of deep-sea shrimps of the family Glyphocrangonidae (Decapoda: Caridea) from the Potiguar Basin, northeastern Brazil. Nauplius 25:1–15

Alves-Júnior FA, Araújo MSLC, Souza-Filho JF (2017c) Occurrence of deep-sea shrimp *Heterocarpus inopinatus* Tavares, 1999 (Crustacea: Decapoda: Caridea) in Potiguar Basin, Northeastern Brazil. Trop Oceanogr 45:38–43

Alves-Júnior FA, de Lemos RJF, Cardoso IA, Araújo MSLC, Bertrand A, Souza-Filho JF (2018a) New records of deep-sea prawn of the genus *Gennadas* Spence Bate, 1881 (Crustacea: Decapoda: Benthesicymidae) from Southwestern Atlantic. Zootaxa 4450:376–384

Alves-Júnior FA, Viana GFS, Araújo MSLC, Souza-Filho JF (2018b) New records of the family Crangonidae (Decapoda: Caridea) from southwestern Atlantic. Zootaxa 4369:46–62

Alves-Júnior FA, Santana JL, Araújo MSLC, Souza-Filho JF (2018c) New records of two deep-sea crabs of the family Lithodidae Samouelle, 1819 (Crustacea: Decapoda) from Southwestern Atlantic. Thalassas 2:1–6

Alves-Júnior FA, Ferreira ÂP, Araújo MSLC (2018d) New records of two species of *Acanthocarpus* Stimpson, 1871 (Crustacea: Decapoda: Calappidae) in northeastern Brazil. PECEN 2:60

Alves-Júnior FA, Araújo MSLC, Cardoso IA, Bertrand A, Souza-Filho JF (2019a) Meso- and Bathypelagic Prawns of the Superfamilies Penaeoidea Rafinesque, 1815 and Sergestoidea Dana, 1852 (Crustacea: Decapoda: Dendrobranchiata) from Southwestern Atlantic: new records and bathymetric distribution. Thalassas 1:1–20

Alves-Júnior FA, Silva ES, Araújo MSLC, Cardoso IA, Bertrand A, Souza-Filho JF (2019b) Taxonomy of deep-sea shrimps of the Superfamily Oplophoroidea Dana 1852 (Decapoda: Caridea) from Southwestern Atlantic. Zootaxa 4613:401–442

Anker A, Pachelle PP, Tavares M (2014) Two new species and two new records of deep-water caridean shrimps from Brazil (Decapoda: Pandalidae, Palaemonidae, Crangonidae). Zootaxa 3815(2):263–278

Baba K, de Saint LM (1996) Crustacea Decapoda: revision of the genus *Bathymunida* Balss, 1914, and description of six new related genera (Galatheidae). In: Crosnier, A (ed) Résultats des Campagnes MUSORSTOM vol 15. Mém Mus Natl Hist Nat 168:433–502

Baba K, Poore GCB (2002) *Munidopsis* (Decapoda, Anomura) from South-Eastern Australia. Crustaceana 75(3–4):231–252

Baba K, Wicksten MK (2017) *Uroptychus nitidus* (A. Milne-Edwards, 1880) and related species (Crustacea: Decapoda: Anomura: Chirostylidae) from the western Atlantic. Zootaxa 4221(3):251–290

Baba K, Macpherson E, Poore CGB, Ahyong ST, Bermudez A, Cabezas P, Lin CW, Nizinski M, Rodrigues C, Schnabel KE (2008) Catalogue of squat lobsters of the world (Crustacea: Decapoda: Anomura families Chirostylidae, Galatheidae and Kiwaidae). Zootaxa 1905:1–220

Balss H (1914) Diagnosen neuer Macruren der Valdivia Expedition. Zool Anz 44:592–599

Balss H (1925) Spanische Süßwasser-Dekapoden. Senckenbergiana 7:206–209

Barea L, Defeo O (1985) Primeros ensayos de captura del crustáceo batial *Geryon quinquedens* Smith en el área común de pesca Argentino Uruguaya. Contrib Dep Oceanog (F.H.C.). Montevideo 2(3):189–203

Barnard KH (1950) Descriptive catalogue of South African Decapod Crustacea. Ann S Afr Mus 38:1–837

Becker EL, Cordes EE, Macko AS, Fisher CR (2009) Importance of seep primary production to Lophelia pertusa and associated fauna in the Gulf of Mexico. Deep-Sea Res 56:786–800

Benedict JE (1901) The anomuran collections made by the Fish Hawk expedition to Porto Rico. Bull U S Fish Comm 2:129–148

Benedict JE (1902) Description of a new genus and forty-six new species of crustaceans of the Family Galatheidae with a list of the known marine species. Proc U S Natl Mus 26(1311):243–334

Bezerra LEA, Ribeiro FB (2015) Primitive decapods from the deep sea: first record of blind lobsters (Crustacea: Decapoda: Polychelidae) in northeastern Brazil. Nauplius 23(2):125–131

Biffar TA (1971) The genus *Callianassa* (Crustacea, Decapoda, Thalassinidea) in South Florida, with keys to the Western Atlantic species. Bull Mar Sci 21(3):637–715

Biffar TA (1973) The taxonomic status of *Callianassa occidentalis* Bate, 1888 and *C. batei* Borradaile, 1903 (Decapoda, Callianassidae). Crustaceana 24:224–230
Birstein JA, Vinogradov LG (1972) Crabdoids (Decapoda, Anomura, Lithodidae) of the Atlantic sector of the Antarctic, South America and South Africa. Zool Zhurnal 5:351–363. (in Russian)
Biscoito M (1993) An account of the shrimps of the family Pandalidae (Crustacea, Decapoda, Caridea) in Madeiran waters. Courier Forsch Inst Senckenberg 159:321–325
Boone L (1927) Scientific results of the First Oceanographic Expedition of the "Pawnee" 1925. Crustacea from tropical East American seas. Bull Bingham Oceanogr Collection 1:1–147
Borradaile LA (1899) On some crustaceans from the South Pacific. Part III. Macrura. Proc Zool Soc London:1000–1015
Borradaile LA (1903) On the classification of the Thalassinidea. Ann Mag Nat Hist 12(71):534–551
Borradaile LA (1915) Notes on Carides. Ann Mag Nat Hist 15(8):205–213
Boschi EE (1976) Nuevos aportes al conocimiento de la distribucion geografica de los crustaceos decapodos del Mar Argentino. Physis Buenos Aires 35A(90):59–68
Bouchet P (2006) The magnitude of marine biodiversity. In: Duarte CM (ed) The exploration of marine biodiversity: scientific and technological challenges. Fundación BBV, Bilbao, pp 31–62
Bouvier E-L (1895) Recherches sur les affinités des *Lithodes* et des *Lomis* avec les Paguridés. Ann Sci Nat Zool 7(8):157–210
Bouvier E-L (1896) Sur la classification des lithodines et sur leur distribution dans les océans. Ann Sci Nat Zool 8(1):1–46
Bouvier E-L (1905a) Sur les Pénéides et les Sténopides recueillis par les expéditions françaises et monégasques dans l'Atlantique oriental. C R Hebd Acad Sci 140:980–983
Bouvier E-L (1905b) Sur les Crustacés Décapodes (abstraction faite des Carides) recueillis par le yacht "Princesse-Alice" au cours de la campagne de 1905. Bull Mus Oceanogr Monaco 55:1–4
Bouvier E-L (1905c) Sur les Palinurides et les Eryonides recueillis dans l'Atlantique oriental par les expéditions françaises et monégasques. C R Acad Sci Paris 140:479–482
Bouvier E-L (1905d) Sur les Crustacés Décapodes (abstraction faite des Carides) recueillis par le yacht "Princesse Alice" au cours de la campagne de 1905. C R Acad Sci Paris 141:644–647
Bouvier E-L (1906a) Suite aux observations sur les *Gennadas* ou Pénéides bathypélagiques. C R Acad Sci Paris 142:746–750
Bouvier E-L (1906b) Sur les *Gennadas* ou Pénéides bathypélagiques. Bull Mus Oceanogr Monaco 80:1–13
Bouvier E-L (1906c) Sur l'*Haliporus androgynus*, Pénéide nouveau provenant des campagnes du Talisman. Bull Mus Hist Nat Paris 12(5):253–256
Bouvier E-L (1907) Quelques impressions d'un naturaliste au cours d'une campagne scientifique de S.A.S. le Prince de Monaco (1905). Bull Inst Oceanogr Monaco 93:1–103
Bouvier E-L (1908) Sur les relations zoologiques des Crevettes de la tribu des Sténopidés. C R Hebd Acad Sci 146:887–891
Bouvier E-L (1925) XLVIII. Les Macroures marcheurs. Reports on the Results of dredging, under the supervision of Alexander Agassiz, in the Gulf of Mexico (1877-78), in the Caribbean Sea (1878-79), and along the Atlantic coast of the United States (1880), by the U.S. Coast Survey Steamer "Blake", Lieut. -Commander CD. Sigsbee U.S.N., and Commander J.R. Bartlett, U.S.N., Commanding. Mem Mus Comp Zool Harvard Coll 47(5):401–472
Brullé A (1839) Crustacés. In: Barker-Webb P, Berthelot S (eds) Histoire naturelle des Iles Canaries: Planche unique. Béthune, Paris, pp 15–18
Bullis HR, Thompson JR (1965) Collections by the exploratory fishing vessels Oregon, Silver Bay, Combat, and Pelican made during 1956-1960 in the southwestern North Atlantic. U S Dept Inter Fish Wild Serv Spec Sci Rep-Fish 510:1–30
Burkenroad MD (1936) The Aristaeinae, Solenocerinae and pelagic Penaeinae of the Bingham Oceanographic collection. Bull Bingham Oceanogr Coll 5:1–151
Burkenroad MD (1938) The Templeton Crocker Expedition. XIII. Penaeidae from the region of Lower California and Clarion Island, with description of four new species. Zoologica 23:55–91
Burkenroad MD (1940) Preliminary descriptions of twenty-one new species of pelagic Penaeidae (Crustacea Decapoda) from the Danish Oceanographical Expeditions. Ann Mag Nat Hist 6:35–54

Burkenroad MD (1963) The evolution of the Eucarida (Crustacea, Eumalacostraca), in relation to the fossil record. Tulane Stud Geol 2:3–17

Burukovsky RN (1977) A new species of the family Pasiphaeidae (Crustacea, Decapoda). Zool Zhurnal 56:473–475

Burukovsky RN (1982) Shrimps of the genus *Parapandalus*: geographical and bathymetrical distribution with the key for the species identification. Byull Moskovsk Obschch Isp Prir Otd Biol 87(3):42–53

Burukovsky RN (1991) Shrimps of the family Nematocarcinidae (Decapoda, Caridea) from the western part of the Indian Ocean. Zool Zhurnal 70:39–46

Burukovsky RN (1996) Shrimps of the genus *Pasiphaea*: systematics and some remarks on new findings (Decapoda, Caridea). Zool Zhurnal 75(6):841–847

Burukovsky RN (2000a) Taxonomy of *Nematocarcinus* (Decapoda: Nematocarcinidae). 8. Structure of distoventral organ in some eastern Atlantic species. Zool Zhurnal 79(12):1–4

Burukovsky RN (2000b) Taxonomy of shrimps from the genus *Nematocarcinus* (Decapoda, Nematocarcinidae). 6. Redescription of species from the groups undulatipes and gracilis with descriptions of two new species. Zool Zhurnal 79:1155–1167

Burukovsky RN (2000c) Taxonomy of shrimps from the genus *Nematocarcinus* (Crustacea: Decapoda: Nematocarcinidae). 1. Description of distoventral organ and revision of *N. productus*, *N. tenuipes*, *N. intermedius*, *N. parvidentatus*, *N. longirostris* and *N. proximatus*. Zool Zhurnal 79(2):161–170

Burukovsky RN (2001) Taxonomy of *Nematocarcinus* (Decapoda, Nematocarcinidae). Description of *Nematocarcinus* from waters of the American continent. Zool Zhurnal 80:1429–1443

Burukovsky RN (2002a) Taxonomy of *Nematocarcinus* (Decapoda, Nematocarcinidae). Results of reinvestigating the collection of the Indo-Pacific *Nematocarcinus* from the National Museum of Natural History (Washington, USA). Zool Zhurnal 81:5–12

Burukovsky RN (2002b) A new shrimp species of the genus *Nikoides* (Decapoda, Processidae). Zool Zhurnal 81:1389–1391

Burukovsky RN (2003) Shrimps of the family Nematocarcinida. Kaliningrad Izdatelistovo KGGU, Kalingrad

Burukovsky RN (2004) Taxonomy of shrimps of the genus *Nematocarcinus* (Decapoda, Nematocarcinidae). A review of taxonomic characteristics and a key to identifying species of the genus. Zool Zhurnal 83(5):549–561

Burukovsky RN (2009) Morphology and biology of shrimp *Nematocarcinus gracilipes* Filhol, 1884 (Decapoda: Nematocarcinidae). Zool Bespozvon 6(2):81–88

Burukovsky RN (2012) Deep Sea shrimps of the family Nematocarcinidae (history of study, systematic, distribution, and biology). Prospekt Nauki, St. Petersburg

Burukovsky RN (2013) Shrimps of the family Nematocarcinidae (Crustacea: Decapoda: Caridea) from Taiwan and the Philippines collected by the TAIWAN, PANGLAO 2005 and AURORA 2007 expeditions in the western Pacific. In: Ahyong ST, Chan T-Y, Corbari L, Ng PKL (eds) Tropical Deep-Sea benthos 27. Muséum National d'Histoire Naturelle, Paris, pp 155–189

Burukovsky RN, Romensky LL (1987) Description of *Pasiphaea balssi* sp. n., a new species of shrimp from South Atlantic (Crustacea, Decapoda, Pasiphaeidae) and polytomous key for identification of the shrimps in the genus. Byull Moskovsk Obschch Isp Prir Otd Biol 92:51–60

Cabral E, Ramos-Porto M, Santos MCF, Acioli FD, Torres MFA, Viana GFS (2000) Shrimps collected in the Northeast of Brazil during the REVIZEE Program (Decapoda: Caridea). Nauplius 8(2):245–248

Calman WT (1925) On macrurous decapod Crustacea collected in South African waters by the S.S. "Pickle". S Afr Fish Mar Biol Survey 4:1–26

Calman WT (1939) Crustacea: Caridea. John Murray Exped 1933-34. Sci Rep 6(4):183–224

Campos N, Navas G, Bermúdez A, Cruz N (2005) Los crustáceos de la franja superior del talud continental (200 y 500 m) del Mar Caribe colombiano. Monografías de Fauna de Colombia. Instituto de Ciencias Naturales, Universidad Nacional de Colombia, Bogotá. Vol. 2

Cardoso IA (2006) New record of *Meningodora vesca* (Smith, 1887) (Caridea, Oplophoridae) to the Southwestern Atlantic. Nauplius 14(1):1–7

Cardoso IA (2009a) Report on some *Plesionika* Bate, 1888 and first record of *Stylopandalus* Coutiére, 1905 (Caridea: Pandalidae) from Brazilian waters. Zootaxa 2120:53–68

Cardoso IA (2009b) First record of *Parapontophilus longirostris* Komai, 2008 (Caridea: Crangonidae) on Brazilian waters. Nauplius 17(1):29–35

Cardoso IA (2010a) First record of family Bathypalaemonellidae (Caridea: Decapoda) on Brazilian deep-sea coral sreefs. Mar Biodivers Rec 3:1–4

Cardoso IA (2010b) Redescription of the sibling species *Plesionika acanthonotus* (Smith, 1882) and *Plesionika holthuisi* Crosnier & Forest, 1968 (Caridea: Pandalidae). Atlantica 32:221–235

Cardoso IA (2011) Three species of the genus *Plesionika* Bate, 1888 (Caridea: Pandalidae) reported for the first time in the southwestern Atlantic. Cah Biol Mar 52:131–146

Cardoso IA (2013) The Southwestern Atlantic Crangonidae (Caridea, Decapoda) including first records of two species. Cah Biol Mar 54:85–92

Cardoso IA, Falkenhaug T, Loureiro LF (2014a) Comparison between the shrimp species richness (Caridea and Dendrobranchiata, Decapoda, Crustacea) of the South and North Mid Atlantic Ridge. Braz J oceanogr 62(1):47–56

Cardoso IA, Burukovsky RN (2014b) *Nematocarcinus* Milne Edwards, 1881 (Crustacea, Decapoda) from Southwestern Atlantic, including the Southern Mid-Atlantic Ridge area. Zootaxa 3887:437–458

Cardoso IA, Serejo CS (2003) Sergestidae (Crustacea, Dendrobranchiata) from the southwestern Atlantic, collected by the Revizee program. Bol Mus Nac Rio de J 512:1–15

Cardoso IA, Serejo CS (2007) Deep sea Caridea (Crustacea, Decapoda) from Campos Basin, RJ, Brazil. Braz J Oceanogr 55:39–50

Cardoso IA, Tavares CR (2006) Three new occurrences of Sergestidae (Dendrobranchiata) for the Southwestern Atlantic slope (13°–22°S). Nauplius 14:9–17

Cardoso IA, Young PS (2005) Deep sea Oplophoridae (Crustacea, Caridea) from the southwestern Brazil. Zootaxa 1031:1–74

Cardoso IA, Falkenhaug T, Fernandes LFL (2014) Comparison between the shrimp species richness (Caridea and Dendrobranchiata: Decapoda: Crustacea) of the south and north mid Atlantic ridge. Braz J Oceanogr 62:47–56

Cardoso IA, Perez JAA, Kitazato H (2017) Deep-sea decapods sampled or observed with a manned submersible at the Rio Grande Rise (SW Atlantic), including the first record of four species from this region. Mar Biodivers 48:695–702

Cardoso IA, Rodríguez-Rey GT, Terossi M, Bartilotti C, Serejo CS (2019) Taxonomic status of the amphi-Atlantic species *Plesionika acanthonotus* (Caridea: Pandalidae): a multi-locus and morphological approach. Zootaxa 4590:457

Carvalho TB, Oliveira RR, Lotufo TMD (2009) Note on the fisheries and biology of the golden crab (Chaceon fenneri) off the northern coast of Brazil. Lat Am J Aquat Res 37:571–576

Caullery M (1896) Crustacés Schizopodes et Décapodes. In: Koehler, R. Résultats Scientifiques de la Campagne du "Caudan" dans le Golfe de Gascogne, Août–Septembre, 1895. Ann Univ Lyon 26:365–419

Chace FA Jr (1936) Revision of the bathypelagic prawns of the family Acanthephyridae, with notes on a new family, Gomphonotidae. J Wash Acad Sci 26:24–31

Chace FA Jr (1939) Reports on the scientific results of the first Atlantis Expedition to the West Indies, under the joint auspices of the University of Havana and Harvard University: preliminary descriptions of one new genus and seventeen new species of decapod and stomatopod Crustacea. Mem Soc Cub Hist Natl 13(1):31–54

Chace FA Jr (1940) Plankton of the Bermuda oceanographic expeditions. IX the bathypelagic caridean Crustacea. Zoologica 25:117–209

Chace FA Jr (1942) Reports on the scientific results of the Atlantis expeditions to the West Indies, under the joint auspices of the University of Havana and Harvard University. The Anomuran Crustacea. I. Galatheidea. Torreia 11:1–106

Chace FA Jr (1947) The deep-sea prawns of the family Oplophoridae in the Bingham Oceanographic collection. Bull Bingham Oceanogr Coll 11(1):1–51

Chace FA Jr (1956) Collections by the Oregon in the Gulf of Mexico. Special Sci Rept Fish 196:1–134

Chace FA Jr (1984) The caridean shrimps (Crustacea: Decapoda) of the Albatross Philippine Expedition, 1907-1910, part 2: families Glyphocrangonidae and Crangonidae. Smith Contrib Zool 397:1–63

Chace FA Jr (1985) The Caridean shrimps (Crustacea: Decapoda) of the Albatross Philippine Expedition, 1907–1910, part 3: families Thalassocarididae and Pandalidae. Smith Contrib Zool 411:1–143

Chace FA Jr (1986) The caridean shrimps (Crustacea: Decapoda) of the Albatross Philippine Expedition, 1907-1910, part 4: families Oplophoridae and Nematocarcinidae. Smith Contrib Zool 432:1–82

Chace FA Jr (1997) The caridean shrimps (Crustacea: Decapoda) of the Albatross Philippine expedition, 1907–1910, part 7: families Atyidae, Eugonatonotidae, Rhynchocinetidae, Bathypalaemonellidae, Processidae, and Hippolytidae. Smith Contrib Zool 587:1–106

Chace FA Jr, Holthuis LB (1978) *Psalidopus*: the scissor-foot shrimps (Crustacea: Decapoda: Caridea). Smith Contrib Zool 277:1–22

Chan TY (1996) Crustacea Decapoda Crangonidae: revision of the three closely related genera *Aegaeon* Agassiz, 1846, *Pontocaris* Bate, 1888 and *Parapontocaris* Alcock, 1901. In: Crosnier A (ed) Résultats des Campagnes MUSORSTOM, vol 15. Mém Mus Natl Hist Natl 168:269–336

Chan TY (2010) Annotated checklist of the world's marine lobsters (Crustacea: Decapoda: Astacidea, Glypheidea, Achelata, Polychelida). Raffles Bull Zool 23:153–181

Chan TY, Crosnier A (1997) Crustacea Decapoda: deep-sea shrimps of the genus *Plesionika* Bate, 1888 (Pandalidae) from French Polynesia, with descriptions of five new species. In: Crosnier A. (ed) Résultats des Campagnes MUSORSTOM, vol 18. Mém Mus Natl Hist Nat 176:187–234

Chan TY, Yu HP (1991) Two similar species: *Plesionika edwardsii* (Brandt, 1851) and *Plesionika crosnieri*, new species (Crustacea: Decapoda: Pandalidae). Proc Biol Soc Wash 104:545–555

Chevaldonne P, Olu K (1996) Occurrence of anomuran crabs (Crustacea: Decapoda) in hydrothermal vent and cold-seep communities: a review. Proc Biol Soc Wash 109(2):286–298

Cleva R (2001) Les Bathypalaemonellidae de Saint-Laurent, 1985 (Crustacea, Decapoda, Caridea) avec description d'une espèce nouvelle et définition d'un genre nouveau. Zoosystema 23:757–782

Cleva R, Guinot D, Albenga L (2007) Annotated catalogue of brachyuran type specimens (Crustacea, Decapoda, Brachyura) deposited in the Muséum national d'Histoire naturelle, Paris. Part I. Podotremata. Zoosystema 29:229–279

Coelho PA (1997) Revisao das espécies de Thalassinidea encontradas em Pernambuco, Brasil (Crustacea, Decapoda). Trab Oceanogr Univ Fed Per 25:137–161

Coelho PA, Coelho-Filho PA (1993) Proposta de classificação da família Xanthidae (Crustacea, Decapoda, Brachyura) através da taxonomia numérica. Rev Bras Zool 10(4):559–580

Coelho PA, Ramos MA (1972) A constituição e a distribuição da fauna de decápodos do litoral leste da América do Sul entre as latitudes de 5°N e 39°S. Trab Oceanogr Univ Fed Per 13:133–236

Coelho PA, Ramos-Porto MA (1985) Sinopse dos crustáceos decápodos brasileiros (famílias Scyllaridae, Palinuridae, Nephropidae, Parastacidae e Axiidae). A Univ Fed Rural Pe 8:41–46

Coelho PA, Ramos-Porto MA (1986) Sinopse dos crustáceos decápodos brasileiros (famílias Callianassidae, Callianideidae, Upogebiidae, Parapaguridae, Paguridae, Diogenidae). Trab Oceanogr Univ Fed Pe 19:27–53

Coelho PA, Ramos-Porto M, Calado TCS (1986) Litoral do Rio Grande do Norte: Decapoda. Cad Omega Univ Fed Rural Per Ser Ciênc Aquáticas 2:79–105

Coelho PA, Ramos-Porto MA, Melo GAS (1990) Crustáceos decápodos do Estado de Alagoas. An Soc Nor Zool 3(3):21–34

Coelho PA, Almeida AO, Souza-Filho JF, Bezerra LEA, Giraldes BW (2006) Diversity and distribution of the marine and estuarine shrimps (Dendrobranchiata: Stenopodidea and Caridea) from North and Northeast Brazil. Zootaxa 1221:4–62

Coelho PA, Almeida AO, Bezerra LEA, Souza-Filho JF (2007) An updated checklist of decapod crustaceans (infraorders Astacidea, Thalassinidea, Polychelida, Palinura, and Anomura) from the northern and northeastern Brazilian coast. Zootaxa 1519:1–16

Coelho PA, Almeida AO, Bezerra LE (2008) Checklist of the marine and estuarine Brachyura (Crustacea: Decapoda) of northern and northeastern Brazil. Zootaxa 1956:1–58

Coelho-Filho PA (2006) Checklist of the Decapods (Crustacea) from the outer continental shelf and seamounts from Northeast of Brazil. REVIZEE Program (NE III). Zootaxa 1184:1–27

Company JB, Maiorano P, Selepides AT, Politou C-Y, Plaity W, Rollant G, Sardà F (2004) Deep-sea decapod crustaceans in the western and central Mediterranean Sea: preliminary aspects of species distribution, biomass and population structure Sci Mar 68:73–86

Coutière H (1905) Sur quelques Crustacés provenant des campagnes de la Princesse-Alice (filet à grande ouverture). CR Hebd Séances Acad Sci 140:1113–1115

Coutière H (1911) Sur les crevettes Eucyphotes recueillies en 1910 au moyen du filet Bourée, par la Princesse-Alice. C R Hebd Séances Acad Sci 152:156–158

Crosnier A (1976) Données sur les Crustacés Décapodes capturés par M. Paul Guézé a l'île de La Réunion lors d'essais de pêche en eau profonde. In: Biologie marine et exploitation des resources de l'océan Indien occidental. Trav Doc ORSTOM 47:225–254

Crosnier A (1978) Crustacés Décapodes Peneides Aristeidae (Benthesicyminae, Aristeinae, Solenocerinae). Faune Madagascar 46:1–197

Crosnier A (1985) Crevettes pénéides d'eau profonde récoltées dans l'océan Indien lors des campagnes Benthedi, Safari I et II, MD 32 Réunion. Bull Mus Natl Hist Nat Paris 7:839–877

Crosnier A (1986) Crevettes de la famille des Pandalidae récoltées durant ces dernières années en Polynésie française. Description de *Plesionika chacei* et *P. carsini* spp. nov. Bull Mus Natl Hist Nat Paris 8:361–377

Crosnier A (1987) Oplophoridae (Crustacea Decapoda) récoltés de 1971 à 1982 par les navires françaises dans l'océan Indien occidental sud. Bull Mus Natl Hist Nat Paris 9:695–726

Crosnier A (1988a) Les *Eupasiphae* (Crustacea, Decapoda, Pasiphaeidae) du sud-ouest de l'océan Indien. description d'*E. paucidentata* sp. nov. Bull Mus Natl Hist Nat Paris 10(4):785–797

Crosnier A (1988b) Sur les *Heterocarpus* (Crustacea, Decapoda, Pandalidae) du sud-ouest de l'océan Indien. Remarques sur d'autres espèces ouest-pacifiques du genre ET description de quatre taxa nouveaux. Bull Mus Natl Hist Nat Paris 1:57–103

Crosnier A, Forest J (1968) Note préliminaires sur les carides receuillis par l "Ombango" au large du plateau continental, du Gabon à l'Angola (Crustacea Decapoda Natantia). Bull Mus Natl Hist Nat Paris 39(2):1123–1147

Crosnier A, Forest J (1969) Note préliminaires sur les pénéides receuillis par l' "Omango" au large du plateau continental, du Gabon à l'Angola (Crustacea Decapoda Natantia). Bull Mus Natl Hist Nat Paris 41:544–554

Crosnier A, Forest J (1973) Les crevettes profondes de l'Atlantique oriental tropical. Faune Trop 19:1–409

Cruz N, Fransen CHJM (2004) Addition of three species of the genus *Plesionika* (Crustacea: Caridea: Pandalidae) to the known Atlantic marine fauna of Colombia. Zool Meded 78(6):131–146

Cruz N, Bermúdez A, Campos NH, Navas G (2002) Los camarones de la familia Crangonidae de la franja superior del talud continental del mar caribe colombiano. Bol Invest Mar Cost 31:183–203

Cunha KMF, Oliveira JEL, Pinheiro AP, Henriques VMC (1999) Estrutura populacional do caranguejo-de-profundidade (*Chaceon fenneri*) no talude continental e bancos oceânicos do nordeste do Brasil. In: Aguilar AET, Malpica ZGC (eds) VIII Congresso Latinoamericano sobre Ciencias del Mar, vol 1. COLACMAR, Trujillo, pp 531–532

Dall W (2001) Australian species of Aristeidae and Benthesicymidae (Penaeoidea: Decapoda). Mem Queensl Mus 46:409–441

Dall'Occo PL, Tavares M (2004) New and additional records of deep-water lobsters from Brazil (Decapoda, Polychelidae). Nauplius 12(2):143–149

Dallagnolo R, Perez JA, Pezzuto PR, Wahrlich R (2009a) The deep-sea shrimp fishery off Brazil (Decapoda: Aristeidae): development and presents status. Lat Am J Aquat Res 37:327–343

Dallagnolo R, Perez JA, Pezzuto PR (2009b) O ordenamento da pescaria de camarões-de-profundidade (Decapoda: Aristeidae) No Sudeste e Sul do Brasil. Braz J Aquat Sci Tech 13(2):31–43

Dawson EW (1997) The deep-sea 'blind' lobsters (Crustacea: Polychelidae) – a species list and bibliographic index. Occas Papers Hutton Foundation, Wallypug Press, Eastbourne, New Zealand (8):1–78

De Grave S, Pentcheff ND, Ahyong ST, Chan T-Y, Crandall KA, Dworschak PC, Felder DL, Feldmann RM, Fransen CHJM, Goulding LYD, Lemaitre R, Low MEY, Martin JW, Ng PKL, Schweitzer SE, Tan S, Tshudy DH, Wetzer R (2009) A classification of living and fossil genera of Decapod Crustaceans. Raffles Bull Zool 21:1–109

De Man JG (1916) Diagnoses of new species of macrurous decapod Crustacea from the Siboga-expedition. Zool Meded 2:147–151

De Man JG (1920) The Decapoda of the Siboga expedition. Part IV: families Pasiphaeidae, Stylodactylidae, Hoplophoridae, Nematocarcinidae, Thalassocaridae, Pandalidae, Psalidopodidae, Gnathophyllidae, Processidae, Glyphgocrangonidae, and Crangonidae. Siboga Exp 39a3:13–18

De Man JG (1928a) The Thalassinidae and Callianassidae collected by the Siboga-Expedition with some remarks on the Laomediidae. Siboga Exp Monog 39a(6):1–187

De Man JG (1928b) A contribution to the knowledge of twenty-two species and three varieties of the genus *Callianassa* Leach. Capita Zool 2(6):1–56

De Saint LM (1966) *Iridopagurus*, genre nouveau de Paguridae (Crustacés: Décapodes) des mers tropicales américaines. Bull Mus Natl Hist Nat Paris 2(38):151–173

De Saint LM (1968) Révision des genres *Catapaguroides* et *Cestopagurus* et description de quatre genres nouveaux. I. *Catapaguroides* A. Milne-Edwards et Bouvier et *Decaphyllus* nov gen. (Crustacés Décapodes Paguridae). Bull Natl Mus Hist Nat Paris 39(5):923–954

Dennell R (1955) Observations on the luminescence of bathypelagic Crustacea Decapoda of the Bermuda area. Zool J Linn Soc London 42:393–406

D'Incao F (1998) Malacostraca. Eucarida. Dendrobranchiata. In: Young P (Ed) Catalogue of Crustacea of Brazil Rio de Janeiro, Museu Nacional, Série Livros:311–321

Doflein F, Balss H (1913) Die Galatheiden der Deutschen Tiefsee-Expedition. Wissenschaftliche Ergebnisse der Deutschen Tiefsee-Expedition auf dem Dampfer "Valvidia" 1898-1899. Jena 20:125–184

Dohrn R (1908) Ueber die Augen einiger Tiefseemacruren: Marburg, pp 1–64

Donaldson H (1975) Vertical distribution and feeding of sergestid shrimps (Decapoda: Natantia) collected near Bermuda. Mar Biol 31:37–50

Fausto-Filho J (1977) Sobre a ocorrência de *Palinustrus truncatus* (H. Milne-Edwards, 1880), no litoral brasileiro e de *Panulirus echinatus* Smith 1860, no litoral do Estado do Ceará, Brasil (Crustacea, Decapoda, Palinuridae). Arq Ciênc Mar Fortaleza 17(1):75–76

Faxon W (1893) Reports on the dredging operations off the west coast of Central America to the Galapagos, to the west coast of Mexico, and in the Gulf of California, in charge of Alexander Agassiz, carried on by the U.S. Fish Commission steamer "Albatross", during 1891, Lieut. Commander Z.L. Tanner, U.S.N., commanding. VI. Preliminary descriptions of new species of Crustacea. Bull Mus Comp Zool 24:149–220

Faxon W (1895) Reports on an exploration off the west coasts of Mexico, Central and South America, and off the Galapagos Islands, in charge of Alexander Agassiz, by the US Fish Commission Steamer "Albatross", during 1891, Lieut.-Commander ZL Tanner, USN, commanding. XV. The Stalkeyed Crustacea. Mem Mus Comp Zool Harvard Coll 18:1–292

Faxon W (1896) Reports on the results of dredging, under the supervision of Alexander Agassiz, in the Gulf of Mexico and the Caribbean Sea, and on the East Coast of the United States, 1877 to 1880, by the US Coast Survey Steamer "Blake", Lieut.-Commander CD Sigsbee, USN, and Commander JR Bartlett, USN., commanding. XXXVII. Supplementary notes on the Crustacea. Bull Mus Comp Zool 30:153–166

Felder DL (1973) An annotated key to crabs and lobsters (Decapoda, Reptantia) from coastal waters of the northwestern Gulf of Mexico. Center for Wetland Resources, Louisiana State University, Baton Rouge

Felder DL, Álvarez F, Goy JW, Lemaitre R (2009) Decapoda (Crustacea) of the Gulf of Mexico, with comments on the Amphionidacea. In: Felder DL, Camp DK (eds) Gulf of Mexico origin, waters, and biota volume 1, biodiversity. Texas A&M University Press, College Station, pp 1019–1104

Fierro Rengifo M, Navas G, Bermúdez A, Campos N (2008) Lista de chequeo de las familias Galatheidae y Chirostylidae (Crustacea: Decapoda: Anomura) del Neotrópico. Biota Colomb 9:1–20

Filhol H (1884) Explorations sous-marines. Voyage du "Talisman". La Nature 12:119–394

Filhol H (1885) La vie au fonds des mers. La Nature 13:355–410

Firth RW, Pequegnat WE (1971) Deep sea lobsters of the families Polychelidae and Nephropidae (Crustacea, Decapoda) in the Gulf of Mexico and Caribbean Sea. Texas A&M University, College Station

Fisher LR, Goldie EH (1961) New records of two deep-sea decapods. Crustaceana 2:78–79

Forest J (1954) Les paguristes des côtes occidentales et méridionales d'Afrique. Ann S Afr Mus 41(4):159–213

Forest J (1964) Sur une crevette recueillie au cours de la campagne de chalutage dans le golfe de Guinée *Plesionika williamsi* sp. nov. Bull Mus Natl Hist Nat Paris 35(6):620–629

Forest J (1974) Les Dromies de l'Atlantique oriental. Description de *Sternodromia* gen. nov. et de deux espèces nouvelles du genre *Dromia* Weber (Crustacea Decapoda Dromiidae). Ann Inst Océanogr Paris 50(1):71–123

Forest J, Holthuis LB (1997) A. Milne Edwards Recueil de figures de Crustacés nouveaux ou peu connus, 1883. New facsimile edition with comments and annotations. Backhuys Publishers, Leiden

Forest J, Saint-Laurent M (1968) Campagne de la Calypso au large des côtes atlantiques de l'Amérique du Sud (1961–1962). 6. Crustacés décapodes: pagurides. Ann Inst Océanogr Monaco 45(2):47–169

Foxton P (1970) The vertical distribution of pelagic decapods [Crustacea: Natantia] collected on the Sond cruise 1965. I. The Caridea. J Mar Biol Assoc UK 50:939–960

Fransen CHJM (1991) Preliminary report on the Crustacea collected in the eastern part of the North Atlantic during the CANCAP and MAURITANIA expeditions of the former Rijksmuseum van Natuurlijke Historie, Leiden. Nationaal Natuurhistorisch Museum, Leiden

Fransen CHJM (2014) True crabs. FAO, Rome

Freitas AJ (1984) The Penaeoidea of Southeast Africa. I. The study area and key to the southeast African species. Oceanogr Res Inst Investig Rep 56:1–31

Galil BS (2000) Crustacea Decapoda: review of the genera and species of the family Polychelidae Wood-Mason, 1874. In: Crosnier A (ed) Résultats des Campagnes MUSORSTOM, volume 20. Mém Mus Natl Hist Nat 184:285–387

Galil BS (2013) On a collection of Polychelidae from Papua New Guinea (Crustacea, Decapoda, Polychelida). Zoosystema 35(4):495–502

García Raso JE (1996) Crustacea Decapoda (excl. Sergestidae) from Ibero-Moroccan waters. Results of Balgim-84 expedition. Bull Mar Sci 58(3):730–752

García-Gómez J (1983) Revision of *Iridopagurus* (Crustacea: Decapoda: Paguridae) with the descriptions of new species from American waters. Bull Mar Sci 33(1):10–54

George RW, Grindley JR (1964) *Projasus*: a new generic name for Parker's crayfish, *Jasus parkeri* Stebbing (Palinuridae: "Silentes"). J R Soc West Aust 3:87–90

Glaessner MF (1969) Decapoda. In: Moore RC (ed) Part R Arthropoda 4(2). Treatise on invertebrate paleontology. The University of Kansas Press and The Geological Society of America, Lawrence, pp R399–R628

Gordan J (1956) A bibliography of pagurid crabs, exclusive of Alcock, 1905. Bull Amer Mus Nat Hist 108(3):253–352

Gordon I (1935) On new and imperfectly known species of Crustacea Macrura. Biol J Linn Soc 39:307–351

Gordon I (1939) A new species of *Sergestes* (Crustacea, Decapoda) from the South Atlantic. Ann Mag Nat Hist 4:498–509

Goy JW (2010) A review of the genus *Engystenopus* (Crustacea: Decapoda: Stenopodidea) *Juxtastenopus*, gen. nov., a new combination for *E. spinulatus* Holthuis, 1946, and transfer of *E. palmipes* Alcock & Anderson, 1894 to the family Spongicolidae Schram, 1986. Zootaxa 2372:263–277

Goy JW, Cardoso IA (2014) Redescription of *Odontozona edwardsi* (Bouvier, 1908) (Decapoda: Stenopodidea: Stenopodidae) and description of a new species of *Odontozona* commensal on the deep-water coral, *Lophelia pertusa* (Linneaus, 1758). Zootaxa 3774(6):552–566

Gray JE (1828) Spicilegia Zoologica; or, original figures and short systematic descriptions of new and unfigured animals 1(1):1–8

Griffin DJG, Stoddart HE (1995) Deep-water decapod Crustacea from eastern Australia: lobsters of the families Nephropidae, Palinuridae, Polychelidae and Scyllaridae. Rec Aust Mus 47:231–263

Griffin DJG, Tranter HA (1986) The Decapoda Brachyura of the Siboga expedition. Part VIII: Majidae. Siboga Exp 39(C4):1–335

Guinot D, Richer de Forges B (1995) Crustacea Decapoda Brachyura: Révision de la famille Homolidae de Haan, 1839. In: Crosnier A (ed) Résultats des Campagnes MUSORSTOM, 1. Mém Mus Natl Hist Nat 163:283–517

Guinot D, Tavares MS (2003) A new subfamilial arrangement for the Dromiidae de Haan, 1833, with diagnoses and descriptions of new genera and species (Crustacea, Decapoda, Brachyura). Zoosystema 25(1):43–129

Gurney R, Lebour M (1940) Larvae of decapod Crustacea. Part VI. The genus *Sergestes*. Discov Rep 20:1–68

Guzmán GG (2008) Camarones pelágicos (Crustacea: Decapoda) en aguas del Pacífico sureste. Contrib Study East Pac Crust 5:27–45

Hanamura Y (1983) Pelagic shrimps (Penaeidae and Caridea) from Baja California and its adjacent region with description of a new species. Bull Biogeog Soc Jpn 38:51–85

Hanamura Y (1994) A new species of *Pasiphaea* Savigny (Crustacea: Caridea: Pasiphaeidae) from north-western Australian waters. The Beagle 11:167–173

Hansen HJ (1896) On the development and the species of the crustaceans of the genus Sergestes. Proc Zool Soc London 1896:936–970

Hansen HJ (1903) On the crustaceans of the genera *Petalidium* and *Sergestes* from the 'Challenger', with an account of luminous organs in *Sergestes challengeri*, n. sp. Proc Zool Soc London 1903:52–79

Hansen HJ (1908) Crustacea Malacostraca. I. Copenhagen. Danish Ingolf-Exp 3:1–120

Hayashi KI (2004) Revision of the *Pasiphaea cristata* Bate, 1888 species group of *Pasiphaea* Savigny, 1816,with descriptions of four new species, and referral of *P. australis* Hanamura, 1989 to *Alainopasiphaea* Hayashi, 1999 (Crustacea: Decapoda: Pasiphaeidae). Trop Deep-Sea Benthos 23(191):319–373

Hayashi KI (2006) Revision of the *Pasiphaea alcocki* species group (Crustacea, Decapoda, Pasiphaeidae). Trop Deep-Sea Benthos 24(193):193–241

Hazlett BA (1966) The behavior of some deep-water hermit crabs (Decapoda: Paguridea) from the straits of Florida. Bull Mar Sci 16(1):76–92

Heller C (1862) Beitrage zur naheren Kenntnis der Macrouren. Sitzungsberichte der Akademie der Wissenschaften in Wien, mathematisch-physikalische. Klasse 45(1):389–426

Henderson JR (1888) Report on the Anomura collected by H.M.S Challenger during the years 1873–1876. In: Murray J (ed) Zoology. Report on the Scientific Results of the Voyage of H.M.S. Challenger During the Years 1873–76 Under the Command of Captain George S Nares, RN, FRS and the Late Captain Frank Tourle Thomson, RN Wyville Thomson, C. and J. Murray (series eds) vol 27. Neill and Company, Edinburgh

Hendrickx ME (1995) Checklist of lobster-like decapod crustaceans (Crustacea: Decapoda: Thalassinidea, Astacidea and Palinuridea) from the eastern tropical Pacific. An Inst Biol Univ Nac Autón México Ser Zool 66(2):151–163

Hendrickx ME (2015) Further records of species of *Gennadas* (Crustacea, Decapoda, Dendrobranchiata, Benthesicymidae) in the Mexican Pacific. Zootaxa 3980:417–426

Hendrickx ME, Estrada-Navarrete FD (1989) A checklist of the species of pelagic shrimps (Penaeoidea and Caridea) from the Eastern Pacific, with notes on their geographic and depth distribution. CalCOFI Rep 30:104–121

Hendrickx ME, Estrada-Navarrete (1996) Los Camarones Pelágicos (Crustacea: Dendrobranchiata y Caridea) del Pacífico Mexicano. Instituto de Ciencias del Mar y Limnología Universidad Nacional Autónoma de Mexico. 157 p

Holthuis LB (1946) Biological results of the Snellius Expedition XIV. The Decapoda Macrura of the Snellius Expedition. I. The Stenopodidae, Nephropsidae, Scyllaridae and Palinuridae. Temminckia 7:1–178

Holthuis LB (1947) Nomenclatorial notes on European macrurous Crustacea Decapoda. Zool Meded 27:312–322

Holthuis LB (1949) The Caridean Crustacea of the Canary Islands. Zool Meded 30:227–255

Holthuis LB (1951) The caridean Crustacea of tropical West Africa. Atlantide Rep 2:7–187

Holthuis LB (1952) Crustacés Décapodes Macrures. Résultats scientifiques de l'Expédition océanographique belge des eaux côtières africaines, Atlantique Sud (1948–1949) 3(2):1–88

Holthuis LB (1955) The recent genera of the caridean and stenopodidean shrimps (class Crustacea, order Decapoda, supersection Natantia) with keys for their determination. Zool Verhand 26:1–157

Holthuis LB (1962) Forty-seven genera of Decapoda (Crustacea); proposed addition to the official list. ZN (S) 1499. Bull Zool Nomencl 19:232–252

Holthuis LB (1969) *Portunus binoculus*, n. sp., a new deep-water swimming crab from the Caribbean region (Crustacea, Decapoda, Brachyura). Bull Mar Sci 19:409–427

Holthuis LB (1971) The Atlantic shrimps of the deep-sea genus *Glyphocrangon* A. Milne Edwards, 1881. Bull Mar Sci 21:267–373

Holthuis LB (1974) The lobsters of the superfamily Nephropidea of the Atlantic Ocean (Crustacea: Decapoda). Bull Mar Sci 24(4):723–884

Holthuis LB (1991) FAO species catalog. Vol. 13. Marine lobsters of the world. An annotated and illustrated catalogue of species of interest to fisheries known to date. FAO Fisheries Synopsis 125. Food and Agriculture Organization of the United Nations, Rome

Illig G (1914) Die Dekapoden der Deutschen Südpolar-Expedition 1901-1903. II. Die Sergestiden. Deutschen Südpolar-Exped 15:349–376

Illig G (1927) Die Sergestiden der Deutschen Tiefsee-Expedition. Wiss Ergebn Valdivia Exped 23:277–354

Ingle RW (1985) *Geryon gordonae* sp. nov. (Decapoda, Brachyura, Geryonidae) from the Northeastern Atlantic Ocean. Crustaceana 48(1):88–98

Ingle RW, Christiansen ME (2004) Lobsters, mud shrimps and anomuran crabs. Keys and notes for the identification of species. Synopses of the British Fauna (New Series). Crothers JH, Hayword PJ (eds) vol 55. Field Studies Council for Linnaean Society of London Estuarine and Coastal Sciences Association, Shrewbury

Iwasaki N (1990) Pasiphaeid shrimps from the eastern North Atlantic and the Caribbean Sea, with the description of a new species of *Pasiphaea* (Crustacea: Decapoda: Pasiphaeidae). Zool Meded 63:187–203

Johnson JY (1867) Descriptions of a new genus and a new species of macrourous decapod crustaceans belonging to the Penaeidae, discovered at Madeira. Proc Zool Soc London 1867:875–901

Juanico M (1973) Hallazgo de *Geryon quinquedens* Smith, 1879 (Crustacea Decapoda), en aguas Uruguayas. Rev Biol Uruguay 1(2):145–149

Judkins DC (2014) Geographical distribution of pelagic decapod shrimp in the Atlantic Ocean. Zootaxa 3895(3):301–345

Judkins DC, Kensley B (2008) New genera in the family Sergestidae (Crustacea: Decapoda: Penaeidae). Proc Biol Soc Wash 121:72–84

Kemp S (1906) On the occurrence of the genus *Acanthephyra* in deep water off the West Coast of Ireland. Scientific Investigations. Dept Agric Tech Instructions Ireland. Fish Branch 1:1–28

Kemp S (1909) The decapods of the genus *Gennadas* collected by H.M.S. 'Challenger'. Proc Zool Soc London 718–730

Kemp S (1910) Notes on Decapoda in the Indian Museum. I. The species of Gennadas. Records of the Indian Museum 5:173–181

Kemp S (1913) The Percy Sladen Trust Expedition to the Indian Ocean in 1905 under the leadership of Mr. J. Stanley Gardiner. Volume 5, no. 5. Pelagic Crustacea Decapoda of the Percy Sladen Expedition in H.M.S. "Sealark". Trans Linn Soc London, Ser 2, Zool 16:53–68

Kemp S (1939) On *Acanthephyra purpurea* and its allies (Crustacea Decapoda: Hoplophoridae). Ann Mag Nat Hist 4:568–579

Kensley BF (1968) Deep sea decapod Crustacea from west of Cape Point, South Africa. Ann S Afr Mus 50:283–323

Kensley BF (1971) The genus *Gennadas* in the waters around Southern Africa. Ann S Afr Mus 57:271–294

Kensley BF (1972) Shrimps and prawns of southern Africa. South African Museum, Cape Town

Kensley BF (1973) A new species of hermit crab from Natal, South Africa (Decapoda, Anomura, Paguridae). Durban Mus Nov 9(19):285–290

Kensley BF (1977) The South African Museum's Meiring Naude cruises. Part 5. Crustacea, Decapoda, Reptantia and Natantia. Ann S Afr Mus 74:13–44

Kensley BF (1981) On the zoogeography of Southern African decapod Crustacea, with a distributional checklist of the species. Smith Contrib Zool 338:1–64

Kensley BF (1987) Deepwater decapod Crustacea from eastern Australia (Penaeidea and Caridea). Rec Austr Mus 39:263–331

Kensley BF, Tranter HA, Griffin DJG (1987) Deepwater decapod Crustacea from eastern Australia (Penaeidea and Caridea). Rec Aust Mus 39:263–331

Kikuchi T (1985) A new species of *Meningodora* (Crustacea, Decapoda, Oplophoridae) from the western North Pacific. Bull Nat Sci Mus 11:191–196

Kikuchi T (1991) Meso- or Bathypelagic Shrimps of the Family Oplophoridae (Crustacea: Decapoda) from the Western North Pacific: Part 1. Genus *Meningodora* Smith, 1882. Science Rep Yokohama Natl Univ II 38:23–40

Kikuchi T, Nemoto T (1986) List of pelagic shrimps (Crustacea, Decapoda) from the western North Pacific. Bull Biogeogr Soc Jpn 41:51–60

Komai T (2004) Deep-Sea Shrimps of the Genus *Glyphocrangon* A. Milne-Edwards (Crustacea, Decapoda, Caridea, Glyphocrangonidae) from off Southeastern Coast of Brazil collected during the REVIZEE Program. Arq Mus Nac Rio de Janeiro 62:31–44

Komai T (2008) A world-wide revision of species of the deep-water crangonid genus *Parapontophilus* Christoffersen, 1988 (Crustacea, Decapoda, Caridea), with descriptions of ten new species. Zoosystema 30:261–332

Komai T, Amaoka K (1993) A new species of the genus *Pasiphaea* (Crustacea Decapoda, Pasiphaeidae) from the North Pacific. Zool Sci 10:367–373

Komai T, Segonzac M (2005) Two new species of *Nematocarcinus* A. Milne Edwards, 1881 (Crustacea, Decapoda, Caridea, Nematocarcinidae) from hydrothermal vents on the North and South East Pacific Rise. Zoosystema 27(2):343–364

Komai T, Wei-Lin C, Chan TY (2012) Bathypelagic shrimp of the genus *Pasiphaea* (Decapoda: Caridea: Pasiphaeidae) from waters around Taiwan, with descriptions of four new species. J Crust Biol 32(2):295–325

Krøyer H (1845) Karcinologiske Bidrag (Fortsaettelse). Naturh Tidsskr New Ser 1:453–538

Krøyer H (1855) Bidrag til Kundskab om Krebsdyrsslaegten *Sergestes* Edw. Og om 11 Arter af same. K Danske Vidensk Selsk Forhand 1855:2–34

Krøyer H (1856) Meddelelser af en Afhandling Ormeslaegten *Sabella* Linn., isaer med Hensyn til dens nordiske Arter [Alternate title: Bidrag til Kundskab af Sabellerne]. K Danske Vidensk Selsk Forhand 1856:1–36

Krøyer H (1859) Forsög til en Monographisk Fremstillung af Kraebsdyrslægten *Sergestes* med Bemærkinger om Dekapodernes Horeredskaber. K Danske Vidensk Selsk Skr 5:217–304

Krygier E, Pearcy W (1981) Vertical distribution and biology of pelagic decapod crustaceans off Oregon. J Crust Biol 1:96–104

Krygier EE, Wasmer RA (1988) Zoogeography of pelagic shrimps (Natantia: Penaeidea and Caridea) in the North Pacific Ocean (with synopses and keys to the species of the Subarcticand Transitional Zones). Bull Ocean Res Inst Univ Tokyo 26:43–98

Lagardère J (1978) Crustacea (adultes pélagiques) Ordre: Decapoda. families: Penaeidae et Sergestidae. Fiches Identif Zoopl 155(156/157):1–15

Latreille PA (1802) Histoire naturelle, générale et particulière des Crustacés et des Insectes. Ouvrage faisant suite à l'histoire naturelle générale et particulière, composée par Leclerc de Buffon, et rédigée par C.S. Sonnini, membre de plusieurs Sociétés savantes, vol. 3, Dufart, Paris

Lemaitre R (1984) Decapod crustaceans from Cay Sal Bank, Bahamas, with notes on their zoogeographic affinities. J Crust Biol 4(3):425–447

Lemaitre R (1989) Revision of the genus *Parapagurus* (Anomura: Paguroidea: Parapaguridae), including redescriptions of the Western Atlantic species. Zool Verhand 21:3–106

Lemaitre R (1990) A review of the eastern Atlantic species of the family Parapaguridae (Decapoda, Anomura, Paguroidea). J Nat Hist 24:219–240

Lemaitre R (1996) Hermit crabs of the family Parapaguridae (Crustacea: Decapoda: Anomura) from Australia: species of *Strobopagurus* Lemaitre, 1989, *Sympagurus* Smith, 1883, and two new genera. Rec Aust Mus 48:163–221

Lemaitre R (2004) A review of *Strobopagurus* Lemaitre, 1989 (Crustacea, Decapoda, Paguroidea, Parapaguridae), with description of a new species. Sci Mar 68(3):355–372

Lemaitre R, McLaughlin PA (1992) Descriptions of megalopa and juveniles of *Sympagurus dimorphus* (Studer, 1883), with an account of the Parapaguridae (Crustacea: Anomura: Paguroidea) from Antarctic and subantarctic waters. J Nat Hist 26:745–768

Lemaitre R, McLaughlin PA (1996) Revision of *Pylopagurus* and *Tomopagurus* (Crustacea: Decapoda: Paguridae), with the descriptions of new genera and species. Part V. *Anisopagurus* McLaughlin, *Manucomplanus* McLaughlin, and *Protoniopagurus* new genus. Bull Mar Sci 59(1):89–141

Lenz H, Strunck K (1914) Die dekapoden der Deutschen Südpolar-Expedition 1901–1903. I. Brachyuren und Macruren mit Ausschluss der Sergestiden. Deutsche Südpolar-Exped 7:257–346

Lessa RP (2006) Recursos pesqueiros da região nordeste. In: Brasil. Ministério do Meio Ambiente (Ed.) Programa REVIZEE: Avaliação do Potencial sustentável de recursos vivos na Zona Econômica Exclusiva do Brasil: relatório executivo. Ministério do Meio Ambiente, Brasília, 153–180

Lessard-Pilon SA, Podowski EL, Cordes EE, Fisher CR (2010) Megafauna community composition associated with *Lophelia pertusa* colonies in the Gulf of Mexico. Deep-Sea Res Part II Top Stud Oceanogr 57(21–23):1882–1890

Lianos L, Mollemberg MC, Lima DJM, Santana W (2017) New records of king crabs (Decapoda: Anomura: Lithodidae) from southern Brazil. Crustaceana 90(7–10):211–218

Lindner MJ, Anderson WW (1941) A new *Solenocera* and notes on the other Atlantic American species. J Wash Acad Sci 31(5):181–187

Lins LSF, Cardoso IA (2010) First record of the deep sea Paguridae *Catapaguroides microps* (Decapoda: Anomura) in the southwestern Atlantic Sea. Cah Biol Mar 51:69–74

Lira C, Bolaños J, Marcano J, Hernández J, Vera-Caripe J, Rodríguez B, López R (2017a) Lista Actualizada de los camarones dendrobranquiados (Crustacea: Decapoda: Dendrobranchiata) marinos y de aguas salobres de Venezuela. Bol Inst Oceanogr Venezuela 56(1):31–45

Lira SMA, Santana CS, Lima CDM, Montes MJF, Schwamborn R (2017b) New records of the larval forms *Cerataspis monstrosa* and *Amphionides reynaudii* (Crustacea: Decapoda) from the western tropical Atlantic. Zootaxa 4237(2):335–346

Lunina AA, Kulagin DN, Vereshchaka AL (2018) Oplophoridae (Decapoda: Crustacea): phylogeny, taxonomy and evolution studied by a combination of morphological and molecular methods. Zool J Linn Soc Lond 039:1–20

Macpherson E (1988) Revision of the family Lithodidae Samouelle, 1819 (Crustacea, Decapoda, Anomura) in the Atlantic Ocean. Monogr Zool Mar 2:9–153

Macpherson E (2007) Species of the genus *Munidopsis* Whiteaves, 1784 from the Indian and Pacific Oceans and reestablishment of the genus *Galacantha* A. Milne-Edwards, 1880 (Crustacea, Decapoda, Galatheidae). Zootaxa 1417:1–135

Macpherson E, Segonzac M (2005) Species of the genus *Munidopsis* (Crustacea, Decapoda, Galatheidae) from the deep Atlantic Ocean, including cold-seep and hydrothermal vent areas. Zootaxa 1095:1–60

Manning RB (1987) Notes on western Atlantic Callianassidae (Crustacea: Decapoda: Thalassinidea). Proc Biol Soc Wash 100(2):386–401

Manning RB, Felder DL (1991) Revision of the American Callianassidae (Crustacea: Decapoda: Thalassinidea). Proc Biol Soc Wash 104(4):764–792

Manning RB, Holthuis LB (1981) West African brachyuran crabs. Smith Contrib Zool 306:1–379

Manning RB, Holthuis LB (1989) Two new genera and nine new species of geryonid crabs (Crustacea, Decapoda, Geryonidae). Proc Biol Soc Wash 102:50–77

Manning RB, Tavares MS, Albuquerque EF (1989) *Chaceon ramosae*, a new deep water crab from Brazil (Crustacea: Decapoda: Geryonidae). Proc Biol Soc Wash 102:646–650

Martin H, Hargreaves P (1991) Shrimps of the Families Pandalidae and Hippolytidae (Crustacea: Decapoda) caught in benthic traps off the Azores. Arquipélago Life Earth Sci 9:47–61

Martin JW, Christiansen JC, Trautwein SE (2001) Crabs of the family Homolodromiidae, VI. *Homolodromia monstrosa* new species (Decapoda: Brachyura) from the western North Atlantic with a redescription of the holotype of *Homolodromia paradoxa* A Milne Edwards, 1880 and comments on sexual dimorphism. Bull Mar Sci 68(2):313–326

Matthews JBL, Pinnoi S (1973) Ecological studies on the deep-water pelagic community of Korsfjorden, western Norway. The species of *Pasiphaea* and *Sergestes* (Crustacea, Decapoda) recorded in 1968 and 1969. Sarsia 52:123–144

McLaughlin PA (1981a) Revision of *Pylopagurus* and *Tomopagurus* (Crustacea: Decapoda: Paguridae) with the descriptions of new genera and species. Part I. Ten new genera of the Paguridae and a redescription of *Tomopagurus* A. Milne Edwards and Bouvier. Bull Mar Sci 31(1):1–30

McLaughlin PA (1981b) Revision of *Pylopagurus* and *Tomopagurus* (Crustacea: Decapoda: Paguridae) with the descriptions of new genera and species. Part II. *Rhodochirus* McLaughlin and *Phimochirus* McLaughlin. Bull Mar Sci 31(2):329–365

McLaughlin PA (2002) A new species of the genus *Catapaguroides* (Decapoda, Anomura, Paguroidea, Paguridae) from Guam, Micronesia. Crustaceana 75:495–504

McLaughlin PA, Lemaitre R (2001) Revision of *Pylopagurus* and *Tomopagurus* (Crustacea: Decapoda: Paguridae), with the descriptions of new genera and species. Part VI. *Pylopagurus* A. Milne-Edwards & Bouvier, 1891, *Haigia* McLaughlin, 1981, and *Pylopaguridium*, a new genus. Proc Biol Soc Wash 114(2):444–483

McLaughlin PA, Camp DK, Angel MV, Bousfield EL, Brunel P, Brusca RC, Cadien D, Cohen AC, Conlan K, Eldredge LG, Felder DL, Goy JW, Haney T, Hann B, Heard RW, Hendrycks EA, Hobbs III HH, Holsinger JR, Kensley B, Laubitz DR, LeCroy SE, Lemaitre R, Maddocks RF, Martin JW, Mikkelsen P, Nelson E, Newman WA, Overstreet RM, Poly WJ, Price WW, Reid JW, Robertson A, Rogers DC, Ross A, Schotte M, Schram FR, Shih C-T, Watling L, Wilson GDF, Turgeon DD (2005) Common and scientific names of aquatic invertebrates from the United States and Canada: crustaceans. American Fisheries Society, Special Publication 31, Bethesda, Maryland

McLaughlin PA, Komai T, Lemaitre R, Rahayu DL (2010) Annotated checklist of anomuran decapod crustaceans of the world (exclusive of the Kiwaoidea and families Chirostylidae and Galatheidae of the Galatheoidea). Part I, Lithodoidea, Lomisoidea and Paguroidea. Raffles Bull Zool 23:5–107

Melo GAS (1996) Manual de Identificação dos Brachyura (Caranguejos e Siris) do litoral brasileiro. Editora Plêiade, São Paulo

Melo GAS (1998) Malacostraca – Eucarida. Brachyura. Oxyrhyncha and Brachyrhyncha. In: Young PS (ed) Catalogue of Crustacea of Brazil. Museu Nacional, Rio de Janeiro, pp 455–515

Melo GAS (1999) Manual de Identificação dos Crustacea Decapoda do litoral brasileiro: Anomura; Thalassinidea; Palinuridea e Astacidea. Plêiade/Fapesp, São Paulo, p 551

Melo GAS, Torres MFA, Campos O (1998) Malacostraca – Eucarida. Brachyura Dromiacea and Oxystomata. In: Young PS (ed) Catalogue of Crustacea of Brazil. Museu Nacional, Rio de Janeiro, pp 439–454

Melo-Filho GAS (1996) Descrição de *Munida victoria* sp. n. e comparação com *M. microphthalma* A. Milne-Edwards 1880, em sua primeira ocorrência na costa brasileira (Crustacea: Decapoda: Galatheidae). Pap Avuls Zool 39(14):271–280

Melo-Filho GAS (1998) Malacostraca - Eucarida. Chirostylidae and Galatheidae. In: Young PS (ed) Catalogue of Crustacea of Brazil. Museu Nacional, Rio de Janeiro, pp 393–397

Melo-Filho GAS (1999) Families Chirostylidae e Galatheidae. In: Buckup L, Bond-Buckup G (eds) Os crustacéos do Rio Grande do Sul. Universidade/UFRGS, Porto Alegre, pp 383–397

Melo-Filho GAS (2006) Reports on the results of the N.Oc. "Prof. W. Besnard" expeditions to the southern coast of Brazil under the Revizee Program: Chirostylidae and Galatheidae (Crustacea: Decapoda: Anomura). Zootaxa 1238:1–22

Melo-Filho GAS (2008) Specimens of Galatheidae (Crustacea, Decapoda) collected by NPq. "Soloncy Moura" off the Southern Coast of Brazil under the Revizee Program. Zootaxa 1841:31–42

Melo-Filho GAS, Melo GAS (1992a) Designation of lectotypes for the species of *Munida* (Crustacea: Anomura: Galatheidae) collected by the U.S. Coast Survey Steamer "Blake" (1877–79) and description of a new species. Proc Biol Soc Wash 105(3):512–520

Melo-Filho GAS, Melo GAS (1992b) Reexamination of the material of *Munida* Leach (Crustacea: Anomura: Galatheidae) collected by the H.M.S. "Challenger" (1872–76) along the Brazilian coast. Proc Biol Soc Wash 105(4):760–774

Melo-Filho GAS, Melo GAS (1997) Redescrição e designação de lectótipos para duas espécies do gênero *Munida* Leach 1820 (Crustacea: Decapoda: Galatheidae) coletadas pelo USFC "Albatross" (1885) no Golfo do México. Pap Avuls Zool 40(12):189–202

Melo-Filho GAS, Melo GAS (2001a) Espécies do gênero *Munida* Leach (Crustacea, Decapoda, Galatheidae) coletadas entre Torres (Rio Grande do Sul, Brasil) e Maldonado (Uruguai) pelo Programa Rio Grande do Sul (PRS I–II, GEDIP). Rev Bras Zool 18(4):1177–1196

Melo-Filho GAS, Melo GAS (2001b) Espécies do gênero *Munida* Leach (Crustacea, Decapoda, Galatheidae), distribuídas na costa do Brasil. Rev Bras Zool 18(4):1135–1176

Melo-Filho GAS, Melo GAS (2001c) Taxonomia e Zoogeografia das espécies do Gênero *Munida* Leach, 1820 (Crustacea: Decapoda: Galatheidae) distribuídas ao longo da costa Temperada-Quente do Atlântico-Sul Ocidental. Trop Oceanogr 29(1):37–57

Miers EJ (1886) Report on the Brachyura collected by H. M. S. Challenger during the years 1873-1876. Report on the scientific results of the H. M. S. Challenger during the years 1873-1876. Zoology 17:362

Milne Edwards H (1837) Note sur le Rhynchocinète, nouveau genre de Crustacé décapode. Ann Sci Nat 2:165–168

Milne-Edwards A (1879) Études sur les Crustacés Podophthalmaires de la región mexicaine. Recherches Zoologiques pour Servir à l'Histoire de la Faune de l'Amérique Centrale et du Mexique 5:185–264

Milne-Edwards A (1880) Reports on the results of dredging under the supervision of Alexander Agassiz, in the Gulf of Mexico and in the Caribbean Sea, 1877, '78, '79, by the US Coast

Survey Steamer "Blake", Lieut-Commander CD Sigsbee, USN, and Commander JR Bartlett, USN commanding VIII. Études préliminaires sur les Crustacés. Bull Mus Comp Zool 8(2):1–68

Milne-Edwards A (1881) Description de quelques crustacés macroures provenant des grandes profondeurs de la Mer des Antilles. Ann Sci Nat 11:1–15

Milne-Edwards A (1883) Recueil de figures de crustacés nouveaux ou peu connus. Plates 1–44:3

Milne-Edwards A (1890) Diagnose d'un crustacé macroure nouveau de la Méditerranée. Bull Soc Zool France 15:163

Milne-Edwards A (1891) Pagurides nouveaux des Açores. Campagnes scientifiques de SA Le Prince de Monaco sur le yacht l'Hirondelle. Bull Soc Zool 16:131–134

Milne-Edwards A, Bouvier E-L (1892) Observations préliminaires sur les paguriens recueillis par les expéditions du Travailleur et du Talisman. Ann Sci Nat 7(13):185–226

Milne-Edwards A, Bouvier E-L (1893) Reports on the results of dredging, under the supervision of Alexander Agassiz, in the Gulf of Mexico (1877–78), in the Caribbean Sea (1878–79), and along the Atlantic coast of the United States (1880), by the US Coast Survey Steamer "Blake", Lieut -Commander C D Sigsbee, USN, and Commander JR Bartlett USN, commanding. XXXIII. Description des Crustacés de la famille des paguriens recueillis pendant l'expédition. Mem Mus Comp Zool Harvard Coll 14(3):1–172

Milne-Edwards A, Bouvier E-L (1894) Considérations générales sur la famille des Galathéidés. Ann Sci Nat Zool 7(16):191–327

Milne-Edwards A, Bouvier E-L (1897) Reports on the results of dredging under the supervision of Alexander Agassiz in the Gulf of Mexico (1877–78), in the Caribbean Sea (1878–79), and along the Atlantic Coast of the United States (1880) by the US Coast Survey Steamer "Blake", Lieut. Commander CD Sigsbee, USN, and Commander JR Bartlett, USN, commanding. XXXV. Description des Crustacés de la famille des Galathéidés recueillis pendant l'expédition. Mem Mus Comp Zool Harvard Coll 19(2):1–141

Milne-Edwards A, Bouvier E-L (1899) Crustacés Décapodes provenant des campagnes de l'Hirondelle et de la Princesse Alice (Brachyoures et Anomoures). Rés Campagnes Sci Albert I 8:63–67

Milne-Edwards A, Bouvier E-L (1900) Crustacés Décapodes (Brachyoures et Anomoures). Exp Sci Travailleur Talisman 1:1–206

Milne-Edwards A, Bouvier E-L (1909) Reports on the results of dredging, under the supervision of Alexander Agassiz, in the Gulf of Mexico (1877-78), in the Caribbean Sea (1878-79), and along the Atlantic coast of the United States (1880), by the U.S.N. Coast Survey Steamer "Blake", Lieut.-Com. C.D. Sigsbee, U.S.N., and Commander J.R. Bartlett, US.N. Commanding. 44. Les Pénéides et Sténopides. Mem Mus Comp Zool Harvard Coll 27:177–274

Milne-Edwards A, Bouvier E-L (1923) Reports on the results of dredging. Under the supervision of Alexander Agassiz, in the Gulf of Mexico (1877–78), in the Caribbean Sea (1878–79), and along the Atlantic coast of the United States (1880), by the U.S. Coast Survey steamer "Blake." Lieut -Com C. Sigsbee, USN, and Commander JR Bartlett, USN, commanding. XLVII: Les Porcellanides et des Brachyures. Mem Mus Comp Zool Harvard Coll 47:283–395

Monod TH (1966) Crevettes et crabes de la côte occidentale d'Afrique. Mem IFAN 77:103–234

Monterossa OE (1988) *Heterocarpus cutressi*, new species and *Plesionika macropoda* Chace, 1939: two caridean shrimps of the family Pandalidae (Crustacea, Decapoda) from Puerto Rico and the U.S. Virgin Islands. Proc Biol Soc Wash 101:633–639

Moreira C (1901) Crustáceos do Brazil. Contribuição para o conhecimento da fauna brasileira. Arq Mus Nac Rio de Janeiro 11:1–151

Murray J, Hjort J (1912) The Depths of the Ocean. A general account of the modern science of oceanography based largely on the scientific researches of the Norwegian Steamer Michael Sars in the North Atlantic. Macmillan, London

Nakazawa K (1932) On three species of *Sergestes* found in Suruga Bay [in Japanese]. Zool Mag (Dobutsugaku Zasshi) 44:31–32

Navas GR, Bermúdez A, Cruz N, Campos NH (2003) Galatéideos (Decapoda, Anomura, Galatheidae) del Caribe colombiano, incluyendo doce primeros registros. Bol Investig Mar Cost 32:183–218

Ng PKL, Guinot D, Davie PJF (2008) Systema Brachyurorum: Part I. An annotated checklist of extant brachyuran crabs of the world. Raffles Bull Zool 17:1–286

Nizinski MS (2003) Annotated checklist of decapod crustaceans of Atlantic coastal and continental shelf waters of the United States. Proc Biol Soc Wash 116(1):96–157

Nucci PR, Melo GAS (2011) Hermit crabs from Brazil: Family Paguridae (Crustacea: Decapoda: Paguroidea), except *Pagurus*. Zootaxa 3104(39):26–41

Nunes DM, Ferreira RCP, Hazin FHV, Travassos P, Souza-Filho JF (2017) Decapod crustaceans of São Pedro and São Paulo Archipelago, Equatorial Atlantic, Brazil. Zootaxa 4324(2):331–347

Oliveira JEL, Pinheiro AP, Freire FAM, Vieira KRR (1999) Potencial de exploração do caranguejo-de-profundidade (*Chaceon fenneri*) no nordeste do Brasil. In: Aguilar AET, Malpica ZGC (eds) VIII Congresso Latinoamericano sobre Ciencias del Mar (COLACMAR), Trujillo, vol 1, pp 50–51

Omori M (1971) Taxonomy and some notes on the biology of a new caridean shrimp, *Plesionika izumiae* (Decapoda, Pandalidae). Crustaceana 20:241–256

Omori M (1974) The biology of pelagic shrimps in the ocean. Adv Mar Biol 12:223–324

Ortmann AE (1893) Decapode nund Schizopoden. Ergebn Plankton-Exped 2:1–120

Palmier G (1993) Crustacés profonds capturés aux casiers aux Antilles françaises. IFREMER, Paris

Pequegnat LH (1970a) Deep-sea caridean shrimps with descriptions of six new species. In: Pequegnat WE, Chace FA Jr (eds) Contributions on the biology of the Gulf of Mexico. Gulf Publishing, Houston, pp 59–123

Pequegnat LH (1970b) Deep-water brachyuran crabs. In: Pequegnat WE, Chace FA Jr (eds) Contributions on biology of the Gulf of Mexico. Texas A&M University Oceanographic Studies, Houston, pp 171–204

Pequegnat LH, Pequegnat WE (1970) Deep-sea anomurans of superfamily Galatheoidea with description of three new species. In: Pequegnat WE, Chace FA Jr (eds) Contributions on the biology of the Gulf of Mexico. Texas A&M University, pp 125–170

Pequegnat WE, Pequegnat LH (1971) New species and new records of Munidopsis (Decapoda: Galatheidae) from the Gulf of Mexico and Caribbean Sea (Suppl to Texas A&M University Oceanographic Studies vol 1) Gulf Publishing Co, Houston

Pequegnat LH, Wicksten MK (2006) Oplophorid shrimps (Decapoda: Caridea: Oplophoridae) in the Gulf of Mexico and Caribbean Sea from the collections of the research vessels Alaminos, Oregon and Oregon II. Crust Res 35:92–107

Pequegnat WE, Pequegnat LH, Firth Jr RW, James BM, Roberts TW (1971) Anomuran Decapoda of the superfamil y Galatheoidea and family Lithodidae. In: Serial Atlas of the Marine Environment, Folio 20. Gulf of Mexico Deep-sea Fauna. Decapoda and Euphausiacea. American Geographical Society, pp 5–8

Perez JAA, Pezzuto PR, Wahrlich R, Soares ALS (2009) Deep-water fisheries in Brazil: history, status and perspectives. Lat Am J Aquat Res 37(3):513–541

Pérez-Farfante I (1973) Morphological study of diagnostic characters in Western Atlantic *Hepomadus* (Crustacea, Decapoda, Penaeidae). Fish B-NOAA 71:441–453

Pérez-Farfante I (1977a) American solenocerid shrimps of the genera *Hymenopenaeus*, *Haliporoides*, *Pleoticus*, *Hadropenaeus* new genus, and *Mesopenaeus* new genus. Fish B-NOAA 75:261–346

Pérez Farfante I (1977b) Penaeopsis eduardoi, a new species of shrimp (Crustacea: penaeidae) from the Indo-West Pacific. Proc Biol Soc Wash 90(1):172–182

Pérez-Farfante I (1980) Revision of the penaeid shrimp genus *Penaeopsis* (Crustacea: Decapoda). Fish B-NOAA 77:721–763

Pérez-Farfante I (1988) Illustrated key to Penaeoid shrimps of commerce in the Americas. NOAA Tech Rep NMFS 64:1–32

Pérez-Farfante I, Bullis HR (1973) Western Atlantic shrimps of the genus *Solenocera* with descriptions of a new species (Crustacea: Decapoda: Penaeidae). Smith Contrib Zool 153:1–33

Pérez-Farfante I, Kensley B (1997) Penaeoid and sergestoid shrimps and prawns of the world. Keys and diagnoses for the families and genera. Mém Mus Natl Hist Nat 175:1–233

Poore GCB (2004) Marine Decapod Crustacea of Southern Australia. A Guide to Identification with Chapter on Stomatopoda, Melbourne, CSIRO Publishing, p 574
Perrier E (1893) Traité de Zoologie. VI Masson, Paris
Poupin J (1994) Faune marine profonde des Antilles françaises: Récoltes du navire Polka faites en 1993. ORSTOM, Paris
Poupin J (2010) Biodiversité de l'Indo-Pacifique tropical français: 2514 espèces de crustacés décapodes et stomatopodes. Rapport Scientifique de l'Institut de Recherche de l'Ecole Navale, Angers/Brest
Poupin J (2018) Les Crustacés décapodes des Petites Antilles: Avec de nouvelles observations pour Saint-Martin, la Guadeloupe et la Martinique. Mus Natl Hist Nat Patrimoines Nat Paris 77:1–264
Powers LW (1977) A catalogue and bibliography to the crabs (Brachyura) of the Gulf of México. Contrib Mar Sci 2:1–190
Rabaud E (1941) Recherches sur l'Adaptation et le Comportement des Pagures. Arch Zool Expérim Générale 82:181–285
Ramos-Porto M, Coelho PA, Souza ST (1987) Sinopse dos crustáceos decápodos brasileiros (Famílias Penaeidae, Solenoceridae, Sicyoniidae). Trabhs Oceanogr Univ Fed Per 20:219–234
Ramos-Porto M, Coelho PA, Souza ST (1989) Sinopse dos crustáceos decápodos brasileiros (Famílias Penaeidae, Solenoceridae, Sicyoniidae). Trabhs Oceanogr Univ Fed Per 20:219–234
Ramos-Porto M, Coelho PA, Young PS (1998) Malacostraca-Eucarida. Caridea (Alpheoidea excluded). In: Catalogue of Crustacea of Brazil. Museu Nacional, Rio de Janeiro, pp 325–350
Ramos-Porto M, Viana GFS, Silva KCA, Cintra IHA, Coelho P (2000) *Stereomastis sculpta* (Smith, 1880) (Decapoda: Polycheloidea: Polychelidae) in Brazilian waters. Nauplius 8:249–251
Ramos-Porto M, Torres MFA, Viana GFS, Santos MCF, Acioli FD, Cabral E (2002) Registers of two species of Crustacea Decapoda Brachyura in Brazilian waters. Nauplius 8(1):169–171
Ramos-Porto M, Muniz APM, Silva KCA, Cintra IHA, Viana GFS (2003) Camarões da Subordem Pleocyemata Burkenroad, 1963 capturados durante pescarias experimentais para o Programa Revizee/Norte (Crustacea, Decapoda) 3(1):77–106
Rathbun MJ (1894) Notes on the crabs of the family Inachidæ in the United States National Museum. Proc U S Natl Mus 17:43–75
Rathbun MJ (1898) The Brachyura of the biological expedition to the Florida Keys and the Bahamas in 1893. Bull Lab Nat Hist State Univ Iowa 4(3):250–294
Rathbun MJ (1901) Investigations of the Aquatic Resources and Fisheries of Porto Rico by the United States Fish Commission Steamer Fish Hawk in 1899. The Brachyura and Macrura of Porto Rico. Bull U S Fish Comm 20:1–127
Rathbun MJ (1904) Decapod crustaceans of the northwest coast of North America. Harriman Alaska Expedition (Harriman Alaska Ser) 10:1–190
Rathbun MJ (1906) The Brachyura and Macrura of the Hawaiian Islands. Bull U S Fish Comm 23:827–930
Rathbun MJ (1918) The grapsid crabs of America. Bull USNM 9:1–461
Rathbun MJ (1925) The spider crabs of America. Bull USNM 129:1–613
Rathbun MJ (1930) The cancroid crabs of America of the families Euryalidae, Portunidae, Atelecyclidae, Cancridae, and Xanthidae. Bull USNM 152:1–609
Rathbun MJ (1937) The oxystomatous and allied crabs of America. Bull USNM 166:1–278
Rego ABB, Cardoso IA (2010) Taxonomy and distribution of shrimps genus *Heterocarpus* A. Milne-Edwards, 1881 collected on the Brazilian slope by the Program REVIZEE – central score. Nauplius 18(2):117–128
Richard J (1905) Campagne scientifique du yacht "Princesse Alice" en 1904: observations sur la faune bathypélagique. Bull Mus Oceanogr Monaco 41:1–30
Richardson LR, Yaldwyn JC (1958) A guide to the natant decapod Crustacea (shrimps and prawns) of New Zealand. Tuatara 7:17–41
Rieger PJ (1998) Malacostraca, Eucarida. Paguroidea. In: Young PS (ed) Catalogue of Crustacea of Brazil. Museu Nacional, Rio de Janeiro, pp 413–429

Riggio G (1905) Contributo alla Carcinologia del Mediterraneo. I. Nota sopra alquanti Crostacei del mare di Messina. Natural Sicil 17:93-96

Risso A (1816) Histoire naturelle des Crustacés des environs de Nice. Librairie Grecque-latine-Allemande, Paris

Risso A (1827) Histoire naturelle des principales productions de l'Europe Méridionale et particulièrement de celles des environs de Nice et des Alpes Maritimes, vol 5. Levrault, Paris

Risso A (1844) Crustacés. In: Nouveau guide du voyageur dans Nice et notices sur l'histoire civile et naturelle de cette ville (2d ed), Nice, pp 93–99

Roberts TW, Pequegnat WE (1970) Deep-water decapod shrimps of the family Penaeidae. Oceanogr Stud Texas A&M Univ 1:21–57

Rodrigues TGA, Cardoso IA (2018) A new species of *Eupasiphae* Wood-Mason & Alcock, 1893 (Crustacea, Decapoda, Pasiphaeidae) from the southwestern Atlantic. Zootaxa 4444(2):189–194

Rodrigues TGA, Cardoso IA (2019) First record of four *Pasiphaea* Savigny, 1816 (Crustacea, Decapoda, Pasiphaeidae) species from the southwestern Atlantic. Zootaxa 4560(2):321–330

Rodrigues C, Serejo C (2010) *Munidopsis transtridens* Pequegnat and Pequegnat, 1971. In: Lavrado HP, Brasil MS (eds) Biodiversidade da região oceânica profunda da Bacia de Campos: Megafauna e ictiofauna demersal. SAG Serv, Rio de Janeiro, pp 216–217

Rodrigues C, Young PS (2003) Duas espécies do gênero *Acanthocarpus* Stimpson, 1871 (Crustacea, Calappidae) para a costa do Brasil. Bol Mus Nac Rio de Janeiro 505:1–8

Rodrigues IB, Cardoso IA, Serejo CS (2017) Catalogue and illustrated key of *Achelous* De Haan, 1833 and *Portunus* Weber, 1795 (Brachyura: Portunidae: Portuninae) species occurring in Brazilian waters. Nauplius 25:1–12

Rodrigues TGA, Alves-Júnior FA, Cardoso IA (2018) A new species of *Pasiphaea* Savigny, 1816 (Crustacea, Decapoda, Pasiphaeidae) from the southwestern Atlantic. Zootaxa 4418(5):93–498

Sakai K (1999) Synopsis of the family Callianassidae, with keys to subfamilies, genera and species, and the description of new taxa (Crustacea: Decapoda: Thalassinidea). Zool Verhand 326:1–152

Sakai K (2005) Callianassoidea of the world (Decapoda, Thalassinidea). Crustaceana Monographs 4:1–285

Sakai K (2011) Axioidea of the world and a reconsideration of the Callianassoidea (Decapoda, Thalassinidea, Callianassida). Crust Monogr 13:1–616

Sankarankutty C, Ferreira AC, Oliveira JEL, Cunha KMF (2001) Occurrence of *Chaceon fenneri* (Manning & Holthuis) (Crustacea, Brachyura, Geryonidae) in the northeast of Brazil. Rev Bras Zool 18:649–652

Santos CF, Michel J, Neves M, Janeiro J, Andrade F, Orbach M (2013) Marine spatial planning and oil spill risk analysis: finding common grounds. Mar Pollut Bull 74:7–381

Scelzo MA, Valentini A (1974) Presencia de *Geryon quinquedens* Smith en aguas del Oceano Atlantico sudoccidental (Decapoda, Brachyura, Geryonidae). Physis, Buenos Aires (A) 33(87):557–567

Schmitt WL (1931) Some carcinological results of the deeper water trawling of the Anton Dohrn, including description of two new species of Crustacea. Carnegie Inst Wash Year Books 30:389–394

Schmitt WL (1935) Mud shrimps of the Atlantic coast of North America. Smithson Misc Contrib 93(2):1–21

Senna A (1903) Nota sui Crostacei. Le esplorazioni abissali nel Mediterraneo del R. Piroscafo "Washigton" nel 1881. II. Boll Soc Entomol Ital 34:235–367

Serejo CS, Cardoso IA (2010) Crustacea. In: Lavrado HP, Brasil ACS (eds) Biodiversidade da região oceânica profunda da Bacia de Campos: Megafauna e Ictiofauna demersal. SAG Serv, Rio de Janeiro, pp 137–231

Serejo CS, Siqueira SGL (2018) Catalogue of the order Amphipoda from Brazil (Crustacea: Peracarida): suborders Amphilochidea, Senticaudata and order Ingolfiellida. Zootaxa 4431(1):1–139

Serejo CS, Young PS, Cardoso IC, Tavares C, Rodrigues C, Almeida TC (2007) Abundância, diversidade e zonação dos crustáceos no talude da costa central do Brasil (11°-22° S) coletados pelo

Programa REVIZEE/Score Central: prospecção pesqueira. In: Costa PAS, Olavo G, Martins AS (eds) Biodiversidade da fauna marinha profunda na costa central brasileira. Série Livros, vol 24. Museu Nacional, Rio de Janeiro, pp 133–162
Silva KCA, Cintra IHA, Ramos-Porto M, Viana GFS (2003) Lagostas capturadas durante pescarias experimentais para o programa REVIZEE/Norte (Crustacea, Nephropoidea, Eryonoidea, Palinuroidea). Bol Tec Cient CEPNOR 3(1):21–35
Sivertsen E, Holthuis LB (1956) Crustacea Decapoda (The Penaeidae and Stenopodidea excepted). Rep Sci Res "Michael Sars" North Atlantic Deep-sea Exped 1910 5(12):1–32
Smith SI (1880) Notice of a new species of the '*Willemoesia* group of Crustacea' (Recent Eryontidae). Proc U S Natl Mus 2:345–353
Smith SI (1881) Preliminary notice of the Crustacea dredged in 64 to 325 fathoms, off the South Coast of New England, by the United States Fish Commission in 1880. Proc U S Natl Mus 3:413–452
Smith SI (1882) Reports on the results of dredging under the supervisión of Alexander Agassiz, on the east coast of the United States during the summer of 1880, by the U.S. Coast Survey Steamer "Blake", Commander J.R. Bartlett, U.S.N., commanding. Bull Mus Comp Zool 10:1–108
Smith SI (1883) Preliminary report on the Brachyura and Anomura dredged in deep water off the south coast of New England by the United States Fish Commission in 1880, 1881, and 1882. Proc U S Natl Mus 6:1–57
Smith SI (1884) Report on the decapod Crustacea of the Albatross dredgings off the East coast of the United States in 1883. Rep U S Fish Comm 10:34–426
Smith SI (1885) On some new or little known decapod Crustacea, from recent Fish Commission dredgings off the east coast of the United States. Proc U S Natl Mus 7:493–511
Smith SI (1886a) The abyssal decapod Crustacea of the 'Albatross' dredgings in the North Atlantic. Ann Mag Nat Hist 17:187–198
Smith SI (1886b) Report on the decapod Crustacea of the Albatross dredgings off the East coast of the United States during the summer and autumn of 1884. Rep U S Fish Comm 13:605–705
Smith SI (1887) Report on the Decapod Crustacea of the Albatross Dredgings off the East coast of the United States, during the Summer and Autumn of 1884. Rep U S Fish Comm 13:605–706
Spence-Bate C (1878a) On the *Willemoesia* Group of Crustacea. Ann Mag Nat Hist 2(5):273–283
Spence-Bate C (1878b) LV. On the *Willemoesia* group of Crustacea. Ann Mag Nat Hist 2(5):484–487
Spence-Bate C (1878c) On the *Willemoesia* group of Crustacea. Rep Br Assoc Adv Sci London 48:561–564
Spence-Bate C (1881) On the Penaeidea. Ann Mag Nat Hist 8:169–196
Spence-Bate C (1888) Report on the Crustacea Macrura collected by the Challenger during the years 1873–76. Report on the Scientific Results of the Voyage of HMS "Challenger"during the years 1873–76 24:1–942
Springer S, Bullis HR (1956) Collections by the Oregon in the Gulf of Mexico. U S Fish Wildl Serv Spec Sci Rep Fish 196:1–134
Squires HJ (1965) Decapod Crustaceans of Newfoundland, Labrador and the Canadian Eastern Arctic. Fisheries Research Board of Canada, Manuscript Report Series, 810:1–212
Squires HJ (1990) Decapoda Crustacea of the Atlantic coast of Canada. Department of Fisheries and Oceans, Ottawa, p 530
Stebbing TRR (1893) A History of Crustacea. Recent Malacostraca, London Kegan Paul, Trench, p 466
Stebbing TRR (1902) South African Crustacea: part II. Maritime Investment South Africa Department Agriculture of Cape Town, pp 1–92
Stebbing TRR (1905) South African Crustacea. Part III. Mar Invest S Afr 4:21–123
Stebbing TRR (1910) General catalogue of South African Crustacea (Part V. of S.A. Crustacea, for the Marine Investigations in South Africa). Ann S Afr Mus 6:281–593
Stebbing TRR (1917) IX. South African Crustacea. Ann S Afr Mus 17(1):23–46
Stimpson W (1860) Notes on North American Crustacea, in the Museum of the Smithsonian Institution. Annals of the Lyceum of Natural History of New York 7:177–246

Stimpson W (1871) Preliminary report on the crustacea dredged in the Gulf Stream in the Straits of Florida by L.F. de Pourtales, Assist. U. S. Coast Survey. Part I. Brachyura. Bull Mus Comp Zool 2:109–160

Studer T (1883) Verzeichnis der während der Crustaceen, welche während der Reise S.M.S. "Gazelle" an der Westküste von Afrika, Ascension un dem Cap der Guten Hoffnung gesammelten Crustaceen. Abh K Akad Wiss Berlin 2:1–32

Sumida PYG (2009) Mar Profundo. In: Pereira R, Soares-Gomes A (eds) Biologia Marinha, 2nd edn. Interciência, Rio de Janeiro, pp 383–398

Sumida PYG, Pires-Vanin AMS (1997) Benthic associations of the shelfbreak and upper slope off Ubatuba-SP, South-Eastern Brazil. Est Coast Shelf Sci 44:779–784

Sund O (1913) The glass shrimps (*Pasiphaea*) in northen waters. Bergens Museums Aarbok 6:1–7

Sund O (1920a) Peneides and stenopides. Rep Sci Res "Michael Sars" North Atlant Deep-sea Exped 1910 3:1–36

Sund O (1920b) The "Challenger" Eryonidea (Crustacea). Ann Mag Nat Hist 6(9):220–226

Takeda M (1983) Crustaceans. In: Takeda M, Okutani T (eds) Crustaceans and Mollusks Trawled off Suriname and French Guiana. Japan Marine Fishery Ressource Research Center, Tokyo, pp 19–180

Takeda M, Okutani T (1983) Crustaceans and mollusks trawled off Surinam and French Guiana. Japan Fisheries Resources Research Center, Tokyo

Tavares MS (1991) The cruise of the "Marion Dufresne" off the Brazilian coast: account of the scientific results and list of stations. Zoosystema 21:597–605

Tavares M (1996) Sur la validité de *Bathyplax typhlus* oculiferus Miers, 1886. The Decapoda Brachyura of the Siboga Expedition. Crustaceana 69(3):413–423

Tavares MS (1999) New species and new records of deep-water caridean shrimps from the South Atlantic Ocean (Crustacea, Decapoda). Zoosystema 21:671–677

Tavares CR (2003) First record of *Bathynectes longispina* Stimpson, 1871 (Crustacea, Brachyura, Portunidae) from southwestern Atlantic. Bol Mus Nac Rio de Janeiro 506:1–6

Tavares CR (2009) New record of *Benthesicymus carinatus* Smith, 1884 (Decapoda: Benthesicymidae), with some notes on its morphology deduced from SEM observations. Cah Biol Mar 50:199–204

Tavares MS, Albuquerque EF (1990) *Paralomis formosa* Henderson, 1888 from off southeastern Brazilian coast (Crustacea, Decapoda, Lithodidae). Bol Mus Nac Rio de Janeiro 340:1–4

Tavares MS, Campinho P (1998) Three new records of deep-sea squat lobsters of the genus *Munidopsis* Witheaves from the southwestern Atlantic Ocean (Decapoda: Galatheidae). Crustacean Research 27:88–100

Tavares CR, Cardoso IA (2006) Deep-sea Pasiphaeidae (Crustacea: Decapoda: Caridea) from off the Brazilian central coast between 11° and 22°S, colleted by the Revizee program. Zootaxa 1174:27–39

Tavares MS, Lemaitre R (2014) New morphological and distributional information on Homolodromiidae and Homolidae (Decapoda: Brachyura) from the Americas, with description of a new species and comments on Western Pacific species. J Crust Biol 34(4):504–524

Tavares MS, Pinheiro AP (2011) A new species of *Chaceon* Manning & Holthuis, 1989, from the southwestern Atlantic, with a key to the western Atlantic species (Crustacea, Decapoda, Geryonidae). Zootaxa 3086:57–68

Tavares MS, Santana W (2018) Refining the genus *Rochinia* A. Milne-Edwards, 1875: reinstatement of *Scyramathia* A. Milne-Edwards, 1880 and *Anamathia* Smith, 1885, and a new genus for *Amathia crassa* A. Milne-Edwards, 1879, with notes on its ontogeny (Crustacea: Brachyura: Epialtidae). Zootaxa 4418:201–227

Tavares CR, Serejo CS (2007) Taxonomy of Aristeidae (Dendrobranchiata: Penaeoidea) from the central coast of Brazil, collected by the Revizee program, between 19° and 22° S. Zootaxa 1585:1–44

Tavares CR, Young PS (2002) Nephropidae (Crustacea, Decapoda) collected by the Revizee Score-Central program from off Bahia to Rio de Janeiro states, Brazil. Arq Mus Nac Rio de Janeiro 60(2):79–88

Tavares MS, Melo-Filho GAS, Melo GAS (2008) The deep-sea squat lobster *Munidopsis transtridens* Pequegnat and Pequegnat, 1971 (Decapoda: Anomura: Galatheidae) from the Southwestern Atlantic. Nauplius 16(2):95–99

Tavares MS, Santana W, Pettan R (2015) *Rochinia confusa*, a junior synonym of *R. umbonata* (Crustacea: Brachyura: Epialtidae) as revealed by ontogenetic changes. J Mar Biol Assoc UK 96(5):1065–1071

Tchesunov AV (1984) A new and rare mesopelagic species of the genus *Pasiphaea* (Crustacea, Decapoda) from the tropic and subtropic Atlantic. Zool Zh 63:993–1003

Thompson JR (1966) The caridean superfamily Bresiloidea (Decapoda Natantia), a revision and a discussion of its validity and affinities. Crustaceana 11:129–140

Tirmizi NM (1969) *Eupasiphae gilesii* (Wood-Mason, 1892) from the Nothern Arabian Sea (Decapoda, Caridea). Crustaceana 16(2):213–218

Torres MFA, Santos MCF, Viana GFS, Ramos-Porto M, Acioli FD, Cabral E (2002) Registro de *Myropsis quinquespinosa* Stimpson, 1871 (Crustacea, Decapoda, Leucosiidae) no nordeste brasileiro. Bol Tec Cient CEPENE 10(1):107–114

Torres MFA, Ramos-Porto M, Viana GFS, Coelho PA, Botter-Carvalho ML, Santos MCF (2006) Redescrição e ocorrência de *Bathynectes longispina* Stimpson, 1871 em águas do Nordeste Brasileiro (Decapoda; Portunidae). Bol Tec Cient CEPNOR 6(1):133–140

Tudge CC, Poore GCB, Lemaitre R (2000) Preliminary phylogenetic analysis of generic relationships within the Callianassidae and Ctenochelidae (Decapoda: Thalassinidea: Callianassoidea). J Crust Biol 20:129–149

Türkay M (1975) Zur Kenntnis der Gattung *Euchirograpsus* mit Bemerkungen zur *Brachygrapsus* and *Litocheira* (Crustacea:Decapoda). Senckenb Biol 52(1/3):103–132

Vázquez-Bader AR, Gracia A (2013) Crangonidae and Glyphocrangonidae (Decapoda; Caridea) of the southern Gulf of Mexico. Zootaxa 3669:367–383

Vázquez-Bader AR, Gracia A (2016) Diversity and distribution of Chirostyloidea and Galatheoidea (Decapoda, Anomura) in the Southern Gulf of Mexico. ZooKeys:1–30

Vereshchaka AL (1990) Pelagic decapods from the seamounts of Nazca and Sala-y-Gómes ridges. Trans P.P.Shirshov Inst Oceanol Moscow 124:129–155

Vereshchaka AL (1994) North Atlantic and Caribbean species of *Sergia* (Crustacea, Decapoda, Sergestidae) and their horizontal and vertical distribution. Steenstrupia 20:73–95

Vereshchaka AL (2000) Revision of the genus *Sergia* (Decapoda: Dendrobranchiata: Sergestidae): taxonomy and distribution. Galathea Rep 18:69–207

Vereshchaka AL (2009) Revision of the genus *Sergestes* (Decapoda: Dendrobranchiata: Sergestidae): taxonomy and distribution. Galathea Rep 22:7–104

Vereshchaka AL, Olesen J, Lunina AA (2014) Global diversity and phylogeny of pelagic shrimps of the former genera *Sergestes* and *Sergia* (Crustacea, Dendrobranchiata, Sergestidae), with definition of eight new genera. PLoS One 9:e112057

Viana GFS, Ramos-Porto M, Torres MFA, Santos MCF (2007) Distribution of *Heterocarpus laevigatus* Bate, 1888 and *Plesionika escatilis* (Stimpson, 1860) along the Atlantic coast of South America (Crustacea: Caridea: Pandalidae). Zootaxa 1577:33–39

Wasmer RA (1989) Supplementary description of *Funchalia taaningi* Burkenroad, 1940 (Crustacea: Decapoda: Penaeidae) from the Central Pacific Ocean and a new record of *F. villosa* (Bouvier, 1905) from the Eastern Indian Ocean. J Nat Hist 23:475–485

Wasmer RA (2005) A remarkable new species of the pelagic shrimp genus *Parapasiphae* Smith, 1884 (Crustacea: Decapoda: Pasiphaeaidae) with double eyes. Proc Biol Soc Wash 118(1):165–175

Webber WR, Booth JD (1988) *Projasus parkeri* (Stebbing, 1902) (Crustacea, Decapoda, Palinuridae) in New Zealand and description of a *Projasus* puerulus from Australia. Natl Mus N Z Rec 3(8):81–92

Wenner EL (1979) Distribution and reproduction of nematocarcinid shrimp (Decapoda: Caridea) from the northwestern North Atlantic. Bull Mar Sci 39(3):380–393

Wicksten MK (2002) Midwater decapods of the Northeastern Pacific. In: M.E. Hendrickx (ed), Contributions to the Study of East Pacific Crustaceans. Instituto de Ciencias del Mar y Limnología, UNAM, Mexico, 1:127–144
Wicksten MK, Méndez MG (1983) *Bathypalaemonella delsolari*, a new species of shrimp from Peru (Decapoda, Caridea, Campylonotidae). Crustaceana 45:225–231
Wicksten MK, Packard JM (2005) A qualitative zoogeographic analysis of decapod crustaceans of the continental slopes and abyssal plain of the Gulf of Mexico. Deep-Sea Res 52:1745–1765
Williams AB (1965) Marine decapod crustaceans of the Carolinas. Fish B-NOAA 65(1):1–298
Williams AB (1974) *Allactaea lithostrota*, a new genus and species of crab (Decapoda: Xanthidae) from North Carolina, U.S.A. Proc Biol Soc Wash 87(3):19–26
Williams AB (1984) Shrimps, lobsters and crabs of the Atlantic coast of the eastern United States, Maine to Florida. Smithsonian Institution Press, Washington, DC
Williams AB (1988) Notes on decapod and euphausiid crustaceans, continental margin, western Atlantic, Georges Bank to Western Florida, USA. Fish B-NOAA 86:67–76
Williams AB, Baba K (1989) New squat lobsters (Galatheidae) from the Pacific Ocean: Mariana Back Arc Basin, East Pacific Rise, and Cascadia Basin. Fish B-NOAA 87:899–910
Wood-Mason J (1892) Illustrations of the Zoology of the Royal Indian Marine Surveying Steamer Investigator under the command of Commander A. Carpenter RN, DSO and Commander RF Hoskyn. RN, Crustacea, Calcutta
Wood-Mason J, Alcock A (1891) Natural history notes from H M Indian marine survey steamer 'Investigator', Commander RF Hoskyn, RN, commanding. Series II, No. 1. On the results of deep-sea dredging during the season 1890-1891. Ann Mag Nat Hist 8(6):268–286
Wood-Mason J, Alcock A (1892) Natural history notes from H.M. Indian marine survey steamer "Investigator", Commander R.F. Hoskyn, R.N., commanding. Series II, No. 1. On the results of deep-sea dredging during the season 1890-1891. Ann Mag Nat Hist 9(6):265–275
Wood-Mason J, Alcock A (1893) Natural history notes from H.M. Indian Marine Survey Steamer "Investigator", Commander R. F. Hoskyn, R. N., commanding. No 22, On the results of the deep-sea dredging during the season 1890-91. Ann Mag Nat Hist 11(6):161–173
Wood-Mason J, Alcock A (1894) Illustrations of the Zoology of the Royal Indian Marine Surveying Steamer "Investigator", under the command of Commander A. Carpenter, RN, DSO, Crustacea Part. II. Pls 6–8. Government Printer, Calcutta
Yaldwyn JC (1957) Deep-water Crustacea of the genus *Sergestes* (Decapoda, Natantia) from Cook Strait, New Zealand. Zoology Publs Vict Univ Coll 22:1–27
Yaldwyn JC, Webber WR (2011) Annotated checklist of New Zealand Decapoda (Arthropoda: Crustacea). Tuhinga 22:171–272
Yokoya Y (1922) A species collected from Suruga Bay. Suissan Gakkai Ho 3:302–303
Yokoya Y (1933) On the distribution of decapod crustaceans inhabiting the continental shelf around Japan, chiefly based upon the materials collected by S. S. Sôyô–Maru, during the year 1923–1930. Journal of the College of Agriculture 12(1):1–226
Yokoya Y, Shibata K (1965) On some shrimps from the Philippine Sea. Bull Fac Fish, Nagasaki Univ 18:1–6
Young CG (1900) The stalk-eyed Crustacea from British Guiana, West Indies and Bermuda. Watkins JM, London
Zarenkov NA (1968) Crustacean Decapoda collected by Soviet Antarctic expeditions in the Antarctic and Antiboreal regions. In: Bykhovskii BE (ed) Biological Reports of the Soviet Antarctic Expedition (1955–1958). Vol. 4. Leningrad, p 153–201
Zariquiey-Álvarez R (1968) Crustáceos Decápodos Ibéricos. Investig Pesq 32:1–510

Chapter 22
An Annotated Checklist and Bibliography of Deep-Water Isopods and Decapod Crustaceans from Chile, Including the Submarine Ridge Salas y Gomez and Nazca Plates

M. A. Retamal, G. Guzmán, and P. De los Ríos-Escalante

Abstract A list of deep-water species (> 200 m depth) of isopods and decapod crustaceans from off Chile is presented, including the areas of South America, the Antarctic, and offshore islands. Also included are the species that were collected in the Nazca and Sala y Gómez seamounts areas during the ex-Soviet Union Expeditions in the 1970s (1973–1978), although not all species found during these expeditions have not yet been recorded for Chilean waters but in international waters. Isopods were represented by nine species only, four in the Perú-Chile Trench, three off Chiloé, and two in the Magellan Strait. In the case of the decapods, 143 species have been reported in Chile, mostly within the suborder Pleocyemata (134 spp.). These 143 species are grouped in 42 families, 5 of these in the Dendrobranchiata.

Keywords Crustacea · Decapoda · Isopoda · Deep-sea · Pacific Ocean · Antarctic Ocean · Chile

---

Marco A. Retamal: Deceased, August 2019.

M. A. Retamal (Deceased)
Facultad de Ciencias Naturales y Oceanográficas, Universidad of Concepción, Concepción, Chile

G. Guzmán
Universidad Arturo Prat, Iquique, Chile

P. De los Ríos-Escalante (✉)
Facultad de Recursos Naturales, Departamento de Ciencias Biológicas y Químicas, Universidad Católica de Temuco, Temuco, Chile

Núcleo de Estudios Ambientales, UC, Temuco, Chile
e-mail: prios@uct.cl

M. E. Hendrickx (ed.), *Deep-Sea Pycnogonids and Crustaceans of the Americas*, https://doi.org/10.1007/978-3-030-58410-8_22

22.1 Introduction

With over 12,000 described species (Appletans et al. 2012; Poore and Bruce 2012), isopods and decapods crustaceans it is the of the most abundant group of Crustacea. In Chilean waters, this group is represented by about 475 species (Retamal and Moyano 2010; Retamal and Ferrada 2016), thus representing about 4% of known species of the world. Chile is a three continents country, and its coasts are under the influence of subantarctic, insular, and continental waters. The continental coast, measured along a straight line, is about 4000 km long. The Chilean maritime territory includes several insular territories, which are the Juan Fernández Archipelago, Desventuradas Islands with two main islands (San Felix and San Ambrosio islands), the isolated territories of Easter Island, and the tiny Sala y Gómez Island. Remarkably, Easter and Sala y Gomez islands are inhabited by both tropical species linked to the Indo-Pacific Ocean and by species with affinities to the Antarctic territory fauna, between 56°W and 93°W. In these islands, the isopods and crustaceans fauna is very scarce. The northern Chilean continental waters have an average temperature of 20 °C, while the southern waters can reach extreme temperatures of 3–4 ° C. Consequently, we find a low variety of isopod and decapod crustaceans in the area, from typical tropical species to subantarctic species.

The list included in this contribution is mostly based on previously published literature and on some unpublished information. Deep-water benthic isopods have been studied first by Beddard (1886), Menzies (1962), and Brandt (1998) in Chiloé islands. Other contributions deal with northern Chile (Brusca and Ninos 1978; Brandt and Wägele 1989; Aydogan et al. 2000) and the Magellan Strait (Park and Wagele 1965; Lorenti and Mariani 1997). Benthic and pelagic deep-water Dendrobranchiata (shrimps) of Chile have been studied mostly by G. Guzmán who produced several contributions on this group of species (e.g., Guzmán and Wicksten 1998, 2000; Guzmán 2004, 2008). Deep-water Pleocyemata (lobsters, hermit crabs, squat lobsters, and true crabs) of Chilean waters has been studied by many authors since the first British expeditions (R/V "Beagle" in 1831–1836; the H.M.S. "Challenger" in 1873–1876) and North American expeditions (U.S.S. "Albatross" in 1888, and U.S.S. "Blake" in 1885–1889) visited the area. More recently, the Swedish Expedition "Lund" (1948–1949) and the Royal Society of Chile Expedition (1958–1959) also made significant contributions to the study of this fauna. In addition, several isolated pioneer contributions were proposed by Molina (1782), Poeppig (1835), d'Orbigny (1842), Philippi (1887), Porter (1901), and Rathbun (1907). Later in the XX century, starting in the 1950s, a long list of contributions was the result of a large group of young Chilean scientists with strong interest for decapod crustaceans (e.g., Guzmán 2003, 2004, 2008).

Between 1950 and 1970, Soviet expeditions have explored the area, collecting and identifying a significant amount of deep-water species of decapods on the Nazca and Salas y Gómez Plates, referred to as "the oceans ridge," near the Juan Fernandez Archipelago. Several species were also collected slightly further than the Chilean EEZ boundary (Parin et al. 1997). In these contributions (i.e., Vereshchaka

1990; Parin et al. 1997; Zhadan 1997), some species collected near the Chilean offshore islands (Pascua and Salas y Gómez) were included because they occur within the limits of Chile Exclusive Economic Zone (EEZ). Other species were captured much further from Chilean territories (> 350 nm offshore), in international waters, and were also included in this contribution as they might also occur closer to the continent. Species of this group have been collected mainly on the submerged ridge Sala y Gómez and Nazca during the 15° cruise of the R/V "Professor Metsayev," the 5° cruise of the R/V "Ikhtiandr" (VNIRO), and the 18° cruise of the R/V "Professor Stockman". They were found in Chilean waters, near the continent but also near the Juan Fernandez Archipelago, and several species occur slightly further than the EEZ boundary (see Parin et al. 1997). Isopod records were obtained by literature reviewing (Beddard 1886; Menzies 1962; Park and Wagele 1965; Brusca and Ninos 1978; Brandt and Wägele 1989; Lorenti and Mariani 1997; Brandt 1998; Aydogan et al. 2000). The list included in this contribution is mostly based on previously published literature and on some unpublished information.

22.2 Material and Methods

Literature dealing with isopods and decapod crustaceans of Chile was reviewed, and a list of species occurring deeper than 200 m was established. Distribution in Chilean waters is provided for each species, either as a geographic range or as more specific localities (latitude and longitude provided when available), together with the depth range recorded in the area. Species originally collected and/or reported by the Soviet expeditions are indicated (*) and are those reported by Burukovsky (1990), Vereshchaka (1990), Parin et al. (1997), Zhadan (1997), and Zarenkov (1990).

Species have been classified according to De Grave et al. (2009), Poore, and Bruce (2012), and recent changes included in WoRMS (WoRMS editorial board, 2020). Within each family and genus, genera and species are classified in alphabetic order. In this contribution we have focused more on records of bentho-demersal or bentho-pelagic species than on pelagic species. For further information on pelagic species, see the contribution in this volume by Guzmán and Olguin (Chap. 12).

22.3 Results

22.3.1 *Systematic Section*

Superorder Peracarida.
Order Isopoda.
Suborder Asellota.

Superfamily Janiroidea.

Family Ischnomesidae Hansen, 1916.

1. *Ischnomesus bacilloides* (Beddard, 1886) [= *Ischnosoma bacilloides* Beddard, 1886, former name basionym]. 42°43′S, 82°11′W, 2610 m (Beddard 1886).

Family Joeropsididae Nordenstam, 1933.

2. *Joeropsis bidens* Menzies, 1962. Gulf of Ancud, 42°46′S, 72°59′W, 250–300 m and intertidal zone (Menzies 1962).

Family Macrostylidae Hansen, 1916.

3. *Macrostylis dellacrocei* Aydogan, Wägele and Park, 2000. 23°15′S, 71°21′W, 7800 m (Aydogan et al. 2000).

Family Munnopsidae Lilljeborg, 1864.

4. *Munneurycope hadalis* Aydogan, Wägele and Park, 2000. 23°15′S, 71°21′W, 7800 m (Aydogan et al. 2000).

Suborder Cymothoida.

Superfamily Cymothooida.

Family Cirolanidae Dana, 1852.

5. *Cirolana diminuta* Menzies, 1962. Peru-Chile (?), 40–2000, dominant at depth > 700 m (Brusca and Ninos 1978).
6. *Natatolana pastorei* (Giambiagi, 1925). Magallanes Strait (no specific locality), 120–515 m. (Lorenti and Mariani 1997).

Suborder Sphaeromatidea.

Superfamily Sphaeromatoidea.

Family Sphaeromatidae Latreille, 1825.

7. *Caecocassidias patagonica* Kussakin, 1967. 43°40′S, 59°35′W, 400–500 m (Brandt 1998).
8. *Exospharoma gigas* (Leach, 1818). Peru-Chile (no specific locality), 0–270 m (Brandt and Wägele 1989).

Suborder Valvifera.

Family Rectarcturidae Poore, 2001.

9. *Rectarcturus tuberculatus* Schultz, 1981. Magallanes Strait (no specific locality), 35–3500 m (Lorenti and Mariani 1997; Park and Wagele 1965).

Superorder Eucarida.

Order Decapoda.

Suborder Dendrobranchiata.

Family Aristeidae Wood-Mason, 1891.

1. *Aristaeomorpha foliacea* (Risso, 1827) (*). Close to seamount Dome Salas y Gómez Ridge (25°04′S, 97°26′W), 200–800 m depth (Burukovsky 1990).

Family Benthesicymidae.

2. *Benthoecetes tanneri* (Faxon, 1893). From 18°45′S to 21°19′S, 505–520 m (Guzmán and Wicksten 2000).
3. *Dalicaris altus* (Spence Bate, 1881). 18°25′S, 485 m, and 20°47′S, 950 m (Guzmán and Wicksten 2000).
4. *Maorrancaris investigatoris* (Alcock and Anderson, 1899) (*). Seamounts Amber, Pearl, and Ichthyologist, Salas y Gomez Ridge (Burukovsly 1990).

Family Solenoceridae Wood-Mason, 1891.

5. *Hadropenaeus lucasii* Bate, 1881 (*). Seamounts Big, Yarala, Pearl, MayDay, Dome, and Cliff of the Salas y Gomez Ridge (Burukovsky 1990).
6. *Haliporoides diomedeae* (Faxon, 1893) Arica (18°25′S, 71°22′W) to Puerto Bueno (50°59′S, 74°13′W), 240–1886 m (Retamal 1994b).
7. *Hymenopenaeus halli* Bruce, 1966 (*). Off Seamount Ichthyologist, Salas y Gómez Ridge (25°07′S, 99°35′W) (Burukovsky 1990).

Family Sicyoniidae Ortmann, 1898.

8. *Sicyonia nasica* Burukovsky, 1990 (*). Off seamounts Dome, New, Dorofeeva, and Yarala, Salas y Gomez Ridge (Burukovsky 1990).

Family Penaeidae Rafinesque, 1815.

9. *Metapenaeopsis stokmani* Burukovsky, 1990 (*). Seamounts Big and Dome, Salas y Gomez Ridge (Burukovsky).

Suborder Pleocyemata.

Infraorder Stenopodoidea.

Family Spongicolidae Schram, 1986.

10. *Spongicoloides* aff. *galapagensis* Goy, 1980. Off Cobquecura (36°00′S), 1000 m (Guzmán and Sellanes 2011).
11. *Spongicola parvispina* Zarenkov, 1990 (*). Seamount Ichthyologist, Salas y Gomez Ridge (Zarenkov 1990).

Infraorder Caridea.

Family Nematocarcinidae Smith, 1884.

12. *Nematocarcinus lanceopes* Bate, 1888. Off Chiloé (42°35′S, 74°48′W), 276–597 m (Guzmán and Quiroga 2005).
13. *Nematocarcinus longirostris* Bate, 1888. SW of Valparaíso (33°02′S, 71°06′W) to SW of Valdivia, (39°48′S, 73°14′W) 2516–2654 m (Retamal 1994b).
14. *Nematocarcinus aff. productus* Bate, 1888. Off Antofagasta (22°48′S, 70°23′S), 1775 m (Guzmán and Quiroga 2005).
15. *Nematocarcinus pseudocursor* Burukovsky, 1990 (*). In seamount "Soldier" of Nazca Ridge and Seamounts Ichthyologist, Pearl, and Amber og Salas y Gomez Ridge (Burukovsky 1990).
16. *Nematocarcinus undulatipes* Bate, 1888 (*). Seamount South Tropic, Nazca Ridge, and Seamount Ichthyologist, Salas y Gomez Ridge (Burukovsky 1990).

Family Processidae Ortmann, 1896.

17. *Processa pygmaea* Burukovsky, 1990 (*). Seamounts Dome and Ichthyologist, Salas y Gomez Ridge (Burukovsky 1990).

Family Campylonotidae Solaud, 1913.

18. *Campylonotus semistriatus* Bate, 1888. Off Iquique (20°12′S, 70°09′W) to Chernuca Harbor (52°45′S, 73°46′W), 276–1424 m (Torti and Boschi 1976).
19. *Campylonotus vagans* Bate, 1888. Valparaíso (33°02′S, 71°06′W) to Wollaston Islands, 18–320 m (Torti and Boschi 1976).

Family Alpheidae Rafinesque, 1815.

20. *Alpheus romensky* Burucovsky, 1990 (*). Seamount Pearl, Salas y Gomez Ridge (Burukovsky 1990).

Family Hippolytidae Bate, 1888.

21. *Chorismus antarcticus* (Pfeffer, 1887). Andvord Bay, Antarctic, 15–915 m (Holthuis 1952).

Family Thoridae Kingsley, 1879.

22. *Eualus dozei* (A. Milne-Edwards, 1891). From off Concepción (36°33′S) to Grey Island (55°6′S, 67°40′W), 15–270 (Retamal 1994a).
23. *Lebbeus antarcticus* Hale, 1941. From off Antofagasta to southern Chile, 1775–2598 m (Guzmán and Quiroga 2005).
24. *Lebbeus bidentatus* Zarenkov, 1976. Peru-Chile Trench, 1680 m (Wicksten and Méndez 1982). Not reported off Chile, but the Peru-Chile Trench connects to the deep water in the SEP.
25. *Lebbeus carinatus* Zarenkov, 1976. Off Chile, 1800 m (Zarenkov 1976).
26. *Lebbeus scripssi* Wicksten and Méndez, 1982. Off Arica (18°40′S, 70°36.00′W), 768–968 m (Wicksten and Méndez 1982).
27. *Leontocaris pacificus* Zarenkov, 1976. Off Chile, 600–700 m (Zarenkov 1976).
28. *Merhippolyte* aff. *americana* Holthuis, 1961. 42°35.35′S, 74°48.33′W, 507 m (Guzmán and Quiroga 2005).

Family Physetocarididae, Chace, 1940.

29. *Physetocaris microphthalma* Chace. Off Arica (18°25′S, 71°43′W), 513 m. (Guzmán 1999).

Family Palaemonidae Rafinesque, 1815.

30. *Bathymenes alcocki* (Kemp, 1922) (*). Seamounts Cliff and Dome, Salas y Gomez Ridge (Burukovsky 1990).

Family Pandalidae Haworth, 1825.

31. *Heterocarpus fenneri* (Crosnier, 1986) (*). Seamounts Dome and Pearl, Salas y Gomez Ridge (Burukovsky 1990).
    Remark. See Burukovsky (1990) for the taxonomic status of this species, referred as *Plesionika fenneri* Crosnier 1986.
32. *Heterocarpus laevigatus* Bate, 1888 (*). Seamounts Amber, Communard, Dome, May Day, Pillar, and Pearl, Salas y Gomez Ridge (Burukovsky 1990).
33. *Heterocarpus reedi* Bahamonde, 1954. Between Tal-Tal and Saavedra Harbor, 200–1000 m (Bahamonde 1955).
34. *Heterocarpus sibogae* de Man, 1917(*). Seamounts Amber, Cliff, Dome, May Day, New, Pearl, Pillar, and Yarala, Salas y Gomez Ridge (Burukovsky 1990).
35. *Pandalina nana* Burukovsky, 1990 (*). Seamounts Dome, New, and Amber, Salas y Gomez Ridge (Burukovsky 1990).
36. *Pandalus amplus* (Bate, 1888). Off Antofagasta, 560–2000 (Zabala and Bahamonde 1998). Off Concepción and off Chiloe (Guzmán & Quiroga 2005).
37. *Plesionika edwardsii* (Brandt, 1851) (*). Seamounts Cliff, Ichthyologist, Long, and New, Sala y Gómez Ridge (Burukovsky 1990).
38. *Plesionika ensis* (A. Milne-Edwards, 1881) (*). Seamounts Amber, Dome Pearl and Yarala, Sala y Gómez Ridge (Burukovsky 1990).
39. *Plesionika martia* (A. Milne-Edwards, 1881) (*). Seamounts Amber, Dome, and Ichthyologist, Sala y Gómez Ridge (Burukovsky 1990).
40. *Plesionika ocellus* (Bate, 1888) (*). Seamount Dome, Sala y Gómez Ridge (Burukovsky 1990).

41. *Plesionika santaecatalinae* (Wicksten & Mendez, 1983). 21°20′S, 70°26′W (Retamal and Soto 1995).
42. *Plesionika* aff. *williamsi* Forest, 1963 (*). Seamount Yarala, Sala y Gómez Ridge (Burukovsky 1990).
43. *Stylopandalus richardi* (Coutière, 1905). Off Caldera (27°04′S, 70°49′W), Valparaíso, Archipélago Juan Fernández (33°38′S, 78°50′W), to former Desventuradas Islands, and near Easter Island, 500–1000 m (Guzmán and Rivera 2002).

Family Stylodactylidae Spence Bate 1888.

44. *Stylodactylus pubescens* Burukovsky, 1990 (*). Seamount May Day, Sala y Gómez Ridge (Burukovsky 1990).

Family Crangonidae Haworth, 1825.

45. *Aegaeon rathbuni* de Man, 1918 (*). Seamount Dome, Sala y Gómez Ridge (Burukovsky 1990).
46. *Metacrangon bahamondei* Retamal and Gorny, 2003. (55°44′S, 66°14′W) 413–746 m (Retamal and Gorny 2003).
47. *Metacrangon procax* Faxon, 1853. Antofagasta to Chiloé, 1351 m (Guzmán and Quiroga 2005).
48. *Notocrangon antarcticus* (Pfeffer, 1887). Foster, Andvord and South Bays, Antarctic Peninsula, 250–760 m (Retamal 1976).
49. *Paracrangon areolata* Faxon, 1893. Between 35°31′S, 73°04′W and 35°43′S, 73°16′W; off Iquique; 580–800 m (Báez and Soto 1997).
50. *Parapontophilus gracilis* (Smith, 1882). 22°48′S, 70°42′W, 1775 m (Guzmán and Quiroga 2005).
51. *Parapontophilus junceus* (Spence Bate, 1888) (*). Seamounts Amber, Ichthyologist, and Pearl, Sala y Gómez Ridge (Burukovsky 1990).
52. *Parapontophilus occidentalis* (Faxon, 1893). 22°48’S, 70°42’W, 1775 m (Guzmán & Quiroga 2005).
53. *Philocheras nikiforovi* (Burukovsky, 1990) (*). Seamounts Dome and Big, Sala y Gómez Ridge (Burukovsky 1990).
54. *Sclerocrangon atrox* Faxon, 1893. Chilean Patagonia (53°23′S to 50°02′S, 76°20′W to 76°20′W), 693–680 m (Bahamonde 1981).

Family Glyphocrangonidae Haworth, 1825.

55. *Glyphocrangon alata* (Faxon, 1893). Between Iquique (20°12′S, 70°09′W) and Lebu, 600–1300 m (Retamal 1994b).
56. *Glyphocrangon loricata* (Faxon, 1895). Off Iquique (20°12′S, 70°09′W) and Algarrobo (33°22′S, 71°40′W), 434–757 m (Retamal 1994b).
57. *Glyphocrangon wagini* Burukovsky, 1990 (*). Seamounts Amber, Ichthyologist, and Pearl, Sala y Gómez Ridge (Burukovsky 1990).

Infraorder Astacidea.

Family Nephropidae Dana, 1852.

58. *Nephropsis occidentalis* Faxon, 1893. Off Iquique (Retamal and Moyano 2010).
59. *Thymops birsteini* (Zarenkov and Semenov, 1972). 56°49′S to 56°48′S, 145–1200 m (Bahamonde 1979).

Infraorder Achelata.

Family Palinuridae Latreille, 1802.

60. *Projasus bahamondei* George, 1976. Off Huasco to Constitución; Juan Fernández Archipelago (33°38′S, 78°50′W), San Félix (26°17′S, 80°05′W), and San Ambrosio (26°20′S, 79°53′W) Islands, possibly at O'Higgins Ridge; 175–550 m (George 1976; Retamal 1994b).
61. *Jasus frontalis* H. Milne Edwards, 1837. Near Robinson Crusoe, Santa Clara, and Alejandro Selkirk Islands (Juan Fernández Archipelago 33°38′S, 78°50′W), San Félix (26°17′S, 80°05′W), and San Ambrosio (26°20′S, 79°53′W) Islands (former Desventuradas Islands), 2–200 m (Retamal and Arana 2000).

Infraorder Axiidea.

Family Axiidae Huxley, 1879.

62. *Calocarides quinqueseriatus* (Rathbun, 1902). 32°08′S, 71°50′W, 320–400 m (Andrade and Báez 1977).

Infraorder Polychelida.

Family Polychelidae Wood-Mason, 1874.

63. *Pentacheles validus* A. Milne-Edwards, 1888. West of Valparaíso (33°42′S7S°17′W), 2500 m (Retamal 1994b).
64. *Stereomastis sculpta* (Smith, 1880). From off Los Vilos (31°56′S, 71°38′ to 71°47′W), 300–500 m, to off Quintero (32°42′S, 71°34′ to 71°45′W), 280–350 m (Báez and Andrade 1979).
65. *Stereomastis pacifica* (Faxon, 1893). Off Arica (18°40.5′S, 70°36.0′W), 768–968 m, and off Valparaíso (30°46′S, 81°31′W), 3000 m (Wicksten & Mendez, 1981).
66. *Stereomastis suhmi* (Bate, 1878). From Coquimbo to the Messier Channel, W of Chilean Patagonia, and near Magellan Strait (53°28′S, 70°47′W), 293–2220 m (Báez and Andrade 1979).
67. *Willemoesia inornata* Faxon, 1893. W of Valparaíso (33°42′S, 78°18′W and 34°7′S, 73°56′W), 2520–4000 m (Wicksten & Mendez, 1981, Retamal 1994b).
68. *Willemoesia pacifica* Sund, 1920. W of Valparaíso (33°42′S, 78°18′W), 2520 m (Retamal 1994b, Wicksten & Mendez, 1981).

Infraorder Anomura.

Superfamily Galatheoidea.

Family Chirostylidae Ortmann, 1892.

69. *Gastroptychus hendersoni* (Alcock and Anderson, 1899). From off Papudo (33°31′S, 71°27′W) to Sarmiento Channel (50°44′S, 74°31′W), Patagonia, 420–750 m (Retamal 1994b).
70. *Uroptychus parvulus* (Henderson, 1885). From off Zapallar (33°32′71°28′W) to Sarmiento Channel (50°44′S, 74°31′W), Magellan Strait (53°28′S, 70°47′W), 300–800 m (Retamal 1994b).

Family Munidopsidae Ortmann, 1892.

71. *Galacantha diomedeae* Faxon, 1893. Off Arica, Iquique, and Antofagasta, associated to Patagonian toothfish fishery (Guzmán and Sellanes 2015).

72. *Galacantha rostrata* (A. Milne-Edwards, 1880). Juan Fernandez Archipelago. 1700–3000 m deep (Retamal 1994b); off Antofagasta, 1175 m (Guzmán and Sellanes 2015).
73. *Munidopsis alfredolaguardai* Hendrickx and Ayon-Parente, 2013. Off Concepción and Chiloe (Guzmán and Sellanes 2015).
74. *Munidopsis agassizii* Faxon, 1893. Off Iquique, associated to Patagonian toothfish fishery (Guzmán and Sellanes 2015).
75. *Munidopsis barrerai* Bahamonde, 1964. 31°44′S to 32°31′S, 300–840 m (Bahamonde 1964).
76. *Munidopsis cochlearis* Khodina, 1973. 23°49′08″S, 71°06′54″W, 4500 m (Khodkina 1973).
77. *Munidopsis follirostris* Khodina, 1973. Near Juan Fernández Island (30°13′09″S, 78°47′03″W), 1280 m (Khodkina 1973).
78. *Munidopsis hamata* Faxon, 1893. Chigualoco Bay (31°44′S, 71°418″W), 400 m (Bahamonde 1973). Off Arica and Iquique (Guzmán and Sellanes 2015).
79. *Munidopsis opalescens* Benedict, 1903. Off Patagonia 700–1000 m (Bahamonde 1973). Off Concepción (Guzmán and Sellanes 2015).
80. *Munidopsis palmatus* Khodina, 1973. 32°11′6″S 71°46′3″W, 660–700 m (Khodkina 1973).
81. *Munidopsis quadrata* Faxon, 1893. Off Arica (Luke 1977). Off Antofagasta and Constitución (Guzmán and Sellanes 2015).
82. *Munidopsis subsquamosa* Henderson, 1888. W of Chiloé Island, 2500–3200 m (Baba et al. 2008). Off Caldera (Guzmán and Sellanes 2015).
83. *Munidopsis tanneri* Faxon, 1893. Arica and Off Valparaiso (Guzmán and Sellanes 2015).
84. *Munidopsis trífida* (Henderson, 1888) Collingwood Strait (Benedict 1902), Aysen (Haig 1955). Off Concepcion and Chiloe (Guzmán and Sellanes 2015).
85. *Munidopsis verrucosus* Khodkina, 1973. 23°47′7″S, 71°03′9″W and 23°15′5″S, 71°39′8″W, 4300–4880 m (Khodkina 1973).
86. *Munidopsis villosa* Faxon, 1893. Off Algarrobo (33°22′S, 71°40′W), 250–800 m (Bahamonde 1964). Off Arica (Luke 1977).

Family Munididae Ahyong, Baba, Macpherson, and Poore, 2010.

87. *Cervimunida johni* Porter, 1903. Coquimbo to Mocha Island (Porter 1903; Haig 1955).
88. *Munida curvipes* Benedict, 1903. Near Chonos Archipelago (45°08′S, 73°14′W), 1890 m (Bahamonde and López 1962); Puerto Otway, Patagonia, 2743 m (Retamal 1994b).
89. *Munida montemaris* Bahamonde and López, 1962. Punta Angeles, Valparaíso (33°02′S, 71°06′W), 280–400 m (Bahamonde and López 1962).
90. *Munida propinqua* (Faxon, 1893). From off Iquique (20°12′S, 70°09′W) to Quintero (32°47′S, 71°42′W), 700–1000 m (Retamal 1994b).
91. *Pleuroncodes monodon* (H. Milne Edwards, 1837). Off Arica (18°25′S, 71°22′W) to Mocha Island (38°22′S, 73°54′W), 20–400 m (unpubl. data).

Family Lithodidae Samouelle, 1819.

92. *Glyptolithodes cristatipes* (Faxon, 1893). Off Arica (18°25′S, 71°22′W), off Quintero (32°47′S, 71°42′W), 245–800 m (Del Solar, 1972).

93. *Lithodes murrayi* (MacPherson, 1988). From off Iquique (20°12′S, 70°09′W) to the Magellan Strait (53°28′S, 70°47′W), 70–581 m (MacPherson 1988).
94. *Lithodes panamensis* Faxon, 1893. Off Arica (18°25′S, 71°22′W), off Iquique (20°12′S, 70°09′W), 760–850 m (Retamal 1992).
95. *Lithodes santolla* (Molina, 1782). Valdivia (39°48′S, 73°14′W) to southern tip of South America, 0–700 m (Retamal 1992).
96. *Lithodes wiracocha* Haig, 1974. From off Iquique (20°12′S, 70°09′W) to Magellan Strait (53°28′S, 70°47′W), 620–800 m (Haig 1974).
97. *Neolithodes diomedeae* (Benedict, 1894). From off Arica (18°25′S, 71°22′W) to Chonos Archipelago, 300–1200 m (Retamal 1992).
98. *Paralomis aspera* Faxon, 1893. Off Iquique (20°12′S, 70°09′W), 560–1270 m (Retamal 1992).
130. *Paralomis chilensis* Andrade, 1980. Off Coquimbo (29°57′S, 71°20′W) and Los Vilos (31°54′S, 71°31′W), 400–1800 m (Andrade 1980).
99. *Paralomis diomedeae* (Faxon, 1896). Off Iquique (20°12′S, 70°09′W), 830–935 m (Del Solar, 1972).
100. *Paralomis longipes* Faxon, 1893. Off Iquique (20°12′S, 70°09′W), 700–800 m (Retamal 1992).
101. *Paralomis otsuae* Wilson, 1988. 22°48′S, 70°42′W, 1740 m (Wilson 1990).
102. *Paralomis papillata* (Benedict, 1895). Off Iquique (20°12′S, 70°09′W), 712–744 m (Haig 1974).
103. *Paralomis sonne* Guzmán, 2009. 22°48.7′S, 70°42.29′W, 1775 m (Guzmán 2009).

Superfamily Paguroidea.

Family Paguridae Latreille, 1802.

104. *Pagurus delsolari* Haig, 1974. Off Iquique (20°12′S, 70°09′W) to Constitución (35°20′S, 72°25′W), 275–650 m (Haig 1974).
105. *Pagurus comptus* White, 1847. Off Coquimbo (29°57′S, 71°20′W) to southern Chile, 16–400 m (Forest and Saint Laurent 1967).

Family Parapaguridae Smith, 1882.

106. *Oncopagurus haigae* (de Saint Laurent, 1972). Valparaíso (33°02′S, 71°06′W), Sala y Gómez (26°28′S, 105°05′W) Plate, from Quisco (33°24′S, 71°42′W) to Taitao Peninsula (46°23′S, 74°41′W), 189–497 m (de Saint Laurent 1972; Olguin et al. 2014).
107. *Oncopagurus mironovi* Zhadan, 1997. Nazca Plate (24°56.5′S, 88°31.6′W), 570–575 m (Zhadan 1997).
108. *Oncopagurus stockmani* Zhadan, 1977 (*). Nazca Plate (27°07′S, 81°18′W), 235 m (Zhadan 1997).
109. *Parapagurus abyssorum* (Filhol, 1885). 39°51′S, 96°52′W, 3603–3621 m; off Valdivia (39°48′S, 73°14′W), 1930 km W of the coast of Chile (Lemaitre 1999)
110. *Parapagurus holthuisi* Lemaitre, 1989. W of Valparaíso (33°42′S, 78°18′W); off Juan Fernández Archipelago, 2115 m (Lemaitre 1999).
111. *Parapagurus jeanette* Lemaitre, 1999. Otway harbor, Gulf of Penas (46°53.15′S, 75°12′W), 82–1485 m (Lemaitre 1999).

112. *Paragiopagurus wallisi* (Lemaitre, 1994) (*). Nazca Plate (25°05.1′S, 97°27.9′W), 260–265 m (Zhadan 1997).
113. *Paragiopagurus boletifer* (de Saint Laurent, 1972) (*). Submerged ridge, Sala y Gómez Ridge (26°28′S, 105°21′W), and Nazca Plate (25°34.0′S, 85°27.0′W), 240–245 m (Zhadan 1997).
114. *Strobopagurus* aff. *gracilipes* (A. Milne-Edwards, 1891) (*). Nazca Plate (24°58.5′S, 88°31.6′W), 570–575 m (Zhadan 1997).
115. *Sympagurus affinis* (Henderson, 1888) (*). Nazca Plate (25°07.8′S, 99°34.0′W), 350–490 m (Parin et al. 1997; station 2023, R/V "Prof. Stockman," 18° cruise), 147–1450 m (Lemaitre 2004).
116. *Sympagurus dimorphus* (Stûder, 1883). 22° S, 57°S, 91–1995 m (Lemaitre 2004; Olguin et al. 2014).
117. *Sympagurus dofleini* (Balss, 1912) (*). Submerged ridge, Sala y Gómez Ridge (26°28′S, 105°05′W), (Lemaitre 2004).
118. *Tylaspis anomala* Henderson, 1885 (*). NE of Easter Island (19°11′S, 102°24′W), 4143 m; 32°36′S, 137°43′W, 4344 m (Lemaitre 1998).

Infraorder Brachyura.
Superfamily Dromioidea.
Family Dromiidae De Haan 1833.
119. *Lauridromia dehaani* (Rathbun, 1923). Seamount Big, Salas y Gomez Ridge (Zarenkov 1990).
Superfamily Homolodromiidea.
Family Homolodromidae De Haan, 1839.
120. *Homolodromia robertsi* (Garth, 1973). 19°03′S, 32°06′W, 560–850 m (Garth 1973; Báez and Martin 1989).
Superfamily Homoloidea.
Family Cymonomidae Bouvier, 1898.
121. *Cymonomus menziesi* Garth, 1971. Chile-Perú Trench, 1000 m (Garth and Haig 1971); off Chiloé (42°35.35′S, 80°37′W), 507 m (Guzmán 2003).
Family Homolidae De Haan, 1839.
122. *Homologenus orientalis* Zarenkov, 1990 (*). Seamount Ichthyologist, Sala y Gómez Ridge (Zarenkov 1990).
123. *Moloha faxoni* (Schmitt, 1921). Off San Félix (26°17′S, 80°05′W) and San Ambrosio (26°20′S, 79°53′W) Islands (unpubl. data).
124. *Paramola japonica* Parisi (*). Seamounts Dorofeeva and New, Sala y Gómez Ridge (Zarenkov 1990).
125. *Paromola rathbuni* Porter, 1908. Juan Fernandez Archipelago (Porter 1927).
Family Latreilliidae Stimpson, 1858.
126. *Eplumula phalangium* (De Haan, 1839 in De Haan, 1833–1850) (*). Seamounts Big and Dome, Sala y Gómez Ridge (Zarenkov 1990).
Superfamily Leucosioidea Samouelle, 1819.
Family Leucosiidae.
127. *Ancylodactyla nana* (Zarenkov, 1990) (*). Seamounts Communard and Dome, Sala y Gómez Ridge (Zarenkov 1990).

128. *Ebalia sculpta* Zarenkov, 1990(*). Seamounts Big, Cliff, and Dome, Sala y Gómez Ridge (Zarenkov 1990).

Superfamily Majoidea.

Family Epialtidae McLeay, 1838.

129. *Libidoclaea granaria* H. Milne Edwards and Lucas, 1842. From off Coquimbo (33°38′S, 78°50′W) to Strait of Magellan, 60–450 m (Rathbun 1925).

130. *Libidoclaea smithi* (Miers, 1886). Canyon off the Bio Bio River (36°49S, 73°17′W), 480 m, to Strait of Magellan (53°28′S, 70°47′W) (Pineda and Retamal 1997).

131. *Lophorochinia parabranchia* Garth, 1969. N of Iquique (20°12′S, 70°09′W) and off Punta Patache (20°48′S, 70°12′W) to Quintero (32°47′S, 71°42′W), 282 m (Retamal 1994b).

Family Inachidae McLeay, 1838.

132. *Cyrtomaia danieli* Zarenkov, 1990 (*). Seamounts Amber, Ichthyologist, May Day, and Yarala, Sala y Gómez Ridge (Zarenkov 1990).

133. *Cyrtomaia platypes* Yokoya, 1933 (*). Seamounts Cliff and Dome, Sala y Gómez Ridge (Zarenkov 1990).

Superfamily Parthenopoidea.

Family Parthenopidae McLeay, 1838.

134. *Zarenkolambrus epibranchialis* (Zarenkov, 1990) (*). Seamounts Cliff and Needle, Sala y Gómez Ridge (Zarenkov 1990).

135. *Hispidolambrus mironovi* (Zarenkov, 1990) (*). Seamounts Big, Dome, and Ichthyologist, Sala y Gómez Ridge (Zarenkov 1990).

Superfamily Calappoidea.

Family Calappidae De Haan, 1833.

136. *Mursia zarenkovi* Galil and Spiridonov, 1998 (*). Seamounts Big, Dome, and Ichthyologist, Sala y Gómez Ridge, and Seamount Eclipse, Nazca Ridge (Zarenkov 1990).

137. *Platymera gaudichaudii* (H. Milne Edwards, 1837). From off Arica (18°25′S, 71°22′W) to Talcahuano (36°42′S, 72°46′W), and off Juan Fernández Archipelago (33°38′S, 78°50′W), 10–450 m (Galil 1993; Retamal et al. 2013).

Superfamily Portunoidea.

Family Geryonidae Colosi, 1923.

138. *Chaceon chilensis* Chirino-Gálvez and Manning, 1989. Near Robinson Crusoe Island (Juan Fernández Archipelago 33°38′S, 78°50′W), 300–1000 m (Chirino-Gálvez and Manning 1989). (*). Seamounts New and Pearl, Sala y Gómez Ridge, and Seamounts Eclipse and Profesor Mesyatsev, Nazca Ridge (Zarenkov 1990).

Family Progeryonidae Stevcic, 2005.

139. *Progeryon mararae* Guinot and Richer de Forges, 1981 (*). Seamount Dome, Sala y Gómez Ridge (Zarenkov 1990).

Superfamily Cancroidea.

Family Atelecyclidae Ortmann, 1893.

140. *Trichopeltarion corallinus* (Faxon, 1893). 18°43′S, 507m, and 36°00.23′S, 922m (Guzmán et al. 2009).

141. *Trichopeltarion hystricosus* (Garth in Garth and Haig, 1971). Off Pisagua (21°19′S, 70°26′W), 605–610 m (Retamal and Soto 1993; Guzmán et al. 2009).

Family Cancridae Latreille, 1802.

142. *Cancer porteri* (Rathbun, 1930). Off continental Chile, 0–500 m (Retamal 1994b).
143. *Platepistoma balssii* (Zarenkov, 1990) (*). Seamounts Big, New, and Yarala, Sala y Gómez Ridge (Zarenkov 1990).

22.4 Discussion

The extended continental Chilean coast has two very well-marked zoogeographic regions: one cold template in the South, from the large island of Chiloé to Cape Horn, and the other warm template in the North, the latter extending, according to some authors (e.g., Brättstrom and Johanssen 1983), from Arica to Talcahuano where a transition zone starts with a mixture of species (Brättstrom and Johanssen 1983). Other authors (e.g., Retamal and Moyano 2010), however, believe that the transition zone starts in Valparaíso and finishes in the large island of Chiloé. The total number of decapods recorded in Chilean waters is 475 (unpubl. data). Some species cited by Vereshchaka (1990), Parin et al.(1997), and Zhadan (1997) from the submerged ridge of Sala y Gómez and Nazca (about 25°S, 75–100°W) have been included herein although these ridges are slightly out of the boundaries of the Chilean EEZ, 200 nm from the coast of continental Chile, and 350 nm from Easter Island and Salas y Gómez Island.

There are only nine species of deep-water isopods on record, four in the Peru-Chile trench, three in Chiloé islands, and two in the Magellan Strait, thus showing a lack of information regarding the abyssal isopods in central Chile and off the oceanic islands. Among the decapods, there exists a clear dominance of shrimps and prawns in Chilean deep water. On the contrary, benthic species of Brachyura are poorly represented.

A total of 143 species of bentho-demersal or bentho-pelagic species of decapod crustaceans occur below 200 m off the coast of Chile (Table 22.1). The Dendrobranchiata are represented by five families that are represented by one to three species. In the Anomura, Parapaguridae is represented by 14 species, Lithodidae by 13, and Munidopsidae by 11. The rest of the deep-water decapods belong to the Brachyura, with 14 families and 12 species. The majority of the Brachyura families have only one to two representatives in deep water. In Chile, including Easter Islands, most records for Brachyura are from shallow water. Brachyuran crabs are also known to be scarce in deep water.

The 46 species discovered on Salas y Gómez and Nazca submerged ridges and around the former Desventuradas Islands live mostly in deep water. However, recent studies in the seamounts surrounding the Desventuradas Island have allowed for the collection of many additional specimens under study. This will probably increase diversity of decapods associated with the seamounts of the SEP (unpubl. data).

Table 22.1 Number of species of deep-water decapod crustaceans recorded for each family present in Chilean waters

Dendrobranchiata		**Astacidea**	
Aristeidae	1	Nephropidae	1
Benthesicymidae	3	**Achelata**	
Solenoceridae	3	Palinuridae	2
Sicyoniidae	1	**Axiidea**	
Penaeidae	1	Axiidae	2
Pleocyemata		**Anomura**	
Caridea		Chirostylidae	2
Nematocarcinidae	4	Munidopsidae	11
Campylonotidae	2	Munididae	5
Hippolytidae	1	Lithodidae	13
Thoridae	6	Paguridae	2
Physetocaridae	1	Parapaguridae	13
Pandalidae	9	**Brachyura**	
Crangonidae	7	Homolodromidae	2
Glyphocrangonidae	2	Cymonomidae	1
Stenopodidea		Calappidae	1
Spongicolidae	1	Atelecyclidae	2
		Cancridae	1
		Epialtidae	3
		Geryonidae	1

While studying deep-water species living around the Juan Fernández Archipelago, Retamal and Arana (2000) found that some of them (i.e., *Jasus frontalis*, *Projasus bahamondei*, *Geryon chilensis*) spend long periods as larval stages, have arrived from distant points, and are present above the submerged ridge, sometime in large numbers.

Acknowledgments The present study was founded by the Universidad de Concepción and project MECESUP UCT 0804. The authors express their gratitude to M.I. and S.M.A., for their valuable comments and suggestions for improving the manuscript.

References

Andrade H (1980) Nueva especie de Paralomis en aguas de Chile: *Paralomis chilensis* n.sp. (Crustacea, Decapoda). Bol Mus Nac Hist Nat Chile 37:269–273

Andrade H, Báez P (1977) *Calastacus rostriserratus* n. sp. *Calocarides quinqueseriatus* (Rathbun, 1902) (Crustacea, Decapoda, Macrura, Axiidae). An Mus Hist Nat Valparaíso 10:65–67

Appletans W, Ahyong ST, Anderson G et al. (2012) The magnitude of global marine species diversity. Curr Biol 22:2189–2202

Arana PM (2000) Pesca exploratoria con trampas alrededor de las islas Robinson Crusoe y Santa Clara, archipiélago de Juan Fernández, Chile. Investig Mar 28:39–52

Arana P, Ahumada M, Guerrero A (2003) Distribución y abundancia de la gamba Haliporoides diomedeae (Crustacea, Decapoda: Penaeidae) frente a la costa central de Chile. Inv Mar Valparaíso 3:57–71

Aydogan A, Wägele JW, Park JY (2000) New deep-sea isopods (Crustacea, Isopoda, Asellota) from the Atacama-trench. Mitt Mus Nat Kd Berlin Zoology Reihe 76:175–194

Baba K, Macpherson E, Poore GCB, Ahyong ST, Bermúdez A, Cabezas P, Lin CW, Nizinski M, Rodrigues C, Schnabel K (2008) Catalogue of squat lobsters of the world (Crustacea: Decapoda: Anomura — families Chirostylidae, Galatheidae and Kiwaidae). Zootaxa 1905: 1–220

Báez P, Andrade H (1979) Crustáceos arquibentónicos frecuentes frente a las costas de Chile central. An Mus Hist Nat Valparaíso 12:219–231

Báez P, Martin J (1989) Crabs of the family Homolodromiidae. I. Description of the male of *Homolodromia robertsi* Garth, 1973 based on specimens from deep waters off the coast of Chile. J Crust Biol 9:492–500

Báez P, Soto R (1997) *Paracrangon areolata* Faxon, 1893 un camarón nuevo para Chile (Decapoda, Crangonidae). Inv Mar Valparaíso 25:277–280

Bahamonde N (1955) Hallazgo de una nueva especie de *Heterocarpus* en aguas chilenas: *H. reedi* n. sp. Inv Zool Chilenas 2:105–114

Bahamonde N (1959) Decápodos chilenos: La familia Homaridae. Inv Zool Chilenas 5:221–227

Bahamonde N, López M (1962) Un Galatheido nuevo para Chile, Munida montemaris n. sp. (Crustacea, Decapoda, Anomura). Rev Chil Hist Nat 55:85–91

Bahamonde N (1964) Dos nuevos *Munidopsis* frente a la costa chilena. (Crustacea, Decapoda, Anomura). Bol Mus Nac Hist Nat Chile 28:157–170

Bahamonde N (1967) Rhynolithodes (Glyptolithodes) cristatipes Faxon frente a la costa chilena (Crustacea, Decapoda, Anomura). Not Mens Mus Nac Hist Nat Chile 136:3–7

Bahamonde N (1973) Munidopsis hamata Faxon en Chile (Crustacea, Decapoda, Anomura). Not Mens Mus Nac Hist Nat Chile 203–204:10–11

Bahamonde N (1979) Registro en Chile de *Thymops birsteini* (Zarenkov and Semenov, 1972) (Crustacea, Decapoda, Nephropidae). Not Mens Mus Nac Hist Nat Chile 24(276):3–5

Bahamonde N (1981) Adiciones a la fauna carcinológica de Chile. *Sclerocrangon atrox* Faxon, 1893. Not Mens Mus Nac Hist Nat Chile 296:3–6

Beddard FE (1886) Report on the Isopoda collected by H.M.S. *Challenger* during the years 1873–1876. Reports of scientific results of the exploring voyage of H.M.S *Challenger* during the years 1873–76. Zoology 11:1–85

Benedict J (1902) Descriptions of a new genus and forty six new species of Crustaceans of the family Galatheidae, with a list of the known marine species. Proc U.S. Nat Mus 26:243–344

Brandt A (1998) Sphaeromatidae (Crustacea, Isopoda) of the Beagle channel and description of *Cymodopsis beageli* n. sp. Beaufortia 48:137–161

Brandt A, Wägele JW (1989) Redescriptions of *Cymodocella tubicauda* Pfeffer, 1987 and *Exosphaeroma gigas* (Leach 1818) (Crustacea, Isopoda, Sphaeromatidae). Antarct Sci 1:205–214

Brättstrom H, Johanssen A (1983) Ecological and regional zoogeography of the marine benthic fauna of Chile. Rep Lund Univ. Chile Exped. 1948–1949. Sarsia 69:289–339

Brusca RC, Ninos M (1978) The status of *Cirolana californiensis* Schultz and *C. deminuta* Menzies and George with a key to the California species of *Cirolana* (Isopoda: Cirolanidae). Proc Biol Soc Wash 91:379–385

Burukovsky RN (1990) Shrimps from Salas-y-Gómez Ridge. T Shirshov Inst Oceanol, Moscow 124:187–217

Chirino-Gálvez LA, Manning R (1989) A new deep sea crab of the genus *Chaceon* from Chile (Crustacea, Decapoda, Geryonidae). Proc Biol Soc Wash 102:401–404

De Grave S, Pentcheff ND, Ahyong ST, Chan T, Crandall KA, Dworschak PC, Felder DL, Feldman RM, CHJM F, LYD G, Lemaitre R, MEY L, Martin JW, PKL N, Schweitzer CE, Tan SH, Tshudy D, Wetzer R (2009) A classification of living and fossil genera of decapod crustaceans. Raffles Bull Zool Suppl 21:1–109

Del Solar F (1972) Addenda al Catálogo de los crustáceos del Perú. Inf Inst Mar Perú-Callao 18:1–21

D'Orbigny A (1842) Coquilles et les échinodermes fossiles recueillis en Colombie, par M. Boussingnault. Voyage dans l'Amérique Méridionale. P. Bertrand, Paris. 64 p

Forest J, de Saint Laurent M (1967) Crustáces Decápodes: Pagurides. Campagne de la Calypso au large des cotes Atlantiques de l'Amerique du Sud (1961-1962). Ann. Inst. Ocean, 45: 47–172

Galil BS (1993) Crustacea Decapoda. A revision of the genus *Mursia* Desmaresti MURSORSTOM 10. Bull Mus Natl Hist Nat Paris 186:347–379

Garth JS (1969) A new genus and species of oxyrhynchous crab from the west coast of South America. Crustacea, Decapoda, Brachyura. Bol Soc Biol Concepción 41:5–8

Garth JS, (1973). The Brachyuran crabs of Easter Island. Proc. Calif. Acad. Sci., 39(17):311–336

Garth JS, Haig J (1971) Decapod Crustacea (Anomura and Brachyura) of the Peru-Chile Trench. Anton Bruun Rep 6:3–20

George RW (1976) A new species of the spiny lobster *Projasus bahamondei* (Palinuridae "Silentes") from the South East Pacific region. Crustaceana 30:27–30

Guzmán G (1999) Nuevos registros de camarones sergéstidos (Crustacea; Decapoda; Dendrobranchiata) en el norte de Chile. Inv Mar Valparaíso 27:87–91

Guzmán G (2003) Primer registro para Chile y descripción de la hembra de *Cymonomus menziesi* Garth, 1971. (Decapoda, Archaeobrachyura, Cymonomidae). Inv Mar Valparaíso 31:91–94

Guzmán G (2004) Oplophoridae (Decapoda, Caridea) in southeastern Pacific Ocean. A revision of Chilean species. Gayana 68:70–75

Guzmán G (2008) Camarones pelágicos (Crustacea: Decapoda) en aguas del Pacífico sureste. Contrib Estudio Crust Pac Este 5:27–45

Guzmán G (2014) Pasiphaea barnardi (Decapoda, Caridea, Pasiphaeidae) en los fiordos australes de Chile. Rev Biol Mar Oceanog 49:1–6

Guzmán G, Quiroga E (2005) New records of shrimps (Decapoda Caridea and Dendrobranchiata) in deep waters off Chile. Gayana 69:285–290

Guzmán G, Rivera J (2002) Stylopandalus richardi (Coutière, 1905) primer registro para aguas del Pacífico sureste (Decapoda, Pandalidae). Inv Mar 30:67–30

Guzmán G, Sellanes J (2015) A review of the Munidopsidae Ortmann, 1898 (Decapoda, Galatheoidea) in Chilean waters, including new records for the Southeastern Pacific. Zootaxa 4021:282–306

Guzmán G, Wicksten MK (1998) Nuevos registros de camarones de la familia Pasiphaeidae (Crustacea, Decapoda) en el Norte de Chile (18° a 22° Lat. S, 70° a 72° Long. W). Gayana 62:203–210

Guzmán G, Wicksten MK (2000) The subfamily Benthesicyminae Bouvier, 1908 (Decapoda, Dendrobranchiata) in Northern Chile (18° to 22°S). Crustaceana 73:925–931

Guzmán G, Báez P, Sellanes J (2009) Primer registro de *Trichopeltarion corallinus* (Faxon, 1893) para el mar de Chile y nuevo registro de *T. hystricosus* (Garth, en Garth & Haig, 1871) (Decapoda: Brachyura; Atelecyclidae). Latin Amer J Aq Res 37:275–279

Haig J (1955) The Crustacea Anomura of Chile. Reports of the Lund University Chile Expedition 1948-1949. Rept Lunds Univ Arskr N F 51:1–68

Haig J (1974) Observations on the Lithodid crabs of Peru, with descriptions of two new species. Bull S Calif Acad Sci 73:152–164

Holthuis L (1952) The Crustacea Decapoda Macrura of Chile. Reports of the Lund University Chile Expediton 1948-1949. Lunds Univ Arsskr N F 247:1–110

Khodkina IV (1973) New species of the genus *Munidopsis* (Decapoda, Anomura) from the East Pacific. Zool Zhurn 52:1156–1167

Lemaitre R (1989) Revision of the genus *Parapagurus* (Anomura, Parapaguridae) including redescriptions of the western Atlantic species. Zool Verhand Leiden 253:3–106

Lemaitre R (1998) Revisiting *Tylaspis anomala* Henderson, 1885 (Parapaguridae) with comments on its relationships and evolution. Zoosystema 20(2):289–305

Lemaitre R (1999) Crustacea Decapoda: a review of the species of the genus *Parapagurus* Smith, 1879 (Parapaguridae) from the Pacific and Indian Oceans. Résultats des Campagnes MURSORSTOM 20. Mém Mus Natl Hist Nat Ser A Zool 180:304–378

Lemaitre R (2003) Revision of *Pylopagurus* and *Tomopagurus* (Crustacea, Decapoda, Paguridae) with descriptions of new genera and species. Addendum and taxonomy summary. Proc Biol Soc Wash 116:464–486

Lemaitre R (2004) A worldwide review of hermit crabs species of the genus *Sympagurus* Smith, 1883. (Crustacea, Decapoda, Parapaguridae). In: Marshall B, Richer de Forbes B (eds) Tropical deep sea benthos, Mémoires du Muséum national d'Histoire naturelle, 1291, vol 23. Muséum national d'Histoire naturelle, Paris, pp 85–149

Lorenti M, Mariani S (1997) Isopod assemblages in the Straits of Magellan: structural and functional aspects. Polar Biol 18:254–259

Luke SR (1977) Catalog of the benthic invertebrate collections. I. Decapod Crustacea and Stomatopoda. University of California. SIO Ref. Ser. 77–9: 72 p.

MacPherson E (1988) Revision of the family Lithodidae Samouelle, 1819 (Crustacea, Decapoda, Anomura) in the Atlantic Ocean. Monog Zool Mar 2:9–153

Menzies RJ (1962) The zoogeography, ecology and systematics of the Chilean marine isopods. Reports of the Lund University Chile Expedition 1948–1949. Lunds Univ Arssk, Lund

Molina JI (1782) Saggio sulla Storia Naturale del Chile. In Bologna MDCCLXXXII. Nella Stamperia de S. Tommaso d' Aquino, 367 p

Olguín N, Guzmán G, Quiroga E, (2014). Familia Parapaguridae (Anomura, Paguroidea) en Chile: Nuevos registros de cangrejos ermitaños de profundidad. Rev Biol Mar Oceanog 49:71–79

Parin NV, Mironov AN, Nesis KN et al. (1997) Biology of the Nazca and Sala y Gómez submarine Ridge, an outpost of the Indo-West Pacific Fauna in the Eastern Pacific Ocean: composition and distribution of the Fauna, its communities and history. Adv Mar Biol 32:147–230

Park J-Y, Wägele W (1965) On small collection of valvifera (Crustacea Isopoda) from the Magellan strait, with description of *Platidotea magellanica* n. gen. n. sp. Bull Zool Mus Univ Amsterdam 14:57–99

Philippi RA (1887) Los Fósiles Terciarios i Cuartarios de Chile. Imprenta de FA Brockhaus, Leipzig, 256 p, 58

Pineda V, Retamal M (1997) Nota sobre la presencia de *Libidoclaea smithi* (Miers, 1888) (Decapoda, Majidae) en los sedimentos del cañón submarino del río Bio Bio. Gayana Oceanol 5:115–118

Poeppig E (1835–36) Reise in Chile, Peru und auf dem Amazoenstrome, wáhrend der Jahre 1827–1832. Vil. 1. Friedrich Fleischer, Leipzig. 464 p

Poore GCB, Bruce NJ (2012) Global diversity of marine isopods (except Asellota and Crustacean Symbionts). PLoS One 7(8):e43529

Porter CE (1901) Instrucciones para la recolección de crustáceos. Rev Chil Hist Nat 5(11–12):260–265

Porter CE (1927) Carcinología Chilena. Nuevo hallazgo de un Homolido. Rev Chil Hist Nat 31: 141–143

Retamal M (1976) Contribución al conocimiento de los Crustacea, Decapoda, Antárticos. Ser Cient Inst Antart Chileno 4(1):89–94

Retamal M (1992) *Percnon pascuensis* una nueva especie de Isla de Pascua (Grapsidae, Plaguisiinae). Invest Mar 30:63–66

Rathbun MJ (1907) South American Crustacea. Rev Chil Hist Nat 11(1):45–50

Rathbun M (1925) The Spider crabs of America. U.S. Nat Mus Bull 129:1–611

Retamal M (1994b) Los Decápodos de Chile. Universidad de Concepción, Dirección de Docencia, Concepción

Retamal M, Arana P (2000) Descripción y distribución de cinco crustáceos decápodos recolectados en aguas profundas en torno a las islas Robinson Crusoe y Santa Clara (Archipiélago Juan Fernández, Chile.). Inv Mar Valparaíso 28:146–163

Retamal M, Moyano H (2010) Zoogeografía de los crustáceos decápodos chilenos marinos y dulceacuícolas. Lat Am J Aquat Res 38: 302–328

Retamal MA, Ferrada P (2016) Catálogo ilustrado de decápodos de la Patagonia chilena. Universidad de Concepción, Programa COPAS Sur-Austral, Concepción

Retamal M, Gorny P (2003) Revisión del género *Metacrangon* y descripción de una nueva especie (Decapoda, Crangonidae). Inv Mar Valparaíso 31:85–90

Retamal M, Moyano H (2010) Zoogeografía de los crustáceos decápodos chilenos marinos y dulceacuícolas. Lat Am J Aquat Res 38:302–328

Retamal M, Soto R (1993) Crustáceos decápodos abisales de la zona de Iquique. Arica Est Oceanol 14:1–4

Retamal M, Soto R (1995) *Plesionika santaecatalinae* Wicksten 1983, en aguas chilenas. Gayana Oceanol 3:1–2

Retamal M, Ernst B, Chamorro J, Manríquez P (2013) Descripción y distribución geográfica de una nueva especie para el archipiélago Juan Fernández, *Platymera gaudichaudii* (H. Milne-Edwards, 1837) (Decapoda, Calappidae). Latin Am J Aq Res 41:1010–1012

Saint Laurent M. de. (1972). Sur la famille des Parapaguridae Smith, 1882. Description de Thyplopagurus foresti gen. nov., sp. nov. et de quinze especes ou sous espèces nouvelles de Parapagurus Smith (Crustacea, Decapoda). Bijd Dierk 42:97–123

Spence Bate C (1888) Report on the Macrura collected by H.M.S challenger during the years 1873–1876. Reports of scientific results of the exploring voyage of H.M.S challenger during the years 1873–76. Zoology 24:1–942

Torti MR, Boschi E (1976) Nuevos aportes al conocimiento de los Crustáceos Decápodos Caridea del género *Campylonotus* Spence Bate, 1888. Physis Sec A 32(84):65–84

Vereshchaka AL (1990) Pelagic decapods from seamounts of the Nazca and Sala-y-Gómez Ridge. In: Mironov AN, Rudjakov JA (eds) Plankton and Benthos from the Nazca and Sala y Gómez submarine Ridge. Academy of the Sciences of the U.R.S.S, pp 129–155. Trans P.P. Shirhov Inst Oceanol

von Tiechenbacher L (1979) Bêitrage zur Taxonomie von *Nematocarcinus longirostris* Spence Bate, 1888 und *Nematocarcinus lanceopes* Spence Bate, 1888, neu fûr die westliche Antarltis. Spixiana 13:229–235

Wicksten M (1980) New records of *Stereomastis sculpta* (Faxon) (Decapoda, Polychelidae) in the Eastern Pacific. Proc Biol Soc Wash 93:914–919

Wicksten M, Méndez M (1981) Alpheus inca a new snapping shrimp (Caridea, Alpheidae) from western South America. J Crust Biol 1:137–142

Wicksten M, Méndez M (1982) New records and new species of the genus *Lebbeus* (Caridea, Hippolytidae) (SIC) in the Easter Pacific Ocean. Bull South Calif Acad Sci 81:106–120

Wicksten M, Méndez M, (1983). Nuevos registros de camarones Carideos en el Perú. Bol Lima 25:1–15

Wilson R (1990) Paralomis otsuae a new species of the Decapoda Anomura from deep waters off the Chilean coast. Crustaceana 58:130–135

Zabala P, Bahamonde N (1998) *Pandalopsis* cf. *ampla* Spence Bate,1888 frente a Antofagasta, Chile (Crustacea, Decapoda, Caridea, Pandalidae). Inv Mar Valparaíso 30:67–70

Zarenkov NA (1976) On the fauna of decapods of the waters adjacent to South America. Byull Moskov Obshch Isp Prir Otd Biol 5:8–18. [in Russian]

Zarenkov NA, Semenov VN (1972) A new species of the genus *Nephropides* from the South West Atlantic. Zool J Moscow 51:599–601

Zarenkov NA (1990) Decapods (Stenopodidea, Brachyura, Anomura) of the Nazca and Sala y Gomez underwater ridges. Trans P.P. Shirshov Inst Oceanol Moscow 124:218–244

Zhadan D (1997) Deep-sea hermit crabs from the submerged Ridge Nazca and Sala-y-Gómez, Southeastern Pacific (Decapoda, Anomura, Parapaguridae). Arthrop Sel 6:55–79

Chapter 23
Deep-Water Stomatopod and Decapod Crustaceans Collected off Central America by the R/V "Miguel Oliver"

M. E. Hendrickx and J. López

Abstract Deep-water stomatopods and decapod crustaceans were collected during an exploratory survey off Central America in 2010 aboard the R/V "Miguel Oliver." A total of 105 trawling operations were performed off Guatemala, El Salvador, Nicaragua, Costa Rica, and Panama. Species captured during these operations are reported, including one species of stomatopod and 25 species of decapods. New geographic records are provided for *Haliporoides diomedeae*, *Solenocera agassizii*, *Pasiphaea emarginata*, and *Portunus xantusii*. Compared to the entire eastern Pacific, the number of species of decapod crustaceans collected off Central America is low and represents only 16% of species recorded below 350 m depth. This percentage increases to 26% when compared to the fauna occurring below 350 off western Mexico.

Keywords Deep sea · Eastern Pacific · Stomatopoda · Decapoda · Distribution

23.1 Introduction

Intensive exploration of deep-water ecosystems initiated in the nineteenth century when the R/V "Challenger," a Royal Navy ship modified for scientific use, navigated about 70,000 nautical miles from 1872 to 1876, circumnavigating the globe, including the southern part of Chile. The bulk of knowledge on decapod crustaceans from the East Pacific, however, came from a research cruise organized in 1891 aboard the

---

Jorge López: Deceased, October 2019.

M. E. Hendrickx (✉)
Laboratorio de Invertebrados Bentónicos, Unidad Académica Mazatlán, Instituto de Ciencias del Mar y Limnología, Universidad Nacional Autónoma de México, Mazatlán, Sinaloa, Mexico
e-mail: michel@ola.icmyl.unam.mx

J. López (Deceased)
OSPESCA, Organización del Sector Pesquero y Acuícola del Istmo Centroamericano, San Salvador, El Salvador

M. E. Hendrickx (ed.), *Deep-Sea Pycnogonids and Crustaceans of the Americas*, https://doi.org/10.1007/978-3-030-58410-8_23

Steamer *Albatross* visiting the west coast of America from Ecuador, including the Galapagos Islands, to Central America and Mexico, including the Gulf of California (see Wicksten 1989; Hendrickx 2012a). As a result of this expedition, one species of stomatopod and 151 species of decapod crustaceans were reported, including a large series of new species and genera collected in deep water (Faxon 1893, 1895).

Since this famous expedition of the *Albatross* took place, until recently interest for deep-water fauna in the East Pacific has been rather incipient, with only incidental sampling being carried out in selected areas along the west coast of America (see Hendrickx 2012a). At the end of the twentieth century and further on within the twentieth-first century, however, some countries decided to dedicate a great deal of efforts to increase their knowledge on the deep-water fauna inhabiting their exclusive economic zone (Wehrtmann et al. 2012), including Mexico (López-Rocha et al. 2006; Hendrickx et al. 2011; Zamorano et al. 2012; Hendrickx 2012a; Cruz-Acevedo et al. 2018; Suárez Mozo et al. 2018), Costa Rica (e.g., Wehrtmann and Nielsen-Munoz 2009; Pedraza and Zapata Padilla 2011; Starr et al. 2012; Villalobos-Rojas et al. 2017; Villalobos-Rojas and Wehrtmann 2018; Sánchez-Jiménez et al. 2018), Colombia (e.g., Puentes and Madrid 1994; Rodríguez et al. 2012; Navas et al. 2013; Beltrán-León et al. 2016; Suárez Mozo et al. 2018), Ecuador (Cornejo Antepara 2010), Peru (Vélez et al. 1992; Kameya et al. 1997; 1998; Arana et al. 2002; Barriga et al. 2009; Aramayo 2016), and Chile (e.g., Retamal 1993; Guzmán and Quiroga 2005; Arana 2003; Arana et al. 2013; Guzmán and Sellanes 2015).

In 2010, an ambitious exploration sampling program was developed within the frame of an international collaboration agreement between MARM (Ministerio de Medio Ambiente y Medio Rural y Marino), España, and OSPESCA (Organización del Sector Pesquero y Acuicola del Istmo Centroamericano). As a result, the R/V "Miguel Oliver" was dispatched to Central America with a view to explore the deep-water ecosystems along the Pacific off Guatemala, El Salvador, Nicaragua, Costa Rica, and Panama (López 2011), between 57 and 1530 m depth.

Part of the results obtained for other groups of species during this cruise were previously presented by Hendrickx and López (2006, 2019) and Robertson et al. (2017; bony fishes). This contribution reports on additional material that was collected on this occasion, specifically decapod crustaceans.

23.2 Material and Methods

The material reported herein was collected between November 10 and December 16, 2010, by the R/V "Miguel Oliver" using a Lofoten fishing gear with 35 mm net aperture. The gear was deployed off Panama, Costa Rica, Nicaragua, El Salvador, and Guatemala, in depths between 57 and 1530 m, with a majority of samples from >200 m depth. In total, 106 trawls were performed. All samples were of at least 20 min. The specimens of species of large crustaceans (stomatopods and decapods) were separated from the catch and preserved for their identification in the laboratory. Although many specimens of each species were occasionally caught in the fishing net, the number of specimens per trawl was not made available for technical reasons. The species were identified using available literature (Faxon 1895; Hendrickx 1995a, b, c, d, e, 2012b, c; Salgado-Barragán and Hendrickx 2010).

23.3 Results

A total of 26 species were collected during the survey: 1 stomatopod and 25 decapod crustaceans (Table 23.1). Among the decapods four species have been reported as pelagic (*Pasiphaea emarginata*, *P. magna*, and *P. tarda*) or occasionally pelagic (*Achelous xantusii*) and are often captured in trawl nets during the ascent of the gear through the water column. During this ascent, the net acts as a mid-water trawl, and

Table 23.1 Number of samples (NS) and countries where the 26 species of stomatopod and decapod crustaceans were captured

Species	NS	Guatemala	El Salvador	Nicaragua	Costa Rica	Panama
Squilla biformis	19	X		X	X	X
Haliporoides diomedeae	28		X		X	X
Solenocera agassizi	17	X	X	X	X	X
Benthesicymus tanneri	1					X
Pasiphaea emarginata	1				X	
Pasiphaea magna	18		X	X	X	X
Pasiphaea tarda	1			X		
Heterocarpus hostilis	43		X	X	X	X
Heterocarpus vicarius	14				X	X
Pandalus amplus	1	X				
Metacrangon procax	1					X
Sclerocrangon atrox	1				X	
Glyphocrangon alata	21				X	X
Glyphocrangon spinulosa	11	X	X	X	X	
Guyanacaris caespitosa	1					X
Nephropsis occidentalis	11		X	X	X	X
Stereomastis pacifica	3				X	
Pleuroncodes planipes	15	X	X	X	X	X
Galacantha diomedeae	2				X	X
Munidopsis depressa	1				X	
Paralomis diomedeae	1					X
Maiopsis panamensis	3					X
Cancer johngarthi	4					X
Platymera gaudichaudii	4					X
Portunus xantusii	1					X
Trichopeltarion corallinum	2			X	X	
Species per country		**5**	**7**	**8**	**16**	**18**

large specimens are retained and mix with truly benthic species (see Hendrickx 2016). The rest of the species reported herein are benthic.

23.4 Systematic Section

Stomatopoda
Family Squillidae

1. *Squilla biformis* Bigelow, 1891

Material Examined. St. 1 (07°24′49.8″N, 78°07′27.0″W), November 12, 2010, 183 m; St. 4 (07°37′36.6″N, 78°41′36.0″W), November 12, 2010, 116 m; St. 38 (07°43′24.6″N, 82°03′25.2″W), November 20, 2010, 144 m; St. 39 (07°44′33.6″N, 82°25′16,8″W), November 20, 2010, 243 m; St. 41 (08°01′57.6″N, 82°33′51.0″W), November 21, 2010, 115 m; St. 45 (08°16′03.0″N, 83°11′21.0″W), November 22, 2010, 1376 m; St. 46 (08°13′54.6″N, 83°10′34.2″W), November 22, 2010, 1260 m; St. 50 (08°48′07.2″N, 84°01′40.8″W), November 23, 2010, 119 m; St. 57 (09°14′25.2″N, 84°34′12.0″W), November 24, 2010, 134 m; St. 61 (09°30′25.2″N, 85°09′01.2″W), November 25, 2010, 140 m; St. 70 (10°41′03.6″N, 86°24′52.8″W), December 4, 2010, 243 m; St. 73 (11°05′03.0″N, 86°44′45.6″W), December 5, 2010, 153 m; St. 81 (11°42′10.8″ N, 87°03′48.6″W), December 7, 2010, 122 m; St. 84 (11°54′34.2″N, 87°38′33.6″W), December 8, 2010, 404 m; St. 85 (11°55′54.6″N, 87°33′30.6″W), December 8, 2010, 181 m; St. 90 (12°17′24.0″N, 88°05′06.6″W), December 9, 2010, 129 m; St. 102 (13°17′33.6″N, 90°24′16.2″W), December 12, 2010, 165 m; St. 103 (13°42′54.4″N, 91°50′06.6″W), December 13, 2010, 117 m; St. 105 (13°28′37.2″N, 91°23′49.8″W), December 13, 2010, 137 m. **Comments** This species was collected in 19 samples (Fig. 23.1), off Guatemala, Nicaragua, Costa Rica, and Panama, in depths between 116 and 1376 m, which is the maximum depth recorded to date for this species. Previous depth records are from 28 to 518 m (Hendrickx and Salgado-Barragán 2002). **Distribution** *Squilla biformis* presents a wide distribution range, from the southern Gulf of California, Mexico, to off Huacho, Peru (Hendrickx and Salgado-Barragán 2002).

Decapoda
Dendrobranchiata
Family Solenoceridae

2. *Haliporoides diomedeae* (Faxon, 1893)

Material Examined. St. 8 (07°32′52.2″N, 79°12′49.8″W), November 11, 2010, 757 m; St. 10 (07°32′55.8″N, 79°11′48.6″W), November 13, 2010, 1020 m; St. 11 (07°23′36.6″N, 79°29′20.4″W), November 14, 2010, 1292 m; St. 12 (07°23′42.0″N, 79°32′20.4″W), November 14, 2010, 1021 m; St. 13 (07°25′55.8″N, 79°33′40.2″W), November 14, 2010, 711 m; St. 16 (07°14′39.6″N, 79°40′03.6″W), November 15, 2010, 1446 m; St. 17 (07°16′16.2″N, 79°39′52.2″W), November 15, 2010, 1141 m;

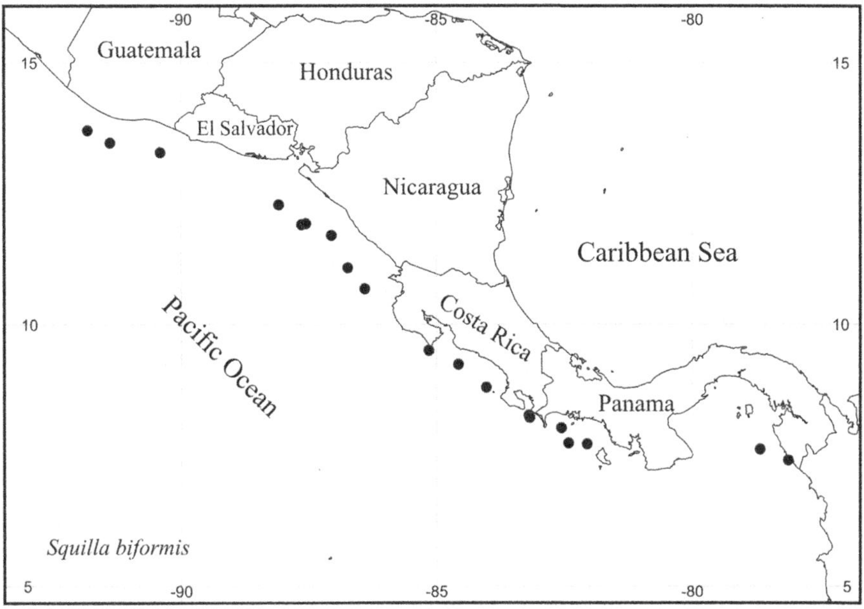

Fig. 23.1 Sampling localities of *Squilla biformis* in the survey area

St. 18 (07°20′57.0″N, 79°40′03.6″W), November 15, 2010, 720 m; St. 19 (07°23′08.4″N, 79°43′51.6″W), November 15, 2010, 480 m; St. 21 (08°09′43.8″N, 80°08′03.0″W), November 16, 2010, 764 m; St. 25 (07°05′20.4″N, 80°55′16.2″W), November 17, 2010, 1113 m; St. 26 (07°08′50.4″N, 81°07′29.4″W), November 17, 2010, 954 m; St. 29 (07°04′48.6″N, 81°22′05.4″W), November 18, 2010, 1126 m; St. 30 (07°05′40.2″N, 81°27′43.8″W), November 18, 2010, 864 m; St. 33 (07°00′37.8″N, 81°44′00.0″W), November 19, 2010, 716 m; St. 34 (06°57′00.6″N, 81°44′43.8″W), November 19, 2010, 1033 m; St. 35 (07°04′14.4″N, 81°42′37.8″W), November 19, 2010, 671 m; St. 36 (07°10′54.0″N, 81°40′00.6″W), November 19, 2010, 143 m; St.40 (07°38′27.0″N, 82°27′04.2″W), November 20, 2010, 1185 m; St. 42 (07°58′55.2″N, 82°44′25.8″W), November 21, 2010, 1115 m; St. 43 (07°56′43.8″N, 82°48′30.0″W), November 21, 2010, 951 m; St. 44 (07°53′27.6″N, 82°50′31.8″W), November 21, 2010, 1093 m; St. 46 (08°13′54.6″N, 83°10′34.2″W), November 22, 2010, 1260 m; St. 52 (08°46′03.0″N, 84°15′43.8″W), November 23, 2010, 836 m; St. 55 (09°08′49.8″N, 84°33′40.8″W), November 24, 2010, 902 m; St. 59 (09°25′40.8″N, 85°09′40.8″W), November 25, 2010, 841 m; St. 65 (09°39′54.0″N, 85°43′54.6″W), December 12, 2010, 1178 m; St. 99 (13°04′58.8″N, 90°36′07.2″W), December 12, 2010, 1054 m. **Comments** Collected in 28 trawls, from off El Salvador, Costa Rica, and Panama (Fig. 23.2), between 143 and 1446 m depth (27 records in water >480 m). The known depth range for this species is from 240 to 1866 m (Peréz-Farfante 1977), but there is an unusual record at 50–95 m off Chile (43°32′S) by Wehrtmann and Carvacho (1997). **Distribution** Previously known from off the Azuero Peninsula (07°30′N), Panama, to off Talcahuano, Chile (Wicksten and Hendrickx 2003). The material from St. 99 (13°04′58.8″N, 90°36′07.2″W) is a new northernmost record for *H. diomedeae* which is now reported from off El Salvador.

3. *Solenocera agassizii* Faxon, 1893

Material Examined. St. 1 (07°24′49.8″N, 78°07′27.0″W), November 12, 2010, 183 m; St. 27 (07°13′09.0″N, 81°05′43.8″W), November 17, 2010, 138 m; St. 38 (07°43′24.6″N, 82°03′25.2″W), November 20, 2010, 144 m; St. 41 (08°01′57.6″N, 82°33′51.0″W), November 21, 2010, 115 m; St. 48 (08°18′38.4″N, 83°07′22.2″W), November 22, 2010, 128 m; St. 49 (08°27′01.2″N, 83°44′10.2″W), November 23, 2010, 128 m; St. 50 (08°48′07.2″N, 84°01′40.8″W), November 23, 2010, 119 m; St. 57 (09°14′25.2″N, 84°34′12.0″W), November 24, 2010, 134 m; St. 61 (09°30′25.2″N, 85°09′01.2″W), November 25, 2010, 140 m; St. 81 (11°42′10.8″N, 87°03′48.6″W), December 7, 2010, 122 m; St. 85 (11°55′54.6″N, 87°33′30.6″W), December 8, 2010, 181 m; St. 90 (12°17′24.0″N, 88°04′28.8″W), December 9, 2010, 131 m; St. 91 (12°44′38.4″N, 88°59′39.6″W), December 10, 2010, 105 m; St. 92 (13°01′11.4″N, 89°27′29.4″W), December 10, 2010, 104 m; St. 98 (13°18′48.0″N, 90°00′11.4″W), December 11, 2010, 114 m; St.104 (13°32′47.4″N, 91°20′52.8″W), December 13, 2010, 114 m; St. 106 (13°39′28.2″N, 91°13′42.0″W), December 13, 2010, 57 m. **Comments** The largest species of the genus in the East Pacific (up to 14 mm long

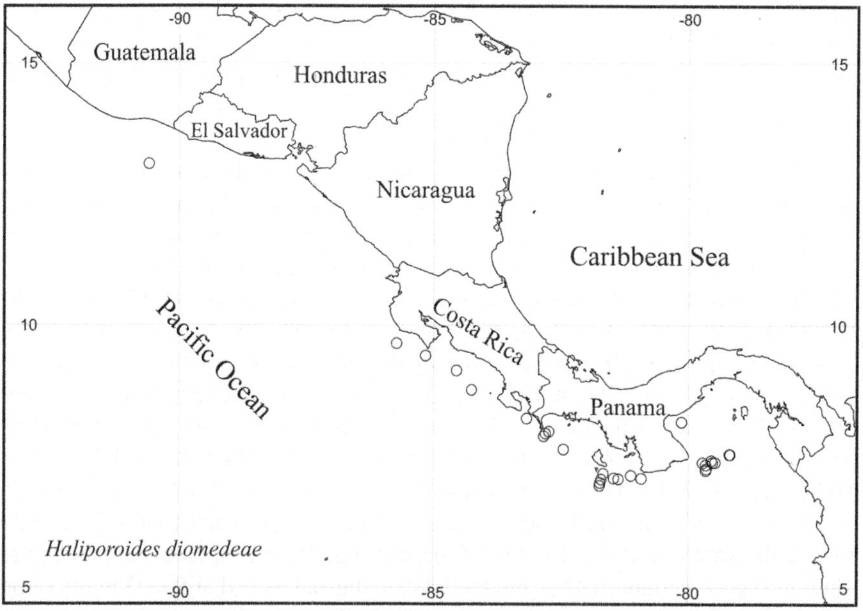

Fig. 23.2 Sampling localities of *Haliporoides diomedeae* in the survey area

in females), *S. agassizii* occurs between 16 and 384 m depth (Hendrickx 1995a). During this survey, it was collected in 17 trawls, off Guatemala, El Salvador, Nicaragua, Costa Rica, and Panama (Fig. 23.3), between 57 and 183 m depth. **Distribution** Previously known from off Cape Blanco, Costa Rica, south to Lobos de Afuera Islands, Peru (Wicksten and Hendrickx 2003). Present records increase the distribution range of *S. agassizii* to off Guatemala (13°32′47.4″N, 91°20′52.8″W).

Family Benthescymidae

4. *Benthesicymus tanneri* Faxon, 1893

Material Examined. St. 18 (07°20′57.0″N, 79°40′36.0″W), November 15, 2010, 720 m. **Comments** According to Hendrickx and Papiol (2015), *B. tanneri* is frequently collected off western Mexico (44 records available) in deep-water nets. It was also abundantly recorded by Faxon (1895) in its distributional range. It occurs between 606 m and 2422 m depth. During this study, it was collected only once, off Panama (Fig. 23.3), in 720 m depth. **Distribution** From San Diego, California, USA, to off Chile (18–22°S), including the Gulf of California to 27°34′N (Guzmán and Wicksten 2000; Hendrickx and Papiol 2015).

Caridea
Family Pasiphaeidae

5. *Pasiphaea emarginata* Rathbun, 1902

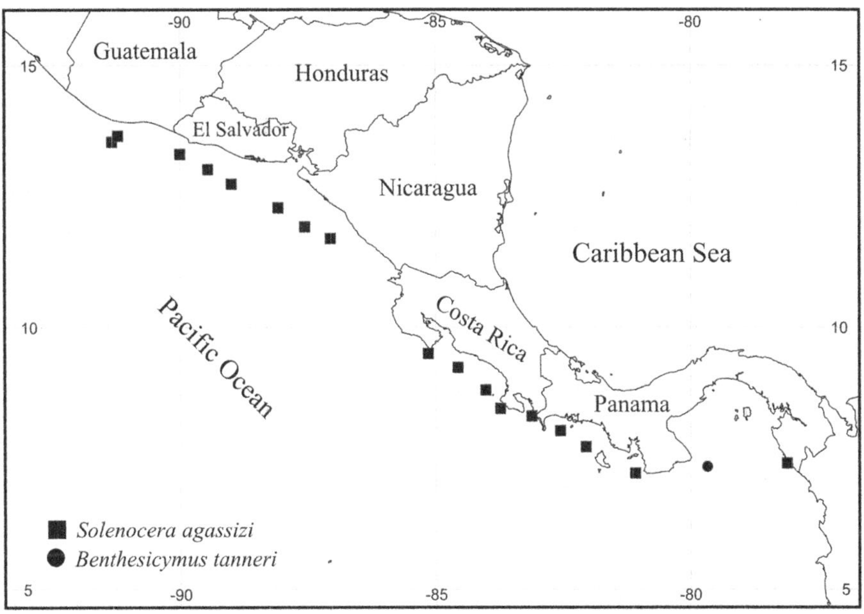

Fig. 23.3 Sampling localities of *Solenocera agassizii* and *Benthesicymus tanneri* in the survey area

Material Examined. St. 32 (07°15′15.0″N, 81°23′44.4″W), November 18, 2010, 108 m; St. 71 (10°41′15.0″N, 86°33′07.8″W), December 5, 2010, 827 m. Comments Found in only one sample, off Costa Rica (Fig. 23.4), *P. emarginata* is a pelagic species incidentally found in bottom trawl nets. Distribution Previously known from off Santa Barbara, California, USA, to SW Mexico (17°36′20″N, 103°57′45″W), including the southern and central Gulf of California (Hendrickx 2013). The southern distribution limit is herein extended to 07°15′15.0″N, off Costa Rica.

6. *Pasiphaea magna* Faxon, 1893

Material Examined. St. 12 (07°23′42.0″N, 79°32′20.4″W), November 14, 2010, 1021 m; St. 24 (07°01′47.4″N, 80°57′57.6″W) November 17, 2010, 1364 m; St. 99 (13°04′58.8″N, 90°36′07.2″W), December 12, 2010, 1054 m; St. 10 (07°30′55.8″N, 79°11′48.6″W), November 13, 2010, 1020 m; St. 26 (07°08′50.4″N, 81°07′29.4″W), November 17, 2010, 954 m; St. 28 (06°54′12.0″N, 81°18′43.2″W), November 18, 2010, 1467 m; St. 29 (07°04′48.6″N, 81°22′05.4″W), November 18, 2010, 1126 m; St. 30 (07°05′40.2″N, 81°27′43.8″W), November 18, 2010, 864 m; St. 42 (07°58′55.2″N, 82°44′25.8″W), November 21, 2010, 1115 m; St. 43 (07°56′43.8″N, 82°48′30.0″W), November 21, 2010, 951 m; St. 44 (07°53′27.6″N, 82°50′31.8″W), November 21, 2010, 1093 m; St. 55 (09°08′49.8″N, 84°33′40.8″W), November 24, 2010, 902 m; St. 62 (09°22′01.2″N, 85°27′13.2″W), November 25, 2010, 1457 m; St. 63 (09°43′47.4″N, 85°49′53.4″W), December 3, 2010, 1530 m; St. 65 (09°39′54.0″N, 85°43′54.6″W), December 3, 2010, 1178 m; St. 78 (11°25′32.4″N, 87°24′04.8″W), December 7, 2010, 1087 m; St. 83 (11° 50′45.6″N, 87°42′32.4″W), December 8, 2010, 1116 m; St. 96 (12°57′39.6″N, 90°19′02.4″W), December 11, 2010, 1100 m. **Comments** A bathypelagic species, *P. magna* is one of the largest pelagic shrimps occurring in the East Pacific (up to 85 mm total length). It was found in 18 samples taken in water with total depth of between 864 and 1530 m, off El Salvador, Nicaragua, Costa Rica, and Panama (Fig. 23.4). **Distribution** From off Oregon, USA, south to Peru (17°08′S) and off Chile (22°13′S), including the SE Gulf of California (Hendrickx and Estrada-Navarrete 1996; Wehrtmann and Carvacho 1997).

7. *Pasiphaea tarda* Krøyer, 1845 (Fig. 23.5a)

Material Examined. St. 80 (11°28′52.2″N, 87°14′57.6″W), December 7, 2010, 460 m. **Comments** A pelagic species incidentally captured in bottom trawls, *P. tarda* was found in one sample only, off Nicaragua (Fig. 23.4). **Distribution** From Unalaska, USA, to off Ecuador. Also known from the Atlantic Ocean (Hendrickx and Estrada-Navarrete 1996).

Family Pandalidae

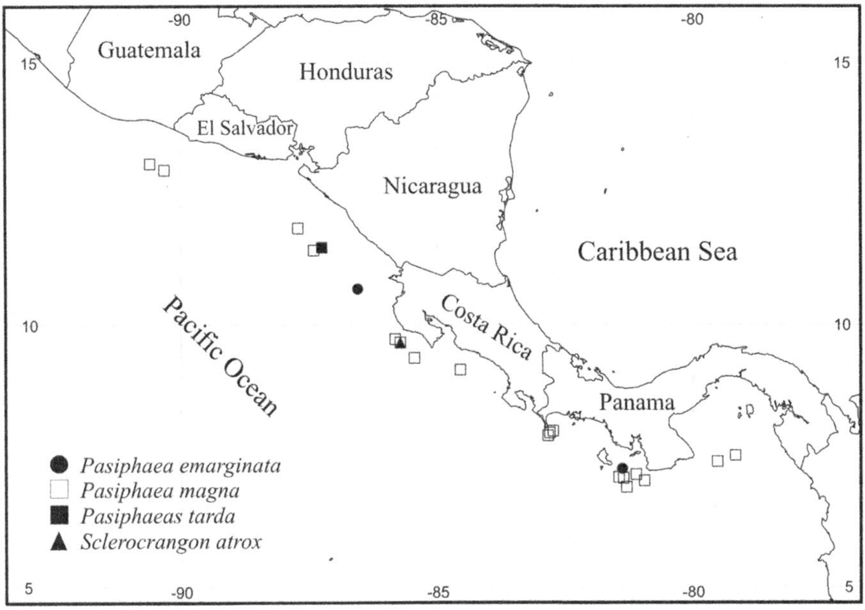

Fig. 23.4 Sampling localities of *Pasiphaea emarginata*, *P. magna*, *P. tarda*, and *Sclerocrangon atrox* in the survey area

8. *Heterocarpus hostilis* Faxon, 1893 (Fig. 23.5b)

Material Examined. St. 11 (07°23′36.6″N, 79°29′20.4″W), November 14, 2010, 1292 m; St. 10 (07°30′55.8″N, 79°11′48.6″w), November 13, 2010, 1020 m; St. 12 (07°23′42.0″N, 79°32′20.4″W), November 14, 2010, 1021 m; St. 13 (07°25′55.8″N, 79°33′40.2″W), November 14, 2010, 711 m; St. 16 (07°14′39.6″N, 79°40′03.6″W), November 15, 2010, 1446 m; St. 17 (07°16′16.2″N, 79°39′52.2″W), November 15, 2010, 1141 m; St. 18 (07°20′57.0″N, 79°40′03.6″W), November 18, 2010, 720 m; St. 24 (07°01′47.4″N, 80°57′57.6″W), November 17, 2010, 1364 m; St. 25 (07°05′20.4″N, 80°55′16.2″W), November 17, 2010, 1113 m; St. 26 (07°08′50.4″N, 81°07′29.4″W), November 17, 2010, 954 m; St. 28 (06°54′12.0″N, 81°18′43.2″W), November 18, 2010, 1467 m; St. 34 (06°57′00.6″N, 81°44′43.8″W), November 19, 2010, 1033 m; St. 40 (07°38′27.00″N, 82°27′04.2″W), November 20, 2010, 1185 m; St. 42 (07°58′55.2″N, 82°44′25.8″W), November 21, 2010, 1115 m; St. 43 (07°56′43.8″N, 82°48′30.0″W), November 21, 2010, 951 m; St. 44 (07°53′27.6″N, 82°50′31.8″W), November 21, 2010, 1093 m; St. 45 (08°16′03.0″N, 83°11′21.0″W), November 22, 2010, 1376 m; St. 46 (08°13′54.6″N, 83°10′34.2″W), November 22, 2010, 1260 m; St. 52 (08°46′03.0″N, 84°15′43.8″W), November 23, 2010, 836 m; St. 53 (09°01′33.6″N, 84°35′34.8″W), November 24, 2010, 1359 m; St. 54 (09°04′52.8″N, 84°34′30.0″W), November 24, 2010, 1188 m; St. 55 (09°08′49.8″N, 84°34′40.8″W), November 24, 2010, 902 m; St. 58 (09°23′30.6″N, 85°10′16.2″W), November 25, 2010, 1270 m; St. 59 (09°25′40.8″N, 85°09′40.8″W), November 25,

2010, 841 m; St. 62 (09°22′01.2″N, 85°27′13.2″W), November 25, 2010, 1457 m; St. 63 (09°43′47.4″N, 85°49′53.4″W), December 3, 2010, 1530 m; St.65 (09°39′54.0″N, 85°43′54.6″W), December 3, 2010, 1178 m; St. 71 (10°41′15.0″N, 86°33′07.8″W), December 5, 2010, 827 m; St. 72 (10°29′28.8″N, 86°27′30.0″W), December 5, 2010, 1212 m; St.75 (11°01′02.4″N, 86°51′54.6″W), December 6, 2010, 784 m; St. 76 (10°50′21.6″N, 86°44′54.0″W), December 6, 2010, 1145 m; St. 77 (11°25′59.4″N, 87°26′34.8″W), December 7, 2010, 1370 m; St. 78 (11°25′32.4″N, 87°24′04.8″W), December 7, 2010, 1087 m; St.81 (11°42′10.8″N, 87°03′48.6″W), December 7, 2010, 122 m; St. 82 (11°43′39.0″N, 87°40′01.8″W), December 8, 2010, 1471 m; St. 83 (1°50′45.6″N, 87°42′32.4″W), December 8, 2010, 1116 m; St. 86 (12°07′09.6″N, 88°16′07.8″W), December 9, 2010, 1389 m; St. 87 (12°10′16.2″ N, 88°14′57.6″W), December 9, 2010, 1100 m; St. 88 (12°12′04.8″N, 88°11′56.4″W), December 9, 2010, 767 m; St. 96 (12°57′39.6″N, 90°19′02.4″W), December 11, 2010, 1100 m; St. 97 (13°02′30.0″N, 90°13′19.8″W), December 9, 2010, 407 m; St. 99 (13°04′58.8″N, 90°36′07.2″W), December 12, 2010, 1054 m; St. 100 (13°06′48.6″N, 90°35′39.6″W), December 12, 2010, 798 m. **Comments** One of the four species of *Heterocarpus* occurring in the East Pacific, *H. hostilis* was one of the most frequently (43 trawls) collected during this survey (Fig. 23.6). It has been reported in depths of 187 to about 1900 m (Hendrickx 1995a). In this study it was collected off El Salvador, Nicaragua, Costa Rica, and Panama, in depths between 122 and 1530 m depth. **Distribution** Previously known from Mexico (23°16′42″N, 110°54′55″W) to off Supe, Peru (Hendrickx and Wicksten 2016).

9. *Heterocarpus vicarius* Faxon, 1893

Material Examined. St. 2 (07°20′36.6″N, 78°06′45.0″W), November 12, 2010, 412 m; St. 5 (07°34′16.8″N, 78°47′40.2″W), November 12, 2010, 430 m; St. 7 (07°37′57.6″N, 79°07′21.0″W), November 13, 2010, 478 m; St. 8 (07°32′52.2″N, 79°12′49.8″W), November 13, 2010, 757 m; St. 14 (07°32′35.4″N, 79°26′52.8″W), November 14, 2010, 435 m; St. 19 (07°23′08.4″N, 79°43′51.6″W), November 15, 2010, 480 m; St. 22 (07°11′21.6″N, 80°10′10.8″W), November 16, 2010, 454 m; St. 23 (07°14′48.0″N, 80°13′50.4″W), November 16, 2010, 103 m; St. 35 (07°04′14.4″N, 81°42′37.8″W), November 19, 2010, 671 m; St. 39 (07°44′33.6″N, 82°25′16.8″W), November 20, 2010, 243 m; St. 40 (07°38′27.0″N, 82°27′04.2″W), November 20, 2010, 1185 m; St. 51 (08°46′19.8″N, 84°05′55.2″W), November 23, 2010, 477 m; St. 52 (08°46′03.0″N, 84°15′43.8″W), November 23, 2010, 836 m; St. 60 (09°27′00.0″N, 85°09′0.00″W), November 25, 2010, 515 m. **Comments** Collected in 14 trawls, between 103 and 1185 m, off Costa Rica and Panama (Fig. 23.7). Reported in trawls between 73 and 760 m and in traps between 329 and 1454 m (Hendrickx 1995a). **Distribution** Gulf of California, Mexico, to off Mollendo, Peru (Wicksten and Hendrickx 2003).

10. *Pandalus amplus* (Bate, 1888) (Fig. 23.5c)

Fig. 23.5 (**a**). *Pasiphaea tarda*. (**b**) *Heterocarpus hostilis*. (**c**) *Pandalus amplus*. (**d**). *Sclerocrangon atrox*. (**e**). *Galacantha diomedeae*. (**f**). *Stereomastis pacifica*

Material Examined. St. 77 (11°25′59.4″N, 87°26′34.8″W), December 7, 2010, 1370 m. **Comments**. Found in one trawl only, at 1350 m depth off Guatemala (Fig. 23.7).

Distribution *Pandalus amplus* is the most widely distributed species of decapod crustaceans of the Americas, ranging from California, USA, to southern Chile and into the SW Atlantic (Boschi 2000; Wicksten and Hendrickx 2003).

Family Crangonidae

11. *Metacrangon procax* (Faxon, 1893)

Material Examined. St. 6 (07°42′28.8″N, 78°57′48.6″W), November 13, 2010, 109 m. Comments. Collected in one sample only, off Panama (Fig. 23.8). This species has been recorded throughout its range between 830 and 1658 m depth (Wicksten 1989), and the depth reported for the specimen examined herein is considered very unlikely. **Distribution** From southern California, USA, to off Atico, Peru, including the central and southern Gulf of California (Hendrickx 2012b, 2014a).

12. *Sclerocrangon atrox* Faxon, 1893 (Fig. 23.5d)

Material Examined. St. 65 (09°39′54.0″N, 85°43′54.6″W), December 3, 2010, 1178 m. Comments Collected in one sample only, off Costa Rica (Fig. 23.4), in 1178 m depth. The bathymetric range for this species is 800–1586 m depth. **Distribution** From western Mexico (27°05′42″N, 114°35′30″W) to off Mollendo, Peru, including the SW Gulf of California (Hendrickx 2014a).

Family Glyphocrangonidae

13. *Glyphocrangon alata* Faxon, 1893

Material Examined. St. 8 (07°32′52.2″N, 79°12′49.8″W), November 13, 2010, 757 m; St. 10 (07°30′55.8″N, 79°11′48.6″W), November 13, 2010, 1020 m; St. 12 (07°23′42.0″N, 79°32′20.4″W), November 14, 2010, 1021 m; St. 13 (07°25′55.8″N, 79°33′40.2″W), November 14, 2010, 711 m; St.14 (07°32′35.4″N, 79°26′52.8″W), November 14, 2010, 435 m; St. 18 (07°20′57.0″N, 79°40′03.6″W), November 15, 2010, 720 m; St. 19 (07°23′08.42″N, 79°43′51.6″W), November 15, 2010, 480 m; St. 21 (07°09′43.8″N, 80°08′03.0″W), November 16, 2010, 764 m; St. 26 (07°08′50.4″N, 81°07′29.4″W), November 17, 2010, 954 m; St. 30 (07°05′40.2″N, 81°27′43.8″W), November 18, 2010, 864 m; St. 33 (07°00′37.8″N, 81°44′00.0″W), November 19, 2010, 716 m; St. 34 (06°57′0.6″N, 81°44′43.8″W), November 19, 2010, 1033 m; St. 35 (07°04′14.4″N, 81°42′37.8″W), November 19, 2010, 671 m; St. 36 (07°10′54.0″N, 81°40′00.6″W), November 19, 2010, 143 m; St. 43 (07°56′43.8″N, 82°48′30.0″W), November 21, 2010, 951 m; St. 44 (07°53′27.6″N, 82°50′31.8″W), November 21, 2010, 1093 m; St. 52 (08°46′03.0″N, 84°15′43.8″W), November 23, 2010, 836 m; St. 53 (09°01′33.6″N, 84°35′34.8″W), November 24, 2010, 1396 m; St. 55 (09°08′49.8″N, 84°33′40.8″W), November 24, 2010, 902 m; St. 58 (09°23′30.6″N, 85°10′16.2″W), November 25, 2010, 1270 m; St. 59 (09°25′40.8″N, 85°09″ 40.8″W), November 25, 2010, 841 m. **Comments** This species appears to be very common off the southern portion of Central America, with 21 samples obtained off Costa Rica and Panama (Fig. 23.8), between 143 and 1369 m depth. The 143 m depth record, however, is doubtful and contrasts with the rest of the samples (20 were deeper than 435 m) and the previously known bathymetric range of this species, i.e., 600–1325 m (Hendrickx 2012b). **Distribution** Known from the SE Gulf of California, Mexico, to Valparaiso, Chile (Hendrickx 2012b).

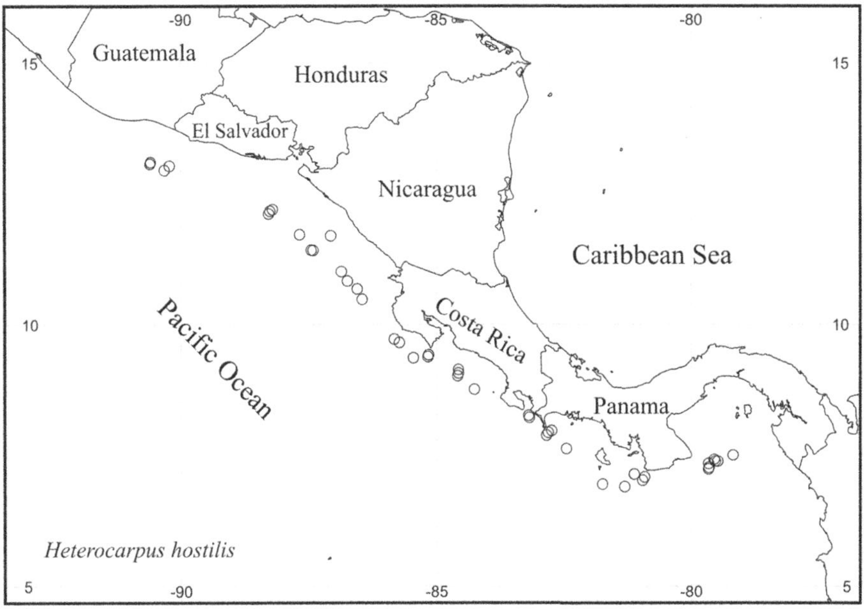

Fig. 23.6 Sampling localities of *Heterocarpus hostilis* in the survey area

14. *Glyphocrangon spinulosa* Faxon, 1893

Material Examined. St. 62 (09°22′01.2″N, 85°27′13.2″W), November 25, 2010, 1457 m; St. 63 (09°43′47.4″N, 85°49′53.4″W), December 3, 2010, 1530 m; St.65 (09°39′54,0″N, 85°43′54.6″W), December 3, 2010, 1178 m; St. 72 (10°29′28.8″N, 86°27′30.0″W), December 5, 2010, 1212 m; St. 77 (11°25′59.4″N, 87°26′34.8″W), December 7, 2010, 1370 m; St. 78 (11°25′32.8″N, 87°24′04.8″W), December 7, 2010, 1087 m; St. 83 (11°50′45.6″N, 87°42′32.4″W), December 8, 2010, 1116 m; St. 86 (12°07′15.6″N, 88°16′07.8″W), December 9, 2010, 1389 m; St. 87 (12°10′16.2″N, 88°14′57.6″W), December 9, 2010, 1100 m; St. 94 (12°53′15.6″N, 90°09′30.6″W), December 11, 2010, 1368 m; St. 99 (13°04′58.8″, 90°36′07.2″W), December 12, 1054 m. **Comments** Collected in 11 samples, off Guatemala, El Salvador, Nicaragua, and Costa Rica (Fig. 23.8), in depths of 1054 to 1530 m, a depth range that fits well with the previously reported depth range, i.e., 956–1586 m (Hendrickx 2012b). **Distribution** Cortez Basin, California, USA, to Panama, including the SE Gulf of California (Hendrickx 2012b.

Axiidea
Family Axiidae

15. *Guyanacaris caespitosa* (Squires, 1979) (Fig. 23.9a)

Material Examined. St. 36 (07°10′54.0″N, 81°40′06.0″W), November 19, 2010, 143 m. Comments Collected in one sample only, off Panama (Fig. 23.10). Known

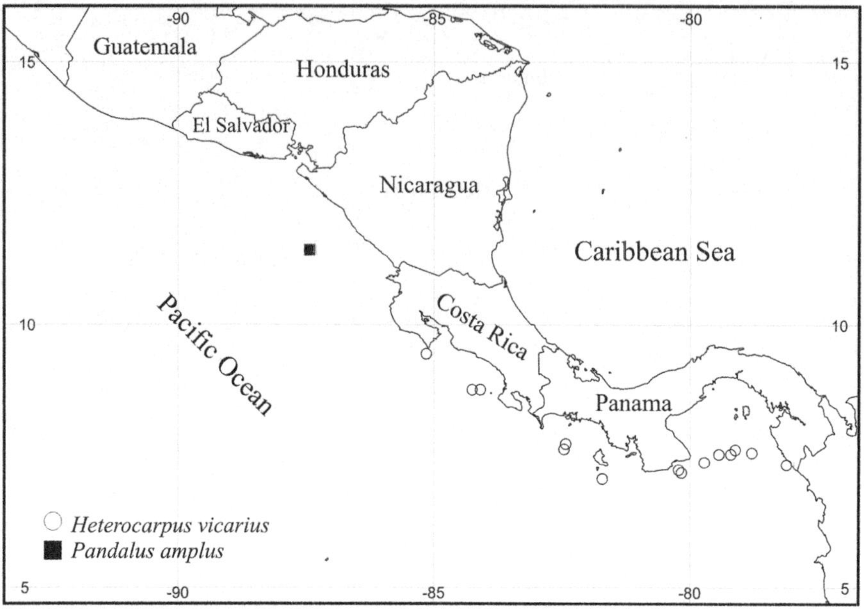

Fig. 23.7 Sampling localities of *Heterocarpus vicarius* and *Pandalus amplus* in the survey area

from depths between 72 and 200 m (Hendrickx 1995b). Distribution East coast of the Gulf of California, Mexico, to Mancora, Peru. In Central America there is only one record in Corinto, Nicaragua (Hendrickx 2005a, b).

Astacidea
Family Nephropidae

16. *Nephropsis occidentalis* Faxon, 1893 (Fig. 23.9b)

Material Examined. St. 26 (07°08′50.4″N, 81°07′29.4″W), November 17, 2010, 954 m; St. 52 (08°46′03.0″N, 84°15′43.8″W), November 23, 2010, 836 m; St. 55 (09°08′49.8″N, 84°33′40.8″W), November 24, 2010, 902 m; St. 65 (09°39′54.0″N, 85°43′54.6″W), December 3, 2010, 1178 m; St. 72 (10°29′28.8″N, 86°27′30.0″W), December 5, 2010, 1212 m; St. 76 (11°01′02.4″N, 86°51′54.6″W), December 6, 2010, 784 m; St. 78 (11°25′32.4″N, 87°24′04.8″W), December 7, 2010, 1087 m; St. 83 (11°50′45.6″N, 67°42′32.4″W), December 8, 2010, 1116 m; St. 87 (12°10′16.2″N, 88°14′57.6″W), December 9, 2010, 1100 m; St. 96 (12°57′39.6″N, 90°19′02.4″W), December 11, 2010, 1100 m; St. 99 (13°04′58.8″N, 90°36′07.2″W), December 12, 2010, 1054 m. **Comments** Collected in 11 samples, off El Salvador, Nicaragua, Costa Rica, and Panama (Fig. 23.11). The specimens were captured between 784 and 1212 m depth that fit well with the known bathymetric distribution of this species which is 270–1310 m (Papiol et al. 2016). **Distribution** *Nephropsis occidentalis* features a very wide distributional range, from off the Baja California Peninsula (27°05′42″N) to southern Chile (Papiol et al. 2016).

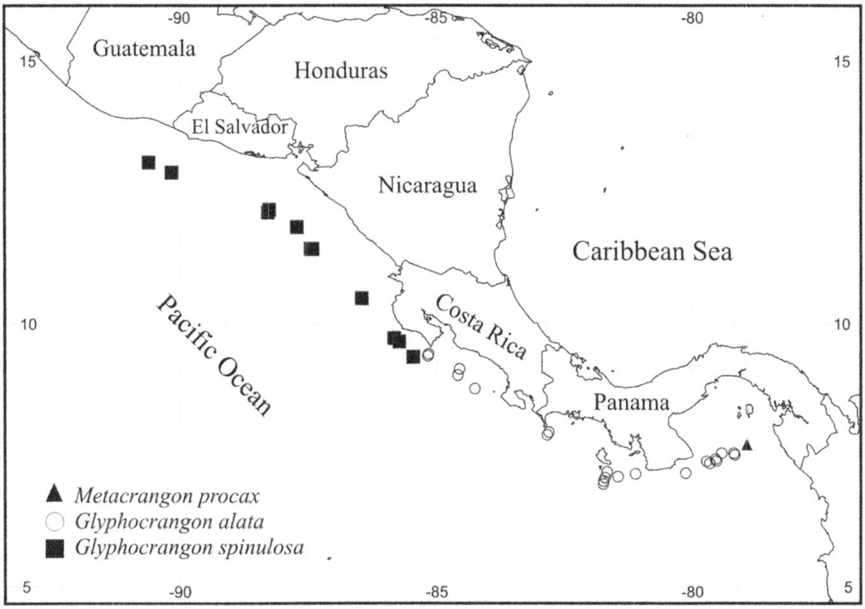

Fig. 23.8 Sampling localities of *Metacrangon procax*, *Glyphocrangon alata*, and *G. spinulosa* in the survey area

Polychelida
Family Polychelidae

17. *Stereomastis pacifica* (Faxon, 1893) (Fig. 23.5f)

Material Examined. St. 63 (09°43′47.4″N, 85°49′53.4″W), December 12, 2010, 1530 m; St. 64 (09°40′49.8″N, 85°43′38.4″W), December 3, 2010, 1105 m; St. 72 (10°39′28.8″N, 86°27′30.0″W), December 5, 2010, 1212 m. **Comments** Collected in three samples, off Costa Rica (Fig. 23.10), between 1105 and 1530 m. This species has been reported between 982 and 3330 depth (Hendrickx). **Distribution** Concepción Point, California, USA, to Chile (25°S, 70°40′W), including the SW Gulf of California, Mexico (Hendrickx 2015a).

Anomura
Family Munididae

18. *Pleuroncodes planipes* (H. Milne Edwards, 1837)

Material Examined. St. 38 (07°43′24.6″N, 82°03′25.2″W), November 20, 2010, 144 m; St. 39 (07°44′33.6″N, 82°25′16.8″W), November 20, 2010, 243 m; St. 57 (09°14′25.2″N, 84°34′12.0″W), November 24, 2010, 134 m; St. 73 (11°05′03.0″N, 86°44′45.6″W), December 5, 2010, 153 m; St. 81 (11°42′10.8″N, 87°03′48.6″W), December 7, 2010, 122 m; St. 83 (11°50′45.6″ N, 87°42′32.4″W), December 8, 2010, 1116 m; St. 85 (11°55′54.6″N, 87°33′30.6″W), December 8, 2010, 181 m; St.

90 (12°17′24.0″N, 88°04′28.8″W), December 9, 2010, 131 m; St. 91 (12°44′38.4″N, 88°59′39.6″W), December 10, 2010, 105 m; St. 92 (13°01′11.4″N, 89°27′29.4″W), December 10, 2010, 104 m; St. 98 (13°18′48.0″N, 90°00′11.4″W), December 11, 2010, 114 m; St. 102 (13°17′33.6″N, 90°24′16.2″W), December 12, 2010, 165 m; St. 103 (13°42′50.4″N, 91°50′06.6″W), December 13, 2010, 117 m; St. 104 (13°32′47.4″N, 91°29′52.8″W), December 13, 2010, 114 m; St. 105 (13°28′37.2″N, 91°23′49.8″W), December 13, 2010, 137 m. Comments A pelagic species with benthic phase, *P. planipes* is often extraordinary abundant in bottom trawls made by fishing boats on the continental shelf (Aurioles-Gamboa and Balart 1995). In this study, it was found in 15 samples in localities with total depth between 104 and 1116 m off Guatemala, El Salvador, Nicaragua, Costa Rica, and Panama (Fig. 23.12). Distribution From California, USA, to off Costa Rica, including the Gulf of California (Hendrickx 2012c).

Family Munidopsidae

19. *Galacantha diomedeae* (Faxon, 1893) (Fig. 23.5e)

Material Examined. St. 30 (07°05′40.2″N, 81°27′43.8″W), November 18, 2010, 864 m; St. 53 (09°01′33.6″N, 84°35′34.8″W), November 24, 2010, 1359 m. **Comments** A widely spread and common species off western Mexico (Hendrickx and Papiol 2019), *G. diomedeae* was found in only two samples, off Costa Rica and Panama (Fig. 23.13), in 864 and 1359 m depth. Its known depth range is 768–3790 m (Baba et al. 2008). **Distribution** From off San Clemente, California, USA, to off Constitución (35°31.48′S), Chile (Baba et al. 2008; Hendrickx and Papiol 2019).

20. *Munidopsis depressa* Faxon, 1893

Material Examined. St. 70 (10°41′03.6″N, 86°24′52.8″W), December 4, 2010, 243 m. **Comments**. Collected in one sample, in 243 m depth, off Costa Rica (Fig. 23.13). This record is surprising as *M. depressa*, a frequently collected species of *Munidopsis* in western Mexico, has been recorded in depths of 780–1300 m (Hendrickx 2012c). **Distribution** *Munidopsis depressa* is known from off Catalina Island, California, USA, to off Guerrero (16°52′N), including the central and southern Gulf of California, Mexico. A closely related species, *M. hamata* Faxon, 1893, ranges south to Chile (26–34°S) (Moscoso 2014).

Family Lithodidae

21. *Paralomis diomedeae* (Faxon, 1893)

Material Examined. St. 18 (07°20′57.0″N, 79°40′03.6″W), November 15, 2010, 720 m. **Comments** Collected in only one station, off Panama (Fig. 23.12), in 720 m depth. The known depth range for this species is 680–935 m (Moscoso 2012). **Distribution** California, USA, Costa Rica, Panama, and to northern Peru (Moscoso 2012).

Brachyura

Family Majidae

22. *Maiopsis panamensis* Faxon, 1893

Material Examined. St. 11 (07°23′36.6″N, 79°29′20.4″W), November 14, 2010, 1292 m; St. 41 (08°01′57.6″N, 82°33′51.0″W), November 21, 2010, 115 m; St. 42 (07°58′55.2″N, 82°44′25.8″W), November 21, 2010, 1115 m.
Comments Collected in three samples, all off Panama (Fig. 23.14), in depths of 115–1292 m. The reported depth range for this species is 15–335 m (maximum depth in the type locality) (Faxon 1895; Hendrickx 1995d). The depth range recorded during this study (115–1292 m) is considerably higher.
Distribution From Abreojos Point (26°42′25″N, 113°34′22″W) to Caleta la Cruz, Peru, including the Gulf of California, Mexico (Hendrickx 2005b).

Fig. 23.9 (**a**). *Guyanacaris caespitosa*. (**b**). *Nephropsis occidentalis*. (**c**). *Achelous xantusii*. (**d**). *Trichopeltarion corallinum*

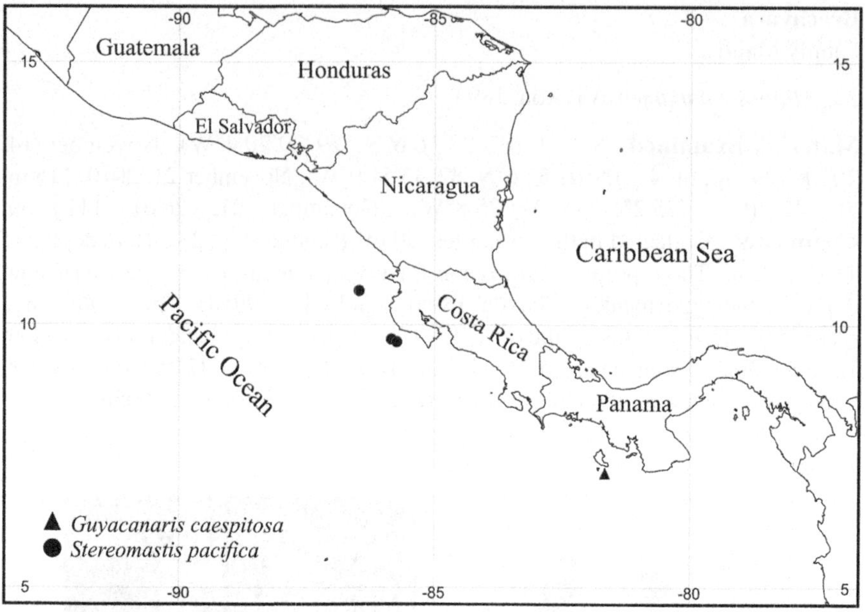

Fig. 23.10 Sampling localities of *Guyanacaris caespitosa* and *Stereomastis pacifica* in the survey area

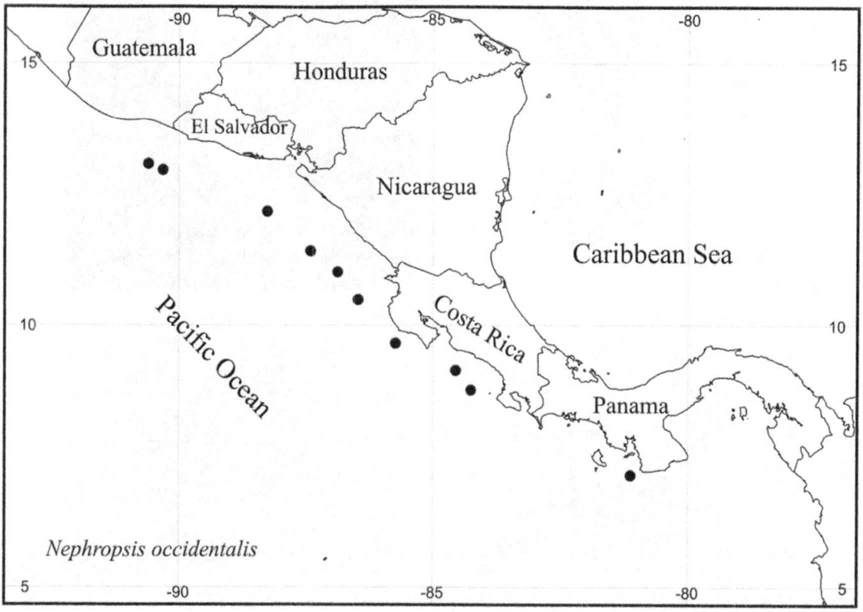

Fig. 23.11 Sampling localities of *Nephropsis occidentalis* in the survey area

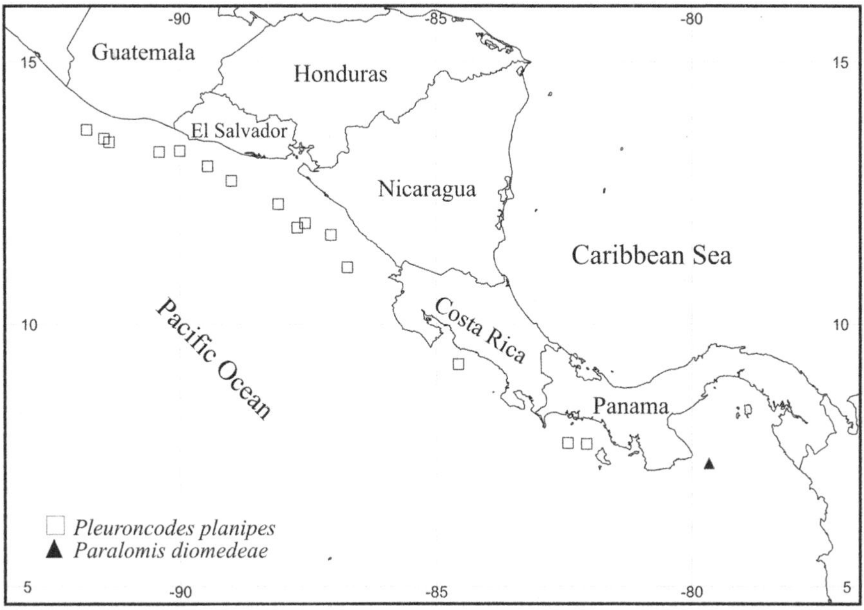

Fig. 23.12 Sampling localities of *Pleuroncodes planipes* and Paralomis diomedeae in the survey area

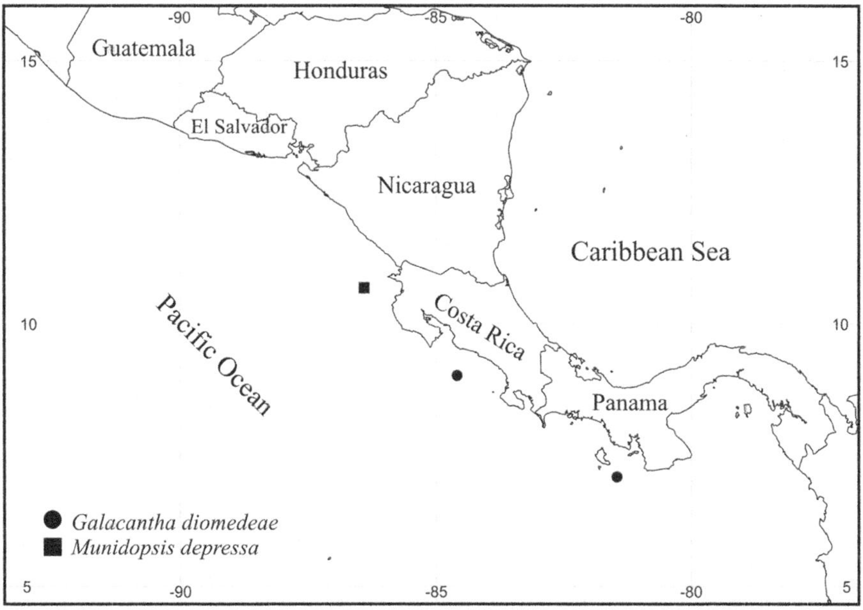

Fig. 23.13 Sampling localities of *Galacantha diomedeae* and *Munidopsis depressa* in the survey area

Family Cancridae

23. *Cancer johngarthi* Carvacho, 1989

Material Examined. St. 7 (07°37′57.6″N, 79°07′21.0″W), November 13, 2010, 478 m; St. 14 (07°32′35.4″N, 79°26′52.8″W), November 14, 2010, 435 m; St. 19 (07°23′08.4″N, 79°43′51.6″W), November 15, 2010, 480 m; St. 22 (07°11′21.6″N, 80°10′10.8″W), November 16, 2010, 454 m. Comments Found in four localities, all off Panama (Fig. 23.14), in 435–480 m depth. The known depth range is 61–523 m (Hendrickx 1995d). Distribution From off Guadalupe Island, Mexico, to the Bay of Panama, including the Gulf of California (Hendrickx 2005b).

Family Calappidae

24. *Platymera gaudichaudii* H. Milne Edwards, 1837

Material Examined. St. 1 (07°24′49.8″N, 78°07′27.0″W), November 12, 2010, 183 m; St. 20 (07°26′52.8″N, 79°50′24.0″W), November 15, 2010, 109 m; St. 32 (07°15′15.0″N, 81°23′44.4″W), November 18, 2010, 108 m; St. 36 (07°10′54.0″N, 81°40′00.6″W), November 19, 2010, 143 m. **Comments** Collected in four samples, off Panama (Fig. 23.15), in depths of 108 to 183 m. The reported depth range for this species is 31–400 m (Hendrickx 1995d). **Distribution** From Queen Charlotte Islands, Canada, to off Talcahuano, Chile (Moscoso 2012, Wicksten 2012).

Family Portunidae

25. *Achelous xantusii* (Stimpson, 1860) (Fig. 23.9c)

Material Examined. St. 22 (07°11′21.6″N, 80°10′10.8″W), November 16, 2010, 454 m. **Comments** Only one sample of this species was collected, off Panama (Fig. 23.15), in depth of 454 m. However, *A. xantusii* presents a pelagic phase and is often seen swimming near surface; thus, it was most certainly captured in the net when the gear was recovered. **Distribution** From Santa Barbara, California, USA, to the central and southern Gulf of California, south to off San Mateo del Mar, Oaxaca, Mexico (Hendrickx et al. 1997; Hendrickx 2005b; as *Portunus xantusii*). The southern distribution limit of *A. xantusii* is herein extended to off Panama.

Family Trichopeltariidae

26. *Trichopeltarion corallinum* (Faxon, 1893) (Fig. 23.9d)

Material Examined. St. 65 (09°39′54.0″N, 85°43′54.6″W), December 3, 2010, 1178 m; St. 76 (10°50′21.6″N, 86°44′54.0″W), December 6, 2010, 1145 m. **Comments** Collected in two samples, off Nicaragua and Costa Rica (Fig. 23.15), in 1145–1178 m. Known depth range is 507 to 1280 m (Guzmán et al. 2009; Moscoso 2012). **Distribution** From the southern Gulf of California, Mexico, to Chiloe (42°35′S), Chile (Hendrickx 1996; Moscoso 2012).

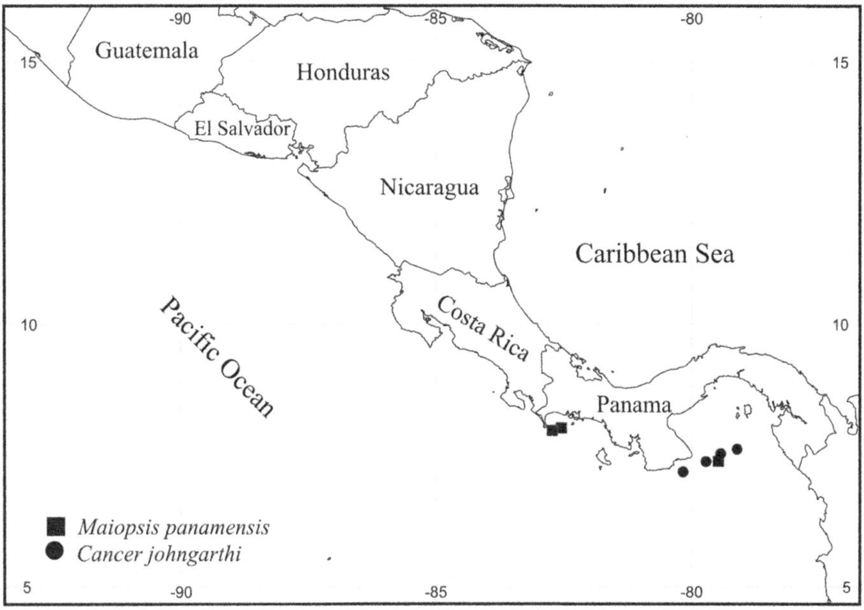

Fig. 23.14 Sampling localities of *Maiopsis panamensis* and *Cancer johngarthi* in the survey area

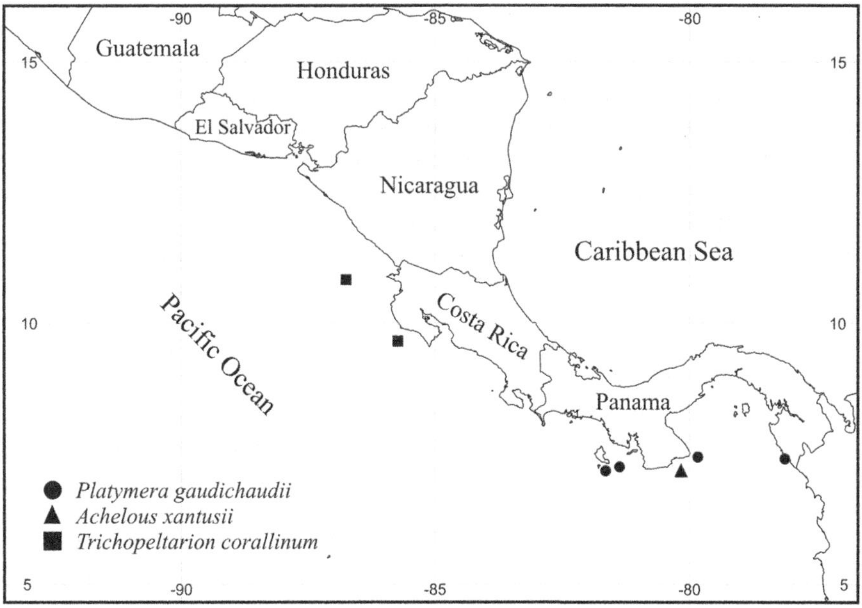

Fig. 23.15 Sampling localities of *Platymera gaudichaudii*, *Achelous xantusii*, and *Trichopeltarion corallinum* in the survey area

23.5 Discussion

In this report, we document the presence of one species of stomatopod and 25 species of decapod crustaceans in deep water off Central America. The number of localities where the species were found during this survey varied considerably, from 1 (10 species) to 43 (Table 23.1). Species with the higher number of records were *Heterocarpus hostilis* (43), *Haliporoides diomedeae* (28), and *Glyphocrangon alata* (21) (Table 23.1), but these 3 species were not those with the widest distribution. Indeed, only two species, *Solenocera agassizii* and *Pleuroncodes planipes*, were collected throughout the survey area, off each of the five countries considered in the study (Table 23.1). Unfortunately, due to adverse circumstances, it was not possible to obtain quantitative data related to capture of the species reported herein.

Records for all species except four correspond to their previously known distribution in the eastern Pacific. Distribution of *Haliporoides diomedeae* was extended to the north, from Panama to off El Salvador. The northernmost distribution limit of *Solenocera agassizii* is extended from off Costa Rica to off Guatemala. In the case of the pelagic shrimp *Pasiphaea emarginata*, the southernmost limit of its distribution range is now set to off Costa Rica, a considerable extension of over 10° of latitude to the south. In a similar way, the swimming crab *Achelous xantusii* is also first recorded off Central America, with an extension of about 9° of latitude to the south. However, the distribution of the species complex *A. xantusii* (formerly reported as *Portunus xantusii*), considered for a long period of time to be composed of three subspecies (see Garth and Stephenson 1966) and recently validated at species rank (Mantelatto et al. 2018), should be carefully reviewed based on re-examination of fresh material.

Some bathymetric ranges observed during this survey are rather surprising and much deeper that previously reported. *Squilla biformis* was found in 116–1376 m depth, much deeper that the previously reported maximum depth, i.e., 518 m. In the East Pacific, stomatopods mostly occur in coastal areas and on the continental shelf (Hendrickx and Salgado-Barragán 2002). Although one particular group of species, the Bathysquilloidea (two families, four species) is known from deep water around the globe (Ahyong 2001), no member of it has so far been recorded off the coast of western America.

Another case is *Metacrangon procax* collected during the "Miguel Oliver" cruise in 109 m depth, almost eight times shallower than the shallowest depth on record (i.e., 830 m depth). One out of 21 samples of *Glyphocrangon alata* was presumably collected in shallow depth, i.e., 143 m vs. > 435 m for the rest of the samples; this is, again, a very unusual depth of residence for this species, and this record is also considered doubtful.

The recognition of *Pleuroncodes planipes* off Central America, one of two species of the genus *Pleuroncodes* which distribution is restricted to the American Pacific, is based on comparison with material from off western Mexico. The two species, *P. planipes* and *P. monodon* (H. Milne Edwards, 1837), are very similar and might co-occur in Central America where *P. monodon* has been reported in deep water (Hernáez and Wehrtmann 2014). Another Galatheoidea, *Munida depressa*, was collected in one sample, in 243 m depth, off Costa Rica. This record is also surprising as *M. depressa*, a frequently collected species of *Munidopsis* in western Mexico, has been recorded in depths >800 m (Hendrickx 2001). The general reported depth range, however, is from 185 to 1255 m (Wicksten 1989). During this survey, the large spider crab *Maiopsis panamensis* was collected twice in depths >1000 m, i.e., 1115 and 1292 m, at least three times deeper than maximum depth on record (15–335 m).

During a previous expedition in the same area aboard the R/V "Dr. F. Nansen" in 1987, from the Gulf of Tehuantepec, Mexico, to Nicaragua, a large amount of specimens were collected from the shelf and the upper slope. The vast majority of the samples, however, were collected in much shallower waters than during the R/V "Miguel Oliver" cruise, from 100 m depth or less (Bianchi 1991). Comparatively, only five species were common to these two cruises or similar: the stomatopod *Squilla biformis* and the decapods *Solenocera agassizii*, *Heterocarpus* sp. (probably *H. vicarius*), *Pleuroncodes monodon* (?*P. planipes*), and *Platymera gaudichaudii* (cited as *Mursia*) (Bianchi 1991).

Deep-water decapods in the eastern Pacific are much more diverse than what can be observed from the results of the R/V "Dr. F. Nansen" (Bianchi 1991) or the R/V "Miguel Oliver" cruises (this study). Based on previous studies by Hendrickx (2011, 2012b, c, 2014a, b, 2019a, b; Hendrickx and Ayón-Parente 2012, 2013, 2014), the number of recorded benthic and benthypelagic species in water deeper than 350 m off western Mexico is 97, not including at least five unpublished new records or undescribed species. While reviewing records of decapod crustaceans occurring mostly in water deeper than 50 m, Wicksten (1989) listed 186 species, of which 161 (87%) were found in depths >350 m. However, only 25 species (13%) showed a bathymetric distribution restricted to water deeper than 1500 m (Wicksten 1989). In a more recent study, Wicksten (this volume) reported 119 species occurring mostly below 700 m depth in the eastern Pacific. Compared to the eastern Pacific, the number of species of decapod crustaceans collected off Central America is low and represents only 16% of species recorded below 350 m depth. This percentage increases to 26% when compared to the fauna occurring below 350 only off western Mexico.

According to published information, it is likely that *Squilla biformis* is the only stomatopod occurring below 200 m depth off Central America. As seen from the comparative data cited previously and analyzing the overall distribution range of decapod crustaceans in the eastern Pacific, the number of species present off Central America is certainly much higher than the one reported herein (i.e., 25). It is therefore necessary to increase the number of surveys in these waters, particularly below the 500 m depth contour, in order to obtain additional, more complete information on these communities that likely occur below the minimum oxygen zone.

From a fishery viewpoint, a few species reported herein are subject to commercial fishery, while some are recognized as presenting a real (high to medium) fishery potential (Table 23.2) due to their size or because similar species are exploited elsewhere. Others have a "low" fishery potential (Table 23.2) for their small size, their

Table 23.2 Fishery potential (high, medium, low) of species collected during this survey and their maximum size on record. TL, total length; CW, carapace width

Species	Potential	Max. size (mm)	Source
Squilla biformis	High	208 (TL)	Hendrickx and Salgado Barragán (1989)
Haliporoides diomedeae	High	215 (TL)	Hendrickx (1995a)
Solenocera agassizi	High	155 (TL)	Rodríguez et al. (2012)
Benthesicymus tanneri	Medium	135 (TL)	Hendrickx (2001)
Pasiphaea emarginata	Low	81 (TL)	Wicksten (2012)
Pasiphaea magna	Medium	145 (TL)	Wicksten (2012)
Pasiphaea tarda	Medium	215 (TL)	Wicksten (2012)
Heterocarpus hostilis	High	140 (TL)	Hendrickx (1995a)
Heterocarpus vicarius	High	115 (TL)	Hendrickx (1995a)
Pandalus amplus	High	170 (TL)	Hendrickx (1995a)
Metacrangon procax	Low	50 (TL)	Hendrickx (1995a)
Sclerocrangon atrox	Medium	165 (TL)	Hendrickx (1995a)
Glyphocrangon alata	Medium	107 (TL)	Hendrickx (1995a)
Glyphocrangon spinulosa	Medium	168 (TL)	Hendrickx (2012b)
Guyanacaris caespitosa	Low	112 (TL)	Hendrickx (2005a)
Nephropsis occidentalis	Medium	130 (TL)	Hendrickx (2003)
Stereomastis pacifica	Low	129 (TL)	Galil (2000)
Pleuroncodes planipes	High	110 (TL)	Aurioles-Gamboa and Balart (1995)
Galacantha diomedeae	Low	32 (CW)	Hendrickx and Papiol (2019)
Munidopsis depressa	Low	38 (TL)	Hendrickx (2001)
Paralomis diomedeae	High	128 (CW)	Haig (1974)
Maiopsis panamensis	High	240 (CW)	Hendrickx (1995d)
Cancer johngarthi	High	140 (CW)	Hendrickx (1995d)
Platymera gaudichaudii	High	162 (CW)	Wicksten (2012)
Achelous xantusii	Low	70 (CW)	Hendrickx (1995d)
Trichopeltarion corallinum	Low	26 (CW)	Tavares and Cleva 2010

low meat-content, their very low abundance, or their bathymetric distribution. The development of intensive fishery activities in deep water, however, should be strongly limited until fundamental aspects linked to their population dynamic are fully understood. Management policy should then be established in order to avoid a rapid depletion of stocks of species that probably experience slow growing process.

Acknowledgments This paper is dedicated to my friend and colleague, the late Jorge A. López (RIP), who took an active part in this contribution and in the gathering of the information reported herein. Always enthusiastic in participating in studies dealing with marine species, Jorge left us when this project was going on. We also acknowledge the "Dirección General Especializada de OSPESCA" and Reinaldo Morales Rodríguez (SICA/OSPESCA) for their permission to use the information included in this contribution. Many thanks to Mercedes Cordero Ruiz for preparing the final version of this manuscript and to two reviewers for improving this manuscript.

References

Ahyong ST (2001) Revision of the Australian Stomatopod Crustacea. Rec Aust Mus Suppl 26:1–326

Aramayo V (2016) Breve síntesis sobre el recurso bacalao de profundidad *Dissostichus eleginoides* en Perú. Rev Biol Mar Oceanogr 51(2):229–239

Arana EP (2003) Experiencia chilena en faenas de pesca en aguas profundas y distantes: Evolución y Perspectivas. In: Yañez RE (ed) Actividad Pesquera y de Acuicultura en Chile. 1. Aspectos globales. Ediciones Universitarias de Valparaiso, Chile, pp 57–79

Arana P, Ahumada M, Guerrero A (2002) Pesca exploratoria de camarones de aguas profundas en las regiones V y VI, año 2002. Informe Final. Estud. Doc, Univ. Católica Valparaíso 20/2002:167 pp

Arana PM, Wehrtmann IS, Orellana JC et al (2013) By-catch associated with fisheries of *Heterocarpus vicarius* (Costa Rica) and *Heterocarpus reedi* (Chile) (Decapoda: Pandalidae): a six-year study (2004–2009). J Crustac Biol 33(2):198–209

Aurioles-Gamboa D, Balart EF (1995) La langostilla: Biología, ecología y aprovechamiento. Centro de Investigaciones Biológicas del Noroeste, S.C. La Paz, BCS, Mexico

Baba K, MacPherson E, Poore GCB et al (2008) Catalogue of squat lobsters of the world (Crustacea: Decapoda: Anomura-families Chirostylidae, Galatheidae and Kiwaidae). Zootaxa 1905:1–220

Barriga E, Salazar C, Palacios J, Romero M, Rodriguez A (2009) Distribución, abundancia y estructura poblacional del langostino rojo de profundidad *Haliporoides diomedeae* (Crustacea: Decapoda: Solenoceridae) frente a la zona norte de Perú (2007–2008). Lat Am J Aquat Res 37(3):371–380

Beltrán-León BS, Ríos Herrera R, Rubio Rincón E (2016) New findings and distribution of some deep sea species of the Colombian Pacific II. *Callionymidae* (Teleostei). Bol Invest Mar Cost – INVEMAR 45(1):155–163

Bianchi G (1991) Demersal assemblages of the continental shelf and slope edge between the Gulf of Tehuantepec (Mexico) and the Gulf of Papagayo (Costa Rica). Mar Ecol Prog Ser 73:121–140

Boschi EE (2000) Species of decapod crustaceans and their distribution in the American marine zoogeographic provinces. Rev Invest Des Pesq 13:3–136

Cornejo Antepara M (2010) Los crustáceos decápodos de aguas profundas del Mar Ecuatoriano. MSc dissertation. Universidad de Guayaquil, Facultad de Ciencias Naturales

Cruz-Acevedo E, Tolimieri N, Aguirre-Villaseñor H (2018) Deep-sea fish assemblages (300–2100 m) in the eastern Pacific off northern Mexico. Mar Ecol Prog Ser 592:225–242

Faxon W (1893) Reports on the dredgings operations off the west coast of Central America to the Galapagos, to the west coast of Mexico, and in the Gulf of California, in charge of Alexander Agassiz, carried on by the U.S. Fish Commision Steamer "Albatross", during 1891, Lieut. Commander Z.L. Tanner, U.S.N., commanding. VI. Preliminary descriptions of new species of Crustacea. Bull Mus Comp Zool Harv Univ 24(7):149–220

Faxon W (1895) Reports on an exploration off the west coast of Mexico, Central and South America, and off the Galapagos Islands, in charge of Alexander Agassiz, carried on by the U.S. Fish Commission steamer "Albatross", during 1891, Lieut. Commander Z.L. Tanner, U.S.N., commanding. XV. The stalk-eyed Crustacea. Mem Mus Comp Zool Harv Coll 18:1–292

Galil BS (2000) Crustacea Decapoda: a revision of the family Polychelidae. In: Crosnier A (ed) Résultats des Campagnes MUSORSTOM, vol 21. Mém Mus ntnl Hist nat Paris Zool 184:285–387

Garth JS, Stephenson W (1966) Brachyura of the Pacific coast of America. Brachyrhyncha: Portunidae. Allan Hancock Monogr Mar Biol 1:1–154

Guzmán G, Quiroga E (2005) New records of shrimps (Decapoda: Caridea and Dendrobranchiata) in deep waters off Chile. Gayana 69(2):285–290

Guzmán GL, Sellanes J (2015) A review of the Munidopsidae Ortmann, 1898 (Decapoda, Galatheoidea) in Chilean waters, including new records for the Southeastern Pacific. Zootaxa 4021(2):282–306

Guzmán G, Wicksten MK (2000) The subfamily Benthesicyminae Bouvier, 1908 (Decapoda, Dendrobranchiata) in northern Chile (18° to 22°S). Crustaceana 73(8):925–932

Guzmán G, Báez P, Sellanes J (2009) Primer registro de *Trichopeltarion corallinus* (Faxon, 1893) para el mar de Chile y nuevo registro de *T. hystricosus* (Garth, en Garth & Haig, 1971) (Decapoda: Brachyura: Atelecyclidae). Lat Am J Aquat Res 37(2):275–279

Haig J (1974) Observations on the lithodid crabs of Peru, with description of two new species. Bull South Calif Acad Sci 73(3):152–164

Hendrickx ME (1995a) CAMARONES. In: Fischer W, Krupp F, Schneider W, Sommer C, Carpenter KE, Niem VH (eds) Guía FAO para la identificación de especies para los fines de la pesca. Pacífico centro-oriental. Vol. I. Plantas e Invertebrados. FAO, Roma, pp 417–537

Hendrickx ME (1995b) LANGOSTAS (Langostas espinosas, bogavantes, cigarras y zapateras, langostas de lodo, etc.). In: Fischer W, Krupp F, Schneider W, Sommer C, Carpenter KE, Niem VH (eds) Guía FAO para la identificación de especies para los fines de la pesca. Pacífico centro-oriental. Vol. I. Plantas e Invertebrados. FAO, Roma, pp 383–416

Hendrickx ME (1995c) ANOMUROS. In: Fischer W, Krupp F, Schneider W, Sommer C, Carpenter KE, Niem VH (eds) Guía FAO para la identificación de especies para los fines de la pesca. Pacífico centro-oriental. Vol. I. Plantas e Invertebrados. FAO, Roma, pp 539–564

Hendrickx ME (1995d) CANGREJOS. In: Fischer W, Krupp F, Schneider W, Sommer C, Carpenter KE, Niem VH (eds) Guía FAO para la identificación de especies para los fines de la pesca. Pacífico centro-oriental. Vol. I. Plantas e Invertebrados. FAO, Roma, pp 565–636

Hendrickx ME (1995e) Checklist of lobsters-like decapod crustaceans (Crustacea: Decapoda: Thalassinidea, Astacidea and Palinuridea) from the Eastern Tropical Pacific. An Inst Bio UNAM 66(2):153–165

Hendrickx ME (1996) New records of deep-water decapod crustaceans in the southeastern Gulf of California, Mexico. Rev Biol Trop 44(2B):945–947

Hendrickx ME (2001) Occurrence of a continental slope decapod crustacean community along the edge of the minimum oxygen zone in the southeastern Gulf of California, Mexico. Belg J Zool 131(Suppl. 2):95–109

Hendrickx ME (2003) Distribution and size of the Pacific deep water lobsterette, *Nephropsis occidentalis* Faxon, 1893 (Decapoda: Astacidea: Nephropidae) in the S.E. Gulf of California, Mexico. Crustaceana 76(2):207–216

Hendrickx ME (2005a) Additional records of *Acanthaxius caespitosa* (Squires, 1979) (Decapoda; Thalassinoidea; Axiidae) from the eastern tropical Pacific. Crustaceana 77(10):1277–1278

Hendrickx ME (2005b) Cap. 14. Crustacea 6. Decapoda: Dendrobranchiata, Caridea, Palinura, Anomura & Brachyura. In: Hendrickx ME, Brusca RC, Findley LT (eds) A Distributional Checklist of the Macrofauna of the Gulf of California, Mexico. Part I. Invertebrates. [Listado y Distribución de la Macrofauna del Golfo de California, México, Parte I. Invertebrados]. Arizona-Sonora Desert Museum, pp 159–194

Hendrickx ME (2011) Distribution and abundance of *Neocrangon resima* (Rathbun, 1902) (Crustacea: Decapoda: Caridea: Crangonidae) collected during the TALUD XIV cruise in the Gulf of California, Mexico. Nauplius 19(2):129–134

Hendrickx ME (2012a) Operaciones oceanográficas en aguas profundas: los retos del pasado, del presente y del proyecto TALUD en el Pacífico mexicano (1989–2009). In: Zamorano P, Hendrickx ME, Caso M (eds) Biodiversidad y comunidades del talud continental del Pacífico mexicano. Secretaría del Medio Ambiente y Recursos Naturales (SEMARNAT). Instituto Nacional de Ecología (INE), México, pp 23–104

Hendrickx ME (2012b) Los Glyphocrangonidae y Crangonidae (Crustacea: Decapoda: Caridea) recolectados durante los cruceros TALUD en el Pacífico mexicano. In: Zamorano P, Hendrickx ME, Caso M (eds) Biodiversidad y comunidades del talud continental del Pacífico mexicano. Secretaría del Medio Ambiente y Recursos Naturales (SEMARNAT). Instituto Nacional de Ecología (INE), México, pp 319–356

Hendrickx ME (2012c) Squat lobsters (Crustacea: Decapoda: Galatheoidea and Chyrostyloidea) collected during the TALUD XIV cruise in the Gulf of California, Mexico, and rediscovery of *Gastropthychus perarmatus* (Haig, 1968) in the eastern Pacific. Zootaxa 3418:28–40

Hendrickx ME (2013) Prevalence and distribution of the dinoflagellate *Thalassomyces californiensis* Collard, 1966 (Ellobiopsidae) on *Pasiphaea emarginata* Rathbun, 1902 (Crustacea: Decapoda: Caridea: Pasiphaeidae), off Western Mexico. Crustaceana 86(6):693–703. https://doi.org/10.1163/15685403-00003194

Hendrickx ME (2014a) Deep-water species of Crangonidae and Glyphocrangonidae (Decapoda, Caridea) collected during the TALUD XV cruise off the west coast of southern Baja California, Mexico. Crustaceana 87(1):83–90. https://doi.org/10.1163/15685403-00003274

Hendrickx ME (2014b) First record of the caridean shrimp genus *Bresilia* Calman, 1896 (Crustacea: Decapoda: Bresiliidae) in the East Pacific and description of a new species. Zootaxa 3878(2):185–195. dx. https://doi.org/10.11646/zootaxa.3878.2.5

Hendrickx ME (2015) Specimens of *Stereomastis pacifica* (Faxon, 1893) collected during the TALUD XVI-B cruise off the west coast of the Baja California Peninsula, Mexico. Crustaceana 88(2):203–207. https://doi.org/10.1163/15685403-00003393

Hendrickx ME (2016) *Plesionika sanctaecatalinae* Wicksten, 1983 (Crustacea Decapoda Caridea Pandalidae) from off the west coast of Pacific Mexico. Zootaxa 4111(4):492–500

Hendrickx ME (2019a) Redescription of the rare shrimp *Heterocarpus nesisi* (Burukovsky, 1986) (Crustacea: Caridea: Pandalidae) rediscovered off western Mexico. Zootaxa 4565(1):49–60. https://doi.org/10.11646/zootaxa.4565.1.3

Hendrickx ME (2019b) Deep-water lithodids of the genus *Paralomis* White, 1856 (Decapoda; Anomura; Lithodidae) off western Mexico. Crustaceana 92(10):1257–1264. https://doi.org/10.1163/15685403-00003942

Hendrickx ME, Ayón-Parente M (2012) First record of *Prionocrangon* Wood Mason & Alcock, 1981 (Crustacea: Decapoda: Caridea: Crangonidae) in the East Pacific and description of a new deep-water species from western Mexico. Zootaxa 3205:63–68

Hendrickx ME, Ayón-Parente M (2013) A new species of *Munidopsis* (Anomura: Galatheoidea: Munidopsidae) from the Gulf of California, western Mexico. Crustaceana 86(11):1306–1317. https://doi.org/10.1163/15685403-00003241

Hendrickx ME, Ayón-Parente M (2014) A new deep-water species of *Odontozona* (Decapoda: Stenopodidea: Stenopodidae) from the East Pacific, and new record of *O. foresti* Hendrickx, 2002. Zootaxa 3835(3):338–348. dx. https://doi.org/10.11646/zootaxa.3835.3.3

Hendrickx ME, Estrada-Navarrete FD (1996) Los camarones Pelágicos (Crustacea: Dendrobranchiata y Caridea) del Pacífico mexicano. Comisión Nacional para el Conocimiento y Uso de la Biodiversidad e Instituto de Ciencias del Mar y Limnología UNAM, México

Hendrickx ME, López J (2006) Geographic and depth distribution of *Bathybembix bairdii* (Dall, 1889) (Mollusca, Gastropoda, Trochidae) in the East Pacific. Oceanides 21(1–2):93–99

Hendrickx ME, López J (2019) The deep-water lophogastrid *Gnathophausia scapularis* Ortmann, 1906 (Crustacea, Lophogastrida) in Central America. Geomare Zool 2:3–5

Hendrickx ME, Papiol V (2015) Distribution of *Benthesicymus tanneri* Faxon, 1893 (Dendrobranchiata, Benthesicymidae) off the west coast of Mexico and notes on its morphology. ZooKeys 473:119–136. https://doi.org/10.3897/zookeys.473.8956

Hendrickx ME, Papiol V (2019) Distribution, relative growth, ecology and fecundity of the deep-water squat lobster *Galacantha diomedeae* (Crustacea: Decapoda: Galatheoidea: Munidopsidae) in the Mexican Pacific. Invertebr Biol. https://doi.org/10.1111/ivb.12248

Hendrickx ME, Salgado Barragán J (1989) Ecology and fishery of stomatopods in the Gulf of California. In: Ferrero EA (ed) (Manning RB, Reaka ML, Wales W (co-eds)) Biology of stomatopods, Collana UZI: Selected Symposia and Monographs. Mucchi Editore, Modena, pp 241–249

Hendrickx ME, Salgado-Barragán J (2002) Stomatopoda. In: Morrone Lupi JJ, Llorente-Bousquets JE, Ponce H (eds) Biodiversidad, taxonomía y biogeografía de artrópodos de México: Hacia una síntesis de su conocimiento, vol III. Universidad Nacional Autónoma de México, pp 373–400

Hendrickx ME, Wicksten MK (2016) New records of decapod crustaceans in the eastern Pacific. Crustaceana 89(5):603–610. https://doi.org/10.1163/15685403-00003541
Hendrickx ME, Demestre M, Esparza-Haro A, Salgado-Barragán J (1997) Stomatopod and decapod crustaceans collected during the CEEMEX P5 and P7 cruises to the Gulf of Tehuantepec, Mexico. Oceanides 11(2):1–28
Hendrickx ME, Mah C, Zárate Montes CM (2011) Deep-water Asteroidea (Echinodermata) collected during the TALUD cruises in the Gulf of California, Mexico. Rev Mex Biodiv 82:798–824
Hernáez P, Wehrtmann IS (2014) Breeding cycle of the red squat lobster *Pleuroncodes monodon* H. Milne Edwards, 1837 (Decapoda, Muninidae) from deepwater Pacific of Costa Rica. Mar Ecol 35(2):204–211
Kameya A, Castillo R, Escudero L et al (1997) Localización, distribución y concentración de langostinos rojos de profundidad. Crucero BIC Humboldt 9607-08 18 de julio a 6 de agosto de 1996. Publ Esp, Instit Mar Perú, 47 p
Kameya A, Moscoso V, Lellish M (1998) Los crustáceos decápodos y estomatópodos del Perú. Inf Inst Mar Perú 136:80–109
López JA (2011) Identificación de especies de crustáceos colectados en el Pacífico centroamericano durante la campaña oceanográfica pesquera del B/O "Miguel Oliver" (Primera parte). SICA-OSPECA-OIRSA Report, San Salvador, 66 p
López-Rocha JA, Arreguin-Sanchez F, Ramirez-Rodriguez M (2006) Stock size estimation of the Mexican crab, *Cancer johngarthi* Carvacho, 1989 on the southwestern peninsula of Baja California, Mexico. Crustaceana 79(9):1033
Mantelatto FL, Robles R, Wehrtmann IS, Schubart CD, Felder DL (2018) New insights into the molecular phylogeny of the swimming crabs of the genera *Portunus* Weber, 1795 and *Achelous* De Haan, 1833 (Brachyura: Portunidae) of the Americas. J Crustac Biol 38(2):190–197
Moscoso V (2012) Catálogo de crustáceos decápodos y estomatópodos del Perú. Bol Inst Mar Perú 27:1–2):1–207
Navas BRS, Bermúdez A, Angel-Yunda C, Campos NH (2013) Afinidades biogeográficas de los galateoideos (Decapoda: Anomura) del Caribe y Pacífico colombiano. Rev MVZ Córdoba 18(2):3501–3511
Papiol V, Hendrickx ME, Serrano D (2016) Distribution and ecology of the Pacific lobsterette *Nephropsis occidentalis* Faxon, 1893 (Crustacea, Decapoda, Astacidea) on the continental slope off western Mexico. In: Riosmena-Rodriguez R (ed) Marine benthos: biology, ecosystem functions and environmental impact. Nova Science Publishers, Inc., New York, pp 197–223
Pedraza MJ, Zapata Padilla LA (2011) Capítulo 5. Síntesis biológico-pesquera de los camarones de profundidad del género *Heterocarpus* (*H. vicarius* y *H. hostilis*), capturados en aguas del Pacífico colombiano. pp 127–156. In: Díaz Merlano JM, Vieira Betancourt C, Melo Sldarriaga G (eds) Diagnóstico de las principales pesquerías del Pacífico colombiano. Fundación Mar Viva, Bogotá, Colombia. cap. 5, 242 pp
Peréz-Farfante I (1977) American solenocerid shrimps of the genera *Hymenopenaeus*, *Haliporoides*, *Hadropenaeus* new genus and *Mesopenaeus* new genus. Fish Bull 75(2):261–346
Puentes V, Madrid N (1994) Evaluación del Camarón de Aguas Profundas del Pacífico Colombiano. Ministerio de Agricultura, Instituto Nacional de Pesca y Acuicultura -INPA-. Informe Final. Septiembre de 1994
Retamal RMA (1993) Crustáceos decápodos abisales de la zona Iquique-Arica. Estud Ocean 12:1–8
Robertson DR, Angúlo A, Baldwin CC et al (2017) Deep-water bony fishes collected by the B/O Miguel Oliver on the shelf edge of Pacific Central America: an annotated, illustrated and DNA-barcoded checklist. Zootaxa 4348(1):1–125
Rodríguez A, Rueda M, Cubillos LA (2012) Agregación reproductiva del camarón de aguas profundas *Solenocera agassizii* (Crustacea: Decapoda) en el Pacífico Colombiano. Bol Invest Mar Cost 41(2):429–446
Salgado-Barragán J, Hendrickx ME (2010) Guía de los estomatópodos del Pacífico oriental. Rev Mex Biodiv 81(Suppl):1–49

Sánchez-Jiménez A, Naranjo-Elizondo B, Rodríguez-Arrieta A et al (2018) Updated catalogue of bony fishes observed in deep waters at Isla del Coco National Park and Las Gemelas seamount, Costa Rica (Eastern Tropical Pacific). Rev Biol Trop 66(5):S1–S113

Starr RM, Green K, Sala E (2012) Deepwater fish assemblages at Isla del Coco National Park and Las Gemelas Seamount, Costa Rica. Rev Biol Trop 60:347–362

Suárez Mozo NY, Valentich-Scott P, Hendrickx ME (2018) Deep-water bivalves from the oxygen minimum zone area off the western Peninsula of Baja California, Mexico. Mol Res 39(2):99–109. https://doi.org/10.1080/13235818.2018.1508550

Tavares M, Cleva R (2010) Trichopeltariidae (Crustacea, Decapoda, Brachyura), a new family and superfamily of eubrachyuran crabs with description of one new genus and five new species. Pap Avulsos Zool (São Paulo) 50(9):97–157

Vélez J, Kameya A, Yamashiro C, Lostaunau N, Valinte O (1992) Investigación del recurso potencial de langostino rojo de profunidad a bordo del BIC “Fridtjof Nansen” (24 de abril - 25 de mayo 1990). Inf Instit Mar Perú C.E.E: 104:3–24

Villalobos-Rojas F, Wehrtmann IS (2018) Reproductive biology of the commercially exploited kolibri shrimp, *Solenocera agassizii* (Decapoda: Solenoceridae), from the Pacific coast of Costa Rica, with considerations for its management. Rev Biol Trop 66(1):S92–S107

Villalobos-Rojas F, Azofeifa-Solano JC, Camacho-García JE, Wehrtmann IS (2017) Gastropods and bivalves taken as by-catch in the deep-water shrimp trawl-fishery along the Pacific coast of Costa Rica, Central America. Mol Res 37(3):175–186

Wehrtmann IS, Carvacho A (1997) New records and distribution ranges of shrimps (Crustacea: Decapoda: Penaeoidea and Caridea) in Chilean waters. Proc Biol Soc Wash 110(1):49–57

Wehrtmann IS, Nielsen-Munoz (2009) The deepwater fishery along the Pacific Costa Rica, Central America. In: Arana P, Perez JAA, Pezzutto PR (eds) Deep-sea fisheries off Latin America. Lat Am J Aquat Res 37(3):543–554

Wehrtmann IS, Arana PM, Barriga E et al (2012) Deep-water shrimp fisheries in Latin America: a review. Lat Am J Aquat Res 40(3):497–535

Wicksten MK (1989) Ranges of offshore decapod crustaceans in the eastern Pacific Ocean. Trans San Diego Soc Nat Hist 21(19):291–316

Wicksten MK (2012) Decapod Crustacea of the Californian and Oregonian zoogeographic provinces. Zootaxa 3371:1–307

Wicksten MK (this volume) Lower slope and abyssal benthic decapods of the eastern Pacific. In: Deep-water Pycnogonida and Crustacea from the Americas. Springer Latin-America, pp xx-ww

Wicksten MK, Hendrickx ME (2003) An updated checklist of benthic marine and brackish water shrimps (Decapoda: Penaoidea, Stenopodidea, Caridea) from the Eastern Tropical Pacific. In: Hendrickx ME (ed) Contributions to the Study of East Pacific Crustaceans 2. [Contribuciones al Estudio de los Crustáceos del Pacífico Este 2]. Instituto de Ciencias del Mar y Limnología, UNAM, Mexico, pp 49–76

Zamorano P, Hendrickx ME, Mendez N et al (2012) Explorando las aguas profundas del Pacífico mexicano: el proyecto TALUD. In: Low Pfeng A, Peters Recagno EM (eds) La frontera final: el océano profundo. Instituto Nacional de Ecología, Secretaria del Medio Ambiente y Recursos Naturales (Semarnat), Mexico, pp 107–151

Chapter 24
Diversity and Biology of Deep-Water Crustaceans in Costa Rica

J. C. Azofeifa-Solano and J. Cortés

Abstract Crustaceans are an important component of deep-sea biodiversity. A brief review of the history of expeditions and studies related to deep-sea crustaceans in Costa Rica is presented. We briefly discuss studies on crustaceans from the Costa Rican deep-sea environments, and we provided an updated list of species recorded for the Pacific and the Caribbean. A total of 147 species has been reported from Costa Rican deep sea; 8 species have been reported from the Caribbean, 138 from the Pacific, and 1 from both basins. Decapoda was the most diverse group with 87 species, followed by Copepoda (23 spp.) and Peracarida (19 spp.). The first deep-sea exploration in Costa Rica began with foreign efforts, with national projects and participation increasing in recent years. Most research dealing with crustaceans has been focused on reproductive biology, in collaboration with the deepwater shrimp fisheries. Future efforts to study the Costa Rican deep-sea will incorporate collaboration with foreign expeditions and private companies since the country does not have enough funding invested in its deep sea. Finally, we discuss the current threats to deep-sea crustaceans, as well as future perspectives for the study of this fascinating group in Costa Rica.

Keywords Biodiversity · Deep water · Benthos · Pelagos · Decapoda · Copepoda Peracarida

J. C. Azofeifa-Solano (✉)
Centro de Investigación en Ciencias del Mar y Limnología (CIMAR), Universidad de Costa Rica, San José, Costa Rica
e-mail: juan.azofeifa@ucr.ac.cr

J. Cortés
Centro de Investigación en Ciencias del Mar y Limnología (CIMAR), Universidad de Costa Rica, San José, Costa Rica

Escuela de Biología, Universidad de Costa Rica, San José, Costa Rica
e-mail: jorge.cortes@ucr.ac.cr

M. E. Hendrickx (ed.), *Deep-Sea Pycnogonids and Crustaceans of the Americas*, https://doi.org/10.1007/978-3-030-58410-8_24

24.1 Introduction

The deep-sea is probably the last frontier of science and is the largest realm on Earth (Roberts et al. 2006; Ramírez-Llodra et al. 2011). Even so, only a small proportion of this habitat has been sampled, but we know that deep-sea diversity is quite remarkable and unique (Woolley et al. 2016). In the last decades, however, depletion of resources on land (e.g., minerals) and onshore waters (e.g., seafood through fishing), coupled with increasing market demands and technological advances, have resulted in major economic, logistical, and political demands to exploit mineral and biological resources in the deep sea (Ramírez-Llodra et al. 2011; Norse et al. 2012; Levin and Le Bris 2015). Nonetheless, the ecosystems and species found in the deep sea are remarkably different from those occurring in shallow waters, and the former are considered highly vulnerable to exploitation and other forms of human-induced pressures, such as plastic pollution and climate change, in part because of their delayed maturity, slow growth, and low or sporadic recruitment (Jennings et al. 1998; Morato et al. 2006; Watson and Morato 2013). Therefore, recording the deep-sea diversity and studying its ecological dynamics are a prevailing requirement to develop conservation and management tools and to properly evaluate the impacts of human activities lurking this deep wilderness (Costello et al. 2013).

The exploration of deep-sea regions in Costa Rica (> 200 m depth) started in the late nineteenth century, and some crustaceans have been recorded from those early explorations (Cortés 2009) (see Sect. 24.3.1). But there has been little crustacean-specific research, and most studies only reported on the presence of "crabs" or "squat lobsters" in the deep-sea habitats, without specifying species or even sampling the organisms for further study. In this contribution we compiled published and unpublished information to reconstruct the history of deep-sea crustacean research, to generate a baseline checklist of recorded species, and to comment on threats and future prospects for Costa Rican deep-sea carcinology.

24.2 Material and Methods

We reviewed the literature on deep-sea habitats from Costa Rica up to May 2020, including expeditions and cruise reports and published articles and books. We followed the widely accepted definition for the deep sea as oceanic waters and seabed below 200 m depth (Gage and Tyler 1991; Greene et al. 1999; Roff and Taylor 2000; UNESCO 2009). We summarized the history of deepwater expeditions and deepwater crustacean records from Costa Rican waters. We also included descriptions of the most commonly studied deep-sea habitats, some references to the general biology of deep-sea crustaceans, and an updated list of the deep-sea crustacean species recorded in Costa Rican waters, both in the Caribbean Sea and in the Pacific Ocean.

24.3 Results and Discussion

24.3.1 *Brief History of Expeditions and Studies Related to Crustaceans*

The Costa Rican deep-sea habitats have been mainly explored by foreign expeditions and cruises in both the Caribbean Sea and the Pacific Ocean (Cortés 2009). The first deepwater expedition took place in 1891, led by Alexander Agassiz on the US Fish Commission Steamer "Albatross" (Agassiz 1892; Townsend 1901; Cortés 2009). They sampled near Isla del Coco from 26 February to 2 March, 1891, with dredging and trawling at 12 stations (#3362–3373), ranging in depth from 95 to 3433 m. Crustaceans were collected and described by Faxon (1893, 1895). There were several expeditions to Costa Rica between the late nineteenth century and the early 1970s, but these took few deepwater samples. In 1925, the "Arcturus" Oceanographic Expedition, led by William Beebe, collected samples from three stations, between 272 and 1636 m depth (Beebe 1926). The New York Zoological Society 1938 expedition to the coast of the Pacific of Mesoamerica, lead again by William Beebe, collected deepwater samples at two stations, between 360 and 910 m depth (Beebe 1938), but we were not able to locate any report of deepwater crustaceans from the stations visited during this expedition. Between 1933 and 1939 the Allan Hancock Pacific Expeditions project collected shallow and shore samples (Fraser 1943). Several species of deep-sea organisms collected during the 1952 "Galathea" expedition have been assigned to Costa Rica, for example, the tanaidacean *Neotanais armiger* described by Wolff (1956), but the station where it was collected was located in international waters (9°23'N, 89°32'W), outside the exclusive economic zone (EEZ) of Costa Rica. This species was again collected during the RV "Vema" expedition of 1958, at a close-by station. The RV "Vema" also collected within the limits of Costa Rican waters (6°21'N, 85°17'W; 1892 m) (Child 1992), but we found no reports on crustaceans. Isolated collections were done by Carl Hubbs and Spencer Luke (Scripps Institution of Oceanography), aboard the RV "Agassiz" in 1973, and some deepwater crustaceans were reported from these collections (Luke 1977; Wicksten 1979).

The Caribbean of Costa Rica has been much less explored, and its crustacean fauna is practically unknown (Cortés 2009). During 1971, the RV "John Elliott Pillsbury" explored offshore waters of the Central American Caribbean, including two stations between 245 and 290 m and another three stations between 715 and 770 m, but no crustacean collection were reported (Voss 1971). In 2011, the fishing vessel RV "Miguel Oliver" sampled along the Caribbean continental slope of Central America, including Costa Rica, in depth from 385 to 1481 m, but only one paper was published (dealing with echinoderms) (Cambronero-Solano et al. 2019).

Between 1987 and 2014, there was a series of geophysical expeditions along the Pacific margin of Costa Rica, including the Ocean Drilling Program; "Alvin" dives in 1994, 2004, and 2014; and a series of German cruises between 1999 and 2003 (Ranero et al. 2008; Sahling et al. 2008; Wheat et al. 2017). An important discovery was the presence of cold methane seeps and hydrothermal vents on the Pacific margin of Costa Rica, with rich biological communities (Sahling et al. 2008; Levin et al. 2009, 2015; Wheat et al. 2019). Since 2006, waters surrounding Isla del Coco have been explored down to 450 m with the "DeepSee" submersible (Cortés and Blum 2008). In 2010, the RV "Miguel Oliver" collected samples along the Pacific coast of Central America, including Costa Rica (Robertson et al. 2017). Crustaceans have been recorded from these explorations, but most studies only report the presence of crabs or squat lobsters in these habitats, without specifying species or sampling the organisms. However, further information on stomatopods and decapod crustaceans collected by this research vessel in 2010 is available in this report (Chap. 23).

The crustaceans inhabiting the Costa Rican deep sea have been poorly studied. Specific studies are limited to species associated with deep-sea fisheries (Wehrtmann and Echeverría-Sáenz 2007; Macpherson and Wehrtmann 2010; Villalobos-Rojas et al. 2020) or remarkable species associated with cold methane seeps such as the yeti crab *Kiwa puravida* (Thurber et al. 2011; Goffredi et al. 2014). Costa Rican scientists have partnered with commercial fisheries as their major source of samples (Wehrtmann and Echeverría-Sáenz 2007; Wehrtmann and Nielsen-Muñoz 2009; Macpherson and Wehrtmann 2010; Villalobos-Rojas et al. 2020).

24.3.2 General Description of Deep-Sea Environments

Costa Rica is a small Central America country, covering 51,100 km^2, with a marine area almost 11 times larger than land area (583,548 km^2). Most marine area (EEZ) corresponds to deep sea: 63% is below 2000 m and 36% below 3000 m (Cortés 2016a, b; Cortés and Benavides-Varela in prep). Along the Pacific margin and in deeper waters, the substrate mostly consists of fine, soft sediments (Townsend 1901, Spinelli and Underwood 2004, Cortés unp data). On the Pacific margin, there are methane seeps where authigenic carbonates are deposited, and a high diversity of benthic organisms is found attached to these rocks (Sahling et al. 2008; Levin et al. 2015). Seamounts and Isla del Coco are rocky environments with a unique biodiversity, including crustaceans (Cortés 2019, Cortés unp. data).

24.3.3 Crustacean Fauna from Costa Rican Deep Waters

24.3.3.1 General Results

A total of 147 deep-sea crustacean species (96 genera, 56 families, 8 orders) has been recorded inhabiting benthic and pelagic habitats off Costa Rica, 138 occurring in the Pacific, 8 in the Caribbean, and 1 reported in both basins (Table 24.1). The group with the most species was the decapods with 87 species. Within these, the

Table 24.1 List of deep-sea crustaceans (>200 m) recorded within Costa Rican waters, in both the Caribbean Sea and the Pacific Ocean

Order	Family	Scientific name	Area	Depth range Costa Rica	Sources
Stomatopoda	Hemisquillidae	*Hemisquilla californiensis* Stephenson, 1967	P	273	Wehrtmann and Echeverría-Sáenz (2007)
	Squillidae	*Squilla biformis* Bigelow, 1891	P	131–350	Camp and Kuck (1990)
		Squilla hancocki Schmitt, 1940	P	220	Camp and Kuck (1990)
		Squilla panamensis Bigelow, 1891	P	206–208	Camp and Kuck (1990)
Euphausiacea	Euphausiidae	*Euphausia diomedeae* Ortmann, 1894	P	500	Azofeifa-Solano and Vargas-Castillo (2020)
		Euphausia distinguenda Hansen, 1908	P	500	Azofeifa-Solano and Vargas-Castillo (2020)
		Euphausia eximia Hansen, 1911	P	200	Azofeifa-Solano et al. (2016)
		Euphausia gibboides Ortmann, 1893	P	500	Azofeifa-Solano and Vargas-Castillo (2020)
		Euphausia lamelligera Hansen, 1911	P	200	Azofeifa-Solano et al. (2016)
		Euphausia tenera Hansen, 1905	P	200	Azofeifa-Solano et al. (2016)
		Nematobrachion flexipes (Ortmann, 1893)	P	500	Azofeifa-Solano and Vargas-Castillo (2020)
		Nematoscelis gracilis Hansen, 1910	P	500	Azofeifa-Solano and Vargas-Castillo (2020)
		Nematoscelis tenella G.O. Sars, 1883	P	500	Azofeifa-Solano and Vargas-Castillo (2020)
		Stylocheiron affine Hansen, 1910	P	200	Azofeifa-Solano et al. (2016)
		Stylocheiron carinatum G.O. Sars, 1883	P	200	Azofeifa-Solano et al. (2016)
		Stylocheiron longicorne G.O. Sars, 1883	P	200	Azofeifa-Solano et al. (2016)
		Stylocheiron maximum Hansen, 1908	P	200	Azofeifa-Solano et al. (2016)

(continued)

Table 24.1 (continued)

Order	Family	Scientific name	Area	Depth range Costa Rica	Sources
		Thysanopoda orientalis Hansen, 1910	P	500	Azofeifa-Solano and Vargas-Castillo (2020)
Decapoda	Acanthephyridae	*Acanthephyra curtirostris* Wood-Mason in Wood-Mason & Alcock, 1891	P	1646	Azofeifa-Solano and Vargas-Castillo (2020)
		Meningodora mollis Smith, 1882	P	1408	Faxon (1893, 1895)
	Aethridae	*Hepatus pudibundus* (Herbst, 1785)	C	14–210	Vargas and Wehrtmann (2009)
		Osachila kaiserae Zimmerman & Martin, 1999	P	2–219	Vargas and Wehrtmann (2009)
	Alpheidae	*Alpheus bellimanus* Lockington, 1877	P	0–300	Vargas and Cortés (1999)
	Alvinocarididae	*Alvinocaris costaricensis* Martin, Wall, Shank, Cha, Seid & Rouse, 2018	P	995–1817	Martin et al. (2018)
	Benthesicymidae	*Bentheogennema burkenroadi* Krygier & Wasmer, 1975	P	500	Azofeifa-Solano and Vargas-Castillo (2020)
		Benthoecetes tanneri (Faxon, 1893)	P	1650–2149	Faxon (1893, 1895)
		Gennadas scutatus Bouvier, 1906	P	1646	Azofeifa-Solano and Vargas-Castillo (2020)
	Calappidae	*Acanthocarpus delsolari* Garth, 1973	P	93–250	Vargas and Wehrtmann (2009)
		Calappula saussurei (Rathbun, 1898)	P	13–275	Vargas and Wehrtmann (2009)
		Cryptosoma bairdii (Stimpson, 1860)	P	2.7–229	Vargas and Wehrtmann (2009)
		Platymera gaudichaudii H. Milne Edwards, 1837	P	22–399	Vargas and Wehrtmann (2009)
	Cancridae	*Cancer johngarthi* Carvacho, 1989	P	90–523	Vargas and Wehrtmann (2009)

(continued)

Table 24.1 (continued)

Order	Family	Scientific name	Area	Depth range Costa Rica	Sources
	Colossendeidae	*Colossendeis* sp.	P	2582	Durkin (2018)
	Crangonidae	*Parapontophilus gracilis* (Smith, 1882)	P	1789–1951	Faxon (1893, 1895)
		Sclerocrangon atrox Faxon, 1893	P	800–1250	Vargas and Wehrtmann (2009)
	Diogenidae	*Paguristes bakeri* Holmes, 1900	P	40–232	Vargas and Wehrtmann (2009)
	Epialtidae	*Stenocionops ovatus* (Bell, 1835)	P	15–275	Vargas and Wehrtmann (2009)
	Ethusidae	*Ethusa ciliatifrons* Faxon, 1893	P	24–410	Vargas and Wehrtmann (2009)
		Ethusina robusta (Miers, 1886)	P	460–3260	Vargas and Wehrtmann (2009)
		Ethusina gracilipes (Miers, 1886)	P	1157–1454	Luke(1977)
		Ethusina smithiana (Faxon, 1893)	P	245	Faxon (1893, 1895)
	Glyphocrangonidae	*Glyphocrangon alata* Faxon, 1893	P	600–1355	Vargas and Wehrtmann (2009)
		Glyphocrangon nobilis A. Milne-Edwards, 1881	P	1408–1951	Faxon (1895)
		Glyphocrangon sicaria Faxon, 1893	P	1866	Wicksten (1979)
		Glyphocrangon vicaria Faxon, 1896	P	1866	Wicksten (1979)
	Inachoididae	*Collodes tenuirostris* Rathbun, 1894	P	5.5–265	Vargas and Wehrtmann (2009)
	Kiwaidae	*Kiwa puravida* Thurber, Jones & Schnabel, 2011	P	1000–1040	Thurber et al. (2011)
	Leucosiidae	*Iliacantha schmitti* Rathbun, 1935	P	18–275	Vargas and Wehrtmann (2009)

(continued)

Table 24.1 (continued)

Order	Family	Scientific name	Area	Depth range Costa Rica	Sources
	Lithodidae	*Neolithodes diomedeae* (Benedict, 1895)	P	640–2450	Vargas and Wehrtmann (2009)
		Lithodes panamensis Faxon, 1893	P	700–1400	Macpherson and Wehrtmann (2010)
		Lithodes wiracocha Haig, 1974	P	700–1400	Macpherson and Wehrtmann (2010)
		Paralomis longipes Faxon, 1893	P	1000–1408	Faxon (1893, 1895); Vargas and Wehrtmann (2009)
		Paralomis papillata (Benedict, 1895)	P	700–1400	Macpherson and Wehrtmann (2010)
		Paralomis diomedeae (Faxon, 1893)	P	770–825	Macpherson and Wehrtmann (2010)
	Majidae	*Maiopsis panamensis* Faxon, 1893	P	7.5–335	Vargas and Wehrtmann (2009)
	Mithracidae	*Pitho lherminieri* (Desbonne in Desbonne & Schramm, 1867)	C	1–220	Vargas and Wehrtmann (2009)
	Munididae	*Munida flinti* Benedict, 1902	C	110–203	Vargas and Cortés (2006)
		Munida gracilipes Faxon, 1893	P	140–320	Wehrtmann et al. (2010)
		Munida microphthalma A. Milne Edwards, 1880	P	245	Faxon (1895)
		Munida obesa Faxon, 1893	P	40–320	Wehrtmann et al. (2010)
		Munida perlata Benedict, 1902	P	190–3292	Vargas and Wehrtmann (2009)
		Munida refulgens Faxon, 1893	P	40–290	Wehrtmann et al. (2010)
	Munidopsidae	*Galacantha diomedeae* Faxon, 1893	P	1157–3433	Faxon (1893, 1895); Luke (1977)

(continued)

Table 24.1 (continued)

Order	Family	Scientific name	Area	Depth range Costa Rica	Sources
		Galacantha rostrata A. Milne Edwards, 1880	P	2149	Faxon (1893, 1895)
		Munidopsis albatrossae W.E. Pequegnat & L.H. Pequegnat, 1973	P	3570	Vargas and Wehrtmann (2009)
		Munidopsis antonii (Filhol, 1884)	P	2519–3676	Vargas and Wehrtmann (2009)
		Munidopsis aspera (Henderson, 1885)	P	245	Faxon (1893, 1895)
		Munidopsis hamata Faxon, 1893	P	1190–1281	Wehrtmann et al. (2010)
		Munidopsis nitida (A. Milne Edwards, 1880)	P	1789	Faxon (1895)
		Munidopsis sp.	P	1000–1040	Levin et al. (2012)
		Munidopsis vicina Faxon, 1893	P	3063–3885	Vargas and Wehrtmann (2009)
		Pleuroncodes monodon (H. Milne Edwards, 1837)	P	150–350	Wehrtmann et al. (2010)
	Nematocarcinidae	*Nematocarcinus agassizii* Faxon, 1893	P	245–1650	Faxon (1893, 1895)
		Nematocarcinus ensifer (Smith, 1882)	P	1789–1951	Faxon (1895)
	Nephropsidae	*Nephropsis occidentalis* Faxon, 1893	P	270–1310	Vargas and Wehrtmann (2009)
	Paguridae	*Solenopagurus diomedeae* (Faxon, 1893)	P	333	Vargas and Wehrtmann (2009)
		Tomopagurus merimaculosus (Glassell, 1937)	P	36–274	Vargas and Wehrtmann (2009)
	Palicidae	*Exopalicus maculatus* (Edmondson, 1930)	P	192–280	Vargas and Wehrtmann (2009)
	Pandalidae	*Heterocarpus affinis* Faxon, 1893	P	900–1244	Luke (1977); Vargas and Cortés (1999)

(continued)

Table 24.1 (continued)

Order	Family	Scientific name	Area	Depth range Costa Rica	Sources
		Heterocarpus hostilis Faxon, 1893	P	933–1789	Faxon (1893, 1895); Luke (1977)
		Heterocarpus vicarius Faxon, 1893	P	62–1454	Vargas and Cortés (1999)
		Plesionika unidens Spence Bate, 1888	P	3–733	Vargas and Cortés (1999)
		Plesionika mexicana Chace, 1937	P	28–258	Vargas and Cortés (1999)
		Plesionika trispinus Squires & Barragan, 1976	P	96–500	Vargas and Cortés (1999)
	Parapaguridae	*Parapagurus holthuisi* Lemaitre, 1989	P	1408–2149	Faxon (1895)
		Probeebei mirabilis Boone, 1926	P	1145	Boone (1926); Lemaitre (1998)
	Penaeidae	*Penaeus brasiliensis* Latreille, 1817	C	365	Tabash (1995)
		Pelagopenaeus balboae (Faxon, 1893)	P	1408	Faxon (1895)
	Pilumnidae	*Pilumnus townsendi* Rathbun, 1923	P	24–288	Vargas and Wehrtmann (2009)
	Polychelidae	*Stereomastis nana* (Smith, 1884)	P	2149	Faxon (1895)
		Stereomastis pacifica (Faxon, 1893)	P	1000–3692	Vargas and Wehrtmann (2009)
	Portunidae	*Achelous gibbesii* (Stimpson, 1859)	C	0–393	Vargas and Wehrtmann (2009)
		Achelous spinicarpus Stimpson, 1871	C	9–550	Vargas and Wehrtmann (2009)
		Achelous iridescens (Rathbun, 1894)	P	13–274	Vargas and Wehrtmann (2009)
		Portunus xantusii (Stimpson, 1860)	P	10–270	Vargas and Wehrtmann (2009)
	Raninidae	*Raninoides lamarcki* A. Milne-Edwards & Bouvier, 1923	C	45–240	Moran and Dittel (1993)

(continued)

Table 24.1 (continued)

Order	Family	Scientific name	Area	Depth range Costa Rica	Sources
	Sergestidae	*Neosergestes consobrinus* (Milne, 1968)	P	500	Azofeifa-Solano and Vargas-Castillo (2020)
		Phorcosergia filicta (Burkenroad, 1940)	P	1646	Azofeifa-Solano and Vargas-Castillo (2020)
		Allosergestes pestafer (Burkenroad, 1937)	P	180–250	Vargas and Cortés (1999)
	Solenoceridae	*Haliporoides diomedeae* (Faxon, 1893)	P	240–1865	Faxon (1895)
		Hymenopenaeus doris (Faxon, 1893)	P	549–4802	Faxon (1895)
		Hymenopenaeus nereus (Faxon, 1893)	P	1951	Faxon (1893, 1895)
		Solenocera agassizii Faxon, 1893	P	16–384	Faxon (1895)
	Xanthidae	*Edwardsium lobipes* (Rathbun, 1898)	P	7–273	Vargas and Wehrtmann (2009)
		Nanocassiope polita (Rathbun, 1894)	P	55–274	Vargas and Wehrtmann (2009)
Amphipoda	Ampeliscidae	*Ampelisca brevisimulata* J.L. Barnard, 1954	P	4–456	Foster et al. (2009)
		Ampelisca hancocki J.L. Barnard, 1954	P	9–210	Foster et al. (2009)
		Ampelisca romigi J.L. Barnard, 1954	P	3–503	Foster et al. (2009)
	Paraphronimidae	*Paraphronima gracilis* Claus, 1879	P	500	Gasca (2009)
	Phronimidae	*Phronima sedentaria* (Forskål, 1775)	P	50–400	Gasca (2009)
Cumacea	Diastylidae	*Diastylis tenebricosa* Jones, 1969	P	3570	Petrescu et al. (2009)
		Makrokylindrus (Adiastylis) menziesi Bacescu, 1962	P	3400–3500	Petrescu et al. (2009)
		Vemakylindrus costaricanus Bacescu, 1961	P	3718	Petrescu et al. (2009)

(continued)

Table 24.1 (continued)

Order	Family	Scientific name	Area	Depth range Costa Rica	Sources
Isopoda	Mesosignidae	*Mesosignum admirandum* Menzies & Frankenberg, 1967	P	1016–1892	Brusca and Wehrtmann (2009)
		Mesosignum asperum Menzies & Frankenberg, 1967	P	3517–3950	Brusca and Wehrtmann (2009)
		Mesosignum macrum Menzies & Frankenberg, 1967	P	3254–3260	Brusca and Wehrtmann (2009)
	Munnopsidae	*Munnopsis longiremus* Richardson, 1912	P	1485–3570	Brusca and Wehrtmann (2009)
		Paropsurus giganteus Wolff, 1962	P	3570–4400	Brusca and Wehrtmann (2009)
		Vanhoeffenura pulchra (Hansen, 1897)	P	2487–3570	Brusca and Wehrtmann (2009)
	Aegidae	*Aega acuminata* Hansen, 1897	P	1353	Brusca and Iverson (1985)
		Aega maxima Hansen, 1897	P	2149	Hansen (1897)
		Aegiochus plebeia (Hansen, 1897)	P	1789	Hansen (1897)
		Rocinela murilloi Brusca & Iverson, 1985	P	1866	Brusca and Iverson (1985)
		Rocinela wetzeri Brusca & France, 1992	P	1157–1454	Brusca and France (1992)
Calanoida	Aetideidae	*Aetideopsis rostrata* Sars G.O., 1903	P	200–600	Morales-Ramírez and Suárez-Morales (2009)
		Aetideus armatus (Boeck, 1872)	P	200–600	Morales-Ramírez and Suárez-Morales (2009)
		Euchirella amoena Giesbrecht, 1888	P	200–600	Morales-Ramírez and Suárez-Morales (2009)

(continued)

Table 24.1 (continued)

Order	Family	Scientific name	Area	Depth range Costa Rica	Sources
		Valdiviella brevicornis Sars G.O., 1905	P	100–1000	Morales-Ramírez and Suárez-Morales (2009)
	Augaptilidae	*Augaptilus longicaudatus* (Claus, 1863)	P	200–600	Morales-Ramírez and Suárez-Morales (2009)
		Haloptilus acutifrons (Giesbrecht, 1893)	P	200–600	Morales-Ramírez and Suárez-Morales (2009)
		Haloptilus longicornis (Claus, 1863)	P-C	200–600	Morales-Ramírez and Suárez-Morales (2009)
		Haloptilus mucronatus (Claus, 1863)	P	200–600	Morales-Ramírez and Suárez-Morales (2009)
		Haloptilus ornatus (Giesbrecht, 1893)	P	200–600	Morales-Ramírez and Suárez-Morales (2009)
		Haloptilus oxycephalus (Giesbrecht, 1889)	P	200–600	Morales-Ramírez and Suárez-Morales (2009)
	Calanidae	*Canthocalanus pauper* (Giesbrecht, 1888)	P	200–600	Morales-Ramírez and Suárez-Morales (2009)
		Undinula vulgaris (Dana, 1849)	C	100–1000	Morales-Ramírez and Suárez-Morales (2009)
	Paracalanidae	*Calocalanus plumulosus* (Claus, 1863)	P	400–600	Morales-Ramírez and Suárez-Morales (2009)
	Euchaetidae	*Euchaeta media* Giesbrecht, 1888	P	600–2000	Morales-Ramírez and Suárez-Morales (2009)

(continued)

Table 24.1 (continued)

Order	Family	Scientific name	Area	Depth range Costa Rica	Sources
	Heterorhabdidae	*Heterorhabdus papilliger* (Claus, 1863)	P	200–600	Morales-Ramírez and Suárez-Morales (2009)
	Lucicutiidae	*Lucicutia bicornuta* Wolfenden, 1905	P	400–500	Morales-Ramírez and Suárez-Morales (2009)
		Lucicutia grandis (Giesbrecht, 1895)	P	300–500	Morales-Ramírez and Suárez-Morales (2009)
	Metridinidae	*Pleuromamma quadrungulata* (Dahl F., 1893)	P	200–600	Morales-Ramírez and Suárez-Morales (2009)
		Pleuromamma robusta robusta (Dahl F., 1893)	P	200–600	Morales-Ramírez and Suárez-Morales (2009)
	Pontellidae	*Pontella agassizii* Giesbrecht, 1895	P	200–600	Morales-Ramírez and Suárez-Morales (2009)
	Scolecitrichidae	*Amallothrix gracilis* (Sars G.O., 1905)	P	200–600	Morales-Ramírez and Suárez-Morales (2009)
		Scolecithricella dentata (Giesbrecht, 1893)	P	200–600	Morales-Ramírez and Suárez-Morales (2009)
Mormonilloida	Mormonillidae	*Neomormonilla minor* (Giesbrecht, 1891)	P	400–1500	Morales-Ramírez and Suárez-Morales (2009)

Areas: P Pacific Ocean, C Caribbean Sea, PC both. Depth range provided for Costa Rica only

most speciose families were Munidopsidae squat lobsters with ten species, followed by Pandalidae shrimps, Lithodidae king crabs, and Munididae squat lobsters, with six species each. Other groups with high number of species were the copepods (23 spp.) and the peracarids (19 spp.) (Table 24.1). The overall depth range of these 147 species occurs from sea surface to 4802 m. Among these, 40 species have depth ranges that include water shallower than 200 m, and most inhabit deep waters down to 400 m. The other 107 species have been recorded exclusively in the deep sea (deeper than 200 m). There are 14 species of decapods and cumaceans that occur below the 3000 m depth, with *Hymenopenaeus doris* (Faxon 1893) and *Paropsurus giganteus* Wolff, 1962, holding the record of maximum depth at 4802 m and 4400 m, respectively (Table 24.1).

24.3.3.2 Stomatopoda

Stomatopods, commonly known as mantis shrimps or sea scorpions (locally known as "alacranes de mar"), are a relatively well-known group in Costa Rica coastal and shallow waters in terms of richness (Reaka and Manning 1980; Vargas 2009; Salas-Moya and Vargas-Castillo 2016). Below 200 m depth, however, only four species have been reported: *Squilla biformis* Bigelow, 1891, *S. hancocki* Schmitt, 1940, *S. panamensis* Bigelow, 1891, and *Hemisquilla californiensis* Stephenson, 1967 (Table 24.1).

The mantis shrimp *S. biformis* was found to be very abundant in the bycatch of the deep-sea shrimp fishery in the Pacific of Costa Rica (Wehrtmann and Echeverría-Sáenz 2007; Wehrtmann and Nielsen-Muñoz 2009), attracting attention as a potential fishing resource considering the alarming decrease of deep-sea shrimp landings, but the project was not attractive to the fishing sector. Nonetheless, this species has been studied to provide information on population demography, spatial distribution, and behavior in order to facilitate implementation of an adequate management policy (Hernáez et al. 2011). Stock of *S. biformis* was composed of two size groups, with large specimens occurring in deeper waters (Hernáez et al. 2011). Large amounts of mantis shrimps were found in the historical fishing grounds where shrimps were trawled in high abundances, suggesting an ecological shift related to the intense fishing pressure (Hernáez et al. 2011).

24.3.3.3 Decapoda

Decapods are probably the most frequently studied crustaceans in the deep-sea habitats of Costa Rica, in part due to their larger sizes compared to other crustaceans and their commercial importance as fishery resources (e.g., shrimps, prawns, king crabs, lobsters) (Wehrtmann and Nielsen-Muñoz 2009; Vargas and Wehrtmann 2009). In addition to the taxonomy and diversity (e.g., Wehrtmann and Echeverría-Sáenz 2007; Vargas and Wehrtmann 2009; Macpherson and Wehrtmann 2010; Thurber et al. 2011; Martin et al. 2018; Azofeifa-Solano and Vargas-Castillo 2020), some research has also focussed on ecology (Wehrtmann et al. 2010; Hernáez et al. 2011; Durkin 2018), reproductive biology (Echeverría-Sáenz and Wehrtmann 2011; Hernáez and Wehrtmann 2011a, b; Villalobos-Rojas and Wehrtmann 2011; Villalobos-Rojas and Wehrtmann 2014), fisheries (Wehrtmann and Nielsen-Muñoz 2009), and microbiome symbiosis (Goffredi et al. 2014).

Many ecological and reproductive biology studies on deep-sea large crustaceans in Costa Rica have resulted from close collaboration between scientists and the deep-sea shrimp fishery, thus offering opportunities to obtain valuable samples taken in deep waters (Wehrtmann and Nielsen-Muñoz 2009). The "camarón fidel," *Solenocera agassizii* Faxon, 1893, and the "camarón camello," *Heterocarpus vicarius* Faxon, 1893, were the two main commercial deepwater shrimps exploited in the Costa Rican Pacific (Wehrtmann and Nielsen-Muñoz 2009). The urgent necessity to develop management plans for these fisheries compelled researchers to study their reproductive biology (Echeverría-Sáenz and Wehrtmann 2011; Villalobos-Rojas and Wehrtmann 2011; Villalobos-Rojas and Wehrtmann 2014).

Anomuran crustaceans are commonly found as bycatch of the deep-sea shrimp fishery, especially squat lobsters and king crabs, also studied (Macpherson and Wehrtmann 2010; Wehrtmann et al. 2010). The red squat lobster, *Pleuroncodes monodon* (H. Milne Edwards, 1837), was very abundant, allowing for the study of their breeding cycle, sexual maturity, and fecundity (Hernáez and Wehrtmann 2011a, b).

The yeti crab (family Kiwaidae) is one of the most emblematic and appealing crustaceans inhabiting the deep-sea vents, not only because of its flamboyant name but also for its no less fantastic biological features (Thurber et al. 2011; Thatje et al. 2015). Such is the case of the endemic yeti crab, *Kiwa puravida*, occurring in the vent seeps in the Costa Rican Pacific (Thurber et al. 2011). These crabs wave their chelipeds close to fluids escaping from methane seeps to promote the chemosynthetic activity of epibiotic bacteria growing in their chelipeds, to be harvested, and to obtain energy in these food-limited environments (Thurber et al. 2011). These bacterial communities are different throughout the yeti crab's body, and ontogenetic changes have been found between different stages of development (Goffredi et al. 2014). However, other vent seeps inhabitants such as shrimps of the genus *Alvinocaris* Williams and Chace, 1982, remain unstudied in Costa Rica.

24.3.3.4 Euphausiacea

There are very few studies regarding this group (commonly known as "krill") in Costa Rican waters (Castellanos et al. 2009; Azofeifa-Solano et al. 2016). A thorough review of the group suggested the presence of 29 species in the area, partly based on offshore studies performed outside the Costa Rican EEZ (Castellanos et al. 2009). During the 2011 and 2012 expeditions to Isla del Coco, seven species were collected from vertical zooplankton samples taken from the surface to 200 m (Azofeifa-Solano et al. 2016). Krill specimens were also collected during the Allan Hancock Foundation and the Los Angeles County Museum expedition in 1973, aboard the MY "VeleroIV" cruises (1244–1247), and seven species were identified from vertical tows between the surface and 500 m (Azofeifa-Solano and Vargas-Castillo 2020). However, known distributions for these species far exceed the 200 m isobath, suggesting that these species also occur in Costa Rican deep sea, but further studies including stratified samples should confirm this assumption.

24.3.3.5 Peracarida

Although peracarids are highly abundant and play a relevant ecological role, they have been poorly studied in Costa Rican waters, and even less information is currently available for deep-sea species (Brusca and Wehrtmann 2009; Heard et al. 2009; Petrescu et al. 2009). A total of three species of deep-sea comma shrimps (Cumacea) of the family Diastylidae were collected between 3400 m and 3718 m in the Pacific: *Vemakylindrus costaricanus* Băcescu, 1961, *Makrokylindrus* (*Adiastylis*) *menziesi* Băcescu, 1962, and *Diastylis tenebricosa* Jones, 1969 (Băcescu 1961, 1962; Jones 1969; Petrescu et al. 2009). In the deep sea, 11 isopods and 5 amphipods have been reported for Costa Rica (Table 24.1). Deep-sea peracarids have not been thoroughly studied, suggesting that their diversity could be far larger than that currently reported in the literature (Brusca and Wehrtmann 2009; Heard et al. 2009; Petrescu et al. 2009).

24.3.3.6 Copepoda

As mentioned earlier, copepods account for 23 deep-sea species recorded for Costa Rica (Morales-Ramírez and Suárez-Morales 2009, Table 24.1). Some studies have focussed on vertical migration and other ecological aspects of deep-sea copepods, such as the contributions by Sameoto (1986), Suárez-Morales and Gasca (1989), and Décima et al. (2016) in the Costa Rican Dome area and by Owre and Foyo (1964a, b) who sampled off the Caribbean coast to 1300 m depth.

24.3.4 Deepwater Fisheries in Costa Rica

Wehrtmann and Nielsen-Muñoz (2009) did an extensive review of the background, current status, and state of knowledge of the fishery and the fishing resources in deep-sea off Costa Rica. Here, we summarize their main results and add research published during the 2010s decade. There are several fisheries and fishing gear types used in Costa Rica fishing fleet, but the semi-industrial shrimp trawling was the only fishery reported to actively target fishing resources deeper than 200 m. The bottom trawl nets were first introduced in Costa Rica in the mid-twentieth century (Wehrtmann and Nielsen-Muñoz 2009) and with them the beginning of a new era of bountiful catches and promising profits for the fishing industry. During the first years of this new fishery, large coastal shrimps were the main targets, but due to stock exploitation, soon enough smaller species, previously discarded, started to be commercially exploited. Fleets started to move deeper as coastal and shallow water resources were depleted and deep-sea shrimps were exploited by the semi-industrial shrimp trawling fishery, e.g., the "camarón fidel" *Solenocera agassizii*, the "camarón camello" *H. vicarius*, and the "camarón real" or "camarón camellón" *Heterocarpus affinis* Faxon, 1893 (Wehrtmann and Nielsen-Muñoz 2009). According to the official landings provided by the Costa Rican Institute of Fishing and Aquaculture (INCOPESCA, by its Spanish acronym), during 1995–2005 the deep-sea shrimps catch accounted for more than half of the total shrimp landings. Deep-sea shrimps accounted for 54.5% of the total catch, divided in *S. agassizii* (27.7%), *H. vicarius* (16.9%), and *H. affinis* (9.9%) (see Wehrtmann and Nielsen-Muñoz 2009). The main lines of shrimp trawlers in Costa Rica have an average length of 823 m, allowing for fishing in both coastal and deep water (Bolaños 2005).

In the 2010s, data showed that shrimp landings were decreasing, while bycatch of noncommercial species was increasing simultaneously, along with increasing costs for the shrimp fleet, driving some companies to anchor their trawling vessels (Wehrtmann and Nielsen-Muñoz 2009; Wehrtmann et al. 2012). A total of ten licenses for deep-sea shrimp fishing were still active in 2014, but during 2017–2018, trawling activities in deep waters were few and sporadic, mainly around 200–400 m, with the deepest trawl recorded at about 874 m depth (Lorna Marchena, pers. comm.). In 2013, the Chamber IV (Sala IV), one of the four chambers of the Supreme Court (Costa Rica's Judicial Branch of the National Government), declared the bottom trawling fishing for shrimp as an "unconstitutional action," and, as a consequence, the INCOPESCA does not grant new fishing permits, authorizations, or licenses, nor does it renew the expired licenses or reactivate the inactive licenses for shrimp fishing with bottom trawling nets, including both semi-industrial and small-scale fishing (sometimes also referred as "artisanal trawling") (Sala IV Sentencia No 2013–10540 2013).

24.3.5 Threats

During the last decades, new technologies have allowed mapping and sampling the deep sea, revealing a great variety of habitats with conservation and economic value. We also have evidence that human activities can easily reach and negatively impact this once unfathomable wilderness (Roberts et al. 2000; Koslow et al. 2001; Baum et al. 2003; Davies et al. 2007; Schlacher et al. 2010). Currently, the main anthropogenic threats to the deep sea are mining (e.g., extraction of oil, gas, and minerals such as rare metals), overfishing (such as trawling and net fishing), pollution and disposal of wastes (structures, radioactive waste, munitions, plastic), and climate change (Roberts 2002; Glover and Smith 2003; Morato et al. 2006; Taylor et al. 2016; Sweetman et al. 2017; Chiba et al. 2018; Heffernan 2019). The deep sea is, indeed, highly vulnerable to human activities, even more in comparison to coastal and shallow waters. Deep-sea species have life-history traits that make them highly vulnerable to extraction. This includes slow growth, delayed maturity, extremely extended longevities, and slow colonization rates (thousands of years in some corals) (Roberts 2002; Clark et al. 2006; Cheung et al. 2007; Heffernan 2019). Deep-sea ecosystems are inherently vulnerable due to the particularities of the species that build these ecosystems. For deep-sea soft bottom habitats, special challenges to conservation include low productivity, low physical energy, low biological rates, and immensity scale; for other habitats having higher productivity, environmental conditions, and diversity (e.g., submarine canyons, seamounts, and vents), challenges include wide spacing and isolation of the deep-sea soft bottoms (Glover and Smith 2003). One of the main concerns is the slow recovery of deep-sea habitats. For example, extraction scars made by the DISCOL experiment on the seafloor are still visible after 26 years (1989–2015), and characteristic organisms such as sponges, soft corals, and sea anemones, once present, have not returned (Heffernan 2019).

24.3.6 Management and Conservation

The deep sea provides vital ecological processes and features such as habitat provision, nutrient cycling, production, trophic support, and carbon sequestration, supporting the health of the ocean and the planet while providing important ecosystem services to humanity (Thurber et al. 2014). Processes occurring at the massive scale of the deep sea, such as organic matter degradation and remineralization, largely contribute to the biogeochemical cycle of carbon, acting as a buffer for ocean acidification (Wenzhöfer et al. 2001). It is clear that the deep-sea environments play a very important role in sustaining the health and functioning of planet Earth (Sweetman et al. 2017). Exploration of pristine deep seas has allowed science to learn more about Earth's interior, ocean's chemistry, and the extreme environments in which life can exist (Van Dover 2011). Deep-sea exploitation, however, should not be pursued before proposing a coherent conservation, management, and mitigation framework (Van Dover 2011).

Despite the fascination that the deep sea inspires and its ecological and economic importance, it is hard for many people to infer connections between the deep-sea processes and their daily life. With this issue in mind, it is not surprising why it has been difficult to obtain politicians' and lawmakers' support. Considering how little we still know about the deep sea, the scarcity of data, and the high costs of exploration this issue will probably not be solved in the near future (Harris et al. 2007; Howell 2010), and scientists likely will keep collaborating with companies pursuing exploitation of the deep-sea resources in order to study these habitats (Brewin et al. 2007; Wehrtmann and Nielsen-Muñoz 2009). The rising concerns about human impacts in the deep sea, however, urge all stakeholders, scientists, conservationists, sectors with interests in exploitation, and the general public to move toward active action and advocate for adequate and science-based management, following a precautionary principle due to the lack of data, because impacts on the deep sea could be irreversible at temporal human scales (Roberts 2002; Waller et al. 2007; Heffernan 2019). Costa Rica should also embrace a precautionary principle and science-based management of its deep-sea resources including of course crustaceans.

24.4 Summary

- Exploration of the deep sea in Costa Rica began with foreign expeditions, and most of the research has been done by foreign scientists, due to the high cost, unbearable for Costa Rican scientists. In recent years (2000 and onward), however, national and foreign resident scientists have aimed for collaborations with international expeditions and deepwater fisheries in order to obtain data and samples from the deep-sea habitats.
- There are several fascinating deep-sea environments known in Costa Rica, including mud volcanoes, cold methane seeps, seamounts, and bathyal soft bottoms.
- There is a total of 147 crustacean species reported inhabiting deep-sea environments in Costa Rica; 8 in the Caribbean Sea, 138 in the Pacific Ocean, and 1 reported in both basins. The most speciose groups are Decapoda (87 spp.), followed by Copepoda (23 spp.), and Peracarida (19 spp.).
- Most studies have focused on reproductive biology of decapods associated with deepwater fisheries, such as shrimps and squat lobsters. In addition, studies on the feeding and microbiome of the endemic yeti crab, *Kiwa puravida*, have been done.
- The deep-water shrimp fisheries started after depletion of stocks of shallow-water species. The absence of adequate management resulted in overexploitation, with decreasing shrimp landings, increasing noncommercial bycatch, and higher costs. Due to public concerns, the Judicial Branch declared the bottom trawling for shrimp as an unconstitutional action, and currently there is no deep-water shrimp trawling fishery in Costa Rica.
- Despite the current absence of commercial activities exploiting deep-sea fishing and mineral resources, the depletion of land and coastal resources is putting pressure on the Costa Rican deep-sea environments. In addition, global threats such as pollution and climate change have also impacted Costa Rica.
- The national and regional agencies do not provide enough funding support to scientists to explore the deep sea. Countries, however, should invest more in research and protection of these habitats. Scientists in Costa Rica will continue to look for collaborations in order to shed light on the nature of the deep sea, the last wilderness.

Acknowledgments We are very grateful to Erik Cordes, who invited us to participate in several expeditions to cold seeps and seamounts in the Costa Rican Pacific, and other participants, especially Lisa Levin, Shana Goffredi, Victoria Orphan, and Greg Rouse. We thank the crew and staff of the RV *Atlantis* and the HOV *Alvin* and the crew and staff of the RV *Falkor* and the ROV *SuBastian*. We are thankful to Beatriz Naranjo, Rita Vargas, Odalisca Breedy, and Arturo Angulo for their collaboration in other projects in which the authors participated and gathered important information for this study. JCAS is grateful to Ingo Wehrtmann and Fresia Villalobos for the information and help provided during many years working in the Unidad de Investigación Pesquera y Acuicultura (UNIP). JCAS is thankful to Conservación Internacional Costa Rica, Lorna Marchena, and Sofía Cortés for their help in gathering legal and administrative information on the Costa Rican deepwater fisheries. Finally, we are very grateful to Michel E. Hendrickx for the invitation to collaborate with a chapter in this book. We are thankful for the comments by Michel E. Hendrickx, Charlotte Seid, and two anonymous reviewers on early drafts of this manuscript.

References

Agassiz A (1892) Reports on the dredging operations off the west coast of Central America to the Galápagos, to the west coast of Mexico, and in the Gulf of California, in charge of Alexander Agassiz, carried on by the U.S. Fish Commission steamer "Albatross," Lieut. Commander Z.L. Tanner U.S.N., commanding. II. General sketch of the expedition of the "Albatross," from February to May, 1891. Bull Mus Comp Zool 23:1–89

Azofeifa-Solano JC, Vargas-Castillo R (2020) Unearthing 20th-century treasures: new records of deep-sea crustaceans from the Costa Rican Pacific. Crustaceana 93:747–757. https://doi.org/10.1163/15685403-bja10017

Azofeifa-Solano JC, Corrales-Ugalde M, Castellano-Osorio I, Morales-Ramírez A (2016) Euphausiids (Crustacea: Euphausiacea) from a hotspot of marine biodiversity, Isla del coco, Costa Rica, eastern tropical Pacific. Rev Biol Trop 64(Suppl 1):S221–S230

Băcescu M (1961) Deux espèces nouvelles de *Makrokylindrus* sous-genre *Vemakylindrus* n. sg. (Crustacés Cumacés) des eaux tropicales du Pacifique (côte américaine). Rev Roum Biol Ser Zool 6:325–333

Băcescu M (1962) Contribution à la connaissance du genre *Makrokylindrus* Stebbing (Crustacea, Cumacea). Espèces nouvelles recueillies au cours des campagnes du Lamont Geological Observatory de New York. In: Abyssal Crustacea. Columbia University, Vema Research Series, pp 209–222

Baum JK, Myers RA, Kehler DG, Worm B, Harley SJ, Doherty PA (2003) Collapse and conservation of shark populations in the Northwest Atlantic. Science 299:389–392

Beebe W (1926) The Arcturus oceanographic expedition. Zoologica 8:1–45

Beebe W (1938) Eastern Pacific expeditions of the New York zoological society, XIV. Introduction, itinerary, list of stations, nets and dredges of the eastern Pacific Zaca expedition, 1937–1938. Zoologica 23:287–298

Bolaños M (2005) Costa Rica. Characterization of the Costa Rican semi-industrial coastal shrimp trawling fishery and the small scale artisanal shrimp trawling fishery. FAO report, EP/GLO/201/GEF, p 1–39

Boone L (1926) Unusual deep-sea Crustacea–some forms secured by the Arcturus oceanographic expedition. Zool Soc Bull 29:69–72

Brewin PE, Stocks KI, Menezes G (2007) A history of seamount research. In: Pitcher TJ, Morato T, Hart PJB, Clark MR, Haggan N, Santos RS (eds) Seamounts: ecology, conservation and management. Blackwell, Oxford, pp 41–61

Brusca RC, France SC (1992) The genus *Rocinela* (Crustacea: Isopoda: Aegidae) in the tropical eastern Pacific. Zool J Linnean Soc 106:231–275

Brusca RC, Iverson EW (1985) A guide to the marine isopod Crustacea of Pacific Costa Rica. Rev Biol Trop 33(Suppl 1):1–77

Brusca RC, Wehrtmann IS (2009) Isopods. In: Wehrtmann IS, Cortés J (eds) Marine biodiversity of Costa Rica, Central America. Springer + Business Media BV, Dordrecht, pp 257–264

Cambronero-Solano S, Benavides R, Solís-Marín FA, Alvarado JJ (2019) New reports of echinoderms on the Caribbean continental slope of Central America. Zoosymposia 15:5–12. https://doi.org/10.11646/zoosymposia.15.1.3

Camp DK, Kuck HG (1990) Additional records of stomatopod crustacean from Isla del coco and Golfo de Papagayo, East Pacific Ocean. Proc Biol Soc Wash 103:847–853

Castellanos I, Suárez-Morales E, Morales-Ramírez A (2009) Euphausiids. In: Wehrtmann IS, Cortés J (eds) Marine biodiversity of Costa Rica, Central America. Springer + Business Media BV, Dordrecht, pp 199–207

Cheung WWL, Watson R, Morato T, Pitcher TJ, Pauly D (2007) Intrinsic vulnerability in the global fish catch. Mar Ecol Prog Ser 333:1–12

Chiba S, Saito H, Fletcher R, Yogi T, Kayo M, Miyagi S, Ogido M, Fujikura K (2018) Human footprint in the abyss: 30 year records of deep-sea plastic debris. Mar Policy 96:204–212. https://doi.org/10.1016/j.marpol.2018.03.022

Child CA (1992) Pycnogonida of the Southeast Pacific biological oceanographic project (SEPBOP). Smithson Contrib Zool 526:1–43

Clark MR, Tittensor D, Rogers AD, Brewin P, Schlacher T, Rowden A, Stocks K, Consalvey M (2006) Seamounts, deep-sea corals and fisheries: vulnerability of deep-sea corals to fishing on seamounts beyond areas of national jurisdiction. UNEPWCMC, Cambridge

Cortés J (2009) A history of marine biodiversity scientific research in Costa Rica. In: Wehrtmann IS, Cortés J (eds) Marine biodiversity of Costa Rica, Central America. Springer + Business Media BV, Dordrecht, pp 47–80

Cortés J (2016a) The Pacific coastal and marine ecosystems. In: Kappelle M (ed) Costa Rican ecosystems. University of Chicago Press, Chicago, pp 97–138

Cortés J (2016b) The Caribbean coastal and marine ecosystems. In: Kappelle M (ed) Costa Rican ecosystems. University of Chicago Press, Chicago, pp 591–617

Cortés J (2019) Chapter 26: Isla del Coco, Costa Rica, Eastern Tropical Pacific. In: Loya Y, Puglise KA, Bridge TCL (eds) Mesophotic coral ecosystems. Springer Nature, Cham, pp 465–474

Cortés J, Benavides-Varela C (in prep) A hypsographic (altitude- and depth-based) view of conservation in Costa Rica. Submitted to Journal for Nature Conservation

Cortés J, Blum S (2008) Life to 450 m depth at Isla del coco, Costa Rica. Rev Biol Trop 56(Suppl. 2):189–206

Costello MJ, Michener WK, Gahegan M, Zhang ZQ, Bourne PE (2013) Biodiversity data should be published, cited, and peer reviewed. Trends Ecol Evol 28:454–461. https://doi.org/10.1016/j.tree.2013.05.002

Davies AJ, Roberts JM, Hall-Spencer J (2007) Preserving deep-sea natural heritage: emerging issues in offshore conservation and management. Biol Conserv 138:299–312

Décima M, Landry MR, Stukel MR, Lopez-Lopez L, Krause JW (2016) Mesozooplankton biomass and grazing in the Costa Rica dome: amplifying variability through the plankton food web. J Plankton Res 38:317–330. https://doi.org/10.1093/plankt/fbv091

Durkin AG (2018) The ecology of deep-sea chemosynthetic habitats from populations to metacommunities. Temple University, Dissertation

Echeverría-Sáenz S, Wehrtmann (2011) Egg production of the commercially exploited Deepwater shrimp, *Heterocarpus vicarius* (Decapoda: Pandalidae), Pacific Costa Rica. J Crustacean Biol 31:434–440. https://doi.org/10.1651/10-3400.1

Faxon W (1893) Reports on the dredging operations off the west coasts of Central America to the Galapagos, to the west coast of Mexico, and in the Gulf of California, in charge of Alexander Agassiz, carried on by the U.S. fish commission steamer "albatross", during 1891, Lieut. Commander Z.L. Tanner, U.S.N., commanding. VI. Preliminary descriptions of new species of Crustacea. Bull Mus Comp Zool 24:149–220

Faxon W (1895) Reports on an exploration off the west coasts of Mexico, central and South America, and off the Galapagos Islands, in charge of Alexander Agassiz, by the U.S. fish commission steamer albatross, during 1891, Lieut. Commander Z.L. Tanner U.S.N., commanding. XV. The stalk-eyed Crustacea. Mem Mus Comp Zool Harvard Coll 18:1–292

Foster JM, LeCroy SE, Heard EW, Vargas R (2009) Gammaridean amphipods. In: Wehrtmann IS, Cortés J (eds) Marine biodiversity of Costa Rica, Central America. Springer + Business Media BV, Dordrecht, pp 265–274

Fraser CM (1943) General account of the scientific work of the *Velero III* in the eastern Pacific, 1931–1941, part III: a ten-year list of the *Velero III* collecting stations (charts 1–115). With an appendix of collecting stations of the Allan Hancock foundation for the year 1942. Allan Hancock Pac Exp 1:259–431

Gage JD, Tyler PA (1991) Deep-sea biology: a natural history of organisms at the deep-sea floor. Cambridge University Press, Cambridge

Gasca R (2009) Hyperiid amphipods. In: Wehrtmann IS, Cortés J (eds) Marine biodiversity of Costa Rica, Central America. Springer + Business Media BV, Dordrecht, pp 275–282

Glover AG, Smith CR (2003) The deep-sea floor ecosystem: current status and prospects of anthropogenic change by the year 2025. Environ Conserv 30:219–241. https://doi.org/10.1017/S0376892903000225

Goffredi SK, Gregory A, Jones WJ, Morella NM, Sakamoto RI (2014) Ontogenic variation in epibiont community structure in the deep-sea yeti crab, *Kiwa puravida*: convergence among crustaceans. Mol Ecol 23:1457–1472. https://doi.org/10.1111/mec.12439

Greene HG, Yoklavich MM, Starr RM, O'Connell VM, Wakefield WW, Sullivan DE, McRea JE, Cailliet GM (1999) A classification scheme for deep seafloor habitats. Oceanol Acta 22:663–678

Hansen HJ (1897) Reports on the dredging operations off the west coast of Central America to the Galapagos, to the west coast of Mexico, and in the Gulf of California, in charge of Alexander Agassiz, carried on by the US. Fish Commission Steamer "Albatross," during 1891, Lieut. Commander Z. L. Tanner, U.S.N., commanding. XXII. The Isopoda. Bull Mus Comp Zool 31:95–129

Harris J, Haward M, Jabour J, Woehler EJ (2007) A new approach to selecting marine protected areas (MPAs) in the Southern Ocean. Antarct Sci 19:189–194

Heard R, Breedy O, Vargas R (2009) Tanaidaceans. In: Wehrtmann IS, Cortés J (eds) Marine biodiversity of Costa Rica, Central America. Springer + Business Media BV, Dordrecht, pp 245–256

Heffernan O (2019) Deep-sea dilemma. Nature 571:465–468

Hernáez P, Wehrtmann IS (2011a) Breeding cycle of the red squat lobster *Pleuroncodes monodon* H. Milne Edwards, 1837 (Decapoda, Muninidae) from Deepwater Pacific of Costa Rica. Mar Biol 35:204–211. https://doi.org/10.1111/maec.12070

Hernáez P, Wehrtmann IS (2011b) Sexual maturity and egg production in an unexloited population of the red squat lobster *Pleuroncodes monodon* (Decapoda, Galatheidae) from Central America. Fish Res 107:276–282. https://doi.org/10.1016/j.fishres.2010.11.010

Hernáez P, Clarke TM, Benavides-Varela C, Villalobos-Rojas F, Nivia-Ruíz J, Wehrtmann IS (2011) Population demography and spatial distribution of the mantis shrimp *Squilla biformis* (Stomatopoda, Squillidae) from Pacific Costa Rica. Mar Ecol Prog Ser 424:157–168

Howell KL (2010) A benthic classification system to aid in the implementation of marine protected area networks in the deep/high seas of the NE Atlantic. Biol Conserv 143:1041–1056

Jennings S, Reynolds JD, Mills SC (1998) Life history correlates of responses to fisheries exploitation. P Roy Soc B-Biol Sci 265:333–339

Jones NS (1969) The systematics and distributions of Cumacea from depths exceeding 200 meters. Galathea Rep 10:99–180

Koslow JA, Gowlett-Holmes K, Lowry JK, O'Hara T, Poore GCB, Williams A (2001) Seamount benthic macrofauna off southern Tasmania: community structure and impacts of trawling. Mar Ecol Prog Ser 213:111–125

Lemaitre R (1998) Revisiting *Tylaspis anomala* Henderson, 1885 (Parapaguridae), with comments on its relationships and evolution. Zoosystema 20:289–305

Levin LA, Le Bris N (2015) The deep ocean under climate change. Science 350:766–768. https://doi.org/10.1126/science.aad0126

Levin LA, Mendoza GF, Konotchick T, Lee R (2009) Macrobenthos community structure and trophic relationships within active and inactive Pacific hydrothermal sediments. Deep-Sea Res Part II 56:1632–1648. https://doi.org/10.1016/j.dsr2.2009.05.010

Levin LA, Orphan VJ, Rouse GW, Rathburn AE, Ussler W, Cook GS, Goffredi SK, Perez EM, Waren A, Grupe BM, Chadwick G, Strickrott B (2012) A hydrothermal seep on the Costa Rica margin: middle ground in a continuum of reducing ecosystems. P Roy Soc Biol Sci 279:2580–2588

Levin LA, Mendoza GF, Grupe BM, Gonzalez JP, Jellison B, Rouse G, Thurber AR, Waren A (2015) Biodiversity on the rocks: macrofauna inhabiting authigenic carbonate at Costa Rica methane seeps. PLoS One 10:e0131080. https://doi.org/10.1371/journal.pone.0131080

Luke SR (1977) Catalog of the benthic invertebrate collections. I Decapod Crustacea and Stomatopoda Scripps Institution of Oceanography Available via UC San Diego: Library – Scripps Digital Collection https://escholarship.org/uc/item/66b0b4k1

Macpherson E, Wehrtmann IS (2010) Occurrence of lithodid crabs (Decapoda, Lithodidae) on the Pacific coast of Costa Rica, Central America. Crustaceana 83:143–151. https://doi.org/10.1163/001121609X12487811051660

Martin JW, Wall AR, Shank T, Cha H, Seid CA, Rouse GW (2018) A new species of *Alvinocaris* (Crustacea: Decapoda: Caridea: Alvinocarididae) from Costa Rican methane seeps. Zootaxa 4504:418–430. https://doi.org/10.11646/zootaxa.4504.3.7

Morales-Ramírez A, Suárez-Morales E (2009) Copepods. In: Wehrtmann IS, Cortés J (eds) Marine biodiversity of Costa Rica, Central America. Springer + Business Media BV, Dordrecht, pp 291–305

Moran DA, Dittel AI (1993) Anomuran and brachyuran crabs of Costa Rica: annotated list of species. Rev Biol Trop 41:599–617

Morato T, Watson R, Pitcher TJ, Pauly D (2006) Fishing down the deep. Fish Fish 7:24–34

Norse EA, Brooke S, Cheung WWL, Clark MR, Ekeland I, Froese R, Gjerde KM, Haedrich RL, Heppell SS, Morato T, Morgan LE, Pauly D, Sumaila RU, Watson R (2012) Sustainability of deep-sea fisheries. Mar Policy 36:307–320. https://doi.org/10.1016/j.marpol.2011.06.008

Owre HB, Foyo M (1964a) Report on a collection of Copepoda from the Caribbean Sea. Bull Mar Sci Gulf Carib 14:359–372

Owre HB, Foyo M (1964b) Plankton of the Florida current. Part IX. Additions of the list of Copepoda, with descriptions of two rare species. Bull Mar Sci Gulf Carib 14:342–358

Petrescu I, Heard R, Vargas R, Breedy O (2009) Cumaceans. In: Wehrtmann IS, Cortés J (eds) Marine biodiversity of Costa Rica, Central America. Springer + Business Media BV, Dordrecht, pp 237–244

Ramírez-Llodra ER, Tyler PA, Baker MC, Bergstad OA, Clark MR, Escobar E, Levin LA, Menot L, Rowden AA, Smith CR, Van Dover CL (2011) Man and the last great wilderness: human impact on the deep sea. PLoS One 6:e22588. https://doi.org/10.1371/journal.pone.0022588

Ranero CR, Grevemeyer I, Sahling H, Barckhausen U, Hensen C, Wallmann K, Weinrebe W, Vannucchi P, von Huene R, McIntosh K (2008) Hydrogeological system of erosional convergent margins and its influence on tectonics and interplate seismogenesis. Geochem Geophys Geosyst 9:Q03S04. https://doi.org/10.1029/2007GC001679

Reaka ML, Manning RB (1980) The distributional ecology and zoogeographical relationships of stomatopod Crustacea from Pacific Costa Rica. Smithson Contr Mar Sci 7:1–29

Roberts CM (2002) Deep impact: the rising toll of fishing in the deep sea. Trends Ecol Evol 17:242–245

Roberts JM, Harvey SM, Lamont PA, Gage JD, Humphery JD (2000) Seabed photography, environmental assessment and evidence for deep-water trawling on the continental margin west of the Hebrides. Hydrobiologia 441:173–183

Roberts JM, Wheeler AJ, Freiwald A (2006) Reefs of the deep: the biology and geology of cold-water coral ecosystems. Science 312:543–547. https://doi.org/10.1126/science.1119861

Robertson DR, Angulo A, Baldwin CC, Pitassy D, Driskell A, Weigt L, Navarro IJF (2017) Deep-water bony fishes collected by the B/O Miguel Oliver on the shelf edge of Pacific Central America: an annotated, illustrated and DNA-barcoded checklist. Zootaxa 4348:1–125. https://doi.org/10.11646/zootaxa.4348.1.1

Roff JC, Taylor ME (2000) National frameworks for marine conservation—a hierarchical geophysical approach. Aquat Conserv Mar Freshw Ecosyst 10:209–223

Sahling H, Masson DG, Ranero CR, Hühnerbach V, Weinrebe W, Klaucke I, Bürk D, Brückmann W, Suess E (2008) Fluid seepage at the continental margin offshore Costa Rica and southern Nicaragua. Geochem Geophys Geosyst 9:Q05S05. https://doi.org/10.1029/2008GC001978

Salas-Moya C, Vargas-Castillo R (2016) First record of Coronida glasselli (Stomatopoda: Coronididae) for Costa Rica (Isla del coco National Park, eastern tropical Pacific). Rev Biol Trop 64:231–233

Sameoto DD (1986) Influence of the biological and physical environment on the vertical distribution of mesozooplankton and micronekton in the eastern tropical Pacific. Mar Biol 93:263–279

Schlacher TA, Rowden AA, Dower JF, Consalvey M (2010) Seamount science scales undersea mountains: new research and outlook. Mar Ecol 31:1–13

Spinelli GA, Underwood MB (2004) Character of sediments entering the Costa Rica subduction zone: implications for partitioning of water along the plate interface. Island Arc 13:432–451

Suárez-Morales E, Gasca R (1989) Copépodos calanoides epiplanctónicos del Domo de Costa Rica (julio-agosto, 1982). Cienc Mar 15:89–102

Sweetman AK, Thurber AR, Smith CR, Levin LA, Mora C, Wei CL, Gooday AJ, Jones DOB, Rex M, Yasuhara M, Ingels J, Ruhl HA, Frieder CA, Danovaro R, Würzberg L, Baco A, Grupe BM, Pasulka A, Meye KS, Dunlop KM, Henry LA, Roberts JM (2017) Major impacts of climate change on deep-sea benthic ecosystems. Elementa-Sci Anthrop 5:4. https://doi.org/10.1525/elementa.203

Tabash FA (1995) An assessment of pink shrimp (*Penaeus brasiliensis*) populations, in three areas of the Caribbean coast of Costa Rica. Rev Biol Trop 43:239–250

Taylor ML, Gwinnett C, Robinson LF, Woodball LC (2016) Plastic microfiber ingestion by deep-sea organisms. Sci Rep 6:33997. https://doi.org/10.1038/srep33997

Thatje S, Marsh L, Roterman CN, Mavrogordato MN, Linse K (2015) Adaptations to hydrothermal vent life in *Kiwa tyleri*, a new species of yeti crab from the east scotia ridge, Antarctica. PLoS One 10:e0127621. https://doi.org/10.1371/journal.pone.0127621

Thurber AR, Jones WJ, Schnabel K (2011) Dancing for food in the deep sea: bacterial farming by a new species of yeti crab. PLoS One 6:e26243. https://doi.org/10.1371/journal.pone.0026243

Thurber AR, Sweetman AK, Narayanaswamy BE, Jones DOB, Ingels J, Hansman RL (2014) Ecosystem functions and services in the deep sea. Biogeosciences 11:3941–3963. https://doi.org/10.5194/bg-11-3941-2014

Townsend CH (1901) Dredging and other records of the United States fish commission steamer albatross, with bibliography relative to the work of the vessel. US Fish Comm Rep 1900:387–562

UNESCO (2009) Global Open Oceans and Deep Seabed (GOODS) – biogeographic classification, IOC technical series, vol 84. UNESCO-IOC, Paris

Van Dover CL (2011) Tighten regulations on deep-sea mining. Nature 470:31–33

Vargas R (2009) Stomatopods. In: Wehrtmann IS, Cortés J (eds) Marine biodiversity of Costa Rica, Central America. Springer + Business Media BV, Dordrecht, pp 193–197

Vargas R, Cortés J (1999) Biodiversidad marina de Costa Rica: Crustacea: Decapoda (Penaeoidea, Sergestoidea, Caridea, Astacidea, Thalassinidea, Palinura) del Pacífico. Rev Biol Trop 47:887–911

Vargas R, Cortés J (2006) Biodiversidad Marina de Costa Rica: Crustacea: Infraorder Anomura. Rev Biol Trop 54:461–488

Vargas R, Wehrtmann IS (2009) Decapod crustaceans. In: Wehrtmann IS, Cortés J (eds) Marine biodiversity of Costa Rica, Central America. Springer + Business Media BV, Dordrecht, pp 209–228

Villalobos-Rojas F, Wehrtmann IS (2011) Gonad development in the commercially exploited Deepwater shrimp *Solenocera agassizii* (Decapoda: Solenoceridae) from the Pacific Costa Rica, Central America. Fish Res 109:150–156. https://doi.org/10.1016/j.fishres.2011.01.032

Villalobos-Rojas F, Wehrtmann IS (2014) Secondary sexual characters and spermatophores of *Solenocera agassizii* (Decapoda: Solenoceridae), including a comparison with other solenocerid shrimp. Sex Early Dev Aquat Org 1:45–55. https://doi.org/10.3354/sedao00006

Villalobos-Rojas F, Azofeifa-Solano JC, Romero-Chavez R, Wehrtmann IS (2020) Hermit crabs associated to the shrimp bottom-trawl fishery along the Pacific coast of Costa Rica, Central America. Mar Fish Sci 33:96–113. https://doi.org/10.47193/mafis.3312020061806

Voss GL (1971) Narrative of R/V John Elliott Pillsbury cruise P-7101 to Central America, 20 January– February 1971. School of Marine and Atmospheric Science-University of Miami, Florida

Waller R, Watling L, Auster P, Shank T (2007) Anthropogenic impacts on the corner rise seamounts, north-West Atlantic Ocean. J Mar Biol Assoc UK 87:1075–1076

Watson RA, Morato T (2013) Fishing down the deep: accounting for within-species changes in depth of fishing. Fish Res 140:63–65

Wehrtmann IS, Echeverría-Sáenz S (2007) Crustacean fauna (Stomatopoda, Decapoda) associated with the Deepwater fishery of *Heterocarpus vicarius* (Decapoda, Pandalidae) along the Pacific coast of Costa Rica. Rev Biol Trop 55:121–130

Wehrtmann IS, Nielsen-Muñoz V (2009) The Deepwater fishery along the Pacific coast of Costa Rica, Central America. Lat Am J Aquat Res 37:543–554. https://doi.org/10.3856/vol37-issue3-fulltext-19

Wehrtmann IS, Herrera-Correal J, Vargas R, Hernáez P (2010) Squat lobsters (Decapoda: Anomura: Galatheidae) from Deepwater Pacific Costa Rica: species diversity, spatial and bathymetric distribution. Nauplius 18:69–77

Wehrtmann IS, Arana PM, Barriga E, Gracia A, Pezzuto PR (2012) Deep-water shrimp fisheries in Latin America: a review. Lat Am J Aquat Res 40:497–535. https://doi.org/10.3856/vol40-issue3-fulltext-2

Wenzhöfer F, Adler M, Kohls O, Hensen C, Strotmann B, Boehme S, Schulz HD (2001) Calcite dissolution driven by benthic mineralization in the deep-sea: in situ measurements of Ca^{2+}, pH, pCO_2 and O_2. Geochim Cosmochim Acta 65:2677–2690. https://doi.org/10.1016/S0016-7037(01)00620-

Wheat CG, McManus J, Fisher AT, Hulme SM, Orcutt BN (2017) Cool seafloor hydrothermal springs reveal global geochemical fluxes. Earth Planet Sci Lett 476:179–188. https://doi.org/10.1016/j.epsl.2017.07.049

Wheat CG, Hartwell AM, McManus J, Fisher AT, Orcutt BN, Schlicht LEM, Niedenzu S, Bach W (2019) Geology and fluid discharge at dorado outcrop, a low temperature ridge-flank hydrothermal system. Geochem Geophys Geosyst 20:487–504. https://doi.org/10.1029/2018GC007933

Wicksten MK (1979) New records of the species of *Glyphocrangon* in the northeastern Pacific Ocean (Caridea: Glyphocrangonidae). Proc Biol Soc Wash 92:217–224

Wolff T (1956) Six new abyssal species of *Neotanais* (Crust. Tanaidacea). Vidensk Meddel Naturhist Foren Kjøbenhavn 118:41–52

Woolley SNC, Tittensor DO, Dunstan PK, Guillera-Arroita G, Lahoz-Monfort JJ, Wintle BA, Worm B, O'Hara TD (2016) Deep-sea diversity patterns are shaped by energy availability. Nature 533:393–396. https://doi.org/10.1038/nature17937

Chapter 25
New Records of Crustaceans Collected Between 670 and 3400 m in the Colombian Caribbean

B. Martínez-Campos, R. A. Plazas-Gómez, H. F. Sáenz, L. F. Dueñas, J. León, and V. Puentes

Abstract As a result of curatorial activites realized in the Museum of Marine Natural History of Colombia, Makuriwa-Invemar, on the material deposited by consulting companies hired by oil and gas operators, 13 new records of deep-water species were recognized within the families Nematocarcinidae, Acanthephyridae, Aristeidae, Oplophoridae, Glyphocrangonidae, and Alvinocarididae (Caridea) and four within amphipods families Eurytheneidae, Scopelocheiridae, and Uristidae. These represent new records for the country, and some are also new records for the Caribbean Sea and the Western Atlantic.

Keywords Deep sea · Decapoda · Shrimps · Amphipoda · Cold seeps Soft bottoms

B. Martínez-Campos (✉)
Instituto de Investigaciones Marinas y Costeras-Invemar, Colombia. Museo de Historia Natural Marina de Colombia Makuriwa-Invemar, Programa de Biodiversidad y Ecosistemas Marinos, Santa Marta D.T.C.H., Colombia
e-mail: bibian.martinez@invemar.org.co

R. A. Plazas-Gómez
Leibniz Centre for Tropical Marine Research (ZMT), Bremen, Germany

H. F. Sáenz
Aquabiósfera SAS Cra 4A N° 6-58 El Rodadero, Santa Marta D.T.C.H., Colombia

L. F. Dueñas
Departamento de Biología, Facultad de Ciencias, Universidad Nacional de Colombia, Bogotá D.C., Colombia

Anadarko Colombia Company-HSE, Bogotá D.C., Colombia

J. León · V. Puentes
Anadarko Colombia Company-HSE, Bogotá D.C., Colombia

M. E. Hendrickx (ed.), *Deep-Sea Pycnogonids and Crustaceans of the Americas*, https://doi.org/10.1007/978-3-030-58410-8_25

25.1 Introduction

Due to a sharp increase in underwater exploration by the hydrocarbon industry in the Colombian Caribbean Sea, covering 658,000 km^2 and bordered by Honduras, Jamaica, Haiti, the Dominican Republic, Nicaragua, Costa Rica, Panama, and Venezuela, a large collection of samples of crustaceans occurring between 670 and 3400 m depth has been obtained. As a result, the holdings of the Makuriwa-Invemar Marine Natural History Museum contain 345 cataloged batches of decapod crustaceans, collected between 2008 and 2018, as well as some amphipods of the deep-water megafaunal component. The amount of material present in the museum's collection contrasts with the few published works dealing with deep-sea megafaunal crustaceans. These include contributions by Zubiría et al. (2016), Ortega-Echavarría et al. (2014), Lemaitre and Tavares (2015), and Martínez-Campos et al. (2017), but there are still many unpublished records and possibly new species.

During the periods of 2016–2018, private consultancy companies and the Scientific Services Coordination (CSC) of Invemar deposited the samples collected during nine exploratory campaigns in this museum. A total of 13 species of shrimp and four species of amphipods were identified and represent new records for Colombia. The specimens were collected in the framework of hydrocarbon exploratory campaigns and monitoring activities of three oil and gas operator companies. Subsequently, some of these specimens were also captured by ROV video on soft sediments.

The Colombian Caribbean is a region with a large number of endemic species. The crustacean megafauna of the deep-sea Colombian Caribbean holds the third place in species diversity, after mollusks and fish. It is also second in terms of abundance, surpassed only by mollusks (Invemar 2010). However, the study of deep-sea crustaceans in Colombia has been focused in the fringe of 200–500 m (Cruz and Fransen 2004; Campos et al. 2005), and families such as Alvinocaridae, Nematocarcinidae, Oplophoridae, Eurytheneidae, Scopelocheiridae, and Uristidae are underrepresented since their lower distribution limit is deeper than 500 m (Pérez-Farfante and Kensley 1997; Komai and Segonzac 2005).

The disitribution patterns and the ecology of the deep-sea crustaceans depend on factors such as the type of sediments, current dynamics, oxygen availability, and temperature. Because crustaceans are mostly secondary consumers, their presence depends largely on the primary production sources and the underlying communities in the substrate (Martin and Haney 2005; Desbruyères et al. 2006; Ramírez-Llorda and Billett 2006; Zamorano et al. 2014).

In deep-sea ecosystems, primary productivity comes from both heterotrophic and chemosynthetic sources. The first corresponds to organic matter falling in the form of microaggregates and marine snow, allowing suspensivore species, such as shrimp, to thrive (Ramírez-Llorda and Billett 2006). Thus, food sources can become ephemeral which drives wandering habits in some species with opportunistic scavenger behavior and a highly sensitive sense of chemoreception, such as the giant amphipods of the genus *Eurythenes* (Koehl 2011).

The second corresponds to chemosynthetic productivity sources, either in hydrothermal vents or cold seep habitats. These have been identified recently in two areas of the Colombian Caribbean Sea: the Sinú-San Jacinto Basin and La Guajira Basin. These habitats show affinities with the communities found off Trinidad and Tobago and the Gulf of Mexico (Digby et al. 2016; Dueñas et al. in review). These ecosystems harbor crustaceans whose presence can be explained by endemism, like the species of the genus *Alvinocaris* that inhabit exclusively these environments, or by species that are opportunistic inhabitants, like the genus *Nematocarcinus*, whose presence is occasional in such ecosystems (Segonzac et al. 1993; Komai and Segonzac 2005; Komai et al. 2005; Martin and Haney 2005; Desbruyères et al. 2006). To date, species of *Nematocarcinus* have been found on muddy plains of the deep sea, and only alvinocarid shrimps have been registered as part of the chemosynthetic communities of Colombia, interacting with mussels of genus *Bathymodiolus* (Dueñas et al. in review).

25.2 Methods

The specimens of this study were collected throughout the course of nine exploratory surveys (Table 25.1). In the case of exploratory wells, several sampling stations (at least nine) were selected in wells vicinity, and nekton or piston core sampling stations were determined based on previous results obtained during other surveys. Specimens were deposited in the museum between 2016 and 2018 and preserved in 70% alcohol. On arrival, the specimens were examined to confirm their identification to the lowest taxonomic level and labeled, and a database recording process was performed.

The material was identified following the contributions and identification keys by Kemp (1939), Holthuis (1955, 1971, 1993), Barnard (1961), Pequegnat (1970), Crosnier and Forest (1973), Crosnier (1978), Williams (1984), Barnard and Karaman (1991), Pérez-Farfante and Kensley (1997), Komai (2004), Stoddart and Lowry (2004), Cardoso and Young (2005), Komai and Segonzac (2005), Cardoso and Serejo (2007), Tavares and Serejo (2007), Zelnio and Hourdez (2009), Vázquez-Bader and Gracia (2013), Cardoso and Burukovsky (2014), D'Udekem D'Acoz and Havermans (2015), Horton and Thurston (2015), Kilgallen and Lowry (2015), and Alves-Júnior et al. (2019). Biological samples were photographed frame by frame in different focal fields, with a Leica Z16 APO-A micro-stereoscope, to generate a high-resolution digital composition using Adobe Photoshop CS6 software. When comparisons between two species were needed, digital illustrations were made using Adobe Illustrator CS6 software.

The following information was provided for each sample: locality, depth, number of individuals, project reference, and catalog number. Recognition characters, general distribution and bathymetry, and some remarks were also provided for each species.

Table 25.1 Collection data of exploration surveys made during this study

Code	Year	Lat N	Long W	Depth (m)	Sampling gear
CA	2014/S/ MBO	10°25′33.8″ to 10°28′49.8″	76°14′35.5″ to 76°16′15.5″	2000–2400	Fish trap
PA	2014/S/ OWC	8°52′58.05″ to 9°54′32.6″	76°52′38.1″ to 76°52′40.09″	200–3000	Fish trap
C5	2015/S/ OWC	9°06′5.8″ to 10°20′37.5″	77°08′41.5″ to 76°30′54.6″	1389–3097	Fish trap
KI	2015/S/ MBO	9°09′53.06″ to 9°10′41.3″	76°49′55.9″ to 76°50′45.8″	1600–1800	Fish trap
C3	2015/S/ OWC	11°18′15.6″ to 11°57′32.2″	74°09′20.0″ to 77°31′12.7″	1200–3400	Fish trap
OR	2015/S/ MBO	12°45′19.4″ to 12°48′34.4″	71°34′59.5″ to 71°35′41.07″	700–840	Fish trap
BR	2016/N/ MBO	12°47′2.5″ to 12°50′18.4″	71°40′19.5″ to 71°42′0.7″	763–854	Fish trap
CAD	2016/ SMBO	10°25′33.8″ to 10°28′49.7″	76°14′35.5″ to 76°16′15.471	2000–2400	Fish trap
C7	2017/N/ CMM	13°01′38.9″ to 13°01′31.7″	73°03′08.8″ to 73°03′16.6″	2812–3229	Piston Core

S, N southern and northern Colombian Caribbean, MBO monitoring of biological communities, OWC oil well characterization, CMM characterization of macro and meiofauna communities

25.3 Results

The family Aristeidae is one of the five families of the superfamily Penaeoidea (recognized by Martin and Davis 2001) recorded in this paper. This family was only previously represented in the area by one species, *Aristaeomorpha foliacea* (Campos et al. 2005); therefore the identification of *Hemipenaeus carpenteri* and *Cerataspis monstrosus* increases the records of the Aristeidae family for the Colombian Caribbean.

Four out of 11 superfamilies recognized in the Caridean infraorder (De Grave and Fransen 2011) were recorded: Bresilioidea, Crangonoidea, Nematocarcinoidea, and Oplophoroidea. This is the first time that the genus *Alvinocaris* is reported in Colombia with the presence of two species: *Alvinocaris muricola* and *A. markensis*. These new reports increase to three the number of records of the Bresilioidea superfamily in the country (Campos et al. 2010). With the reports of *Glyphocrangon aurantiaca* and *G. longirostris*, the number of species of the superfamily Crangonoidea in the Colombian Caribbean increases to 12 (Bermúdez et al. 2002; Campos et al. 2010). Finally, in this study we report for the first time a species of the superfamily Nematocarcinoidea, with the presence of *Nematocarcinus ensifer* and *N. rotundus*.

Only one species of the superfamily Oplophoroidea had been previously recorded in the country; however, in the frame of this review, two families and five species are presented herein as new reports. The new records include *Systellaspis debilis* belonging to the Oplophoridae family, and four species of the Acanthephyridae family: *Acanthephyra acutifrons*, *A. quadrispinosa*, *A. curtirostris*, and *A. stylorostratis*.

The deep-sea amphipods here recorded belong to the superfamily Lysianassoidea, a group that has been poorly studied in the Colombian Caribbean Sea. The amphipods registered here are scavengers, larger than 500 μm, with great capacity for active swimming. They are opportunistic consumers of decaying corpses, and they are commonly captured in fish traps (Stoddart and Lowry 2004; Horton and Thurston 2015). These are the first records of amphipods for the deep-sea Colombian Caribbean.

Eurythenes gryllus and *E. thurstoni* are the first records of the Eurytheneidae family, *Paracallisoma alberti* Chevreux, 1903 is the first record of the family Scopelocheiridae, and *Stephonyx biscayensis* is the first record of the family Uristidae.

25.3.1 Systematic Section

Order Decapoda Latreille, 1802.
Suborder Dendrobranchiata Bate, 1888.
Infraorder Caridea Dana, 1852.
Superfamily Penaeoidea Rafinesque, 1815.
Family Aristeidae Wood-Mason in Wood-Mason & Alcock, 1891.

1. *Hemipenaeus carpenteri* Wood-Mason in Wood-Mason & Alcock, 1891 (Fig. 25.1a).

Hemipenaeus carpenteri Wood-Mason, 1891 (in Wood-Mason & Alcock):189.
Aristaeus (*Hemipenaeus*) *carpenteri*.— Alcock, 1901: 32–33.
Hemipenaeus carpenteri.— Crosnier 1978:76–80, Figs. 27c–d, 28a–b, 29a; Gore 1985:122–124; Crosnier 1994:369, Fig. 5; Pérez-Farfante and Kensley 1997: 46; Dall 2001:418, Fig. 6; Tavares and Serejo 2007:23–27, Figs. 14–16; Felder et al. 2009:1049.

Material Examined. Off Moñitos, Córdoba, 9°16′5.5092″-76°46′56.657″, 1800 m, 1 org, PA-E03. INV CRU8459 **Recognition characters**. Rostrum short, with three dorsal spines. Gastrorbital carina present. Cervical carina reaching the dorsum of carapace. Abdominal somites 4–6 with a dorsal carina, third with a large dorsal spine curved downward, sixth with a small terminal spine (Tavares and Serejo 2007). **Distribution**. Northwestern Atlantic Ocean: off Bahamas. Gulf of Mexico. Caribbean Sea: Colombia (Southwestern Colombian Caribbean: off Moñitos, Córdoba). Southwestern Atlantic: Brazil. Indian Ocean: Madagascar; Arabian Sea; Bay of Bengal. West Pacific Ocean: Japan; Wallis and Futuna Islands; Western and Northeastern Australia. East Pacific Ocean: off Gulf of Panama, Galapagos Islands

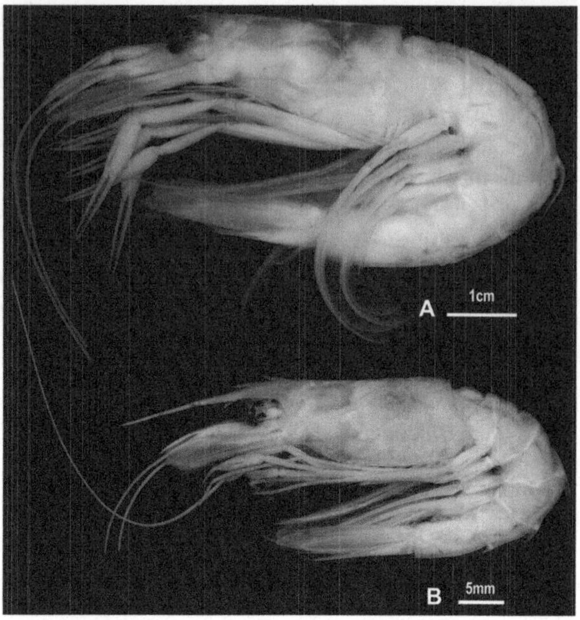

Fig. 25.1 (**a**) *Hemipenaeus carpenteri* PA-E03. (**b**) *Cerataspis monstrosus* C5-E13

(Tavares and Serejo 2007; this paper). **Bathymetric distribution**. 900–3900 m (Tavares and Serejo 2007). **Remarks**. This is the first record of this species for the Colombian Caribbean and for the Caribbean Sea. No differences from the description of Tavares and Serejo (2007).

2. *Cerataspis monstrosus* Gray, 1828 (Fig. 25.1b).

Aristaeus armatus Bate, 1881:188; 1888:312–317, pls. 45–46, Figs. 1–2.
Aristaeus (*Aristaeopsis*) *armatus*.— Alcock 1901: 41.
Plesiopenaeus armatus.— Crosnier and Forest 1973:294–296, Fig. 99c–d; Crosnier 1978:92–94, Figs. 31d–e, 32d–f, 33b; Pérez-Farfante and Kensley 1997:50–52, Figs. 19–20; Dall 2001:421–422, Fig. 9; Tavares and Serejo 2007:33–37, Figs. 20–22; Felder et al. 2009:1049.

Material Examined. Off Cartagena, 10°27′19.64″-76°30′54.68″, 2873 m, 1 org, C5-E13 INV CRU8485. **Recognition characters**. Rostrum slightly curved upward, with three dorsal spines. Cervical carina reaches the middle of carapace. Abdominal pleurae with small terminal spines. Abdominal somites 3–6 with a median dorsal carina, and with dorsal spines (Tavares and Serejo 2007). **Distribution**. Eastern Atlantic Ocean: off Azores; Madeira; Canary Islands; Cape Verde. Gulf of Mexico. Caribbean Sea: Colombia (off Cartagena). Indian Ocean: Zanzibar; Madagascar; Maldives Islands; Bay of Bengal. West Pacific Ocean: the Philippines; Japan; Tuamotu Islands; Wallis and Futuna Islands; northeast of Australia; Hawaii (Tavares and Serejo 2007; present sudy). **Bathymetric distribution**. 752–5413 m (Tavares and Serejo 2007). **Remarks.** First record for the Colombian Caribbean and for the Caribbean Sea. No differences from the description of Tavares and Serejo (2007).

Infraorder Caridea Dana, 1852.
Superfamily Bresilioidea Calman, 1896.
Family Alvinocarididae Christoffersen, 1986.

3. *Alvinocaris markensis* Williams, 1988 (Figs. 25.2b, 25.3a, c and 25.4c, d).

Alvinocaris markensis Williams, 1988:264, Figs. 1, 2, 7.
Alvinocaris muricola.— Shank et al. 1999:246 (not *Alvinocaris muricola* Williams, 1988).
Alvinocaris aff. *markensis*.— Desbruyères et al. 2001:1335.
Alvinocaris markensis.— Dixon and Dixon 1996:9, Figs. 1–3; Vereshchaka 1996:577; Shank 1997:192; Shank et al. 1998:89; Shank et al. 1999:246, 247, Fig. 2; Kikuchi and Hashimoto 2000:146–148 (key); Desbruyères et al. 2000: 209; Komai and Segonzac 2005:1123, 1124, 1128–1132, Figs. 6, 7, 14a, 29; Martin and Haney 2005:448; Zelnio and Hourdez 2009:67, 68 (key).

Material Examined. Off La Guajira, 13°15′55.9″-73°03′43.2″, 2980 m, 1 org, INV CRU8988. **Recognition characters**. Rostrum directed forward or slightly descending, usually reaching to second segment of antennular peduncle and armed with

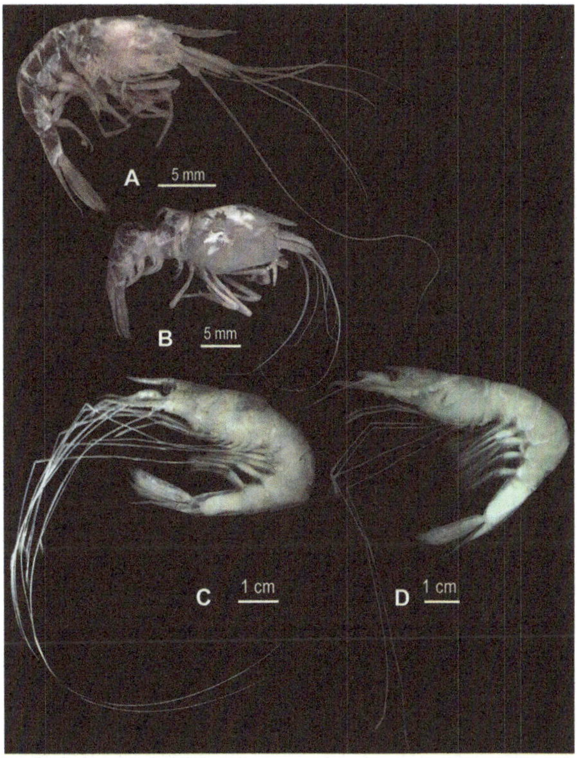

Fig. 25.2 (**a**) *Alvinocaris muricola* INV CRU_8989. (**b**) *Alvinocaris markensis* INV CRU_8988. (**c**) *Nematocarcinus ensifer* CD-NA3. (**d**) *Nematocarcinus rotundus* C3-E03

Fig. 25.3 (**a**, **c**) Cold seeps in which *Alvinocaris* sp. specimens were collected in association with mussels of the genus *Bathymodiolus* (Photo: CSA Ocean Science Inc. – Anadarko Colombia Company©). (**b**) *Nematocarcinus ensifer* captured by trawl camera on deep soft bottoms. (Photo: CSA Ocean Science Inc. – Anadarko Colombia Company©)

14–19 teeth (5–7 on posterior carapace). Posterior most tooth arising from 0.24 to 0.31 of carapace length. Ventral margin with six to nine small teeth. Carapace 0.52–0.65 times as wide as long. Pterygostomian tooth distinctly exceeding antennal tooth; post-antennal groove shallow. Branchial region not particularly inflated. Telson not reaching or reaching posterior margin of uropodal endopod, length less than 2.90 times anterior width armed with six to eight dorsolateral spines. Posterior margin convex, armed with two pairs of lateral spines and 12–14 plumose setae (Komai and Segonzac 2005). **Distribution**. Caribbean Sea: Colombia (La Guajira). Mid-Atlantic Ridge: Snake Pit, Lucky Strike, and Rainbow. Hydrothermal vents of Broken Spur, TAG, and Logatchev (Komai and Segonzac 2005; present study). **Bathymetric distribution**. 2292–3650 m (Komai and Segonzac 2005). **Remarks**. First record for the Colombian Caribbean and for the Caribbean Sea. The overlapping of key characters between *A. markensis* and *A. muricola*, widely discussed by Komai and Segonzac (2005), makes difficult the determination in immature specimens. In this case, of the 13 specimens of the *Alvinocaris* genus reviewed, 4 specimens were found in adult stage, and only 1 of them corresponds with *A. markensis* (Fig. 25.4). However, this specimen was collected with seven immature specimens with overlapping characters between *A. markensis* and *A. muricola*. Consequently,

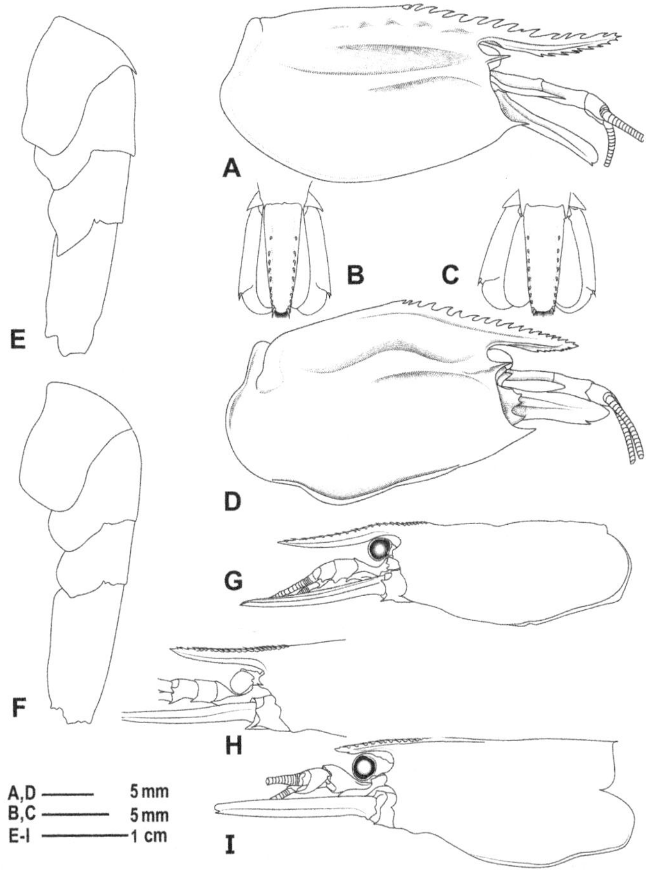

Fig. 25.4 Differences in the carapace and telson in the specimens collected of *Alvinocaris* genus. (**a**) Carapace. (**b**) Telson of *Alvinocaris muricola*. (**c**) Telson. (**d**) Carapace of *Alvinocaris markensis*. (**e**) Abdominal segments of *Nematocarcinus ensifer*. (**f**) Abdominal segments of *Nematocarcinus rotundus* (note the dorsal projection on third and fourth dorsal segments in *N. ensifer* differentiating this species from *N. cursor* and *N. exilis*. (**g**), (**h**) Length and tip of rostrum of *Nematocarcinus ensifer*. (**i**) Length and tip of rostrum of *Nematocarcinus rotundus*

we report only the adult specimen, because there is no certainty on the identity of the immature specimens.

4. *Alvinocaris muricola* Williams, 1988 (Figs. 25.2a, 25.3a, c and 25.4a, b).

Alvinocaris muricola Williams, 1988:268, Figs. 3, 4, 7; Shank et al. 1999:246, Fig. 2; Kikuchi and Hashimoto 2000:146, 148 (key); Komai and Segonzac 2005:1123–1124, 1132–1143, Figs. 2, 3, 8–14, 29; Martin and Haney 2005:448; Felder et al. 2009:1053.

Material Examined. Off La Guajira, 13°15′55.9″-73°03′43.2″, 2980 m, 4 org, INV CRU8986; off La Guajira, 13°05′03.4″-73°00′48.5″, 2154 m, 1 org, INV CRU8989; off La Guajira, 13°15′55.9″-73°03′43.2″, 1 org, 2980 m, INV CRU8990. **Recognition characters.** Rostrum directed forward, weakly curved dorsally or straight, usually reaching to the second segment of antennular peduncle in females, occasionally overreaching in males. Dorsal margin with 10–17 teeth (4–6 on carapace) posterior most tooth arising from 0.34 to 0.40 of carapace length; ventral margin with 3–13 small teeth. Second segment of antennular peduncle 1.9–2.1 times longer than wide; antennal scale 1.9–2.1 times longer than wide. Carapace 0.69–0.83 times as wide as long. Pterygostomian tooth strongly produced anteriorly far beyond tip of antennal tooth. Post-antennal groove deep, almost parallel to horizontal plane of carapace. Branchial region notably convex. Telson nearly reaching to slightly overreaching posterior margin of uropodal endopod; length more than 2.90 times anterior width, armed with 6–8 dorsolateral spines; posterior margin moderately convex, with two pairs of spines at lateral angles and 12–14 plumose setae (Komai and Segonzac 2005; Zelnio and Hourdez 2009). **Distribution**. Gulf of Mexico: West Florida Escarpment. Caribbean Sea: Barbados, Colombia (off La Guajira). Eastern Atlantic: West Equatorial Africa (Komai and Segonzac 2005; present study). **Bathymetric distribution**. 1697–3277 m (Komai and Segonzac 2005). **Remarks**. First record for the Colombian Caribbean Sea, based on three adult and three immature specimens. All of them presented the diagnostic characteristics of the species, including the juveniles.

Superfamily Crangonoidea Haworth, 1825.
Family Glyphocrangonidae Smith, 1884.

5. *Glyphocrangon aurantiaca* Holthuis, 1971 (Fig. 25.5a, b).

   Glyphocrangon aurantiaca Holthuis, 1971:303, Fig. 8; Takeda and Okutani 1983:68, textfig; Komai 2004:35, Fig. 2a, b; Vázquez-Bader and Gracia 2013:381 (key).

Material Examined. Off La Guajira, 12° 45′ 20.33″-71° 36′ 39.83″, 700–800 m, 1 org OR-E02 INV CRU8881; off La Guajira, 12° 45′ 19.40″-71° 35′ 00.47″, 700 m, 3 org, OR-E11 INV CRU8891; off La Guajira, 12° 47′ 2.504″-71° 40′ 21.402″, 763 m, 1 org, BR- E06 INV CRU8429. **Recognition characters**. Rostrum shorter than carapace with two pairs of lateral spines; dorsal surface lacking corrugations. Carapace with the submedian carina composed of low, blunt tubercles; anterior-most tubercle of anterior intermediate carina obsolete; antennal carina not extending to hepatic region; posterior third carina ending in a blunt tubercle; anterior lateral carina not forming acute lamina, armed with two spines. Antennal spine strong, far more strongly divergent than branchiostegal spine. Abdomen with low, broad median carinae; and low dorsolateral carinae (Komai 2004). **Distribution**. Caribbean Sea: Colombia (La Guajira), Tobago. Southwestern Atlantic: French Guiana, Suriname, and Brazil (Holthuis 1971; Komai 2004; present study). **Bathymetric distribution**. 394–761 m (Holthuis; 1971, Komai 2004). **Remarks**.

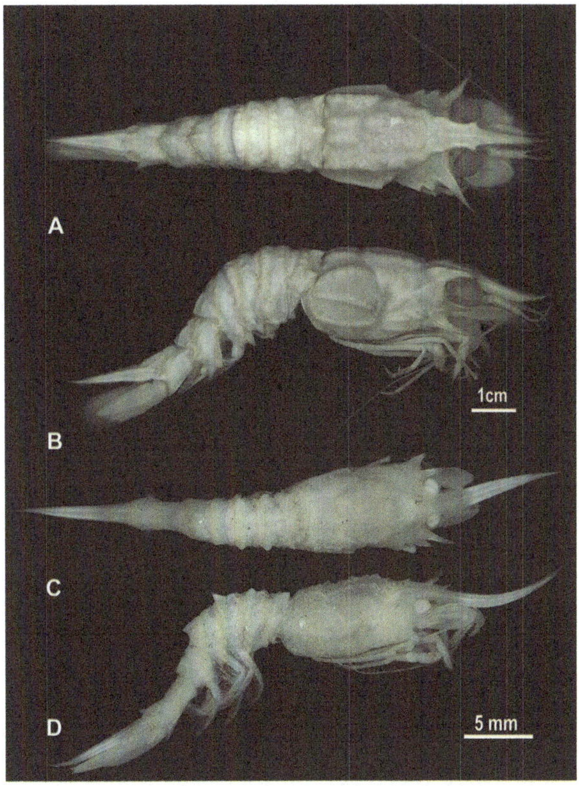

Fig. 25.5 (**a**) *Glyphocrangon aurantiaca* (BR-E06). (**b**) *Glyphocrangon longirostris* C5-E02

First record of *G. aurantiaca* for the Colombian Caribbean. Previous records indicate the presence of *G. longleyi* in Colombia. This species shows affinity with *G. aurantiaca*, but it can be differentiated by the presence of a long, strongly projecting antennal spine and a posterior antennal carina devoid of spine. In turn, *G. longleyi* features a straight posterior lateral carina and lacks medium carina in the second and third abdominal somite (Holthuis 1971).

6. *Glyphocrangon longirostris* (Smith, 1882) (Fig. 25.5c, d).

Rhachocaris longirostris Smith, 1882:51, pl. 5, Fig. 1, pl. 6, Fig. 1.
Glyphocrangon longirostris.— Pequegnat 1970:106.— Holthuis 1971: 330, Figs. 11–13.— Crosnier and Forest 1973:230, Fig. 73a, b.— Chace 1984:8 (in key).— D'Udekem D'Acoz 1999:138.— Komai 2004:35, Fig. 2c, d.— Cardoso and Serejo 2007:40, Fig. 1. — Serejo et al. 2007:139.— Felder et al. 2009:1061.— Vázquez-Bader and Gracia 2013:381 (key).

Material Examined. Off Córdoba, 9°2′19.12″-76°59′40.04″, 1389 m, 1 org, C5-E02 INV CRU8481. **Recognition characters**. Rostrum shorter than carapace with two pairs of lateral spines. Carapace with cervical and lateral grooves. Anterior and posterior submedian carina composed of tubercles; intermediate anterior carina with anterior tubercle spiniform, posterior carina composed of small tubercles. Anterior antennal carina short, posterior well developed. Anterior lateral carina not expanded, not aligned with anterior antennal carina, with an anterior spine. Branchiostegal and antennal spines strong (based on Cardoso and Serejo 2007). **Distribution**. Northwestern Atlantic: the USA (Massachusetts). Gulf of Mexico. Caribbean Sea: Colombia (off Córdoba); Soutwestern Atlantic: Brazil (Ceará and Rio Grande do Norte; Potiguar Basin, Bahia, Espírito Santo, Rio de Janeiro). Eastern Atlantic: Ireland to South Africa (Cardoso and Serejo 2007; present study). **Bathymetric distribution**. 908–2094 m (Cardoso and Serejo 2007). **Remarks**. First record for the Colombian Caribbean and for the Caribbean Sea. No differences from the description of Cardoso and Serejo (2007).

Superfamily Nematocarcinoidea Smith, 1884.
Family Nematocarcinidae Smith, 1884.

7. *Nematocarcinus ensifer* (Smith, 1882) (Figs. 25.2c, 25.3b and 25.4g, h, e).

Eumiersia ensifera Smith 1882:77.
Nematocarcinus ensiferus.— Smith 1884:368; 1886: 188; 1887:665; Caullery, 1896:377.
Nematocarcinus ensifer.— Sivertsen and Holthuis 1956: 19 (in part); Pequegnat 1970: 75; Crosnier and Forest 1973:98, 116–123, Figs. 32a–c, 33a–c; Cardoso and Serejo 2007:41; Felder et al. 2009:1053; Cardoso and Burukovsky 2014:556 (key).

Material Examined. Off Barranquilla, 11°43′21.6″-74°45′04.3″, 2015 m, 2 org, C3-E08 INV CRU8434; off Cartagena, 10°27′11.798″-76°15′25.406″, 2300 m, 1 org, CD-NA3 INV CRU8901; off Cartagena, 10°27′19.64″-76°30′54.68″, 2873 m, 2 org, C5-E13 INV CRU8484. **Recognition characters.** Rostrum directed obliquely upward, armed with more than 20 teeth on dorsal margin; ventral margin unarmed. Carapace with antennal and branchiostegal spines. Stylocerite with broad base and acute tip. Scaphocerite elongate, with distal tooth overreaching blade. Posterior edge of abdominal segments 3–4 forming an obtuse dorsal projection above the next segment. Fifth abdominal somite pleura with a stout spine; distoventral organ at sixth abdominal somite formed by two pairs of parallel rows of long plumose setae, inner row less developed (Crosnier and Forest 1973; Cardoso and Serejo 2007; Cardoso and Burukovsky 2014). **Distribution**. Northwestern Atlantic: the USA (Philadelphia, Maryland, Delaware). Gulf of Mexico. Caribbean Sea: Colombia (off Barranquilla, Cartagena). Southwestern Atlantic: Brazil. Eastern Atlantic: Portugal, Gulf of Gascoigne, Azores, Morocco, Cape Verde. Mediterranean Sea (Crosnier and Forest 1973; Cardoso and Serejo 2007; Felder et al. 2009; present study). **Bathymetric distribution**. 1430–3549 m (Felder et al. 2009). **Remarks**. First record for the Colombian Caribbean and for the Caribbean Sea. The rostrum of the specimens examined does not overreach the antennular peduncle or overreaches it with less than one third of its length (although it has more than 20 teeth), and specimens can be confused with *N. cursor*, following the keys proposed by Cardoso and

Burukovsky (2014). Nevertheless, *N. ensifer* can be separated from *N. cursor* by the dorsal projection on abdominal segments 3–4 (Fig. 25.4).

8. *Nematocarcinus rotundus* Crosnier & Forest, 1973 (Figs. 25.2c and 25.4i, f).

Nematocarcinus rotundus Crosnier and Forest, 1973:98, 103–116, Fig. 29c, 30f–i, 31e, f.
Nematocarcinus cursor.— Smith 1886 b: pp. 189, 192, 196, 198 (in part); Smith 1887:665 (in part), pl. 17, Fig. 1–a; Agassiz 1888:46 (in part); Thompson 1966 b:138, Fig. 4–5; Pequegnat 1970:73 (in part).
Nematocarcinus rotundus.— Wenner 1979:380; Takeda and Okutani 1983:53; Squires 1990:105; Burukovsky 2001:1440; 2003:144; 2004:558; 2012:173; Cardoso and Burukovsky 2014:445–449, 556 Figs. 5–7.

Material Examined. Off Barranquilla, 11°24′26.2″-74°53′59.2″, 1300 m, 3 org, C3-E11 INV CRU8440; off Barranquilla, 11°56′05.0″-74°09′20.0″, 1800 m, 1 org, C3-E03 INV CRU8446. **Recognition characters**. Rostrum straight, usually reaching the distal margin of second segment antennular peduncle, armed with one well-defined ventrodistal tooth and 10–16 (usually 12–15) dorsal teeth. Distance between apex and dorsodistal tooth equaling distance between the remaining rostrum and the dorsal teeth. Somite 3 with a posterodorsal margin poorly developed, rounded. Pleura of somite 5 widely rounded, without spine or with a poorly developed spine, its sides intersecting at an angle of almost 120°. Telson with 6–8 pairs of dorsolateral spines and one accessory spine (Cardoso and Burukovsky 2014). **Distribution**. Northwestern Atlantic: between Long Island and Chesapeake Bay, between the Bahamas archipelago and the south of Florida. Gulf of Mexico. Caribbean Sea: Lesser Antilles, Colombia (off Barranquilla, off Santa Marta). Southwestern Atlantic: French Guiana; Brazil (Bahia to Rio de Janeiro) (Cardoso and Burukovsky 2014; present study). **Bathymetric distribution**. 421–1875 m (Cardoso and Burukovsky 2014). **Remarks**. First record for the Colombian Caribbean Sea. No differences from the description of Cardoso and Burukovsky (2014).

Superfamily Oplophoroidea Dana, 1852.
Family Acanthephyridae Spence Bate, 1888.

9. *Acanthephyra acutifrons* Spence Bate, 1888 (Fig. 25.6a).

Acanthephyra acutifrons Spence Bate 1888:749, pl. 126: Fig. 3; Chace Jr 1940:146, Fig. 23; 1986:10, Figs. 2b, 4b, 5b; Crosnier 1987:696; Kensley et al. 1987:283; Cardoso and Young 2005:8, Figs. 3a–d, 4a–g, 5a–d, 6a; Pequegnat and Wicksten 2006:95; Judkins 2014:304; Alves-Júnior et al. 2019:403, 404, Figs. 1a–b, 2.

Material Examined. Off Barranquilla, 11°55′17.0″-74°54′00.0″, 3100 m, 1 org, C3-E06 INV CRU8448. **Recognition characters**. Rostrum reaching 2/3 of scaphocerite, with 6–8 dorsal teeth, one ventral. Antennal spine absent; branchiostegal spine present, without distinct carina. All abdominal somites dorsally carinate, somites 3–6 with posteromesial tooth, distinctly strong in the third. Male pleopod 1

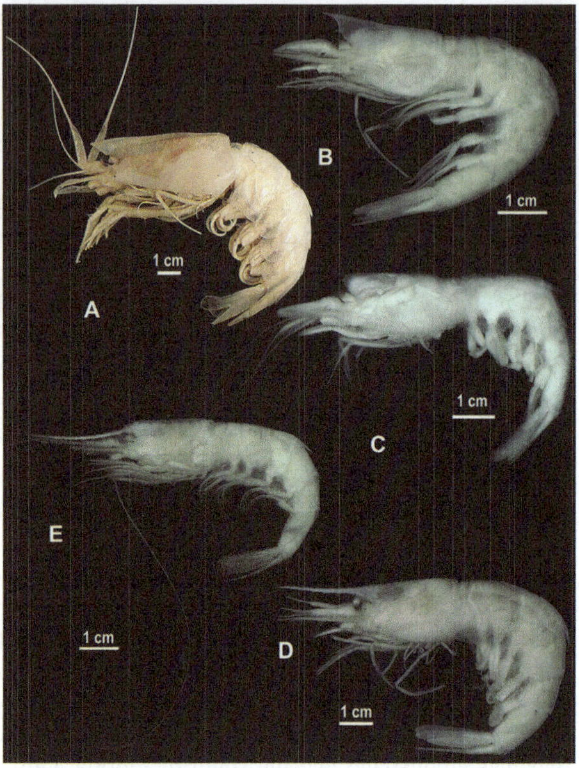

Fig. 25.6 (**a**) *Acanthephyra acutifrons* C3-E06. (**b**) *Acanthephyra curtirostris* (C5-E13). (**c**) *Acanthephyra stylorostratis* (PA-E01). (**d**) *Acanthephyra quadrispinosa* (KI-FSE03). (**e**) *Systellaspis debilis* (CD-NA3)

with endopod rounded, lateral margins with densely plumose setae, distal lobe with numerous distal spines (Cardoso and Young 2005). **Distribution**. Northwestern Atlantic: Cuba, Bahamas, Tortuga Island. Gulf of Mexico: Mexico. Caribbean Sea: Colombia (off La Guajira); Soutwestern Atlantic: Guiana, Suriname, French Guiana, Brazil (Rio Grande do Norte, Rocas Atoll, Pernambuco, Fernando de Noronha, off Alagoas, Bahia, and Espírito Santo). Meso-Atlantic Ridge. Eastern Atlantic: off Western Sahara. Indo-Pacific Oceans: West Indian Ocean, Madagascar, Philippines, Indonesia (Sumatra), Australia, Japan (Alves-Júnior et al. 2019; present study). **Bathymetric distribution**. 650–4200 m (Alves-Júnior et al. 2019). **Remarks**. First record for the Colombian Caribbean and for the Caribbean Sea. Cardoso and Young (2005) report carapace lengths of 41–51 mm in specimens from Brazilian waters. The specimen examined presented a remarkably long carapace: 60 mm.

10. *Acanthephyra curtirostris* Wood-Mason in Wood-Mason & Alcock, 1891 (Fig. 25.6b).

Acanthephyra acutifrons Spence Bate 1888:749 (in part); Kemp 1906: 22.
Acanthephyra curtirostris Wood Mason, 1891 (in Wood-Mason & Alcock):195; Wood-Mason and Alcock 1892:364, pl. 3, Fig. 5; Faxon 1895:164, pl. 43,

Figs. 2–5; Anderson, 1896:94; Alcock 1899:76; 1901:81; Kemp 1906:22; De Man 1920:44; Balss 1925:261, Fig. 30; Chace Jr 1936:26; 1937:111; 1947:17; Calman 1939:194; Springer and Bullis Jr 1956:11; Figueira 1957:28, pl. 2, Fig. 1; Pearcy and Forss 1966:1137; Crosnier and Forest 1968:1129; 1973:39, Fig. 8a; Kensley 1981a:57; 1981b:21; Kensley et al. 1987:283; Vereshchaka 1990:139; Pequegnat and Wicksten 2006:96; Alves-Júnior et al. 2019:404–406, Figs. 3a–b, 4.

Material Examined. Off Cartagena, 10°27′19.64″-76°30′54.68″, 2873 m, 1 org, C5-E13 INV CRU8474. **Recognition characters**. Integument firm. Rostrum not reaching beyond antennular peduncle with 10–14 dorsal teeth, one ventral. Carapace not dorsally carinate posteriorly. Branchiostegal spine and carina present; the last extends back to posterior part of branchial region. Abdominal somites 2–6 dorsally carinate, somites 3–6 with a median posterior spine. Telson dorsally sulcate on proximal half with 6–15 dorsolateral spines (Alves-Júnior et al. 2019). **Distribution.** Northwestern Atlantic: the USA (Oregon), Bermuda, Bahamas. Gulf of Mexico. Caribbean Sea: Antilles, Panamá Basin. Colombia (Soutwestern Colombian Caribbean, off Arboletes, Antioquia). Southwestern Atlantic: British Guiana, Brazil (Pear, Rocas Atoll, Fernando de Noronha, and off Pernambuco). Eastern Atlantic: Portugal (Madeira Island). Indo-Pacific Ocean: East coast of Africa, Aldabra Atoll, Arabian Sea, Maldives Islands, Gulf of Bengal, Andaman Sea. Eastern Pacific: the USA (Northern California, USA, to Baja California, Mexico) and Peru (Alves-Júnior et al. 2019; present study). **Bathymetric distribution:** 550–5900 m (Alves-Júnior et al. 2019). **Remarks**: First record for the Colombian Caribbean Sea. No differences from the description of Alves-Júnior et al. (2019).

11. *Acanthephyra quadrispinosa* Kemp, 1939 (Fig. 25.6d).

Acanthephyra batei Stebbing 1905:107, pl. 24B.
Acanthephyra quadrispinosa Kemp 1939:576; Barnard 1950:668, Fig. 124 g; Kensley 1968:311; 1972:40 (key), Fig. 18c, d; 1981a:57; 1981b:21; 1987:284; Chace Jr 1986:26, Figs. 3 h, 4 t, 7 g, 10c, 14; Crosnier 1987:697; Kensley et al. 1987:284; Cardoso and Young 2005:21, Figs. 14–18; Judkins 2014:304; Alves-Júnior et al. 2019:410–411, Figs. 9 a–b, 10, 39e.

Material Examined. Off Gulf of Morrosquillo, 10°6′58.07″-76°47′57.36″, 3097 m, 1 org, C5-E11 INV CRU8473; off Arboletes, Antioquia, 9°11′30.2849″-76°50′45.879″, 1600–1800 m, 1 org, KI-E01 INV CRU8470; off San Juan de Urabá, 9°2′35.03400″-76°52′38.157″, 1200 m, 1 org, PA-E02 INV CRU8451. **Recognition characters**. Rostrum overreaching scaphocerite, ventral margin with 3–4 teeth; antennal spine present; carapace with branchiostegal spine and carina present. All the abdominal somites except the first strongly carinate dorsally; 3–6 segments armed with posteromesial tooth. Third distinctly strong. Telson sulcate on dorsal midline, with four pairs of dorsolateral stout setae (Cardoso and Young 2005, Alves-Júnior et al. 2019). **Distribution**. Caribbean Sea: Colombia (Gulf of Morrosquillo, Arboletes, San Juan de Urabá). South Western Atlantic: Brazil (off Pernambuco, Espírito Santo, and Rio de Janeiro), off Uruguay.

Indo-Pacific Oceans: from South Africa, Madagascar, Arabian Sea, Sumatra, Indonesia, Australia, Papua New Guinea, Japan, Canada (Alves-Júnior et al. 2019; present study). **Bathymetric distribution**. 250–3716 m (Alves-Júnior et al. 2019). **Remarks**. First record for the Colombian Caribbean and for the Caribbean Sea. It considerably increases the range of latitudinal distribution in the Atlantic Ocean. According to the description of Alves-Júnior et al. (2019), *A. quadrispinosa* is armed with 3–7 dorsal teeth on the rostrum, while the specimens examined here exhibit 10 spines. The specimens showed similarities with *Acanthephyra purpurea*, but the presence of a tooth at the distal end of the dorsal carina of the fourth abdominal somite indicates that they belong to *A. quadrispinosa*.

12. *Acanthephyra stylorostratis* (Spence Bate, 1888) (Fig. 25.6c).

Bentheocaris stylorostratis Spence Bate, 1888: p. 726, pl. 123, Fig. 4.
Acanthephyra stylorostratis.— Calman 1925:14; Chace Jr 1936:30; 1940:144, Fig. 22; 1986:10; Barnard 1950:666; Cardoso and Young 2005:27, Figs. 14 a–e, 15 a–g, 16 a–d, 17a; Pequegnat and Wicksten 2006:97; Serejo and Cardoso 2010:194, Fig. 1a–c; Alves-Júnior et al. 2019:412–413, Figs. 11a–c, 12, 39 f.

Material Examined. Off Barranquilla, 11°55′17.0″-74°54′00.0″, 3100 m, 1 org, C3-E06 INV CRU8445; off San Juan de Urabá, 8°52′58.0512″-76°52′40.090″, 200 m, 1 org, PA-E01 INV CRU8453. **Recognition characters**. Rostrum projecting like a rounded crest, dorsally armed with six spiniform teeth and one styliform and long tooth on ventral margin. Carapace with antennal spine, and branchiostegal spine and carina. Somites 2–6 carinate and 3–6 with posteromesial tooth, third distinctly strong (Cardoso and Young 2005). **Distribution**. Northwestern Atlantic: the USA (off New Jersey, Florida), off Bermuda. Gulf of Mexico, Caribbean Sea: Honduras, Nicaraguan Rise, Colombia (Southern and central Colombian Caribbean, off Barranquilla, San Juan de Urabá), Venezuela (Isla la Tortuga). Southwestern Atlantic: Brazil (Ceará Chain, Rio Grande do Norte, Alagoas, and off Espírito Santo). Eastern Atlantic: Portugal (Canary, Madeira, and Cape Verde Islands). Sahara Occidental. Indo-Pacific Oceans: South Africa (Natal), Tuamotu Archipelago (Pequegnat and Wicksten 2006; Alves-Júnior et al. 2019; present study). **Bathymetric distribution**. 700–3548 m (Alves-Júnior et al. 2019). **Remarks**. First record for the Colombian Caribbean. These records extend the bathymetric distribution from shallow waters to 200 m.

Family Oplophoridae Dana, 1852.

13. *Systellaspis debilis* (A. Milne-Edwards, 1881) (Fig. 25.6e).

Acanthephyra debilis A. Milne-Edwards 1881: p. 13.
Miersia gracilis Smith 1882: p. 70.
Systellaspis bouvieri Coutière 1905:8, Fig. 3.
Systellaspis debilis.— Holthuis 1951:32; Chace Jr 1940:181, Fig. 51; 1986:67, Figs. 34 m–o, 35 g, h; Kensley 1972:38, Fig. 17b–c; Crosnier and Forest 1973:87, Figs. 26b, 27b; Vereshchaka 1990:140; Poupin 1996:6; Cardoso

and Young 2005:64, Figs. 49–53; Pequegnat and Wicksten 2006:102; Felder et al. 2009:1053; Poupin 2010:73; Alves-Júnior et al. 2019:431–432, Figs. 35 a–b, 36.

Material Examined. Off Córdoba, 9°21′12.47″-77°1′4.32″ 2423 m1 org C5-E05 INV CRU8486. (Off Arboletes, Antioquia) 10°27′11.798″-76°15′25.406″ 2300 m 1 org CD-NA3 INV CRU8901. **Recognition characters**. Rostrum overreaching scaphocerite and armed with 10 dorsal and 15 ventral teeth, carapace with branchiostegal spine. Abdominal somites not dorsally carinate, 3–5 with posteromesial tooth, third distinctly strong. Tergum of somites 4 and 5 with markedly spinulose posteromesial margin (Cardoso and Young 2005). **Distribution**. Northwestern Atlantic: Bahamas. Gulf of Mexico. Caribbean Sea: Antilles, Colombia (off Córdoba, off Arboletes, Antioquia). Southwestern Atlantic: Brazil (Bahia). Eastern Atlantic: West African coast, from Cape Verde, Guinea to Gabon. Indian and Pacific Oceans: Madagascar, Western Indian Ocean, South China Sea, Zanzibar, the Philippines, Indonesia (Cardoso and Young, 2005; present study). **Bathymetric distribution**. 25–4594 m frequently between 300 and 600 m (Alves-Júnior et al. 2019). **Remarks**. First record for the Colombian Caribbean. No differences from the description of Cardoso and Young (2005).

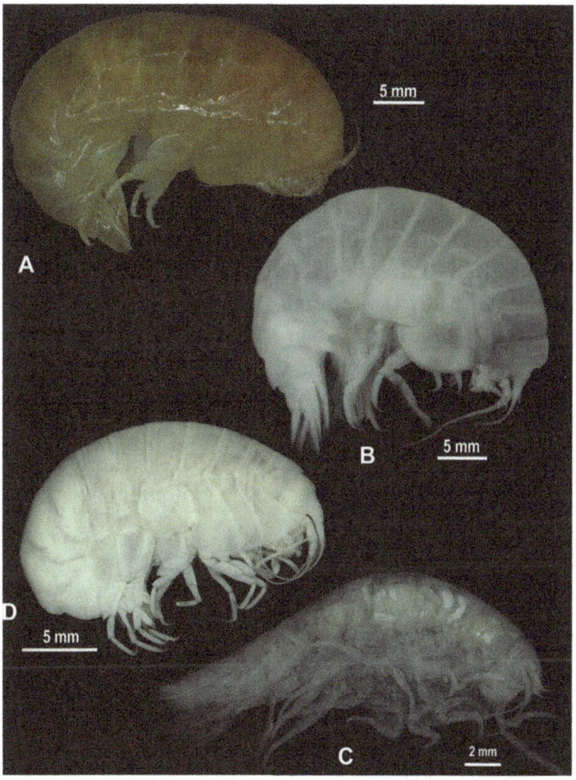

Fig. 25.7 (**a**) *Eurythenes gryllus* (KI-FSN04). (**b**) *Eurythenes thurstoni* (BR-E03). (**c**) *Paracallisoma alberti* (CD-NA3). (**d**) *Stephonyx biscayensis* (KI-FSE02)

Order Amphipoda Latreille 1816.
Suborder Amphilochidea Boeck, 1871.
Superfamily Lysianassoidea Dana, 1849.
Family Eurytheneidae Stoddart & Lowry, 2004.

14. *Eurythenes gryllus* (Lichtenstein in Mandt, 1822) (Fig. 25.7a).

Gammarus gryllus Lichtenstein in Mandt, 1822:34.
Eurythenes gryllus.— Stoddart and Lowry, 2004: 429–445, Figs. 1–11 (complete synonymy); D'Udekem D'Acoz and Havermans 2015:23–41, Figs. 12–26.

Material Examined. Off Morrosquillo, 10°27′19.64″-76°30′54.68″, 2873 m, 7 org, C5-E13 INV CRU8479; off Córdoba, 3097 m, 10°6′58,07″-76°47′57,36″, 63 org, C5-E11 INV CRU8480; off isla San Bernardo, 10°2′22.9379″-76°37′32.584″, 3000 m, 4 org, PA-E08 INV CRU8454; off isla San Bernardo, 10°9′32.8859″-76°40′5.9483″, 2800 m, 14 org, PA-E09 INV CRU8455; off Cartagena, 10°20′51.824″-76°27′28.591″, 2600 m, 4 org, PA-E11 INV CRU8457; off Cartagena, 10°28′49.093″-76°16′15.471″, 2000–2400 m, 5 org, CA-E01 INV CRU8461; off Arboletes, 9°8′15.11330″-76°50′44.324″, 1600–1800 m, 2 org, KI-FSN04 INV CRU8465. **Recognition characters**. Anterodorsal margin of head smooth. Gnathopod 1 parachelate, length of basis 2–2.5 the width; propodus slightly tapering distally. Pereopods 3–7 dactyl short, about one third as long as article six, combined length of articles 3–7 of third pereopod twice as long as second article. Pereopod 4 coxa as deep as wide. Basis of pereopod 7 as long as width or equaling; posteroventral margin straight. Third pleonite with anterodorsal notch. Epimeron 3 posteroventral corner broadly rounded. Urosomite 1 not dorsodistally produced over urosomite 2. Uropod of third peduncle devoid of robust setae (Barnard 1961; Stoddart and Lowry 2004). **Distribution**. Cosmopolitan (D'Udekem D'Acoz and Havermans 2015). In the Colombian Caribbean (off Arboletes, Morrosquillo, Cartagena). **Bathymetric distribution**. 839–3000 m (D'Udekem D'Acoz and Havermans 2015). **Remarks**. Despite a cosmopolitan condition, according to the map in Stoddart and Lowry (2004), this is the first record of the species in the Colombian Caribbean Sea. Escobar-Briones et al. (2010) mentioned the presence of the *E. gryllus* in the Caribbean waters but did not clarify the localities. Moreover, they are not included in the listing of amphipods that are recognized for the Caribbean, based on Martín et al. (2013). Therefore we consider it relevant to confirm the presence of the species in Colombian waters.

15. *Eurythenes thurstoni* Stoddart & Lowry, 2004 (Fig. 25.7b).

Eurythenes thurstoni Stoddart and Lowry, 2004:451–460, Figs. 16–20.
Eurythenes gryllus.– Hurley 1957:2 (in part); Barnard 1961:35 (in part), Figs. 6, 7 (material from South Pacific); Lowry and Stoddart 1993:72; France and Kocher 1996a:633 (in part).
Eurythenes cf. *gryllus*.— Lowry and Stoddart 1994:158 (in part).
Eurythenes sp.— France and Kocher 1996b:306.

Eurythenes thurstoni.— Senna 2009:86 (table); Quadra et al. 2014:376, Figs. 2–3; D'Udekem D'Acoz and Havermans 2015:74–75.

Material Examined. Colombia (La Guajira) 12°50′17.578″-71° 40′19.550″ 840 m 2 org BR-E03 INV CRU8427. **Recognition characters**. Anterodorsal margin of head forming an upturned ridge. Gnathopod 1 subchelate; basis three times longer than wide; propodus margins subparallel; palm of gnathopod 2 very protruding. Dactyl of pereopods 3–6 short. Coxa of pereopod 4 deeper than wide. Posterodistal lobe of basis of pereopod 7 very long (1.5 times longer than width), posteroventral margin straight but angled. Third pleonite without anterodorsal notch. Third epimeron with posteroventral corner subquadrate. Urosomite 1 dorsodistally produced over urosomite 2. Peduncle of uropod 3 with robust setae on medial face (Stoddart and Lowry, 2004). **Distribution**. Northwestern Atlantic: Bahamas. Gulf of Mexico. Caribbean Sea: Colombia (La Guajira). Southwestern Pacific Ocean: Indonesia, eastern and southern Australia, New Zealand, Loyalty Islands Basin, Wallis and Futuna Islands, Tonga (Stoddart and Lowry 2004; present study). **Bathymetric distribution**. 128–4670 m (Stoddart and Lowry 2004). **Remarks**. First record for the Colombian Caribbean and for the Caribbean Sea. No differences from the description of Stoddart and Lowry (2004).

Family Scopelocheiridae Lowry & Stoddart, 1997.

16. *Paracallisoma alberti* Chevreux, 1903 (Fig. 25.7d).

Paracallisoma alberti Chevreux, 1903:84, Figs. 2, 3; Stebbing 1906:719; Chevreux 1935:39, pl. 1, Fig. 3, pl. 9, Fig. 2, pl. 16, Fig. 5; Birstein and Vinogradov 1955:223, 279 (in part); Schellenberg 1955:191 (in part); Barnard 1958:97 (list); Belloc 1960:4; Lampitt et al. 1983:76; Desbruyères et al. 1985:236, Fig. 1, 237; Thurston 1990:266; Barnard and Karaman 1991:511; Palerud and Vader 1991:41; Lopes et al. 1993:209; Dauvin and Bellan-Santini 2002:316; Horton 2006:20; Horton et al. 2013:355. Kilgallen and Lowry 2015:22–24, Figs. 22–23; Horton and Thurston 2015:94–101, Figs. 1–5.

Material Examined. Off Cartagena, 10°27′11.798″-76°15′25.406″, 2300 m, 2 org, CD-NA3 INV CRU8904; off Cartagena, 10°28′49.795″-76°14′36.848″, 2400 m, 2 org, CD-NA2 INV CRU8903; off Cartagena, 10°25′33.851″-76°16′14.057″, 2400 m, 3 org, CD-NA4 INV CRU8905 **Recognition characters**. Head deeper than long, lateral cephalic lobe large, narrow, subacute. Coxae 1–2 subrectangular, not tapering distally. Gnathopod 1 with coxa slightly longer than wide, margins subparallel; basis not expanded; propodus subrectangular, dactylus anterior margin with rows of long slender setae. Gnathopod 2 with propodus broadened distally, dactyl shorter than palm, inserted at the anterodistal angle of propodus. Pereopods 3–7 not prehensile. Carpus of pereopods 3–4 distinctly longer than wide. Coxa of fourth pereopod subacutely produced on posteroventral lobe. Basis of fifth pereopod pyriform. Urosomite 1 concave anteriorly with a rounded smooth boss posteriorly. Uropod 1 peduncle 1.2 times inner ramus; rami subequal. Uropod 2 peduncle as long as inner ramus. Uropod 3 peduncle shorter than inner ramus; outer ramus, two-articulate. Telson weakly tapering and deeply cleft (Kilgallen and Lowry 2015; Horton and Thurston 2015). **Distribution**. Northeastern Atlantic: Madeiran and Iberian Basins; Porcupine Seabight, Bay of Biscay; Faroe-Shetland Channel. Caribbean Sea, Colombia (off Cartagena) (Horton and Thurston, 2015; present study). **Bathymetric distribution**. 1661–2628 m (Horton and Thurston 2015). **Remarks**. First record for the Colombian Caribbean, the Caribbean Sea, and the northwestern Atlantic.

17. *Stephonyx biscayensis* (Chevreux, 1908) (Fig. 25.7c).

Euonyx biscayensis Chevreux, 1908:1, Fig. 1; Barnard 1916:110; Stephensen, 1923:42; Schellenberg 1926:200; Chevreux 1927:47; Pirlot 1933:120 (key); Chevreux, 1935:7, pl. 5 Fig. 2; Barnard 1958:91; Belloc 1960:6; Desbruyères et al. 1985:236, 237; Andres 1987:96; Costello et al. 1989:32; Barnard and Ingram 1990:2, 3 (key); Barnard and Karaman 1991:485; Holmes et al. 1997:186; Dauvin and Bellan-Santini 2002:315.

Stephonyx biscayensis.— Lowry and Stoddart 1989:522, Figs. 2, 3; Palerud and Vader 1991:43; Poupin 1994:16; Lowry and Stoddart 1997:129, Fig. 63; Escobar-Briones and Winfield 2003:42; Ortiz et al. 2007:516; Diffenthal and Horton 2007:40 (key); Senna and Serejo 2007:13 (key); Brown and Thatje 2011:1, Figs. 1–3; Narahara et al. 2012:1486, 1506 (key), Figs. 7–11; Corrigan et al. 2014:10, Fig. 5; Lowry and Kilgallen 2014:66–67.

Material Examined. Off Antioquia, 9°2′19.12″-76°59′40.04″, 1389 m, 15 org, C5-E02 INV CRU8478; off Arboletes, Antioquia, 9°13′31.0605″-76°49′7.6814″, 1600–1800 m, 2 org, KI-E02 INV CRU8464. **Recognition characters**. Antenna 1 shorter than antenna 2; flagellum article 1 distinctly elongate. Antenna 2 as long as or longer than body. Lateral cephalic lobes rounded. Gnathopod 1 chelate; coxa subquadrate and reduced. Ischium extremely elongate; carpus very long, of subequal length to propodus. Propodus margins subparallel, palm obtuse. Gnathopod 2 subchelate; carpus at least twice as long as propodus, dactyl minute. Coxa of fifth pereopod equilobate, basis distinctly broader than long. Coxa 6 produced into a posterior lobe. First epimeron with the anterodistal margin obtuse or broadly rounded; third epimeron with posterodistal margin slightly acute. Telson more than twice as long as broad, deeply cleft (Kilgallen 2006). **Distribution**. Northwestern Atlantic: off Florida, the USA. Gulf of Mexico. Caribbean Sea: Puerto Rico Basse Terre, Guadeloupe, Colombia (off Antioquia). Northeastern Atlantic: Bay of Biscay; Faeroes; Ireland; Hatton-Rockall Basin; northwest Africa. Southeastern Atlantic: South Africa (Cape Point). Indian Ocean. East China Sea: Okinawa Trough and off Aguni-Jima Island (Lowry and Kilgallen 2014; present study). **Bathymetric distribution**. 494–1510 m (Lowry and Kilgallen 2014). **Remarks**. First record for the Colombian Caribbean. No differences from the description of Kilgallen (2006).

25.3.2 Distribution Considerations

The specimens collected and captured with the ROV coincide mostly with those found in other regions of the Western Atlantic deep marine ecosystems. Actually, several species presented a wide latitudinal distribution. However, it is remarkable that only nine of the 17 species had been previously recorded in Caribbean waters. It is also important to emphasize that *Cerataspis monstrosus*, *Alvinocaris muricola*, *Eurythenes thurstoni*, and *Stephonyx biscayensis* are distributed from the northwestern Atlantic to the Caribbean Sea, and *Alvinocaris markensis* and *Acanthephyra quadrispinosa* from the southwestern Atlantic to the Caribbean Sea. This indicates that the Colombian Caribbean is a borderland for distribution patterns and, at the same time, a point of convergence for species that belongs to two different biogeographic regions.

Locally, the specimens presented a pattern of distribution that coincides with what has been reported in literature, indicating that there are at least two different faunal groups, which allow for a clear differentiation between the northern and southern sectors of the Colombian Caribbean (Invemar 2010). *Alvinocaris markensis*, *A. muricola*, *G. aurantiaca*, *Acanthephyra acutifrons*, and *Eurythenes thurstoni* are representatives of the northern Colombian area, while the remaining species recorded herein are representatives of the southern Colombian area. However, it is imperative to enrich the database and complete the information related to the distribution of the deep-sea species occurring off Colombia in order to elucidate whether this emerging pattern is due to ecological factors or to knowledge gaps about the distribution of the regional deep fauna.

Acknowledgments We would like to express our gratitude to the oil and gas operator companies Shell, Petrobras, and Anadarko Colombia Company for sending the samples of their offshore exploratory projects, whether it was mandatory or not, to the Makuriwa-Invemar Marine Natural History Museum. This was also possible, thanks to the environmental consulting company Aquabiósfera SAS and the Scientific Services Coordination (CSC) of Invemar, which collected, preliminary processed, properly stored, and sent the samples to the Museum, on behalf of the oil and gas companies mentioned. Finally thanks to the Instituto de Investigaciones Marinas y Costeras-Invemar and the Museum of Marine Natural History of Colombia Makuriwa-Invemar for the taxonomic identification. Contribution No. 1280 of Invemar.

References

Agassiz A (1888) Three cruise of the Blake. Bul Mus Comp Zool 15:1–220

Alcock A (1899) A summary of the Deep-sea Zoological work of the Royal Indian Marine Survey Ship Investigator from 1884 to 1897. Sci Mem Med Officers Army India 11:1–49

Alcock A (1901) Materials for a carcinological fauna of India. No. 1. The Brachyura Osyrhyncha. J Asiatic Soc Bengal 64(2):157–291

Alves-Júnior FA, Dos Santos SE, De Araújo MLC, Cardoso I, Bertrand A, Souza-Filho JF (2019) Taxonomy of deep-sea shrimps of the Superfamily Oplophoroidea Dana 1852 (Decapoda: Caridea) from Southwestern Atlantic. Zootaxa 4613(3):401–442

Anderson ARS (1896) An Account of the deep-sea Crustacea collected during the season 1894–95. Natural History Notes from the R. I. M. Survey Steamer "Investigator", Commander CF Oldham, RN commanding. Series II No 21. J Asiat Soc Bengal 65(2):88–106

Andres HG (1987) Die Gammaridea der 76. Reise von FFS Walther Herwig mit Beschreibung von *Parachevreuxiella lobata* gen n und sp n (Crustacea: Amphipoda). Mitt Hamb Zool Mus Inst 84:95–103

Balss H (1925) Macrura der deutschen Tiefsee-Expedition. 2. Natantia, Teil A Wiss Ergebnisse Deutschen Tiefsee-Exped Valdivia 1898–1899 20:217–315

Barnard KH (1916) Contributions to the crustacean fauna of South Africa. 5. -The Amphipoda. Ann South African Mus 15, 105–302, pls. 126–128

Barnard KH (1950) Descriptive catalogue of South African decapod Crustacea. Ann South African Mus 38:1–837

Barnard JL (1958) Index to the families, genera, and species of the gammaridean Amphipoda (Crustacea). Allan Hancock Found Publ, Occ Paper 19:1–145

Barnard JL (1961) Gammaridean Amphipoda from depths of 400–6000 meters. Ekaudette Foundation. Solvang. California, 106 p

Barnard JL, Ingram C (1990) Lysianassoid Amphipoda (Crustacea) from Deep-Sea Thermal Vents. Smith Contrib Zool 499:1–80

Barnard JL, Karaman GS (1991) The families and genera of marine gammaridean Amphipoda (except marine gammaroids). Part 1. Rec Austr Mus 13(1):1–418

Bate CS (1881) On the Penaeidae. Ann Mag Nat Hist 5(8):169–196

Bate CS (1888) Report on the Crustacea Macrura collected by the Challenger during the years 1873–76. Report on the scientific results of the Voyage of HMS Challenger during the years 1873–76. Zoology 24 (52)

Belloc G (1960) Catalogue des types d'amphipodes du Musée océanographique de Monaco. Bull Inst Océanog (Monaco) 57(1170):1–28

Bermúdez A, Campos NH, Navas GR (2002) Los camarones de la familia Crangonidae de la franja superior del talud continental del mar Caribe colombiano. Bol Inv Mar Cos 31:183–203

Birstein JA, Vinogradov ME (1955) Pelagic gammarideans (Amphipoda-Gammaridea) from the Kurile-Kamchatka Trench. Akademiya Nauk SSSR, Trudy Inst Okeanol 12: 210–287 (in Russian)

Brown A, Thatje S (2011) Respiratory response of the deep-sea amphipod *Stephonyx biscayensis* indicates bathymetric range limitation by temperature and hydrostatic pressure. PLoS One 6(12):e28562

Burukovsky RN (2001) Taxonomy of *Nematocarcinus* (Decapoda, Nematocarcinidae). Description of *Nematocarcinus* from waters of the American continent. Zool Zhurn 80(12):1429–1443

Burukovsky RN (2003) Shrimps of the family Nematocarcinidae. Kaliningrad State Technical University, Kaliningrad p 250

Burukovsky RN (2004) Taxonomy of shrimps of the genus *Nematocarcinus* (Decapoda, Nematocarcinidae). A review of taxonomic characteristics and a key to identifying species of the genus. Zool Zhurn 83(5):549–561

Calman WT (1896) On deep-sea Crustacea from the south west of Ireland. Trans Royal Irish Acad 31:1–22

Calman WT (1925) On macrurous decapod Crustacea collected in South African waters by the S.S. "Pickle". Rep Fish Mar Biol Surveys South Africa 4:1–26

Calman WT (1939) Crustacea: Caridea. Scientific Rep John Murray Exped 6:183–224

Campos NH, Navas GR, Bermúdez A, Cruz N (2005) Los crustáceos decápodos de la franja superior del talud continental (300-500 m) del mar Caribe colombiano. UNAL, Unibiblos, Bogotá p 272

Campos NH, Rodríguez-Salcedo N, Bermúdez A (2010) Crustáceos decápodos del Caribe colombiano In: INVEMAR (ed) Biodiversidad del margen continental del Caribe colombiano. Publ Espec Invemar pp. 229-253

Cardoso I, Burukovsky R (2014) *Nematocarcinus* Milne Edwards, 1881 (Crustacea, Decapoda) from Southwestern Atlantic, including the Southern Mid-Atlantic Ridge area. Zootaxa 3887(3):437–458

Cardoso I, Serejo CS (2007) Deep Sea Caridea (Crustacea, Decapoda) from Campos Basin, RJ, Brazil. Braz J Oceanogr 55(1):39–50

Cardoso I, Young P (2005) Deep-sea Oplophoridae (Crustacea Caridea) from the southwestern Brazil. Zootaxa 1031:1–76

Caullery M (1896) Crustacés Schizopodes et Decapoda. In: Koehler, R., Résultats scientifiques de la campagne du Caudan das le Golfe de Gascogne, aout-septembre, 1895. Ann Univ Lyon 26: 365–419 plates 13–17

Chace FA Jr (1936) Revision of the bathypelagic prawns of the family Acanthephyridae, with notes on a new family, Gomphonotidae. J Wash Acad Sci 26:24–31

Chace FA Jr (1940) Plankton of the Bermuda Ocaenographic expeditions. IX. The bathypelagic Caridean Crustacea. Zoologica 25 2(11):17–209

Chace FA Jr (1984) The Caridean shrimps (Crustacea: Decapoda) of the Albatross Philippine Expedition, 1907-1910, Part 2: Families Glyphocrangonidae and Crangonidae. Smithson Contrib Zool 397:1–63

Chace FA Jr (1986) The Caridean shrimps (Crustacea: Decapoda) of the Albatross Philippine Expedition, 1907-1910. Part 4: Families Oplophoridae and Nematocarcinidae. Smithson Contrib Zool 432:5–57

Chevreux E (1903) Campagnes scientifiques de S.A. le P. Albert Ier de Monaco. Note préliminaire sur les amphipodes de la famille des Lysianassidae recueillis par la "Princesse Alice" dans les eaux profondes de l'Atlantique et de la Méditerranée. Bull Soc Zool Fr 28:81–97

Chevreux E (1908) Diagnoses d'amphipodes nouveaux provenant des campagnes de la "Princesse-Alice" dans l'Atlantique nord. Bull Inst Océanogr 117:1–13

Chevreux E (1927) Malacostracés (suite). III. Crustacés Amphipodes. Expéditions Scientifiques du "Travailleur" et du "Talisman" pendant les années 1880, 1881, 1882, 1883 9:41–152

Chevreux E (1935) Amphipodes provenant des campagnes du Prince Albert Ier de Monaco. Résultats des Campagnes scientifiques accomplies sur son Yacht par Albert Ier Prince Souverain de Monaco 90: 1–214, pls 1–16

Christoffersen ML (1986) Phylogenetic relationships between Oplophoridae, Atyidae, Pasiphaeidae, Alvinocarididae fam. n. Bresiliidae, Psalidopodidae and Disciadidae (Crustacea Caridea Atyoidea). Bol Zool Univ São Paulo 10:273–281

Corrigan LJ, Horton T, Fotherby H, White TA, Hoelzel AR (2014) Adaptive evolution of the deep-sea amphipods from the superfamily Lysianassoidea [sic] in the North Atlantic. Evol Biol 41(1):154–165

Costello MJ, Holmes JMC, McGrath D, Myers A (1989) A review and catalogue of the Amphipoda (Crustacea) in Ireland. Irish Fish Invest, Ser B Marine 33:1–70

Coutière H (1905) Note préliminaire sur les Eucyphotes recueillis par S.A.S. le Prince de Monaco à l'aide du filet à grande ouverture. Bull Inst Océanogr 48:1–35

Crosnier A (1978) Crustacés decápodes pénéides Aristeidae (Benthesicyminae, Aristeinae, Solenocerinae). Faune Madagascar 46:1–197

Crosnier A (1987) Oplophoridae (Crustacea Decapoda) récoltés de 1971 à 1982 par les navires français dans l´ócean Indien occidental sud. Ann Mus Hist Nat Paris 4:695–726

Crosnier A (1994) Crustacea Decapoda: Observations complémentaires sur les *Metapenaeopsis* indo-ouest pacifiques sans appareil stridulant (Penaeidae). Description de deux espèces nouvelles. In: Crosnier A (ed) Résultats des Campagnes MUSORSTOM, Vol 12. Mém Mus Natl Hist Nat 161:339–349

Crosnier A, Forest J (1968) Note préliminaire sur les Carides recueillis par l'Ombango au large du plateau continental, du Gabon à l'Angola (Crustacea Decapoda Natantia). Bull Mus Natl Hist Nat 2(39):1123–1147

Crosnier A, Forest J (1973) Les crevettes profondes de l'Atlantique Oriental Tropical. ORSTOM, Paris, p 409

Cruz N, Fransen CHJM (2004) Addition of three species of the genus *Plesionika* (Crustacea: Caridea: Pandalidae) to the known Atlantic marine fauna of Colombia. Zool Med Leiden 78:131–146

D'Udekem D'Acoz C (1999) Inventaire et distribution des crustacés décapodes de l'Atlantique nord-oriental, de la Méditerranée et des eaux continentales adjacentes au nord de 25°. Colle Patrimoines Nat 40:1–383

D'Udekem D'Acoz C, Havermans C (2015) Contribution to the systematics of the genus *Eurythenes* S.I. Smith in Scudder, 1882 (Crustacea: Amphipoda: Lysianassoidea: Eurytheneidae). Zootaxa 1:001–080

Dall W (2001) Australian species of Aristeidae and Benthesicymidae (Penaeoidea: Decapoda). Mem Queensland Mus 46(2):409–441

Dana JD (1849) Synopsis of the genera of Gammaridea. Am J Sci Arts 8(2):135–140

Dana JD (1852) On the classification of the Chrystoidea, Paguridea, etc. Amer J Sci Arts 13(37):119–124

Dauvin JC, Bellan-Santini D (2002) Les crustacés amphipodes Gammaridea benthiques des côtes françaises métropolitaines: Bilan des connaissances. Crustaceana 75:299–340

De Grave S, Fransen CHJM (2011) Carideorum catalogus: the recent species of the Dendrobranchiate, Stenopodidean, Procarididean and Caridean shrimps (Crustacea: Decapoda). Zool Med Leiden 85(9):195–589

De Man JG (1920) The Decapoda of the Siboga expedition, Vol. 23, Part 4. Families Pasiphaeidae, Stylodactylidae, Oplophoridae, Nematocarcinidae, Thalassocaridae, Pandalidae, Gnathophyllidae, Processidae, Glyphocrangonidae, and Crangonidae. Siboga Exped 23(39):1–318

Desbruyères D, Geistdorfer P, Ingram CL, Khripoudoff A, Lagardère JP (1985) Répartition des populations de l'épibenthos carnivore. In: Laubier L, Monniot C (eds) Peuplements profonds du golfe de Gascogne. Ifremer, Paris, pp 234–251

Desbruyères D, Almeida A, Biscoito M, Comtet T, Le Bris N, Sarradin PM, Segonzac M (2000) A review of the distribution of hydrothermal vent communities along the northern Mid-Atlantic Ridge: dispersal vs. environmental controls. Hydrobiologia 440:201–216

Desbruyères D, Biscoito M, Caprais JC, Colaco A, Comtet T, Crassous P, Fouquet Y, Khripounoff A, Le Bris N, Olu K, Riso R, Sarradin PM, Segonzac M, Vangriesheim A (2001) Variations in deep-sea hydrothermal vent communities on the Mid-Atlantic Ridge near the Azores Plateau. Deep-Sea Res Part I Oceanog Res Papers 48:1325–1346

Desbruyères D, Segonzac M, Bright M (2006) Handbook of deep-sea hydrothermal vent fauna. Denisia 18:1–544

Diffenthal M, Horton T (2007) *Stephonyx arabiensis* (Crustacea: Amphipoda: Lysianassoidea: Uristidae), a new deep-water scavenger species from the Indian Ocean, with a key to the genus *Stephonyx*. Zootaxa 1665:31–41

Digby A, Puentes V, León J (2016) Cold seeps associated with structured benthic communities: more accurate identification and evaluation using a new multibeam survey methodology in the offshore Southern Colombian Caribbean. Int J Geosci 7:761–774

Dixon DR, Dixon LRJ (1996) Results of DNA analyses conducted on vent-shrimp post larvae collected above the Broken Spur vent field during the CD95 cruise, August 1995. BRIDGE News-letter 11:9–15

Dueñas L, Herrera S, León J, Puentes V (In review) Characterization of deep-sea chemosynthetic communities in the Colombian Caribbean. Front Mar Sci

Escobar-Briones E, Winfield I (2003) Checklist of the Benthic Gammaridea and Caprellidea (Crustacea: Peracarida: Amphipoda) from the Gulf of Mexico Continental Shelf and Slope. Belg J Zool 133(1):37–44

Escobar-Briones E, Nájera-Hillman E, Álvarez F (2010) Unique 16S rRNA sequences of *Eurythenes gryllus* (Crustacea: Amphipoda: Lysianassidae) from the Gulf of Mexico abyssal plain. Rev Mex Biodiv 81:177–185

Faxon W (1895) Reports on an exploration off the West coasts of Mexico, Central and South America, and off the Galapagos Island, in charge of Alexander Agassiz, by the U. S. Fish Commission steamer Albatross during 1891. XV. The stalk eyed Crustacea Mem Mus Comp Zool 18:1–292

Felder DL, Álvarez F, Goy JW, Lemaitre R (2009) Decapoda (Crustacea) of the Gulf of Mexico, with comments on the Amphionidacea. In: Felder DL, Camp DK (eds) Gulf of Mexico–origins, waters, and biota. Biodiversity. Texas A&M University Press, College Station, Texas, pp 1019–1104

Figueira AJG (1957) Madeiran Decapod crustaceans in the collection of the Museum Municipal do Funchal I. On some interesting deep-sea prawns of the families Pasiphaeidae, Oplophoridae and Pandalidae. Bol Mus Municip Funchal 10:22–51

France SC, Kocher TD (1996a) Geographic and bathymetric patterns of mitochondrial 16s rRNA sequence divergence among deep-sea amphipods, *Eurythenes gryllus*. Mar Biol 126:633–643

France SC, Kocher TD (1996b) DNA sequencing of formalin-fixed crustaceans from archival research collections. Mol Mar Biol Biotechnol 5(4):304–313

Gore RH (1985) Some rare species of abyssobenthic shrimps (Families Crangonidae, Glyphocrangonidae and Nematocarcinidae) from the Venezuela Basin, Caribbean Sea (Decapoda, Caridea). Crustaceana 48:269–285

Gray JE (1828) Spicilegia Zoologica; or, original figures and short systematic descriptions of new and unfigured animals. Treuttel, Würtz & Co 1(1):1–8

Haworth AH (1825) A new binary arrangement of the macrurous Crustacea. Philos Mag J 65:183–184

Holmes JMC, Costello MJ, Connor DW (1997) Crustacea. In: Howson CM, Picton BE (eds) The species directory of the marine fauna and flora of the British Isles and surrounding seas. Ulster Museum Publication, Ulster Museum, Belfast, pp 142–209

Holthuis LB (1951) The caridean crustacea of tropical West Africa. Atlantide Rep 2:7–187

Holthuis LB (1955) The recent genera of the caridean and stenopodidean shrimps (class Crustacea, order Decapoda, supersection Natantia) with keys for their determination. Zool Verh 26:1–157

Holthuis LB (1971) The Atlantic shrimps of the deep-sea genus *Glyphocrangon* A. Milne Edwards, 1881. Biological Results of the University of Miami Deep-Sea Exped 21(1):267–373

Holthuis LB (1993) The recent genera of the caridean and stenopodidean shrimps (Crustacea, Decapoda): with an appendix on the order Amphionidacea. Nationaal Natuurhistorisch Museum, Leiden

Horton T (2006) Deep-sea scavenging amphipods from the Faroe-Shetland Channel. Porcup Mar Nat Hist Soc Newsl 19:17–20

Horton T, Thurston MH (2015) A revision of the genus *Paracallisoma* Chevreux, 1903 (Crustacea: Amphipoda: Scopelocheiridae: Paracallisominae) with a redescription of the type species of the genus *Paracallisoma*, the description of two new genera and two new species from the Atlantic Ocean. Zootaxa 3995(1):91–132

Horton T, Thurston MH, Duffy GA (2013) Community composition of scavenging amphipods at bathyal depths on the Mid-Atlantic Ridge. Deep-Sea Res Part II Top Stud Oceanogr 98:352–359

Hurley DE (1957) Some Amphipoda, Isopoda and Tanaidacea from Cook Strait. Zool Public Victoria Univ College 21:1–20

Invemar (2010) Biodiversidad del margen continental del Caribe colombiano. Public Espec Invemar 20:458

Judkins DC (2014) Geographical distribution of pelagic decapod shrimp in the Atlantic Ocean. Zootaxa 3895(3):301–345

Kemp SW (1906) On the occurrence of the genus *Acanthephyra* in deep water off the West coast of Ireland. Sci Inv Fish Branch, Ireland 1:1–28

Kemp S (1939) On *Acanthephyra purpurea* and its allies (Crustacea Decapoda: Oplophoridae). Ann Mag Nat Hist 11(4):568–579

Kensley BF (1968) Deep sea decapod Crustacea from West of Cape Point, South Africa. Ann S African Mus 50(12):283–323

Kensley BF (1972) Shrimps and prawns of Southern Africa. S African Mus 1:1–65

Kensley BF (1981a) On the Zoogeography of Southern African Decapod Crustacea, with a distributional checklist of the species. Smithson Contrib Zool 338:1–66

Kensley BF (1981b) The South African museum's Meiring Naude cruises Part 12 Crustacea Decapoda of the 1977, 1978, 1979 cruises. Ann S African Mus 83:49–78

Kensley B (1987) Deepwater Decapod Crustacea from Eastern Australia (Penaeidea and Caridea). Rec Aus Mus 39:263–331

Kensley B, Tranter HA, Griffin DJG (1987) Deepwater Decapod Crustacea from eastern Australia (Penaeidea and Caridea). Rec Aus Mus 39(5):263–331

Kikuchi T, Hashimoto J (2000) Two new Caridean shrimps of the family Alvinocarididae (Crustacea, Decapoda) from a hydrothermal field at the Minami-Ensei Knoll in the Mid-Okinawa Trough, Japan. Species Div 5:135–148

Kilgallen NM (2006) Taxonomy of the Lysianassoidea of the Northeast Atlantic and Mediterranean: An interactive identification key and studies on problematic groups. PhD Thesis GMIT 306 p

Kilgallen NM, Lowry J (2015) A review of the scopelocheirid amphipods (Crustacea, Amphipoda, Lysianassoidea), with the description of new taxa from Australian waters. Zoosyst Evol 91(1):1–43

Koehl MAR (2011) Hydrodynamics of sniffing by crustaceans. In: Breithaupt T, Theil M (eds) Chemical Communication in Crustaceans. Springer Verlag, p 85–102

Komai T (2004) Deep-sea shrimps of the genus *Glyphocrangon* A. Milne-Edwards (Crustacea, Decapoda, Caridea, Glyphocrangonidae) from off Southeastern coast of Brazil collected during the revizee program1. Arqu Mus Nac Rio de Janeiro 62(1):31–44

Komai T, Segonzac M (2005) A revision of the genus *Alvinocaris* Williams & Chace (Crustacea: Decapoda: Caridea: Alvinocarididae), with descriptions of a new genus and a new species of *Alvinocaris*. J Nat Hist 39(15):1111–1175

Komai T, Shank TM, Van Dover CL (2005) A new species of *Alvinocaris* (Crustacea: Decapoda: Caridea: Alvinocarididae) and a new record of *A. muricola* from methane seeps on the Blake Ridge Diapir, northwestern Atlantic. Zootaxa 1019:27–42

Lampitt RS, Merrett NR, Thurston MH (1983) Inter-relations of necrophagous amphipods, a fish predator, and tidal currents in the deep sea. Mar Biol 74:73–78

Latreille PA (1802) Histoire naturelle, générale et particulière, des crustacés et des insectes, vol 3. F. Dufart, Paris

Latreille PA (1816) Amphipodes. In: Biot JB, Bosc LAG, Chaptal JA, Desmarest AG, Dutour M, Huzard JB, Lamarck, JBPA de Monet de, Latreille PA, Lucas JAH, Olivier GA, Palisot de Beauvois AMFJ, Parmentier AG, Patrin EML, Richard LC, Sonini CS, Thouin A, Tollard C, Vieillot LP, Virey JJ, Yvart JAV (eds) Nouveau dictionnaire d'histoire naturelle, appliqué aux arts, à l'agriculture, à l'économie rurale et domestique, à la médecine, etc. 2nd ed Vol. 1. Librairie Deterville, Imprimerie d'Abel Lanoë, Paris, p 467–469

Lemaitre R, Tavares M (2015) New taxonomic y distributional information on hermit crabs (Crustacea: Anomura: Paguroidea) from the Gulf of Mexico, Caribbean Sea, and Atlantic coast of South America. Zootaxa 3994(4):451–506

Lopes MFR, Marques JC, Bellan-Santini D (1993) The benthic amphipod fauna of the Azores (Portugal): an up-to-date annotated list of species, and some biogeographic considerations. Crustaceana 65(2):204–217

Lowry JK, Kilgallen NM (2014) A generic review of the lysianassoid family Uristidae and descriptions of new taxa from Australian waters (Crustacea, Amphipoda, Uristidae). Zootaxa 3867(1):1–92

Lowry JK, Stoddart HE (1989) The scopelocheirid genus *Aroui* (Crustacea: Amphipoda: Lysianassoidea) with notes on the association between scopelocheirid amphipods, cassid gastropods and spatangoid echinoids. Rec Aust Mus 41:111–120

Lowry JK, Stoddart HE (1993) Crustacea Amphipoda: Lysianassoids from Philippine and Indonesian waters. Mém Mus Natl Hist Nat A Zool 156:55–109

Lowry JK, Stoddart HE (1994) Crustacea Amphipoda: lysianassoids from the tropical Western South Pacific Ocean. In: Crosnier A (ed) Résultats des campagnes MUSORSTOM vol. 12. Mém Mus Nat Hist Natur Ser A 161:127–223

Lowry JK, Stoddart HE (1997) Amphipoda Crustacea IV. Families Aristiidae, Cyphocarididae, Endevouridae, Lysianassidae, Scopelocheiridae, Uristidae. Mem Hourglass Cru 10:1–148

Mandt MW (1822) Observationes in historiam naturalem et anatomiam comparatam in itinere Groenlandico factae. Dissertation inauguralis quam consensu et auctoritate gratiosi micorum ordinis in universitate literaria berolinensi ut summi in medicina et chirurgia honores rite, p 40

Martin JW, Davis GE (2001) An updated classification of the recent Crustacea. Science Series. Natl His Mus Los Angeles County 39:1–124

Martin JW, Haney TA (2005) Decapod crustaceans from hydrothermal vents and cold seeps: a review through 2005. Zool J Linnean Soc 145:445–522

Martín A, Díaz Y, Miloslavich P, Escobar-Briones E, Guerra-García JM, Ortiz M, Valencia B, Giraldo A, Klein E (2013) Diversidad regional de Amphipoda en el Mar Caribe. Rev Biol Trop 61(4):1682–1720

Martínez-Campos B, Campos NH, Lemaitre R (2017) Catálogo de los cangrejos ermitaños del Caribe colombiano/Catalog of hermit crabs from Colombian Caribbean. Instituto de Investigaciones Marinas y Costeras-Invemar. Public Espec Invemar 32:1–440

Milne-Edwards A (1881) Description de quelques crustacés macroures provenant des grandes profondeurs de la Mer des Antilles. Ann Sc Nat 11(6):1–15

Narahara Y, Tomikawa K, Torigoe T (2012) Four species of the genus *Stephonyx* (Crustacea: Amphipoda: Uristidae) from Japan, with description of a new species. J Nat Hist 46(23–24):1477–1507

Ortega-Echavarría C, Bermúdez A, Navas G (2014) Presencia de *Munidopsis robusta* en el Caribe colombiano (A. Milne-Edwards, 1880) (Crustacea: Anomura: Munidopsidae). Univ Sci 19(1):43–49

Ortiz M, Martín A, Díaz YJ (2007) Lista y referencias de los crustáceos anfípodos (Amphipoda: Gammaridea) del Atlántico occidental tropical. Rev Biol Trop 55(2):479–498

Palerud R, Vader W (1991) Marine Amphipoda Gammaridea in North-East Atlantic and Norwegian Arctic. Tromura, Naturvit 68:1–97

Pearcy WG, Forss CA (1966) Depth distribution of oceanic shrimps (Decapoda Natantia) off Oregon. J Fish Res Board Can 23(8):1135–1143

Pequegnat LH (1970) Deep-Sea Caridean Shrimps with descriptions of six new species. In: Pequegnat and Chace (ed) Contribution on the Biology of the Gulf of Mexico. Texas A&M Univ Ocean Stud 1:59–123

Pequegnat LH, Wicksten MK (2006) Oplophorid shrimps (Decapoda: Caridea: Oplophoridae) in the Gulf of Mexico and Caribbean Sea from the collections of the research vessels Alaminos, Oregon and Oregon II. Crust Res 35:92–107

Pérez-Farfante I, Kensley B (1997) Penaeoid and sergestoid shrimps and prawns of the world, keys and diagnoses for the families and genera. Min Mus Natl Hist Nat Sér A Zool 5:1–233

Pirlot JM (1936) Les amphipodes de l'expédition du Siboga. Deuxième partie: Les amphipodes gammarides, II. Les amphipodes de la mer profonde. 3: Addendum et partie générale. III. - Les amphipodes littoraux. 1: Lysianassidae, Ampeliscidae, Leucothoidae, Stenothoidae, Phliantidae, Colomastigidae, Ochlesidae, Liljeborgiidae, Oedicerotidae, Synopiidae, Eusiridae, Gammaridae. Siboga-Exped Monog 33:237–328

Poupin J (1994) Faune marine profonde des Antilles françaises. Récoltes du navire Polka faites en 1993. ORSTOM, Paris

Poupin J (1996) Atlas des crustacés marins profonds de Polynésie française. Récoltes du navire Marara (1986/1996). Service mixte de Surveillance radiologique et biologique, Montlhéry

Poupin J (2010) Biodiversité de l'Indopacifique tropical français: 2514 espèces de crustacés décapodes et stomatopodes. Rapport Scientifique de l'Institut de Recherche de l'Ecole Navale, Angers/Brest

Quadra A, Sorrentino R, Senna AR, Serejo CS (2014) First record of *Eurythenes thurstoni* Stoddart & Lowry (2004). (Crustacea: Amphipoda: Lysianassoidea) from the South Mid-Atlantic Ridge. Lat Am J Aquat Res 42(2):376–380

Rafinesque CS (1815) Analyse de la nature ou tableau de l'univers et des corps organisés. Aux Dépends de l' Auteur, Palerme

Ramírez-Llorda E, Billett DSM (2006) Ecosistemas de las profundidades marinas: Reservorio privilegiado de la biodiversidad y desafíos tecnológicos. In: Duarte CM (ed) La exploración de la biodiversidad marina. Desafíos científicos y tecnológicos. Fundación BBVA 3:65–94

Schellenberg A (1926) Amphipoda 3: Die Gammariden der Deutschen Tiefsee-Expedition. Wiss Ergeb Deutschen Tiefsee-Exped Dampfer Valdivia 1898–1899 23(5):193–243

Schellenberg A (1955) Amphipoda. Rep Swedish Deep-Sea Exped ser. 2. Zoology 14:181–195

Segonzac M, de Saint LM, Casanova B (1993) L'énigme du comportement trophique des crevettes Alvinocarididae des sites hydrothermaux de la dorsale medio- atlantique. Cah Biol Mar 34:535–571

Senna AR (2009) The giant deep-sea amphipods (Lysianassoidea: Eurytheneidae) from Brazilian waters. Nauplius 17(2):86–96

Senna AR, Serejo CS (2007) A new deep-sea species of *Stephonyx* (Lysianassoidea) from off the central coast of Brazil. Nauplius 15(1):7–14

Serejo CS, Cardoso IA (2010) Crustacea. In: Lavrado HP, Brasil ACS (eds) Biodiversidade da região oceânica profunda da Bacia de Campos: Megafauna e ictiofauna demersal. SAG Serv, Rio de Janeiro

Serejo CS, Young PS, Cardoso IC, Tavares C, Rodrigues C, Almeida TC (2007) Abundância, diversidade e zonação de Crustacea no talude da costa central do Brasil (11° – 22°S) coletado pelo Programa REVIZEE Pesca – Score Central. In: Costa PAS, Martins AS, Olavo G (eds) Pesca e potenciais de exploração de recursos vivos da região central da Zona Econômica Exclusiva brasileira. Rio de Janeiro: Museu Nacional 24:133–162

Shank T (1997) *Alvinocaris lusca* Williams & Chace Jr, 1982, and *A. markensis* Williams, 1988. In: Desbruyères D, Segonzac M (eds) Handbook of deep-sea hydrothermal vent fauna. Ifremer, Brest, pp 191–192

Shank TM, Lutz RA, Vrijenhoek RC (1998) Molecular systematics of shrimp (Decapoda: Bresiliidae) from deep-sea hydrothermal vents I: Enigmatic "small orange" shrimp from the Mid-Atlantic Ridge are juvenile *Rimicaris exoculata*. Molec Mar Biol Biotech 7:88–96

Shank TM, Black MB, Halanych KM, Lutz RA, Vrijenhoek RC (1999) Miocene radiation of deep-sea hydrothermal vent shrimp (Caridea: Bresiliidae): evidence from mitochondrial cytochrome oxidase subunit I. Mol Phylogenet Evol 13:244–254

Sivertsen E, Holthuis LB (1956) Crustacea Decapoda (The Penaeidae and Stenopodidae accepted). Report on the Scientific Results of the "Michael Sars" North Atlantic Deep-sea Exped 1910 5(12):1–54

Smith SI (1882) Reports on the results of dredging under the supervisión of Alexander Agassiz, on the east coast of the United States during the summer of 1880, by the U.S. Coast Survey Steamer "Blake", Commander J.R. Bartlett, U.S.N., commanding. Bull Mus Comp Zool 10(1):1–108

Smith SI (1884) XV Report on the decapod Crustacea of the Albatross dredging off the east coast of the United States In: Unites States Commission of Fish and Fisheries. Part X. Report of the Commission for 1882, p 345–424

Smith SI (1886) XXI report on the decapod Crustacea of the albatross dredging off the east coast of the United States during the summer and autumn of 1884. In: Unites States Commission of Fish and Fisheries Part XIII Report of the Commission for 1885, p 605–705

Smith SI (1887) Report on the decapod Crustacea of the Albatross dredging off the east coast of the United States during the summer and autumn of 1884. In: Unites States Commission of Fish and Fisheries. Part XIII. Report of the Commission for 1885, p. 605–706

Springer S, Bullis HR Jr (1956) Collections by the Oregon in the Gulf of Mexico. U.S States Department of the Interior, Fish and Wildlife Service. Service Sci Rep-Fish 196:1–134

Squires HJ (1990) Decapod Crustacea of the Atlantic coast of Canada. Can J Fash Aquat Sci 221:1–532
Stebbing TRR (1905) South African Crustacea. Part III. Mar Invest S Afr 4:21–123
Stebbing TRR (1906) Amphipoda. I. Gammaridea. Das Tierreich 21:1–806
Stephensen K (1923) Crustacea Malacostraca, V: (Amphipoda, I). Danish Ingolf-Exped 3:1–100
Stoddart HE, Lowry JK (2004) The deep-sea lysianassoid genus *Eurythenes* (Crustacea, Amphipoda, Eurytheneidae n fam). Zoosystema 26(9):425–468
Takeda M, Okutani T (1983) Crustaceans and mollusks trawled off Suriname and French Guiana. Japan Marine Fishery Reseach Center, Tokyo
Tavares CR, Serejo CS (2007) Taxonomy of Aristeidae (Dendrobranchiata: Penaeoidea) from the central coast of Brazil, collected by the Revizee program, between 19° and 22°S. Zootaxa 1585:1–44
Thompson JR (1966) The caridean superfamily Bresilioidea (Decapoda Natantia). A revision and a discussion of its validity and affinities. Crustaceana 11(2):129–140
Thurston MH (1990) Abyssal necrophagous amphipods (Crustacea: Amphipoda) in the northeast and tropical Atlantic Ocean. Prog Oceanog 4:257–274
Vázquez-Bader AR, Gracia A (2013) Crangonidae and Glyphocrangonidae (Decapoda; Caridea) of the Southern Gulf of Mexico. Zootaxa 3669(3):367–383
Vereshchaka AL (1990) Pelagic decapods from the seamounts of Nazca and Sala-y-Gómez ridges. Tr PP Shirshov Inst Okeanol, Moscow 124:129–155
Vereshchaka AL (1996) Comparative analysis of taxonomic composition of shrimps as edificators of hydrothermal communities in the Mid-Atlantic Ridge. Dokl Biol Sci 351:576–578
Wenner EL (1979) Distribution and reproducction of Nematocarcinid shrimp (Decapoda: Caridea) from the Northwestern North Atlantic. Bull Mar Sci 29(3):380–393
Williams AB (1984) Shrimps, lobsters and crabs of the Atlantic coast of the eastern United States, Maine to Florida. Smithsonian Institution Press, Washington, DC
Williams AB (1988) New marine decapod crustaceans from waters influenced by hydrothermal discharge, brine, and hydrocarbon seepage. Fish Bull 86:263–287
Wood-Mason J, Alcock A (1891) Natural history notes from H.M. Indian marine survey steamer 'Investigator', Commander RF Hoskyn RN commanding. No. 21. Note on the results of the last season's deep-sea dredging (cont.). Ann Mag Nat Hist 38(6):186–202
Wood-Mason J, Alcock A (1892) Natural history notes from H.M. Indian marine survey Steamer "Investigator", Commander RF, Hoskyn RN, commanding. Ser II, no.1. On the results of deep-sea dredging during the season 1890– 91. Ann Mag Nat Hist 9(6):265–275
Zamorano P, Hendrickx ME, Méndez N, Gómez S, Serrano D, Aguirre H, Madrid J, Morales-Serna FM (2014) La exploración de las aguas profundas del Pacífico mexicano: Proyecto TALUD. In: Zamorano P, Hendrickx ME, Caso M (eds) La Frontera Final: El Océano Profundo. Instituto Nacional de Ecología, Mexico DF
Zelnio K, Hourdez S (2009) A new species of *Alvinocaris* (Crustacea: Decapoda: Caridea: Alvinocarididae) from hydrothermal vents at the Lau Basin, Southwest Pacific, and a key to the species of Alvinocarididae. Proc Biol Soc Wash 122(1):52–71
Zubiría W, Campos NH, Bermúdez A (2016) Presencia de *Munidopsis bermudezi* en el mar Caribe colombiano (Crustacea: Anomura: Munidopsidae). Rev Mar Cost 8(2):67–74

Index

M. E. Hendrickx (ed.), *Deep-Sea Pycnogonids and Crustaceans of the Americas*, https://doi.org/10.1007/978-3-030-58410-8

B

F

G

H

O

The manufacturer's authorised representative in the EU is Springer Nature Customer Service Centre GmbH, Europaplatz 3, 69115 Heidelberg, Germany. If you have any concerns regarding our products, please contact ProductSafety@springernature.com

Printed and bound by CPI Group (UK) Ltd, Croydon, CR0 4YY
28/09/2026
02224067-0012